Cotton Production and Uses

Shakeel Ahmad • Mirza Hasanuzzaman
Editors

Cotton Production and Uses

Agronomy, Crop Protection, and Postharvest Technologies

Springer

Editors
Shakeel Ahmad
Department of Agronomy, Faculty of Agricultural Sciences and Technology
Bahauddin Zakariya University
Multan, Pakistan

Mirza Hasanuzzaman
Department of Agronomy, Faculty of Agriculture
Sher-e-Bangla Agricultural University
Dhaka, Bangladesh

ISBN 978-981-15-1471-5 ISBN 978-981-15-1472-2 (eBook)
https://doi.org/10.1007/978-981-15-1472-2

This Springer imprint is published by the registered company Springer Nature Singapore Pte Ltd.
The registered company address is: 152 Beach Road, #21-01/04 Gateway East, Singapore 189721, Singapore

Preface

Cotton is a multifaceted crop of which wholesome or all parts individually can be used for their by-products in addition to their domestic or economic uses. It provides lint raw material to the ever-increasing textile industry, cottonseed oil for culinary purposes, and edible oil and protein-rich oil cake residue for livestock. Cotton can benefit human being through its sticks, fibers, seed, and oil as the primary products, whereas several secondary products are manufactured by utilizing these components of cotton. The cotton fiber is unique in generating a host of products that sustain and make life more comfortable and aesthetically appealing. It is one of the ancient crops, but still, many aspects of its production and processing are still under research. On the threatening issues, cotton consumes more pesticide than any other crop; it is estimated that 25% of the worldwide use of insecticide and 10% of pesticide use are accounted for by cotton cultivation. Pesticides sprayed across cotton fields easily run off and pollute freshwater sources. Therefore, numerous research works have been carried out in the past couple of decades to invent an eco-friendly integrated pest management approach for cotton production. In recent decades, organic production has drawn much attention to the growers and users which does not simply mean replacing synthetic fertilizers and pesticides with organic ones. Organic cultivation methods are based more on knowledge of agronomic processes than input-based conventional production. Unlike the agronomic crops, cotton needs special postharvest technologies. Several articles were published dealing with cotton production and processing. Some of the genetic approaches, such as GM cotton for pest resistance, have also faced extreme debate in the last decades. In the era of climate changes, cotton is facing diverse abiotic stresses such as salt, drought, toxic metals, and environmental pollutants. Scientists are trying to develop stress tolerance cultivars using agronomic, genetic, and molecular approaches. Although there are many papers on these developments, there is no comprehensive book where readers can find all information ready. Therefore, this book will be the first comprehensive volume of its kind. It presents the recent development of cotton production and processing in an organized way.

We, the editors, would like to give special thanks to the authors for their outstanding and timely work in producing such fine chapters. Our profound thanks also to Dr. Kamrun Nahar and Dr. Md. Mahabub Alam for their critical review and valuable support in formatting and incorporating all editorial changes in the manuscripts. We are highly thankful to Ms. Mei Hann Lee, Editor (Editor, *Life Sciences*), Springer, Japan, for her prompt responses during the acquisition. We are also thankful to Sivachandran Ravanan, Project Coordinator of this book, and all other editorial staffs for their precious help in formatting and incorporating editorial changes in the manuscripts.

Multan, Pakistan Shakeel Ahmad
Dhaka, Bangladesh Mirza Hasanuzzaman

Contents

Editors and Contributors

About the Editors

Shakeel Ahmad is Professor of Agronomy at Bahauddin Zakariya University, Multan, Pakistan. In 2006, he received his Ph.D. from the University of Agriculture, Faisalabad, Pakistan. Later, he completed his postdoctoral research from the University of Georgia, USA. He joined as a Lecturer in the Department of Agronomy, Bahauddin Zakariya University, in October 2002. He was promoted to Professor in 2016. He has been devoting himself in teaching and researching the field of arable crops, especially focused on crop modeling, climate change impact assessment, and adaptation strategies since 2004. He continuously earned Research Productivity Award (RPA) for 5 years from the Pakistan Council for Science and Technology (PCST) through the Ministry of Science and Technology, Government of Pakistan, Islamabad. He has published over 125 articles and 50 book chapters. His publications received over 1500 citations with an h-index of 20 (according to Scopus). He is Editor and Reviewer of many peer-reviewed international journals. He is also an Active and Life Member of professional societies like Pakistan Society of Agronomy and Pakistan Botanical Society. He has attended and presented papers and posters in national and international conferences in different countries.

Department of Agronomy, Faculty of Agricultural Sciences and Technology, Bahauddin Zakariya University, Multan, Pakistan

Mirza Hasanuzzaman is Professor of Agronomy at Sher-e-Bangla Agricultural University in Dhaka. He received his Ph.D. on "Plant Stress Physiology and Antioxidant Metabolism" from Ehime University, Japan. Later, he completed his postdoctoral research at the Center of Molecular Biosciences, University of the Ryukyus, Japan. He was also the Recipient of the Australian Government's Endeavour Research Fellowship for postdoctoral research as an Adjunct Senior Researcher at the University of Tasmania, Australia. His current work is focused on the physiological and molecular mechanisms of environmental stress tolerance. He has published over 80 articles in peer-reviewed journals, edited 6 books, and written 30 book chapters. According to Scopus®, his publications have received roughly 3600 citations with an h-index of 30. He is an Editor and Reviewer of more than 50 peer-reviewed international journals and was a Recipient of the "Publons Peer Review Award 2017 and 2018." He has been honored by different authorities for his outstanding performance in different fields, like research and education, and has received the World Academy of Sciences Young Scientist Award (2014).

Department of Agronomy, Faculty of Agriculture, Sher-e-Bangla Agricultural University, Dhaka, Bangladesh

Contributors

Ghulam Abbas Department of Agronomy, Bahauddin Zakariya University, Multan, Pakistan

Khalid Abdullah Ministry of National Food Security and Research, Islamabad, Pakistan

Muhammad Abid Institute of Plant Protection and Agro-products Safety, Anhui Academy of Agricultural Sciences, Hefei, China

Department of Plant Pathology, Bahauddin Zakariya University, Multan, Pakistan

Irfan Afzal Department of Agronomy, University of Agriculture, Faisalabad, Pakistan

Muhammad Naveed Afzal Central Cotton Research Institute, Multan, Pakistan

Ashfaq Ahmad Climate Change, US.-Pakistan Centre for Advanced Studies in Agriculture and Food Security, University of Agriculture Faisalabad, Faisalabad, Pakistan

Burhan Ahmad Pakistan Meteorological Department, Islamabad, Pakistan

Fiaz Ahmad Central Cotton Research Institute Multan, Multan, Pakistan

Ijaz Ahmad Agriculture Pest Warning & Quality Control of Pesticides, Government of Punjab, Layyah, Pakistan

Ishfaq Ahmad Centre for Climate Research and Development, COMSATS University, Islamabad, Pakistan

Maqsood Ahmad Department of Environmental Sciences, Bahauddin Zakariya University, Multan, Pakistan

Munir Ahmad Department of Entomology, Pir Mehr Ali Shah, Arid Agriculture University Rawalpindi, Rawalpindi, Pakistan

Niaz Ahmed Department of Soil Science, Bahauddin Zakariya University, Multan, Pakistan

Muhammad Ahmed Central Cotton Research Institute, Multan, Pakistan

Mukhtar Ahmed Department of Agronomy, Pir Mehr Ali Shah, Arid Agriculture University, Rawalpindi, Pakistan

Saeed Ahmad Department of Agronomy, Muhammad Nawaz Shareef University of Agriculture, Multan, Pakistan

Muhammad Naeem Akhtar Department of Soil and Environmental Sciences, MNS University of Agriculture, Multan, Pakistan
Pesticide Laboratory, Multan, Pakistan

Salim Akhter Institute of Pure and Applied Biology, Bahauddin Zakariya University, Multan, Pakistan

Rida Akram Department of Environmental Sciences, COMSATS Institute of Information Technology, Vehari, Pakistan

Amjed Ali University College of Agriculture, University of Sargodha, Sargodha, Pakistan

Hafiz Rizwan Ali Department of Agronomy, Muhammad Nawaz Shareef University of Agriculture, Multan, Pakistan

Hakoomat Ali Department of Agronomy, Bahauddin Zakariya University, Multan, Pakistan

Muhammad Arif Ali Department of Soil Science, Bahauddin Zakariya University, Multan, Pakistan

Nawab Ali Department of Soil Science, Bahauddin Zakariya University, Multan, Pakistan

Khizer Amanet Department of Environmental Sciences, COMSATS Institute of Information Technology, Vehari, Pakistan

Asad Amin Department of Environmental Sciences, COMSATS Institute of Information Technology, Vehari, Pakistan
Queensland Alliance for Agriculture and Food Innovation (QAAFI), The University of Queensland, Brisbane, Australia

Muhammad Waqas Amjid Department of Agriculture, Bacha Khan University, Khyber Pakhtunkhwa, Pakistan

Muhammad Anees Department of Entomology, Bahauddin Zakariya University, Multan, Pakistan

Muhammad Arif Department of Agronomy, University of Agriculture Peshawar, Khyber Pakhtunkhwa, Pakistan

Muhammad Arshad Institute of Environmental Sciences and Engineering, National University of Science and Technology, Islamabad, Pakistan

Farzana Ashraf Central Cotton Research Institute, Multan, Pakistan

Farhena Aslam Department of Agronomy, Bahauddin Zakariya University, Multan, Pakistan

Muhammad Aslam Department of Agriculture (Extension Wing), Government of Punjab, Multan, Punjab, Pakistan

Ateeq-ur-Rehman Department of Plant Pathology, Bahauddin Zakariya University, Multan, Pakistan

Atique-ur-Rehman Department of Agronomy, Bahauddin Zakariya University, Multan, Pakistan

Muhammad Awais Department of Agronomy, The Islamia University, Bahawalpur, Pakistan

Salman Ayub Department of Agronomy, University of Agriculture, Faisalabad, Pakistan

Muhammad Azim Khan Department of Weed Science, Agriculture University, Peshawar, Pakistan

Adem Bardak Department of Agricultural Biotechnology, Kahramanmaras Sutcu Imam University, Kahramanmaras, Turkey

Muhammad Mohsin Alam Bhatti Department of Plant Pathology, Bahauddin Zakariya University, Multan, Pakistan

Usman Khalid Chaudhry Department of Agricultural Genetic Engineering, Ayhan Sahenk Faculty of Agricultural Sciences and Technologies, Nigde Omer Halisdemir University, Nigde, Turkey

Sobia Chohan Department of Plant Pathology, Bahauddin Zakariya University, Multan, Pakistan

Subhan Danish Department of Soil Science, Bahauddin Zakariya University, Multan, Pakistan

Shah Fahad Department of Agriculture, University of Swabi, Swabi, Pakistan
College of Plant Science and Technology, Huazhong Agricultural University, Wuhan, P.R. China

Sumrina Faiz Institute of Pure and Applied Biology, Bahauddin Zakariya University, Multan, Pakistan

Hafiz Umar Farid Department of Agricultural Engineering, Bahauddin Zakariya University, Multan, Pakistan

Omer Farooq Department of Agronomy, Bahauddin Zakariya University, Multan, Pakistan

Zartash Fatima Department of Agronomy, Bahauddin Zakariya University, Multan, Pakistan

Abdul Ghaffar Department of Agronomy, Muhammad Nawaz Shareef University of Agriculture, Multan, Pakistan

Muhammad Habib ur Rahman Department of Agronomy, Muhammad Nawaz Shareef University of Agriculture, Multan, Pakistan
Institute of Crop Science and Resource Conservation (INRES) Crop Science Group, University Bonn, Bonn, Germany

Ghulam Haider Department of Agronomy, Muhammad Nawaz Shareef University of Agriculture, Multan, Pakistan

Hafiz Mohkum Hammad Department of Environmental Sciences, COMSATS Institute of Information Technology, Vehari, Pakistan

Hafiz Abdul Haq Central Cotton Research Institute, Multan, Pakistan

Waseem Hassan Department of Soil and Environmental Sciences, Muhammad Nawaz Shareef University of Agriculture, Multan, Multan, Pakistan

Khezir Hayat Central Cotton Research Institute, Multan, Pakistan

Gerrit Hoogenboom Agricultural and Biological Engineering Department, Institute for Sustainable Food Systems (ISFS), University of Florida, Gainesville, FL, USA

Babar Hussain Faculty of Life Sciences, University of Central Punjab, Lahore, Pakistan
Faculty of Engineering and Natural Sciences, Sabanci University, Istanbul, Turkey

Mubshar Hussain Department of Agronomy, Bahauddin Zakariya University, Multan, Pakistan
School of Veterinary and Life Sciences, Murdoch University, Murdoch, WA, Australia

Nazim Hussain Department of Agronomy, Bahauddin Zakariya University, Multan, Pakistan

Sajjad Hussain Department of Horticulture, Bahauddin Zakariya University, Multan, Pakistan

Sajjad Hussain Department of Environmental Sciences, COMSATS Institute of Information Technology, Vehari, Pakistan

Fariha Ilyas Department of Soil Science, Bahauddin Zakariya University, Multan, Pakistan

Hafiz Muhammad Imran Central Cotton Research Institute, Multan, Pakistan

Muhammad Imran Department of Environmental Sciences, COMSATS Institute of Information Technology, Vehari, Pakistan

Danish Iqbal Fibre Technology Section, Central Cotton Research Institute, Multan, Pakistan

Pakeeza Iqbal Department of Botany, University of Agriculture, Faisalabad, Pakistan

Khawar Jabran Department of Plant Production and Technologies, Faculty of Agricultural Sciences and Technologies, Niğde Ömer Halisdemir University, Niğde, Turkey

Muhammad Kamran Department of Agronomy, University of Agriculture, Faisalabad, Pakistan

Abdul Khaliq Department of Agronomy, University of Agriculture Faisalabad, Faisalabad, Pakistan

Tasneem Khaliq Agro-Climatology Lab, Department of Agronomy, University of Agriculture Faisalabad, Faisalabad, Pakistan

Amna Khan Department of Agronomy, Bahauddin Zakariya University, Multan, Pakistan
Department of Agronomy, University College of Agriculture, University of Sargodha, Sargodha, Pakistan

Azhar Abbas Khan College of Agriculture, Bahauddin Zakariya University, Layyah, Pakistan

Muhammad Azam Khan In-Service Agricultural Training Institute, Sargodha, Pakistan

Seema Mahmood Institute of Pure and Applied Biology, Bahauddin Zakariya University, Multan, Pakistan

Sultan Mahmood Department of Plant Breeding and Genetics, Bahauddin Zakariya University, Multan, Pakistan

Waqas Malik Department of Plant Breeding and Genetics, Bahauddin Zakariya University, Multan, Pakistan

Amar Matloob Department of Agronomy, MNS-University of Agriculture, Multan, Pakistan

Zahid Mehmood Central Cotton Research Institute, Multan, Pakistan

Muhammad Azam Mian Central Cotton Research Institute, Multan, Pakistan

Noor Mohammad Central Cotton Research Institute Multan, Multan, Pakistan

Khuram Mubeen Muhammad Nawaz Shareef University of Agriculture Multan, Multan, Punjab, Pakistan

Muhammad Mubeen Department of Environmental Sciences, COMSATS Institute of Information Technology, Vehari, Pakistan

Wali Muhammad Agriculture Pest Warning & Quality Control of Pesticides, Government of Punjab, Layyah, Pakistan

Hassan Munir Department of Agronomy, University of Agriculture, Faisalabad, Pakistan

Ghulam Mustafa Department of Soil Science, Bahauddin Zakariya University, Multan, Pakistan

Kamrun Nahar Department of Agricultural Botany, Sher-e-Bangla Agricultural University, Dhaka, Bangladesh

Syed Atif Hasan Naqvi Department of Plant Pathology, Bahauddin Zakariya University, Multan, Pakistan

Wajid Nasim Department of Agronomy, University College of Agriculture and Environmental Sciences, The Islamia University of Bahawalpur (IUB), Bahawalpur, Punjab, Pakistan

Ahmad Nawaz College of Agriculture, BZU, Layyah, Pakistan

Sibgha Noreen Institute of Pure and Applied Biology, Bahauddin Zakariya University, Multan, Pakistan

Dony Parlak Kahramanmaras Sutcu Imam University, Kahramanmaras, Turkey

Asia Perveen Central Cotton Research Institute Multan, Multan, Pakistan

Rashida Perveen Department of Plant Pathology, Bahauddin Zakariya University, Multan, Pakistan

Abdul Qayyum Department of Plant Breeding and Genetics, Bahauddin Zakariya University, Multan, Pakistan

Fahd Rasul Department of Agronomy, University of Agriculture, Faisalabad, Pakistan

Ghulam Rasul Pakistan Meteorological Department, Islamabad, Pakistan
International Center for Integrated Mountain Development, Kathmandu, Nepal

Haseeb Ur Rehman Department of Agronomy, Bahauddin Zakariya University, Multan, Pakistan

Muhammad Sajid Department of Plant Pathology, Bahauddin Zakariya University, Multan, Pakistan

Muhammad Sajid Department of Agronomy, University of Agriculture, Faisalabad, Pakistan

Asif Sajjad Department of Entomology, University College of Agriculture and Environmental Sciences, The Islamia University of Bahawalpur, Bahawalpur, Pakistan

Muhammad Asif Saleem Department of Plant Breeding and Genetics, Bahauddin Zakariya University, Multan, Pakistan

Muhammad Sarfraz Soil Salinity Research Institute, Pindi Bhattian, Pakistan

Muhammad Ilyas Sarwar Fibre Technology Section, Central Cotton Research Institute, Multan, Pakistan

Naeem Sarwar Department of Agronomy, Bahauddin Zakariya University, Multan, Pakistan

Sarfraz Ali Shad Department of Entomology, Bahauddin Zakariya University, Multan, Pakistan

Khurram Shahzad Department of Soil Science, Bahauddin Zakariya University, Multan, Pakistan

Muhammad Nouman Tahir Department of Plant Pathology, Bahauddin Zakariya University, Multan, Pakistan

Muhammad Tayyab Tahir Department of Agri. Business and Marketing, Bahauddin Zakariya University, Multan, Pakistan

Muhammad Tariq Central Cotton Research Institute, Multan, Pakistan

Ahmet Uludag Plant Protection Department, Faculty of Agriculture, Canakkale Onsekiz Mart University, Canakkale, Turkey

Ummad-ud din Umar Department of Plant Pathology, Bahauddin Zakariya University, Multan, Pakistan

Abdul Wahid Department of Environmental Sciences, Bahauddin Zakariya University, Multan, Pakistan

Aftab Wajid Agro-Climatology Lab, Department of Agronomy, University of Agriculture Faisalabad, Faisalabad, Pakistan

Azra Yasmeen Department of Agronomy, Bahauddin Zakariya University, Multan, Pakistan

Iqra Zakir Department of Agronomy, Bahauddin Zakariya University, Multan, Pakistan

Chapter 1
World Cotton Production and Consumption: An Overview

Muhammad Azam Khan, Abdul Wahid, Maqsood Ahmad, Muhammad Tayab Tahir, Mukhtar Ahmed, Shakeel Ahmad, and Mirza Hasanuzzaman

Abstract Agriculture contributes mainly to national economies specifically in developing countries, and cotton is an important cash crop. In certain countries, it is recognized as "white gold" since it is earning foreign exchange. In the world, cotton fiber is a distinguished fiber that serves as a raw material for textile industries having a yearly significant economic impact of at least $600 billion. Genetic diversity and its usage in getting sustainability of lint cotton and cotton yield, and usage of bio-based substitute such as procession and change in various biochemical, physiological, morphological and genetically significant traits. Nearly 25 M tons of total cotton is produced worldwide annually. Top ten cotton-producing countries are India, China, the United States, Pakistan, Brazil, Australia, Uzbekistan, Turkey, Turkmenistan, and Burkina Faso.

Keywords *Gossypium hirsutum* · Productivity · Consumption · Export · Import

The word cotton, derived from the Arabic word "quotn" (Lee and Fang 2015), has a place with *Gossypium* variety, which was additionally derived from the Arabic word "goz" (Gledhill 2008), meaning a delicate material. Cotton is a delicate, soft staple

M. A. Khan
In-Service Agricultural Training Institute, Sargodha, Pakistan

A. Wahid · M. Ahmad
Department of Environmental Sciences, Bahauddin Zakariya University, Multan, Pakistan

M. T. Tahir
Department of Agri. Business and Marketing, Bahauddin Zakariya University, Multan, Pakistan

M. Ahmed
Department of Agronomy, PMAS Arid Agriculture University, Rawalpindi, Pakistan

S. Ahmad (✉)
Department of Agronomy, Bahauddin Zakariya University, Multan, Pakistan
e-mail: shakeelahmad@bzu.edu.pk

M. Hasanuzzaman
Department of Agronomy, Sher-e-BanglaAgricultural University, Dhaka, Bangladesh

S. Ahmad, M. Hasanuzzaman (eds.), *Cotton Production and Uses*,
https://doi.org/10.1007/978-981-15-1472-2_1

fiber that develops in a boll or defensive case, around the seeds of the plants of the genus *Gossypium* in the mallow family Malvaceae (Cobley 1956). Cotton species *G. arboretum* and *G. herbaceum* were previously used as shrubs (Iqbal et al. 2001).

Agriculture is the main contributor to most of the country's economy especially in developing countries, and cotton is one of the important crops in agriculture (Ahmad et al. 2014, 2017; Ahmad and Raza 2014; Abbas and Ahmad 2018). In some countries, it is known as "white gold" because it is producing so much revenue (Ali et al. 2011, 2013a, 2014a). Cotton is the world's best preeminent fiber and natural crop extending one of the biggest textile industries having a yearly economic impact of at least $600 billion worldwide. Genetic diversity and its usage in getting sustainability of lint cotton and cotton yield, and usage of bio-based substitute such as procession and change in various biochemical, physiological, morphological and genetically significant traits (Tariq et al. 2017; Amin et al. 2017). Enormous event in narrow and broad genetic base of cotton cultivars. It is the most widely used fiber in every cloth we can think of. About 25 million tons of total cotton is produced worldwide per year, and its worth is about 12 billion dollars. Cotton plant requires plenty of sunshine and 60–120 cm rain (Khan et al. 2004; Usman et al. 2009; Rahman et al. 2018). Due to genetic engineering, different varieties of cotton have been developed like *Bacillus thuringiensis* (Bt) cotton, which resulted in dramatic increase in cotton production (The Daily Records January 2, 2019) (Sawan 2018). Around the globe, more than 100 countries (Fig. 1.1) are producing cotton, and total

Fig. 1.1 Cotton-producing areas around the globe. (Source: https://www.picswe.com/pics/world-cotton-e1.html)

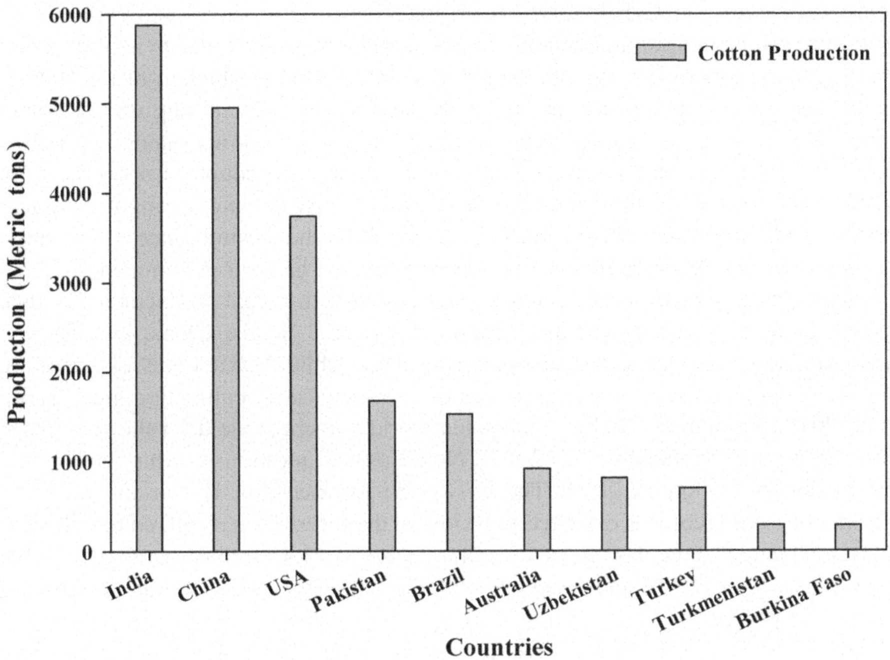

Fig. 1.2 Top ten cotton-producing countries around the globe. (Source: The Daily Records 02 January 2019)

worldwide yearly planted area is 33 M ha for the production during the year 2014 (Bremen Cotton Exchange 2014). Among these countries, the top ten cotton-producing countries are India, China, the United States, Pakistan, Brazil, Australia, Uzbekistan, Turkey, Turkmenistan, and Burkina Faso, and their per year production is given in Fig. 1.2. Although India is at number one and is producing 26% of the world's total cotton, its yield per acre is very low. In China, cotton planting is in 24 provinces out of 35, it is primary crop of China, and 99.5% of total cultivated area has been used for cotton plantation. The United States of America is leading in cotton exports; its major portion of the cotton production is in southern states including Mississippi, Louisiana, and Arkansas. Pakistan is also a major cotton-producing and -consuming country. Indus Valley Civilization is the place where the oldest cotton plantation is traced so far (Ahmad et al. 2018; Ali et al. 2013b, 2014b). Fifteen percent of the country's land is used to grow cotton. In Brazil, a major portion of the cultivated land is used for cotton farming, and the country is the fourth largest exporter of cotton in the world. Australia is using 100% local seed technology, and the country is the third largest exporter of cotton products, and the country is known for contaminant-free cotton, having good fiber length and color. The fifth largest exporter of cotton product is Uzbekistan. They are exporting 17% of their total production, but Uzbek cotton is facing issues due to slaves and child labor. Turkey is producing premium quality of cotton in the world, but due to raging war in the region, export of the product dented, and since 2002 production of the "white gold" was facing steady decline; but now production is increasing due to some remedial measures (using technology and quality seeds). Turkmenistan is at number nine among the top ten producers. Cotton production in the country has declined up to 50% due to scarcity of water. Burkina Faso is at last number among the top ten cotton-producing countries. Production is gradually increasing since 1980; only during 2017–2018, about 20% increase was estimated by their government.

As major cotton-producing countries were increased, global production was also raised about 14% during the year 2018. The United States, China, and Turkey approximately were expected to rise up to 20%, while Mexico was expected to double. This increase in production was due to increase in cultivation area (Tariq et al. 2018; Amin et al. 2018). Anticipated world's average yield for the year 2018 was 792 kg ha^{-1}(Johnson et al. 2013). Now the key question is: what will be the production of cotton in 2019. The answer to the question is that the price of competitive crop (corn) is expected to be low in the future; as a result, cotton planted area is expected to increase from 14 million acres to 14.45 million acres during the year 2019 (2019/2020 Fundamentals, Outlook, and Caveats), so rise in production graph is expected.

Historical data of the world about production, export, and import of cotton from the year 1995–1996 to 2017–2018 is given in Table 1.1. It shows that although increase in production year to year is not constant, there is an overall 75.86% increase during the last 23 years, and same trend can be seen in trade (import and export). It is worth mentioning that the use of cotton is 70.10% higher in 2017–2018 than the year 1995–1996 (Table 1.1).

According to a survey, worldwide purchases of cotton during the year 2017–2018 were US$49.9 billion. From a region prospective, two third (65.5%) of global cotton was imported by Asian countries. The remaining portion was purchased by Europe (16%), Africa (7.8%), Latin America including Caribbean but excluding Mexico (6.1%), North America (4.1%), and Oceania (16%), and the rest was purchased by Australia and New Zealand.

Among the top 15 importer countries, China is at number one; it imported cotton of worth US$8.6 billion which was 17.3% of total cotton import globally. Bangladesh was at number two with US$5.3 billion (10.7%), Vietnam was at number three with US$4.2, Turkey was at the fourth position with import of US$3 billion, Indonesia was at the fifth position with US$2.1 billion, Hong Kong was at the sixth position with US$1.5 billion, Italy was at seventh with US$1.3, South Korea was at the eighth position with US$1.2 billion, Germany was at ninth, and Mexico was at the tenth position with import of US$1 billion; India imports cotton of US$991.4 million and was at the eleventh position, Pakistan is importing cotton of worth US$975 and was at the twelfth position, the United States was at the thirteenth position with import of cotton worth US$940.6, Thailand was at the fourteenth

Table 1.1 Historical data of the world about production, export, and import of cotton from the year 1995–1996 to 2017–2018

Million 480 lb. bales

Marketing year	Beginning stocks	Production	Imports	Mill use	Exports	Ending stocks	S-U-Rn ratio (%)
1995–1996	32.02	93.90	27.00	85.94	27.40	40.14	46.7
1996–1997	40.14	90.05	28.58	87.94	26.78	44.64	50.8
1997–1998	44.64	92.37	25.93	87.27	26.78	49.47	56.7
1998–1999	49.47	86.07	24.48	84.77	23.52	52.86	62.4
1999–2000	52.86	87.91	27.99	91.09	27.13	51.14	56.1
2000–2001	51.14	89.09	26.21	92.15	26.16	49.56	53.8
2001–2002	49.56	98.50	29.30	94.38	29.08	54.68	57.9
2002–2003	54.68	91.02	30.19	98.41	30.40	47.88	48.7
2003–2004	47.88	96.68	34.15	98.09	33.15	48.38	49.3
2004–2005	48.38	121.55	33.97	109.21	34.95	60.98	55.8
2005–2006	60.98	116.36	44.67	116.97	44.92	61.91	52.9
2006–2007	61.91	122.69	38.31	124.21	37.42	62.99	50.7
2007–2008	62.99	120.05	39.45	123.84	38.87	61.88	50.0
2008–2009	61.88	108.07	30.57	110.30	30.21	61.45	55.7
2009–2010	61.45	103.08	36.93	119.49	35.80	46.18	38.6
2010–2011	46.18	117.30	36.30	115.49	34.90	49.26	42.7
2011–2012	49.26	127.24	45.42	104.12	45.87	72.15	69.3
2012–2013	72.15	123.89	47.63	108.24	46.44	89.33	82.5
2013–2014	89.33	120.36	41.20	109.91	40.84	100.05	91.0
2014–2015	100.05	119.22	36.07	112.23	35.51	107.32	95.6
2015–2016	107.32	96.16	35.44	113.24	34.87	90.55	80.0
2016–2017	90.34	106.66	37.70	116.18	37.91	80.40	69.2
2017–2018	80.40	123.78	40.93	122.58	40.92	81.14	66.2

(Source: National Cotton Council of America)

position with US$777.7, and Honduras was at the last position (US$768.9) among the top 15 cotton importer countries. Percent contribution of the world's import of top 15 countries is given in Fig. 1.3 (World's Top Importers 2018).

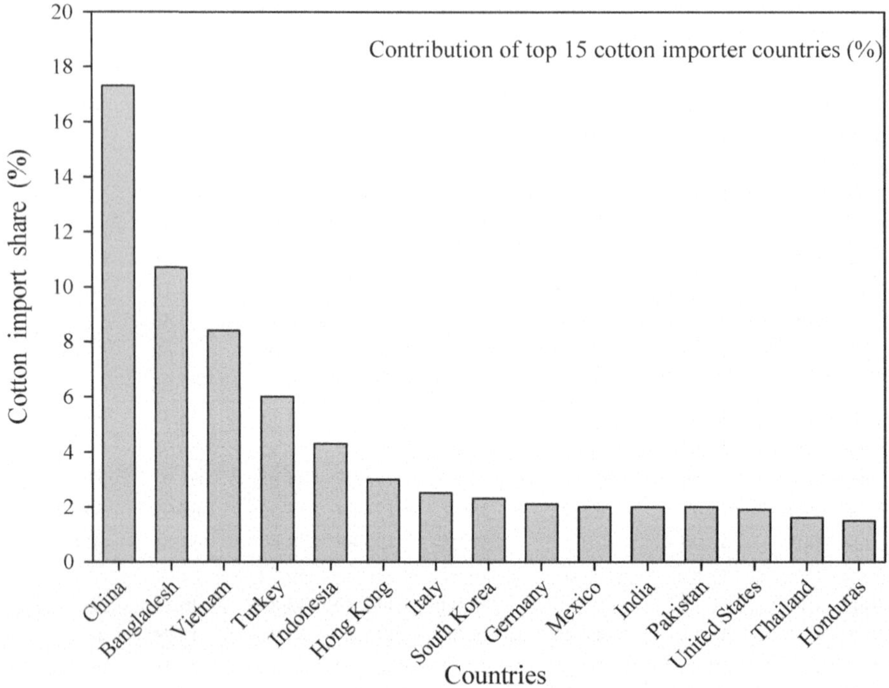

Fig. 1.3 Percent contribution of top 15 countries in world total import. (Source: World's Top Importers 2018)

References

Abbas Q, Ahmad S (2018) Effect of different sowing times and cultivars on cotton fiber quality under stable cotton-wheat cropping system in southern Punjab, Pakistan. Pak J Life Soc Sci 16:77–84

Ahmad S, Raza I (2014) Optimization of management practices to improve cotton fiber quality under irrigated arid environment. J Food Agri Environ 2(2):609–613

Ahmad S, Raza I, Ali H, Shahzad AN, Atiq-ur-Rehman, Sarwar N (2014) Response of cotton crop to exogenous application of glycinebetaine under sufficient and scarce water conditions. Braz J Bot 37(4):407–415

Ahmad S, Abbas Q, Abbas G, Fatima Z, Atique-ur-Rehman, Naz S, Younis H, Khan RJ, Nasim W, Habib ur Rehman M, Ahmad A, Rasul G, Khan MA, Hasanuzzaman M (2017) Quantification of climate warming and crop management impacts on cotton phenology. Plan Theory 6, 7:1–16

Ahmad S, Iqbal M, Muhammad T, Mehmood A, Ahmad S, Hasanuzzaman M (2018) Cotton productivity enhanced through transplanting and early sowing. Acta Sci Biol Sci 40:e34610

Ali H, Afzal MN, Ahmad F, Ahmad S, Akhtar M, Atif R (2011) Effect of sowing dates, plant spacing and nitrogen application on growth and productivity on cotton crop. Int J Sci Eng Res 2 (9):1–6

Ali H, Abid SA, Ahmad S, Sarwar N, Arooj M, Mahmood A, Shahzad AN (2013a) Integrated weed management in cotton cultivated in the alternate-furrow planting system. J Food Agri Environ 11(3&4):1664–1669

Ali H, Abid SA, Ahmad S, Sarwar N, Arooj M, Mahmood A, Shahzad AN (2013b) Impact of integrated weed management on flat-sown cotton (*Gossypium hirsutum* L.). J Anim Plant Sci 23 (4):1185–1192

Ali H, Hameed RA, Ahmad S, Shahzad AN, Sarwar N (2014a) Efficacy of different techniques of nitrogen application on American cotton under semi-arid conditions. J Food Agri Environ 12 (1):157–160

Ali H, Hussain GS, Hussain S, Shahzad AN, Ahmad S, Javeed HMR, Sarwar N (2014b) Early sowing reduces cotton leaf curl virus occurrence and improves cotton productivity. Cer Agron Moldova XLVII(4):71–81

Amin A, Nasim W, Mubeen M, Nadeem M, Ali L, Hammad HM, Sultana SR, Jabran K, Habib ur Rehman M, Ahmad S, Awais M, Rasool A, Fahad S, Saud S, Shah AN, Ihsan Z, Ali S, Bajwa AA, Hakeem KR, Ameen A, Amanullah, Rehman HU, Alghabar F, Jatoi GH, Akram M, Khan A, Islam F, Ata-Ul-Karim ST, Rehmani MIA, Hussain S, Razaq M, Fathi A (2017) Optimizing the phosphorus use in cotton by using CSM-CROPGRO-cotton model for semi-arid climate of Vehari-Punjab, Pakistan. Environ Sci Pollut Res 24(6):5811–5823

Amin A, Nasim W, Mubeen M, Ahmad A, Nadeem M, Urich P, Fahad S, Ahmad S, Wajid A, Tabassum F, Hammad HM, Sultana SR, Anwar S, Baloch SK, Wahid A, Wilkerson CJ, Hoogenboom G (2018) Simulated CSM-CROPGRO-cotton yield under projected future climate by SimCLIM for southern Punjab, Pakistan. Agric Syst 167:213–222

Bremen Cotton Exchange (2014). https://cottonaustralia.com.au/cotton-library/fact-sheets/cotton-fact-file-the-world-cotton-market

Cobley L (1956) An introduction to the botany of tropical crops. Longman Green and Co, London

Fundamentals, Outlook, and Caveats (2019). https://cottonmarketing.tamu.edu/newcrop-fundamentals-outlook-and-caveats/

Gledhill D (2008) The names of plants. Cambridge University Press, Cambridge

Iqbal M, Reddy O, El-Zik K, Pepper A (2001) A genetic bottleneck in the 'evolution under domestication' of upland cotton *Gossypium hirsutum* L. examined using DNA fingerprinting. Theor Appl Genet 103(4):547–554

Johnson JD, Kiawu J, MacDonald S, Meyer LA, Skelly C (2013) The world and United States cotton outlook.

Khan MB, Khaliq A, Ahmad S (2004) Performance of mashbean intercropped in cotton planted in different planting patterns. J Res (Sci) 15(2):191–197

Lee JA, Fang DD (2015) Cotton as a world crop: origin, history, and current status. Cotton (agronmonogr57):1–24

Rahman MH, Ahmad A, Wang X, Wajid A, Nasim W, Hussain M, Ahmad B, Ahmad I, Ali Z, Ishaque W, Awais M, Shelia V, Ahmad S, Fahad S, Alam M, Ullah H, Hoogenboom G (2018) Multi-model projections of future climate and climate change impacts uncertainty assessment for cotton production in Pakistan. Agric For Meteorol 253-254:94–113

Sawan ZM (2018) Climatic variables: Evaporation, sunshine, relative humidity, soil and air temperature and its adverse effects on cotton production. Inform Process Agric 5(1):134–148

Tariq M, Yasmeen A, Ahmad S, Hussain N, Afzal MN, Hasanuzzaman M (2017) Shedding of fruiting structures in cotton: factors, compensation and prevention. Trop Subtrop Agroecosyst 20(2):251–262

Tariq M, Afzal MN, Muhammad D, Ahmad S, Shahzad AN, Kiran A, Wakeel A (2018) Relationship of tissue potassium content with yield and fiber quality components of Bt cotton as influenced by potassium application methods. Field Crops Res 229:37–43

The Daily Records (2019). http://www.thedailyrecords.com/2018-2019-2020-2021/world-famous-top-10-list/world/largest-cotton-producing-countries-world/12785/#1_India_5879_thousand_metric_ton

Usman M, Ahmad A, Ahmad S, Irshad M, Khaliq T, Wajid A, Hussain K, Nasim W, Chattha TM, Trethowan R, Hoogenboom G (2009) Development and application of crop water stress index for scheduling irrigation in cotton (*Gossypium hirsutum* L.) under semiarid environment. J Food Agri Environ 7(3&4):386–391

World's Top Importers (2018). http://www.worldstopexports.com/cotton-imports-by-country/

Chapter 2
Soil Management and Tillage Practices for Growing Cotton Crop

Muhammad Arif Ali, Fariha Ilyas, Subhan Danish, Ghulam Mustafa, Niaz Ahmed, Sajjad Hussain, Muhammad Arshad, and Shakeel Ahmad

Abstract For cultivation of crops, among all soil adaptive practices, tillage has been considered a fundamental crop-growing practice for centuries to clear and soften the soil. Due to changing climatic conditions and perturbation of resources, there is a need to implement soil adaptive practices and improve tillage practices to ensure security of food and guaranteed fiber production to achieve zero hunger. This chapter covers the influences of climate change and important soil adaptation and tillage practices especially for cotton crop. Our goal was to provide a framework regarding factors responsible for low cotton yield and soil adaptations that can improve cotton productivity. We attempt to highlight possible negative effects of climate change, i.e., high temperature, greenhouse gas emission, drought stress, salinity stress, insect/pest/disease attack, and primary techniques to mitigate climatic adverse effects on cotton crop. Keeping the current scenario, we suggest that advance research is still required to address the adverse effects of climate through better implementation of soil adaptations.

Keywords Tillage · Field capacity · Conventional · Water use efficiency · Soil water

Abbreviations

AFP Air-filled porosity
AW Available water

M. A. Ali (✉) · F. Ilyas · S. Danish · G. Mustafa · N. Ahmed
Department of Soil Science, Bahauddin Zakariya University, Multan, Pakistan
e-mail: arif1056@bzu.edu.pk

S. Hussain
Department of Horticulture, Bahauddin Zakariya University, Multan, Pakistan

M. Arshad
Institute of Environmental Sciences and Engineering, National University of Science and Technology, Islamabad, Pakistan

S. Ahmad (✉)
Department of Agronomy, Bahauddin Zakariya University, Multan, Pakistan
e-mail: shakeelahmad@bzu.edu.pk

S. Ahmad, M. Hasanuzzaman (eds.), *Cotton Production and Uses*,
https://doi.org/10.1007/978-981-15-1472-2_2

CT	Conventional tillage
FC	Field capacity
GHGs	Greenhouse gases
ICAC	International Cotton Advisory Committee
ISR	Induced systemic resistance
N	Nitrogen
NT	No tillage
NUE	Nitrogen use efficiency
P	Phosphorus
PWP	Permanent wilting point
SEEP	Social, Environmental and Economic Performance
WUE	Water use efficiency
WHC	Water holding capacity

2.1 Introduction

Cotton (*Gossypium*) is a tropical shrub and belongs to Malvaceae family which is cultivated worldwide especially from north of latitude 30°N, including the USA, the Union of Soviet Socialist Republics, and China being the major producers of cotton. Outside the tropical belt, cotton can only be grown during the summer season (Tariq et al. 2017, 2018). The cotton plant has erect branching stems, alternate leaves on it, and large and showy flowers with five petals mainly white or cream purplish. Cotton has a tap root system and can grow up to 60 cm depth with adequate moisture and good soil conditions. The fruit is like a capsule with three to five leathery valves. Ovoid shaped cotton seed is embodied in fruit with a coating of long hair like threads/fibers. Seeds of cotton can have dimensions as 10 × 6 mm and up to 80 mg of weight. It has a hard seed coat covered by cuticles. There are 50 known species of cotton discovered, among which only 4 are cultivated globally and the remaining species grow wild in tropical and sub-tropical areas (Gotmare et al. 2000). The four common cultivated cotton species are *Gossypium hirsutum*, *G. herbaceum*, *G. barbadense*, and *G. arboreum*. The species are different from their fiber length, maturity, strength, and micronaire. Climatic condition and crop genetic makeup of certain species proved some species suitable over the others, for a specific area. *G. hirsutum* species produce about 90% of total cotton and have high-quality cotton (Brown 2002; Liu et al. 2013). Thus, it is widely grown due to higher productivity and wide adaptability in various agro-environmental conditions (Avci et al. 2013). However, high temperature can cause sterility and boll shedding of cotton (Sawan et al. 2002) in certain regions (Amin et al. 2017, 2018). In Asia and Africa, other species of cotton are often predominant (Wu et al. 2005) because these species are unable to adapt variations in climatic conditions and low yields (Avci et al. 2013; Amin et al. 2017, 2018).

The seed cotton is the most demanded part of the plant and is used in various industries as a raw material like textiles, edible oil, paper, and animal feed, besides medicinal products (Hegde et al. 2004; Aluri et al. 2008; Ezuruike and Prieto 2014). The fiber of cotton is used in many products because of its numerous positive characteristics like comfort, color retention, absorbency, and strength (Hegde et al. 2004). In 2013–2014, cotton global cultivation yield has increased to estimated production of over 23 M tons. The entire plant is a source of some important compounds including terpenes, phenolics, fatty acids, lipids, and carbohydrates, besides proteins (Bell 1986; Hu et al. 2011; Essien et al. 2011). Site-specific and crop-specific management is necessary for obtaining maximum economic profit and quality within limited resources (Ali et al. 2011, 2013a, b, 2014a, b). Site management includes management of soil, water, tillage, insect and pest attack, and soil's physical and chemical properties (Fig. 2.1). The aim of these managements is to get precision, increase profitability and crop yield and productivity, use environment-friendly approaches, and sustain water-plant-soil relationship (Atherton et al. 1999; Ahmad et al. 2014, 2017, 2018; Ahmad and Raza 2014; Abbas and Ahmad 2018). Topography is one of the important factors that must be kept under consideration. Land should not be completely leveled, but it has to be smooth, i.e., there is no gully formation, so that mechanization should be performed easily. The crop is recommended to be cultivated in rows for hassle-free mechanical operations. Over

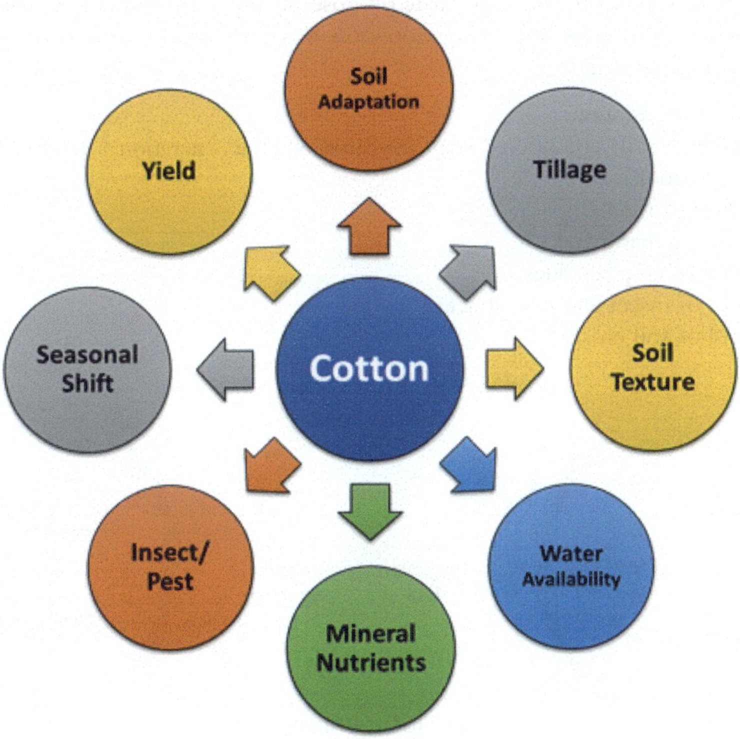

Fig. 2.1 Factors affecting cotton growth and yield

a period of time, soil management and cultural practices have remained to change for better productivity (Khan et al. 2004; Usman et al. 2009; Rahman et al. 2018). Changes or adaptations thus brought for better productivity of cotton belonged to soil adaptations directly or indirectly. Other adaptations include on-farm management practices (insects, pests, water requirements, cropping season, and pattern) legislative or combination of many approaches. In this chapter, some of the practices are explained for better crop productivity.

2.2 Soil Adaptations and Tillage Practices for Growing Cotton

Soil tilling is one of the most fundamental phenomena in crop husbandry. In order to prepare a good seedbed for the sowing of cotton, some kind of tillage is an essential practice (Figs. 2.2 and 2.3). History of tillage practice is as old as human's life. Tillage practices started from hand pulling to scratching, to log dragging, to animal-drawn plows, to modern steel plows using tractors and machines with different strength of tillage. Nowadays, there is a vast variety of tillage instruments and tractors with varying power. The basic purpose of tilling the soil is to boost up the natural soil condition for improving crop growth. Tillage practices are performed for the following purpose:

- Seedbed preparation.
- Improving soil conditions (e.g., infiltration rate, aeration, organic matter decomposition).
- Removal of weeds.
- Breaking of hardpans.
- Burying of crop residues.
- Control of insect and pest attacks.
- Control of soil erosion.

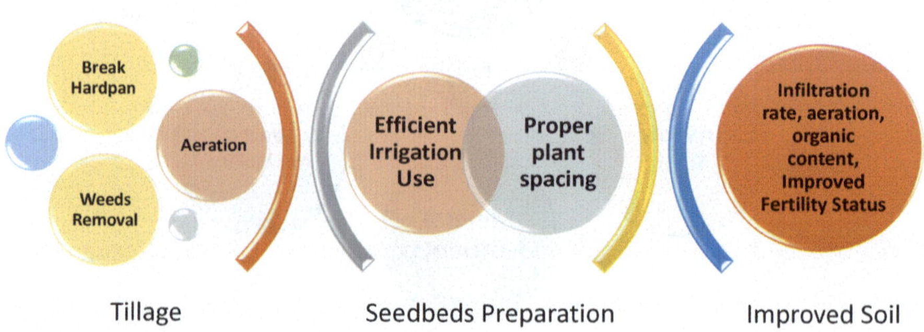

Fig. 2.2 Potential benefits of tillage and seedbed preparation for soil and better cotton yield

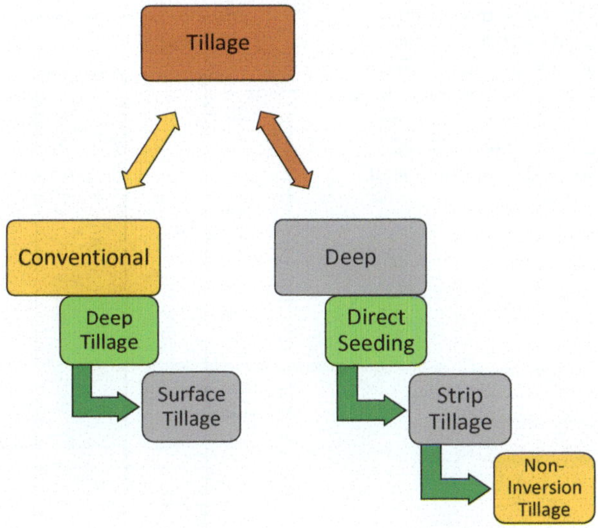

Fig. 2.3 Tillage practices classification for cotton

Tillage does not control plant growth directly; it imparts an effect on soil moisture, temperature, aeration, organic carbon content, bulk density, and structure which directly effects on plant growth (Bama et al. 2017). Excessive tillage speeds up the removal of soil moisture by exposing the soil pores into the atmosphere, enhancing aeration, and disrupting macroaggregates, whereas reduced tillage saves the moisture (Shu et al. 2015; Kabiri et al. 2015). Tilling the soil during spring season imparts a heating effect on the soil as it removes the weeds, which shaded the soil and break hardpans or compactness of soil which allow the exchange of gases in between pedosphere and atmosphere. Bulk density is reduced by tillage practices up to <1 g cm^{-3}, which is optimum for efficient seeding operation. The bulk density of soil also has a direct impact on soil aeration, temperature, moisture retention, and erosion. Disturbing the surface soil rapidly changes the bulk density of soil temporarily and sometimes permanently. Tillage intensity varies from the minimum or no tillage to conservation tillage. The CT alters the physical properties of soil by exposing more organic matter for microbial attacks, whereas minimum or no tillage enhances the formation of macroaggregates; both phenomena improve the organic matter use efficiency (Yang et al. 2005) (Fig. 2.4). Organic matter decomposition improves soil structure which alternatively improves water retention in macroaggregates (Tisdall and Oades 1982). These irregular macroaggregates are loosely held together, producing macropores. These pores ameliorate the soil structure and improve crop growth by affecting water infiltration rate and enhance drainage and rapid gaseous exchange between pedosphere and atmosphere (Magdoff and Van Es 2000). Whereas, soil organic matter also influences on chemical properties of soils as its effects on nutrients availability and pH. Nutrients are released on mineralization/decomposition of organic matter in available forms for plant absorption. The decomposition rate of organic matter is more in a tropical climate than temperate

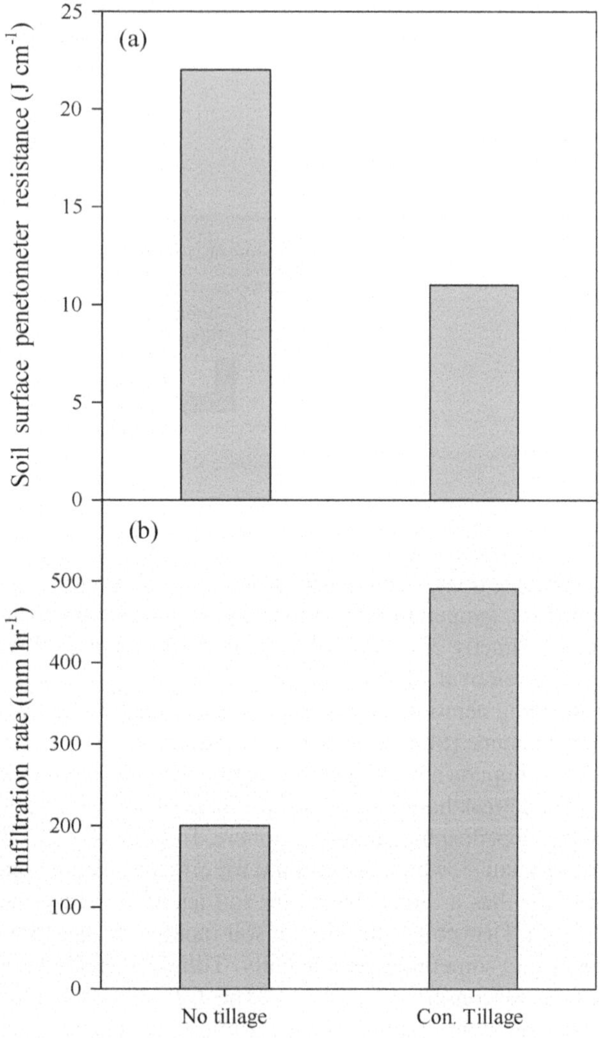

Fig. 2.4 Comparison of penetrometer resistance (**a**) and infiltration rate (**b**) in conventional and no tillage practices with cotton crop. (Source: Modified and adapted from Foster et al. 2018)

region because of more microbial activity in the warmer region. Type of tillage used depends upon the soil type, crop to be grown, and climate of that region. Some features of tillage can be improved by excessive tillage but some can be destroyed by this. So, there is a need of less or minimum tillage for the sustainable farming system (Iqbal 2005). Tillage can be conventional, conservation, and no till. Within the United Kingdom, current CT systems are comprised of both primary and secondary cultivations. Primary tillage involves inverting soil employing mouldboard plow (Gajri et al. 2002). Secondary tillage is comprised of using a single or double pass of a cultivator to produce a fine tilth or seedbed for sowing (Bell 1996).

2.2.1 Conventional Tillage (CT)

The CT is a dominant practice across the world; however, its long-term feasibility is not researched. It is defined as reduced tillage which left 30% residual crops on surface (Baker et al. 2002). This technique develops a good seedbed by burying all crop/plant residues, weeds, and pest giving the crop ideal conditions for seed germination and root penetration (Gajri et al. 2002). The soil is usually plowed with disc after harvesting (22 cm depth) and sweep-plowed (50 cm sweep) in early winter, rows are bedded during winter (15 cm height), and the seedbed is prepared with a bed topper (Foster et al. 2018). Conventional cultivation is mostly practiced on soils with drainage issues, e.g., poor structure of the soil (Fig. 2.4). Farmers often routinely use this practice for better yield security, ideal seedbed preparation, poor tillage conditions, and for ease of drilling. However, plowing often causes problems like soil erosion (El Titi 2003) and soil compaction besides lower work rates. The soil is exposed to rainfall and high wind, thus accelerating the erosion aside from soil degradation processes that reduced crop yield along with the quality of soil (Lal et al. 2007; Abdullah 2014).

2.2.2 Minimum Tillage

In minimum tillage practice, there are only 35% residues left on the surface. In no tillage (NT), there is much less disturbance of soil and crop residues by the use of the crop. Heavy machinery causes compaction problems in the subsoil; a subsoiler is often practiced in the United Kingdom for softening the compacted soil (Batey 2009). In non-inversion tillage, the soil is not inverted; instead, tine besides disc implements are used, which lift and mix straw residues and break the clods to leave a fine tilth (Christian 1994). The seedbed is not prepared in no tillage technique; the soil is only disturbed during crop sowing and harvesting with tines or disc (Foster et al. 2018). They summarized that no tillage proves efficient during drought conditions as it has the ability to retain moisture. They suggested that NT system is beneficial for cotton-sorghum rotation in an arid and semiarid region. Cotton produces the same yield under both tillage systems. However, NT provides greater cotton yield during drought as compared to CT and proves economical for dryland farming. Bulk density of both treatments remained the same for 0–15 cm depth, but it was high in CT as compared to NT at 15–30 cm depth.

Strip tillage consists of planting cotton on 40 in. row spacing. This technique includes strip operation apart from planting into surface residues from previous crop besides cover crop (Torbert et al. 2015). It has several benefits over CT as it improves physical properties and biodiversity in soil (Reeder 2000). It conserves soil moisture by using crop residue which protects the soil and prevents erosion besides increasing environmental benefits for soil wildlife (Reeder 2002).

2.3 Soil Adaptations in Relation to Soil Texture and Water Adaptability

Globally, freshwater resources are becoming insufficient due to increasing population. In the agriculture sector, high priority must be given to the lesser use of water resources and the efficient use of water by enhancing crop productivity (Jalota et al. 2006). In most of the developing countries like Pakistan, India, etc., groundwater is the main source for irrigation which is declining at a faster rate during the past few years (Hira et al. 2004). Irrigation water is one of the important limiting factors for crops of summer season or in tropical areas. Cotton (*Gossypium hirsutum* L.) crop requires deficit irrigation (lesser number of irrigations) as compared to wheat. At present farmers grow cotton-wheat in only 10% of total cultivated area with four to five irrigations to both crops through border method of 75 mm depth per irrigation. Knowledge about irrigation requirement at a specific stage of plant growth is essential for aiming to develop deficit irrigation techniques (Yazar et al. 2002a). Moisture stress during the growing season reduced lint yield. Cotton lint yield is dependent on production besides retention of bolls which are reduced by water stress (Yazar et al. 2002b).

For cotton topography of the area is an essential factor that has to be taken under consideration. In rain-fed areas, land must be formed in a way which allows proper drainage of excess water. In the case of irrigated or insufficient rainfall areas, there is some slope in land which must provide inlet and outlet channels. Over- and under-irrigation prove detrimental to the cotton plant, but different varieties respond differently on different soils.

Table 2.1 indicates field capacity (FC) of soils and water available in different soils, for the growth of plants. Sandy loam soils contain 65 mm available water per feet depth of soil, and clayey soils have 49 mm available water per feet depth of soil. These values provide us with the idea that clayey soil requires more water for wetting up as compared to sandy soils and sandy soils retain less plant available water as compared to clayey soils. In mid-summer irrigation water must be 4 in./102 mm water depth in each irrigation for best and desired cotton yield. Irrigation must be applied after a small interval of time because of fast leaching property of sandy soils. Cotton in sandy soils shows an active growth after irrigation and suffers from

Table 2.1 Textural properties of soil which impact on water availability to plants

Soil texture	Field capacity (%)	PWP (%)	Available water (mm/ft.)
Sand	6.7	1.8	20
Fine sandy loam	25.6	1.3	65
Loamy	32.2	1	63
Silt loam	35	1	71.6
Silt clay loam	31.4	1.1	48
Clay loamy	30	1	42
Clay	39.4	22.1	49

Source: Salter and Williams (1965)

drought stress in the next week. Fine-textured soils' physical conditions often limit root penetration and thus effect on water translocation due to the development of hardpans. Cotton roots become unable to take advantage of high WHC of fine-textured soils. Such soils required deep tillage for breaking of hardpans below the surface of soils (Longenecker and Eire 1966).

Fine-textured soils contain more available water as compared to sandy- or coarse-textured soils, but water molecules are strictly adsorbed on clayey particles; due to this reason, roots required high energy for absorbing this water. For obtaining this water plants must have to attain higher osmotic potential which takes several weeks or months during which these plants exhibit symptom of moisture stress. This situation is not true for sandy- or coarse-textured soils. The following practices can be applied to conserve the soil and moisture (Delgado et al. 2011).

- Mulching from crop residues protects the soil from erosion and increases susceptibility.
- Addition of organic matter and green manures enhances C content of soil. Soil C sequestration is a healthy cycle for soil health. On average increase in 1% of soil organic C (10 g C/kg soil) increases moisture content at saturation, FC, permanent wilting point (PWP), as well as available water (AW) of soil at 2.95, 1.61, 0.17, and 1.16 mm H_2O 100 mm/kg soil, respectively. Sandy soils are affected to a much greater extent as compared to loams and clayey soil (Minasny and Mcbratney 2018).
- Harvesting of plant residues should be avoided as it aids in maintaining soil moisture and organic matter. It also enhances the microbial biomass of soil and dehydrogenase activity and high carbon stock (Bama et al. 2017).
- Grasses can contribute to C sequestration besides better protect environment than arable crops used for energy. These effects are likely to vary greatly by region and by field within field, besides type of crop (Bakshi et al. 2015).
- Use of agricultural management practices such as ridge-furrow planting technique proves beneficial for improving soil moisture in silty loam soils (Sun et al. 2017).

2.4 Soil Adaptation in Relation to Mineral Nutrient Status

At the farm level, global mineral resources and environmental hazards (by nutrient losses) are usually ignored. However, nutrient management is an essential component in sustainable crop production. It not only effects on crop productivity but also has a strong impact on water and soil management (Pietrzak 2013). Many techniques have been developed to improve nutrient management especially N, P, and K management and to minimize the nutrient losses according to the chemistry of the soil. Soil nutrients can be managed by organic amendments and inorganic chemical fertilizers. Application of chemical fertilizers resulted in reducing microbial richness and causes an imbalance of soil microbial biomass (Sun et al. 2015). Multiple crops have varying rates at which they uptake nutrients and have differences in ways by which they distribute element spatially within the plant. Few ions have an intense effect on other ions' mobility and on the distribution within certain plant organs, e.g., sodium ion effect on calcium ion.

Animal manure (including cow, sheep, and poultry manure, etc.) and green manure contain a high ratio of nutrients including N, P, and K and other macro- and micro- nutrients and, due to their safe nature, are widely used in the pasture and crop fertilization. Regular practice of applying manures in fields results in stabilized high productivity of cotton crop (Blaise et al. 2006). Use of inorganic fertilizer for fulfilling the nutrient demand of crop often proves costly. Upon decomposition in soil, organic manures mineralize nutrients by microbial activity, particularly nitrogen (N) and phosphorus (P) and other micronutrients, in plant available forms. Their application to croplands is also an environment-friendly technique of disposing waste produced in large quantities (Nyakatawa et al. 2000). Mineralization or decomposition of organic matter highly depends upon C:N ratio, so the addition of chemical fertilizer often increases the mineralization rate. Although the use of chemical fertilizer degrades the natural soil environment, there is a need to develop a sustainable management system which enhances crop output. Application of manures in addition to chemical amendment significantly improves the yield of cotton (Reddy et al. 2017).

Under poor quality of irrigation water, application of zinc along with fertilizer and organic amendment proves beneficial in a cotton-wheat rotation system (Buttar et al. 2017). Uses of multiple organic amendments (FYM, vermicompost, biofertilizers, green manure) along with the chemical inorganic fertilizers enhance nutrient use efficiency and the available nutrient in the soil. It not only increases the crop yields but also increases the efficiency of added fertilizers as well as the fertility status of the soil (Verma et al. 2018)

Soil adaptation to mineral nutrients for growing cotton using tillage practices is a very important component for management of cotton crop and its productivity (Fig. 2.5). Low aeration is very common in clayey soil that is caused due to heavy and frequent irrigations, waterlogging, and soil compaction due to heavy machinery that restricts the root proliferation and optimal nutrient uptake. Multiple strategies related to modifying physical properties, applying nutrients, adding organic amendments, microbial inoculation, and/or integration of management practices for soil adaptation can affect mineral nutrients.

Soil aeration is significantly affected in potting or soil media by air-filled porosity (AFP). The acceptable range of % AFP for growing cotton crop enhances the nutrient uptake. Using suitable tillage practices in compact soil, the aeration level in the soil is improved that increases the available nutrient contents and nutrient uptake and plant nutrient uptake, and hence cotton plant growth is enhanced. Optimum soil aeration affects the soil enzymatic activities and soil microbial activities for optimal nutrient supply (Li et al. 2016). Soil conservation practices are found as better environmental protection technologies than conventional farmer's practices (Aggarwal et al. 2017). High residue in conservation tillage systems for cotton (*Gossypium hirsutum* L.) productivity has been suggested as

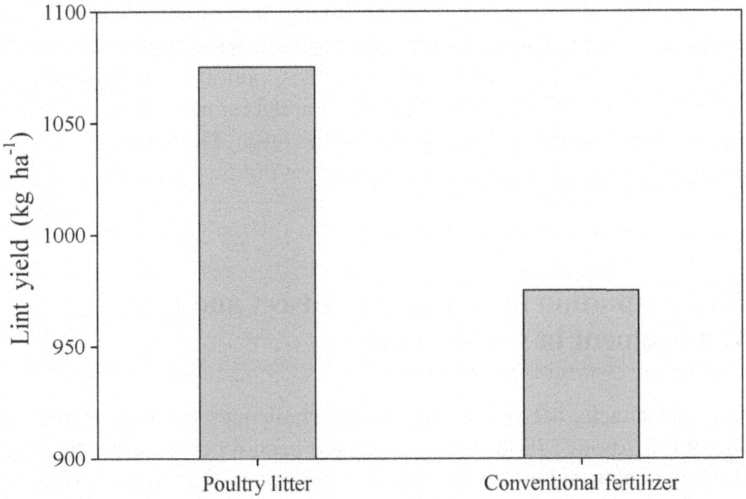

Fig. 2.5 Average lint yield obtained in kg ha^{-1} by poultry litter and fertilizer application. (Source: Modified and adapted from Endale et al. (2002))

having potential to be both economically and environmentally sustainable, as research has indicated many advantages for conservation tillage systems compared to CT (Torbert et al. 2015).

The intensive cultivation in the cotton growing area needs frequent irrigation that needs a lot of irrigation water. On the other hand, canal irrigation water is lacking. Continuous use of tube well irrigation creates salinity and fertility problems. Many kinds of cultural practices are adopted to tackle different issues of mineral nutrients; however, they are situation specific, like burying of stubbles for building up soil fertility and organic matter and application of farmyard manure for enhancing organic matter and nutrient contents and maintaining or increasing beneficial bacterial community. Soil compaction on the other hand significantly decreases cotton productivity because of its deep-rooted nature. Soil compaction can be reduced by deep plowing either annually or biannually to make nutrients available and soil suitable for cotton growth. It can also be improved by cultivating deep-rooted cover crops, which penetrate compacted soil zone besides creating channels. Such cover crops can reduce the need for annual deep tillage prior to planting and increase soil OM that provides greater water infiltration besides available WHC. Significantly reduced soil compaction, increased cotton LY besides soil moisture content, reduced nematode population densities, and increased soil available P, K, Mn, as well as OM content compared to conventional without cover crop (Marshall et al. 2016) have been observed.

Among nutrients, N is the most affected nutrient in the soil due to its speciation, mobility, and availability to plants. Nitrogen use efficiency (NUE) varies under different soil and climatic conditions. Better NUE is a better indicator of cotton yield that is possible by maintaining the optimal concentration of N in the cotton crop root zone. The NUE helps to decide fertilizer rates, optimal N uptake, and crop yield (Rochester 2011). The irrigation and fertilizer adjustments depending upon precipitation and groundwater level will be the best option to optimize nitrate in the cotton root zone (Jiao et al. 2017). Soil can be adapted for mineral nutrient deficiency especially nitrogen by intercropping pulses with cotton. Furthermore, they add a lot of organic matter that enhances water and nutrient holding capacity (Lal 2017).

2.5 Soil Adaptation in Relation to Insect and Pest Management in Cotton Crop

Insect and pest attacks are one of the major challenges throughout the cropping season of cotton. About 249 insect species have been reported so far that can cause severe damage to cotton. Different types of major insect and pest attacks include sap sucking, leaf chewing, nematodes, and diseases caused by viruses and fungi. However, their presence and intensity are variable in different areas. Similarly, soil adaptations are also varied from area to area and are pest species specific. It has been observed that intensive cultivation, lack of zoning area for cotton, cultivation of susceptible varieties, malpractices, and resistance of biotypes against pesticides are critical aspects that are contributing toward intensive damage caused by insects and pests in cotton crops. Soil temperature affects the cotton bollworm pupae. The opening of soil due to increase in soil temperature significantly decreases pupae of *Aphis gossypii*, *Phenacoccus solenopsis*, *Dysdercus cingulatus*, and *Helicoverpa armigera* by 28, 70, 29, and 50%, respectively. Hence, soil temperature played an imperative role in the reduction of pest attack on crop (Huang 2016; Hansen et al. 2011).

The role of soil health for controlling the insect and pest attack is also a very important aspect that needs attention in recent times. In healthy soils, most of beneficial organisms, i.e., microbes (e.g., root endophytic fungi, mycorrhizal fungi, plant growth-promoting fungi, and rhizobacteria rhizobia) (Bezemer and van Dam 2005; Gehring and Bennett 2009), induced systemic resistance (ISR) against diseases (Schulz 2006; Weyens et al. 2009). The ISR saves the crops against a wide range of diseases and can be initiated by a wide variety of plant growth-promoting microbes (Van Wees et al. 2008; Van der Ent et al. 2009). Plant growth-promoting microbes, i.e., *Acremonium alternatum* in cabbage (Raps and Vidal 1998) and *Bacillus pumilus* in tomato and cucumber (Zehnder et al. 1997; Disi et al. 2018), interact with aboveground insects, i.e., herbivores, natural enemies, and pollinators through plant-mediated mechanisms, often leading to negative effects on the insect herbivore. Augmentation of microbes in soil depends on type of soil, nature of soil problem, pest type, or specific area. Furthermore, there is a need for research to add such microbial communities that have potential to suppress the pests and induce positive influence on plant health (Hodson and Lewis 2016) (Fig. 2.6).

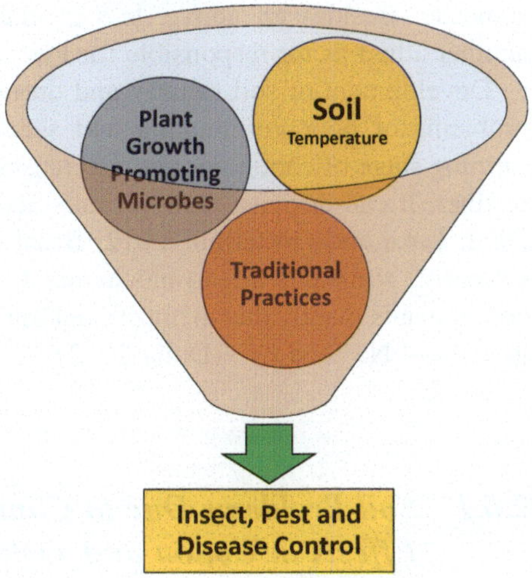

Fig. 2.6 Major soil adaptation to control insect, pest, and diseases in cotton

Many traditional methods such as application of organic materials in soil (Chau and Heong, 2005), tillage practice (Gaylor 1989), and cover cropping (Ludwig and Meyhofer 2016) can also positively control the diseases (blast and sheath blight) and insect/pest (brown plant hopper, stem borer, leaf folder) attack (Chau and Heong 2005). Moreover, there is a need for integrated soil health tests that should be validated for specific area and support the soil adaption for the insect/pests of that area.

2.6 Soil Adaptation in Relation to Climatic Changes and Seasonal Shifts

Rapid change in climatic conditions, i.e., higher temperature due to emission of greenhouse gases (GHGs), and low precipitation, as well as extensive use of groundwater, are critical factors that are negatively affecting productivity of cotton (Smith et al. 2007; Karl et al. 2009; Karmakar et al. 2016). It is expected that till 2030, the emission of GHGs through agriculture will be 40% higher as compared to current emission (Smith et al. 2007). The International Cotton Advisory Committee (ICAC) Panel on Social, Environmental and Economic Performance (SEEP 2009) estimates the GHG emissions in cotton production system ranged from 0.15 to 4 tons CO_2 eq. ha^{-1}. However, an increase in atmospheric temperature due to elevated level of GHG is a major factor that is directly involved in the reduction of cotton yield in terms of small size of cotton bolls besides less maturation (Reddy et al. 1999). Furthermore, several studies have showed that global warming due to GHG can influence the pest's metabolism as well due to which pest population rate is increased (Karl et al. 2009). Karl et al. (2009) argued that elevated temperatures of environment reduced effectiveness of many pesticides, i.e., pyrethroids and spinosad. Besides all above, less availability of good-quality irrigation water is another allied factor responsible for low yield of cotton (Karl et al. 2009).

Development of soil salinity and drought, due to high temperatures, creates the problems of poor germination and seedling growth. Drought stress at the boll-forming stage of cotton not only decreases the yield but also deteriorates the quality of fiber. It can decrease 6–60% bolls and 2–40% boll weight in cotton (Ma et al. 2002; Loka and Oosterhuis 2012; Basal et al. 2009; Lokhande and Reddy 2014). Excessive soil salt leads to biochemical, physiological, and metabolic disorders in cotton plants mainly due to nutritional imbalance, osmotic effects (dehydration), and toxicity of Na^+ and Cl^- (Dong 2012).

2.6.1 Soil Problems Due to Climate Changes and Their Effects in Cotton and Yield Decrease and Soil Adaptation

Burying straw and straw mulch increases the soil moisture enhancing soil aggregate formation rather than dispersion. It also reduces the soil salinity (Zhao et al. 2014). The NT has advantages over tillage practice for maximum carbon, nitrogen, and microbial community shifts (Mbuthia et al. 2015). By growing cotton nursery in the plug trays in various growing media and under a plastic cover in a small area to make the temperature and humidity suitable for cotton crop growth. The soilless media are rich in nutrient contents and organic matter that not only supplies suitable nutrient contents but also boosts the cotton growth.

Sustainable agricultural practices for mitigating climate change having potential for improving pest management including plant species diversification, cover cropping, tillage practices that retain crop residue, application of organic fertilizers, as well as water management practices like irrigation and sustainable crop intensification must be promoted. However, further research is needed that explicitly tests pest besides predator responses to agricultural practices under climate uncertainty if these practices are to be effectively promoted and implemented as agricultural pest management strategies (Murrell 2017). Cover crop enables the soil to adapt the soil degradation and soil dispersion by soil carbon sequestration and soil aggregation (Lal 2015). Climatic changes and seasonal drifts are continuously declining the yield of cotton; a tender crop. The crop sowing time is delayed and that leads to yield reduction. By 2050, the cotton yield is expected to decrease up to 17%. Moreover, by 2050, there will be a need of 50% more irrigations to get optimum moisture (Williams et al. 2015). Climate change has direct or indirect effects on soil microbes and plants' geographical movement (Classen et al. 2015). Climate change affects the extracellular enzymatic activities that affect the SOM turnover and fate of ecosystems (Puissant et al. 2015). Water use efficiency (WUE) changes with climate change (Luo et al. 2017). Elevated CO_2 and temp affect nutrient uptake and hence affect cotton production. Soil responses to climate change should be considered while developing a strategy for climate change (Osanai et al. 2017). Soil conservation practices play a significant role in soil adaptations to environmental conditions and mitigating seasonal shifts and changing fertilizers, irrigation practices, and crop diversification of crop varieties (Abid et al. 2015). Pakistan ranks 12th in the world with respect to global climate change and severe conditions (Kreft et al. 2013). The critical temperature for cotton growth is 28 °C (Reddy et al. 1991). Critical temperature at which cotton reproductive stage stops is 35 °C (Hatfield and Prueger 2015). Planting cotton, 6 weeks earlier than normal season is a better strategy to help the soil to adapt the climate change and enhance cotton yield (Anapalli et al. 2016). Prediction of water demand helps the soil to adapt to the climate change scenarios (Yang et al. 2014). Black carbon increases cation exchange capacity of soil (Liang et al. 2006). Climate change affects the soil erosion that can be managed by soil conservation using suitable crop rotations and decreasing soil loss by continuously changing the soil conservation strategies (O'Neal et al. 2005). Climate change include the elevation of temperature and CO_2 levels (Osanai et al. 2017). Using climate-smart agriculture, the carbon sequestration and reducing greenhouse gas emissions (Thierfelder et al. 2017) could be a great strategy for sustaining agriculture. Tillage practices altered the carbon contents; community of nematodes and bacterial feeders and soil adapt to climate change (Ito et al. 2015). The coupling of meteorological and crop growth models helps the soil to adapt to climate change (Tsakmakis et al. 2017).

Waterlogging and salinization have direct and indirect adverse effects on plant growth and yield (Singh 2015). Development of soil salinity and drought, due to high temperatures, creates the problems of poor germination and seedling growth. Drought stress at the boll-forming stage of cotton not only decreases the yield but also deteriorates the quality of fiber. It can decrease 6–60% bolls and 2–40% boll weight in cotton (Ma et al. 2002; Loka and Oosterhuis 2012; Basal et al. 2009; Lokhande and Reddy 2014). Excessive salt in soil leads to physiological besides biochemical and metabolic disorders in cotton mainly as a result of osmotic effects, nutritional imbalance, as well as toxicity of salt ions (Na^+ and Cl^-) (Dong 2012).

However, the sensitivity of soil toward higher organic matter degradation is adapted by carbon quality and soil nutrients (Sihi et al. 2016). Presence of balance soil carbon might enable the soil to adapt current changes in the climate (Bradford et al. 2016). Therefore, it is suggested that for improvement in cotton yield, optimization of irrigation and soil fertility status (Fig. 2.7) is the necessity of time under changing climatic conditions (Jiao et al. 2017) that can be achieved through:

- Minimizing soil erosion and burning of cotton residues in field.
- Including cover crops in cropping systems.
- Revising sowing season to prevent moisture loss and pest attacks.
- Including organic matter addition in production technology along with inorganic fertilizers.
- Improving water and nutrient use efficiency.
- Minimizing soil tillage to save soil organic residues.
- Cultivating resistant varieties of cotton under changing climate scenario.

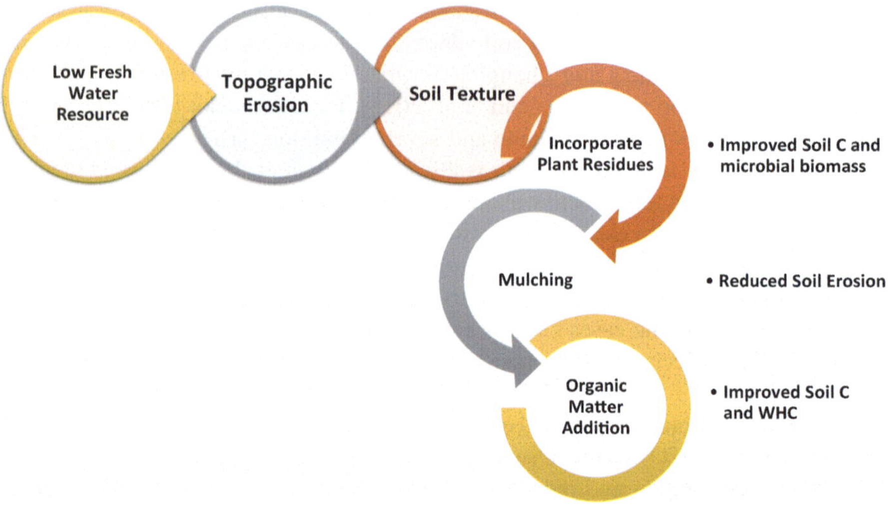

Fig. 2.7 Major constrains associated with low cotton yield and possible adaptation for improvement under climatic scenario

2.7 Conclusion

Changing climatic conditions are inducing negative impacts on the growth and yield of cotton. Increasing temperature, drought, salinity, and emission of greenhouse gases are making the condition more adverse for cotton production. Under such climatic conditions and perturbation of resources, implementation of soil adaptive and tillage practices is the necessity of time to ensure better food and fiber production. There are certain soil adaptations, i.e., use of organic matter, proper removal of crop residues, and mulching, that could be effective to mitigate environmental stresses to some extent. Besides that, the role of tillage, cropping system, and sowing season are also vital for improving the growth and yield of cotton under the current scenario.

References

Abbas Q, Ahmad S (2018) Effect of different sowing times and cultivars on cotton fiber quality under stable cotton-wheat cropping system in southern Punjab, Pakistan. Pak J Life Soc Sci 16:77–84

Abdullah AS (2014) Minimum tillage and residue management increase soil water content, soil organic matter and canola seed yield and seed oil content in the semiarid areas of Northern Iraq. Soil Tillage Res 144:150–155

Abid MEA, Scheffran J, Schneider UA, Ashfaq M (2015) Farmers' perceptions of and adaptation strategies to climate change and their determinants: the case of Punjab province, Pakistan. Earth Syst Dynam 6(1):225–243

Aggarwal P, Bhattacharyya R, Mishra AK, Das TK, Šimůnek J, Pramanik P, Sudhishri S, Vashisth A, Krishnan P, Chakraborty D, Kamble KH (2017) Modelling soil water balance and root water uptake in cotton grown under different soil conservation practices in the Indo-Gangetic Plain. Agric Ecosyst Environ 240:287–299

Ahmad S, Abbas Q, Abbas G, Fatima Z, Atique-ur-Rehman, Naz S, Younis H, Khan RJ, Nasim W, Habib ur Rehman M, Ahmad A, Rasul G, Khan MA, Hasanuzzaman M (2017) Quantification of climate warming and crop management impacts on cotton phenology. Plan Theory 6(7):1–16

Ahmad S, Iqbal M, Muhammad T, Mehmood A, Ahmad S, Hasanuzzaman M (2018) Cotton productivity enhanced through transplanting and early sowing. Acta Sci Biol Sci 40:e34610

Ahmad S, Raza I (2014) Optimization of management practices to improve cotton fiber quality under irrigated arid environment. J Food Agri Environ 2(2):609–613

Ahmad S, Raza I, Ali H, Shahzad AN, Atiq-ur-Rehman, Sarwar N (2014) Response of cotton crop to exogenous application of glycinebetaine under sufficient and scarce water conditions. Braz J Bot 37(4):407–415

Ali H, Abid SA, Ahmad S, Sarwar N, Arooj M, Mahmood A, Shahzad AN (2013a) Integrated weed management in cotton cultivated in the alternate-furrow planting system. J Food Agri Environ 11(3&4):1664–1669

Ali H, Abid SA, Ahmad S, Sarwar N, Arooj M, Mahmood A, Shahzad AN (2013b) Impact of integrated weed management on flat-sown cotton (*Gossypium hirsutum* L.). J Anim Plant Sci 23 (4):1185–1192

Ali H, Afzal MN, Ahmad F, Ahmad S, Akhtar M, Atif R (2011) Effect of sowing dates, plant spacing and nitrogen application on growth and productivity on cotton crop. Int J Sci Eng Res 2 (9):1–6

Ali H, Hameed RA, Ahmad S, Shahzad AN, Sarwar N (2014a) Efficacy of different techniques of nitrogen application on American cotton under semi-arid conditions. J Food Agri Environ 12 (1):157–160

Ali H, Hussain GS, Hussain S, Shahzad AN, Ahmad S, Javeed HMR, Sarwar N (2014b) Early sowing reduces cotton leaf curl virus occurrence and improves cotton productivity. Cer Agron Moldova XLVII(4):71–81

Aluri JSR, Karyamsetty HJ, Rao SP (2008) Traditional extraction of bark tannin from the mangrove tree *Ceriops decantra* (Griff.) Ding Hou and its use in treating cotton fishing nets. Nat Prod Radiance 7:173–175

Amin A, Nasim W, Mubeen M, Ahmad A, Nadeem M, Urich P, Fahad S, Ahmad S, Wajid A, Tabassum F, Hammad HM, Sultana SR, Anwar S, Baloch SK, Wahid A, Wilkerson CJ, Hoogenboom G (2018) Simulated CSM-CROPGRO-cotton yield under projected future climate by SimCLIM for southern Punjab, Pakistan. Agr Syst 167:213–222

Amin A, Nasim W, Mubeen M, Nadeem M, Ali L, Hammad HM, Sultana SR, Jabran K, Habib ur Rehman M, Ahmad S, Awais M, Rasool A, Fahad S, Saud S, Shah AN, Ihsan Z, Ali S, Bajwa AA, Hakeem KR, Ameen A, Amanullah, Rehman HU, Alghabar F, Jatoi GH, Akram M, Khan A, Islam F, Ata-Ul-Karim ST, Rehmani MIA, Hussain S, Razaq M, Fathi A (2017) Optimizing the phosphorus use in cotton by using CSM-CROPGRO-cotton model for semi-arid climate of Vehari-Punjab, Pakistan. Environ Sci Pollut Res 24(6):5811–5823

Anapalli S, Pettigrew W, Reddy K, Ma L, Fisher D, Sui R (2016) Climate-optimized planting windows for cotton in the lower Mississippi delta region. Agronomy 6(4):46. https://doi.org/10.3390/agronomy6040046

Atherton BC, Morgan MT, Shearere SA, Stombawgh TS, Ward AD (1999) Site-specific farming: a perspective on information needs, benefits and limitations. J Soil Water Conserv 54:455–461

Avci U, Pattathil S, Singh B, Brown VL, Hahn MG, Haigler CH (2013) Cotton fiber cell walls of *Gossypium hirsutum* and *Gossypium barbadense* have differences related to loosely-bound xyloglucan. PLoS One 8(2):e56315

Baker CJ, Saxton KE, Ritchie WR (2002) No-tillage seeding: science and practice, 2nd edn. CAB International, Wallingford, p 258

Bakshi P, Wali VK, Iqbal M, Amit J, Kour K, Rafiq A, Bakshi M (2015) Sustainable fruit production by soil moisture conservation with different mulches: a review. Afr J Agric Res 10:4718–4729

Bama KS, Somasundaram E, Thiageshwari S (2017) Influence of tillage practices on soil physical chemical and biological properties under cotton maize cropping sequence. Int J Chem Studies 5:480–485

Basal H, Dagdelen N, Unay A, Yilmaz E (2009) Effects of deficit drip irrigation ratios on cotton (*Gossypium hirsutum* L.) yield and fibre quality. J Agron Crop Sci 195:19–29

Batey T (2009) Soil compaction and soil management—a review. Soil Use Manag 25:335–345

Bell AA (1986) Physiology of secondary products. In: Mauney JR, Stewart JMD (eds) Cotton Physiology. The Cotton Foundation, Memphis, TN, pp 597–621

Bell B (1996) Farm machinery, 4th edn. Farming Press Books, Ipswich, p 320

Bezemer TM, van Dam NM (2005) Linking aboveground and belowground interactions via induced plant defenses. Trends Ecol Evol 20:617–624

Blaise D, Ravindran CD, Singh JV (2006) Trends and stability analysis to interpret results of long-term effects of application of fertilizers and manure to cotton grown on rainfed vertisols. J Agron Crop Sci 192:319–330

Bradford MA, Wieder WR, Bonan GB, Fierer N, Raymond PA, Crowther TW (2016) Managing uncertainty in soil carbon feedbacks to climate change. Nat Clim Chang 6:751–758

Brown HB (2002) Cotton history, species, varieties, morphology, breeding, culture, diseases, marketing and uses, 2nd edn. Biotech Books, Delhi, pp 27–40

Buttar GS, Thind HS, Sekhon KS, Kaur A, Gill RS, Sidhu BS, Aujla MS (2017) Management of saline-sodic water in cotton-wheat cropping system. J Agr Sci Tech 19:465–474

Chau LM, Heong KL (2005) Effects of organic fertilizers on insect pest and diseases of rice. Omonrice 13:26–33

Christian DG (1994) Experience with direct drilling and reduced cultivation in England. In: Tebrugge F (ed) EC workshop: experience with the applicability of no-tillage crop production in the West-European countries, 27–28 June 1994, Giessen, Germany. Wissenschaftlicher Fachverlag, Langgons, pp 25–31

Classen AT, Sundqvist MK, Henning JA, Newman GS, Moore JA, Cregger MA, Moorhead LC, Patterson CM (2015) Direct and indirect effects of climate change on soil microbial and soil microbial-plant interactions: what lies ahead? Ecosphere 6(8):1–21

Delgado JA, Groffman PM, Nearing MA, Goddard T, Reicosky D, Lal R, Kitchen NR, Rice CW, Towery D, Salon P (2011) Conservation practices to mitigate and adapt to climate change. J Soil Water Conser 66:118A–129A

Disi JO, Kloepper JW, Fadamiro HY (2018) Seed treatment of maize with Bacillus pumilus strain INR-7 affects host location and feeding by Western corn rootworm, *Diabrotica virgifera virgifera*. J Pest Sci 91:515–522

Dong H (2012) Technology and field management for controlling soil salinity effects on cotton. Aus J Crop Sci 6:333–341

El Titi A (2003) Soil tillage in agroecosystems. CRC Press, Boca Raton, p 384

Endale DM, Carbrera ML, Steiner JL, Radcliffe DE, Vencill WK, Schomberg HH, Lohr L (2002) Impact of conservation tillage and nutrient management on soil water and yield of cotton fertilized with poultry litter or ammonium nitrate in the Georgia Piedmont. Soil Till Res 66:55–68

Essien EE, Aboaba SA, Ogunwande IA (2011) Constituents and antimicrobial properties of the leaf essential oil of *Gossypium barbadense* (Linn.). J Med Plant Res 5:702–705

Ezuruike UF, Prieto JM (2014) The use of plants in the traditional management of diabetes in Nigeria: pharmacological and toxicological considerations. J Ethnopharmacol 155:857–924

Foster JL, Bean ME, Morgan C, Morgan G, Mohtar R, Landivar J, Young M (2018) Comparison of two tillage practices in a semi-arid cotton–grain sorghum rotation. Agron J 110:1572–1579

Gajri PR, Arora VK, Prihar SS (2002) Tillage for sustainable cropping. Food Products Press, New York, p 195

Gaylor MJ (1989) Direct and indirect effects of conservation tillage on the management of insect pests of cotton. In Proceeding Department of Entomology Alabama Agricultural Experiment Station, Auburn University Alabama, 36849, pp 10–12

Gehring C, Bennett A (2009) Mycorrhizal fungal-plant-insect interactions: the importance of a community approach. Environ Entomol 38:93–102

Gotmare V, Singh P, Tule B (2000) Wild and cultivated species of cotton. In: Technical bulletin, vol 5. Central Institute for Cotton Research, Nagpur

Hansen JD, Johnson JA, Winter DA (2011) History and use of heat in pest control: a review. Inter J Pest Manag 54:267–289

Hatfield JL, Prueger JH (2015) Temperature extremes: effect on plant growth and development. Weather and Climate Extremes 10:4–10

Hegde RR, Dahiya A, Kamath MG, Gao X, Jangala PK (2004) Cotton fibres. Tickle College of Engineering, University of Tennessee, Knoxville, TN

Hira GS, Jalota SK, Arora VK (2004) Efficient management of water resources for sustainable cropping in Punjab. Res Bull

Hodson A, Lewis EE (2016) Managing for soil health can suppress pests. Calif Agr 70:137–141

Hu G, Houston NL, Pathak D, Schmidt L, Thelen JJ, Wendel JF (2011) Genomically biased accumulation of seed storage proteins in allopolyploid cotton. Genetics 189:1103–1115

Huang J (2016) Effects of soil temperature and snow cover on the mortality of overwintering pupae of the cotton bollworm, *Helicoverpa armigera* (Hübner) (Lepidoptera: Noctuidae). Int J Biometeorol 60:977–989

Iqbal M, Anwar-Ul-Hassan, Ali A, Rizwanullah M (2005) Residual effect of tillage and farm manure on some soil physical properties and growth of wheat (*Triticum aestivum* L). Int J Agri Biol 7:54–57

Ito T, Araki M, Higashi T, Komatsuzaki M, Kaneko N, Ohta H (2015) Responses of soil nematode community structure to soil carbon changes due to different tillage and cover crop management practices over a nine-year period in Kanto, Japan. Appl Soil Ecol 89:50–58

Jalota SK, Sood A, Chahal GBS, Choudhury BU (2006) Crop water productivity of cotton (*Gossypium hirsutum* L.)—wheat (*Triticum aestivum* L.) system as influenced by deficit irrigation, soil texture and precipitation. Agr Water Manage 84:137–146

Jiao X, Maimaitiyiming A, Salahou MK, Liu K, Guo W (2017) Impact of groundwater level on nitrate nitrogen accumulation in the vadose zone beneath a cotton field. Water 9:171

Kabiri V, Raiesi F, Ghazav MA (2015) Six years of different tillage systems affected aggregate-associated SOM in a semi-arid loam soil from Central Iran. Soil Till Res 154:114–125

Karl TR, Melillo JM, Peterson TC (2009) Global climate change impacts in the United States. Cambridge University Press, New York, p 196

Karmakar R, Das I, Dutta D, Rakshit A (2016) Potential effects of climate change on soil properties: a review. Sci Int 4:51–73

Khan MB, Khaliq A, Ahmad S (2004) Performance of mashbean intercropped in cotton planted in different planting patterns. J Res (Sci) 15(2):191–197

Kreft S, Eckstein D, Junghans L, Kerestan C, Hagen U (2013) Global climate risk index 2014. In: Who suffers most from extreme weather events. Weather-related loss events in 2012 and 1994–2012. German watch, Bonn

Lal R (2015) Soil carbon sequestration and aggregation by cover cropping. J Soil Water Conserv 70:329–339

Lal R (2017) Improving soil health and human protein nutrition by pulses-based cropping systems. Adv Agron:167–204

Lal R, Reicosky DC, Hanson JD (2007) Evolution of the plow over 10,000 years and the rationale for no-till farming. Soil Till Res 93:1–12

Li Y, Niu W, Wang J, Liu L, Zhang M, Xu J (2016) Effects of artificial soil aeration volume and frequency on soil enzyme activity and microbial abundance when cultivating greenhouse tomato. Soil Sci Soc Am J 80:1208–1221

Liang B, Lehmann J, Solomon D, Kinyangi J, Grossman J, O'neill B, Skjemstad JO, Thies J, Luizao FJ, Petersen J, Neves EG (2006) Black carbon increases cation exchange capacity in soils. Soil Sci Soc Am J 70(5):1719–1730

Liu C, Yuan D, Zhang X, Lin Z (2013) Isolation, characterization and mapping of genes differentially expressed during fibre development between *Gossypium hirsutum* and *G. barbadense* by cDNA-SRAP. J Genet 92:175–181

Loka DA, Oosterhuis DM (2012) Water stress and reproductive development in cotton. In: Oosterhuis DM, Cothren JT (eds) Flowering and fruiting in cotton. The Cotton Foundation, Candova, TN, pp 51–58

Lokhande S, Reddy KR (2014) Reproductive and fiber quality responses of upland cotton to moisture deficiency. Agron J 106:1060–1069

Longenecker DE, Eire LJ (1966) Irrigation water management. In: Eilliot FC, Hoover M, Porter WK (eds) Advances in production and utilization of cotton: principles and practices. Iowa State University Press, Ames, pp 321–346

Ludwig M, Meyhofer R (2016) Efficacy of crop cover netting against cabbage pests and their natural enemies and relevance of oilseed rape. J Plant Dis Protect 123(6):331–338

Luo Z, Feng W, Luo Y, Baldock J, Wang E (2017) Soil organic carbon dynamics jointly controlled by climate, carbon inputs, soil properties and soil carbon fractions. Glob Change Biol 23:4430–4439

Ma FY, Li MC, Yang JR, Ji XJ, Shentu XD, Tao HJ (2002) A study of effect of water deficit of three periods during cotton anthesis on canopy apparent photosynthesis and WUE. Sci Agric Sin 12:1467–1472

Magdoff F, van Es H (2000) Building soils for better crops. Sustainable agriculture network, 2nd edn. University of Nebraska Press, Lincoln, NE, p 240

Marshall MW, Williams P, Nafchi AM, Maja JM, Payero J, Mueller J, Khalilian A (2016) Influence of tillage and deep rooted winter cover crops on soil properties, pests, and yield responses in cotton. Open J Soil Sci 6:149–158

Mbuthia LW, Acosta-Martínez V, DeBruyn J, Schaeffer S, Tyler D, Odoi E, Mpheshea M, Walker F, Eash N (2015) Long term tillage, cover crop, and fertilization effects on microbial community structure, activity: implications for soil quality. Soil Biol Biochem 89:24–34

Minasny B, McBratney AB (2018) Limited effect of organic matter on soil available water capacity. Eur J Soil Sci 69:39–47

Murrell EG (2017) Can agricultural practices that mitigate or improve crop resilience to climate change also manage crop pests? Curr Opin Insect Sci 23:81–88

Nyakatawa EZ, Reddy KC, Mays DA (2000) Tillage, cover cropping and poultry litter effects on cotton: II. Growth and yield parameters. Agron J 92:1000–1007

O'Neal MR, Nearing MA, Vining RC, Southworth J, Pfeifer RA (2005) Climate change impacts on soil erosion in Midwest United States with changes in crop management. Catena 61 (2–3):165–184

Osanai Y, Tissue DT, Bange MP, Anderson IC, Braunack MV, Singh BK (2017) Plant-soil interactions and nutrient availability determine the impact of elevated CO_2 and temperature on cotton productivity. Plant Soil 410:87–102

Pietrzak S (2013) Sporządzanie bilansow skladnikow nawozowych metodą "u bramy gospodarstwa". W: Samoocena gospodarstw w zakresie zarządzania składnikami nawozowymi i oceny warunków środowiskowych [Making balance of fertilizers by using "the-farm gate" method]. In: Ulen L, Pietrzak S, Tonderski K (eds) Self-assessment of farms in the management of fertilizers and environmental conditions. Wydaw, IMUZ, Falenty, pp 7–33

Puissant J, Cécillon L, Mills RT, Robroek BJ, Gavazov K, De Danieli S, Spiegelberger T, Buttler A, Brun JJ (2015) Seasonal influence of climate manipulation on microbial community structure and function in mountain soils. Soil Biol Biochem 80:296–305

Rahman MH, Ahmad A, Wang X, Wajid A, Nasim W, Hussain M, Ahmad B, Ahmad I, Ali Z, Ishaque W, Awais M, Shelia V, Ahmad S, Fahad S, Alam M, Ullah H, Hoogenboom G (2018)

Multi-model projections of future climate and climate change impacts uncertainty assessment for cotton production in Pakistan. Agric For Meteorol 253-254:94–113

Raps A, Vidal S (1998) Indirect effects of an unspecialized endophytic fungus on specialized plant—herbivorous insect interactions. Oecologia 114:541–547

Reddy DD, Blaise D, Kumrawat B, Singh AK (2017) Evaluation of integrated nutrient management interventions for cotton (*Gossypium hirsutum*) on a Vertisol in central India. Commun Soil Sci Plan 48(4):469–475

Reddy KR, Davidonis GH, Johnson AS, Vinyard BT (1999) Temperature regime and carbon dioxide enrichment alter cotton boll development and fiber properties. Agron J 91:851–858

Reddy VR, Reddy KR, Baker DN (1991) Temperature effect on growth and development of cotton during the fruiting period. Agron J 83(1):211–217

Reeder R (2000) Conservation tillage systems and management. Mid-West Plan Service, Ames

Reeder R (2002) Consider strip tillage to alleviate soil compaction [online]. Meredith Corporation, Des Moines. http://www.agriculture.com/default.sph/AgNews.class?FNC=DetailNewsAsearch listAg-News.html48797

Rochester IJ (2011) Assessing internal crop nitrogen use efficiency in high-yielding irrigated cotton. Nutr Cyc Agroecosyst 90:147–156

Salter PJ, Williams JB (1965) The influence of texture on the moisture characteristics of soils. Eur J Soil Sci 16:1–15

Sawan ZM, Hanna LI, Gad El Karim GA, McCuistion WL (2002) Relationships between climatic factors and flower and boll production in Egyptian cotton (*Gossypium barbadense*). J Arid Environ 52:499–516

Schulz BJE (2006) Mutualistic interactions with fungal root endophytes. In: Schulz BJE, Boyle CJC, Sieber TN (eds) Microbial root endophytes. Springer-Verlag, Berlin, pp 261–279

SEEP (2009) Report from the expert panel on social, environmental and economic performance. Performance (SEEP) of cotton production. Executive summary to the 68th ICAC plenary meeting, September 2009. p 6

Shu X, Zhu A-N, Zhang J-B, Yang W-L, Xin X-L, Zhang X-F (2015) Changes in soil organic carbon and aggregate stability after conversion to conservation tillage for seven years in the Huang-Huai-Hai Plain of China. J Integr Agr 14:1202–1211

Sihi D, Inglett PW, Inglett KS (2016) Carbon quality and nutrient status drive the temperature sensitivity of organic matter decomposition in subtropical peat soils. Biogeochemistry 131:103–119

Singh A (2015) Soil salinization and waterlogging: a threat to environment and agricultural sustainability. Ecol Indic 57:128–130

Smith P, Martino D, Cai Z, Gwary D, Janzen H, Kumar P, McCarl B, Ogle S, O'Mara F, Rice C, Scholes B, Sirotenko O (2007) Agriculture. In: IPCC. Climate change 2007: mitigation. Contribution of working group III to the fourth assessment report of the intergovernmental panel on climate change. pp 498–540

Sun C, Feng D, Mi Z, Li C, Zhang J, Gao Y, Sun J (2017) Impacts of ridge-furrow planting on salt stress and cotton yield under drip irrigation. Water 9:49

Sun R, Zhang X-X, Guo X, Wang D, Chu H (2015) Bacterial diversity in soils subjected to long-term chemical fertilization can be more stably maintained with the addition of livestock manure than wheat straw. Soil Biol Biochem 88:9–18

Tariq M, Afzal MN, Muhammad D, Ahmad S, Shahzad AN, Kiran A, Wakeel A (2018) Relationship of tissue potassium content with yield and fiber quality components of Bt cotton as influenced by potassium application methods. Field Crops Res 229:37–43

Tariq M, Yasmeen A, Ahmad S, Hussain N, Afzal MN, Hasanuzzaman M (2017) Shedding of fruiting structures in cotton: factors, compensation and prevention. Trop Subtrop Agroecosyst 20(2):251–262

Thierfelder C, Chivenge P, Mupangwa W, Rosenstock TS, Lamanna C, Eyre JX (2017) How climate-smart is conservation agriculture (CA)?–its potential to deliver on adaptation, mitigation and productivity on smallholder farms in southern Africa. Food Secur 9(3):537–560

Tisdall JM, Oades JM (1982) Organic-matter and water-stable aggregates in soil. Eur J Soil Sci 33:141–163

Torbert HA, Ingram JT, Prior SA (2015) High residue conservation tillage system for cotton production: a farmer's perspective. Eur Agrophy J 2:1–14

Tsakmakis I, Kokkos N, Pisinaras V, Papaevangelou V, Hatzigiannakis E, Arampatzis G, Gikas GD, Linker R, Zoras S, Evagelopoulos V, Tsihrintzis VA, Battilani A, Sylaios G (2017) Operational precise irrigation for cotton cultivation through the coupling of meteorological and crop growth models. Water Resour Manag 31(1):563–580

Usman M, Ahmad A, Ahmad S, Irshad M, Khaliq T, Wajid A, Hussain K, Nasim W, Chattha TM, Trethowan R, Hoogenboom G (2009) Development and application of crop water stress index for scheduling irrigation in cotton (*Gossypium hirsutum* L.) under semiarid environment. J Food Agri Environ 7(3&4):386–391

Van der Ent S, Van Wees SCM, Pieterse CMJ (2009) Jasmonate signaling in plant interactions with resistance-inducing beneficial microbes. Phytochemistry 70:1581–1588

Van Wees SCM, Van d, EntCorné S, Pieterse MJ (2008) Plant immune responses triggered by beneficial microbes. Curr Opin Plant Biol 11:443–448

Verma N, Chaudhary S, Goyal S (2018) Long term effects of inorganic fertilizers and organic amendments on ammonification and nitrification activity of soils under cotton-wheat cropping system. Int J Curr Microbiol App Sci 7:718–724

Weyens N, van der Lelie D, Taghavi S, Newman L, Vangronsveld J (2009) Exploiting plant microbe partnerships to improve biomass production and remediation. Trends Biotechnol 27:591–598

Williams A, White N, Mushtaq S, Cockfield G, Power B, Kouadio L (2015) Quantifying the response of cotton production in eastern Australia to climate change. Clim Chang 129:183–196

Wu Z, Soliman KM, Zipf A, Saha S, Sharma GC, Jenkins JN (2005) Isolation and characterization of genes differentially expressed in fiber of *Gossypium barbadense* L. J Cotton Sci 9:166–174

Yang C, Yang L, Ouyang Z (2005) Organic carbon and its fractions in paddy soil as affected by different nutrient and water regimes. Geoderma 124:133–142

Yang H, Qi J, Xu X, Yang D, Lv H (2014) The regional variation in climate elasticity and climate contribution to runoff across China. J Hydrol 517:607–616

Yazar A, Sezen SM, Gencel B (2002a) Drip irrigation of corn in the Southeast Anatolia Project (GAP) area in Turkey. Irrig Drain 51:293–300

Yazar A, Sezen SM, Sesveren S (2002b) LEPA and trickle irrigation of cotton in the Southeast Anatolia Project (GAP) area in Turkey. Agr Water Manage 54:189–203

Zehnder G, Kloepper J, Tuzun S, Yao C, Wei G, Chambliss O, Shelby R (1997) Insect feeding on cucumber mediated by rhizobacteria-induced plant resistance. Entomol Exp Appl 83:81–85

Zhao Y, Pang H, Wang J, Huo L, Li Y (2014) Effects of straw mulch and buried straw on soil moisture and salinity in relation to sunflower growth and yield. Field Crop Res 161:16–25

Chapter 3
Managing Planting Time for Cotton Production

Muhammad Naveed Afzal, Muhammad Tariq, Muhammad Ahmed, Ghulam Abbas, and Zahid Mehmood

Abstract Since cotton is responsive to its surrounding environments, an appropriate planting time is very important to realize its optimum yield potential. The temperature is an important environmental factor for which planting time varies across the world. It may be from few days to weeks, called as planting window. It must be optimized according to yield target, forthcoming weed, pest, disease, and abiotic stress problems. The planting time is considered optimum if it permits the completion of the crop vegetative and reproductive growth stages in an efficient way. There is an optimum planting time and further delay drastically reduced seed cotton yield. It determines the duration of various phenological stages such as appearance of first square, flower, and boll split on 50% population and total crop duration. Generally, early planting results in higher yield by extending growth period, and late planting reduces yield because of shortened season. The planting time has a major role for fiber and seed quality optimization to ensure good-quality yarn and better seed germination. The planting time is strictly monitored and legally implemented in certain regions of the world to prevent disease and pests. The climate change is threatening cotton production throughout the world, and adjustment of planting dates may play an effective role as an adaptation strategy to match fruiting phase to favorable climate.

Keywords Planting window · Yield · Seed · Lint quality · Climate change

Abbreviations

BB	Bacterial blight
CCRI	Central Cotton Research Institute
CIM	Cotton Institute Multan
CLCuD	Cotton leaf curl virus disease

M. N. Afzal · M. Tariq (✉) · M. Ahmed · Z. Mehmood
Central Cotton Research Institute, Multan, Pakistan
e-mail: mtariq131@ccri.gov.pk

G. Abbas
Department of Agronomy, Bahauddin Zakariya University, Multan, Pakistan

S. Ahmad, M. Hasanuzzaman (eds.), *Cotton Production and Uses*,
https://doi.org/10.1007/978-981-15-1472-2_3

ENSO El Niño-Southern Oscillation
NIAB Nuclear Institute for Agriculture and Biology
RLS *Ramularia* leaf spot

3.1 Introduction

The cotton crop possesses indeterminate growth habits which take advantage of duration of growing season (Ali et al. 2011, 2013a, b, 2014a, b). The growth and development of cotton are highly influenced by prevailing weather variables at various growth stages (Ahmad et al. 2014, 2017, 2018; Abbas and Ahmad 2018; Ahmad and Raza 2014). Planting at various dates creates variable weather conditions which affect the growth and yield. The planting time is optimized in such a way that the developmental stage may coincide with desirable environmental conditions (Amin et al. 2017, 2018; Tariq et al. 2017, 2018). The optimum planting time ensures uniform stand establishment, growth, and improved seed cotton yield (Khan et al. 2004; Rahman et al. 2018; Usman et al. 2009). The decision of best planting time is based on soil temperature; availability of canal or rainwater; incidence of diseases, pests, and weeds; and harvesting time of crops in rotation with cotton. The early planting prolongs the growth season which translates into higher yield. It is preferred in regions where low heat units are accumulated. In this way, the crop will be mature before the onset of the fall. The late planting is often left with no chance of successful replanting in case of germination failure. The selection of planting date is even more important than irrigation for obtaining higher yield (Hallikeri et al. 2009). The cotton yield is highly influenced by planting time due to its relevance with heat unit accumulation. Therefore, the planting time varies across cotton-growing countries which have been summarized in Table 3.1.

The advent and duration of various phenological stages were also affected by prevailing temperature which depends upon the planting time. The planting time affects the growing degree days' accumulation for completion of various stages. The

Table 3.1 Time of planting of various cotton-growing countries

Regions	Time of planting
USA	March, April
Greece	April
Central Asia	April
China	April
Middle East and Egypt	May
South Asia	June
Argentina	September, October, November
Australia	September, October, November
Brazil	October, November

Source: US Department of Agriculture [Major World Crop Areas and Climatic Profiles]

days taken to start emergence, 50% emergence, and mean emergence time were gradually decreased with delay in planting from 14th February till 15th May (Shah et al. 2017). The days taken to square, flower, and first boll split were reduced with delay in sowing (Sohair et al. 2015; Shah et al. 2017; Iqbal et al. 2018). The cotton seed quality is highly influenced by time of planting and picking. It was revealed that emergence of seed harvested from early sowing was delayed (Kamran et al. 2017).

The significantly higher seed cotton, seed index, and dry matter yield from early planting have been witnessed in literature (Arshad et al. 2007; Ali et al. 2009; Huang 2016). However, oil contents were improved with delay in sowing (Mohamed et al. 2016). The fiber traits like fiber length, uniformity, and spinning consistency index showed non-significant relationship with planting time. However, micronaire, fiber strength, and fiber elongation were decreased, and short fiber contents were increased with the delay in sowing (Killi and Bolek 2014).

The early sown crop can better withstand with weeds than late sown. The lower weed competition, density, dry matter, and yield losses were reported in early sown crop (Webster et al. 2009; Tauseef et al. 2012; Hariharasudhan et al. 2017). The planting time has been used for disease and insect management. The early planting has been suggested for tarnished plant bug management in Mississippi (MS) Delta (O'Berry et al. 2008) and cotton leaf curl disease management in Pakistan (Iqbal and Khan 2010). On the other hand, the early planting has been banned in Pakistan to tackle the problem of pink bollworm infestation.

3.2 Goals of Planting Time Optimization

3.2.1 High Yield and Better Lint Quality

Optimum planting provides proper growing season to best utilize the natural sources for higher yield. The planting time determines the crop growth season, duration of phenophase, and dry matter partitioning to various plant organs (Ali et al. 2009). The yield increases of about 10% in Mississippi Delta, USA (Pettigrew and Adamczyk 2006); 60.1% in Dera Ismail Khan, Pakistan (Usman et al. 2016); and 51.4% at Sakha Agriculture Research Experimental Station, Egypt (Sohair et al. 2015), have been reported by adjusting planting time. Although boll weight and ginning out turn vary under the influence of planting time, the number of bolls is more affected. The low yield in delayed planting in comparison with early planting is the result of short reproductive period, heat stress during early development, and sub-optimal temperature, photo period, and solar radiations during reproductive stages (Rahman et al. 2016).

The fiber length, strength, micronaire values, and uniformity index are some of the key attributes of fiber quality. In general, cultivar selection and growing environment matter the fiber quality, while the impact of crop management remains inconsistent. The ginning out turn, fiber length, strength, fiber elongation, uniformity index, and micronaire values decreased with late sowing (Ali et al. 2009; Awan et al. 2011; Sohair et al. 2015; Usman et al. 2016). The lower micronaire values in late sown conditions suggested that late sowing of genotypes with high micronaire value may offer an opportunity for reducing its value. However, in certain environments, the mid planting produced higher ginning out turn and superior fibers over early and delayed planting (Awan et al. 2011; Shah et al. 2017).

3.2.2 Insect, Disease, and Weed Management

The selection of planting time is an important tool for the management of insects, diseases, and weeds. The early planting was suggested for the management of tarnished plant bug in MS Delta (O'Berry et al. 2008). Likewise, the early sowing is being discouraged in Pakistan for the management of pink bollworm. The diseases' infestation level also varies in response to planting time. The early planted cotton during the third week of December in Brazilian conditions was more infested with *Ramularia* leaf spot disease than the crop planted during the third week of January (Ascari et al. 2016). Similarly, early cotton planting is a better management strategy to minimize the cotton leaf curl virus incidence in Pakistan (Iqbal and Khan 2010). The bacterial blight is the most common disease of cotton in Sudan which flourishes in warm humid weather. The results of experiment conducted by Mohammed et al. (2003) suggested late planting is safer for the disease incidence in comparison with early planting. For such diseases, particular planting window is defined and implemented on large scale by the government agencies.

The weeds' competitive ability varied with planting time. Generally, the weeds compete less in early sown crops than late sown crop. The lowest weed density and dry matter were recorded in first August planting as compared to 15th August, first September, and 15th September planting (Hariharasudhan et al. 2017). The damp and moist weather in the month of July favors growth and development of narrow leaves weeds. The problem is more severe for late sown crop with poor plant canopy development at very early stage (Tauseef et al. 2012). The competitive ability of Benghal dayflower was assessed in Georgia cotton, sown at various times. Its presence resulted in yield loss of about 21–30% in May-planted crop, while 40–60% more yield losses were recorded in June planting (Webster et al. 2009).

3.2.3 Heat and Drought Stress Management

Adjusting planting time is helpful to offset the effects of bad weather events, particularly heat and drought. In the areas where planting window is wider, the farmers may prefer different planting time to avoid unpredicted weather events. It may reduce the bad effects of the drought events, pest attack, boll rot and other seedling disease, etc. Mitigating heat stress through planting time adjustment may be less important for irrigated than rainfed cotton (Luo et al. 2016). Shifting planting time may avoid the period of high evaporation demand to efficiently use available water. The concept of climate zones thus has been used to decide planting time according to soil water availability and to match subsequent crop growth with optimum climatic conditions (da Silva et al. 2007).

3.2.4 *Seed Quality*

The seed quality determines the success of planting particularly in stressful environment. The planting time affects the seed quality mainly when boll opening coincides with rainy season. The seed quality reduced with delay in sowing from 25th April to 10th June. The second (25th May) and third sowing dates (10th June) produced 1.74 and 3.35%, 3.90% and 14.99%, 3.40% and 8.76%, and 1.75% and 5.00% less seed weight, embryo weight, embryo oil, and protein contents, respectively. The differences in seed quality between planting time are due to variations in daily mean, maximum and minimum temperature, and solar radiations. The oil and protein accumulation is faster in early sown crop than late sown crops, mainly from temperature effect (Wei et al. 2017). The delayed sowing mostly pushes the crop to cold environment and poor light penetration which reduces the sucrose translocation from leaves to developing boll (Liu et al. 2013).

3.3 Factors Affecting Choice of Planting Time

3.3.1 *Soil Temperature and Precipitation*

The soil temperature is an important environmental factor which affects the date of planting. The cotton planting is not recommended if the soil temperature is below 16 °C (Luo et al. 2016). It must be consistent about three consecutive days because short changes in temperature may be misleading. Such variations in required temperature make the cotton planting time different which has been listed in Table 3.1. The lower temperature delays the germination process and produces poor stand. Conventionally, the calendar dates are used to define regional planting windows based on long-term farming experience. The weather projections may be used to estimate the approximate date with the right temperature to make all the necessary arrangements for planting. The occurrence of rainfall during sowing determines the planting time of rainfed cotton. The knowledge of future weather conditions is required to coincide peak irrigation requirement with rainy season. The planting time should be adjusted in such a way that boll opening coincides with less rainy season to avoid seed and fiber quality deterioration from boll rot disease.

3.3.2 Genotypes

The prevailing weather conditions at critical growth have a strong impact on the yield formation process. There are two means to adjust particular growth stages with specific environment which include shifting planting time and selection of genotypes. The potential yield of cultivar can be harvested only if planted at optimum time. The continuous experimentation is being carried out to find out the optimum planting time for different cotton varieties in local environment. In fact, different genotypes have been identified for specific planting windows, and yield decline occurs in case of delayed planting. The crop duration is reduced in case of late planting for which short-duration genotypes should be preferred, while long-duration genotypes are suitable for early planting to harvest maximum yield benefits. The earliness character also plays an important role for deciding planting time in areas where boll opening may coincide with rainy season. Therefore, decision of planting time is based according to genotypes in hand for cultivation.

3.3.3 Cropping Sequence

The cropping sequence affects the planting time by making land fallow. Harvesting of previous crop and cultivation of successive crop determines the planting time. In some cropping sequence, a lot of time span is available between harvesting and planting, and planting window is wider for such sequence. However, the harvesting and planting time overlap in certain cropping sequence, e.g., wheat-cotton and planting window become narrow. The cotton crops sown in wheat-cotton, sunflower-cotton, and maize-cotton sequence in Pakistan often get late (Ahmad et al. 2017; Tariq et al. 2018). Hence, cropping season moves with planting time. In cropping systems, where planting time overlaps the harvesting of the preceding crop, the relay cropping technology is preferred for timely planting.

3.3.4 Availability of Inputs and Labor

The labor availability is an important factor to decide the planting and picking time. If two crops mature simultaneously at local level, there should be delayed or early sowing of one of the crops to avoid labor at crises at maturity. For example, the maturity time of cotton and peanut matches in South Georgia where early planting time of short maturity varieties may be used for early picking to overcome the labor crises.

3.3.5 Diseases and Pests

The disease incidence requires specific environmental conditions in addition to host availability and likely to prevail at particular time. The farmers are usually well familiar with the disease appearance time and critical sensitive growth stages. In this regard, shifting planting dates aims to avoid host availability or enable host plant to better tolerate the diseases. Such examples include *Ramularia* leaf spot (RLS) in Brazil, cotton leaf curl virus disease (CLCuD) in Pakistan, bacterial blight (BB) in Sudan, and *Fusarium* wilt in Australia (Ascari et al. 2016; Iqbal and Khan 2010; Mohammed et al. 2003; Allen 2005). The early sowing is preferred in areas which are very prone to CLCuD incidence. It enables the crop to attain enough vigor to tolerate the disease problem.

3.3.6 Technological Advancement

Planting time also varies with the advancement in production technology. For example, planting time in China has been advanced about 7–10 days by plastic film mulching (Dai and Dong 2014; CRI 2013). Similarly, the area under early planting was significantly increased in Pakistan with the arrival of transgenic (*Bacillus thuringiensis* (Bt)) cotton cultivars. This technology provides the built-in protection against selected bollworms which were a major threat for conventional cultivars under early planting system. The mechanical picking is gaining popularity which would require the development of compact cotton cultivation. The planting time is a very important tool to keep the crop compact to make it harvestable through mechanical picker. The results of experiments conducted at Yellow River valley, China, indicate that late planting should be preferred over seasonal planting for mechanical picking (Wang et al. 2016).

3.4 Planting Time Vital for Cultivar Selection

The cotton cultivars are responsive to environmental conditions in terms of phenological development like time for initiation of squaring, flowering, and boll maturation (Singh et al. 2007). The cotton cultivar NIAB-112 was regarded as an early maturing cultivar because it completes phenological development earlier about 6–8 days in comparison with AA-802 and IR-3701, the long-duration cultivar. The boll opening of 21st June planting was reduced by 18 days over 10th March planting which drastically affected the boll development in the last planting and produced poor yield (Rahman et al. 2016). Therefore, different cultivars are sown at their proper time to coincide growth stages with optimum weather conditions. The genotype yield and growth response are modified by various planting dates, and such association indicates that genotypes had differential weather requirement. The response of CIM-602 was different than CIM-598, CIM-599, and Ali-Akber 703 because CIM-602 produced the highest bolls and seed cotton yield with 04th May planting, while 19th April was the optimum planting time for the rest of the genotypes (Usman et al. 2016). The short-duration cultivar will perform better for late planting. The early planting of long-duration cultivars is required in areas where the risk of end season frost is present. However, early maturing varieties can be sown in normal sowing time to avoid the peak fruiting period with frost.

3.5 Shifting Planting Time for Climate Change

The climate change is a continuous process and has become the major challenge for cotton production. The year-to-year variations among various planting date experiments are due to results of climate impacts. The impact of climate change is not the same in various regions but depends on prevailing temperature. For instance, the yield increase of 782.6 kg ha^{-1} was expected in Northwest China with 1 °C rise in diurnal temperature range from peak bloom to maturity, whereas a mean temperature increase of 1 °C during whole growing season may result a yield increase of 4765.9 kg ha^{-1}. The sunshine hour's value from budding to anthesis is important for leaf area index. The yield may be improved by 0.0121 kg ha^{-1} per sunshine hour from budding to anthesis (Huang 2016), while it has a negative relation with yield in hot countries because a yield decline of 110 kg ha^{-1} is estimated with every increase in ambient temperature above optimal value (Singh et al. 2007). Hence, better understanding of climate change impact is necessary to adjust planting time. The shedding of fruiting structure is accelerated with rise in temperature (Tariq et al. 2017, 2018). The global warming in addition to yield reduction would accelerate the occurrence of phenological events, thus reducing the duration of phenophases. Generally, sowing occurs earlier with increasing temperature trends during planting season. The cotton sowing has been advanced by 5.35 and 0.24 days decade^{-1} in Pakistan and North China Plain from 1980 to 2015 and 1981 to 2012, respectively (Ahmad et al. 2017; Wang et al. 2017). In such circumstances, the regional sowing time should be critically reviewed and revised to take the full advantage of heat sources and other climatic attributes. Various environmental conditions are imposed at different growth stages through planting time adjustment. The shifting planting time is being taken as adaptation strategies to minimize the impact of various weather vagaries. The ENSO (El Niño-Southern Oscillation) influences the cotton production in Georgia, USA. One of the suggested key management to minimize the impact of ENSO is shifting planting time (Paz et al. 2012). The analysis of 33-year yield data revealed that planting time should be 20 days back than 10th May under Faisalabad, Pakistan, environment to minimize the impact of adverse weather on yield (Rahman et al. 2017). The results of multiyear sowing time trials indicated the year-to-year yield variability (Pettigrew and Dowd 2011; Pettigrew and Meredith 2009), suggesting the role of climatic variability for planting time decision. The short-term field experiments of about 2–4 years do not produce reliable results for deciding planting windows for long-term climate variability. About three decade field experimentation would be required to account variations in climatic variables (Anapalli et al. 2016). Indeed, the approach is very expensive and time-consuming; however, the crop modelling approach enables us to explore the climatic variable impacts on a long-term basis from short period field experiments.

3.6 Planting Time Adjustment for Cotton Leaf Curl Disease Management

The cotton leaf curl disease is the most important disease of the cotton in Pakistan, India, and Africa which is characterized by upward leaf curling, vein thickening, leaf enation, and stunted plant growth. It appeared as a minor disease in Pakistan during 1967; however, it became a serious disease during the early 1990s. It caused huge yield losses of about one million bales by reducing productivity from 9.05 to 8.04 million bales during 1992–1993 (Mahmood et al. 2003). Various disease management approaches including development of resistant strains were tested. But no single approach provided desirable results on a long-term basis, and integrated strategy was outlined for its management. The disease appearance depends upon environment where day and night temperature difference and level of relative humidity are very important. The incidence is favored by small differences in day and night temperature and higher levels of relative humidity. Therefore, the presence of younger plants along with such favorable environment promotes the disease infestation. The planting time may be adjusted to provide enough growth season prior to disease arrival. Three-year field experiments (2015–2017) were conducted at Central Cotton Research Institute (CCRI), Multan, Pakistan, to monitor the disease incidence on transgenic and non-transgenic genotypes at 15-day interval for various planting times. The transgenic genotypes were planted on 01st March till 15th May, and non-transgenic genotypes were planted from 15th April till 15th June. The results revealed that March planting of transgenic genotypes avails about 105-day

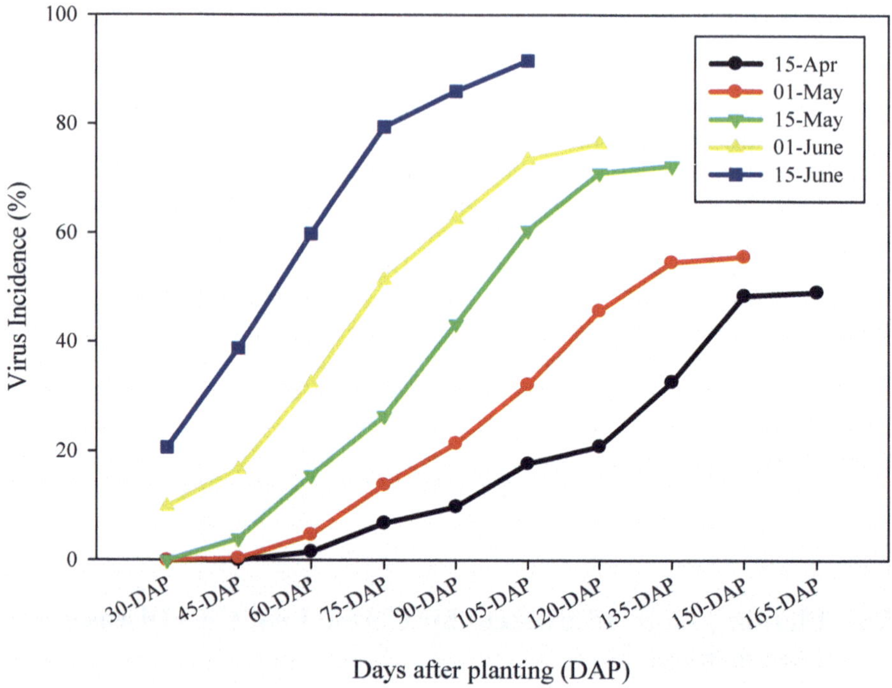

Fig. 3.1 Cotton leaf curl incidence (%) trend on transgenic genotypes across various planting time

virus-free period. The length of disease-free period gradually decreased with delay in sowing, and disease infestation appears at very early growth stages (Fig. 3.1). The March planting showed virus incidence below 50% at 180 days, while 100% disease incidence was observed at 105 days for May planting. The disease progression rate remains low for early planting and high for late planting. Therefore, March planting is a safer planting window to minimize the risk of disease infestation. The same trend appears for non-transgenic cotton where disease appears at 60 days after planting for April planting, while June planting is infested with disease at very early stages (Fig. 3.2). The impact of disease on yield reduction is more if the crop is infested during early growth season. In case of late planting, the virus-tolerant genotypes should be preferred to secure yield from reducing the virus infestation level.

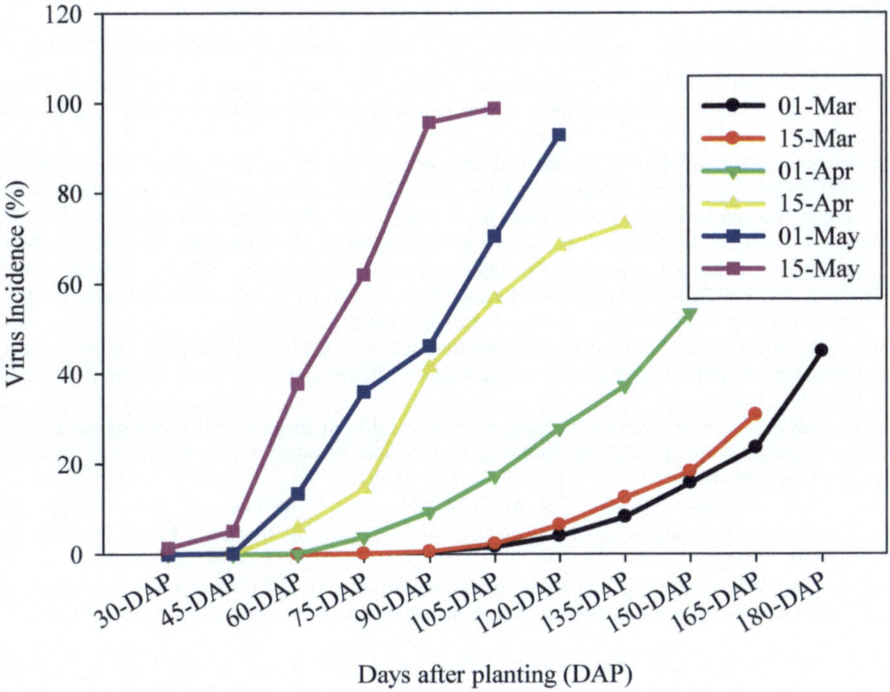

Fig. 3.2 Cotton leaf curl incidence (%) trend on non-transgenic genotypes across various planting time

3.7 Conclusion

Because shifting planting times alters the crop growth season, the impact of climatic conditions would vary at various planting time. Such variations have a deep impact on yield formation process. The planting time decides yield potential because of its relation with crop phenology and seasonal crop duration. The decision regarding selection of planting time is even more important than cultivar selection in cotton. The selection of planting time is objective oriented which may be adjusted for higher yield; better fiber quality; management tool for weeds, diseases, and insects; and drought and heat stress. The impact of planting time is significant for yield and yield components and remains variable for fiber quality in various environments.

References

Abbas Q, Ahmad S (2018) Effect of different sowing times and cultivars on cotton fiber quality under stable cotton-wheat cropping system in southern Punjab, Pakistan. Pak J Life SocSci 16:77–84

Ahmad S, Abbas Q, Abbas G, Fatima Z, Rehman AU, Naz S, Younis H, Khan RJ, Nasim W, Rehman MHU, Ahmad A, Rasul G, Khan MA, Hasanuzzaman M (2017) Quantification of climate warming and crop management impacts on cotton phenology. Plan Theory 6:1–16

Ahmad S, Iqbal M, Muhammad T, Mehmood A, Ahmad S, Hasanuzzaman M (2018) Cotton productivity enhanced through transplanting and early sowing. Acta Sci Biol Sci 40:e34610

Ahmad S, Raza I (2014) Optimization of management practices to improve cotton fiber quality under irrigated arid environment. J Food Agri Environ 2(2):609–613

Ahmad S, Raza I, Ali H, Shahzad AN, Atiq-ur-Rehman, Sarwar N (2014) Response of cotton crop to exogenous application of glycinebetaine under sufficient and scarce water conditions. Braz J Bot 37(4):407–415

Ali H, Abid SA, Ahmad S, Sarwar N, Arooj M, Mahmood A, Shahzad AN (2013a) Integrated weed management in cotton cultivated in the alternate-furrow planting system. J Food Agri Environ 11(3&4):1664–1669

Ali H, Abid SA, Ahmad S, Sarwar N, Arooj M, Mahmood A, Shahzad AN (2013b) Impact of integrated weed management on flat-sown cotton (*Gossypium hirsutum* L.). J Anim Plant Sci 23 (4):1185–1192

Ali H, Afzal MN, Ahmad F, Ahmad S, Akhtar M, Atif R (2011) Effect of sowing dates, plant spacing and nitrogen application on growth and productivity on cotton crop. Int J Sci Eng Res 2 (9):1–6

Ali H, Afzal MN, Ahmad S, Muhammad D (2009) Effect of cultivars and sowing dates on yield and quality of *Gossypium hirsutum* L. crop. J Food Agric Environ 7(3&4):244–247

Ali H, Hameed RA, Ahmad S, Shahzad AN, Sarwar N (2014a) Efficacy of different techniques of nitrogen application on American cotton under semi-arid conditions. J Food Agri Environ 12 (1):157–160

Ali H, Hussain GS, Hussain S, Shahzad AN, Ahmad S, Javeed HMR, Sarwar N (2014b) Early sowing reduces cotton leaf curl virus occurrence and improves cotton productivity. Cer Agron Moldova XLVII(4):71–81

Allen SJ (2005) Delayed planting as a control strategy for Fusarium Wilt of cotton in Australia. In: Proc Beltwide Cotton Conf. New Orleans, LA, 4–7 January 2005, vol 1. National Cotton Council Beltwide Cotton Conference, Memphis, TN, pp 131–135

Amin A, Nasim W, Mubeen M, Ahmad A, Nadeem M, Urich P, Fahad S, Ahmad S, Wajid A, Tabassum F, Hammad HM, Sultana SR, Anwar S, Baloch SK, Wahid A, Wilkerson CJ, Hoogenboom G (2018) Simulated CSM-CROPGRO-cotton yield under projected future climate by SimCLIM for southern Punjab, Pakistan. Agric Syst 167:213–222

Amin A, Nasim W, Mubeen M, Nadeem M, Ali L, Hammad HM, Sultana SR, Jabran K, Habib urRehman M, Ahmad S, Awais M, Rasool A, Fahad S, Saud S, Shah AN, Ihsan Z, Ali S, Bajwa AA, Hakeem KR, Ameen A, Amanullah, Rehman HU, Alghabar F, Jatoi GH, Akram M, Khan A, Islam F, Ata-Ul-Karim ST, Rehmani MIA, Hussain S, Razaq M, Fathi A (2017) Optimizing the phosphorus use in cotton by using CSM-CROPGRO-cotton model for semi-arid climate of Vehari-Punjab, Pakistan. Environ Sci Pollut Res 24(6):5811–5823

Anapalli SS, Pettigrew WT, Reddy KN, Ma L, Fisher DK, Sui R (2016) Climate-optimized planting windows for cotton in the lower Mississippi Delta region. Agronomy 6:46. https://doi.org/10.3390/agronomy6040046

Arshad M, Wajid A, Maqsood M, Hussain K, Aslam M, Ibrahim M (2007) Response of growth, and quality of different cotton cultivars to sowing dates. Pak J Agric 44(2):208–212

Ascari JP, De Araújo DV, Dias LDE, Bagatini GJ, Mendes IRN (2016) Severity of Ramularia leaf spot and seed cotton yield in different sowing times. Rev Caatinga Mossoró 29(3):603–610

Awan H, Awan I, Mansoor M, Khan EA, Khan MA (2011) Effect of sowing time and plant spacing on fiber quality and seed cotton yield. Sarhad J Agric 27(3):411–413

Cotton Research Institute, Chinese Academy of Agricultural Sciences (CRI) (2013) Cultivation of cotton in China. Shanghai Science and Technology Press, Shanghai, China. (in Chinese)

da Silva JC, Heldwein A, Martins FB, Streck NA, Guse FI (2007) Risco de stressetérmico para o feijoeiroem Santa Maria RS. Ciência Rural. Santa Maria 37(3):643–648

Dai J, Dong H (2014) Intensive cotton farming technologies in China: achievements, challenges and countermeasures. Field Crop Res 155:99–110

Hallikeri SS, Halemani HL, Patil VC, Palled YB, Patil BC, Katageri IS (2009) Influence of sowing time and moisture regimes on growth, seed cotton yield and fibre quality of Bt-cotton. Karnataka J Agric Sci 22(5):985–991

Hariharasudhan V, Chinnusamy C, Prabhakaran NK (2017) Optimum time of sowing and weed management methods on weeds interference and productivity and profitability of Bt cotton hybrid in western zone of Tamil Nadu. Int J Chem Stud 5(3):793–796

Huang J (2016) Different sowing dates affected cotton yield and yield components. Int J Plant Prod 10(1):63–83

Iqbal M, Khan MA (2010) Management of cotton leaf curl virus by planting time and plant spacing. Adv Agri Botanics 2(1):25–34

Iqbal M, Ul-Allah S, Naeem M, Hussain M, Ijaz M, Wasaya A, Ahmad MQ (2018) Reproductive development and seed cotton yield of *Gossypium hirsutum* L. affected by genotype and planting time. Int J AgricBiol 20(7):1590–1596

Kamran M, Afzal I, Basra SMA, Khan SHU, Mahmood A (2017) Improvement of cotton crop performance by estimating optimum sowing and picking time. Int J Agric Biol 19:241–247

Khan MB, Khaliq A, Ahmad S (2004) Performance of mashbean intercropped in cotton planted in different planting patterns. J Res (Sci) 15(2):191–197

Killi F, Bolek Y (2014) Timing of planting is crucial for cotton yield. Acta Agr Scand B-S P 56 (2):155–160

Liu J, Ma Y, Lv F, Chen J, Zhou Z, Wang Y, Abudurezike A, Oosterhuis DM (2013) Changes of sucrose metabolism in leaf subtending to cotton boll under cool temperature due to late planting. Field Crops Res 144:200–211

Luo Q, Bange M, Braunack M, Johnston D (2016) Effectiveness of agronomic practices in dealing with climate change impacts in the Australian cotton industry—a simulation study. Agric Syst 147:1–9

Mahmood T, Arshad M, Gill MI, Mahmood HT, Tahir M, Hussain S (2003) Burewala strain of cotton leaf curl virus: a threat to CLCuV cotton resistance varieties. Asian J Plant Sci 2:968–970

Mohamed KA, Yagoub SO, Elsalam AEKA, Abuali AI (2016) Response of sowing dates, cultivars and nitrogen application on growth, yield and oil contents of cotton crop (*Gossypium hirsutum* L.) growth at Nuba Mountain. Scholar J Agric Vet Sci 3(5):351–357

Mohammed OE, Ahmed NE, Eneji AE, Ma YQ, Ali E, Inanaga S, Sugimoto Y (2003) Effect of sowing dates on the incidence of bacterial blight and yield of cotton. Basic Appl Ecol 4:433–440

O'Berry NB, Faircloth JC, Edmisten KL, Collins GD, Stewart AM, Abaye AO, Herbert DA Jr, Haygood RA Jr (2008) Plant population and planting date effects on cotton (*Gossypium hirsutum* L.) growth and yield. J Cotton Sci 12:178–187

Paz JO, Woli P, Garcia y Garcia A, Hoogenboom G (2012) Cotton yields as influenced by ENSO at different planting dates and spatial aggregation levels. Agric Syst 111:45–52

Pettigrew WT, Adamczyk JJ Jr (2006) Nitrogen fertility and planting date effects on lint yield and cry1ac (Bt) endotoxin production. Agron J 98:691–697

Pettigrew WT, Dowd MK (2011) Varying planting dates or irrigation regimes alters cotton seed composition. Crop Sci 51:2155–2164

Pettigrew WT, Meredith WR Jr (2009) Seed quality and planting date effects on cotton lint yield, yield components and fiber quality. J Cotton Sci 13:37–47

Rahman MH, Ahmad A, Wang X, Wajid A, Nasim W, Hussain M, Ahmad B, Ahmad I, Ali Z, Ishaque W, Awais M, Shelia V, Ahmad S, Fahad S, Alam M, Ullah H, Hoogenboom G (2018)

Multi-model projections of future climate and climate change impacts uncertainty assessment for cotton production in Pakistan. Agric For Meteorol 253-254:94–113

Rahman MH, Ahmad A, Wajid A, Hussain M, Akhtar J, Hoogenboom G (2016) Estimation of temporal variation resilience in cotton varieties using statistical models. Pak J Agri Sci 53 (4):787–807

Rahman MH, Ahmad A, Wajid A, Hussain M, Rasul F, Ishaque W, Islam MA, Shelia V, Awais M, Ullah A, Wahid A, Sultana SR, Saud S, Khan S, Fahad S, Hussain M, Hussain S, Nasim W (2017) Application of CSM-CROPGRO-Cotton model for cultivars and optimum planting dates: evaluation in changing semi-arid climate. Field Crops Res 238(2019):139–152. https://doi.org/10.1016/j.fcr.2017.07.007

Shah MA, Farooq M, Shahzad M, Khan MB, Hussain M (2017) Yield and phenological responses of BT cotton to different sowing dates in semi-arid climate. Pak J Agri Sci 54(2):233–239

Singh RP, Prasad PVV, Sunita K, Giri SN, Reddy KR (2007) Influence of high temperature and breeding for heat tolerance in cotton: a review. Adv Agron 93:313–385

Sohair EDE, Abdalla AMA, Abdel-Gawad SDN, Wageda AEF (2015) Effect of delaying planting date on yield, fiber and yarn quality properties in some cultivars and promising crosses of Egyptian cotton. American-Eurasian J Agric Environ Sci 15(5):754–763

Tariq M, Afzal MN, Muhammad D, Ahmad S, Shahzad AN, Kiran A, Wakeel A (2018) Relationship of tissue potassium content with yield and fiber quality components of Bt cotton as influenced by potassium application methods. Field Crops Res 229:37–43

Tariq M, Yasmeen A, Ahmad S, Hussain N, Afzal MN, Hasanuzzaman M (2017) Shedding of fruiting structures in cotton: factors, compensation and prevention. Trop Subtrop Agroecosys 20:251–262

Tauseef M, Ihsan F, Nazir W, Farooq J (2012) Weed flora and importance value index (ivi) of the weeds in cotton crop fields in the region of Khanewal, Pakistan. Pak J Weed Sci Res 18 (3):319–330

Usman K, Ayatullah KN, Khan S (2016) Genotype-by-sowing date interaction effects on cotton yield and quality in irrigated condition of Dera Ismail Khan, Pakistan. Pak J Bot 48 (5):1933–1944

Usman M, Ahmad A, Ahmad S, Irshad M, Khaliq T, Wajid A, Hussain K, Nasim W, Chattha TM, Trethowan R, Hoogenboom G (2009) Development and application of crop water stress index for scheduling irrigation in cotton (*Gossypium hirsutum* L.) under semiarid environment. J Food Agri Environ 7(3&4):386–391

Wang X, Hou Y, Du M, Xu D, Lu H, Tian X, Li Z (2016) Effect of planting date and plant density on cotton traits as relating to mechanical harvesting in the Yellow River valley region of China. Field Crops Res 198:112–121

Wang Z, Chen J, Xing F, Han Y, Chen F, Zhang L, Li Y, Li C (2017) Response of cotton phenology to climate change on the North China Plain from 1981 to 2012. Sci Rep 7:6628

Webster TM, Grey TL, Flanders JT, Culpepper AS (2009) Cotton planting date affects the critical period of Benghal Dayflower (*Commelina benghalensis* L.) control. Weed Sci 57:81–86

Wei HU, Chen M, Zhao WQ, Chen BL, Wang YH, Wang SS, Zhao ZG (2017) The effects of sowing date on cotton seed properties at different fruiting-branch positions. J Integr Agr 16:1322–1330

Chapter 4
Sowing Methods for Cotton Production

Omer Farooq, Khuram Mubeen, Azhar Abbas Khan, and Shakeel Ahmad

Abstract Among responsible factors for getting maximum yield of cotton, appropriate sowing method has prime importance, as it plays a decisive role in proper crop stand and root penetration and development. Furthermore, suitable plant-to-plant and row-to-row adjustment accelerates the utilization of available resources. Cotton is an important cash crop besides being a big earning source for many leading cotton-producing countries of the world. Various planting techniques including flat planting by drilling in differently spaced lines, ridge sowing and raised bed planting are being practised in different countries of the world. Among these methods, each has associated merits and demerits under prevailing soil and climatic conditions. Studies have revealed that higher yield targets can be achieved through exploiting all aforementioned planting techniques, but management factor makes a real difference under all conditions. Due to specific climatic requirement, cotton crop has to deal with a variety of pests including insect, disease-causing pathogens and weeds, so, through good management practices, we can save this white gold from yield losses and get targeted yield.

Keyword Cotton sowing methods · Drilling of cotton · Ridge plantation · Bed sowing

O. Farooq (✉) · S. Ahmad (✉)
Department of Agronomy, Bahauddin Zakariya University, Multan, Pakistan
e-mail: shakeelahmad@bzu.edu.pk

K. Mubeen
Muhammad Nawaz Shareef University of Agriculture Multan, Multan, Punjab, Pakistan

A. A. Khan
College of Agriculture, Bahauddin Zakariya University, Layyah, Pakistan

S. Ahmad, M. Hasanuzzaman (eds.), *Cotton Production and Uses*,
https://doi.org/10.1007/978-981-15-1472-2_4

4.1 Introduction

There are many factors responsible for the good growth and yield of cotton crop; following those we can get potential yield of all cultivars (Ahmad et al. 2014, 2017, 2018; Abbas and Ahmad 2018; Ahmad and Raza 2014; Ali et al. 2011, 2013a, b, 2014a, b). Among these factors, selection of good planting method is a prerequisite in overcoming the yield challenges of this important crop. Moreover it plays a crucial role in getting desired crop stand and proper root penetration and development and also accelerates the utilization of available resources through maintaining proper plant-to-plant besides row-to-row distances.

Plant growth along with development is mainly influenced by favourable soil conditions i.e. appropriate moisture contents and temperature and minimum root penetration resistance. Consequently, suitable sowing technique is required which may utilize a tillage system that improves physical characteristics, germination, growth and yield (Krause et al. 2009).

However, each sowing method has its own associated merits and demerits under different conditions; therefore, studies are required to determine suitability of different sowing techniques on site-specific basis. Cotton is a big earning source for many countries of the world (Amin et al. 2017, 2018; Khan et al. 2004; Rahman et al. 2018; Tariq et al. 2017, 2018; Usman et al. 2009). World top exporters of cotton are the USA, India, Brazil and Australia (Statista 2018). Cotton is usually planted by various methods depending upon the available resources and objectives of the growers/farmers and researchers. Primarily, main focuses should be available seed type, soil conditions/soil physical properties, irrigation facilities available/irrigation methods to be followed and tools to deal with the associated cotton pests (weeds, insects and diseases).

Among all aspects, cotton planting technique is a dynamic feature that has momentous influence on growth and development of crop and eventually its final yield. Several sowing techniques such as flat planting by drilling in differently spaced lines, ridge sowing and raised bed planting are being followed for cultivation in Pakistan and rest of the world. Moreover, considering all the factors responsible for development, growth and lint yield of cotton, sowing technique has crucial importance as it not only supports in instituting proper crop stand but also assists in getting maximum yield contributing traits (Ali et al. 2012). Furthermore, planting geometry plays a significant role in the expression of plant growth and yield characteristics of cotton crop (Kaggwa-Asiimwe et al. 2013; Mao et al. 2014). Canopy arrangement, light interception and ultimately fruit development are all highly affected by cotton population density. It has also been well studied that plant-to-plant distance modifies plant height, photosynthetic efficiency of plant leaves, plant architecture and finally fruiting boll size (Hussain et al. 2000; Siebert et al. 2006). Increasing plant density may increase the lint yield and leaf area index, but the individual leaf area can be reduced (Gwathmey and Clement 2010). Similarly comparison among different planting densities provides that dry matter partitioning towards fruit was higher at 11 plants m^{-2} compared to 6 or 15 plants m^{-2} (Wang et al. 2011). Such parameters like change in canopy size, leaf area and shoot area determine light harvesting and also light-driven plasticity, e.g. foliage inclination angle distribution and spatial aggregation (Niinemets 1998).

Additionally, better planting techniques not only support in maintaining ideal plant population with improved emergence but also empower plants to uniformly and efficiently utilize resources like land, light and other applied inputs. It is therefore necessary to develop improved planting techniques to avoid unnecessary plant crowding and thereby to facilitate plants' more efficient and effective exploitation of aforementioned resources (Quanqi et al. 2008). Currently, success of modern agriculture is totally dependent upon the plentifulness of quality irrigation water; unluckily, continuous supply of good-quality irrigation water is a big challenge, and the focus of the world is water-wise cultivation (Deng et al. 2006). Accordingly, water-wise cultivation emphasizes on improving water use efficiency (WUE) by harvesting higher economic returns from our crop lands with least water demand (Ali and Ehsanullah 2007; Farooq et al. 2009). Enhancing WUE is an important tool to economize full use of precipitations besides skillful management of irrigation network by choosing an appropriate planting technique (Hussain et al. 2003). In addition to enhancing the WUE, different planting techniques also improve water application efficiency.

4.2 Different Planting Methods for Cotton Production

4.2.1 Bed Planting of Cotton

Among all the planting techniques of cotton, this method is the most advanced and profitable method (Table 4.1). Raised seedbeds are prepared mechanically with the help of a raised bed planter/shaper (Fig. 4.1), while the shape of the bed is shown in Fig. 4.2. Seeds are manually planted to a depth of 1 inch, whereas the planting geometry is as follows: row-to-row distance on beds is 75 cm, while plants are spaced at 25 cm. Sowing deeper than this depth may cause germination problem.

Planting of seed on raised beds enhances germination percentage and also eliminates the formation of crust that sometimes reduces seed germination (Ahmad et al. 2009). Furthermore, bed planting technique also improves crop stand, ensures better plant population and hence the final yield, and in comparison to flat planting, yield obtained by bed planting is much better that sometimes may be up to 33% (Iftikhar et al. 2010; Anwar et al. 2003).

Worldwide bed planting technique is considered as one of the best planting techniques due to several benefits like good crop stand, drainage of surplus water after rainfall, less delta of water required, better nutrient use efficiency, better water distribution efficiency, less lodging in comparison to flat sowing, better weed control and increased yield (Hobbs and Gupta 2003; Sayre 2000). Bed planting resulted in considerable water saving, which may be up to 30% (Fahong et al. 2003) compared

Table 4.1 Yield comparison of cotton planted by different sowing techniques

Sr. #	Maximum yield producing sowing method	Comparison with other sowing method/s	References
	Bed sowing $\frac{R\times R-55\ \text{cm}}{P\times P-30\ \text{cm}}$	Bed sowing $\frac{R\times R-70\ \text{cm}}{P\times P-30\ \text{cm}}$	Bakhsh et al. (2018)
1	Bed sowing $\frac{R\times R-75\ \text{cm}}{P\times P-22.5\ \text{cm}}$	Ridge sowing and flat sowing	Irfan et al. (2014)
2	Flat sowing $\frac{R\times R-25\ \text{cm}}{P\times P-30\ \text{cm}}$	Bed sowing and ridge sowing	Ehsanullah et al. (2017)
3	Ridge sowing $\frac{R\times R-75\ \text{cm}}{P\times P-30\ \text{cm}}$	Flat sowing	Nadeem et al. (2013)
4	Bed sowing $\frac{R\times R-75\ \text{cm}}{P\times P-22.5\ \text{cm}}$	Ridge sowing and flat sowing	Ali et al. (2010)
5	Ridge sowing $\frac{R\times R-75\ \text{cm}}{P\times P-30\ \text{cm}}$	Flat sowing	Buttar et al. (2005)
6	Paired row sowing $\frac{R\times R-35/100\ \text{cm}}{P\times P-30\ \text{cm}}$	Single row	Aujla et al. (2005)
7	Row spacing $\frac{R\times R-100\ \text{cm}}{P\times P-40\ \text{cm}}$	Row spacings $\frac{R\times R-100\ \text{cm}}{P\times P-50\&60\ \text{cm}}$	Jat et al. (2014)
8	Row spacing $\frac{R\times R-100\ \text{cm}}{P\times P-40\ \text{cm}}$	Row spacings $\frac{R\times R-100\ \text{cm}}{P\times P-50\&60\ \text{cm}}$	Narkhede et al. (2000)
9	Row spacing $\frac{R\times R-120\ \text{cm}}{P\times P-45\ \text{cm}}$	Row spacings $\frac{R\times R-120\&150\ \text{cm}}{P\times P-60, 90\&45, 60\ \text{cm}}$	Narkhede et al. (2000)
10	Row spacing $\frac{R\times R-75\ \text{cm}}{P\times P-15\ \text{cm}}$	Row spacings $\frac{R\times R-75\ \text{cm}}{P\times P-22\&30\ \text{cm}}$	Nehra and Kumawat (2003)
11	Ridge sowing $\frac{R\times R-75\ \text{cm}}{P\times P-30\ \text{cm}}$	Flat sowing	Maqsood et al. (2006)
12	Yield comparison between flat and ridge sowing was NS[a]		Maqbool et al. (2001)

[a]Non-significant

Fig. 4.1 Raised bed planter

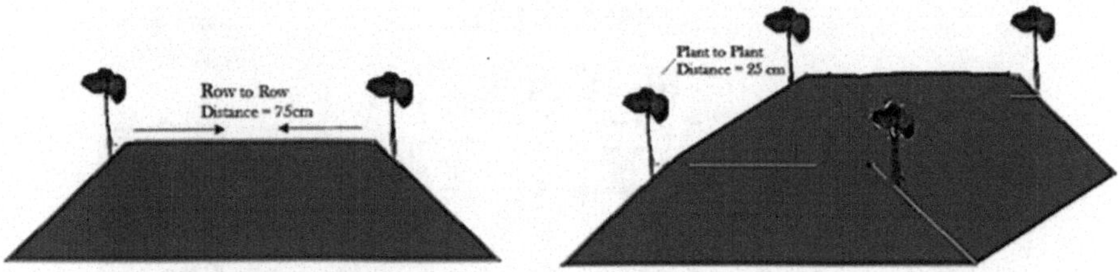

Fig. 4.2 Sketch of cotton raised bed and planting geometry

to conventional sowing technique. Better crop stand besides enhanced yield is due to enhancement in root proliferation in bed planting (Ahmad and Mahmood 2005).

Practically, bed sowing technology has the following benefits that motivate the growers to follow this system of planting:

- Less irrigation water is required and water use efficiency is also improved.
- Weed control is easier and less competition between cotton plants and weeds; further mechanical weed control is possible with good efficacy.
- Plant growth is much improved and better yield is estimated.
- There is low insect and disease pressure in comparison to other planting techniques.

4.2.2 Ridge Planting of Cotton

Ridge planting is another important sowing technique of cotton dominantly practised in all the cotton-producing countries of the world and gave higher yield in comparison to other sowing techniques (Table 4.1). Ridges are also prepared mechanically with the help of ridger (Fig. 4.3), while the shape of the ridges is shown in Fig. 4.4.

Ridge-furrow planting of cotton, where ridge and furrows are formed with a ridger while cotton is planted manually on the side of the ridge, augmented germination and yields of cotton compared to flat sowing and is widely practised in Pakistan and other cotton-producing countries of the world. Ridge-sown cotton has been found to increase physical characteristics like increased moisture contents and decrease root penetration resistance (Gürsoy et al. 2011). The ridge planting offered high lint productivity along with more earliness compared to conventional planting methods. However, cotton sown on beds produced high SCY (seed cotton yield) compared to ridge and flat sowing (Ali et al. 2012). Hussain et al. (2003) found that ridge-sown cotton significantly produced higher yield as compared to flat-sown crop.

Fig. 4.3 Ridge planter

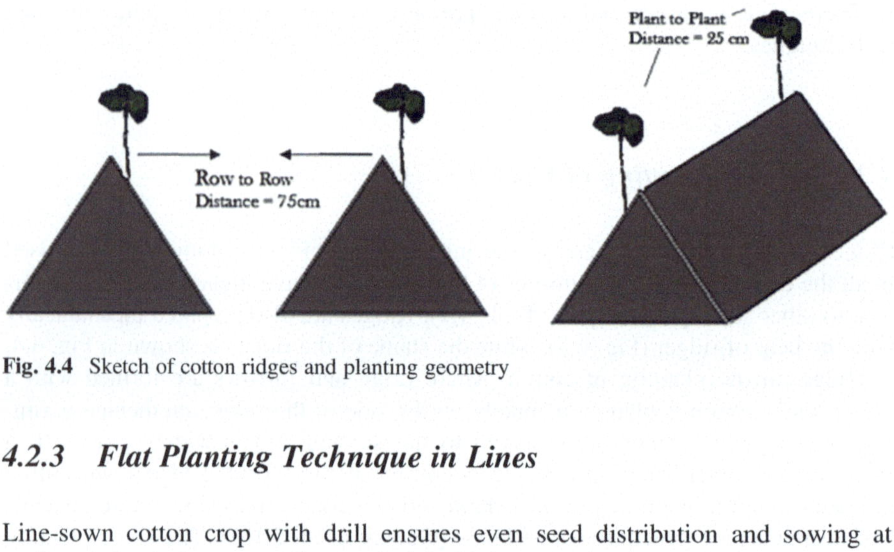

Fig. 4.4 Sketch of cotton ridges and planting geometry

4.2.3 *Flat Planting Technique in Lines*

Line-sown cotton crop with drill ensures even seed distribution and sowing at preferred depth that generally resulted in higher germination and even crop stand (Tariq et al. 2001). This method of planting cotton is also widely adopted in all cotton-growing areas and yields best results (Table 4.1). Usually the same planting geometry is followed as followed by ridge planting technique, i.e. R-R, 75 cm, and P-P, 25 cm (Fig. 4.5). Row spacing is a key management component in improving crop productivity (Saleem et al. 2009), and different experiment has been performed by changing P-P and R-R distances (Table 4.1). Cotton sowing in narrow rows might be a feasible strategy than conventional grown to increase lint harvest (Jahedi et al. 2013). Whereas, number of bolls plant^{-1}, boll weight and lint productivity increase

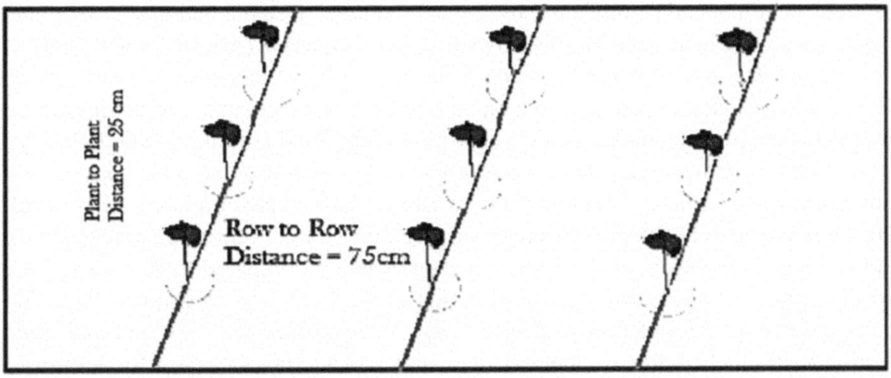

Fig. 4.5 Sketch of cotton sowing on flat surface and planting geometry

at wider plant spacings (Boquet 2005), it has also been witnessed that higher plant density compensates for losses in yield (Kasap and Killi 2004).

4.3 Weeds' Behaviour Under Different Sowing Methods

Growth behaviour of weeds varies under diverse planting techniques. In flat sowing, weeds are present everywhere, while in ridge sowing, weeds grow at specific places. Herbicides' action also differs in diverse sowing techniques (Maqbool et al. 2001). Weeds incur potential damages to crops and ecosystems. Interference with crops; reduction in growth, yield and quality of crops; serving as alternate host for many insect pests and pathogens; and affecting humans and livestock are some of the drawbacks due to weeds. Sowing methods which favour narrow row widths and increased plant population per unit area will reduce the sunlight available for weeds resulting in reduced weed dry mass emerging later under the canopy of the crop. Rapid canopy closure reduced soil weed seed bank, and reduced growth of weed seedlings can be achieved by growing crops in closer rows. Under such situation weeds' growth will reduce resulting in effective weed management and improved crop productivity. Sowing patterns can improve crop competitiveness, reduce weed suppressive ability and ultimately improve crop production (Sumathi et al. 2010). Sowing methods change the microclimate and structure of crop canopy; hence weed competition can be reduced (Singh 2014).

Weed behaviour changes with change in community behaviour in response to change in sowing method of cotton. Demographic quantitative models lay the foundation of predicting future weed behaviour. Change in sowing method of cotton may bring changes in sowing time, plant population and plant spacing which brings changes in microclimate, physiology of crop plants, pest damage, weed dynamics, damages incurred by weeds, etc. Weeds behave differently in response to variation in the method of sowing cotton by bringing change in weed population, and weed biomass is influenced significantly by sowing methods. Bed plantation may facilitate better seedling emergence and avoids soil crust formation (Ahmad et al. 2009); it may result in weed seedling emergence on beds with little weeds in the furrows. Hence it explains why cotton grown on beds gave better yield than flat sowing where weeds have equal chance to grow everywhere in the field (Iftikhar et al. 2010).

Cotton sown on ridges also reveals improved emergence and seed cotton yield due to improved soil moisture and less resistance to root penetration (Gürsoy et al. 2011). It may also be attributed to reduced weed density per unit area compared with flat sowing. Relocation of weed seeds in ridge sowing compared to flat sowing at a specific place will vary for weed positioning in the field, and subsequent herbicide application method and dose will also vary. As compared with flat bed and ridge furrow method, bed furrow method of cotton sowing favours crop for improved growth and development and reduces weed infestation (Jha and Soni 2013). Studies on pit plantation of cotton have been undertaken in the past that resulted in more fruiting potential and boll size (Nazir et al. 2001). However it still needs to evaluate existing sowing methods of cotton including its impact on weed dynamics. Line-sown cotton through drill results in uniform seed distribution with desired seeding depth and even crop stand (Tariq et al. 2001) and leaves reduced space and liberty for weeds to grow when compared with flat sowing of cotton.

4.4 Insects' Behaviour Under Different Sowing Methods

Insects and weeds are the most common factors which cause severe damages to field crops by 37% and 34%, respectively. Additionally viruses and pathogens have share of 11% loss if kept untreated (Oerke and Dehne 2004). Surveys indicate that a massive loss of cotton produce (up to 70%) may occur due to insect pests' infestation in the absence of any compatible control measure (Rehman et al. 2017). Many insect pests are responsible for decreasing profit and output of cotton crop including whitefly, mealy bug, jassid, dusky cotton bug and a large number of lepidopteron species. About 18.78% drop in cotton yield has been observed due to infestation of jassid (Ali 1992), and huge yield losses (40–50%) in cotton crop have been recorded during severe attack of thrips and jassid in cotton field plots (Naqvi 1976). The CLCuV causal agent carrier "whitefly" (Malik et al. 1999) disturbs photosynthesis process of crop in addition to direct losses through sucking cell sap of plant, and in Pakistan ~38.7% income reduction has been recorded in year 1993 (Khan and Khan 1995). Spacing alters the behaviour of insects by reducing egg laying site and shelter leading to plant growth and lower risk of pest outbreak and diseases. Intercropping is considered as vital constituent of cropping systems in developing nations (Sodiya et al. 2010). Experiments conducted for cotton thrips in onion field plots showed that onion plus clover intercropping causes lessening in thrips besides higher productivity (Trdan et al. 2006; Hildenhagen et al. 1995). Narrow-row or ultra-narrow-row cotton cultivation is another well-known procedure to cultivate the cotton crop escape of harmful insects and diseases. The population of whitefly and jassid has been recorded more in field plots where plant spacing was less (22.5 cm) and vice versa for wider planting space (30 cm). Increasing plant spacing from 23 cm to 30 cm, sucking insect pests (whitefly) population is toward a decreasing trend; similarly higher population of jassid (39/plant) has also been observed in least spacing compared to wider plant spacing. This decreased number of insect pests is due to sufficient aeration in plants and may have effect on pest infestation, their growth and their development (Arif et al. 2006; Hussain 2017). At the same time, vigorous growth of plant may facilitate strong resistance mechanisms to tolerate, escape or prevent pest attack.

Generally it is thought that tillage does not have any adverse effect on insects population in cotton fields. However, few studies indicated that ridge plantation of cotton lowers down the population of *Empoasca* spp. and *Bemisia tabaci* when compared with conventional sowing methods (Gencsoylu and Yalcin 2004). At the same time it has also been recorded that some predacious beetles, i.e. *Pterostichus* spp., were found more (0.27/trap) in conventional plots as compared to conservation tillage system. However, in proceeding year, more numbers (3.77/trap) of *C. sexguttata* were observed in plots where conservation tillage was applied. The predacious ground beetles richness in conservation tillage plots showed that field conditions, i.e. sowing methods, tillage and plantation space, have significant effect on population of predatory insects that ultimately keep check on soft bodied insect pests and help to keep the pests population below economic threshold level (ETL) (Shrestha and Parajulee 2010).

4.5 Conclusion

In conclusion, we can say that each planting technique has comparative advantages for the cotton grower. There are many factors like soil type, availability of irrigation water and insect pest management, responsible in selecting the best choice for cotton sowing. However bed sowing can be considered as a best choice for cotton plantation in getting higher net benefits and good economic returns to the farmers. Furthermore, management factor should always be given the highest priority after selecting any of the planting techniques of cotton, as it has major role in getting high yield targets of this crop.

References

Abbas Q, Ahmad S (2018) Effect of different sowing times and cultivars on cotton fiber quality under stable cotton-wheat cropping system in southern Punjab, Pakistan. Pak J Life Soc Sci 16:77–84

Ahmad N, Arshad M, Shahid MA (2009) Bed-furrow system to replace conventional flood irrigation in Pakistan. In: Proceedings of 59 IEC meeting and 20 ICID conference held at New Delhi, India, pp 6–11

Ahmad N, Mahmood N (2005) Impact of raised bed technology on water productivity and lodging of wheat. Pak J Water Res 9:1–9

Ahmad S, Abbas Q, Abbas G, Fatima Z, Atique-ur-Rehman, Naz S, Younis H, Khan RJ, Nasim W, Habib urRehman M, Ahmad A, Rasul G, Khan MA, Hasanuzzaman M (2017) Quantification of climate warming and crop management impacts on cotton phenology. Plan Theory 6(7):1–16

Ahmad S, Iqbal M, Muhammad T, Mehmood A, Ahmad S, Hasanuzzaman M (2018) Cotton productivity enhanced through transplanting and early sowing. Acta Sci Biol Sci 40, e34610

Ahmad S, Raza I (2014) Optimization of management practices to improve cotton fiber quality under irrigated arid environment. J Food Agri Environ 2(2):609–613

Ahmad S, Raza I, Ali H, Shahzad AN, Atiq-ur-Rehman, Sarwar N (2014) Response of cotton crop to exogenous application of glycinebetaine under sufficient and scarce water conditions. Braz J Bot 37(4):407–415

Ali A (1992) Physio-chemical factors affecting resistance in cotton against jassid, *Amrasca devastans* (Dist.) and thrips, *Thrips tabaci* (Lind.) in Punjab, Pakistan. Ph.D. dissert. Dept. Entomol., Univ. Agric. Faisalabad, Pakistan, pp 430

Ali M, Ali L, Sattar M, Ali MA (2010) Response of seed cotton yield to various plant populations and planting methods. J Agric Res 48:134–137

Ali H, Abid SA, Ahmad S, Sarwar N, Arooj M, Mahmood A, Shahzad AN (2013a) Integrated weed management in cotton cultivated in the alternate-furrow planting system. J Food Agri Environ 11(3&4):1664–1669

Ali H, Abid SA, Ahmad S, Sarwar N, Arooj M, Mahmood A, Shahzad AN (2013b) Impact of integrated weed management on flat-sown cotton (*Gossypium hirsutum* L.). J Anim Plant Sci 23 (4):1185–1192

Ali H, Afzal MN, Ahmad F, Ahmad S, Akhtar M, Atif R (2011) Effect of sowing dates, plant spacing and nitrogen application on growth and productivity on cotton crop. Int J Sci Eng Res 2 (9):1–6

Ali H, Hameed RA, Ahmad S, Shahzad AN, Sarwar N (2014a) Efficacy of different techniques of nitrogen application on American cotton under semi-arid conditions. J Food Agri Environ 12 (1):157–160

Ali H, Hussain GS, Hussain S, Shahzad AN, Ahmad S, Javeed HMR, Sarwar N (2014b) Early sowing reduces cotton leaf curl virus occurrence and improves cotton productivity. Cer Agron Moldova XLVII(4):71–81

Ali L, Ehsanullah (2007) Seed cotton yield and water use efficiency of cotton (*Gossypium hirsutum* L.) sown in different planting methods. Pak J Agri Sci 44(4):571–574

Ali M, Ali L, Sattar M, Ali MA (2012) Response of seed cotton yield to various plant populations and planting methods. J Agric Res 48:134–137

Amin A, Nasim W, Mubeen M, Ahmad A, Nadeem M, Urich P, Fahad S, Ahmad S, Wajid A, Tabassum F, Hammad HM, Sultana SR, Anwar S, Baloch SK, Wahid A, Wilkerson CJ, Hoogenboom G (2018) Simulated CSM-CROPGRO-cotton yield under projected future climate by SimCLIM for southern Punjab, Pakistan. Agric Syst 167:213–222

Amin A, Nasim W, Mubeen M, Nadeem M, Ali L, Hammad HM, Sultana SR, Jabran K, Habib urRehman M, Ahmad S, Awais M, Rasool A, Fahad S, Saud S, Shah AN, Ihsan Z, Ali S, Bajwa AA, Hakeem KR, Ameen A, Amanullah, Rehman HU, Alghabar F, Jatoi GH, Akram M, Khan A, Islam F, Ata-Ul-Karim ST, Rehmani MIA, Hussain S, Razaq M, Fathi A (2017) Optimizing the phosphorus use in cotton by using CSM-CROPGRO-cotton model for semi-arid climate of Vehari-Punjab, Pakistan. Environ Sci Pollut Res 24(6):5811–5823

Anwar MM, Gill MI, Zaki MS (2003) Effect of bed-furrow planting on cotton crop. The Pak Cottons 47:41–46

Arif MJ, Gogi MD, Mirza M, Zia K, Hafeez F (2006) Impact of plant spacing and abiotic factors on population dynamics of sucking insect pests of cotton. Pak J Biol Sci 9:1364–1369

Aujla MS, Thind HS, Buttar GS (2005) Cotton yield and water use efficiency at various levels of water and N through drip irrigation under two methods of planting. Agric Water Manag 71:167–179

Bakhsh A, Chauhdary JN, Ahmad N (2018) Improving crop water productivity of major crops by adopting bed planting in Rechna Doab Pakistan. Pak J Agri Sci 55:965–972
Boquet DJ (2005) Cotton in ultra- narrow row spacing: plant density and nitrogen fertilizer rates. Agron J 97:279–287
Buttar GS, Mahey RK, Aggarwal N (2005) Effect of sowing dates, planting methods and irrigation scheduling on the growth and yield of American cotton (*Gossypium hirsutum*). J Cotton Res Dev 19:213–215
Deng XP, Shan L, Zhang H, Turner N (2006) Improving agricultural water use efficiency in arid and semiarid areas of China. Agr Water Manage 80:23–40
Ehsanullah, Shahzad MA, Anjum SA, Zohaib A, Ishfaq M, Warraich EA (2017) Effect of different sowing methods and planting densities on growth, yield, fiber quality and economic efficacy of cotton. Pak J Agric Res 30(1):67–74
Fahong W, Xuquing W, Sayre KD (2003) Comparison study on two different planting methods for winter wheat in China, bed planting course. CIMMYT, Mexico city
Farooq M, Kobayashi N, Wahid A, Ito O, Basra SMA (2009) Strategies for producing more rice with less water. Adv Agron 101:351–388
Gencsoylu I, Yalcin I (2004) The effect of different tillage systems on cotton pests and predators in cotton fields. Asian J Plant Sci 3(1):39–44
Gürsoy S, Sessiz A, Karademir E, Karademir C, Kolay B, Urgun M, Malhi SS (2011) Effects of ridge and conventional tillage systems on soil properties and cotton growth. Int J Plant Prod 5 (3):227–236
Gwathmey CO, Clement JD (2010) Alteration of cotton source–sink relations with plant population density and mepiquat chloride. Field Crops Res 116:101–107
Hildenhagen R, Richter E, Hommes M (1995) Vorkommen und gezielteBekämpfung von *Thrips tabaci* an Porree und Zwiebeln. Mitt Dtsch Ges Allgangew Ent 10:183–187
Hobbs PR, Gupta RK (2003) Resource conserving technologies for wheat in rice–wheat systems. In: Ladha JK, Hill J, Gupta RK, Duxbury J, Buresh RJ (eds) Improving the productivity and sustainability of rice–wheat systems: issues and impact, vol. 65, paper 7, ASA special publications. ASA, Madison, WI, pp 149–171
Hussain F (2017) Impact of various agronomic practices on the incidence of cotton leaf curl virus. PhD. thesis, Department of Agronomy, Faculty of Agriculture, University of Agriculture, Faisalabad, Pakistan, pp 129
Hussain S, Farid ZS, Anwar M, Gill MI, Baugh MD (2000) Effect of plant density and nitrogen on the yield of seed cotton variety CIM-443. Sarhad J Agri 16:143–147
Hussain T, Jehanzeb, Tariq M, Siddiqui BN (2003) Effect of different irrigation levels on the yield and yield components of cotton (*Gossypium hirsutum* L.) under two sowing methods. J Biol Sci 3:655–659
Iftikhar T, Babar LK, Zahoor S, Khan NG (2010) Impact of land pattern and hydrological properties of soil on cotton yield. Pak J Bot 42:3023–3028
Jahedi MB, Vazin F, Ramezani MR (2013) Effect of row spacing on the yield of cotton cultivars. Cer Agron Moldova 46:31–38
Jat RD, Nanwal RK, Kumar P, Shivran AC (2014) Productivity and nutrient uptake of *Bt* cotton (*Gossypium hirsutum* L) under different spacing and nutrient levels. J Cotton Res Dev 28:260–262
Jha AK, Soni M (2013) Weed management by sowing methods and herbicides in soybean. Indian J Weed Sci 45:250–252
Kaggwa-Asiimwe R, Andrade-Sanchez P, Wang G (2013) Plant architecture influences growth and yield response of upland cotton to population density. Field Crops Res 145:52–59
Kasap Y, Killi F (2004) Effect of row space and nitrogen interaction on seed cotton (*Gossypium hirsutum* L.) yield under irrigated conditions of Turkey. Ind J Agron 49:64–67
Khan MB, Khaliq A, Ahmad S (2004) Performance of mashbean intercropped in cotton planted in different planting patterns. J Res (Sci) 15(2):191–197

Khan WS, Khan AG (1995) Cotton situation in the Punjab, an overview. National Seminar on Strategies for Increasing Cotton Production in Agriculture

Krause U, Koch HJ, Maerlaender B (2009) Soil properties effecting yield formation in sugar beet under ridge and flat cultivation. Eur J Agron 31(1):20–28

Malik AK, Mansoor S, Saeed NA, Asad S, Zafar Y, Stanley J, Markham (1999) Development of CLCV resistance cotton varieties through genetic engineering. Mongr. Directorate Agric Inform Punjab Pakistan 3

Mao L, Zhang L, Zhao X, Liu S, van der Werf S, Zhang S, Spiertz H, Li Z (2014) Crop growth, light utilization and yield of relay intercropped cotton as affected by plant density and a plant growth regulator. Field Crops Res 155:67–76

Maqbool MM, Tanveer A, Ali A, Ahmad R (2001) Effect of sowing methods and herbicides on weeds and yield of cotton (*Gossypium hirsutum* L.). Pak J Bot 33:383–387

Maqsood M, Hussain T, Tayyab M, Ibrahim M (2006) Effect of different irrigation levels on the yield and radiation use efficiency of cotton (*Gossypium hirsutum*) under two sowing methods. Pak J Agri Sci 43:21–24

Nadeem MA, Idrees M, Ayub M, Tanveer A, Mubeen K (2013) Effect of different weed control practices and sowing methods on weeds and yield of cotton. Pak J Bot 45:1321–1328

Naqvi KM (1976) Crop protection to boost up cotton production. In: Seminar organized by ESSO. Fertilizer Company Limited Pakistan 74

Narkhede WN, Jadhao JK, Bhatade SS, Dhoble MV (2000) Response of upland cotton varieties (*Gossypium hirsutum*) to plant densities and nitrogen levels on vertisol under rainfed condition. J Cotton Res Dev 14:30–32

Nazir MS, Jabbar A, Gardezi TR, Bhatti IH, Nawaz S (2001) Effect of bunch and flat plantation in different spatial arrangements on the yield and staple length of upland cotton. J Bio Sci 1:565–567

Nehra PL, Kumawat PD (2003) Response of *G. hirsutum* cotton varieties to spacing and nitrogen levels. J Cotton Res Dev 17:41–42

Niinemets U (1998) Adjustment of foliage structure and function to a canopy light gradient in two co-existing deciduous trees. Variability in leaf inclination angles in relation to petiole morphology. Trees 12:446–451

Oerke EC, Dehne HW (2004) Safeguarding production-losses in major crops and the role of crop protection. Crop Prot 23:275–285

Quanqi L, Yuhai C, Mengyu L, Xunbo Z, Baodi D, Songlie Y (2008) Water potential characteristics and yield of summer maize in different planting patterns. Plant Soil Environ 54:14–19

Rahman MH, Ahmad A, Wang X, Wajid A, Nasim W, Hussain M, Ahmad B, Ahmad I, Ali Z, Ishaque W, Awais M, Shelia V, Ahmad S, Fahad S, Alam M, Ullah H, Hoogenboom G (2018) Multi-model projections of future climate and climate change impacts uncertainty assessment for cotton production in Pakistan. Agric For Meteorol 253-254:94–113

Rehman A, Jingdong L, Chandio AA, Hussain I, Wagan SA, Memon QA (2017) Economic perspectives of cotton crop in Pakistan: a time series analysis (1970–2015) (Part 1). J Saudi Soci Agric Sci 18(1):49–54

Saleem MF, Anjum SA, Shakeel A, Ashraf MY, Khan HZ (2009) Effect of row spacing on earliness and yield in cotton. Pak J Bot 41:2179–2188

Sayre KD (2000) Effects of tillage, crop residue retention and nitrogen management on the performance of bed-planted, furrow irrigated spring wheat in northwest Mexico. Presented at the 15th conference of the international soil tillage research organization, 2–7 July 2000, Fort Worth, TX

Shrestha RB, Parajulee MN (2010) Effect of tillage and planting date on seasonal abundance and diversity of predacious ground beetles in cotton. J Ins Sci 10(174):1–17

Siebert JD, Stewart AM, Leonard BR (2006) Comparative growth and yield of cotton planted at various densities and configurations. Agron J 98:562–568

Singh R (2014) Weed management in major kharif and rabi crops. Nat Training Adv Weed Manage:31–40

Sodiya AS, Akinwale AT, Okeleye KA, Emmanuel JA (2010) An integrated decision support system for intercropping. In J Decis Suppo Syst Tech 2:51–66

Statista (2018) Leading cotton exporting countries in 2017/2018. US Department of Agriculture; Economic Research Service

Sumathi V, Subramanyam D, Rao DSK, Reddy DS (2010) Effect of planting pattern and weed management on weed flora and yield of rabi sunflower. Indian J Weed Sci 42:212–216

Tariq M, Afzal MN, Muhammad D, Ahmad S, Shahzad AN, Kiran A, Wakeel A (2018) Relationship of tissue potassium content with yield and fiber quality components of Bt cotton as influenced by potassium application methods. Field Crops Res 229:37–43

Tariq M, Ali H, Jan A (2001) Influence of sowing methods and mulching on yield and yield components of wheat. Pak J Biol Sci 4:657–659

Tariq M, Yasmeen A, Ahmad S, Hussain N, Afzal MN, Hasanuzzaman M (2017) Shedding of fruiting structures in cotton: factors, compensation and prevention. Trop Subtrop Agroecosyst 20(2):251–262

Trdan S, Znicdarcic D, Valic N, Rozman L, Vidrih M (2006) Intercropping against onion thrips, *Thrips tabaci* Lindeman (Thysanoptera: Thripidae) in onion production: on the suitability of orchard grass, *Lacy phacelia*, and buckwheat as alternatives for white clover. J Plant Dis Protect 113:24–30

Usman M, Ahmad A, Ahmad S, Irshad M, Khaliq T, Wajid A, Hussain K, Nasim W, Chattha TM, Trethowan R, Hoogenboom G (2009) Development and application of crop water stress index for scheduling irrigation in cotton (*Gossypium hirsutum* L.) under semiarid environment. J Food Agri Environ 7(3&4):386–391

Wang G, Kaggwa-Asiimwe R, Andrade SP (2011) Growth and yield response to plant population of two cotton varieties with different growth habits. Arizona Cotton Report, August 2011

Chapter 5
Irrigation Scheduling for Cotton Cultivation

Sajjad Hussain, Ashfaq Ahmad, Aftab Wajid, Tasneem Khaliq, Nazim Hussain, Muhammad Mubeen, Hafiz Umar Farid, Muhammad Imran, Hafiz Mohkum Hammad, Muhammad Awais, Amjed Ali, Muhammad Aslam, Asad Amin, Rida Akram, Khizer Amanet, and Wajid Nasim

S. Hussain · M. Mubeen (✉) · M. Imran · H. M. Hammad · R. Akram · K. Amanet
Department of Environmental Sciences, COMSATS Institute of Information Technology, Vehari, Pakistan
e-mail: muhammadmubeen@ciitvehari.edu.pk

A. Ahmad
Program Chair, Climate Change, US.-Pakistan Centre for Advanced Studies in Agriculture and Food Security, University of Agriculture Faisalabad, Faisalabad, Punjab, Pakistan

A. Wajid · T. Khaliq
Agro-Climatology Lab, Department of Agronomy, University of Agriculture Faisalabad, Faisalabad, Punjab, Pakistan

N. Hussain
Department of Agronomy, Bahauddin Zakariya University, Multan, Punjab, Pakistan

H. U. Farid
Department of Agricultural Engineering, Bahauddin Zakariya University, Multan, Punjab, Pakistan

M. Awais
Department of Agronomy, University College of Agriculture and Environmental Sciences, The Islamia University of Bahawalpur (IUB), Bahawalpur, Punjab, Pakistan

A. Ali
University College of Agriculture, University of Sargodha, Sargodha, Punjab, Pakistan

M. Aslam
Department of Agriculture (Extension Wing), Government of Punjab, Lahore, Punjab, Pakistan

A. Amin
Department of Environmental Sciences, COMSATS Institute of Information Technology, Vehari, Pakistan

Queensland Alliance for Agriculture and Food Innovation (QAAFI), The University of Queensland, Brisbane, QLD, Australia

W. Nasim
Department of Agronomy, University College of Agriculture and Environmental Sciences, Islamia University of Bahawalpur, Bahawalpur, Pakistan

S. Ahmad, M. Hasanuzzaman (eds.), *Cotton Production and Uses*,
https://doi.org/10.1007/978-981-15-1472-2_5

Abstract Crops need water for evaporation (E) and transpiration (T), known as evapotranspiration (ET). However, too much water is not good for various crops. Crop water need depends on growth stage, climate, and crop type. Approximately 40% cotton is produced under irrigated conditions. Water for irrigation is becoming limited in many cotton-growing regions and competition for water is increasing speedily in areas normally having plentiful water resources. So, many cotton producers and the associations representing cotton producers are interested in the scheduling of irrigation strategies that increase water use efficiency (WUE). Responses of cotton under water stress depend on stage of growth, duration, time, and extent of stress. Cotton is a drought-tolerant crop; however, it performs better under optimum water conditions. The water requirement of cotton is 27–51 acre inches depending upon growing duration and prevailing climatic conditions. However, it is essential to apply uniform and accurate amount of water at proper time for maximum cotton yield. Normally, cotton uses less water from sowing to emergence. However, pre-sowing irrigation is mandatory to ensure good cotton seed germination. After germination, crop water demand increases from 0.2 to 0.44 in. per day. Lack of water can reduce plant growth, the number of fruiting sites because of shedding of young bolls, and boll size, consequently resulting in loss of yield potential. There are various irrigation scheduling tools, the main purpose of which is to supply water according to the need of the plant. Water balance method, estimating crop water use, and sensor-based scheduling are a few important tools to maintain irrigation scheduling in cotton.

Keywords Water use efficiency · Evapotranspiration · Irrigation methods · Irrigation techniques

Abbreviations

CO_2	Carbon dioxide
CT	Canopy temperature
ET	Evapotranspiration
FDR	Frequency domain reflectometry
GMS	Granular matrix sensors
MIST	Mississippi irrigation scheduling tool
SDI	Subsurface drip irrigation
TDR	Time domain reflectometry
VMC	Volumetric moisture content
WUE	Water use efficiency

5.1 Introduction

Cotton is the major irrigated crop worldwide (Ahmad et al. 2014, 2017, 2018; Abbas and Ahmad 2018; Ahmad and Raza 2014; Ali et al. 2011, 2013a, b, 2014a, b), and precise application of irrigation is important for optimum yield. Currently, the irrigation requirement is changed due to unfavorable climatic conditions. Usually no irrigation is required during the rainy season until and unless rainfall is very erratic. *Kharif* cotton is usually rain-fed and is sown in the second week of June with the start of the monsoon, while summer crop is sown in the second week of May (Amin et al. 2017, 2018; Khan et al. 2004; Rahman et al. 2018; Tariq et al. 2017, 2018; Usman et al. 2009). If there is no optimum rainfall, preplanting irrigation is vital for ensuring excellent germination. Cotton needs sufficient water during pre-budding and through bud formation. The seasonal length of cotton varies from 150 to 195 days. Two or three subsequent irrigations are required after 1 month of monsoon. Region-wise irrigation requirements are changing constantly over a period of time (Wang et al. 2015). Among the different techniques, furrow irrigation is the dominant practice. However, drip irrigation method can potentially improve irrigation efficiency. Irrigation scheduling in cotton plays a major role in water saving, achieving higher irrigation performance and managing percolation resulting from a lot of water supplied as irrigation. Irrigation is a key cultural practice along with fertilizer application or tillage, which provides support in arid and semiarid regions and acts as a supplement to rainfall in other regions. Generally, temporal distribution of rainfall does not coincide with the crop water demands. Therefore, supplemental irrigation is of supreme importance for successful crop production. Supplemental irrigation (to reduce soil water deficit stress) can improve emergence, incorporate herbicides, improve nutrient uptake, support plant growth, prevent water stress, reduce square and boll shedding, maintain fiber quality, and support maximum yield potential (Nehra and Nehra 2004).

5.2 Why Plants Need Water?

Water is an essential component of all living cells. Just like all other living organisms, plants need water for different physiological processes. Plants require water for the absorption of nutrients from soil, for transpiration, and photosynthesis. Water is essential for nutrient transportation from the root to the other parts of the plant and for the transfer of photo-assimilates from source to sink. Thus, the plants' roots absorb water from the soil for proper growth. A considerable amount of the absorbed water is lost through transpiration. Water from groundwater surface is lost in the form of vapors to the atmosphere during the daytime.

During sunshine hours, the leaves of plants obtain energy from the light and also open their stomata to absorb carbon dioxide (CO_2) for metabolism (this procedure is called photosynthesis).

$$6CO_2 + 12H_2O - - - - - - - - - - - - \rightarrow C_2H_{12}O_6 + 6H_2O + 6O_2$$

While the stomata open to absorb CO_2, the plant transpires water from the leaves, which causes the cooling of the plant surface (Yazar et al. 2002).

Fig. 5.1 Risk of too much water

5.3 The Risk of Too Much Water

Without irrigation water, crops cannot be produced. Too much water is not suitable for many crops either. Apart from rice, only a few crop plants can grow "with their feet in the water." If there is too much rain, the soil will be full of water, and there will be no sufficient air (Fig. 5.1). The extra water must be removed. The removal of extra water either from the ground or from the root zone is called drainage.

If there is too slight rain, water must be provided from other resources; irrigation is required. The quantity of irrigation water depends not only on the quantity of water previously available from rainfall but also on the overall quantity of water required for the different crops. In addition, the irrigation water is the difference between the total water required for the crops and the quantity of rainfall available to the crops (Zwart and Bastiaanssen 2004).

5.4 Why Irrigation Scheduling?

All field crops require air, soil, light (sunshine), and water to grow. Besides providing anchorage, soil serves as a reservoir of water and nutrients, which plants absorb through their roots. The sun supplies the energy that is essential for plant growth. Water is needed for ET and is an important physiological component of plant body playing its role in photosynthesis and other metabolic processes. Water is provided either through rainfall or irrigation. In many places, rainfall is not sufficient to grow plants, so the plants need irrigation water. Making up-to-date management of irrigation decisions is an important element for excellent irrigation management performance and it regularly plays an important role in profitability and yield. Irrigation scheduling refers to the number of irrigations for a crop and their timing. This depends on several factors including the type of crop, its stage of development, extent of the root system, rate of ET, and the water holding capacity of the soil (Leib et al. 2003).

The positive relationship between the cotton yield produced by unit ET and the water used is called WUE. Only a small amount of water is being retained in the root zone from the total water supplied from the sources (river, well). The rest of the water is lost during the conveyance and distribution or field application loses (Fig. 5.2; Table 5.1) (Vories et al. 2007, 2011).

Fig. 5.2 Increased water use efficiency through furrow irrigation

Table 5.1 Function of efficiencies designed for irrigation structure

Irrigation method	Function of efficiency
Land irrigation	
Furrow	50–70%
Stage of basin	60–80%
Border	60–75%
Sub-irrigation	50–80%
Sprinkler irrigation	60–85%
Drip irrigation	80–90%

5.5 Irrigation Methods

5.5.1 *Drip Irrigation*

Drip irrigation is an efficient water application technique for various types of soils to ensure high WUE. Although a bit expensive, drip irrigation is very effective under limited water availability. It saturates the root zone slowly but continuously, which results in less water loss due to leaching and ET. Through drip irrigation, we can also reduce the conveyance and application losses, thereby improving irrigation efficiency (Fig. 5.3).

5.5.2 *Sprinkler Irrigation*

Sprinkler irrigation is more effective for sandy soils of rough topography, where leveling is not feasible or where labor and water are limited. Sprinklers are beneficial compared to surface application methods because water can be delivered at a preferred and controlled rate, to ensure a uniform allocation of water as well as high efficiency of water use. It is an expensive method and is used only in advanced countries; therefore, it is not accepted due to the high initial cost. Now a day, sprinklers should be utilized for a few types of soil and crop except extremely fine-textured soils (Bhattarai et al. 2006).

5.5.3 *Subsurface Irrigation*

Subsurface irrigation is a common technique in which the water table is raised and held close to the plant root zone by the use of channels (ditches) and also by subsurface drains. Subsurface irrigation is not frequently used for arid and semiarid irrigated regions, in which constant irrigation is required to germinate crops. Thus, it

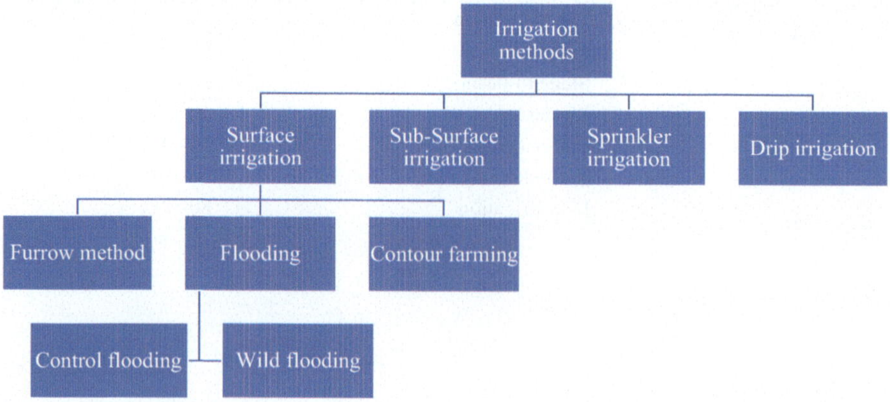

Fig. 5.3 Irrigation methods

is characteristically used in combination with subsurface drainage and controlled drainage (Nehra and Nehra 2004).

5.5.4 Surface Irrigation

5.5.4.1 Furrow Irrigation

Furrow irrigation is the most widely used method of irrigation. It saves almost 50% of irrigation water. It is well adopted in deep and almost level soils having regular and reasonable slopes. Furrow slope can be restricted to 0.3% and the length of the slope of furrow might be decreased to control erosion and give adequate drainage. Loam soils, along with the clay loam, are well suitable for furrow irrigation, as the total water holding ability is high in these soils, while it is not suitable for sandy soils.

5.5.4.2 Skip Furrow

The row-to-row distance in this method is maintained at 60 cm and water is applied alternatively into furrows. It saves almost 50% of irrigation water. Alternate furrows should be skipped and may be converted to ridges having a wide bed formation. The space accessible in the skipped furrow might be intercropped by pulses. The method may be modified to 2:1 skipped furrow, while under severe water scarcity, 3:1 skipped furrow may be used (Howell et al. 2004).

5.5.4.3 Alternate Furrow Irrigation

In alternate furrow irrigation method, the water is being applied in alternate furrows. The frequency of irrigation is high in this method compared to conventional methods of irrigation. The subsequent irrigation is applied in leftover furrows, and this sequence is followed till crop maturity. This system of irrigation is more suitable for clay and loamy soils. Irrigation scheduling according to this system recorded the maximum seed yield of cotton as compared to flood irrigation (Pereira et al. 2007).

5.5.4.4 Irrigation Under Paired Row Planting

In paired row planting, instead of making ridges and furrows 75 cm apart, furrows are made 150 cm apart and seeds are sown (at 60 cm apart) on the both sides of furrows and irrigated. This technique gives equal number of rows and optimal population of the plants in a unit area as in conventional planting. In addition, it permits every row to obtain additional light on one side than the second side. Cotton plants respond well to the sunlight, and 70–80% of bolls might be produced on the open side. Moreover, less water would be lost by evaporation in the two inner sides because of more rapid and complete shading by the plant (resulting in less weeds). Moreover weed control is easier on the external side because of more available space. More aeration of soil is obtained due to the dry inter space between the paired rows. Pereira et al. (2007) obtained the highest yield of cotton using paired row planting technique.

5.5.5 *Micro-Irrigation*

Fresh water has many applications like irrigation, industry, power production, livestock, and is used for domestic purposes in rural as well as in urban regions. Because of the increasing cost of irrigation projects and limited supply of good quality fresh water, water becomes a high-value commodity and is called liquid gold. Micro-irrigation is a well-known method for enhancing the WUE of crops (Duchemin et al. 2006).

5.6 Seasonal Water Needs Vary by Climate

Like other crops, ET or seasonal water requirement of cotton varies with prevailing climatic conditions. Under dry and hot climate, plants transpire more water for cooling and better biomass production. Weather elements like air temperature, cloudiness, humidity, solar radiations, and speed of wind determine the requirement of water for plants known as **evaporative demand**. Soil moisture content indicates the amount of water to be supplied to the rhizosphere to fulfill plants' evaporative demand. During moisture scarcity, the soil is not capable of providing enough water to plants to meet their evaporative demand (Evett et al. 2016).

5.7 Targeted Schedule Irrigation

Targeted irrigation scheduling permits the cotton grower to precisely manage the crop by managing the duration and amount of water stress, which, in turn, might impact the overall crop performance. Climate and soil water availability mainly determine the optimum time and proper amount of irrigation water for effective crop production. Understanding the interaction of the two for their effect on crop water status is very much needed for good water management practices.

Different schedule policies are generally required depending on the grower's objectives. For example, attaining the highest yield through surplus irrigation would not essentially lead to maximum productivity of water and net profit. Actually, maximum crop water productivity may be achieved through irrigation levels below the full water requirements of the crop.

When water availability is limited, the conventional practice of highest irrigation for highest land productivity may no longer be a good strategy. In this situation, irrigation should be optimized, (which may be more profitable), with attempts to obtain the greatest of crops per drop instead of most crops per unit soil. This needs management like deficit irrigation application on all the land instead of full irrigation on a part of land (Vories et al. 2011).

5.8 The Science and Art of Irrigation Scheduling

Irrigation scheduling in cotton is basically the science of determining when to irrigate, how much to apply, where to apply, and for what purpose to apply. The main objective of irrigation is to make sure that reasonably high crop water use is sustained during the crop season, which permits nonlimiting moisture in the soil for good plant growth and development. However, sound irrigation practices also maintain ample moisture stress between irrigation events to effectively manage the vegetative growth depending upon the specific needs of plants. With these complexities which impact irrigation decision, there is no doubt that various successful farmers regard scheduling of irrigation for cotton equally as a science and an art (De Jonge and Thorp 2017).

5.9 Cotton Water Requirements

5.9.1 Evapotranspiration

The quantity of water evaporating from wet soils along with transpiration from plant surfaces is called ET. Its rate is mainly determined by climatic features such as solar radiation, temperature, humidity, and wind. Water requirements by crops include the total quantity of water used in ET (Fig. 5.4). To successfully plan irrigation, growers need to plan according to ET. Evapotranspiration is the full amount of water use for a crop, with evaporation from the soil surface and transpiration from the plant.

- Cotton irrigation scheduling is a challenge as there is much variation in rainfall, temperature, and cloudiness throughout the growing period.
- The characteristics of cotton crop, in addition to prevailing ecological situation, are crucial in formative water use.
- Water use rises regularly from the early period (mainly by loss of water as evaporating surface) to the developmental period, lastly peaking on the mid-period.
- This period of highest water use matches the period of maximum canopy and highest boll load of cotton, which occurs usually in August.

Fig. 5.4 Evapotranspiration from cotton field

Cotton utilizes water during its whole lifecycle through the joint procedures of evaporation and transpiration, commonly referred to as ET. The ET is a result of climate variables (mostly solar emission, atmospheric temperature, and moisture), with soil, crop characteristics, and cultural operations (Zwart and Bastiaanssen 2004).

5.9.2 Measuring Evapotranspiration

An excellent method to evaluate ETc (after determining the coefficient of cotton crop) is through a lysimeter, which measures the loss and gain of water during daytime (Fig. 5.5). The variation between ecological and cultural factors needs the determination of local ET_O, ET_c with coefficients K_c for a scheduled irrigation for cotton (Fortes et al. 2005).

To calculate and monitor the crop ET, the following information is needed:

1. A native weather station should provide the approximate cotton crop ET.
2. A rainfall gauge located in every field and group of adjacent cotton field.
3. A good approximation of the allowable water depletion for the soil.

5.9.3 Water Use

Water use in cotton at crucial periods (early, mid-season, and developmental) is presented in Table 5.2. At the early period of cotton crop (approximately 0–25 days after planting), the everyday ETc range is 0.03–0.20 in. per day, having an average 0.09 in. per day and 0.63 in. per week. Use of water on average is 0.22 in. per day

Fig. 5.5 Measuring evapotranspiration

Table 5.2 Cotton growth period with corresponding every day ET_c and K_c values

Period of cotton growth	Length of period (days)	Every day ET_c (in.)	Average ET_c (in.)	Average K_c
Early	25	0.03–0.20	0.09	0.48
Crop developmental	35	0.09–0.36	0.22	1.02
Mid-season	50	0.18–0.44	0.30	1.44

(roughly 1.5 in. per week) on the developmental period of the crop and 0.28 in. per day (roughly 2.0 in. per week) in the mid-season. Equivalent K_c values are 0.48, 1.02, and 1.44 for early, developmental, and mid-season period, correspondingly. Use of crop water in Stoneville, Mississippi, was related to lower K_c values. The range of water use was about 0.05 in. per day initially to a peak of 0.28 in. per day, and then further decreased to 0.12 in. per day at opening boll. Cotton crop coefficient value ranged from 0.4 during early stage to 1.2 during mid-season, and then reduced to 0.6 at the ending (Hunsaker and Elshikha 2017) Table 5.3.

5.10 Water Use and Crop Coefficients

Cotton water requirement changes with respect to cultivar, duration of growing period, temperature of atmosphere, daily sunshine hours, quantity and distribution of precipitation, and soil characteristics. This water use dynamics for the cotton field (during 160 days' period) shows that through progression in the growth, the use of water raises progressively from 2.5 mm per day in seedling period, 2.5–6.2 mm per day as of squaring to initial bloom, to the maximum of 6.2–10 mm per day in full flowering, and then decreases to 5.1 mm per day during boll formation; it falls well below 2.0 mm per day through the boll opening period. When we talk of percentage

Table 5.3 Benefits and detriments of irrigation at various cotton growth stages

Growth stage	Sensitivity to deficit water	Irrigation benefits	Detriments of irrigation
Emergence	Reasonable	Stimulates herbicides; hydrates germinates; cools soil land	Seedling virus if cool; generates soil crust; herbicide damage
Pressuring	Slight	Little unless extreme	Shallow root structure
Squaring	Reasonable	Keeps away from loss of subsoil water; desired size of plant; enhances fertilizers in the solution	Soil diffusion; excessive vegetation
Initial bloom	Maximum	Maintains bolls; original fruit or leaves; recompense for nematodes, compaction or herbicide damage; maintains fiber excellence	Little
Late bloom	Reasonable	Reserves bolls; healthy leaves; maintains fiber	Little
Cutout	Slight	Vigorous leaves	Delayed maturity
Opening	Not	Some	Boll rot; delayed harvest

of complete use of water in a season, crop water need is 20% till initial flower, 40% at the time of initial flower toward the peak flower, 30% through peak flower to bursting of bolls, and just 10% till maturity.

Cotton is basically a water scarcity-tolerant crop; however, it responds well to enough water. Depending on the weather and the length of the overall growing stage, cotton water requirements range from 27 to 51 in. (700–1300 mm). During the initial vegetative stage, the water requirement of the crop is 10% of the total (Bhatti and Patel 2012).

5.11 Water-Sensitivity Stages of Cotton Growth

Irrigation improves the harmful effect of the soil water shortage stress through two diversified physiology-related processes in the plants (while they cannot obtain enough water). The main physiological method to estimate water shortage in plants is growth of cells. From root tips expanding through the soil to fibers elongating on seed coats, the ability of individual cells within a plant to expand is largely determined by the availability of soil water. In addition to decreasing growth, soil water shortage activates hormonal variation in reproductive growth, which results in shedding of fruiting parts (squares as well as bolls). Management of irrigation should have the prime objective to reduce stress at crucial stages so that plants have the maximum capacity to begin, mature and retain bolls.

5.11.1 Planting to Emergence

Water is important for the germination, and irrigation during this period is principally required for seedling establishment. On the other hand, irrigation after planting would cool the soil and can decrease seedling growth. Root establishment is rather fast; taproots grow up to 2.4 in. per day after emergence from the seed. Cotton water use at planting and through emergence is low. If the seedbed is too dry to support germination, preplanting irrigation is preferable to irrigating following planting. Irrigating shortly after planting would cool the soil and might facilitate the development of seedling diseases. Once seeds germinate, young cotton plants put much of their energy into developing roots. Without very low soil moisture during this time, irrigation contribution to yield is very little. Few water shortages during initial periods can be the cause of a deeper root system (Modala et al. 2015).

5.11.2 Emergence to Initial Square

After stand establishment, early period of water deficit is normally not a problem, if there is enough water for emergence as well as initial development of seedlings. Water need during this time is very low and early plants of cotton do partitioning of appreciable resources to the roots. With very low soil moisture during this time, irrigation contribution to yield is very little. As a matter of fact, an early mild water shortage can encourage production of roots, particularly encouraging deeper roots. Primed Acclimation (PA) is an irrigation idea that uses deliberate mild drought stress at early vegetative growth to induce physiological variations in the plant and make it a bit drought-tolerant in mid-season, while unfavorable effects of drought stress are maximal. As the name shows, a time of mild, organized deficit of water acclimates plants of water-limited conditions; thus beginning a start of plant to increase WUE. A few of the changes may be enhanced root growth, decreased water use, variability in pattern of fruiting, and elicited molecular and enzymatic reaction. PA can conserve yield with appreciable water decrease. In cotton, the period of PA lasts for 35 days, starting on complete seedling establishment (approximately 14 days of planting) to late squaring and initial bloom. Throughout this growing stage, water utilization may be lowered by as low as 30%, with no yield loss in a few production areas in the South. An additional advantage to appropriately applied PA is the decreased plant growth hormones needed shortly for growing period and further regular maturity (Thorp et al. 2014).

5.11.3 Initial Square to Initial Flower

Once cotton plants emerge and grow, irrigation should be used to supplement rainfall and prevent depleting the soil moisture profile prior to bloom. For approximately 21 days from initial square to initial bloom, cotton vegetative expansion is extremely rapid; the fruiting locations number is predetermined. This is the stage during which plants very quickly absorb soil phosphorus along with potassium due to fast growth of roots. The evidence based on ground imaging and cotton roots system measurement shows that the highest rooting depth achievable may be as high as 36 in. After 40–60 days of sowing, the maximum depth may be achieved. Severe water shortage at this stage is particularly damaging for the cotton crop in environments of short-season (De Jonge and Thorp 2017).

5.11.4 Initial Flower to Peak Bloom

Crop water demand increases to 0.2–0.44 in. per day during this period. Lack of water can reduce plant growth, reduce the number of fruiting sites, cause the shedding of young bolls, reduce boll size, and result in loss of yield potential. Severe water deficit at this stage can also lead to shorter staple, higher micronaire, and lower fiber quality. Lack of water does not typically result in square shed between first flower and peak bloom. If shedding of square is observed during this time, other reasons must be explored. Water shortage stress during this stage also has impact on yield by lowering the size of existing bolls. During this time, the greatest root deepening is attained; however, lateral roots carry on growing throughout the rooting zone; therefore, the maximum size of the roots may not be achieved till 90 days of sowing (De Jonge et al. 2012).

5.11.5 Peak Bloom to Open Bolls

Water demand begins to decrease as bolls begin to mature and open. Water shortage stress at this growth period is not so critical than during squaring and initial flowering stages. Water deficit at this stage can have an effect on square as well as little boll shedding. Yet, these late fruit losses exert little effect on yield as compared to initial losses of season bolls. The parameters of fiber quality influenced by water shortage are length of fiber and micronaire, mainly in little bolls. Farmers should decrease or stop irrigation and allow plants to experience some water stress. After the start of boll opening, water stress should be imposed to plants to allow for better harvesting conditions. This stress reduces regrowth, makes defoliation easier, and encourages boll opening (Phene et al. 1992).

5.12 Irrigation Scheduling Tools

Weather information is used to evaluate the quantity of water that can be transpired in a reference crop, for example, grass; a crop coefficient is used to set this reference value for the particular crop.

5.12.1 Method of Water Balance

Irrigation scheduling with a balance of water and checkbook method depends on the water available in soil. However, a checkbook, inputs are added (credited) in the totality of soil water, withdrawals are deducted (debited) from the soil water. The everyday total balance of water in this equation is then:

$$\text{soil water (today)} = \text{soilwater (yesterday)} + \text{inputs} - \text{withdrawals}$$

Inputs (contribution) in soil moisture include irrigation and rainfall water. Withdrawals comprise transpiration by the plant, evaporation out of the soil, and deep percolation into the layers of soil. Throughout the overall growing period, evaporation or transpiration is a very crucial method by which water escapes the soil surface. The share of deep percolation is very small in terms of withdrawals throughout the growing period, and therefore is implicit to be insignificant (Yang et al. 2010).

5.12.2 Soil and Water Content

Equation of water balance needs information for available soil water. Determination of following level of soil water depends on the early soil moisture content. Soil moisture depends on the soil texture. Sandy soil has coarse particles with pores which retain less water, lowering the soil water accessible to the plants. On the other hand, clay soils have various, extremely little pores. The particles of clay attach to the soil water extra firmly. Although clay soils retain higher water compared to sandy soils, a little water is obtainable by the plants due to the tight binding. Loamy soils contain reasonable pores for holding moisture and also do not attach soil moisture strongly as to prevent plants in obtaining the water. Such soils have relatively more water accessible to the plants (Hunsaker and Elshikha 2017).

5.12.3 Maintaining Water and Soil

Scheduled irrigation is way of maintaining available soil moisture for the plants between field capacity and permanent wilting point. For the soil water balance approach, the early soil moisture is approximated on the basis of texture of the soil. Water present in the soil is traced using everyday variation in the use of water and water inputs.

Inputs (contribution) to crop may be precisely determined. Rainfall is determined by rain gauges placed in the field, and by tipping bucket rain gauges in the weather observatory. Irrigation of cotton should be measured directly or can be calculated based on the total quantity of water given in a specified region. The outputs (deductions) from this system, however, are a bit complicated to determine exactly and are, therefore, expected (Grismer 2002).

5.12.4 Estimating the Use of Crop Water

Water use in crop can be approximated through different means. A typical procedure which has been devised approximates a reference ET with the help of climate parameters. Then the reference ET is converted to cotton ET by using its crop coefficients. As an alternate approach, procedures have been devised that approximate ET from the curves developed over the years of measurements in the flied. These basic approaches can be accurate; however, they are location specific (from where the information was gathered).

5.12.5 Mississippi Irrigation Scheduling Tool (MIST)

The Mississippi Irrigation Scheduling Tool (MIST) depends on quite recent technical information of water use by crop in order to support producers during making decisions for irrigation. The method is intended to be simple for access and use. The user does not need to take readings from field or input data, as the MIST collects data automatically from national and local databases, and constantly determines use of crop water. Soil hydrology and texture information can be downloaded by Natural Resource and Conservation System, which is established according to spatial locations. Data for cotton field can be entered through user, or it may be downloaded automatically from data collected. The MIST uses NRCS runoff equations in order to handle field runoff differences. This provides extra correct field-soil moisture of cotton after a rain event. No plant and soil measurements are needed to run this scheduler. Automatically downloadable data from such databases permit growers to employ the MIST, with no requirement of extensive information collection and input to model. With the MIST, we need no program installation or maintenance. This method is accessed by the internet, and is obtainable on many more platforms (smartphones, desktop computers, tablet computers, and laptops). So, it enables the user to estimate the needs of crop water from anywhere, and tells instantly about the cotton crop water need. By the use of a daily time-step calculations and by weather updates, it becomes easy to get a more reliable determination of available soil water. Daily time-step calculations also help determine crop water needs for the next many days, thus allowing farmers to manage their water resources in a better way (Dursun and Suzden 2014).

5.12.5.1 Combining Soil Moisture Monitoring and Water Balancing

Combining soil moisture monitoring and a water balance approach helps to avoid the water shortfalls. For example, during early growing period, while irrigation is scheduled for a poor drained soil, a water balance method may specify the need to irrigate and the water sensor might indicate plenty of water available. On the other hand, following an intense rainfall, a balance of water can specify that irrigation is not necessary as it does not automatically identify how much rainfall ran off, whereas a soil sensor might show drier than estimated situation leading to an early irrigation following rainfall. It is apparent that sensors may be the answer to such limitations; however, it must be kept in mind that measurement of sensors is only for a very small location of field and it may not represent whole field. Hence, we recommend the combination of soil sensors and water balance (Hunsaker et al. 2005).

5.12.6 Sensor-Based Scheduling

For more than 60 years, sensors have been in use for observing soil water situation and supplying information to help conclude irrigation timing. A major challenge in practical utilization of these sensors in commercial farming is the time taken by sensors to reach the field for recording sensor output This challenge becomes worse when the number of fields of a person increases and also keeping in mind that during peak water demand, such systems may be needed to monitor water after at least 3 days. The current accessibility of different systems of sensors incorporated with practically reasonable wireless information communication capabilities have made sensor-based scheduling further useful. These modern tools are encouraging, as Cotton Incorporated's 2008 Natural Resource Survey showed that only 10% of cotton producers employed weather-based scheduling method and soil and crop monitoring systems (Shah and Das 2012).

5.12.6.1 Different Measurements Types

There are three physical properties that may be measured through sensors for scheduling irrigation:

Soil matric potential, which is a measure of water binding to soil; the higher this potential, the more stress the plant experiences. Sensors that determine matric potential comprise electronic sensors and tensiometers, for example, "WaterMark" sensor of Irrometer. Volumetric moisture content (VMC) is used to determine water volume for each soil volume. Many different types of sensors are used to determine these properties, which include capacitance sensors, neutron probes, and time domain reflectometry (TDR) sensors.

Canopy temperature (CT) is a determination of the surface temperature of leaves. Transpiration has a cooling effect on leaves and an increase in water stress results with the decrease in transpiration; therefore, the canopy will be at a higher temperature. The CT can be calculated through insertion of thermocouples straight on top of the leaves, but it is generally measured through an infrared thermometer.

5.12.6.2 Different Types of Sensors

The type of method that a sensor employs to measure soil water content or soil moisture tension has important implications for knowing the performance characteristics of sensors in cotton production. Tensiometer makes use of a porous ceramic tip for direct contact to soil to evaluate soil water tension directly. Granular Matrix Sensors (GMS) determine the transformation of electrical resistance, which is because soil water goes inside and away from sensor in reaction to neighboring soil moisture; this electrical resistance calculation is also connected with soil water tension. Neutron probe calculates the number of neutrons that collide with hydrogen within water and also is generally connected through the volumetric content of water (Hunsaker et al. 2005).

Tensiometers, neutron scattering, and GMS have a long history of use in cotton schedule irrigation. For more than 20 years, very latest sensor types are available in market which determines soil's capacitance and dielectric constant (capacity of a matter to store electricity). The quantity of water (compared to air) within soil pores is the main factor affecting the dielectric constant of soil. One of the methods to find out this electrical property is the measurement of any change in a frequency of radio wave while it passes through the soil, which is called Frequency Domain Reflectometry (FDR).

An alternative method is to determine the reflectance pattern of the voltage pulse that is applied to a cable guide positioned in the soil, which is called Time Domain Reflectometry (TDR). Sensors to compute dielectric stable are regularly connected to the volumetric content of soil water (Yang et al. 2010).

5.13 Methods and Costs of Obtaining Soil and Water Data

There are many strategies for obtaining soil water stress data from the sensor-based scheduling. The easiest process, we will identify "in-field data collection" in which sensors are set up in a field of cotton through wire leads approaching to surface. For this installation, a farmer will enter a field of cotton having a hand reader attached with wire leads. In this position, the information is collected by hand. This information may require graphing and formatting to enhance the understanding of these results. This method is of very low cost. Other locations will have less expense as hand reader can easily be transported to other places. Keep in mind, to add time needed and cost of transfer anybody for reading of sensors. This method is useful for obtaining decisions about irrigation on reading time but is not considered good for recording soil water tension and content. It is hard to make time for many readings every week, and in a humid area the sensor does not obtain any reading during the rainy season or during a prolonged dry period. Lastly, locations of sensor position can be misplaced with the crop growth and nobody enters the field of wet cotton for taking readings.

The second method we will identify is "edge of field logging," in which the sensor leads are wired to logger or to a radio source which transmits this reading wirelessly to data logger on the edge of the cotton field, which is simple to approach. During this situation, somebody has to go the field of cotton to download logger readings as well as uploading these readings to software; however, nobody needs to enter the cotton field. A few loggers have displays which are onboard and do not need these steps of download or upload. This resulted in continuing information set which can be easily connected to rainfall as well as scheduling irrigation. In a number of instances, the total cost requires to be revised for every new sensor position; however, some other sensor positions can be linked by the original logger and wireless receiver or data logger; hence may be used in a lot of positions.

A third method is "office computer or smartphone access" in which case, data logger can be situated along with sensors in cotton field and on field edge; that logger sends out the reading of the sensor by internet through long-range radio, satellite and cell phone. Lastly, we have "portable sensors and data loggers"; such a sensor is lowered to a PVC access pipe at every observing position and readings are taken at different soil depths within access pipe. Equipment price per measurement position quickly decreases since number of locations increases. Hence, this method has been mainly used by a large number of producers and irrigation advisors. It must be documented that a portable sensor or logger needs more travel toward field of cotton and inside crop. Although, if a counselor is in traveling, you get a different set of eyes surveillance your crop of cotton and every week delivery of set schedule irrigation information (Aqeel-ur-Rehman et al. 2014).

5.14 Conclusions

In order to improve irrigation system efficiency and avoid nonoptimal water use, it is vital to devise an intelligent irrigation scheduling that ensures optimized water use along with irrigation application when and where needed. If there is variation in temperature or humidity in environment, the sensors sense the variations and provide signal for irrigation. Sensor technologies are found appropriate for collecting data pertaining to soil, crop, and weather and help in devising solutions for major agricultural processes associated with irrigation, etc. The increase of wireless sensors in the agricultural industry makes it possible to boost productivity besides profitability of farm operations and eventually boosting resource use efficiency. Field research that testifies effects of variable irrigation regimes besides timings will assist in quantifying the most appropriate decision criteria for using crop models as irrigation scheduling tools.

References

Abbas Q, Ahmad S (2018) Effect of different sowing times and cultivars on cotton fiber quality under stable cotton-wheat cropping system in southern Punjab, Pakistan. Pak J Life Soc Sci 16:77–84

Ahmad S, Abbas Q, Abbas G, Fatima Z, Atique-ur-Rehman NS, Younis H, Khan RJ, Nasim W, Habib ur Rehman M, Ahmad A, Rasul G, Khan MA, Hasanuzzaman M (2017) Quantification of climate warming and crop management impacts on cotton phenology. Plan Theory 6(7):1–16

Ahmad S, Iqbal M, Muhammad T, Mehmood A, Ahmad S, Hasanuzzaman M (2018) Cotton productivity enhanced through transplanting and early sowing. Acta Sci Biol Sci 40:e34610

Ahmad S, Raza I (2014) Optimization of management practices to improve cotton fiber quality under irrigated arid environment. J Food Agric Environ 2(2):609–613

Ahmad S, Raza I, Ali H, Shahzad AN, Atiq-ur-Rehman, Sarwar N (2014) Response of cotton crop to exogenous application of glycinebetaine under sufficient and scarce water conditions. Braz J Bot 37(4):407–415

Ali H, Abid SA, Ahmad S, Sarwar N, Arooj M, Mahmood A, Shahzad AN (2013a) Integrated weed management in cotton cultivated in the alternate-furrow planting system. J Food Agric Environ 11(3&4):1664–1669

Ali H, Abid SA, Ahmad S, Sarwar N, Arooj M, Mahmood A, Shahzad AN (2013b) Impact of integrated weed management on flat-sown cotton (*Gossypium hirsutum* L.). J Anim Plant Sci 23 (4):1185–1192

Ali H, Afzal MN, Ahmad F, Ahmad S, Akhtar M, Atif R (2011) Effect of sowing dates, plant spacing and nitrogen application on growth and productivity on cotton crop. Int J Sci Eng Res 2 (9):1–6

Ali H, Hameed RA, Ahmad S, Shahzad AN, Sarwar N (2014a) Efficacy of different techniques of nitrogen application on American cotton under semi-arid conditions. J Food Agric Environ 12 (1):157–160

Ali H, Hussain GS, Hussain S, Shahzad AN, Ahmad S, Javeed HMR, Sarwar N (2014b) Early sowing reduces cotton leaf curl virus occurrence and improves cotton productivity. Cercetări Agronomice în Moldova XLVII(4):71–81

Amin A, Nasim W, Mubeen M, Ahmad A, Nadeem M, Urich P, Fahad S, Ahmad S, Wajid A, Tabassum F, Hammad HM, Sultana SR, Anwar S, Baloch SK, Wahid A, Wilkerson CJ, Hoogenboom G (2018) Simulated CSM-CROPGRO-cotton yield under projected future climate by SimCLIM for southern Punjab, Pakistan. Agric Syst 167:213–222

Amin A, Nasim W, Mubeen M, Nadeem M, Ali L, Hammad HM, Sultana SR, Jabran K, Habib urRehman M, Ahmad S, Awais M, Rasool A, Fahad S, Saud S, Shah AN, Ihsan Z, Ali S, Bajwa AA, Hakeem KR, Ameen A, Amanullah, Rehman HU, Alghabar F, Jatoi GH, Akram M, Khan A, Islam F, Ata-Ul-Karim ST, Rehmani MIA, Hussain S, Razaq M, Fathi A (2017) Optimizing the phosphorus use in cotton by using CSM-CROPGRO-cotton model for semi-arid climate of Vehari-Punjab, Pakistan. Environ Sci Pollut Res 24(6):5811–5823

Aqeel-Ur-Rehman, Abbasi AZ, Islam N, Shaikh ZA (2014) A review of wireless sensors and networks applications in agriculture. Comput Stand Inter 36:263–270

Bhattarai SP, McHugh AD, Lotz G, Midmore DJ (2006) The response of cotton to subsurface drip and furrow irrigation in a vertisol. Exp Agr 42(1):29–49

Bhatti GH, Patel HM (2012) Estimation of evapotranspiration for four major crops using dual crop coeffcient method in SardarSarovar command area. Hydro–2012, Conference on Hydraulics, Water Resources, Coastal and Environmental Engineering. Mumbai: Indian Institute of Technology

DeJonge KC, Ascough JC, Andales AA, Hansen NC, Garcia LA, Arabi M (2012) Improving evapotranspiration simulations in the CERES-maize model under limited irrigation. Agric Water Manag 115:92–103

DeJonge KC, Thorp KR (2017) Standardized reference evapotranspiration and dual crop coefficient approach in the DSSAT cropping system model. Trans ASABE 60(6):1965–1981

Duchemin B, Hadria R, Erraki S, Boulet G, Maisongrande P, Chehbouni A, Escadafal R, Ezzahar J, Hoedjes JCB, Kharrou MH, Khabba S, Mougenot B, Olioso A, Rodriguez JC, Simonneaus V (2006) Monitoring wheat phenology and irrigation in Central Morocco: on the use of relationships between evapotranspiration, crops coefficients, leaf area index and remotely sensed vegetation indices. Agric Water Manag 79:1–27

Dursun M, Özden S (2014) An efficient improved photovoltaic irrigation system with artificial neural network based modeling of soil moisture distribution - a case study in Turkey. Comput Electron Agric 102:120–126

Evett SR, Howell STA, Schneider AD, Copeland KS, Dusek DA, Brauer DK, Gowda PH (2016) The Bushland weighing lysimeters: a quarter century of crop ET investigations to advance sustainable irrigation. Trans ASABE 59(1):163–179

Fortes PS, Platonov AE, Pereira LS (2005) GISAREG—A GIS based irrigation scheduling simulation model to support improved water use. Agric Water Manag 77:159–179

Grismer ME (2002) Regional cotton lint yield, ET_c and water value in Arizona and California. Agric Water Manag 54:227–242

Howell TA, Evett SR, Tolk JA, Schneider AD (2004) Evapotranspiration of full-, deficit-irrigated, and dryland cotton on the northern Texas High Plains. J Irrig Drain Eng 130(4):277–285

Hunsaker DJ, Barnes EM, Clarke TR, Fitzgerald GJ, Pinter JPJ (2005) Cotton irrigation scheduling using remotely sensed and FAO-56 basal crop coefficients. Trans ASAE 48(4):1395–1407

Hunsaker DJ, Elshikha DM (2017) Surface irrigation management for guayule rubber production in the U.S. desert southwest. Agric Water Manag 185:43–57

Khan MB, Khaliq A, Ahmad S (2004) Performance of mashbean intercropped in cotton planted in different planting patterns. J Res (Sci) 15(2):191–197

Leib BG, Jabro JD, Matthews GR (2003) Field evaluation and performance comparison of soil moisture sensors. Soil Sci 168(6):396–408

Modala NR, Ale S, Rajan N, Munster CL, DeLaune PB, Thorp KR, Barnes EM (2015) Evaluation of the CSM-CROPGRO-cotton model for the Texas Rolling Plains region and simulation of deficit irrigation strategies for increasing water use efficiency. Trans ASABE 58(3):685–696

Nehra IPL, Nehra KC (2004) Effect of FIRB (Furrow Irrigated Raised Bed) system on productivity of cotton–wheat cropping system. In: Proceeding of International Symposium on Strategies for Sustainable Cotton Production- A global vision, 23–25th November 2004

Pereira LS, Gonc¸alves JM, Dong B, Mao Z, Fang SX (2007) Assessing basin irrigation and scheduling strategies for saving irrigation water and controlling salinity in the upper Yellow River Basin, China. Agric Water Manag 93(3):109–122

Phene CJ, Detar WR, Clark DA (1992) Real-time irrigation scheduling of cotton with an automated pan evaporation system. Am Soc Agric Eng 8(6):787–793

Rahman MH, Ahmad A, Wang X, Wajid A, Nasim W, Hussain M, Ahmad B, Ahmad I, Ali Z, Ishaque W, Awais M, Shelia V, Ahmad S, Fahad S, Alam M, Ullah H, Hoogenboom G (2018) Multi-model projections of future climate and climate change impacts uncertainty assessment for cotton production in Pakistan. Agric For Meteorol 253-254:94–113

Shah NG, Das I (2012) Precision irrigation sensor network based irrigation, problems, perspectives and challenges of agricultural water management, IIT Bombay, India, pp 217–232

Tariq M, Afzal MN, Muhammad D, Ahmad S, Shahzad AN, Kiran A, Wakeel A (2018) Relationship of tissue potassium content with yield and fiber quality components of Bt cotton as influenced by potassium application methods. Field Crop Res 229:37–43

Tariq M, Yasmeen A, Ahmad S, Hussain N, Afzal MN, Hasanuzzaman M (2017) Shedding of fruiting structures in cotton: factors, compensation and prevention. Trop Subtrop Agroecosyst 20(2):251–262

Thorp KR, Ale S, Bange MP, Barnes EM, Hoogenboom G, Lascano RJ, White JW (2014) Development and application of process-based simulation models for cotton production: a review of past, present, and future directions. J Cotton Sci 18(1):10–47

Usman M, Ahmad A, Ahmad S, Irshad M, Khaliq T, Wajid A, Hussain K, Nasim W, Chattha TM, Trethowan R, Hoogenboom G (2009) Development and application of crop water stress index for scheduling irrigation in cotton (*Gossypium hirsutum* L.) under semiarid environment. J Food Agric Environ 7(3&4):386–391

Vories ED, Greene JK, Teague TG, Stewart JH, Phipps BJ, Pringle HC, Clawson EL, Hogan RJ, O'Leary PF, Griffin TW (2011) Determining the optimum timing for the final furrow irrigation on mid-south cotton. Trans ASABE 27(5):737–745

Vories ED, Hogan R, Tacker PL, Glover RE, Lancaster SW (2007) Estimating the impact of delaying irrigation for midsouth cotton on clay soil. Trans ASABE 50(3):929–937

Wang J, Wang H, He J, Li L, Shen M, Tan X, Min H, Zheng L (2015) Wireless sensor network for real-time perishable food supply chain management. Comput Electron Agric 110:196–207
Yang Y, Yang Y, Moiwo JP, Hu Y (2010) Estimation of irrigation requirement for sustainable water resources reallocation in North China. Agric Water Manag 97(11):1711–1721
Yazar A, Sezen SM, Sesveren S (2002) LEPA and trickle irrigation of cotton in the Southeast Anatolia project (GAP) area in Turkey. Agric Water Manag 54:189–203
Zwart SJ, Bastiaanssen WGM (2004) Review of measured crop water productivity values for irrigated wheat, rice, cotton, and maize. Agric Water Manag 69:115–133

Chapter 6
Role of Macronutrients in Cotton Production

Niaz Ahmed, Muhammad Arif Ali, Subhan Danish, Usman Khalid Chaudhry, Sajjad Hussain, Waseem Hassan, Fiaz Ahmad, and Nawab Ali

Abstract Sound nutrition plays a key role in enhancing cotton yield. As cotton undergoes vegetative and reproductive growth at the same time, its nutritional requirements are dissimilar, compared to other field crops. Cotton is grown as an annual crop with an indeterminate growth pattern. The vegetative branching provides a potential fruiting place except under abiotic and biotic stresses. Moreover, cotton has a deep root system with low density of roots in the surface layer of soils where availability of nutrients is high. The rooting system makes cotton crop more dependent on the subsoil for nutrition. A continuous supply of nutrients is required to sustain morphogenesis. The rate of both nutrients absorption and dry matter production increases progressively during the seedling, vegetative, and fruiting periods and peaks near the end of the bloom period. Nitrogen, phosphorus, and potassium are required in large quantities and are limited in many soils. The deficiencies of macro- and micronutrients decrease plant growth and development, and consequently seed cotton yield is reduced. The deficiency of phosphorous (P), calcium (Ca), potassium (K), boron (B), magnesium (Mg), and zinc (Zn) affects fruit production in cotton than vegetative growth, while the deficiencies of nitrogen (N), sulfur (S), molybdenum (Mo), and manganese (Mn) affect equally vegetative and reproductive growth of cotton. A bevy of literature concerning the role of macronutrients in growth and development is presented in the following paragraphs.

N. Ahmed · M. A. Ali (✉) · S. Danish · N. Ali
Department of Soil Science, Bahauddin Zakariya University, Multan, Pakistan
e-mail: arif1056@bzu.edu.pk

U. K. Chaudhry
Department of Agricultural Genetic Engineering, Ayhan Sahenk Faculty of Agricultural Sciences and Technologies, Nigde Omer Haliesdimer University, Nigde, Turkey

S. Hussain
Department of Horticulture, Bahauddin Zakariya University, Multan, Pakistan

W. Hassan
Department of Soil and Environmental Sciences, MNS University of Agriculture, Multan, Pakistan

F. Ahmad
Central Cotton Research Institute Multan, Multan, Pakistan

S. Ahmad, M. Hasanuzzaman (eds.), *Cotton Production and Uses*,
https://doi.org/10.1007/978-981-15-1472-2_6

Keywords Phosphorus · Nitrogen · Potassium · Calcium · Magnesium · Zinc · Boron

Abbreviations

B	Boron
Ca	Calcium
CEC	Cation exchange capacity
CO_2	Carbon dioxide
GHGs	Greenhouse gases
K	Potassium
Mg	Magnesium
Mn	Manganese
Mo	Molybdenum
N	Nitrogen
NH_4^+	Ammonium
NO_3^-	Nitrate
P	Phosphorous
S	Sulfur
Zn	Zinc

6.1 Introduction

For good development and yield, cotton crop wants uninterrupted availability of essential nutrient (Ahmad et al. 2014, 2017, 2018; Abbas and Ahmad 2018; Ahmad and Raza 2014; Ali et al. 2011, 2013a, b, 2014a, b). Sufficient availability as well as deficiency of nutrient results in the decrease of crops' growth and yield (Table 6.1). It is set up certainly that ideal dietary prerequisite of cotton is of essential significance to help up its production. Various studies to observe the response of cotton to different fertilizer doses can help us understand the nutrient requirements of cotton among the cotton developing zones, and their supply is basically needed for enhancing cotton yield per hectare (Amin et al. 2017, 2018; Rahman et al. 2018; Tariq et al. 2017, 2018; Usman et al. 2009). The mineral nourishment of cotton relies upon both the cotton root's capacity to investigate the soil and on the soil's capacity to supply N, P, and K nutrients (Bisson et al. 1994). Varshney (1979) recommended the optimum dose of nitrogen as 131 kg ha^{-1} and reported that further increase in application resulted in increase of seed cotton yield. Colakoglu (1980) suggested the ideal dose of NPK (80–120 kg, 60–90 kg, and 100–200 kg ha^{-1}) for enhancing

Table 6.1 Critical and adequate concentrations of macronutrients in cotton at various growth stages

Nutrients	Growth stage	Index tissues	Concentration		Reference
			Critical	Adequate	
N as (NO_3)	Early square	Petiole	15,000–25,000	–	–
	First bloom	Petiole	10,000–18,000	–	–
	Peak bloom	Petiole	3000–10,000	–	–
	First ppen boll	Petiole	1000–4000	–	–
	Late season	Petiole	–	<2000	–
P as (PO_4)	First bloom	Petiole	–	>1500	–
	Peak bloom	Petiole	–	>1200	–
	First open boll	Petiole	–	>1000	–
K	First bloom	Petiole	40,000–45,000	–	–
	50% bloom	Leaf	18,000	–	–
	Peak bloom	Petiole	30,000	–	–
	First open boll	Petiole	15,000	–	–
	Late season	Petiole	10,000	–	–
Ca	Seasonal	Petiole	–	6000–25,000	–
	Early bloom	Leaf	–	17,000–25,000	–
Mg	Seasonal	Petiole	–	2000–8000	–
	Peak bloom	Leaf blade	–	>3000	–
S	Early square	Shoot	–	>2000	Mahler (1989)
	Seasonal	Leaf blade	2400–4000	>4000	–

Source: Adapted from Cassman (1993)

cotton yield by considering the specific cotton-growing regions of Turkey. However as we know, nutrient requirements vary from region to region and mainly are dependent upon climatic conditions, thereby in Sindh province of Pakistan, Suhag et al. (1981) recommended that addition of fertilizer 112 kg N + 50 kg P ha^{-1} is better for achieving good yield returns. In another study conducted by Mithaiwala et al. (1981), they observed the influence of combined NPK fertilizers and alone on growth and yield of cotton, and it was inferred that sole application of P was not significant though; addition of N alone even with its combination with K was fruitful for cotton crop. Soil analysis done in Pakistan elaborated the deficiency of N and P with occasional deficiency of K (Wahhab 1985). Khan et al. (1987) described and strengthened the aforementioned importance of nitrogen for cotton and inferred that P application was not helpful for improving morphological and yield traits of cotton. Moreover another study conducted by the same group to highlight nitrogen significance showed that nitrogen alone at 100 kg ha^{-1} was also economical and beneficial rather than higher levels of other nutrient application to cotton in Sakrand conditions. A 12-year field study was conducted with 56 fertilizer experiments on cotton with a conclusion that nitrogen requirement is essential for cotton yield; contrarily phosphorus did not show positive response, while potassium was negligible (Setatou and Simonis 1994).

6.2 Nitrogen

6.2.1 Nitrogen and Its Function

Nitrogen (N) is very important nutrient that play imperative parts in enhancing yield of lint and photosynthesis in cotton (Khan et al. 2017a, b). Therefore, pre-stored assimilates, N uptake, remobilization, and gain in cotton biomass are correlated linearly with addition of N (Guarda et al. 2004; Gibson et al. 2007; Yang et al. 2011). Unnecessary or minimum N addition rate can minimize yield, nitrogen-use efficiency, crop development, and N plant recovery (NPR) in cotton (Amanullah 2014; Lokhande and Reddy 2015). Addition of N as fertilizer, its intake, and consumption are absolutely associated with canopy and dry matter buildup in crop (Olesen et al. 2002). Ideal N applications are influenced by multifarious factors, i.e., soil fertility, yield potential, and traditional management practices (Bange and Milroy 2004; Clawson et al. 2008; Dai and Dong 2014). Higher N addition is usually adopted for cotton cultivation (Guarda et al. 2004), though higher N addition may enhance environmental pollution and decrease lint yield (Bundy and Andraski 2004; Hou et al. 2007).

6.2.2 Nitrogen Availability

Nitrogen gaining and passage in soils are dependent on architecture of the root; cellular membrane; available forms, i.e., ammonium (NH_4^+) and nitrate (NO_3^-), of nitrogen in soil solution; carbon; N metabolites; and soil properties (Jackson et al. 2012). Ammonium and nitrate are the major forms of inorganic N (Xu et al. 2012). Roots of crops uptake ammonium and nitrate forms with variable affinities. Both low and high affinities of NO_3^- in transport systems consist of constitutive and NO_3-inducible elements (Miller et al. 2007). Many membrane proteins are involved in translocation, compartmentation, NO_3^- intake, and remobilization (Dechorgnat et al. 2011). The architecture and activities of ammonium (NH_4^+) and nitrate (NO_3^-) transporters in roots are controlled by N-available form and content, temperature, and diurnal variations that affect the nitrogen intake of roots (Garnett et al. 2009). Higher contents of NO_3^- in soil can modify the NO_3^- buildup in plant in spite of higher NR movement during early stages of plant growth. In addition, NO_3^- acquisition is affected by nutrient as well as water supply, temperature variations, and radiation (Aliyan 2013).

6.2.3 Cotton Need for Nitrogen

Cotton (*Gossypium hirsutum* L.) is a major seed oil- and fiber-producing crop that cannot be neglected worldwide (Constable and Bange 2015). Cotton plants have high N demand, i.e., 50–412 kg N ha^{-1} that play important role in emissions of GHGs as well as acceleration of denitrification. Overuse of N fertilizer interrupts the natural ecosystem as well as biodiversity and accelerates environmental pollution (Hirel et al. 2007; Zhang et al. 2008; Yang et al. 2013a). Fiber production has intensified globally leading to environmental problems especially polluting air, soil, and surface and underground water. Improving crop cultivation with efficient N use is a concern (Yang et al. 2013b) to escape from environmental risks and economic losses, and N fertilizers should be used more competently in crop cultivation. A significant influence to releases and flux of GHGs from cotton field across the world has been documented because of nitrogenous fertilization. Numerous management choices have been documented to decrease GHG emissions in atmosphere and enhance nutrients-use efficiency for cotton crop.

6.2.4 Nitrogen Application and GHGs

Ideal use of nitrogenous manures to harvests may expand the uptake effectiveness of crops and lessen the N misfortunes through draining. There is immediate connection between soil biota development and action, expanding the N sum, which promotes the development and incitement of microorganisms with increments in oxygen utilization and coming about denitrification (Velthof et al. 2011). N_2O emanation increments with the application measure of the creature fertilizer by 0.025% and 0.85% were because of greater denitrification (Paul et al. 1993). However, data related to addition, optimal addition of organic as well as inorganic nutrient, will be beneficial for effective uptake of N by plants and decreasing the GHG emissions.

6.2.5 Nitrogen Management in Cotton

Nitrogen is the most problematic supplement to manage in cotton under irrigation; it affects cotton yield. N can start from a scope of sources, i.e., barometrical obsession, mineralization, and connected N as compost. N improves photosynthetic limit in leaves; root morphology, i.e., root length, thickness, volume, and mass; and physiological movement (Zhang et al. 2017a, b). In nature, N changes by starting with one structure then onto the next through a progression of responses and lastly comes back to the environment (Fig. 6.1). In cotton creation frameworks where N is connected, its administration is unwieldy because of its reactivity and high portability in the soil (Robert et al. 2009). N is fundamental for covering improvement and

Fig. 6.1 Nitrogen deficiency—cotton | Yara United States (Crop Nutrition n.d.)

better light capture, use, and vitality exchange (Barraclough et al. 2010), and subsequently, crops depend intensely on N preparation. A suitable supply of N prompts intelligible biomass aggregation and yield development in cotton, making it the absolute most significant development-constraining element.

6.2.6 Nitrogen Effects on Yield

To fulfill the need of fiber from a developing total populace, half higher N compost application is required every year for harvest creation. Out of 85 to 90 metric huge amounts of every year connected N to horticultural harvests around the world, 40–70% is decreased through water invasion, volatilization, denitrification, and overflow with waste water (Wu and Liu 2008), though a little part is consumed by harvests (Carranca 2012). Consolidated use of NH_4^+ and NO_3^- compared with single N can possibly build biomass amassing and absolute N substance of cotton (Cao et al. 2007). In any case, crops provided with just NO_3^- had inferior development, poor root advancement, and littler leaves with limited chlorophyll-contrasted and NH_4^+-provided plants. More noteworthy misfortune and natural contamination as outcomes of N are issues for agronomists, mechanical associations, government organizations, and nongovernment areas (Yang et al. 2013b).

6.3 Phosphorus

6.3.1 Phosphorus and Its Functions

Phosphate (Pi) availability is inadequate in soil covering an area of more than 30% of agricultural lands, while the figures are higher for acidic soil covering up to 70% of arable lands which limits the growth of crop (Lenka and Lal 2012). Phosphate is a part of nucleic acids and cell layers and basic for metabolic procedures. Phosphate insufficiency diminishes plant development and photosynthesis and along these lines biomass collection and yield. The harvest interest for supplements, for example, Pi, has expanded because of the presentation of high-yielding cultivars, and this necessity might be much higher because of expanded plant development under rising environmental CO_2 focuses (Rogers et al. 1993; Lenka and Lal 2012). The current environmental CO_2 centralization of around 394 mmol mol^{-1} is anticipated to be multiplied before the twenty-first century is over (IPCC 2007).

Cotton development, photosynthesis, and buildup yield react emphatically to raised CO_2 and accessibility of supplements including Pi (Rogers et al. 1993). It may be, the positive reaction of raised CO_2 is frequently diminished when cotton is developed in combination with low supply of supplements such as phosphorus (Rogers et al. 1993; Barrett and Gifford 1995).

Phosphorus inadequacy (Fig. 6.2) hinders cotton development and advancement by diminishing photosynthetic limit leaf extension, biomass gathering, and yield (Barrett and Gifford 1995).

Fig. 6.2 Phosphorus deficiency—cotton I Yara United States (Crop Nutrition n.d.)

6.3.2 Causes of Phosphorus Deficiency

1. Soil pH (low or high).
2. Organic matter (low).
3. Temperature and humidity.
4. Root development.
5. Low P availability.
6. Soils with a high phosphate capacity.
7. High Fe content.

6.3.3 Phosphorus Is Important for

Improved early root development.
Improved yield.
Earlier boll set and maturity.

Rogers et al. (1993) recommended that basic convergence of inorganic phosphorus required for greatest harvest efficiency may increase in a CO_2-advanced condition. In this way, plants right now that are not constrained by Pi may progress toward becoming restricted because of anticipated increments in air CO_2 (Lenka and Lal 2012). Studies assessing cotton development and physiology under fluctuating dimensions of Pi supply and CO_2 are restricted. The greater part of the earlier examinations researching crop reactions under fluctuating supplement supply has concentrated on supplements other than Pi, and few have considered significant line crops including cotton and their connection with CO_2 focuses (Jacob and Lawlor 1991; Campbell and Sage 2006; Jin et al. 2011).

6.4 Potassium

6.4.1 Potassium and Its Function

Untimely senescence of cotton brought about by potassium (K) inadequacy has been watched worldwide for over two decades (Brouder and Cassman 1990; Wright 1999; Dong et al. 2005). This issue essentially decreases the term and viability of cotton leaf photosynthetic limit (Wright 1999), which may thusly lessen cotton yields by as much as 20% (Cassman et al. 1989) and could quantifiably weaken fiber quality (Pettigrew 2003; Clement-Bailey and Gwathmey 2007). Intriguingly, untimely senescence has likewise been seen in fields with moderately abnormal amounts of accessible K^+, on which different harvests were not influenced (Wright 1999). As far as we could possibly know, this inconsistency might be brought about by awkwardness between the substantial boll load and the root's absorptive limit of K^{+2}. Also,

Fig. 6.3 Potassium deficiency—cotton I Yara United States (Crop Nutrition n.d.)

high temperature stress (Reddy et al. 1999; Dai et al. 2015) and lack of nitrogen (Tewolde et al. 1994) could influence cotton yield and fiber properties by advancing untimely senescence.

Cotton fiber is a fundamental crude material for the material business. Fiber length and quality decide yarn quality, as it was (Young 1990; Zhao et al. 2012). Past investigations have affirmed that K inadequacy could fundamentally diminish fiber length (Yang et al. 2016a, b; Bauer et al. 1998), quality (Cassman et al. 1990; Minton and Ebelhar 1991), and micronaire increase (Minton and Ebelhar 1991). Fiber cells start from cotton ovule epidermal cells. After quick lengthening for roughly 16 days, extension moderates, and concentrated cellulose amalgamation happens until development. By development, more noteworthy than 94% of the fiber dry weight is unadulterated cellulose (Ruan et al. 2001). The point of cellulose stores in the cell divider framework (Davidonis et al. 2004), and the attributes of cellulose aggregation enormously decide the last fiber quality (Shu et al. 2007). Be that as it may, the components by which K insufficiency represses fiber properties and fiber advancement stay indistinct. The greater part of starch for fiber cellulose union originates from the leaves subtending the cotton bolls (Lunn and Hatch 1995; Liu et al. 2013). Past investigations have affirmed that K lack influences cotton leaf photosynthetic limit by restraining sugar amalgamation and the starch transportation rate (Bednarz and Oosterhuis 1999; Pettigrew et al. 2005). Along these lines, the speculation that K insufficiency influences cotton fiber properties by inciting a starch procurement trouble speaks to a decent section point for investigating the systems underlying this process (Fig. 6.3).

6.4.2 *Causes of Potassium Deficiency*

Soil pH (low).
Leaching (sandy soils).
Low water-induced stress.
Heavy rainfall or irrigation.
Heavy soils (clay).

Low K reserves in soil.
Magnesium enrich soils.

6.4.3 Potassium Is Important for

Robust and enhancing yield.
Improving fiber-associated traits (fiber maturity and length).
Reducing early senescence.

Calcium (Ca), among the secondary macronutrients, is required in a larger quantity by cotton than magnesium and sulfur. Ca gives strength to the cell wall in the plants by forming calcium pectate of the middle lamella and thus helps in developing strong plants. The growing point of the plant such as roots and shoots particularly meristematic cell division and elongation is accelerated in the presence of Ca. Calcium promotes root length and diameter and thus invigorates early plant stand. Adequate availability of Ca favorably influences the enzyme activity within the plant cells. Calcium accumulation is one of the salient features of cotton which triggers synthesis of organic acids within the plant.

6.5 Calcium

6.5.1 Sources of Calcium

Limestone is the primary source which contains the highest percentage of calcium. The other sources are gypsum, basic slag, hydrated lime, burnt lime, ground oyster shells, and dried, crushed eggshells. Hydrated and burnt lime provides readily available Ca than gypsum and basic slag. Eggshells are a rich source and may contain up to 95% calcium as CaO. Finely ground eggshells and oyster shells are used to neutralize acidic soils.

6.5.2 Calcium Nutrition

Calcium (Ca) deficiency is comparatively occasional in cotton plants; nevertheless it has been documented in upland soils of the old cotton belt (Donald 1964). This mechanism develops aluminum toxicity comparative to low concentration of calcium (Soileau et al. 1969). Under low pH soils (<7), the availability of Ca to cotton crop is dependent on cation exchange capacity (CEC) of the soil (Hinkle and Brown 1968). To remediate such conditions, application of lime can play an imperative role in neutralizing the acidic conditions of soils (Hinkle and Brown 1968; Mason and

Table 6.2 Materials containing calcium as a main nutrient

Material	Calcium (CaO)	Magnesium (MgO)	Sulfur (SO_4^{2-})
	Percentage (%)		
Calcitic lime	31.7	3.4	0.10
Dolomitic lime	21.5	11.4	0.30
Gypsum	22.5	0.4	0.30
Basic slag	29.0	3.4	0.30
Ground oyster shells	47.5	0.62	0.40
Dried, crushed eggshells	95.0	0.90	

Maskell 1931). About 60% of calcium is stored in leaves as chloroplast (Mason and Maskell 1931).

The concentration of Ca is variable among various parts of plants. Calcium concentration in cotton plant varies from 90 to 160 kg ha^{-1} of Ca in mature cotton plants (Bassett et al. 1970; Olson and Bledsoe 1942) but low in seeds and fibers (Donald 1964; Cooper et al. 1947). The ideal Ca content in cotton crop is analyzed by temperature and salinity interaction (Gerard and Hinojosa 1973).

Calcium exists in enormous quantity in Earth's crust ranging up to 1.0%. Plants absorb calcium in the form of Ca^{2+} ions from the soil. Calcium is required by meristematic and differentiating tissues of the plant and is necessary for cell membrane functions. Ca is required by plant during mitotic spindle formation and in activating enzymes and has important role in regulating metabolic activities. Ca also acts as a regulator of micronutrients and facilitates uptake of iron, manganese, zinc, and copper by plants and also maintains a healthy balance of nutrients within plants. Ca plays positive correlation for root growth, and stunted cotton root growth was observed with a limited supply of Ca for cotton (Adams and Moore 1983). In cotton the uptake of Ca is, however, dependent on the target yields. For instance, the total Ca uptake was found to be 71, 94, and 155 kg ha^{-1} at the respective lint yield levels of 1000, 1800, and 2400 kg ha^{-1} (Rochester 2007) (Table 6.2).

6.5.3 *Factor Affecting Ca Availability to Plants*

Calcium has the tendency to form insoluble compounds with phosphorus, thus limiting its availability to plants (Fig. 6.4). Being positively charged, Ca^{2+} is adsorbed on the negatively charged surfaces such as clay and organic matter particles, referred to as "exchangeable Ca." The exchangeable Ca can be replaced by other positively competing ions like Na^+, K^+, and Mg^{2+} (Fageria 2001). The abundance of competing ions lowers down Ca availability to plants and may also damage soil structure in case Ca^{2+} is replaced by Na^+. The presence of high concentrations of Al^{3+} also limit Ca^{2+} availability to plants (Foy et al. 1972), often observed in acidic soils. Ca availability to plants increases with the increase in pH

Fig. 6.4 Calcium deficiency—cotton I Yara United States (Crop Nutrition n.d.)

and CEC as higher CEC of a soil reflects the greater capacity of the soil to absorb and hold Ca.

6.5.4 Functions of Ca in Plants

A large number of scientists have reviewed the role of Ca in cotton in the past (Bangerth 1979; Christensen and Foy 1979; Millaway and Wiersholm 1979). It has been found that Ca plays a key role in maintaining the integrity of membrane in cotton (Jones and Lunt 1967; Marinos 1962). The primary function of Ca in plant growth is to provide structural support by strengthening cell walls through production of calcium pectate compounds which stabilize cell walls and cement cells together. Ca promotes cell elongation and is involved in enzymatic and hormonal processes, formation of proteins, and carbohydrate movement in plants. Ca protects plants against heat stress by regulating and improving stomatal functions and participating in heat shock protein induction. Ca helps in plant's defense against diseases by strengthening cell walls which are protected from impairment by the enzymes produced by numerous bacteria and fungi. Adequate Ca nutrition improves fruit quality. Calcium is essential in the germination of pollens and pollen tube growth (Ge et al. 2009) and subsequent fertilization of ovule (Digonnet et al. 1997).

Calcium concentration impacts the leakiness of membranes and facilitates selective transport of ions in plasmalemma (Jones and Lunt 1967; Epstein 1961). The enhancement of K uptake by enhanced levels of Ca is an established fact (Viets

1944). Calcium deficiency is directly associated with reduction in uptake of other nutrients especially potassium, boron, nitrate, and chloride (Foote and Hanson 1964). Adequate Ca may also be significant in Ca-dependent processes (Poovaiah and Leopold 1973; Sharpies and Johnson 1977). Structurally, Ca pectate is a constituent of the cell wall (Jones and Lunt 1967). Tissues that have low Ca contents remained considerably disorganized and weak in their wall. Calcium's important role is the differentiation of cell organelles (Burström 1968). Calcium has the capability to offset the opposing effects of amplified acidity on ion absorption and aids in coping with heavy metal toxicity (Rains et al. 1964; Foy et al. 1969).

6.5.5 Ca Disorders in Cotton Plant

Ca is lost from the soil through a number of ways like removal by drainage water, plant removal, immobilization by soil organisms, leaching down by rain water, and absorption onto the clay particles. Calcium deficiencies are not common but may occur in sandy soils facing drought conditions. The deficiency of Ca is normally observed at the early stage of crop with a stunted and reduced number of developments of young shoots. Firstly leaves exhibit patches of chlorotic stains; later the severe symptoms cause curling of leaf margins with a blackening of leaves (Donald 1964; Stocking and Ongun 1962; Jones and Lunt 1967). Moreover calcium deficiency results in tissue deterioration (Jones and Lunt 1967). Germination is also reduced by low calcium levels (Hinkle and Brown 1968). Ca-deficient plants show abortion of roots tips and terminal growing points and develop exceptionally dark green color of the foliage and weakened stems and increased abscission of fruiting bodies (Addicott and Lyon 1973) or even complete failure of the crop to set bolls (Joham 1957) due to disturbance in translocation of carbohydrates. Healthy root growth and development have been attributed to adequate Ca supply (Jones and Lunt 1967; Epstein 1961). The deficiency of Ca has been reported to decrease the basipetal shift of auxin (dela Fuente and Leopold 1973) and thus promotes abscission. Generally low calcium in soil causes formation of larger plants but fewer fruiting although the case is opposite with sufficient supply of calcium which ultimately results in abundant fruiting (Donald 1964). Excessive calcium causes toxicity which is not an issue for cotton growth, but alkaline soils might induce deficiencies of micronutrients (Jones and Lunt 1967).

6.5.6 Ca^{2+} in Abiotic Stress Tolerance

Climatic changes has triggered various environmental stresses known as abiotic stress for crops which significantly reduced the growth and in severe circumstances might result in complete crop failure (Schenkel 2010; Gregory and George 2011). So it is a dire need to alleviate the damages caused due to environmental pressure as plants are being immutable to move toward favorable growing conditions (Shahzad et al. 2016). Abiotic stresses are categorized as excessive accumulation of salt, drought conditions, and heat and cold stress (Zhu 2016). Anthropogenic activities directly influence our planet with excessive usage and uncontrolled measures; especially in developing and underdeveloped countries, this situation is worse. Several mechanisms and approaches have been proposed until now to counteract the severity of environmental stresses, and the role of appropriate nutrient supply is also one of the mechanisms for alleviating stress conditions (Wani et al. 2016; Srivastava et al. 2016; Latif et al. 2015).

Currently we are illuminating the specific role of Ca^{2+} ion against abiotic stresses. It is a crucial element for regulation of cellular responses against harsh conditions (Sanders et al. 2002; Dodd et al. 2010). With the onset of stress faced by plant, the first organ to give signals to upper plant parts is the root with an immediate increase in Ca^{2+} concentration in several cell compartments such as mitochondria, nucleus, and cytosol. Ca^{2+} transient stimuli differ in their spatiotemporal patterns, being therefore considered as specific stress signals (McAinsh and Hetherington 1998). Each Ca^{2+} stimulus contributes to a pool in plants, and the most significant role of calcium is the closure of the stomata to reduce transpiration (Allen et al. 2001; Miwa et al. 2006). Calcium-dependent specific proteins are involved for the synthesis of calcium ions or controlling regulations. Calcium sensor proteins were observed to be more than 250 in *Arabidopsis* a model plant (Day et al. 2002) which are further classified as calcium-dependent protein kinases (CPKs) and the calcium and calmodulin-dependent protein kinase (CCaMK) (Cheng et al. 2002; Wang et al. 2015).

Calcium is involved in regulating plant mechanism and shifting plants to coping strategy by adjusting to external harsh climatic conditions such as stress signaling against drought response (Shao et al. 2008). Higher Ca^{2+} signaling and accumulation confer plants' drought resistance (Cousson 2009). External application of calcium under drought conditions improved the chlorophyll content and improved antioxidant enzyme activities.

Soil salinity is one of the emerging problems across the globe due to poor supply of irrigation and improper management of lands (Dassanayake and Larkin 2017). Salinity impedes crop growth and causes yield losses every year (Shabala and Cuin 2008) and thereby a major abiotic stress to crop yield and threat for sustainable productivity (Wan et al. 2017). Salinity can be reduced by external supplementation of nutrients, and calcium (Ca) is a chief macronutrient in this regard and is also a stress stimulus as mentioned earlier (Khan et al. 2010). It is well known that calcium signaling prepares bonds with phospholipid bilayer with a stabilization and promotion of structural integrity of membranes of plants against salt stress.

Temperature fluctuation from the optimal plant growth requirements is a stress either high or low. Generally it is perceived that the Earth temperature is increasing every year, so it is a point of interest for many researchers to adopt their future crops with changing climatic conditions (Lobell and Asner 2003). Heat stress is mostly studied to elaborate or highlight the ongoing changes in plants the most significant is that it severely disrupts the negatively affects photosynthetic machinery (Sharkey

2005; Allakhverdiev et al. 2008). However calcium was applied as foliar application, and fruitful results were reported for alleviating heat stress with reduced oxidative damages and improved photosynthesis (Gong et al. 1998a; Larkindale and Huang 2004). Ca^{2+} binds directly with the cell membrane and provides protection against heat injury with minimal fluidity of cell membrane (Toprover and Glinka 1976; Cooke et al. 1986). Moreover, calcium moves to pool in the cytosol with the influx of heat stress (Gong et al. 1998b; Wang et al. 2009), which resultantly activates numerous Ca^{2+}-dependent proteins, such as calmodulins, and mitigates heat stress responses (Saidi et al. 2009; Kudla et al. 2010).

6.6 Magnesium

Magnesium (Mg^{2+}) occurs at the center of the chlorophyll ($C_{55}H_{72}O_5N_4Mg$) molecules. Therefore it is essential for photosynthesis. Mg is also an activator for many enzymes required in growth processes. It activates photophosphorylation converting ADP to ATP which is the energy currency in different developmental processes. So Mg is important in P utilization. It also activates the enzyme ribulose diphosphate carboxylase which plays an important role in N metabolism. Mg plays a role in nucleic acid synthesis. It also stabilizes the ribosomal structure. Mg mainly assists in translocating cellulose and controls fiber quality of cotton. Mg enhances resistance of tissues against degradation. Mg compliments or antagonizes other minerals which affect defense mechanism against different disease (Buchanan 1980; Marschner 1995; Tanoi and Kobayashi 2015).

6.6.1 Magnesium in Plants and Soil

Magnesium is the eighth abundant element in the Earth's crust with a range of more than 20,000 ppm (Tanoi and Kobayashi 2015). In plant magnesium is the second most abundant with a range of 0.15–0.35% of the dry weight (Tanoi and Kobayashi 2015; Marschner 1995). In cotton Mg is found to be 801.6 mg/kg, 765.4 mg/kg, and 649.5 mg/kg in leaf, stem, and roots, respectively.

6.6.2 Magnesium Deficiency Causes and Deficiency Symptom

In spite of Mg abundance, it is deficient in field due to soil acidity and its competitor nutrient cations, i.e., its nitrogen (N^+), potassium (K^+), hydrogen (H^+), and calcium (Ca^{2+}) in soil (Uchida 2000). In plants its deficiency symptom is "reddening" of leaves. In addition to reddening of leaves, chlorosis, necrosis and droop also occur.

This reddening symptom is due to the reduction in photosynthesis of plant (Tanoi and Kobayashi 2015).

6.6.3 *Effects of Mg Deficiency on Physiological Processes*

In Mg deficiency photosynthates, sugars, and starches begin to accumulate in addition to reduce photosynthesis, which causes the reddening of leaves, chlorosis, and necrosis. Transpiration also decrease (Tanoi and Kobayashi 2015).

6.7 Sulfur

The need for sulfur in plant nutrition was demonstrated in the early nineteenth century. Sulfur is the fourth major plant nutrient. Plants require sulfur in amount equal to or slightly less than phosphorus.

6.7.1 *Deficiency Symptoms of S*

Often, coarse-textured soils with low organic matter content and good drainage are S deficient (Waddoups 2011). Sulfur deficiencies in crops including cotton (Younge 1942; Harris et al. 1945) are occurring due to minimal/no application of S to the soil, use of high analysis fertilizer, and more removal of S by high yielding crop varieties and high cropping intensity (Scherer 2001). Hill (2000) and Havlin et al. (2005) reported crop response to S application under low temperature conditions owing to reduced mineralization of organic S to SO_4.

Sulfur is an immobile nutrient in cotton plant; therefore its obvious deficiency symptoms are younger leaves compared to older leaves. Deficient leaves become pale green and with the severity yellowish green, but the leaf veins tend to remain somewhat green than interveinal tissue. Plants suffering from S deficiency showed short stature, stunted growth, and lesser vegetative branches with a fewer bolls per plant and are short and having few vegetative branches with small bolls (Cassman 1993). Sulfur deficiency is not observable on older leaves as they remain green.

6.7.2 *Cotton Response to Sulfur*

Cotton crop showed responses to S application in cotton belt and elsewhere in the world (Mahler 1989; Messick 1992; Mullins and Burmester 1993; Delgado and Amacher 1995; Mullins 1998).

6.8 Conclusion

Balance uptake of macronutrients is basic necessity for getting optimum yield of cotton. For that the level of soil fertility is a very important aspect as cotton has deep root system and is dependent on subsoil nutrients. Among these nutrients, deficiency of macronutrients not only induced adverse effects on cotton vegetative growth but also caused decline in reproductive growth as well. Keeping in mind the deficiency symptoms of macronutrients in cotton, an efficacious plan can be developed to overcome this problem. However, the importance of soil testing to assess the fertility status of soil for management of macronutrients cannot be denied.

References

Abbas Q, Ahmad S (2018) Effect of different sowing times and cultivars on cotton fiber quality under stable cotton-wheat cropping system in southern Punjab, Pakistan. Pak J Life Soc Sci 16:77–84

Adams F, Moore BL (1983) Chemical factors affecting root growth in subsoil horizons of coastal plain soils. Soil Sci Soc Am J 47:99–102

Addicott FT, Lyon JL (1973) Physiological ecology of abscission. In: Kozlowski TT (ed) Shedding of plant parts. Academic Press, New York/London, pp 85–124

Ahmad S, Raza I (2014) Optimization of management practices to improve cotton fiber quality under irrigated arid environment. J Food Agri Environ 2(2):609–613

Ahmad S, Raza I, Ali H, Shahzad AN, Atiq-ur-Rehman, Sarwar N (2014) Response of cotton crop to exogenous application of glycine betaine under sufficient and scarce water conditions. Braz J Bot 37(4):407–415

Ahmad S, Abbas Q, Abbas G, Fatima Z, Atique-ur-Rehman Z, Naz S, Younis H, Khan RJ, Nasim W, Habib urRehman M, Ahmad A, Rasul G, Khan MA, Hasanuzzaman M (2017) Quantification of climate warming and crop management impacts on cotton phenology. Plan Theory 6(7):1–16

Ahmad S, Iqbal M, Muhammad T, Mehmood A, Ahmad S, Hasanuzzaman M (2018) Cotton productivity enhanced through transplanting and early sowing. Acta Sci Biol Sci 40:e34610

Ali H, Afzal MN, Ahmad F, Ahmad S, Akhtar M, Atif R (2011) Effect of sowing dates, plant spacing and nitrogen application on growth and productivity on cotton crop. Int J Sci Eng Res 2 (9):1–6

Ali H, Abid SA, Ahmad S, Sarwar N, Arooj M, Mahmood A, Shahzad AN (2013a) Integrated weed management in cotton cultivated in the alternate-furrow planting system. J Food Agri Environ 11(3&4):1664–1669

Ali H, Abid SA, Ahmad S, Sarwar N, Arooj M, Mahmood A, Shahzad AN (2013b) Impact of integrated weed management on flat-sown cotton (*Gossypium hirsutum* L.). J Anim Plant Sci 23 (4):1185–1192

Ali H, Hameed RA, Ahmad S, Shahzad AN, Sarwar N (2014a) Efficacy of different techniques of nitrogen application on American cotton under semi-arid conditions. J Food Agri Environ 12 (1):157–160

Ali H, Hussain GS, Hussain S, Shahzad AN, Ahmad S, Javeed HMR, Sarwar N (2014b) Early sowing reduces cotton leaf curl virus occurrence and improves cotton productivity. Cercetări Agronomice în Moldova XLVII(4):71–81

Aliyan R (2013) Effects of different nitrogen fertilization rate on morphological and physiological characteristics of cotton root in Jujube cotton intercropping system. Acta Agriculturae Boreali-occidentalis Sinica 22:96–102

Allakhverdiev SI, Kreslavski VD, Klimov VV, Los DA, Carpentier R, Mohanty P (2008) Heat stress: an overview of molecular responses in photosynthesis. Photosynth Res 98:541–550

Allen GJ, Chu SP, Harrington CL, Schumacher K, Hoffmann T, Tang YY, Grill E, Schroeder JI (2001) A defined range of guard cell calcium oscillation parameters encodes stomatal movements. Nature 411:1053–1057

Amanullah (2014) Source and rate of nitrogen application influence agronomic N-use efficiency and harvest index in maize (*Zea mays* L) genotypes. Maydica 59:80–89

Amin A, Nasim W, Mubeen M, Nadeem M, Ali L, Hammad HM, Sultana SR, Jabran K, Habib urRehman M, Ahmad S, Awais M, Rasool A, Fahad S, Saud S, Shah AN, Ihsan Z, Ali S, Bajwa AA, Hakeem KR, Ameen A, Amanullah, Hafeez Ur Rehman, Alghabar F, Jatoi GH, Akram M, Khan A, Islam F, Ata-Ul-Karim ST, Rehmani MIA, Hussain S, Razaq M, Fathi A (2017) Optimizing the phosphorus use in cotton by using CSM-CROPGRO-cotton model for semi-arid climate of Vehari-Punjab, Pakistan. Environ Sci Pollut Res 24(6):5811–5823

Amin A, Nasim W, Mubeen M, Ahmad A, Nadeem M, Urich P, Fahad S, Ahmad S, Wajid A, Tabassum F, Hammad HM, Sultana SR, Anwar S, Baloch SK, Wahid A, Wilkerson CJ, Hoogenboom G (2018) Simulated CSM-CROPGRO-cotton yield under projected future climate by SimCLIM for southern Punjab, Pakistan. Agric Syst 167:213–222

Bange MP, Milroy SP (2004) Growth and dry matter partitioning of diverse cotton genotypes. Field Crops Res 87:73–87

Bangerth F (1979) Calcium-related physiological disorders of plants. Arm Rev Phytopathol 17:97–122

Barraclough PB, Howarth JR, Jones J, Lopez-Bellido R, Parmar S, Shepherd C, Hawkesford MJ (2010) Nitrogen efficiency of wheat: genotypic and environmental variation and prospects for improvement. Eur J Agron 33:1–11

Barrett DI, Gifford RM (1995) Photosynthetic acclimation to elevated CO_2 on relation to biomass allocation in cotton. J Biogeogr 2:331–339

Bassett DM, Anderson WD, Werkhoven CHE (1970) Dry matter production and nutrient uptake in irrigated cotton (*Gossypium hirsutum*). Agron J 62:299–303

Bauer PJ, May OL, Camberato JJ (1998) Planting date and potassium fertility effects on cotton yield and fiber properties. J Prod Agric 11(4):415–420

Bednarz C, Oosterhuis D (1999) Physiological changes associated with potassium deficiency in cotton. J Plant Nutr 22:303–313

Bisson P, Cretenet M, Jallas E (1994) Nitrogen phosphorus and potassium availability in the soil physiology of the assimilation and use of the nutrients by the plant. In: Proceedings of the world cotton research conference, 14–17 Feb 1994, Brisbane, Australia, pp 115–124

Brouder S, Cassman K (1990) Root development of two cotton cultivars in relation to potassium uptake and plant growth in a vermiculitic soil. Field Crop Res 23:187–203

Buchanan BB (1980) Role of light in the regulation of chloroplast enzymes. Annu Rev Plant Physiol 31:341–374

Bundy LG, Andraski TW (2004) Diagnostic tests for site-specific nitrogen recommendations for winter wheat. Agron J 96:608–614

Burström H (1968) Calcium and plant growth. Biol Rev 43:287–316

Campbell CD, Sage RF (2006) Interactions between the effects of atmospheric CO_2 content and P nutrition on photosynthesis in white lupin (*Lupinus albus* L.). Plant Cell Environ 29:844–853

Cao Y, Fan XR, Sun SB, Xu G, Shen Q, Di T (2007) Effect of partial replacement of NH_4^+ by NO_3^- on nitrate reductase activity and their genetic expression patterns in rice. Plant Nutr Fertil Sci 13:99–105

Carranca C (2012) Nitrogen use efficiency by annual and perennial crops. In: Lichtfouse E (ed) Farming for food and water security. Springer Science+Business Media, Dordrecht, pp 57–82

Cassman KG (1993) Cotton. In: Bennett WF (ed) Nutrient deficiencies in crop plants. Am Phytopathol Soc Press, St Paul, MN, pp 111–119
Cassman KG, Kerby TA, Roberts B, Bryant D, Brouder S (1989) Differential response of two cotton cultivars to fertilizer and soil potassium. Agron J 81:870–876
Cassman KG, Kerby TA, Roberts BA, Bryant DC, Higashi SL (1990) Potassium nutrition effects on lint yield and fiber quality of Acala cotton. Crop Sci 30(3):672–677
Cheng SH, Willmann MR, Chen HC, Sheen J (2002) Calcium signaling through protein kinases. The Arabidopsis calcium-dependent protein kinase gene family. Plant Physiol 129:469–485
Christensen MN, Foy CD (1979) Fate and function of calcium in tissue. Commun Soil Sci Plant Anal 10:427–442
Clawson EL, Cothren JT, Blouin DC, Satterwhite JL (2008) Timing of maturity in ultra-narrow and conventional row cotton as affected by nitrogen fertilizer rate. Agron J 100:421–431
Clement-Bailey J, Gwathmey CO (2007) Potassium effects on partitioning, yield, and earliness of contrasting cotton cultivars. Am Soc Agron 99:1130–1136
Colakoglu H (1980) Nutrient requirement and fertilization of cotton in Turkey. Potash Rev 12:3–9
Constable GA, Bange MP (2015) The yield potential of cotton (*Gossypium hirsutum* L.). Field Crops Res 182:98–106
Cooke A, Cookson A, Earnshaw MJ (1986) The mechanism of action on calcium in the inhibition on high temperature-induced leakage of betacyanin from beet root discs. New Phytol 102:491–497
Cooper HP, Paden WR, Garman WH (1947) Some factors influencing the availability of magnesium in soil and the magnesium content of certain crop plants. Soil Sci 63:27–41
Cousson A (2009) Involvement of phospholipase C-independent calcium-mediated abscisic acid signaling during Arabidopsis response to drought. Biol Plant 53:53–62
Crop Nutrition (n.d.) Nutrients deficiencies-Cotton. Yara's Podcast. https://www.yara.us/crop-nutrition/cotton/nutrient-deficiencies/potassium-deficiency-cotton/
Dai J, Dong H (2014) Intensive cotton farming technologies in China: achievements, challenges and countermeasures. Field Crop Res 155:99–110
Dai Y, Chen B, Meng Y, Zhao W, Zhou Z, Oosterhuis DM, Wang Y (2015) Effects of elevated temperature on sucrose metabolism and cellulose synthesis in cotton fibre during secondary cell wall development. Funct Plant Biol 42:909–919
Dassanayake M, Larkin JC (2017) Making plants break a sweat: the structure, function, and evolution of plant salt glands. Front Plant Sci 8:1–20
Davidonis GH, Johnson AS, Landivar JA, Fernandez CJ (2004) Cotton fiber quality is related to boll location and planting date. Agron J 96(1):42–47
Day IS, Reddy VS, Shad AG, Reddy ASN (2002) Analysis of EF-hand containing proteins in Arabidopsis. Genome Biol 3:RESEARCH0056. https://doi.org/10.1186/gb-2002-3-10-research0056
Dechorgnat J, Nguyen CT, Armengaud P, Jossier M, Diatloff E, Filleur S, Daniel-Vedele F (2011) From the soil to the seeds: the long journey of nitrate in plants. J Exp Bot 62:1349–1359
dela Fuente RK, Leopold AC (1973) A role for calcium in Auxin transport. Plant Physiol 51:845–847
Delgado JA, Amacher MC (1995) Modeling the uptake of sulfur by crops on two alluvial soils of Louisiana: cotton. Commun Soil Sci Plant Anal 26:1921–1934
Digonnet C, Aldon D, Leduc N, Dumas C, Rougier M (1997) First evidence of a calcium transient in flowering plants at fertilization. Development 122:2867–2874
Dodd AN, Kudla J, Sanders D (2010) The language of calcium signaling. Annu Rev Plant Biol 61:593–620
Donald L (1964) Nutrient deficiencies in cotton. In: Sprague HB (ed) Hunger signs in crops. David McKay Company, New York, pp 59–98
Dong H, Li W, Tang W, Zhang D (2005) Research progress in physiological premature senescence in cotton. Cotton Sci 17:56–60

Epstein E (1961) The essential role of calcium in selective cation transport by plant cells. Plant Physiol 36:437–444

Fageria VD (2001) Nutrient interactions in crop plants. J Plant Nutr 24:1269–1290

Foote BD, Hanson JB (1964) Ion uptake by soybean root tissue depleted of calcium by ethylene diamine-tetraacetic acid. Plant Physiol 39:450–460

Foy CD, Fleming AJ, Arminger WH (1969) Aluminum tolerance of soybean varieties in relation to calcium nutrition. Agron J 61:505–511

Foy CD, Fleming AL, Gerloff GC (1972) Differential aluminum tolerance in two snap bean varieties. Agron J 64:815–818

Garnett T, Conn V, Kaiser BN (2009) Root based approaches to improving nitrogen use efficiency in plants. Plant Cell Environ 32:1272–1283

Ge LL, Xie CT, Tian HQ, Russell SD (2009) Distribution of calcium in the stigma and style of tobacco during pollen germination and tube elongation. Sex Plant Reprod 22:87–96

Gerard CJ, Hinojosa E (1973) Cell wall properties of cotton roots as influenced by calcium and salinity. Agron J 65:556–560

Gibson LR, Nance CD, Karlen DL (2007) Winter triticale response to nitrogen fertilization when grown after corn or soybean. Agron J 99:49–58

Gong M, Li YJ, Chen SZ (1998a) Abscisic acid-induced thermotolerance in maize seedlings is mediated by calcium and associated with antioxidant enzymes. J Plant Physiol 153:488–496

Gong M, van der Luit AH, Knight MR, Trewavas AJ (1998b) Heat-shock-induced changes in intracellular Ca2+ level in tobacco seedlings in relation to thermotolerance. Plant Physiol 116:429–437

Gregory PJ, George TS (2011) Feeding nine billion: the challenge to sustainable crop production. J Exp Bot 62:5233–5239

Guarda G, Padovan S, Delogu G (2004) Grain yield, nitrogen-use efficiency and baking quality of old and modern Italian bread-wheat cultivars grown at different nitrogen levels. Eur J Agron 2 (1):181–192

Harris HC, Bledsoe RW, Calhoun PW (1945) Responses of cotton to sulfur fertilization. J Am Soc Agron 37:323–329

Havlin JL, Beaton JD, Tisdale SL, Nelson WL (2005) Sulfur, calcium, and magnesium. In: Havlin JL et al (eds) Soil fertility and fertilizers, 7th edn. Pearson Prentice Hall, Upper Saddle River, NJ, pp 219–243

Hill P (2000) Crop response to tillage systems. In: Reeder R (ed) Conservation tillage systems and management, vol 45. Midwest Plan Service Publications/Iowa State University, Ames, IA, pp 47–60

Hinkle DA, Brown AL (1968) Secondary nutrients and micronutrients. In: Elliot FC, Hoover M, Porter WK Jr (eds) Advances in production and utilization of quality cotton: principles and practices. Iowa State University Press, Ames, IA, pp 282–310

Hirel B, Le Gouis J, Ney B, Gallais A (2007) The challenge of improving nitrogen use efficiency in crop plants: towards a more central role for genetic variability and quantitative genetics within integrated approaches. J Exp Bot 58:2369–2387

Hou Z, Li P, Li B, Gong Z, Wang Y (2007) Effects of fertigation scheme on N uptake and N use efficiency in cotton. Plant Soil 290:115–126

IPCC (2007) Climate change 2007: the physical science basis. In: Solomon S, Qin D, Manning M, Chen Z, Marquis M, Averyt KB, Tignor M, Miller HL (eds) Contribution of working group I to the fourth assessment report of the intergovernmental panel on climate change. Cambridge University Press, Cambridge/New York, NY

Jackson MR, Melideo SL, Jorns MS (2012) Human sulfide: Quinone oxidoreductase catalyzes the first step in hydrogen sulfide metabolism and produces a sulfane sulfur metabolite. Biochemistry 51:6804–6815

Jacob J, Lawlor DW (1991) Stomatal and mesophyll limitations of photosynthesis in phosphate deficient sunflower, maize and wheat plants. J Exp Bot 42:1003–1011

Jin SH, Huang JQ, Li XQ, Zheng BS, Wu JS, Wang ZJ, Liu GH, Chen M (2011) Effects of potassium supply on limitations of photosynthesis by mesophyll diffusion conductance in *Carya cathayensis*. Tree Physiol 31:1142–1151

Joham HE (1957) Carbohydrate distribution as affected by calcium in cotton. Plant Physiol 32:113–117

Jones RGW, Lunt OR (1967) The function of calcium in plants. Bot Rev 33:407–426

Khan MA, Soomro AW, Khan RU, Arain AS (1987) Response of cotton to phosphorus fertilizer and its time of application at Sakrand Sindh. Pak Cottons 31:55–62

Khan MN, Siddiqui MH, Mohammad F, Naeem M, Khan MMA (2010) Calcium chloride and gibberellic acid protect linseed (*Linum usitatissimum* L.) from NaCl stress by inducing antioxidative defence system and osmoprotectant accumulation. Acta Physiol Plant 32:121–132

Khan A, Najeeb U, Wang L, Tan DKY, Yang G, Munsif F, Ali S, Hafeez A (2017a) Planting density and sowing date strongly influence growth and lint yield of cotton crops. Field Crops Res 209:129–135. https://doi.org/10.1016/j.fcr.2017.04.019

Khan A, Tan DKY, Afridi MZ, Luo H, Tung SA, Ajab M, Fahad S (2017b) Nitrogen fertility and abiotic stresses management in cotton crop: a review. Environ Sci Pollut Res 24:1–16

Kudla J, Batistic O, Hashimoto K (2010) Calcium signals: the lead currency of plant information processing. Plant Cell 22:541–563

Larkindale J, Huang B (2004) Thermotolerance and antioxidant systems in Agrostis stolonifera: involvement of salicylic acid, abscisic acid, calcium, hydrogen peroxide, and ethylene. J Plant Physiol 161:405–413

Latif F, Ullah F, Mehmood S, Khattak A, KhanAU KS, Hussain I (2015) Effects of salicylic acid on growth and accumulation of Phenolics in *Zea mays* L. under drought stress. Acta Agric Scand Sect B Soil Plant Sci 66:325–332

Lenka NK, Lal R (2012) Soil-related constraints to the carbon dioxide fertilization effect. Crit Rev Plant Sci 31:342–357

Liu J, Ma Y, LvF CJ, Zhou Z, Wang Y, Abudurezike A, Oosterhuis DM (2013) Changes of sucrose metabolism in leaf subtending to cotton boll under cool temperature due to late planting. Field Crops Res 144:200–211

Lobell DB, Asner GP (2003) Climate and management contributions to recent trends in U.S. agricultural yields. Science 299:1032

Lokhande S, Reddy KR (2015) Cotton reproductive and fiber quality responses to nitrogen nutrition. Int J Plant Prod 9:191–210

Lunn JE, Hatch MD (1995) Primary partitioning and storage of photosynthate in sucrose and starch in leaves of C_4 plants. Planta 197:385–391

Mahler RJ (1989) Sulfur effects on cotton cultivars grown in a greenhouse. J Plant Nutr 12:187–206

Marinos NG (1962) Studies on submicroscopic aspects of mineral deficiencies. 1. Calcium deficiency in the shoot apex of barley. Am J Bot 49:834–841

Marschner H (1995) Mineral Nutrition in higher plants, 2nd edn. Academic Press, London

Mason TG, Maskell EJ (1931) Further studies on transport in the cotton plant. Ann Bot 45:125–173

McAinsh MR, Hetherington AM (1998) Encoding specificity in Ca^{2+}signalling systems. Trends Plant Sci 3:32–36

Messick DL (1992) Soil test interpretation for Sulphur in the United States an overview. Sulphur in Agric 16:24–25

Millaway RM, Wiersholm L (1979) Calcium and metabolic disorders. Comm Soil Sci Plant Anal 10:1–28

Miller AJ, Fan XR, Orsel M, Smith SJ, Wells DM (2007) Nitrate transport and signalling. J Exp Bot 58:2297–2306

Minton EB, Ebelhar MW (1991) Potassium and aldicarb-disulfoton effects on verticillium wilt, yield, and quality of cotton. Crop Sci 31(1):209–212

Mithaiwala IK, Mirbahar MJ, Channa AA, Arain MH (1981) Effect of fertilizers on the yield of long staple variety K-68/9 in Guddu Barrage area. Pak Cottons 25:73–79

Miwa H, Sun J, Oldroyd GE, Downie JA (2006) Analysis of calcium spiking using a cameleon calcium sensor reveals that nodulation gene expression is regulated by calcium spike number and the developmental status of the cell. Plant J 48:883–894
Mullins GL (1998) Cotton response to the rate and source of sulfur on a sandy coastal plain soil. J Prod Agric 11:214–218
Mullins GL, Burmester CH (1993) Uptake of sulfur by four cotton cultivars grown under field conditions. J Plant Nutr 16:1071–1081
Olesen JE, Berntsen J, Hansen EM, Petersen BM, Petersen J (2002) Crop nitrogen demand and canopy area expansion in winter wheat during vegetative growth. Eur J Agron 16:279–294
Olson LC, Bledsoe RP (1942) The chemical composition of the cotton plant and the uptake of nutrients at different growth stages. Georgia Agr Exp Sta Bul 222
Paul JW, Beauchamp EG, Zhang X (1993) Nitrous and nitric oxide emissions during nitrification and denitrification from manure-amended soil in the laboratory. Can J Soil Sci 73:539–553
Pettigrew W (2003) Relationships between insufficient potassium and crop maturity in cotton. Agron J 95(5):1323–1329
Pettigrew W, Meredith W, Young L (2005) Potassium fertilization effects on cotton lint yield, yield components, and reniform nematode populations. Agron J 97:1245–1251
Poovaiah BW, Leopold AC (1973) Deferral of leaf senescence with calcium. Plant Physiol 52:236–239
Rahman MH, Ahmad A, Wang X, Wajid A, Nasim W, Hussain M, Ahmad B, Ahmad I, Ali Z, Ishaque W, Awais M, Shelia V, Ahmad S, Fahad S, Alam M, Ullah H, Hoogenboom G (2018) Multi-model projections of future climate and climate change impacts uncertainty assessment for cotton production in Pakistan. Agric For Meteorol 253-254:94–113
Rains DW, Schmid WE, Epstein E (1964) Absorption of cations by roots. Effect of hydrogen ions and essential role of calcium. Plant Physiol 39:274–278
Reddy KR, Davidonis GH, Johnson AS, Vinyard BT (1999) Temperature regime and carbon dioxide enrichment alter cotton boll development and fiber properties. Agron J 91:851–858
Robert L, Boman R, McFarl M, Bean B, Provin T, Hons F (2009) Nitrogen management in cotton. J Agric Life Ext 1–9
Rochester IJ (2007) Nutrient uptake and export from an Australian cotton field. Nut Cyc Agroecosys 77:213–223
Rogers GS, Payne L, Milham P, Conroy J (1993) Nitrogen and phosphorus requirements of cotton and wheat under changing atmospheric CO_2 concentrations. Plant Soil 155–56:231–234
Ruan YL, Llewellyn DJ, Furbank RT (2001) The control of single-celled cotton fiber elongation by developmentally reversible gating of plasmodesmata and coordinated expression of sucrose and K+ transporters and expansin. Plant Cell 13(1):47–60
Saidi Y, Finka A, Muriset M, Bromberg Z, Weiss YG, Maathuis FJ, Goloubinoff P (2009) The heat shock response in moss plants is regulated by specific calcium-permeable channels in the plasma membrane. Plant Cell 21:2829–2843
Sanders D, Pelloux J, Brownlee C, Harper JF (2002) Calcium at the crossroads of signaling. Plant Cell 14:S401–S417
Schenkel R (2010) Science policy. The challenge of feeding scientific advice into policy-making. Science 330:1749–1751
Scherer HW (2001) Sulphur in crop production. Eur J Agron 14:81–111
Setatou HB, Simonis AD (1994) Response of cotton to NPK fertilization B the Greek experience. Proc World Cotton Res Conf 1:147–151
Shabala S, Cuin TA (2008) Potassium transport and plant salt tolerance. Physiol Plant 133:651–669
Shahzad S, Chaudhry UK, Anwar B, Saboor A, Yousaf MF, Saeed F, Yaqoob S (2016) Drought stress effect on morphological and physiological characteristics of different varieties of annual verbena (Verbena hybrida). J Bio Env Sci 9:32–46
Shao HB, Song WY, Chu LY (2008) Advances of calcium signals involved in plant anti-drought. C R Biol 331:587–596

Sharkey TD (2005) Effects of moderate heat stress on photosynthesis: importance of thylakoid reactions, rubisco deactivation, reactive oxygen species, and thermotolerance provided by isoprene. Plant Cell Environ 28:269–277

Sharpies RO, Johnson DS (1977) The influence of calcium on senescence changes in apples. Ann Appl Biol 85:450–453

Shu H, Wang Y, Chen B, Hu H, Zhang W, Zhou Z (2007) Genotypic differences in cellulose accumulation of cotton fiber and its relationship with fiber strength. Acta Agron Sin 33:921–926

Soileau JM, Engelstad OP, Martin JB Jr (1969) Cotton growth in an acid fragipan subsoil: 11. Effects of soluble calcium, magnesium, and aluminum on roots and tops. Soil Sci Soc Am Proc 33:919–924

Srivastava AK, Penna S, Nguyen DV, Tran LS (2016) Multifaceted roles of aquaporins as molecular conduits in plant responses to abiotic stresses. Crit Rev Biotechnol 36:389–398

Stocking CR, Ongun A (1962) The intracellular distribution of some metallic elements in leaves. Am J Bot 49:284–289

Suhag SM, Memon AB, Jhatyal NA (1981) Fertilizer requirement of cotton for two series in Sindh. Pak Cottons 25:161–168

Tanoi K, Kobayashi NI (2015) Leaf senescence by magnesium deficiency. Plan Theory 4:756–772

Tariq M, Yasmeen A, Ahmad S, Hussain N, Afzal MN, Hasanuzzaman M (2017) Shedding of fruiting structures in cotton: factors, compensation and prevention. Trop Subtrop Agroecosyst 20(2):251–262

Tariq M, Afzal MN, Muhammad D, Ahmad S, Shahzad AN, Kiran A, Wakeel A (2018) Relationship of tissue potassium content with yield and fiber quality components of Bt cotton as influenced by potassium application methods. Field Crops Res 229:37–43

Tewolde H, Fernandez C, Foss D (1994) Maturity and lint yield of nitrogen-and phosphorus-deficient Pima cotton. Agron J 86:303–309

Toprover Y, Glinka Z (1976) Calcium ions protect beet root cell membranes against thermally induced changes. Physiol Plant 37:131–134

Uchida R (2000) Essential nutrients for plant growth: nutrient functions and deficiency symptoms. In: Silva JA, Uchida R (eds) Plant nutrient management in Hawaii's soils, approaches for tropical and subtropical agriculture. College of Tropical Agriculture and Human Resources, University of Hawaii at Manoa, Hawaii, HI

Usman M, Ahmad A, Ahmad S, Irshad M, Khaliq T, Wajid A, Hussain K, Nasim W, Chattha TM, Trethowan R, Hoogenboom G (2009) Development and application of crop water stress index for scheduling irrigation in cotton (*Gossypium hirsutum* L.) under semiarid environment. J Food Agri Environ 7(3&4):386–391

Varshney OP (1979) Effect of NPK alone and in various combinations on the development yield of cotton (*Gossypium hirsutum* L.). Soils Fertil 42:273–285

Velthof G, van Bruggen C, Groenestein C, De Haan B, Hoogeveen M, Huijsmans J (2011) A model for inventory of ammonia emissions from agriculture in the Netherlands. Atmos Environ 46:248–255

Viets FG Jr (1944) Calcium and other polyvalent cations as accelerators of ion accumulation by excised barley roots. Plant Physiol 19:466–480

Waddoups M (2011) Interpreting soil & plant tissue tests. Northwest Agricultural Consultants, Kennewick, WA. http://www.nwag.com/interpre.shtml. Accessed 21 Apr 2011; verified 15 Sept 2011

Wahhab A (1985) Crop response to fertilizer and soil data interpretation. FAO, project report NEDC/069/Pak

Wan H, Chen L, Guo J, Li Q, Wen J, Yi B, Ma C, Tu J, Fu T, Shen J (2017) Genome-wide association study reveals the genetic architecture underlying salt tolerance-related traits in rapeseed (*Brassica napus* L.). Front Plant Sci 8:1–15. https://doi.org/10.3389/fpls.2017.00593

Wang Y, Yu Q, Tang X, Wang L (2009) Calcium pretreatment increases thermotolerance of Laminaria japonica sporophytes. Prog Nat Sci 19:435–442

Wang JP, Munyampundu JP, Xu YP, Cai XZ (2015) Phylogeny of plant calcium and calmodulin-dependent protein kinases (CCaMKs) and functional analyses of tomato CCaMK in disease resistance. Front Plant Sci 6:1075

Wani SH, Kumar V, Shriram V, Sah SK (2016) Phytohormones and their metabolic engineering for abiotic stress tolerance in crop plants. Crop J 4:162–176

Wright P (1999) Premature senescence of cotton (*Gossypium hirsutum* L.)–predominantly a potassium disorder caused by an imbalance of source and sink. Plant Soil 211:231–239

Wu L, Liu M (2008) Preparation and properties of chitosan-coated NPK compound fertilizer with controlled-release and water-retention. Carbohydr Polym 72(2):240–224

Xu WF, Jia L, Baluška F, Ding G, Shi W, Ye N, Zhang J (2012) PIN2 is required for the adaptation of Arabidopsis roots to alkaline stress by modulating proton secretion. J Exp Bot 63:6105–6114

Yang G, Tang H, Nie Y, Zhang X (2011) Responses of cotton growth, yield, and biomass to nitrogen split application ratio. Eur J Agron 35:164–170

Yang G, Chu K, Tang H, Nie Y, Zhang X (2013a) Fertilizer 15N accumulation, recovery and distribution in cotton plant as affected by N rate and split. J Integr Agric 12(6):999–1007

Yang S, Wang F, Guo F, Meng JJ, Li XG, Dong ST, Wan SB (2013b) Exogenous calcium alleviates photoinhibition of PSII by improving the xanthophyll cycle in peanut (*Arachis hypogaea*) leaves during heat stress under high irradiance. PLoS One 8:e71214

Yang J, Hu W, Zhao W, Chen B, Wang Y, Zhou Z, Meng Y (2016a) Fruiting branch K+ level affects cotton fiber elongation through osmoregulation. Front Plant Sci 7:13

Yang JS, Hu W, Zhao W, Meng Y, Chen B, Wang Y, Zhou Z (2016b) Soil potassium deficiency reduces cotton fiber strength by accelerating and shortening fiber development. Sci Rep 6:28856

Young RJ (1990) In: Kroschwitz JI (ed) Polymers: fibers and textiles, a compendium. Wiley, New York

Younge OR (1942) Sulfur deficiency and its effect on cotton production on Coastal Plain soils. Soil Sci Soc Am Proc 6:215–218

Zhang L, Spiertz JHJ, Li Z, van derWerf B (2008) Nitrogen economy in relay intercropping systems of wheat and cotton. Plant Soil 303:55–68

Zhang H, Khan A, Tan DKY, Luo H (2017a) Rational water and nitrogen management improves root growth, increases yield and maintains water use efficiency of cotton under mulch drip irrigation. Front Plant Sci 8:912

Zhang H, Khan A, Tan DKY, Luo H (2017b) Rational water and management improves root growth, increases yield and maintains water use efficiency of cotton under mulch drip irrigation. Front Plant Sci 8:912

Zhao W, Meng Y, Li W, Chen B, Xu N, Wang Y, Zhou Z, Oosterhuis DM (2012) A model for cotton (*Gossypium hirsutum* L.) fiber length and strength formation considering temperature-radiation and N nutrient effects. Ecol Model 243:112–122

Zhu JK (2016) Abiotic stress signaling and responses in plants. Cell 167:313–324

Chapter 7
Essential Micronutrients for Cotton Production

Niaz Ahmed, Muhammad Arif Ali, Sajjad Hussain, Waseem Hassan, Fiaz Ahmad, and Subhan Danish

Abstract Micronutrients are vital for cotton growth and development. Fundamental micronutrients like boron, zinc, iron, manganese, and copper. Deficiency of these micronutrients increases under calcareous soil and high soil pH conditions which make them unavailable for plant. Currently cotton yield is falling due to nonavailability of micronutrients, so it is need of the hour to ensure sufficient supply of these micronutrients throughout cotton growth period. Micronutrients are required for various growth functions in the cotton plant. Soil application of micronutrients often results in binding with soil particles due to unfavorable soil conditions making them unavailable. Foliar application of micronutrients in such conditions is an economically viable option and also beneficial because of their application and relatively quick absorption by the leaves. In this chapter we discussed about the role of different micronutrients for cotton growth and development. We also elaborated their deficiency symptoms and possible remedial measures to correct the deficiencies.

Keywords Micronutrients · Nutrient supply · Cotton · Growth · Yield

Abbreviations

Ca	Calcium
CGR	Crop growth rate
Co	Cobalt

N. Ahmed · M. A. Ali (✉) · S. Danish
Department of Soil Science, Bahauddin Zakariya University, Multan, Pakistan
e-mail: arif1056@bzu.edu.pk

S. Hussain
Department of Horticulture, Bahauddin Zakariya University, Multan, Pakistan

W. Hassan
Department of Soil and Environmental Sciences, Muhammad Nawaz Sharif University of Agriculture, Multan, Pakistan

F. Ahmad
Central Cotton Research Institute Multan, Multan, Pakistan

S. Ahmad, M. Hasanuzzaman (eds.), *Cotton Production and Uses*,
https://doi.org/10.1007/978-981-15-1472-2_7

Cu	Copper
DAS	Days after sowing
EDTA	Ethylenediaminetetraacetic acid
Fe	Iron
GA	Gibberellic acid
IAA	Indole acetic acid
$MgSO_4$	Magnesium sulfate
Mn	Manganese
NAR	Net assimilation rate
RGR	Relative growth rate
Zn	Zinc
$ZnSO_4$	Zinc sulfate

7.1 Introduction

Cotton is a cash and industrial crop in the world that provides 60% exchange to the national exchequer (Ahmad et al. 2014, 2017, 2018a, b; Abbas and Ahmad 2018; Ahmad and Raza 2014; Ali et al. 2011a, b, 2013a, b, 2014a, b). Researchers are making efforts to improve the per hectare yield of cotton crop via improvement in the production technology (Soomro et al. 2000; Amin et al. 2017, 2018; Khan et al. 2004; Rahman et al. 2018; Tariq et al. 2017, 2018; Usman et al. 2009). It is documented fact that balance use of fertilizer holds a key role in production technology for significant improvement in the yield of crops (Ebeed and Abd El-migeed 2005; Hasina et al. 2011; Zou et al. 2012; Ahmad et al. 2018a, b). Micronutrients control numerous physiological and biochemical processes in the growth and development of crops (Liu et al. 2000; Rana et al. 2012; Das et al. 2013; Tahir et al. 2018). Essential micronutrients such as boron (B), copper (Cu), iron (Fe), manganese (Mn), zinc (Zn), and molybdenum (Mo) offer a dynamic part in crop physiology by regulating enzymes and enzymatic reactions (Mendel and Hänsch 2002). However, reduction in crop yields due to unavailability of these micronutrients, mainly Zn, B, and Fe, in calcareous soils is considerable. Significant reduction in yields of cottonseed (or seed cotton) grown on calcareous soils has been shown under slight imbalance of these nutrient elements. Localized nutrient deficiencies are common based on cropping system (Memon et al. 2012); however extensive farming of crops with high micronutrients demand on calcareous soils with low OM has made 90% of Pakistan's soils deficient in Zn, B, and Fe (Jiskani 2011; Yaseen et al. 2013). These field-scale deficiencies have halted cotton yield in the last decade (Ali 2012; Soleymani and Shahrajabian 2012); if not managed properly it is predicted to further decline in the future.

7.2 Essentiality of Micronutrients

Micronutrients are required in small quantities and are essential for crop growth, development, and yield. Micronutrients not only affect yield and quality of agricultural products but also have indirect impact on human and animal health. Each essential micronutrient offers a specific role in plant physiology and development; therefore balanced ratios of these elements are required for optimum plant nutrition (Millaleo et al. 2010; Mousavi et al. 2011; Sharifianpour et al. 2013). Cheng et al. (1997) found that foliar application of $CaCl_2$ (0.01 mol $CaCl_2$/l) on young cotton seedlings promoted the water holding capacity of cotton leaves and strengthened physiological activities by maintaining structural stability of cytoplasmic membranes, thereby improving drought resistance in cotton. Combined foliar application of chelated Ca at 50 mg/l and Cu, Zn, or Co at 12.5 mg/l and Fe or Mn at 25 mg/l, all applied 3 times as foliar spray at 70, 85, and 100 DAS, increased the SCY, viability, and seedling vigor with increasing N rate and application of Ca, Cu, Zn, Fe, or Mn. Increased seed index as well as increased length of the hypocotyls and the entire seedlings with increasing N and with the application of Cu, Fe, or Mn was found with no significant effect on germination rate index (Sawan et al. 1989). Kamalanathan et al. stated that foliar application of Zn, Mo, Fe, B, and Mn at 2.0, 0.5, 5.0, 0.5, and 2.0 lb. acre^{-1}, respectively, on the 60th DAS increased SCY, ginning, % and fiber quality. Foliar application of 1% Micnelf MS-16 on 30, 65, and 90 DAS produced greater yields than foliar application of Micnelf MS-16 supplemented with $MgSO_4$ in *Kharif* season in Khandwa (India). The highest seed cotton yield of 8.12 tons per ha was obtained from the co-application of the prescribed NPK rate (120:60:60 kg/ha) + 1% Micnelf MS-16 at 30, 65, and 90 DAS (Namdeo et al. 1992). Heitholt (1994) studied the application of 0.32 kg boron per hectare combined with 16.6 kg N as triazone-N (24% of N as S-tetrahydro triazone). Application of boron alone decreased lint yield (1.4 t ha^{-1}) than control (1.56 t ha^{-1}). There was no N × B interaction on most of the fiber properties. Yaseen et al. (2013) also reported improved leaves nutrient status and increased flowers and number of bolls and thus higher seed cotton yields when sprayed with a mix of Zn, B, Mn, Cu, and Fe along with NPK on calcareous soils. Thus augmented yield ensued 20–30% additional financial benefit compared to NPK-alone application.

7.2.1 Boron

Burkalov (1969) reported 14.4 kg or higher cottonseed (means seed) and seed cotton (means phutti) yields as a result of spraying 0.5% boron. Koegh and Mapples (1969) showed higher yields of cotton in response to soil application of Bat 116 lb. (very high rate) per acre. Braud et al. also recorded increased bolls in cultivar D-9 grown in culture solution containing 0.21–3.4 mg of boron. Foliar application of boron in combination with insecticides improved all growth and yield attributes such as boll weight and plant height and yields ranging from 4.1 to 4.4 g, 1.1 to 1.4 m, and 1.4 t/ha, respectively. Application of boron at 45–80 days after sowing (DAS) increased the B concentration in leaves but not always related with yield increase. Micronutrient spray from seedling stage to boll formation stage (Guertal et al. 1996) or up to 100 DAS (Cheng et al. 1997) was found to influence cotton yields positively. Lower doses (0.15 kg boron/ha) of B as a foliar spray at the early stages improved both the fiber quality and yield. However when applied at the later stages (flowering), better effects were observed only at the higher rates (Carvalho et al. 1996). In a greenhouse study, there was a difference in tissue B content in cotton (8–11 mg kg^{-1}) by foliar application of different sources of B like boric acid (17.5%), sodium borate (Solubor® 20.5% B), and Traco Liquid B (10% B); however these did not affect tissue B content in soybean (Guertal et al. 1996). Dong (1995) reported that three rounds of foliar application of 0.2% B at seedling, flowering, or boll formation stages increased kapas yields by 15%. The yield increase was 13.1% and 8.6% for two and one round of spray, respectively. The average yield increase was 16.1%. The CGR, RGR, and NAR of cotton during vegetative growth were unaffected by foliar spray of either Cu (15 g 10% Cu-EDTA per liter water) or B application (0.2 g boric acid per liter). Leaf area ratio was generally increased by B and decreased by Cu. Boll weight was unaffected by B or Cu application, while kapas yield (1468 kg/ha) and open bolls $plant^{-1}$ (19.13%) were increased (Hosny et al. 1984). Sun and Xu (1986) observed that when soil-available B ranged from 0.05 to 4.48 ppm with soil pH of 8.0, foliar application of 0.2% boric acid at square formation, early, and peak flowering increased the seed cotton yield by 0.03–17.3% and staple length by 0.17 mm and also reported that the critical value of B application was 0.55 ppm.

7.2.2 Zinc

Zinc is an essential micronutrient that is a key component of many enzymes and plays regulatory role in crop development and yields. Zinc is taken up by the plants in its divalent form (Zn^{2+}) through xylem routes. In the phloem, sap zinc makes complexes with low molecular weight organic acids and increases its concentration. However like other micronutrients, Zn becomes immobile and unavailable to plants in soil solutions of eroded, calcareous, and acidic soils. Zinc deficiencies are usually accompanied with Fe deficiency in calcareous soils due to the low solubility of minerals containing iron in alkaline soils of arid regions (Tavakoli et al. 2014). Rajendran et al. (2010) concluded that either three sprays of $MgSO_4$ (1%) + $ZnSO_4$ (0.5%) at squaring, flowering, and boll developmental stage were recommended to obtain increased seed cotton yield. Application of mepiquat chloride alone and in combination with 0.15 l/ha^{-1}B at flowering stage and 2 weeks after resulted in by far the highest open bolls and yield (Gormus 2006). Sharma et al. (1988) observed 14.6q/ha as a result of 0.5% Zn on 50 and 65 DAS compared to 11.82q/ha cotton yields without Zn. Zinc increased the dry matter production and weight of leaves and stems at different growth stages, bolls plant^{-1}, and boll weight but had no effect on plant height. Biozyme (GA3 + IAA + Zeatin + micronutrients) application at 600 ml/ha increased the plant height, sympodial branches plant^{-1}, bolls plant^{-1}, and SCY as compared with water-sprayed control (Thakar Singh and Sidhu 1997). Gomaa (1991) reported that in cotton cv. Giza 70 foliar spray of 1.0 kg $ZnSO_4$ ha^{-1} in soil applied with 125 kg P_2O_5 ha^{-1} resulted in the highest yield in Alexandria University (Egypt). Total and open boll numbers were increased only by the foliar Zn application. Zn foliar application at 0.1–0.3% at the third leaf stage was found to increase SCY by 22.7% in coastal areas of Dafeng County (China) where the soils contained 0.12–0.93 ppm Zn (Lu and Shi 1986). Biswas (1984) reported that the two foliar sprays of 0.02% boric acid +0.1% $ZnSO_4$ in 1:1 ratio at start and full flowering markedly augmented the B and Zn contents in different plant organs.

7.2.3 Manganese

Manganese is a constituent of chlorophyll and therefore plays an essential role in the lifelong processes of plants, and its deficiency causes growth retardation and reduced yields (Millaleo et al. 2010). Manganese ions in redox reactions readily oxidize from Mn^{2+} to Mn^{3+} and Mn^{4+} and play a key role in electron transport chain in photosynthesis. Its deficiency affects plants in a variety of ways by disrupting photosynthesis rates and efficiency and has very serious effects on carbohydrate synthesis particularly carbohydrate content of roots (Mousavi et al. 2011). Limited availability of manganese results in low fertility and reduced carbohydrate content during grain filling and thus crop quality and yield decreases (Sharifianpour et al. 2013). Cotton yields were 4.48 and 3.68 tons per hectare with early and late Mn applications, respectively, compared with 1.96 tons per hectare without foliar Mn application (Levin 1992). In a fine sandy loam soil, cotton was supplied with 0 or 16 kg Mn ha^{-1} by foliar as $MnSO_4$ and 40 kg/ha N or 45 kg/ha N + 45 kg/ha P_2O_5 + 45 kg/ha K_2O as basal dose as well as 57 kg N + 57 kg K_2O as a side dressing. Foliar Mn application was found to produce darker green leaves but had no effect on yields.

7.2.4 Copper

Application of Cu at 37.5 ppm with Mn at 50 ppm through foliar spray produced premier seed index, oil content, and yield. The highest oil and protein yields ensued from applying 25 ppm Cu or Mn, while oil refractive index and iodine value increased with application of Cu and/or Mn (Sawan et al. 1993).

Fig. 7.1 Micronutrient deficiency symptoms in cotton: (**a**) boron deficiency showing abortion of terminal growth, (**b**) Manganese deficiency showing chlorotic leaf, (**c**) zinc deficiency showing interveinal chlorosis, and (**d**) iron deficiency showing bleached and discolored leaf (Source: IPNI 2013)

7.3 Diagnosis and Remedial Measures

Micronutrient deficiency symptoms can be easily observed on cotton leaves (Fig. 7.1). Deficiency has detrimental effect on cotton growth and yield; therefore it is necessary to understand which micronutrient is deficient in the plant by observing the phenotypical appearance. Boron deficiency symptoms include distortion of leaves with stunted terminal resulting in abnormal growth of upper leaves and aborted flowering (Zhao and Oosterhuis 2002). Manganese deficiency results in chlorosis on younger leaves because its mobility is limited within plants, so it is deposited more in older leaves (Dordas 2009). Zinc causes leathery appearance and upturning of leaves, and it is mostly caused by excessive application of phosphorus (Cakmak and Marschner 1986). Iron-deficient leaves look like chlorotic and bleached except for veins (IPNI 2013). Micronutrient deficiency can be corrected by exploiting several practices. First option is to supply chemical fertilizer to supplement micronutrients to the soil (Abid et al. 2002; Ahmad et al. 2013). Soil conditioners are also beneficial such as organic matter to improve the soil characteristics and provide favorable conditions for micronutrients uptake for plants (Li et al. 2007).

7.4 Improving Fertilizer-Use Efficiency

Supplementation, priming, and food fortification are commonly practiced strategies to tackle micronutrient-deficient soils (Gibson 2011). These strategies do not address the primary cause of meager micronutrient status, insufficient dietary intake, and condition of food security and may not be the most appropriate strategy for rural and underprivileged households. Complementary approaches are needed in this context to support community- and agriculture-based interventions, with concurrent struggles to improve capacity and reach of supplementation and fortification (Gibson 2011; Girard et al. 2012). Soil application of micronutrients compared to the foliar application on calcareous soils having higher soil pH is less efficient, and their availability decreases to plant roots (Rashid and Ryan 2004). Foliar application is the most efficient way for alleviation of nutrient deficiency and readily available for cotton after absorption via cotton leaves (Sawan et al. 2008).

7.4.1 Foliar Application of Micronutrients

Priming of soil with micronutrients is found to be less effective since they get immobile and remain inaccessible to plant roots (Rashid and Ryan 2004; Ali et al. 2008). Instead, foliar application of micronutrients not only decreases environmental pollution but also increases fertilizer-use efficiency. Particularly in cotton crop, foliar application of micronutrients during reproductive stages enhances the efficient utilization of nutrients. Resultantly, it reduces boll shedding and increases the seed cotton yield (Ali et al. 2011a, b). Under certain circumstances plant is unable to uptake micronutrients to its vegetative growth parts due to which visible symptoms of nutrient deficiency can be easily noticed. However foliar-applied nutrients are directly absorbed by the vegetative organs of cotton which ultimately results in a good crop performance against micronutrient-deficient soil as well as provides vigorous growth stands and cotton yield (Yaseen et al. 2013). It also protects the crop from excessive amount of micronutrient application to soil which ultimately causes micronutrient toxicity to cotton (Ahmed et al. 2008). Foliar application is a quick way of fixing nutrient deficiency and supplementing micronutrient in cotton (Fig. 7.2).

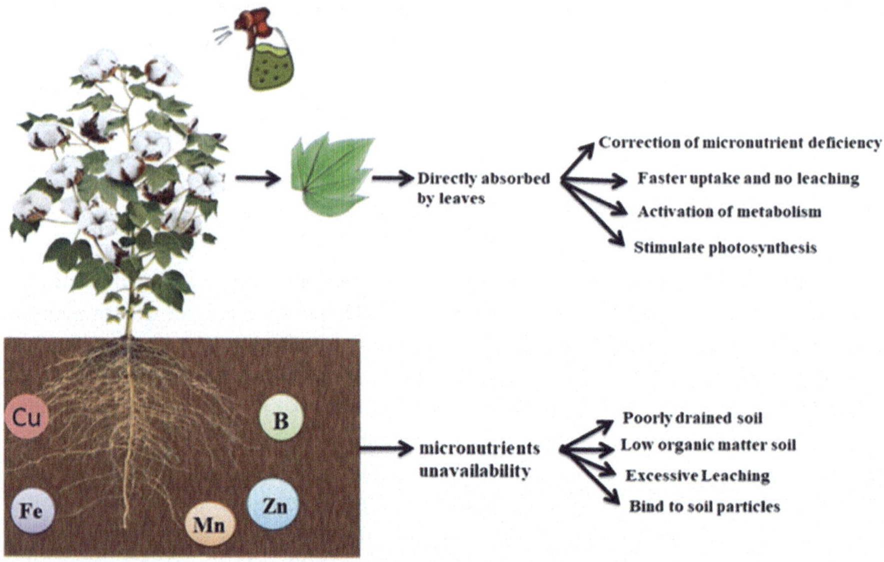

Fig. 7.2 Benefits of foliar application over soil application for cotton

7.5 Micronutrient Dosage and Benefits for Cotton

Cotton is deemed to require high supply of micronutrients for its growth and development as compared to other crops (Oosterhuis 2001). It is sensitive to deficiency of boron and zinc where boron is required in enormous amount, while zinc is required in nominal amount (Shorrocks 1992; Rashid 1995). Cotton genotypes are differing for being susceptible and tolerant to micronutrient-deficient conditions (Soomro et al. 2000). Proper dose of boron is essential for cotton yield and availing higher lint quality (Stewart 1986). Among all the micronutrient groups of fertilizer, the demand of boron is relatively higher and necessary for its growth (Howard et al. 2000). Improvement in cotton fruiting per plant and higher seed yield were achieved by boron fertilizer; basal application of 2.0 kg ha^{-1} boron conferred better performance for cotton starting from its overall growth stages, i.e., stem height, to number of nodes with a lesser number of boll shedding which resulted in higher fiber yield (Abid et al. 2007). However crop also performed well with the application of 1.0 kg B ha^{-1} (Rashid 1995). Boron application benefit is far more for cotton as it was reported in a study on cotton to improve boll weight and seeds per boll (Malik et al. 1990). Boron application to calcareous soil conditions improved boron supply and assisted in supply of other essential nutrients with higher growth returns in cotton (Ahmed et al. 2011). Contrary to this, boron deficiency has devastating effect on cotton via reduction in translocation of photosynthates and halts its growth with abnormal development of reproductive organs (Wang and Zhou 1992). Zinc and boron application to cotton improved its yield by 14% as compared to cotton that was deprived of these nutrients (Rashid and Rafique 2000). Boron and zinc dosage at 2 kg ha^{-1} and 5 kg ha^{-1}, respectively, provided encouraging

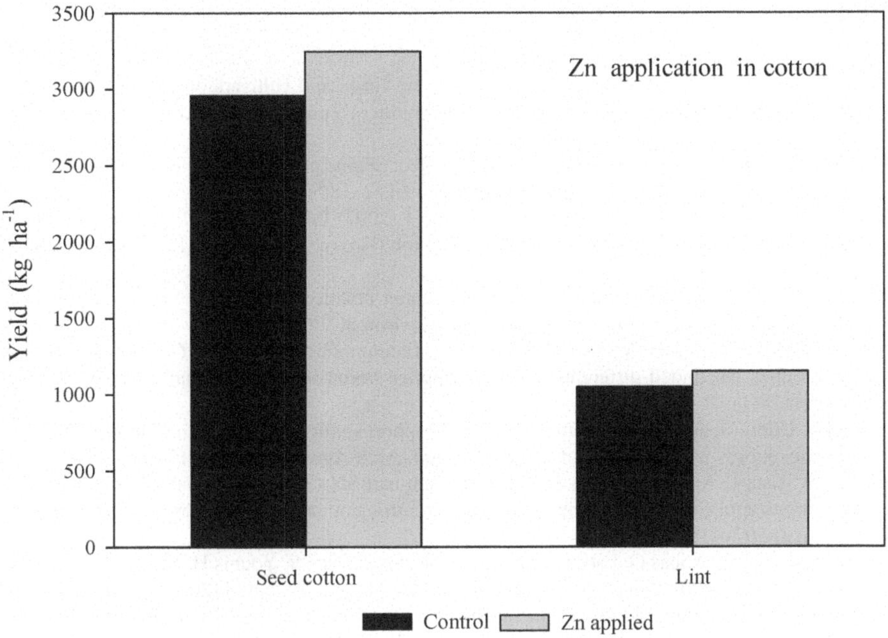

Fig. 7.3 Increase in cotton yield with zinc application

response toward enhancing cottonseed yield (Soomro et al. 2000). Boron improved morphological characteristics of cotton and fiber quality traits (Ahmad et al. 2009). These two nutrients have positive correlation toward increase in cotton yield. Application of zinc improved cotton yield as illustrated in Fig. 7.3 (Sawan et al. 2008). Foliar application of various micronutrients was tested on cotton growth and yield; it was concluded that all of the micronutrients are vital for growth and some have crucial role in triggering plant enzymatic activities assisting in improving the photosynthesis of cotton (Oosterhuis and Weir 2010).

7.6 Conclusion

Micronutrients are equally important as that of macronutrients for the proper growth of cotton. Physiochemical properties of soils are major factors that directly influenced the bioavailability of micronutrients in cotton plants. As change in soil chemical attributes is not an easy task due to high buffering ability, foliar application of micronutrients can overcome this problem to many extents. In that regard, identification of micronutrient deficiency in cotton could be very helpful for balance and timely application of micronutrients. That's why this chapter elaborates the deficiency symptoms and possible remedial measures to overcome the micronutrient deficiency.

References

Abbas Q, Ahmad S (2018) Effect of different sowing times and cultivars on cotton fiber quality under stable cotton-wheat cropping system in southern Punjab, Pakistan. Pak J Life Soc Sci 16:77–84

Abid M, Ahmad N, Jahangir M, Ahmad I (2002) Effect of zinc, iron and manganese on growth and yield of rice (*Oryza sativa* L.). Pak J Agric Sci 39:177–180

Abid M, Ahmad N, Ali A, Chaudhry MA, Hussain J (2007) Influence of soil – applied boron on yield, fiber quality and leaf boron contents of cotton (*Gossypium hirsutum* L.). J Agri Soc Sci 3 (1):7–10

Ahmad S, Raza I (2014) Optimization of management practices to improve cotton fiber quality under irrigated arid environment. J Food Agri Environ 2(2):609–613

Ahmad S, Akhtar LH, Iqbal N, Nasim M (2009) Short communication cotton (*Gossypium hirsutum* L.) varieties responded differently to foliar applied boron in terms of quality and yield. Soil Environ 28(1):88–92

Ahmad F, Uddin S, Ahmad N, Islam R (2013) Phosphorus–microbes interaction on growth, yield and phosphorus-use efficiency of irrigated cotton. Arch Agron Soil Sci 59(3):341–351

Ahmad S, Raza I, Ali H, Shahzad AN, Atiq-ur-Rehman SN (2014) Response of cotton crop to exogenous application of glycine betaine under sufficient and scarce water conditions. Braz J Bot 37(4):407–415

Ahmad S, Abbas Q, Abbas G, Fatima Z, Atique-ur-Rehman, Naz S, Younis H, Khan RJ, Nasim W, Habib ur Rehman M, Ahmad A, Rasul G, Khan MA, Hasanuzzaman M (2017) Quantification of climate warming and crop management impacts on cotton phenology. Plan Theory 6(7):1–16

Ahmad I, Bibi F, Bakhsh A, Ullah H, Danish S, Asif-ur-Rehman (2018a) Assessment of various levels of potassium citrate and sucrose along with boric acid on quality and yield of SufaidChaunsa. Int J Biosci 13(1):188–195

Ahmad S, Iqbal M, Muhammad T, Mehmood A, Ahmad S, Hasanuzzaman M (2018b) Cotton productivity enhanced through transplanting and early sowing. Acta Sci Biol Sci 40:e34610

Ahmed N, Abid M, Ahmad F (2008) Boron toxicity in irrigated cotton (*Gossypium hirsutum* L.). Pak J Bot 40(6):2443–2452

Ahmed N, Abid M, Ahmad F, Ullah MA, Javaid Q, Ali MA (2011) Impact of boron fertilization on dry matter production and mineral constitution of irrigated cotton. Pak J Bot 43(6):2903–2910

Ali EA (2012) Effect of Iron nutrient care sprayed on foliage at different physiological growth stages on yield and quality of some durum wheat (*Triticum durum* L.) varieties in sandy soi. Asian J Crop Sci 4:139–149

Ali S, Khan AR, Mairaj G, Arif M, Fida M, Bibi S (2008) Assessment of different crop nutrient management practices for yield improvement. Aust J Crop Sci 2:150–157

Ali H, Afzal MN, Ahmad F, Ahmad S, Akhtar M, Atif R (2011a) Effect of sowing dates, plant spacing and nitrogen application on growth and productivity on cotton crop. Int J Sci Eng Res 2 (9):1–6

Ali L, Ali M, Mohyuddin Q (2011b) Effect of foliar application of zinc and boron on seed cotton yield and economics in cotton-wheat cropping pattern. J Agric Res 49(2):173–180

Ali H, Abid SA, Ahmad S, Sarwar N, Arooj M, Mahmood A, Shahzad AN (2013a) Integrated weed management in cotton cultivated in the alternate-furrow planting system. J Food Agri Environ 11(3&4):1664–1669

Ali H, Abid SA, Ahmad S, Sarwar N, Arooj M, Mahmood A, Shahzad AN (2013b) Impact of integrated weed management on flat-sown cotton (*Gossypium hirsutum* L.). J Anim Plant Sci 23 (4):1185–1192

Ali H, Hameed RA, Ahmad S, Shahzad AN, Sarwar N (2014a) Efficacy of different techniques of nitrogen application on American cotton under semi-arid conditions. J Food Agri Environ 12 (1):157–160

Ali H, Hussain GS, Hussain S, Shahzad AN, Ahmad S, Javeed HMR, Sarwar N (2014b) Early sowing reduces cotton leaf curl virus occurrence and improves cotton productivity. CercetăriAgronomiceîn Moldova XLVII(4):71–81

Amin A, Nasim W, Mubeen M, Nadeem M, Ali L, Hammad HM, Sultana SR, Jabran K, Habib urRehman M, Ahmad S, Awais M, Rasool A, Fahad S, Saud S, Shah AN, Ihsan Z, Ali S, Bajwa AA, Hakeem KR, Ameen A, Amanullah, Hafeez Ur Rehman, Alghabar F, Jatoi GH, Akram M, Khan A, Islam F, Ata-Ul-Karim ST, Rehmani MIA, Hussain S, Razaq M, Fathi A (2017) Optimizing the phosphorus use in cotton by using CSM-CROPGRO-cotton model for semi-arid climate of Vehari-Punjab, Pakistan. Environ Sci Pollut Res 24(6):5811–5823

Amin A, Nasim W, Mubeen M, Ahmad A, Nadeem M, Urich P, Fahad S, Ahmad S, Wajid A, Tabassum F, Hammad HM, Sultana SR, Anwar S, Baloch SK, Wahid A, Wilkerson CJ, Hoogenboom G (2018) Simulated CSM-CROPGRO-cotton yield under projected future climate by SimCLIM for southern Punjab, Pakistan. Agric Syst 167:213–222

Biswas CR (1984) A note on mode of application of urea to un-irrigated cotton in Sundarbans, West Bengal. J Indian Soc Coastal Agric Res 2:49–50

Burkalov N (1969) Foliar application of B, Co, Zn, Mo and on cotton plants. Nauchi Tr. Selkostop Institute, Kolar 18(1):17–22

Cakmak I, Marschner H (1986) Mechanism of phosphorus-induced zinc deficiency in cotton. I. Zinc deficiency-enhanced uptake rate of phosphorus. Physiol Plant 68(3):483–490

Carvalho LH, Silva NM, BrasilSobrinho MOC, Kondo JI, Chiavegato EJ (1996) Aplicação de boro no algodoeiro, emcobertura e empulverização foliar. RevistaBrasileira de Ciência do Solo, Campinas. Revesta- Brasileria de Ciencia do solo 20:265–269

Cheng LH, Tang LS, Zhang YG, Yan JY, Zhang H (1997) The effect of $CaCl_2$ on the drought resistance of young cotton seedlings. China Cottons 24:17–18

Das JK, Salam RA, Kumar R, Bhutta ZA (2013) Micronutrient fortification of food and its impact on woman and child health: a systematic review. Syst Rev 2:67

Dong JF (1995) The yield increasing ability of spraying cotton with boron. Henan NongyeKexue 3:6

Dordas C (2009) Foliar application of manganese increases seed yield and improves seed quality of cotton grown on calcareous soils. J Plant Nutr 32(1):160–176

Ebeed S, Abd El-migeed MMM (2005) Effect of spraying sucrose and some nutrient elements on Fagri Kalan mango trees. J Appl Sci Res 1(5):341–346

Gibson RS (2011) Strategies for preventing multi-micronutrient deficiencies: a review of experiences with food-based approaches in developing countries. In: Thompson B, Amoroso L (eds) Combating micronutrient deficiencies: food-based approaches. CAB International and FAO, Rome, pp 7–27

Girard AW, Self JL, McAuliffe C, Olude O (2012) The effects of household food production strategies on the health and nutrition outcomes of women and young children: a systematic review. Paediatr Perinat Epidemiol 26(Suppl 1):205–222

Gomaa MA (1991) Response of cotton plant to phosphatic and zinc fertilization. Ann Agric Sci, Moshtohor (Egypt)

Gormus O (2006) Effect of mepiquat chloride and boron on irrigated cotton (Gossypiumhirsutum) in Turkey. Indian J Agron 51(2):149–151

Guertal EA, Abaye AO, Lippert BM, Miner GS, Gascho GJ (1996) Sources of boron for foliar fertilization of cotton and soybean. Commun Soil Sci Plan 27(15–17):2815–2828

Hasina G, Said A, Saeed B, Mohammad F, Ahmad I (2011) Effect of foliar application of nitrogen, potassium and zinc on wheat growth. ARPN J Agric Biol Sci 6:56–58

Heitholt JJ (1994) Comparison of adjuvant effects on cotton leaf potassium concentration and lint yield. J Plant Nutr 17(12):2221–2233

Hosny AA, Kadry W, Mohamad HMH (1984) Effect of foliar application of boron and copper on growth yield and yield components of Giza 75 cotton variety. Ann Agric Sci 21:25–35

Howard DD, Essington ME, Gwathmey CO, Percell WM (2000) Buffering of foliar potassium and boron solutions for no-tillage cotton production. J Cotton Sci 4:237–244

International Plant Nutrition Institute (IPNI) (2013) Nutrient Deficiencies & Toxicities in Cotton. http://www.ipni.net/article/IPNI-3327
IPNI (2013) Nutrient deficiencies & toxicities in cotton. http://www.ipni.net/article/IPNI-3327
Jiskani M (2011) Imbalanced nutrition causes a-biotic and increases biotic disease in crops. http://www.pakissan.com/%0Aenglish/allabout/farminputs/fertilizers/imbalanced.nutrition.%0Acauses.shtml
Khan MB, Khaliq A, Ahmad S (2004) Performance of mashbean intercropped in cotton planted in different planting patterns. J Res (Sci) 15(2):191–197
Koegh JL, Mapples R (1969) Boron for cotton and soybeans on hoersial plains soil. Bull Agric Exp Stud 7(1):11–19
Levin I (1992) Hassadeh 72(11):1390–1392
Li BY, Zhou DM, Cang L, Zhang HL, Fan XH, Qin SW (2007) Soil micronutrient availability to crops as affected by long-term inorganic and organic fertilizer applications. Soil Tillage Res 96 (1–2):166–173
Liu A, Hamel C, Hamilton RI, Ma BL, Smith DL (2000) Acquisition of Cu, Zn, Mn and Fe by mycorrhizal maize (*Zea mays* L.) grown in soil at different P and micronutrient levels. Mycorrhiza 9(6):331–336
Lu FM, Shi JF (1986) Yield increasing effect of zinc sulphate to cotton. China Cottons 4–31
Malik MN, Iqrar HS, Shahabuddin (1990) Effect of phosphorus fertilization and seed cotton yield and phosphorus availability in clay loam soils. Pak Cottons 34:115–121
Memon M, Jamro GM, Memon NUN, Memon KS, Akhtar MS (2012) Micronutrient availability assessment of tomato grown in Taluka Badin. Sindh Pak J Bot 44:649–654
Mendel RR, Hänsch R (2002) Molybdoenzymes and molybdenum cofactor in plants. J Exp Bot 53 (375):1689–1698
Millaleo R, Reyes DM, Ivanov AG, Mora ML, Alberdi M (2010) Manganese as essential and toxic element for plants transport, accumulation and resistance mechanisms. J Soil Sci Plant Nutr 10:470–481
Mousavi SR, Shahsavari M, Rezaei M (2011) A general overview on manganese (Mn) importance for crops production. Aust J Basic Appl Sci 5(9):1799–1803
Namdeo KN, Sharma JK, Mandoli KC (1992) Effect of foliar feeding of micro-nutrients on rainfed hybrid cotton (JK Hy. 1). Crop Res 5:451–455
Oosterhuis D (2001) Physiology and nutrition of high yielding cotton in the USA. InformaçõesAgronômicas 95:18–24
Oosterhuis DM, Weir BL (2010) Foliar fertilization of cotton. In: Stewart JMD (ed) Springer physiology of cotton. Science + Business Media B.V., California, pp 225–237
Rahman MH, Ahmad A, Wang X, Wajid A, Nasim W, Hussain M, Ahmad B, Ahmad I, Ali Z, Ishaque W, Awais M, Shelia V, Ahmad S, Fahad S, Alam M, Ullah H, Hoogenboom G (2018) Multi-model projections of future climate and climate change impacts uncertainty assessment for cotton production in Pakistan. Agric For Meteorol 253-254:94–113
Rajendran K, Amanullah MM, Vaiyapuri K (2010) Foliar nutrition in cotton-a review. Agric Rev 31 (2):120–126
Rana A, Saharan B, Nain L, Prasanna R, Shivay YS (2012) Enhancing micronutrient uptake and yield of wheat through bacterial PGPR consortia. Soil Sci Plant Nutr 58:573–582
Rashid A (1995) Nutrient indexing of cotton and micronutrient requirement of crops. Micronutrient project annual report 1993–94. NARC, Islamabad, p 91
Rashid A, Rafique I (2000) Boron and zinc fertilizer use in cotton: importance and recommendations. Pakistan Agriculture Research Council, Islamabad
Rashid A, Ryan J (2004) Micronutrient constraints to crop production in soils with mediterranean-type characteristics: a review. J Plant Nutr 27:959–975
Sawan ZM, Maddah El Din MS, Gregg B (1989) Effect of nitrogen fertilisation and foliar application of calcium and micro-elements on cotton seed yield, viability and seedling vigour. In: Proceedings of the International Seed Test Association

Sawan ZM, Basyony AE, McCuistion WL, El Farra AHA (1993) Direct and residual effect of foliar applications of copper and manganese on cottonseed yield and quality. J Agric Food Chem 41 (10):1756–1759

Sawan ZM, Mahmoud MH, El-Guibali AH (2008) Influence of potassium fertilization and foliar application of zinc and phosphorus on growth, yield components, yield and fiber properties of Egyptian cotton (*Gossypium barbadense* L.). J Plant Ecol 1(4):259–270

Sharifianpour G, Zaharah AR, Ishak CF, Hanafi MH, Nejat N, Sahebi M, Sharifkhani A, Azizi P (2013) Elucidating the expression of zinc transporters involved in zinc uptake by upland rice landraces in Malaysia. Adv Environ Biol 7(14):4854–4857

Sharma JC et al (1988) Agric Sci Digest 8:165–169

Shorrocks VM (1992) Boron-A global appraisal of the occurrence, diagnosis and correction of boron deficiency. In: Portech S (ed) Proceedings of international symposium on the role of sulphur, magnesium and micronutrients in balanced plant nutrition. The Sulphur Institute, Washington, DC, pp 39–53

Singh T, Sidhu AS (1997) Ann Biol 132:317–318

Soleymani A, Shahrajabian MH (2012) The effects of Fe, Mn and Zn foliar application on yield, ash and protein percentage of forage sorghum in climatic condition of Esfahan. Int J Biol 4:3–7

Soomro AW, Soomro AR, Leghari AB, Chang MS, Soomro AH, Tunio GH (2000) Effect of boron and zinc micronutrients on seed cotton yield and its components. Pak J Biol Sci 3 (12):2008–2009

Stewart JMD (1986) Integrated events in the flower and fruit. In: Murphy JR, Stewart JMD (eds) Cotton physiology, The cotton foundation reference book series No. 1. National Cotton Council America, Memphis, TN, pp 261–297

Sun ZY, Xu CJ (1986) Soil available B in the south of Hebei Province and application of boron to cotton. J Soil Sci 17(3):130–132

Tahir FA, Ahamad N, Rasheed MK, Danish S (2018) Effect of various application rate of zinc fertilizer with and without fruit waste biochar on the growth and Zn uptake in maize. Int J Biosci 13(1):159–166

Tariq M, Yasmeen A, Ahmad S, Hussain N, Afzal MN, Hasanuzzaman M (2017) Shedding of fruiting structures in cotton: factors, compensation and prevention. Trop Subtrop Agroecosyst 20(2):251–262

Tariq M, Afzal MN, Muhammad D, Ahmad S, Shahzad AN, Kiran A, Wakeel A (2018) Relationship of tissue potassium content with yield and fiber quality components of Bt cotton as influenced by potassium application methods. Field Crops Res 229:37–43

Tavakoli MT, Chenari AI, Rezaie M, Tavakoli A, Shahsavari M, Mousavi SR (2014) The importance of micronutrients in agricultural production. Adv Environ Biol 8(10):31–35

Usman M, Ahmad A, Ahmad S, Irshad M, Khaliq T, Wajid A, Hussain K, Nasim W, Chattha TM, Trethowan R, Hoogenboom G (2009) Development and application of crop water stress index for scheduling irrigation in cotton (*Gossypium hirsutum* L.) under semiarid environment. J Food Agri Environ 7(3&4):386–391

Wang YH, Zhou XF (1992) The effects of boron on the anatomical structures of cotton petioles. In: Portech S (ed) Proceedings of international symposium on the role of sulphur, magnesium and micronutrients in balanced plant nutrition. The Sulphur Institute, Washington, DC, pp 70–80

Yaseen M, Ahmed W, Shahbaz M (2013) Role of foliar feeding of micronutrients in yield maximization of cotton in Punjab. Turk J Agric For 37(4):420–426

Zhao D, Oosterhuis DM (2002) Cotton carbon exchange, nonstructural carbohydrates, and boron distribution in tissues during development of boron deficiency. Field Crops Res 78(1):75–87

Zou CQ, Zhang YQ, Rashid A, Ram H, Savasli E, Arisoy RZ, Ortiz-Monasterio I, Simunji S, Wang ZH, Sohu V, Hassan M, Kaya Y, Onder O, Lungu O, Mujahid MY, Joshi AK, Zelenskiy Y, Zhang FS, Cakmak I (2012) Biofortification of wheat with zinc through zinc fertilization in seven countries. Plant Soil 361(1–2):119–130

Chapter 8
Plant Growth Regulators for Cotton Production in Changing Environment

Sibgha Noreen, Seema Mahmood, Sumrina Faiz, and Salim Akhter

Abstract In the recent past, much interest has been generated for the commercial application of plant growth regulators in cotton production, especially in the eventual threat to climate change. Since the 1950s, a number of plant growth regulators are in use to improve quantity and quality of cotton produce. These substances are organic ones, unlike inorganic nutrients, which are applied in low concentrations to mainstream various physiological, biochemical, and anatomical processes during the life span. These organic substances constitute auxins, gibberellins, cytokinins, abscisic acid, and ethylene that are applied to leverage seed germination and seedling vigor; upscale photosynthetic machinery, fruitfulness, and greater divergence of photoassimilates from source to sink; and improve tolerance to external vagaries and timely harvesting of crop ahead of weather onslaught. Other than these, brassinosteroids, jasmonates, and salicylic acid are also applied. The journey of application of various substances starts from seed treatment to final harvest. The growth and development of cotton crop are greatly interacted with ecological environments and other farm management practices. Therefore, rates and timings of application of various substances are adjusted in concurrence with its architectural growth and developmental habit in its growth cycle. The usage of plant growth regulators has been termed "shotgun" approach in the realm of climate-smart cotton production. This management tool is not only cost-effective but also ready to use during the season. The advancing footprints of climate change are to be mitigated by adaptation of strategies considering region-specific scenarios.

Keywords Indole butyric acid · Abscisic acid · Jasmonic acid · Salicylic acid

Abbreviations

1-MCP	1-methylcyclopropeng
ABA	Abscisic acid
BAP	6-benzylaminopirine
BRs	Brassinosteroids

S. Noreen (✉) · S. Mahmood · S. Faiz · S. Akhter
Institute of Pure and Applied Biology, Bahauddin Zakariya University, Multan, Pakistan

S. Ahmad, M. Hasanuzzaman (eds.), *Cotton Production and Uses*,
https://doi.org/10.1007/978-981-15-1472-2_8

Bt	*Bacillus thuringiensis*
BT	Glycine betaine
eCO_2	Elevated carbon dioxide
GA	Gibberellic acid
IBA	Indole butyric acid
JA	Jasmonic acid
PGRs	Plant growth regulators
PUT	Putrescine
SA	Salicylic acid
SPD	Spermidine
SPM	Spermine

8.1 Introduction

Cotton plant belongs to genus *Gossypium* (*G. hirsutum*, *G. barbadense*, *G. herbaceum*, and *G. arboreum*) of which 95% comes from *Gossypium hirsutum* species. This is a perennial woody shrub, which originated from tropical climate countries. Presently, it is cultivated in an annual cycle and follows an orderly manner with monopodial vegetative main stem and late monopodial fruiting branches (Usman et al. 2009; Ahmad et al. 2014; Abbas and Ahmad 2018; Ahmad and Raza 2014; Ali et al. 2011, 2013a, b, 2014a, b). Cotton plant having an indeterminate growth habit, thereby, has no expansive limit to plant development (Hearn and Constable 1984). The production of fruiting bodies continues to progress, until the supply of photosynthates turns to equilibrium with demand from developing bolls (Guinn 1985). The productivity is limited due to reduction and production of successive nodes on main stem, loss in fruiting sites, and accelerated fruit shedding, ultimately leading to reduction in boll size and quantity of produce (Pettigrew et al. 1992). The plant diagrams are used to "map" the position of bolls to assess the impact of input and expected yield outcome (Jenkins et al. 1990; Pervez et al. 2005). Cotton is mainly grown in warm areas; therefore it is better adapted having inbuilt thermotolerance mechanism (Riaz et al. 2013; Ahmad et al. 2017, 2018; Amin et al. 2017, 2018; Rahman et al. 2018; Tariq et al. 2017, 2018). The kinetic window is 30/22 °C (day/night) temperature (Reddy et al. 1992); however, temperatures higher than 45/33 °C exhibit pollen sterility in cotton crop (Singh et al. 2010; Taha and Noakes 1982).

It is extensively grown on an area of 33–35 million hectares in arid, semiarid, and humid regions of the world (Wakelyn and Chaudhry 2010). The production of lint and cottonseed is estimated 26.0 and 50.0 million tons, respectively. The value of lint is more than US $ 40.0 billion, while cottonseed shares 15–20% of cotton produce (Wakelyn and Chaudhry 2010). It is projected that the demand for food, fiber, feed, and fuel will increase by more than 70% over current quantum of agricultural production to meet the demand of 9.0 billion humans as well as livestock production by the middle of the twenty-first century. On the other hand, advancing footprints of climate change are being witnessed in the developing economies (Bellard et al. 2012). The productivity of cotton is greatly impacted by rising temperature, elevated carbon dioxide (eCO_2), and greater variation in precipitation (Luo et al. 2015). Maximum yield potential can be exploited by growing such cotton cultivars that are highly resilient and adaptable under various ecological conditions. However, breeding of cultivars having wider adaptability traits is a long-term strategy, but it also holds a little success because cultivars are bred akin to specific ecology, rather than for varied ecological conditions. The sustained cotton production rests with adoption of best management practices, improvement of water and fertilizer-use efficiency, reduction of carbon footprints, and resilience to climate change. Among other agronomic practices, application of plant growth regulators (PGRs) holds a key input in manipulating cotton crop for harvesting potential. PGRs are comprised of broad array of organic compounds, viz., growth retardants (mepiquat chloride, Pix Plus, Cycocel) and growth or yield prompters (PGR-IV, cytokinins, tricontanol, NAA, atonik, and cytozyme), and are used in low concentrations. These products are produced naturally by plant or though synthetic process (Zhao and Oosterhuis 2000).

Since the dawn of discovery of plant hormones/PGRs in the 1930s, many advances have been made in the arena of development and their application in agricultural, horticultural, and viticultural crops across the globe. In plant system, PGRs hold integral part in stimulating enzymatic battery network, mitigating the biotic and abiotic stresses, accelerating/reducing the physiological, biochemical and anatomical phenomena, enhancing quantity and quality of produce, and also lengthening shelf life (Abeles et al. 1992). These products are also used to reduce pest infestation, because of defoliation and dropping of younger/green bolls close to maturity (Graham et al. 1987). The crop treated with mepiquat chloride contains higher gossypol content in leaves and square; thereby population of bollworm is reduced upon eating these parts (Hedin and McCarty 1991; McCarty and Hedin 1994).

In most of the cases, quantum of endogenous hormones is insufficient in the plant system to ward off the vagaries of external threats (Noreen 2010). The breeding of crop varieties having traits of greater tolerance/resistance to changing climatic conditions is a long-term strategy and costs heavy investment in capital and human resources (Noreen 2010). The exogenous application of PGRs is not only cost-effective but also environment friendly. Majority of cotton farmers have already necessary spray machinery and skill to protect cotton crop. The use of PGRs is considered as "shotgun" approach to gather "white gold" by manipulating cotton crop under imminent climate change. The efficiency of PGRs does not depend only on concentration of active ingredient but also chemistry of formulation (solvents, surfactants, water quality), stage of growth, genetic makeup of varieties, canopy temperature, and structure of epidermis (Rademacher 2015). In the days to come, there is a greater urgency to make advancement in knowledge and application of PGRs for gathering the potential harvest in the wake of advancing footprints of climate change. The purpose of this book chapter is to review the usage of plant growth regulators to mainstream the growth and development of cotton crop to harvest potential yield under imminent climatic changes.

8.2 Climate Change and Cotton Production

Climate is an essential input in agricultural production and will remain so in the times to come. Cotton production systems are very much dynamic and effected greatly in response to even small fluctuations in the environment (Hardaker and Lien 2010). The increase in mean temperature by 1.0 °C may shorten the appearance of first square by 7 days (from 63 to 56 days). The more fruitfulness occurring in cotton grown under high atmospheric CO_2 (720 μmol mol^{-1}) leads to higher translocation of assimilates to metabolic sink (Reddy et al. 1999), while water-use efficiency is altered (Ephrath et al. 2011). The cumulative effects of elevated carbon dioxide (eCO_2), extreme temperatures, and erratic precipitation produce heavy toll on growth and development of cotton crop (IPCC 2014). The prevalence of high temperatures and eCO_2 content affects the feeding pattern and rate of survival of insects (Sankaranarayanan et al. 2010a), reduces expression of *Bacillus thuringiensis* (Bt) toxin, increases resistance to *Helicoverpa* species (Wu 2007; Chen et al. 2007), increases weed infestation (Sankaranarayanan et al. 2010b), increases tolerance to glyphosate (Ziska et al. 1999), and controls certain diseases in cotton crop.

The cotton water prints show that irrigation water usage has declined from 1214 L to 644 L to produce 1.0 kg lint +2.0 kg seeds. Water productivity can be enhanced by water harvesting optimizing water delivery through alternate furrows, sprinklers, or subsurface irrigation technology (www.icac.org). The opportunities do exist to mitigate the foot prints of extreme weather conditions that could be overridden by integration of certain agronomic practices (Bange et al. 2016). The adaptation strategies include crop rotation and tillage system; cultivation of transgenic (Bt) cultivars (Campbell et al. 1991); disease resistance (Liu et al. 2014a, b); drought tolerance (Stiller et al. 2005); heat stress tolerance (Bibi et al. 2008); water logging (Conaty et al. 2008); farm management practices, differing in maturity period, choice of best suited PGRs (Constable 1995); and less indeterminate growth habit cultivars (Sankaranarayanan et al. 2010a, b).

8.3 Potential Application of PGRs

The application of PGRs induces tolerance to imminent weather vagaries starting from seed germination and seedling development to maturity, which pave the way toward quantum of yield. The occurrence of rank growth due to overuse of nitrogenous fertilizers and irrigation practices can be controlled to the breaking point, where the plant system tends to transport its greater amount of assimilates from

Table 8.1 Plant hormones and their mode of action

Group	Class	Biological active form	Specific role
Growth promoters	Auxins	Indole acetic acid	Cell elongation and growth
	Gibberellins	Gibberellic acid	Stem elongation, braking dormancy
	Cytokinin	Zeatin	Cell division
Growth inhibitors	Ethylene	Ethylene	Senescence, fruit ripening
	Abscisic acid	Abscisic acid	Stomatal regulation
Recent discoveries	Jasmonate	Harmonic acid	Herbivore defense
Proteinase inhibitors	Salicylic acid	Salicylic acid	Pathogen defense, thermotolerance
Systemic-acquired resistance	Brassinosteroid	Brassinolide	Light regulation of genes

source to sink for fair response to yield (Fernandez et al. 1991). These chemicals augment, in retention of fruit on the first three positions along sympodia (from where two-thirds yield) is gathered at maturity. These substances maintain progressive vegetative and reproductive growth, which causes stress to absorb greater quantum of water and nutrients from the rhizosphere (Klepper and Taylor 1979). However, the uptake of nutrients is greatly reduced close to maturity of cotton crop.

The efficacy and efficiency of PGRs are greatly interacted with various agronomic practices and environmental factors (Guinn 1986). The use of mepiquat chloride at earlier stages of growth may cause reduction in yield (Cathey and Meredith 1988), enhance tolerance to abiotic stress (Huang and Gausman 1982), shorten the internodal length (Zhang et al. 1991), increase uptake of nutrients from root tissues (Zhang et al. 1991), and cause differential response to accumulate osmolytes (Boquet and Coco 1993). There are instances that the use of some insecticides, viz., hydrocarbons, organophosphates, chlordimeform, (N′-4-chloro-o-tolyl)-N, and N-dimethyl formamidine, may enhance yield by mainstreaming and upregulating various physiological phenomena (Youngman et al. 1990).

In cotton, the various phases of growth and development are potentiated and manipulated by different PGRs (Cathey and Thomas 1986) (Tables 8.1, 8.2, and 8.3).

8.4 Mode of Action of Plant Growth Regulators

In the contemporary world of PGRs, their potential usage in cotton production is evidenced from continuous demand from the cotton farmers. These products start their journey from planting through maturity. They have specific mode of action and their application is time-bound in plant's life. Among PGRs, mepiquat chloride is effective in mitigating the chilling effects at seedling stage and tolerance to high

Table 8.2 Response of cotton to plant growth regulators (PGRs)

Compound	Response	References
1. Germination and seedling development		
Oleic acid D-a-tocopherol cottonseed oil	Reduced herbicide damage caused by trifluralin	Christiansen and Hilton (1974)
Indole-3-acetic acid (IAA) kinetin	Overcome trifluralin damage	Hassawy and Hamilton (1971)
Gibberellic acid	Induced more rapid emergence and taller seedlings	Ergle (1958)
	Reduced chilling injury in germinating cotton	Cole and Wheeler (1974)
Adenosine-3′-5′-cyclic (AMP) monophosphate	Reduced chilling injury in germinating cotton	Cole and Wheeler (1974)
Antitranspirants	Prevented chilling injury to seedling cotton	Christiansen and Ashworth (1978)
Mepiquat chloride (1,1-dimethylpiperidinium chloride)	Reduced wind-blown sand damage to cotton	Gausman et al. (1980a)
	Increased drought resistance of seedling cotton	Xu and Taylor (1992)
	Seed treatment reduced chlorosis and necrosis of cotton	Corbin and Frans (1991)
Indole acetic acid	Growth and development, fruitfulness, seed cotton yield	Sheng et al. (1988)
Indole butyric acid	Growth and development, yield	Sheng et al. (1988)
2,4, 6-trichlorophenoxy	Malformation of leaves decline in yield	Sheng et al. (1988)
Mepiquat chloride	Increase in number of bolls, boll weight, and cotton yield	York (1983)
PGR-IV	Root growth, boll retention, and yield	Livingston et al. (1992)
2. Vegetative development		
Chlormequat (CCC) (2-chloroethyl)-trimethylammonium chloride	Reduced plant height and yield	Thomas (1964)
	Reduced plant height but not yield	De Silva (1971)
	Promoted drought resistance	Singh (1975)
	Increased fruit loss	Kittock and Arle (1977)
	Seed treatment reduced	Thomas (1967)
	Chlorosis and necrosis of cotton treated with fluometuron	Corbin and Frans (1991)
	Decreased growth rate	Marani and Levi (1973)
	Increased yields	Singh (1970)
N-dimethyl-N-B-chlorethyl hydrazonium chloride (CMH)	Decreased growth rate	Marani and Levi (1973)
2,3,5-Triiodobenzoic acid (TIBA)	Reduced plant size and dry weight; reduced seed cotton yield	Thomas (1967)

(continued)

Table 8.2 (continued)

Compound	Response	References
Mepiquat chloride	Darker green leaves	Gausman et al. (1980a); Walter et al. (1980)
	Height reduction under luxuriant growth conditions	Heilman and Brown (1981)
	Reduction in canopy width Reduced internode length	Walter et al. (1980)
	Reduced bolls/plant; individual boll weight increased	Walter et al. (1980)
	Smaller bolls; increased flower production	York (1983)
	Erratic yield response	
	Thicker leaves with reduced surface area	Heilman and Brown (1981)
	Increased chlorophyll a/b ratio	Gausman et al. (1979)
	Stimulation of CO_2 uptake	Gausman et al. (1980b)
	Rate of increase in root density significantly less	Cappy and Cothren (1980)
	Increased gross canopy photosynthesis	Hodges et al. (1991)
	Lack of effect on carbon-use efficiency in well-watered and water-deficient plants; helped maintain leaf turgor potential in water-stressed plants	Fernandez et al. (1991, 1992)
	Increase in fine root hairs	Fernandez et al. (1991)
	More favorable water status; higher leaf water potential	Stuart et al. (1980)
	Reduced boll rot	Snow et al. (1981)
	Mitigates symptom expressions of *Verticillium* wilt	Erwin et al. (1979)
Indole butyric acid (IBA)	Increased NADH oxidase activity in treated roots, indicating increase in root respiration	Urwiler and Oosterhuis (1986)
Thidiazuron	Timely harvest, sustain yield, improve fruit quality	Suttle (1988)
3. Reproductive development		
Chlordimeform [N-(chloro-O-tolyl)-N, N-dimethylformamide]	Increased initiation and retention of fruit	Phillips et al. (1977)
	No differences in fruiting response under insect-free environment	Cathey and Bailey (1987)

(continued)

Table 8.2 (continued)

Compound	Response	References
	Yield increase	Cathey and Bailey (1987)
	Cytokinin-like activity	Bauer and Cothren (1990)
Acephate (O,S-dimethyl acetyl-phosphoramidothioate)	Increased flowering rate, boll production and size, and yield; no effect on previously mentioned parameters	Cathey (1979)
Guthion [O,)-dimethyl S-(4-oxo-1, 2, 3, -benzo triazin -3 (4H-yl) mehtyl) phosphorodithioate]	Produced more flowers and more bolls; longer maturation period	Hacskaylo and Scales (1959)
Gibberellic acid (GA3)	Significant increase in boll set; erratic yield results	Walhood (1958)
	Taller plants; slightly less yield	Lane (1958)
	Flower bud and boll abscission reduced; yields not affected	Subbiah and Mariakulandai (1972)
Naphthalene acetic acid (NAA)	Increase in boll numbers; 8 to 10% increase in yield at first harvest	Negi and Singh (1956)
	Decreased boll shedding; increased yields	Murty et al. (1976)
	Increased fiber length	Patel (1992)
2,3,5-Triiodobenzoic acid (TIBA)	Slight increase in boll set and slightly earlier termination; no significant yield effect	Thomas (1967)
	Yield increases; increased boll size and number of bolls per plant; lowered position of first fruiting branch	Freytag and Coleman (1973)
Cytozyme crop⁺ (bacterial enzyme plus micronutrients)	Significant decrease in transpiration and volatile nitrogen loss; trends, but non-significant increases in yield	Cothren and Cotterman (1980)
Chlormequat	Low concentrations, coarser fiber; higher concentrations, increased length and fineness; decreased strength maturity and yield	Bhatt et al. (1972)
	Significant yield increase	Nageswara et al. (1980)
Indole acetic acid (IAA)	Improved length and fineness of fiber	Bhatt et al. (1972)
Naphthalene acetic acid (NAA)	Low concentrations increased fiber fineness; high concentration decreased fiber fineness	Bhatt et al. (1972)

(continued)

Table 8.2 (continued)

Compound	Response	References
Gibberellic acid (GA3)	Increased fiber length; no effect on fiber length	Bhatt and Ramanujam (1971); Sitaram and Abraham (1973)
Mepiquat chloride	Increased boll weight, seed and lint index, lint yield, and earliness	Sawan et al. (1984)
	Increased levels of petiole nitrates compared to controls	Stedman and Taylor (1982)
Paclobutrazol ([2RS-3RS]-1-[4-chlorophenyl]-4-4-dimethy1-2-1,4-triazol-y1-pentan-3-ol)	Reduced plant height, number of mainstem nodes, and dry weight of stems and leaves	Ben-Porath and Baker (1990)
Cutechin	Applied at flowering to seeds soaked in chlormequat chloride r succinic acid, increased drought tolerance, increased root volume, higher seed cotton yield	Shanmugham (1992)
Triacontanol	Increased lint yield; increased fiber length	Patel (1992)
Polyhydroxycarboxylic acid (PHCA)	Yield increase	Munoz (1994)
1-Methylcyclopropene (1-MCP)	Ethylene inhibitor, stomatal resistance to drought stress	Binder and Bleecker (2003)
Methanol	Improves water-use efficiency, plant structure, and cotton yield	Makhdum et al. (2002)
Methanol plus urea	Tolerance to drought improvement in photosynthetic rate, improve yield	Makhdum et al. (2002)

Source: Adapted after Cothren and Oosterhuis (2010) and updated

temperatures at advanced stages of growth within the kinetic window (<10 °C to <45 °C), by mediating the plasma membrane leakage and greater storage of starch in the chloroplast (Zhang et al. 1991; Reddy et al. 1999). The growth retardants, viz., mepiquat chloride, kinetin, and pentaborate, reduce rank growth and augment the cotton plant to retain greater number of shooting bodies along the sympodia through partitioning of photoassimilates for their greater movement from source to sink (Gwathmey and Clement 2010). Other products, e.g., mefluidide [N-(2-4-dimethyl-5-{(trifluoromethyl) sulfonyl] amino} phenyl) acetamide], maleic hydrazide (1, 2-dehydro-3, 6-Pyridazinediene), and dikegulac, may be used with caution to avoid toxicity to the growing tissues (Bocion et al. 1975).

PGRs sensitize the plant system to divert greater quantum of organic substances to the developing bolls. For example, mepiquat chloride respond fairly to short-season conventional varieties (Boquet and Coco 1993), while having a little effect on transgenic (*Bacillus thuringiensis*) under ultra-narrow cotton (Underbrink 1999; Gerik et al. 1998). The absorption and translocation of nutrients, namely, nitrogen, phosphorus, potassium, and micronutrients, from roots to upper parts are mediated

Table 8.3 Efficacy of plant hormones under abiotic stresses

Type of stress	Hormone	Type of application	Dose	Parameters enhanced	Reference
Heat stress	Putrescine	Foliar application	0 and 10 mM	Flower and fruit development	Bibi et al. (2010)
Drought	Jasmonic acid	Foliar application	0 and 2.5 μM	Fiber length	Tan et al. (2012)
Drought	Putrescine	Foliar application	0, 200, 400, and 600 ppm	Antioxidant defense and yield	Shallan et al. (2012)
Drought	Salicylic acid	Foliar application	0 and 100 mg L^{-1}	Leaf area index and yield	Noreen et al. (2013)
Drought	Cytokinin	Endogenous synthesis	Internal concentration (according to level of stress)	Yield components	Kuppu et al. (2013)
Salinity	Putrescine	Foliar application	0, 1, and 2 ppm	Growth, yield, and enzymatic antioxidants	Hanafy Ahmed et al. (2013)
Drought	Jasmonic acid	Foliar application	0 and 200 ppm	Antioxidants and yield	Chen et al. (2013)
Salinity	Putrescine (Put) and humic acid (HA)	Foliar application	Put (0, 1, and 2 ppm) and HA (0, 1, and 2%)	Nutrient uptake and proteins	Alobaidy (2013)
Salinity	Salicylic acid	Foliar application	0 and 0.1 mM	Photosynthesis and antioxidant defense	Liu et al. (2014a, b)
Drought	Abscisic acid	Endogenous synthesis	Synthesize according to level of stress	Photosynthesis and water-use efficiency	Mittal et al. (2014)
Heat stress	Auxin	Endogenous synthesis	Synthesize according to level of stress	Starch synthesis and photosynthesis	Min et al. (2014)
Salinity	Nitric oxide	Rooting medium	0, 0.1, and 1.0 mM	Antioxidant defense	Dong et al. (2014)
Lead toxicity	Hydrogen sulfide	Rooting medium	0 and 200 μM	Photosynthesis and antioxidants	Bharwana et al. (2014a, b)

Drought	Salicylic acid	Foliar application	0 and 100 mg/L	Yield and agronomic parameters	Zafar et al. (2015)
Drought	Jasmonic acid	Rooting medium	0 and 10 mM	Regulation of stomatal closure by regulation of ABA	Riemann et al. (2015)
Salinity	Auxin	Rooting medium	0, 0.01, and 0.001 μg/mL	Increase in yield and disease resistance	Egamberdieva et al. (2015)
Salinity	Brassinolide	Rooting medium	0 and 100 nM	Antioxidant machinery and nutrient uptake	Shu et al. (2015)
Copper-induced oxidative stress	Salicylic acid	Seed priming	0 and 100 μM	Nutrient uptake and antioxidant defense system	Mei et al. (2015)
Salinity	Nitric oxide	Foliar application	0 and 300 μM	Nutrient uptake	Kong et al. (2016)
Heat Stress	Salicylic acid	Foliar application	0, 0.5, 1.0, and 1.5 mM	Antioxidant defense system	Galani et al. (2016)
Drought	Putrescine (Put) and 24-epibrassinolide (EBL)	Foliar application	Put (0, 1, and 2 ppm) and EBL (0, 10^{-9}, and 10^{-7} M)	Nutrient uptake and antioxidants	Hanafy Ahmed et al. (2016)
Drought	Putrescine (Put) and 24-epibrassinolide (EBL)	Foliar application	Put (0, 1, and 2 ppm) and EBL (0, 10^{-9}, and 10^{-7} M)	Cotton fiber and antioxidants	Ahmed et al. (2017)
Salinity	Salicylic acid (SA) and potassium citrate (PC)	Foliar application	SA (0 and 200 ppm) and PC (0 and 2.5 g/L)	Yield and antioxidants	El-Beltagi et al. (2017)
Toxin-induced stress	24-Epibrassinolide	Seed priming	0 and 10 nM	Callus growth	Bibi et al. (2017)
Heat stress	Salicylic acid	Foliar application	0 and 50 ppm	Enzymatic antioxidants	Sarwar et al. (2018)
Salinity	Melatonin	Seed priming	0, 10, 20, 50, 100, and 200 μM	Seed germination	Xiao et al. (2019)
Drought stress	Glycine betaine	Foliar spray	0, 1, 3, and 6 kg/ha	Growth and development	Gorham (1995)

(continued)

Table 8.3 (continued)

Type of stress	Hormone	Type of application	Dose	Parameters enhanced	Reference
Drought stress	Glycine betaine	Foliar spray	0, 1, 3, and 6 kg/ha	Increase in number of bolls and yield in cotton	Makhdum and Shababuddin (2006)
Drought stress	Glycine betaine	Foliar spray	0, 1, 3, and 6 kg/ha	Increase in net CO_2 assimilation rate transpiration rate, stomatal conductance, and water-use efficiency	Makhdum and Shababuddin (2006)
Drought stress	Glycine betaine	Foliar spray	0, 1, 3, and 6 kg/ha	Increase in pressure potential, reduction in leaf water potential, and osmotic potential	Makhdum et al. (2006)
Drought stress	Salicylic acid	Foliar spray	50–100 mg kg^{-1}	Number of bolls, increase in yield, reduced fruit shed	Noreen et al. (2013)
Drought stress	Proline	Foliar spray	100 mg kg^{-1}	Increase in boll number and weight, seed cotton yield	Noreen et al. (2013)
Drought stress	Salicylic acid	Foliar spray	100 mg Kg^{-1}	Increase in biological yield, fruit production, and yield	Noreen et al. (2015)
Drought stress	Proline	Foliar spray	100 mg kg^{-1}	Increase in biological yield, fruit production, and yield	Noreen et al. (2015)
Drought stress	Glycine betaine	Foliar spray	100 mg kg^{-1}	Increase in biological yield, fruit production, and yield	Noreen et al. (2015)

by application of PGRs (Han et al. 1991; Oosterhuis and Zhao 1994). Some of the harvest-aid chemicals, viz., ethephon, ethephon plus cyclanilide, thidiazuron, and diuron, are employed for early opening of bolls and uniform harvesting process (Scott 1990).

8.4.1 Auxins

The auxins (indole acetic acid) are produced in apical meristem and travel to the base. These endogenous hormones work in phototropic and gravitropic reactions and enhance the accumulation of starch in seeds of cereal crops (Abeles et al. 1992). These organic substances stimulate cell elongation at the cellular level (Salisbury and Ross 1992), enzyme and protein synthesis (Marre 1979), simulation of gene activity (Key 1989), leaf senescence, and wound healing (Salisbury and Ross 1992; Lau et al. 2008). The other auxins comprised of indole-3-acetic acid and 1-naphthylacetic acids are also being used in tissue culture technology for proliferation of root system (Lüthen 2015). The mixing of cyclanilide (1-[(2, 4-dichlorophenyl) carbonyl] cyclopropane carboxylic acid) with ethephon is used as harvest-aid chemicals in cotton (Fritz 1997). Spraying of indole acetic acid results in improvement in growth and development, greater fruit formation, and increase in cotton yield (Shah et al. 1998).

8.4.2 Multiple Entity PGRs

The PGR-IV containing auxins, indole butyric acid (IBA), and gibberellic acid (GA) blended nutrient solutions and fermented broth and are being marketed for improving photosynthesis (Hickey 1996), greater allocation of osmolytes from production to consumption organs and increasing yield and quality of cotton (Livingston and Parker 1994), increase in boll weight and size, and high reproduction to vegetative dry weight ratios (Zhao and Oosterhuis 2000)

8.4.3 Cytokinins and Cytokinin-Like

The cytokinins are found abundantly in root and shoot apical meristems (Taiz and Zeiger 1998). They are involved in cell division and elongation, embryo development, anthocyanin production, chloroplast biogenesis (Taiz and Zeiger 1998), nucleic acid metabolism, mobilization of nutrients, and inducing tolerance to drought and heat stress (Javid et al. 2011). The adenine-type cytokinins comprising of [N-(2 – furylmethyl)-3H-purin-6-amine] and 6-benzyladenine and B-benzyladenine are used in tissue culture technology. The diphenylurea-type cytokinins, thidiazuron, and forchlorfenuron are used as cotton defoliant and dropping of immature fruit in apple trees (Greene 1999). Other than these, cytokinin-like compounds (burst, cytozyme, cytokine, triggrr) are also used to enhance fruit production, retention of fruits along sympodia, and partitioning of assimilates for treasuring sink pool capacity in cotton plant (Parker and Salk 1990).

8.4.4 Gibberellins

The gibberellins, GA3 (gibberellic acid), GA4, and GA7, are phytohormones that are produced by fermentation of fungus (*Gibberella fujikuroi*). These organic substances work in longitudinal development and production of fruit (Noggle and Fritz 1983), induction of enzymes in germinating seed, cell division in the apical dominance, and cell wall plasticity. These GAs promote internodal length, size of leaves, and structural development in cotton plant (Taylor and Cosgrove 1989; Taiz and Zeiger 1998). The mixture of GAs is used in producing seedless grapes, improving quality of citrus, and reducing fruit russeting in apples (Rademacher 2015). Out of 100 tons of GAs, two-thirds of GA3 is used in agricultural production (Hornsey 2003).

8.4.5 Glycine Betaine (GB)

Glycine betaine (N-methyl substantiated by glycine derivatives), a member of quaternary ammonium compounds, is abundantly formed in high plant (Karabudak et al. 2014). The accumulation of GB in the plant system helps in higher retention of water, maintaining photosynthetic activity, and membrane integrity for alleviating external threats (Ashraf and Foolad 2007). The growth, development, and yield productivity by foliar application of GB in cotton (Makhdum and Shababuddin 2006; Makhdum et al. 2002) is enhanced significantly.

8.4.6 Proline

Proline, an amino acid, works in stabilization of subcellular structures. It is involved in osmo-protection as well as progressive growth under drought stress (Gill et al. 2014) and increased tolerance to salinity. The application of 100 mg L^{-1} caused an increase in cotton yield under water stress (Noreen et al. 2013).

8.4.7 Ethylene/Harvest Aid

Ethylene as a plant hormone in the form of gaseous molecules works in a number of physiological phenomena, i.e., inducing flower initiation, accelerating/limiting growth of cells (Bartholomew and Sanewski 2018), leaf abscission, and dehiscence of bolls before mechanical harvesting in cotton (Taiz and Zeiger 1998; Abeles et al. 1992). The occurrence of loss due to spraying of ethephon during the season may not be compensated by most cotton cultivars (Pettigrew et al. 1992). The ethylene synthesis is increased with rise in canopy temperature, resulting in fruit abscission (Reddy et al. 1992). The synthesis could be suppressed by the use of 1-methylcyclopropene (1- MCP) by reducing membrane electrolyte leakage (Blankenship and Dole 2003). The generation of reactive oxygen species could reduce by foliar spray of chaperons (comprised of nitrophenolates, sodium 5-nitroguaiacolate, sodium otho-nitrophenolate, and sodium para-nitrophenolate) for alerting abscission of fruiting bodies.

The categories of harvest-aid chemicals include (a) defoliants (thidiazuran, diuron, dimethipin) that increases ethylene concentration in leaves and reduces auxin hormones; (b) boll openers/conditioners (ethephon, cyclanilide, aminomethane dihydrogen tetraoxosulfate) that induce ethylene for separation of boll walls; and (c) desiccants and herbicides (sodium chloride, magnesium chloride, glyphosate, diquat, paraquat, carfentrazone-ethyl) that disrupt the leaf membrane integrity for rapid desiccation of leaves. The product 1-methylcyclopropene (1-MCP) is used as liquid spray to improve tolerance in cotton, maize, wheat, and rice in response to heat and drought stresses (Watkins 2015). Cotton crop is sprayed with PGR-IV, glyphosate (N-phosphonomethyl glycine), 2-chloroethyl phosphoric ethephon, and thidiazuran for accelerating dehiscence of late-season green bolls to avoid diapausing population of pink bollworm [*Pectinophora gossypiella* (Saunders)] and larvae of *Heliothis* species for their overwintering (Bariola et al. 1990). The use of these products facilitates mechanical picking without any loss in yield (Taylor and Cosgrove 1989; Jones et al. 2001; Scott 1990).

8.4.8 Abscisic Acid

Abscisic acid (ABA) is considered as stress tolerance hormone and works as signaling mediator for tolerance to abiotic stresses (Raghavendra et al. 2010). Naturally, it is present as (+)—cis-trans—isomer (S-ABA) and is produced from fungi *Botrytis cinerea* containing S-ABA (Marumo et al. 1982). ABA is involved is a wide array of functions, i.e., inhibition of progressive growth by interfering nucleic acid synthesis (Wright and Hiron 1972). The stomata are closed due to greater transportation of ABA from roots to apical portion (Horton 1971). The hydraulic conductivity also increases due to greater accumulation of ABA (Taiz and Zeiger 1998). The chemicals, polyamines (putrescine, PUT), spermidine (SPD), and spermine (SPM) suppress the production of ABA and potentiate early flowering and fruit development.

The hormones, viz., BAP (6-benzylaminopirine) and benzyl adenine, help in main streaming turgor pressure and stomatal closure in cotton grown under water-logged and drought stress.

8.4.9 *Jasmonic Acid*

Jasmonic acid (JA) is a synthetic prohydrojasmon, and its methyl ester works as signaling molecule and is involved in inducing tolerance to salinity and drought (Kondo et al. 2011; Kazan and Manners 2012), promoting senescence (Gross and Parthier 1994), enhancing fruit formation and ripening, pigment formation (Davies 2010, 2013), and also ethylene production (van Loon and Glick 2004). The reduction in chlorophyll content and chloroplast proteomics is accelerated by application of jasmonates (Creelman and Mullet 1995). Jasmonates work as defense mechanism to pathogen attack (Creelman and Mullet 1995). The ethylene production is also accelerated due to spray of jasmonates (van Loon and Glick 2004).

8.4.10 *Salicylic Acid*

Salicylic acid (SA) or ortho-hydroxy benzoic acid is comprised of a diverse group of plant phenolics and fairly found in the plant kingdom (Radin et al. 1978). It is involved in resistance to biotic stresses (van Loon and Glick 2004), lengthening the flowering period, inhibiting ethylene production, antagonizing negative effects of abscisic acid, and improving tolerance to low temperatures. Under the stress environment, SA regulates plant metabolism and is a signaling molecule for defense against biotic and abiotic stresses (Noreen et al. 2017). Salicylic acid produces pathogenesis -related proteins, which help in the tolerance to pathogen attack (Sticher et al. 1997).

8.4.11 *Brassinosteroids*

The brassinosteroids (BRs) belonging to the group of steroids are involved in a variety of physiological processes, i.e., cell division and cell elongation (Clouse and Sasse 1998), light-regulated development (Li and Chory 1999), reproductive development and leaf senescence (Vriet et al. 2012), and inducing tolerance to salinity and drought stresses and pathogen infection (Manavalan et al. 2012). The application of 24-epibrassiolide or 28- homobrassinolide improves antioxidant enzymatic system proline content and other growth and developmental processes in response to aluminum toxicity.

8.5 Future Perspectives

Since the 1930s, plant growth regulators have been putting feathers in their caps to mainstream the productivity and quality of agricultural, horticultural, and viticultural crops. Among arable crops, cotton crop shares its greatest proportion in the consumption of various products, the reason being that they constitute an integrated network of stimulus and augment various enzymatic processes and improve tolerance capacity to ward off the external threats. In view of the imminent changes in the climate, R&D endeavors are to be continued to look for new formulation, which would be more stable in their efficacy, and to be resilient under the footprints of climate change. In the foreseeable future, new entrants could be multi-entity PGRs, polyhydroxycarboxylic acid alone, and/or a mixture containing micronutrients, apart from other potential products, viz., jasmonate, salicylic acid, and brassinosteroids. Furthermore, extension services are to be reinforced to advocate the cotton farmers about practical usage of various PGRs during crop cycle in response to changing weather conditions. There is an urgency that R&D efforts are to be stepped up to integrate effects of climate change (temperature, humidity, CO_2, and water stress) for sustainable cotton production. Region-specific adaptation strategies need to be developed. Simulation models will be helpful in assessing the impact and adaptation options for future climate change.

References

Abbas Q, Ahmad S (2018) Effect of different sowing times and cultivars on cotton fiber quality under stable cotton-wheat cropping system in southern Punjab, Pakistan. Pak J Life Soc Sci 16:77–84

Abeles FB, Morgan PW, Saltveit ME Jr (1992) Ethylene in Plant Biology, 2nd edn. Academic Press, Inc., San Diego, pp 1–13

Ahmad S, Raza I (2014) Optimization of management practices to improve cotton fiber quality under irrigated arid environment. J Food Agri Environ 2(2):609–613

Ahmad S, Raza I, Ali H, Shahzad AN, Atiq-ur-Rehman, Sarwar N (2014) Response of cotton crop to exogenous application of glycinebetaine under sufficient and scarce water conditions. Braz J Bot 37(4):407–415

Ahmad S, Abbas Q, Abbas G, Fatima Z, Atique-ur-Rehman, Naz S, Younis H, Khan RJ, Nasim W, Habib ur Rehman M, Ahmad A, Rasul G, Khan MA, Hasanuzzaman M (2017) Quantification of climate warming and crop management impacts on cotton phenology. Plants 6(7):1–16

Ahmad S, Iqbal M, Muhammad T, Mehmood A, Ahmad S, Hasanuzzaman M (2018) Cotton productivity enhanced through transplanting and early sowing. Acta Sci Biol Sci 40:e34610

Ahmed AHH, Darwish E, Alobaidy MG (2017) Impact of putrescine and 24-epibrassinolide on growth, yield and chemical constituents of cotton (*Gossypium barbadense* L.) plant grown under drought stress conditions. Sciences 16:9–23

Ali H, Afzal MN, Ahmad F, Ahmad S, Akhtar M, Atif R (2011) Effect of sowing dates, plant spacing and nitrogen application on growth and productivity on cotton crop. Int J Sci Eng Res 2 (9):1–6

Ali H, Abid SA, Ahmad S, Sarwar N, Arooj M, Mahmood A, Shahzad AN (2013a) Integrated weed management in cotton cultivated in the alternate-furrow planting system. J Food Agri Environ 11(3&4):1664–1669

Ali H, Abid SA, Ahmad S, Sarwar N, Arooj M, Mahmood A, Shahzad AN (2013b) Impact of integrated weed management on flat-sown cotton (*Gossypium hirsutum* L.). J Anim Plant Sci 23 (4):1185–1192

Ali H, Hameed RA, Ahmad S, Shahzad AN, Sarwar N (2014a) Efficacy of different techniques of nitrogen application on American cotton under semi-arid conditions. J Food Agri Environ 12 (1):157–160

Ali H, Hussain GS, Hussain S, Shahzad AN, Ahmad S, Javeed HMR, Sarwar N (2014b) Early sowing reduces cotton leaf curl virus occurrence and improves cotton productivity. Cercetări Agronomice în Moldova XLVII(4):71–81

Alobaidy M (2013) Effect of putrescine and humic acid on cotton plant growing under salinity stress conditions

Amin A, Nasim W, Mubeen M, Nadeem M, Ali L, Hammad HM, Sultana SR, Jabran K, Habib ur Rehman M, Ahmad S, Awais M, Rasool A, Fahad S, Saud S, Shah AN, Ihsan Z, Ali S, Bajwa AA, Hakeem KR, Ameen A, Amanullah, Hafeez Ur Rehman, Alghabar F, Jatoi GH, Akram M, Khan A, Islam F, Ata-Ul-Karim ST, Rehmani MIA, Hussain S, Razaq M, Fathi A (2017) Optimizing the phosphorus use in cotton by using CSM-CROPGRO-cotton model for semi-arid climate of Vehari-Punjab, Pakistan. Environ Sci Pollut Res 24(6):5811–5823

Amin A, Nasim W, Mubeen M, Ahmad A, Nadeem M, Urich P, Fahad S, Ahmad S, Wajid A, Tabassum F, Hammad HM, Sultana SR, Anwar S, Baloch SK, Wahid A, Wilkerson CJ, Hoogenboom G (2018) Simulated CSM-CROPGRO-cotton yield under projected future climate by SimCLIM for southern Punjab, Pakistan. Agr Syst 167:213–222

Ashraf M, Foolad M (2007) Roles of glycine betaine and proline in improving plant abiotic stress resistance. Environ Exp Bot 59:206–216

Bange MP, Baker JT, Bauer PJ, Broughton KJ, Constable GA, Luo Q, Oosterhuis DM, Osanai Y, Payton P, Tissue DT (2016) Climate change and cotton production in modern farming systems. CABI, Wallingford

Bariola L, Chu C, Henneberry T (1990) Timing applications of plant growth regulators and last irrigation for pink bollworm (Lepidoptera: Gelechiidae) control. J Econ Entomol 83:1074–1079

Bartholomew DP, Sanewski GM (2018) 11 Inflorescence and fruit development and yield. In: The pineapple: botany, production and uses. CABI, Wallingford, p 233

Bauer P, Cothren J (1990) Growth-promoting activity of chlordimeform. Agron J 82:73–75

Bellard C, Bertelsmeier C, Leadley P, Thuiller W, Courchamp F (2012) Impacts of climate change on the future of biodiversity. Ecol Lett 15:365–377

Ben-Porath A, Baker D (1990) Taproot restriction effects on growth, earliness, and dry weight partitioning of cotton. Crop Sci 30:809–814

Bharwana S, Ali S, Farooq MA, Iqbal N, Hameed A, Abbas F, Ahmad MSA (2014a) Glycine betaine-induced lead toxicity tolerance related to elevated photosynthesis, antioxidant enzymes suppressed lead uptake and oxidative stress in cotton

Bharwana SA, Ali S, Farooq MA, Ali B, Iqbal N, Abbas F, Ahmad MSA (2014b) Hydrogen sulfide ameliorates lead-induced morphological, photosynthetic, oxidative damages and biochemical changes in cotton. Environ Sci Pollut Res 21:717–731

Bhatt J, Ramanujam T (1971) Some responses of a short-branch cotton variety to gibberellin. Cotton Growing Rev 48:136

Bhatt J, Raman C, Sankaranarayanan T, Iyer S (1972) Changes in lint characters of cotton varieties by growth regulators. Cotton Growing Rev 49:160

Bibi A, Oosterhuis D, Gonias E (2008) Photosynthesis, quantum yield of photosystem II and membrane leakage as affected by high temperatures in cotton genotypes. J Cotton Sci 12:150–159

Bibi AC, Oosterhuis DM, Gonias ED (2010) Exogenous application of putrescine ameliorates the effect of high temperature in *Gossypium hirsutum* L. flowers and fruit development. J Agron Crop Sci 196:205–211

Bibi N, Ahmed IM, Fan K, Dawood M, Li F, Yuan S, Wang X (2017) Role of brassinosteroids in alleviating toxin-induced stress of Verticillium dahliae on cotton callus growth. Environ Sci Pollut Res 24:12281–12292

Binder BM, Bleecker A (2003) A model for ethylene receptor function and 1-methylcyclopropene action. ACTA Hort 628:177–187

Blankenship SM, Dole JM (2003) 1-Methylcyclopropene: a review. Postharvest Biology and Technology 28:1–25

Bocion P, De Silva W, Hüppi G, Szkrybalo W (1975) Group of new chemicals with plant growth regulatory activity. Nature 258:142

Boquet D, Coco A (1993) Cotton yield and growth interactions among cultivars, row spacings and soil types under two levels of Pix. In: Proceedings, Beltwide Cotton Conferences, 10-- 14 January, New Orleans, Louisiana, USA. National Cotton Council, Memphis, USA, pp 1370–1372

Campbell C, Zentner R, Bowren K, Townley-Smith L, Schnitzer M (1991) Effect of crop rotations and fertilization on soil organic matter and some biochemical properties of a thick Black Chernozem. Can J Soil Sci 71:377–387

Cappy J, Cothren J (1980) Root growth and development of Pix-treated cotton plants

Cathey GW (1979) Acceleration of ball dehiscence with desiccant chemicals. Agron J 71:505–510

Cathey G, Bailey J (1987) Evaluation of chlordimeform for cotton yield enhancement. J Econ Entomol 80:670–674

Cathey GW, Meredith WR (1988) Cotton response to planting date and mepiquat chloride. Agron J 80:463–466

Cathey GW, Thomas RO (1986) Use of plant growth regulators for crop modification. In: Mauney JR, Stewart JMD (eds) Cotton physiology. The Cotton Foundation, Memphis, TN

Chen ZJ, Scheffler BE, Dennis E, Triplett BA, Zhang T, Guo W, Chen X, Stelly DM, Rabinowicz PD, Town CD (2007) Toward sequencing cotton (Gossypium) genomes. Plant Physiol 145:1303–1310

Chen Y, Liu Z-H, Feng L, Zheng Y, Li D-D, Li X-B (2013) Genome-wide functional analysis of cotton (*Gossypium hirsutum*) in response to drought. PLoS One 8:e80879

Christiansen M, Ashworth E (1978) Prevention of chilling injury to Seedling cotton with anti-transpirants 1. Crop Sci 18:907–908

Christiansen M, Hilton J (1974) Prevention of trifluralin effect on cotton with soil applied lipids 1. Crop Sci 14:489–490

Clouse SD, Sasse JM (1998) Brassinosteroids: essential regulators of plant growth and development. Annu Rev Plant Biol 49:427–451

Cole D, Wheeler J (1974) Effect of pregermination treatments on germination and growth of cottonseed at suboptimal temperatures 1. Crop Sci 14:451–454

Conaty W, Tan D, Constable G, Sutton B, Field D, Mamum E (2008) Genetic variation for waterlogging tolerance in cotton. J Cotton Sci 12:53–61

Constable GA (1995) Predicting yield responses of cotton to growth regulator. In: Constable GA, Forrester NW (eds) Challenging the future. Proceedings of the World Cotton Research Conference-1, CSIRO, Brisbane, Australia, pp 3–5

Corbin BR, Frans RE (1991) Protecting cotton (*Gossypium hirsutum*) from Fluometuron injury with seed protectants. Weed Sci 39:408–411

Cothren J, Cotterman C (1980) Evaluation of Cytozyme Crop+ as a foliar application to enhance cotton yields. Arkansas Farm Res 29(6):2

Cothren JT, Oosterhuis D (2010) Use of growth regulators in cotton production. In: Stewart JM, Oosterhuis DM, Heitholt JJ, Mauney JR (eds) Physiology of cotton. Springer, pp 289–303

Creelman RA, Mullet JE (1995) Jasmonic acid distribution and action in plants: regulation during development and response to biotic and abiotic stress. Proc Natl Acad Sci 92:4114–4119

Davies PJ (2010) The plant hormones: their nature, occurrence, and functions. In: Plant hormones. Springer, Dordrecht, pp 1–15

Davies PJ (2013) Plant hormones: physiology, biochemistry and molecular biology. Springer Science & Business Media, Dordrecht

De Silva W (1971) Some effects of the growth retardant chemical CCC on cotton in Uganda. In: Cotton Growing Rev

Dong Y, Jinc SS, Liu S, Xu LL, Kong J (2014) Effects of exogenous nitric oxide on growth of cotton seedlings under NaCl stress

Egamberdieva D, Jabborova D, Hashem A (2015) Pseudomonas induces salinity tolerance in cotton (*Gossypium hirsutum*) and resistance to Fusarium root rot through the modulation of indole-3-acetic acid. Saudi J Biol Sci 22:773–779

El-Beltagi HS, Ahmed SH, Namich AAM, Abdel-Sattar PR (2017) Effect of salicylic acid and potassium citrate on cotton plant under salt stress. Fresen Environ Bull 26:1091–1100

Ephrath J, Timlin D, Reddy V, Baker J (2011) Irrigation and elevated carbon dioxide effects on whole canopy photosynthesis and water use efficiency in cotton (*Gossypium hirsutum* L.). Plant Biosyst 145:202–215

Ergle DR (1958) Compositional factors associated with the growth responses of young cotton plants to gibberellic acid. Plant Physiol 33:344

Erwin D, Tsai S, Khan R (1979) Growth retardants mitigate Verticillium wilt and increase yield of cotton. California Agric 33:8–10

Fernandez C, Cothren J, McINNES K (1991) Partitioning of biomass in well-watered and water-stressed cotton plants treated with mepiquat chloride. Crop Sci 31:1224–1228

Fernandez C, Cothren J, McInnes K (1992) Carbon and water economies of well-watered and water-deficient cotton plants treated with mepiquat chloride. Crop Sci 32:175–180

Freytag A, Coleman E (1973) Effect of multiple applications of 2, 3, 5-triiodobenzoic acid (TIBA) on yields of stormproof and nonstorm proof cotton. Agron J 65:610–612

Fritz CD (1997) Finish® cotton harvest-aid EUP results. In: Davis TD (ed) Proceedings of the 23rd Annual Meeting of the Plant Growth Regulation Society of America. Plant Growth Regulation Society of America, Research Triangle Park, pp 224–229

Galani S, Hameed S, Ali MK (2016) Exogenous application of salicylic acid: inducing thermotolerance in cotton (*Gossypium hirsutum* L.) seedlings

Gausman H, Walter H, Stein E, Rittig F, Leamer R, Escobar D, Rodriguez R (1979) Leaf CO_2 uptake and chlorophyll ratios of PIX-treated cotton. In: Proceedings-Plant Growth Regulator Working Group

Gausman H, Stabenow J, Rittig, F, Escobar D, Garza M (1980a) Mepiquat chloride effects on cotton leaf anatomy. In: Proceedings of the Plant Growth Regulator Working Group, pp 8–14 (03366982565)

Gausman H, Walter H, Rittig F, Escobar D, Rodriguez R (1980b) Effect of mepiquat chloride (Pix) on CO_2 uptake of cotton plant leaves. In: Proceedings of the Plant Growth Regulator Working Group, pp 1–6

Gerik TJ, Oosterhuis DM, Torbert HA (1998) Managing cotton nitrogen. Adv Agron 64:115–147

Gill SS, Gill R, Anjum NA (2014) Target osmoprotectants for abiotic stress tolerance in crop plants—glycine betaine and proline. In: Anjum NA, Gill SS, Gill R (eds) Plant adaptation to environmental change: significance of amino acids and their derivatives. CAB International, Wallingford, CT, pp 97–108

Gorham J (1995) Betaines in higher plants-biosynthesis and role in stress metabolism. Amino acids and their derivatives in higher plants, pp 171-203

Graham CT, Jenkins JN, McCarty JC, Parrott WL (1987) Effects of mepiquat chloride on natural plant resistance to tobacco budworm in cotton 1. Crop Sci 27:360–361

Greene DW (1999) Tree growth management and fruit quality of apple trees treated with prohexadione-calcium (BAS 125). Hort Sci 34:1209–1212

Gross D, Parthier B (1994) Novel natural substances acting in plant growth regulation. J Plant Growth Regul 13:93
Guinn G (1985) Fruiting of cotton. III. Nutritional stress and cutout 1. Crop Sci 25:981–985
Guinn G (1986) Hormonal relations during reproduction. In: Cotton physiology. The Cotton Foundation, Memphis, TN, pp 113–136
Gwathmey C, Clement J (2010) Alteration of cotton source–sink relations with plant population density and mepiquat chloride. Field Crops Res 116:101–107
Hacskaylo J, Scales A (1959) Some effects of guthion alone and in combination with DDT and of a dieldrin-DDT mixture on growth and fruiting of the cotton plant. J Econ Entomol 52:396–398
Han I-S, Jongewaard I, Fosket DE (1991) Limited expression of a diverged 0-tubulin gene during soybean (Glycine ma% [L.] Merr.) development. Plant Mol Biol l6:225–234
Hanafy Ahmed AH, Darwish E, Hamoda SAF, Alobaidy MG (2013) Effect of putrescine and humic acid on growth, yield and chemical composition of cotton plants grown under saline soil conditions. American-Eurasian J Agric Environ Sci 13:479–497
Hanafy Ahmed A, Darwish E, Alobaidy M (2016) Impact of putrescine and 24-epibrassinolide on growth, yield and chemical constituents of cotton (*Gossypium barbadense* L.) plant grown under drought stress conditions
Hardaker JB, Lien G (2010) Probabilities for decision analysis in agriculture and rural resource economics: The need for a paradigm change. Agricultural Systems 103:345–350
Hassawy GS, Hamilton K (1971) Effects of trifluralin and organophosphorus compounds on cotton seedlings. Weed Sci 19:166–169
Hearn A, Constable G (1984) Irrigation for crops in a sub-humid environment VII. Evaluation of irrigation strategies for cotton. Irri Sci 5:75–94
Hedin PA, McCarty JC Jr (1991) Effects of kinetin formulations on allelochemicals and agronomic traits of cotton. J Agric Food Chem 39:549–553
Heilman M, Brown J (1981) Interactions of nitrogen with Pix on the growth and yield of cotton. In: Proceedings of Beltwide Cotton Production Research Conferences, New Orleans, LA, 4–8 Jan 1981, National Cotton Council of America, Memphis, TN
Hickey J (1996) Farm verification of active bloom applications of PGR-IV to enhance yield and maturity. In: Dugger P, Richter DA (eds) Proceedings of Beltwide Cotton Production Research Conferences. National Cotton Council of America, Memphis, TN
Hodges H, Reddy V, Reddy K (1991) Mepiquat chloride and temperature effects on photosynthesis and respiration of fruiting cotton. Crop Sci 31:1302–1308
Hornsey IS (2003) A history of beer and brewing. Royal Society of Chemistry, Cambridge
Horton RF (1971) Stomatal opening: the role of abscisic acid. Can J Bot 49:583–585
Huang S, Gausman H (1982) Ultrastructural observations on temperature resistance increase in cotton plants (*Gossypium hirsutum* L.) by mepiquat chloride. In: Wilson WC (ed) Proceedings annual meeting Plant Growth Regulator Society of America Conference. Plant Growth Regulator Society of America, Lake Alfred, FL
IPCC (2014) Climate change 2014: Synthesis report. Contribution of Working Groups I, II and III to the Fifth Assessment Report of the Intergovernmental Panel on Climate Change [core writing team, RK Pachauri, and LA Meyer, editors]. IPCC, Geneva
Javid MG, Sorooshzadeh A, Moradi F, Modarres Sanavy SAM, Allahdadi I (2011) The role of phytohormones in alleviating salt stress in crop plants. Aust J Crop Sci 5:726–734
Jenkins JN, McCarty J, Parrott W (1990) Fruiting efficiency in cotton: boll size and boll set percentage. Crop Sci 30:857–860
Jones ML, Kim ES, Newman SE (2001) Role of ethylene and 1-MCP in flower development and petal abscission in zonal geraniums. Hort Sci 36:1305–1309
Karabudak T, Bor M, Özdemir F, Türkan İ (2014) Glycine betaine protects tomato (Solanum lycopersicum) plants at low temperature by inducing fatty acid desaturase7 and lipoxygenase gene expression. Mol Biol Reports 41:1401–1410
Kazan K, Manners JM (2012) JAZ repressors and the orchestration of phytohormone crosstalk. Trends Plant Sci 17:22–31

Key JL (1989) Modulation of gene expression by auxin. Bioessays 11:52–58
Kittock D, Arle HF (1977) Termination of late season cotton fruiting with plant growth regulators 1. Crop Sci 17:320–324
Klepper B, Taylor H (1979) Limitations to current models describing water uptake by plant root systems. In: The soil–root interface. Elsevier, pp 51–65
Kondo S, Fiebig A, Okawa K, Ohara H, Kowitcharoen L, Nimitkeatkai H, Kittikorn M, Kim M (2011) Jasmonic acid, polyamine, and antioxidant levels in apple seedlings as affected by ultraviolet-C irradiation. Plant Growth Regul 64:83–89
Kong X, Wang T, Li W, Tang W, Zhang D, Dong H (2016) Exogenous nitric oxide delays salt-induced leaf senescence in cotton (*Gossypium hirsutum* L.). Acta Physiologiae Plantarum 38:61
Kuppu S, Mishra N, Hu R, Sun L, Zhu X, Shen G, Blumwald E, Payton P, Zhang H (2013) Water-deficit inducible expression of a cytokinin biosynthetic gene IPT improves drought tolerance in cotton. PLoS One 8:e64190
Lane H (1958) Response of cotton on the Texas high plains to foliar treatments of gibberellc acid. In: Proceedings of Beltwide Cotton Production Research Conferences. National Cotton Council of America, Memphis, TN
Lau S, Jürgens G, De Smet I (2008) The evolving complexity of the auxin pathway. Plant Cell 20:1738–1746
Li J, Chory J (1999) Brassinosteroid actions in plants. J Exp Bot 50:275–282
Liu G, Li X, Jin S, Liu X, Zhu L, Nie Y, Zhang X (2014a) Overexpression of rice NAC gene SNAC1 improves drought and salt tolerance by enhancing root development and reducing transpiration rate in transgenic cotton. PLoS One 9:e86895
Liu S, Dong Y, Xu L, Kong J (2014b) Effects of foliar applications of nitric oxide and salicylic acid on salt-induced changes in photosynthesis and antioxidative metabolism of cotton seedlings. Plant Growth Regul 73:67–78
Livingston S, Parker R (1994) Lint yield responses to applications of PGR-IV and mepiquat chloride applied to five cotton varieties in South Texas. In: Proceedings Beltwide Cotton Conferences. National Cotton Council of America, Memphis, TN
Livingston S, Anderson D, Wilde Jr L, Hickey J (1992) Use of foliar application of pix PGR, and PHCA in low rate multiple applications for cotton improvement under irrigated and dryland conditions. In: Proceedings of Beltwide Cotton Production Research Conferences. National Cotton Council of America, Memphis, TN
Luo Q, Bange M, Johnston D, Braunack M (2015) Cotton crop water use and water use efficiency in a changing climate. Agric Ecosyst Environ 202:126–134
Lüthen H (2015) What we can learn from old auxinology. J Plant Growth Regul 34:702–707
Makhdum I, Shababuddin M (2006) Effect of different doses of glycine betaine and time of spray application on yield of cotton (*Gossypium hirsutum* L.). J Res (Sci) 17:241–245
Makhdum I, Nawaz A, Shabab M, Ahmad F, Illahi F (2002) Physiological response of cotton to methanol foliar application. J Res (Sci) 13:37–43
Makhdum M, Shababuddin I, Ahmad F (2006) Effect of exogenous application of glycinebetaine on water relations in cotton (Gossypium hirsutum L.) in aridisols. J Sci Technol 30:11–14
Manavalan LP, Chen X, Clarke J, Salmeron J, Nguyen HT (2012) RNAi-mediated disruption of squalene synthase improves drought tolerance and yield in rice. J Exp Bot 63:163–175
Marani A, Levi D (1973) Effect of soil moisture during early stages of development on growth and yield of cotton plants 1. Agron J 65:637–641
Marre E (1979) Fusicoccin: a tool in plant physiology. Ann Rev Plant Physiol 30:273–288
Marumo S, Katayama M, Komori E, Ozaki Y, Natsume M, Kondo S (1982) Microbial production of abscisic acid by Botrytis cinerea. Agric Biol Chem 46:1967–1968
McCarty JC Jr, Hedin PA (1994) Effects of 1, 1-dimethylpiperidinium chloride on the yields, agronomic traits, and allelochemicals of cotton (*Gossypium hirsutum* L.), a nine year study. J Agric Food Chem 42:2302–2304
Mei L, Daud MK, Ullah N, Ali S, Khan M, Malik Z, Zhu SJ (2015) Pretreatment with salicylic acid and ascorbic acid significantly mitigate oxidative stress induced by copper in cotton genotypes. Environ Sci Pollut Res 22:9922–9931

Min L, Li Y, Hu Q, Zhu L, Gao W, Wu Y, Ding Y, Liu S, Yang X, Zhang X (2014) Sugar and auxin signaling pathways respond to high-temperature stress during anther development as revealed by transcript profiling analysis in cotton. Plant Physiol 164:1293–1308

Mittal A, Gampala SSL, Ritchie GL, Payton P, Burke JJ, Rock CD (2014) Related to ABA-Insensitive3(ABI3)/Viviparous1 and AtABI5 transcription factor coexpression in cotton enhances drought stress adaptation. Plant Biotechnol J 12:578–589

Munoz S (1994) Nutrient uptake and plant physiology enhancement by PHCA treatments on control plants. In: Proceedings of Beltwide Cotton Production Research Conferences. National Cotton Council of America, Memphis, TN

Murty P, Raju D, Rao G (1976) Effect of plant growth regulators on flower and boll drop in cotton. Food Farming and Agriculture

Nageswara Rao P, Prabhakara Rao J, Amariah M (1980) Research note on effect of Cycocel on MCU-5. Cotton Development

Negi L, Singh A (1956) A preliminary study on the effect of some hormones on yield of cotton. Indian Cotton Gr Rev 10:153

Noggle GR, Fritz GJ (1983) Introductory plant physiology. Prentice-Hall Inc, Englewood Cliffs

Noreen S (2010) Influence of exogenous application of salicylic acid on physiological and biochemical attributes of sunflower (*Helianthus annuus* L.) under salinity stress. University of Agriculture Faisalabad, Pakistan

Noreen S, Athar HUR, Ashraf M (2013) Interactive effects of watering regimes and exogenously applied osmoprotectants on earliness indices and leaf area index in cotton (*Gossypium hirsutum* L.) crop. Pak J Bot 45:1873–1881

Noreen S, Zafar Z, Shah KH, Athar H, Ashraf M (2015) Assessment of economic benefits of foliarly applied osmoprotectants in alleviating the adverse effects of water stress on growth and yield of cotton (*Gossypium hirsutum* L.). Pak J.Bot 47:2223–2230

Noreen S, Siddiq A, Hussain K, Ahmad S, Hasanuzzaman M (2017) Foliar application of salicylic acid with salinity stress on physiological and biochemical attributes of sunflower (*Helianthus annuus* L.) crop. Acta Scientiarum Polonorum-Hortorum Cultus 16:57–74

Oosterhuis DM, Zhao D (1994) Increased root length and branching in cotton by soil application of the plant growth regulator PGR-IV. Plant Soil 167:51–56

Parker L, Salk P (1990) Foliar TRIGGER on cotton: a comprehensive statistical review. In: Proceedings – Beltwide Cotton Production Research Conferences USA, pp 659-660

Patel JK (1992) Effect of triacontanol and naphthalene acetic acid on lint yield, fibre quality, and nitrogen, phosphorus and potash uptake in cotton (Gossypium species). Indian J Agron 37:332–337

Pervez H, Ashraf M, Makhdum M (2005) Influence of potassium rates and sources on seed cotton yield and yield components of some elite cotton cultivars. J Plant Nutr 27:1295–1317

Pettigrew W, Heitholt J, Meredith W (1992) Early season floral bud removal and cotton growth, yield, and fiber quality. Agron J 84:209–214

Phillips J, Herzog G, Nicholson W (1977) Effect of chlordimeform on fruiting characteristics and yield of cotton [Heliothis]. Arkansas Farm Res 26(2):4

Rademacher W (2015) Plant growth regulators: backgrounds and uses in plant production. J Plant Growth Regul 34:845–872

Radin JW, Parker LL, Sell CR (1978) Partitioning of sugar between growth and nitrate reduction in cotton roots. Plant Physiol 62:550–553

Raghavendra AS, Gonugunta VK, Christmann A, Grill E (2010) ABA perception and signalling. Trends Plant Sci 15:395–401

Rahman MH, Ahmad A, Wang X, Wajid A, Nasim W, Hussain M, Ahmad B, Ahmad I, Ali Z, Ishaque W, Awais M, Shelia V, Ahmad S, Fahad S, Alam M, Ullah H, Hoogenboom G (2018) Multi-model projections of future climate and climate change impacts uncertainty assessment for cotton production in Pakistan. Agr Forest Meteorol 253-254:94–113

Reddy K, Reddy VR, Hodges HF (1992) Temperature effects on early season cotton growth and development. Agron J 84:229–237

Reddy KR, Davidonis GH, Johnson AS, Vinyard BT (1999) Temperature regime and carbon dioxide enrichment alter cotton boll development and fiber properties. Agron J 91:851–858

Riaz M, Farooq J, Sakhawat G, Mahmood A, Sadiq MA, Yaseen M (2013) Genotypic variability for root/shoot parameters under water stress in some advanced lines of cotton (*Gossypium hirsutum* L.). Genet Mol Res 12:552–561

Riemann M, Dhakarey R, Hazman M, Miro B, Kohli A, Nick P (2015) Exploring jasmonates in the hormonal network of drought and salinity responses. Front Plant Sci 6:1077

Salisbury F, Ross C (1992) Hormones and growth regulators: auxins and gibberellins. Plant Physiol 4:401–405

Sankaranarayanan K, Praharaj CS, Nalayini P, Bandyopadhyay K, Gopalakrishnan N (2010a) Climate change and its impact on cotton (Gossypium sp.). Indian J Agric Sci 80:561–575

Sankaranarayanan K, Praharaj CS, Nalayini P, Bandyopadhyay K, Gopalakrishnan N (2010b) Effect of magnesium, zinc, iron and boron application on yield and quality of cotton (*Gossypium hirsutum*). Indian J Agric Sci 80:699–703

Sarwar M, Saleem MF, Ullah N, Rizwan M, Ali S, Shahid MR, Alamri SA, Alyemeni MN, Ahmad P (2018) Exogenously applied growth regulators protect the cotton crop from heat-induced injury by modulating plant defense mechanism. Sci Rep 8:17086

Sawan Z, Sakr R, El-Kady M (1984) Effect of ethrel treatment on the yield components and fiber properties of the Egyptian cotton. Zeitschrift fur Acker-und Pflanzenbau 153(1):72–78

Scott JC (1990) Domination and the arts of resistance: hidden transcripts. Yale University Press, New Haven, p 269

Shah S, Li J, Moffatt BA, Glick BR (1998) Isolation and characterization of ACC deaminase genes from two different plant growth-promoting rhizobacteria. Can J Microbiol 44:833–843

Shallan MA, Hassan HMM, Namich AAM, Ibrahim AA (2012) Effect of sodium nitroprusside, putrescine and glycine betaine on alleviation of drought stress in cotton plant. American-Eurasian J Agric Environ Sci 12:1252–1265

Shanmugham K (1992) Seed soaking and foliar application of growth-regulants and anti-transpirant chemicals for increasing drought resistance in rainfed upland cotton (*Gossypium hirsutum*). Indian J Agric Sci 62:744–750

Sheng C, Hopper K, Meredith W, King E, Ma S (1988) Cotton development, yield, and quality after early square removal with ethephon

Shu H, Ni W, Guo S, Gong Y, Shen X, Zhang X, Xu P, Guo Q (2015) Root-applied brassinolide can alleviate the NaCl injuries on cotton. Acta Physiologiae Plantarum 37:75

Singh S (1970) Revolution in cotton yield with CCC. Indian Farming 20:5–6

Singh S (1975) Studies on the effects of soil moisture stress on the yield of cotton. Indian J Plant Physiol 18:49

Singh S, Singh VV, Choudhary AD (2010) Combining ability estimates for oil content, yield components and fibre quality traits in cotton (G. hirsutum) using an 8× 8 diallel mating design. Trop Subtrop Agroecosyst 12:161–166

Sitaram MS, Abraham ES (1973) Note on effect of gibberellic acid on quality of Laxmi cotton. Cotton Growing Rev 50:150–151

Snow JP, Crawford SH, Berggren GT, Marshall JG (1981) Growth regulator tested for cotton boll rot control [Pix (NN-Dimethylpiperidinium chloride), Fusarium]. Louisiana Agriculture (USA) 24(3):324

Stedman S, Taylor B (1982) The effect of Pix® applications on cotton treated with four levels of nitrogen. In: Brown JM (ed) Proceedings Beltwide Cotton Production Research Conferences. National Cotton Council of America, Memphis, TN

Sticher L, Mauch-Mani B, Métraux JP (1997) Systemic acquired resistance. Annu Rev Phytopathol 35:235–270

Stiller WN, Read JJ, Constable GA, Reid PE (2005) Selection for water use efficiency traits in a cotton breeding program: cultivar differences. Crop Sci 45:1107–1113

Stuart B, Wendt C, Abernathy J (1980) The influence of mepiquat chloride on the plant water status of cotton

Subbiah KK, Mariakulandai A (1972) Application of gibberellic acid and naphthalene acetic acid in preventing bud and boll shedding in Cambodia cotton. Madras Agr J 59:350–352

Suttle JC (1988) Disruption of the polar auxin transport system in cotton seedlings following treatment with the defoliant thidiazuron. Plant Physiol 86:241–245

Taha MB, Noakes DE (1982) The effect of age and season of the year on testicular function in the dog, as determined by histological examination of the seminiferous tubules and the estimation of peripheral plasma testosterone concentrations. J Small Anim Pract 23:351–357

Taiz L, Zeiger E (1998) Plant physiology, vol 998. Sinauer Associates, Sunderland, p 79

Tan J, Tu L, Deng F, Wu R, Zhang X (2012) Exogenous jasmonic acid inhibits cotton fiber elongation. J Plant Growth Regul 31:599–605

Tariq M, Yasmeen A, Ahmad S, Hussain N, Afzal MN, Hasanuzzaman M (2017) Shedding of fruiting structures in cotton: factors, compensation and prevention. Trop Subtrop Agroecosyst 20(2):251–262

Tariq M, Afzal MN, Muhammad D, Ahmad S, Shahzad AN, Kiran A, Wakeel A (2018) Relationship of tissue potassium content with yield and fiber quality components of Bt cotton as influenced by potassium application methods. Field Crops Res 229:37–43

Taylor A, Cosgrove DJ (1989) Gibberellic acid stimulation of cucumber hypocotyl elongation: effects on growth, turgor, osmotic pressure, and cell wall properties. Plant Physiol 90:1335–1340

Thomas RO (1964) Effects of application timing and concentration of 2-Chloroethyl trimethylammonium chloride on plant size and fruiting responses of cotton. Crop Sci 4:403–406

Thomas R (1967) Effects of two growth retardants on flowering and boll production of greenhouse cotton plants

Underbrink SM (1999) Agronomic differences in growth and yield between BT and conventional cotton treated with mepiquat chloride. Texas A&M University

Urwiler MJ, Oosterhuis DM (1986) The effect of the growth regulators Pix and IBA on cotton root growth. Arkansas Farm Res 36(6):5

Usman M, Ahmad A, Ahmad S, Irshad M, Khaliq T, Wajid A, Hussain K, Nasim W, Chattha TM, Trethowan R, Hoogenboom G (2009) Development and application of crop water stress index for scheduling irrigation in cotton (*Gossypium hirsutum* L.) under semiarid environment. J Food Agri Environ 7(3&4):386–391

van Loon LC, Glick BR (2004) Increased plant fitness by rhizobacteria. In: Sandermann H (ed) Molecular ecotoxicology of plants. Springer, Berlin, pp 177–205

Vriet C, Russinova E, Reuzeau C (2012) Boosting crop yields with plant steroids. Plant Cell 24:842–857

Wakelyn PJ, Chaudhry MR (2010) Cotton: technology for the 21st century. International Cotton Advisory Committee, 2010, pp 475

Walhood V (1958) Effect of gibberellins on yield and growth of cotton

Walter H, Gausman H, Rittig F, Namken L, Escobar D, Rodriguez R (1980) Effect of mepiquat chloride on cotton plant leaf and canopy structure and dry weights of its components. In: Proceedings of Beltwide Cotton Production Research

Watkins CB (2015) Advances in the use of 1-MCP Advances in postharvest fruit and vegetable technology. CRC Press, Boca Raton, FL, pp 117–145

Wright S, Hiron R (1972) The accumulation of abscisic acid in plants during wilting and under other stress conditions. In: Carr DJ (ed) Plant growth substances 1970. Springer, Berlin, pp 291–298

Wu K (2007) Monitoring and management strategy for Helicoverpa armigera resistance to Bt cotton in China. J Invertebr Pathol 95:220–223

Xiao S, Liu L, Wang H, Li D, Bai Z, Zhang Y, Sun H, Zhang K, Li C (2019) Exogenous melatonin accelerates seed germination in cotton (*Gossypium hirsutum* L.). PLoS One 14(6):e0216575

Xu X, Taylor H (1992) Increase in drought resistance of cotton seedlings treated with mepiquat chloride. Agron J 84:569–574

York A (1983) Cotton cultivar response to mepiquat chloride. Agron J 75:663–667

Youngman RR, Leigh TF, Kerby TA, Toscano NC, Jackson CE (1990) Pesticides and cotton: effect on photosynthesis, growth, and fruiting. J Econ Entomol 83:1549–1557
Zafar ZU, Hussain K, Athar HUR, Noreen S, Ashraf M (2015) Assessment of economic benefits of foliarly applied osmoprotectants in alleviating the adverse effects of water stress on growth and yield of cotton (*Gossypium hirsutum* L). Pak J Bot 47:2223–2230
Zhang F, Romheld V, Marschner H (1991) Release of zinc mobilizing root exudates in different plant species as affected by zinc nutritional status. J Plant Nutr 14:675–686
Zhao D, Oosterhuis DM (2000) Pix plus and mepiquat chloride effects on physiology, growth, and yield of field-grown cotton. J Plant Growth Regul 19:415–422
Ziska LH, Teasdale JR, Bunce JA (1999) Future atmospheric carbon dioxide may increase tolerance to glyphosate. Weed Sci 47(5):608–615

Chapter 9
Weed Management in Cotton

Muhammad Tariq, Khalid Abdullah, Shakeel Ahmad, Ghulam Abbas, Muhammad Habib ur Rahman, and Muhammad Azim Khan

Abstract The weeds are becoming a major challenge for cotton production across the world, and the crop is infested by broad leaves, grasses, and sedges. Although the presence of weeds is deleterious for potential yield, the extent of losses depends on weed type, density, duration, and crop growth stage. Some weeds like *Datura stramonium* L., *Amaranthus palmeri* L., *Amaranthus retroflexus* L., and *Ambrosia trifida* L. cause significant yield losses at very low densities in comparison with *Cucumis melo* L. and *Eleusine indica* L. The cotton crop is very sensitive at early growth stages where weed presence during the first 2 months of growth may reduce yield from 10% to 90%. The most common effects of weeds on cotton quality are higher trash contents and lint staining problem. The various weed control options have been used for weed management in cotton; however, the efficacy of weed control methods remains low, probably by practicing weed control out of the critical period of weed-crop competition and selection of inappropriate method. The cost of weed control ranges from 54.5 to 320.6 US\$ ha^{-1} in various cotton-growing countries of the world. The use of herbicides for weed control is common in the world; however, excessive and non-judicious use of herbicides led to the evolution of resistant weed biotypes. It provided the basis for development of glyphosate-tolerant cotton genotypes. The big change has been made for weed management with the advent of genetically modified glyphosate-tolerant

M. Tariq (✉)
Agronomy Section, Central Cotton Research Institute, Multan, Pakistan

K. Abdullah
Ministry of National Food Security and Research, Islamabad, Pakistan

S. Ahmad
Department of Agronomy, Faculty of Agricultural Sciences and Technology, Bahauddin Zakariya University, Multan, Pakistan

G. Abbas
Department of Agronomy, Bahauddin Zakariya University, Multan, Pakistan

M. H. ur. Rahman
Department of Agronomy, Muhammad Nawaz Shareef University of Agriculture, Multan, Pakistan

M. A. Khan
Department of Weed Science, Agriculture University, Peshawar, Pakistan

S. Ahmad, M. Hasanuzzaman (eds.), *Cotton Production and Uses*,
https://doi.org/10.1007/978-981-15-1472-2_9

cotton. It offers flexibility in herbicide use and time of application and improves weed control efficiency and economic returns. The evolution of herbicide resistance has been reported in *Amaranthus palmeri* L., *Commelina benghalensis* L., *Conyza bonariensis* L., *Conyza canadensis* L., and *Sorghum halepense* L. Some new technologies like Roundup Ready Xtend cotton and Enlist technology are being developed to cope to the challenge of glyphosate resistance.

Keywords Weed flora · Control · Resistance · Integrated weed management

Abbreviations

CPWR	Critical period for weed removal
CWFP	Critical weed-free period
DAE	Days after emergence
GDD	Growing degree days
IWM	Integrated weed management
WAE	Weeks after emergence

9.1 The Importance and Scope of Weed Control

The prevailing weed flora and emerging weed problems are great challenges for cotton production throughout the world. Almost all types of weeds like grasses, sedges, and broad leaf weeds are associated with cotton crop. An overview of weed infestation in cotton at early stage is presented in Fig. 9.1. The weeds compete with the cotton crop because of its slow growth and wider spaces available between the rows. The cotton is also very prone to various insect pests which also make the crop less competitive than weeds. Weeds continue to emerge throughout the season;

Fig. 9.1 Overview of weeds infestation in cotton

Table 9.1 Weed density effect on cotton yield losses

Weed species	Density	Yield losses (%)	Reference
Amaranthus retroflexus L.	0.20–0.33 plants/ 1 m row	50	Ma et al. (2015)
Abutilon theophrasti L.	1.0 plant/1 m row	42	Bailey et al. (1999)
Amaranthus palmeri	1.0 plant/1 m row	66	Rowland et al. (1999)
Ambrosia trifida L.	0.26 plants/1 m row	50	Barnett and Steckel (2013)
Anoda cristata L.	8.0 plants/12 m row	30	Chandler and Oliver (1979)
Commelina benghalensis L.	10 plants/m^2	40–62	Ahanchede (1996)
Cucumis melo L.	10.0 plants/4 m row	34	Tingle et al. (2003)
Datura stramonium L.	1.0 plant/1 m row	65	Scott et al. (2000)
Eleusine indica L.	4.0 plants/1 m row	20–27	Xiao-yan et al. (2015)
Sida spinosa L.	1.0 plant/30 ft. row	0.26	Buchanan et al. (1977)
Sorghum halepense L.	1.0 plant/15 m row	5.5	Wood et al. (2002)

however, weeds emerging at very early stage are more deleterious for yield. The weeds like redroot pigweed grow taller than cotton and adversely affect the canopy development (Ma et al. 2015). Besides yield and fiber quality deterioration, weeds also increase the cost of production and serve as alternate host for disease and pests. The possible mechanism of yield reduction from weeds may be the result of weed allelopathy, parasitism, and competition. However, it is very difficult under field conditions to state that whether the yield reduction is due to competition or allelopathic impact. Generally, yield losses in cotton may be up to 30–40% due to weed infestation. However, grassy weeds cause 10% more yield losses than broad leaf weeds (Khan and Khan 2003). The weeds are the second most important constraint of cotton followed by insects which are responsible for about 34% losses at global level (Oerke and Dehne 2004). However, magnitudes of yield losses vary from field to field because of weeds' competitive ability which depends upon weed type, weed infestation duration, and density. The weeds like *Ambrosia trifida* L., *Amaranthus retroflexus* L., *Amaranthus palmeri* L., and *Datura stramonium* L. are highly competitive. However, *Eleusine indica* L., *Anoda cristata* L., and *Cucumis melo* L. are less competitive (Table 9.1). The examples of weeds inhibiting cotton growth through exertion of allelopathic compounds include unicorn and smooth pigweed (Riffle et al. 1987; Munger et al. 1984). The impact of weeds on fiber quality is variable because some weeds like goose grass have no adverse effects on fiber quality (Xiao-yan et al. 2015). However, Palmer amaranth did not influence fiber properties except micronaire (Rowland et al. 1999). The weeds also reduce the efficiency of insecticides and growth regulators through direct interception by the weeds. Certain weeds remain green at the time of cotton picking. Such weeds get mixed with seed cotton during picking operations and contaminate with green color. The mixed trash would therefore require additional operations during ginning.

Although season long weed management practices are adopted to avoid yield losses, initial weed control is very important because yield penalties during this phase are highly significant (Ahmad et al. 2014, 2017, 2018; Abbas and Ahmad 2018; Ahmad and Raza 2014; Ali et al. 2011, 2013a, b, 2014a, b; Amin et al. 2017, 2018; Khan et al. 2004; Rahman et al. 2018; Tariq et al. 2017, Tariq et al. 2018a, b; Usman et al. 2009). A weed-free period of about 20 days after sowing produced 73.4% higher seed cotton yield over unweeded plots (Bishnoi et al. 1993). Therefore, more efforts should be concentrated for early season weed management. The weeds like *Echinochloa crus-galli* (L.) Beauv must be controlled during first 9 weeks of planting (Keeley and Thullen 1991). However, the presence of *Amaranthus palmeri* during late season growth also reduces the yield (Ward et al. 2013). Because of the prime importance of weed management in cotton, genetic approaches are also being utilized for development of glyphosate-resistant cotton. The yield reduction potential of weeds at various densities is given in Table 9.1.

9.2 Weed Flora of Cotton

The cotton crop is rich in weeds ranging from narrow leaves, broad leaves, to sedges. The common weeds of cotton in various cotton-growing countries have been reported in the literature (Table 9.2).

9.3 Critical Period of Weed Control

It refers to the period of crop growth during which weeds must be controlled to avoid yield losses (Bukun 2004). Therefore, it is often called as time span, time interval, and window. The knowledge of critical weed-crop competition is required for deciding the time and need of weed control. The concept of CPWC was introduced by Nieto et al. (1968). It has two components, i.e., critical period for weed removal (CPWR) and critical weed-free period (CWFP). CPWR represents the maximum weed competition period before it results to irrevocable crop yield losses, while CWFP is the minimum period from planting that must be kept weed free to avoid significant yield losses. Hence, it starts after when weed presence did not reduce yield and ends up after which weeds are no longer able to cause yield losses. The weeds, crop, and environmental characters determine the start and duration of CPWC (Hall et al. 1992). It showed that CPWC may vary among crops, growing environment, and weed composition of even the same crop. The concept has practical application for annual weeds; however, it may not be applicable for perennial weeds that must be controlled before reaching unmanageable levels. The cotton crop should be kept weed free during the first 4 to 7 weeks (Douti 1997). The weed-free period of about 11 weeks is necessary to avoid yield losses (Papamichail et al. 2002). The weeds that grow after critical weed-crop competition can be controlled before seed set to prevent next-year infestation. The critical period of weed-crop competition must be kept in mind for deciding the time of weed control

Table 9.2 Certain weeds of cotton crop in various cotton-growing countries

Sr. no.	Weed type	Global distribution	Reference
1.	*Alternanthera tenella* Colla	Brazil	Silva et al. (2009)
2.	*Amaranthus* spp.	Greece, Turkey	Economou et al. (2005), Özaslan et al. (2015)
3.	*Commelina benghalensis* L.	USA, Brazil, Botswana	Webster et al. (2005), Silva et al. (2016), Abdullahi (2004)
4.	*Convolvulus arvensis* L.	Greece, Pakistan, Turkey, Spain	Economou et al. (2005), Khan and Khan (2003), Bükün (2005), García-Ruiz et al. (2018)
5.	*Cyperus rotundus* L.	Greece, India, Pakistan, Turkey	Economou et al. (2005), Jadhav et al. (1995), Khan and Khan (2003), Bükün (2005)
6.	*Digera arvensis* L.	India, Pakistan	Giri and Bhosle (1997), Oad et al. (2007)
7.	*Echinochloa colona* L.	Turkey, Pakistan	Bükün (2005), Tauseef et al. (2012)
8.	*Eleusine indica* L.	China, Greece, India	Xiao-yan et al. (2015), Papamichail et al. (2002), Nalini et al. (2015)
9.	*Leptochloa chinensis* L. Nees	China	Zhang (2003)
10.	*Setaria viridis* L.P. Beauv	China, India	Zhang (2003), Nalini et al. (2015)
11.	*Solanum nigrum* L.	Greece, Spain, Turkey	Economou et al. (2005), García-Ruiz et al. (2018), Pala and Mennan (2017)
12.	*Sorghum halepense* L. Pers.	Greece, Turkey, India	Economou et al. (2005), Özaslan et al. (2015), Nalini et al. (2015)
13.	*Trianthema portulacastrum* L.	India, Pakistan, America, and Africa	Panwar et al. (1998), Ara et al. (2015)
14.	*Tribulus terrestris* L.	Turkey, India	Bükün (2005), Singh and Pathania (2017)

because weed control outside this period does not result significant in yield benefits. The weed control practices must be optimized during this span to improve efficiency. The weed-crop competition is related to time of emergence and duration of presence of weeds. The competition may be very severe when growth of both the partners continues simultaneously. It is represented in days after emergence (DAE), weeks after emergence (WAE), and growing degree days (GDD). It is also an important component of integrated weed management.

9.4 The Share of Weed Control in Cost of Production

The data presented in Table 9.3 showed the comparison of weed control cost of various cotton-growing countries. The weed control is a significant part of cost of production which ranges about 2.5–14.8% in various cotton-growing countries. The

Table 9.3 Cost of weed control for 1 ha (US$) for year 2016

Sr. no.	Name of countries	Herbicide	Weeding	Hoeing	Total cost of production	Weed control share (%)
1.	Greece	113.9	56.95	45.7	3220.13	6.7
2.	China	–	45.8	274.8	3765.36	8.5
3.	India	–	99.6	60.9	1225.6	13.1
4.	Pakistan	–	64.4	–	1447.88	4.4
5.	Turkey	11.8	185.8	–	3790.4	5.2
6.	Brazil	90.2	–	31.3	1606.92	7.6
7.	Australia	35.9	15.0	3.57	2219.11	2.5
8.	Argentina	36.8	–	22.7	850.92	7.0
9.	Tanzania	–	111.6	–	755.61	14.8
10.	Egypt	56.3	30.4	61.9	2203.27	6.7

Source: Cost of production of raw cotton. ICAC (2016)

maximum weed control cost was reported from China followed by Turkey, India, and Egypt. Australia, Argentina, and Pakistan are among countries investing lower amount for weed control. The proportion of cost of weed control relative to total cost is higher for Tanzania and India. The herbicide cost is maximum for Greece followed by Brazil and minimum for Turkey. However, the cost of other weeding option is high in Turkey than other countries (ICAC 2016).

9.5 Weeds as Alternate Host for Insects and Diseases

The weeds support biodiversity and provide a range of ecological services for many insects. The weeds act as alternate host plant for different pests when main crop is absent. Sometimes pests are also present in weeds even though main crop is present, and it may be due to relative host plant preference. The whitefly presence was noted on common ragweed and piemarker about 10 days before the cotton in Langfang, Hebei province, China (Zhang et al. 2014). Similarly, the mealybug (*Phenacoccus solenopsis* L.), a major cotton pest, has also been found in ragweed (*Parthenium hysterophorus* L.) around fields (Shabbir 2014). Also, Bharathi and Muthukrishnan (2017) identified 26 weeds, as host plant of mealybug. The thorn apple (*Datura ferox* L.) is host for *Heliothis* and mites (Charles and Roberts 2013).

9.6 Weed Control Methods

It refers to the elimination of existing weed population in the field. The cultural, mechanical, and biological means has been an important weed management tool before the advent of organic herbicides during the twentieth century. All the weed control methods may not be equally effective for all growing areas, and success varies according to climate, soil type, prevailing weed flora, and other agronomic management.

9.6.1 Preventive Control

The preventive measures aim to avoid spread of weed propagation sources like seed, rhizomes, and tubers. These include clean cultivation, use of weed-free seed and water, fully decomposed farmyard manures, and restricted movement of animals from weed-infested field to cotton field. The post-cultivation cleaning of tillage implements is advised to reduce the shift of weeds to new fields. The weed removal prior to seed setting is also an important preventive measure to reduce the weed pressure for the next growing season. It is very effective for freshly cultivated fields and newly appearing weeds in local areas.

9.6.2 Cultural Control

It is important weed control method following preventive control and ignored in most cases. The agronomic practices more favorable for cotton and less favorable for weeds are preferred. The time and frequency of cultural operations change the weed growth and development pattern. It protects the environment and generally requires no additional cost. The cultural control did not replace the herbicides, but it improves efficiency of in-season-applied herbicides. The crop rotation, planting configuration, row orientation, use of competitive cultivars, planting time adjustment, and stale bed technique are some important cultural weed control practices for cotton.

The difference in time for row closure, plant height, early growth vigor, and branching habits resulted variations in cultivars' competitive ability. The differences in cultivars' competitive ability against various weeds have been reported in literature (Chandler and Meredith 1983; Rezakhanlou et al. 2013). The row space is adjusted to reach the row closure stage at the earliest. It reduces weed emergence and subsequent growth by limiting light penetration in interrow spaces. The superior weed control of *Amaranthus retroflexus* L. from narrow spacing has been witnessed in literature (Gwathmey et al. 2008). The chances of new flushes of weeds are reduced at full canopy cover which is achieved about 100–120 days after sowing. Therefore, increasing plant density from either reducing interrow or plant spacing is a better strategy for early season weed management. The weed flora changes according to sowing time. Both broad and narrow leaf density increases with delay in sowing. The damp and moist weather in the month of July favors growth and development of narrow leaf weeds. The problem is more severe for late-sown crop with poor plant canopy development at very early stage (Tauseef et al. 2012). The weeds become abundant in continuous cropping through adding up seeds in soil.

Fig. 9.2 Mechanical weed control

Rotating cotton with high biomass crops such as sorghum, millet, and maize is very helpful to reduce the weed pressure in cotton. Irrigation followed by cultivation provides the favorable environmental condition for weed seed; hence, post-irrigation interrow cultivation at specific interval may help out to reduce weed infestation. It often bears no additional cost or modest cost; however, managerial skills are very important for cultural control of weeds.

9.6.3 Mechanical Weed Control

In this technique, the weeds are controlled through various mechanical tools. It is nonselective weed control method where hoe, disks, rotary weeders, and other interculturing implements are used. These tools incorporate, chop, cover, and uproot the weeds. The moldboard plow incorporates the weeds and buried the seed in deeper soil where these are unable to germinate. Such practices are also beneficial for improving root aeration and earthing up. However, weeds in the interplant spaces are controlled through manual means. The weed control through interculturing is more successful when weeds are small and cotton is low in height. Shallow interrow cultivation destroys the weeds (Fig. 9.2) and also stimulates the weed seed germination. However, bringing weed seed to surface from deeper soil, spreading of reproductive weeds organs throughout the field, and soil degradation are certain demerits of weed control through tillage. Hand hoeing had been an important and effective weed control means before the introduction of herbicides, while now it is considered secondary weed management option. The mechanical weed control starts from primary and secondary tillage and continues during crop season in the form of interculturing. However, repeated operations are carried out until and unless crop does not suffer. The weeds resistant to herbicides are easily handled through mechanical means. It can be easily implemented without any risk of herbicide residue problem. However, ineffective weed control in the close proximity of cotton plant and promotion of spread of weed propagating material are some key concerns of mechanical weed control.

9.6.4 Allelopathy

It involves production and release of various allelopathic compounds from plant parts. The weed control through allelopathy provides natural control and reduces the risks linked with synthetic herbicides. The allelopathic crops such as sorghum, soybean, sunflower, and sesame have the potential for managing weeds in cotton. These crops can be intercropped and rotated and has residue incorporation, aqueous extracts, and mulching to minimize the weed pressure. Spreading residues of allelopathic crops within cotton rows also reduces the weed buildup. The allelopathic extracts are effective alone and in combination with herbicides. One of such example includes purple nut sedge where sorghum allelopathic extract (Sorgaab) along with lower dose of S-metolachlor resulted about 62–92% control (Iqbal and Cheema 2008). Furthermore, the horse purslane was also successfully controlled in cotton through sorghum, mulberry, sunflower, and brassica allelopathic extracts along with lower doses of pendimethalin and paraquat (Luqman et al. 2017).

9.6.5 Chemical Weed Control

The chemical weed control is accomplished with the use of herbicides. The farmers adopted chemical weed control with the expectation to overcome the issues related to other methods. It starts with the herbicide application prior to sowing and proceeds through in season-selective and season-directed spray of nonselective herbicides. The better weed control through herbicides reduced and even replaced other weed management tools in some regions. However, the expert opinion considered it not a good practice to say bye to other available weed control options for herbicides. Although herbicides are very effective, these do not provide season-long control and nor are equally effective for all weeds. The use of preemergence herbicides is very common in cotton; however, very few selective postemergence herbicides are available. In most cases, the preemergence herbicides are banded over seeding zone, while avoiding application on drier space of beds. The preemergence herbicides reduce the weed population or delay the growth to make the weeds less

Fig. 9.3 Initial weed control through application of preemergence herbicides

competitive than cotton. Figure 9.3 indicates the initial weed control from preemergence herbicides. The pendimethalin, S-metolachlor, and trifluralin are important preemergence herbicides (Singh et al. 2016; Tariq et al. 2018a, b), while pyrithiobac-sodium, quizalofop-ethyl, and haloxyfop-P-methyl are postemergence herbicides (Veeraputhiran and Srinivasan 2015; Tariq et al. 2018a, b). The evolution of glyphosate-resistant cotton further extended the scope of herbicides which permits the in-season application of nonselective herbicides (Glyphosate). Proper dose and application time should be ensured to avoid development of herbicide-resistant weeds which is major challenge for chemical weed control.

9.6.6 Biological Weed Control

The concept of biological weed control is old where various organisms were considered for specific weed control. The biological agents including pathogens, geese, nematodes, insects, etc. are used for the purpose. But success was very low in practical term for crops especially cotton. These agents require special rearing laboratories, supplemental feed, protection from predators, and alternate habitat during insecticidal spray. The use of *Orrinaphyllobia* (a parasitic nematode) for control of silverleaf nightshade (*Solanum elaeagnifolium* Cav.) has been practiced in the southwestern USA (Orr et al. 1975). However, the results were not satisfactory. The sicklepod (*Cassia obtusifolia*) is an important cotton weed in southeastern areas of the USA and was controlled through a fungus, *Alternaria cassiae* (Charudattan et al. 1986). However, the scope of biological control seems to be attractive with the introduction of Bollgard II due to low use of insecticides. The major drawback of biological control is high development cost and nonuniversal weed control for all weeds.

9.7 Prospective of Glyphosate-Resistant Cotton

The development of herbicide-resistant cotton was started through genetic engineering where bromoxynil-resistant cotton was an initial product in 1995. However, the efforts were diverted toward development of glyphosate-resistant crops because broad-spectrum weed control (Reddy and Nandula 2012). Glyphosate [N-(phosphonomethyl) glycine] is a widely used herbicide in the world due to low cost, broad-spectrum weed control, and presence of tolerance in the transgenic crops (Latif et al. 2015). Therefore, glyphosate has been considered as "a-once-in-a-century herbicide" (Duke and Powles 2008). The glyphosate-tolerant cotton was made commercially available during 1997 and covered significant area (Latif et al. 2015). Later on glufosinate-resistant cotton was also introduced at commercial level during 2004, but it produced lower yield than glyphosate-resistant cotton (Reddy and Nandula 2012).

The herbicide tolerance enabled the crop to survive which were previously damaged with herbicides along with targeted weeds. The prospective of herbicide-resistant cotton is quite clear from a significant area under cultivation. It makes flexible in-season postemergence (glyphosate) application along with preemergence application which was otherwise impossible in crops lacking glyphosate tolerance. However, the volunteer herbicide-tolerant crops may act as resistant weeds. The evolution of glyphosate-resistant weed biotype is another concern of the use of herbicide-tolerant crops. The shifts in weed flora have been observed in cotton fields which were continuously treated with glyphosate (Culpepper 2006). The population of *Conyza bonariensis* (L.) Cronquist., *Sonchus oleraceus* L., and *Chloris virgata* Sw. in Australia was significantly increased in 2011 over past data recorded in 2008 (Werth et al. 2013). The changes in weed composition from 3 years of consecutive use of glyphosate have been witnessed in Spain. The *Portulaca oleracea* L. became the most abundant weed in continuously glyphosate-treated plots (García-Ruiz et al. 2018).

9.8 Herbicide Resistance and Its Management

The herbicide resistance refers to the inherit ability of the weed to survive and reproduce to herbicide dose that has been previously effective. The herbicide application is an easy and effective means of weed control which favored the evolution of herbicide-resistant biotypes. The main reasons of development of herbicide resistance may be low dose, delayed time of application over recommended one, improper spray coverage, poor mixing, and sprayer calibration (McCloskey 2012). However, the phenomenon of herbicide resistance led the crop scientist to develop herbicide-tolerant crops such as cotton, maize, and soybean (Reddy and Nandula 2012). The genetic (gene mutation rate, gene flow, inheritance, etc.) and generation turnover are some important species-related characteristics which determine the resistance evolution rate (Jasieniuk et al. 1996; Stanton et al. 2008). The rate of resistance development is less in weeds with less number of generations round the year and vice versa. For example, *Eleusine indica* L. in Malaysia have four generations and developed resistance in 3 years of glyphosate use (Lee and Ngim 2000). Whereas the *Lolium rigidum* L. has one generation in Australia, it took 15 years to develop resistance (Powles et al. 1998).

The evolution of glyphosate-resistant weeds such as *Commelina benghalensis* L., *Amaranthus palmeri* L., *Conyza bonariensis* L., *Eleusine indica* L., *Echinochloa colona* L., *Lolium multiflorum* L., *Amaranthus tuberculatus* L., *Conyza canadensis* L., and *Sorghum halepense* L. in the USA and Australia has been reported (Norsworthy et al. 2007; Webster et al. 2005; Werth et al. 2011; Heap 2015). However, the *Palmer amaranth* has been regarded a highly harmful weed for cotton in the USA. It infested about 254,921 ha cotton field during 2008 in various states of the USA (Norsworthy et al. 2016).

The herbicide substitution with variable mode of action, optimizing dose and timing of herbicide application, diversifying weed management options, adopting cultural ways of weed management like deep tillage, narrowing row space, shifting planting time, and crop rotation are some important tools of herbicide resistance management. The changing cropping sequence, for example, cultivating Roundup Ready Flex with LibertyLink cotton, can be helpful to avoid delay in development of resistance. The growers must be taught about such opportunities for herbicide resistance management. Besides these, some new technologies are in process which would permit the overtop application of some additional herbicides rather than to rely on glyphosate. These include Roundup Ready Xtend cotton by Monsanto and Enlist technology by Dow Agro Sciences. The former has flexibility for in-season application of glyphosate, glufosinate, and dicamba and later allows the overtop use of 2,4-D in addition to glyphosate and glufosinate (Norsworthy et al. 2016). Such technologies will aid up for glyphosate resistance management in weeds.

9.9 Integrated Weed Management in Cotton

The integrated weed management (IWM) emphasizes the combined application of crop management and science in such a way that also considers the causes rather to respond on present weed infestation (Buhler 2002). It integrates the preventive, cultural, mechanical, chemical, and biological weed control options that make the survival of weeds difficult. It should focus on weed propagation, survival, and transition from propagation to seedlings. No single weed control tool should be taken more beyond the component of IWM. These methods may be applied alone or in conjunction with each other. However, the timing of application of these operations determines the effectiveness of integrated plan. Despite the significant yield losses from weeds, the IWM lagged behind the integrated pest management. Because weed flora and weed biology are continuously changing due to many factors, it is very essential to consider the weed management not as a product but a continuous process. Knowledge of critical period of weed-crop competition is very important for designing effective IWM strategy. It uses all available weed control methods from sowing to maturity and also includes fallow season weed management. There are no single and specific approaches which ensure equal weed control for all field and locations. Although weed control with herbicides is more common, the problems associated with herbicides like environmental pollution, herbicide resistance, and weeds shift require integration of all available nonchemical weed control methods. The IWM for annual weeds aims to reduce the seed bank. However, some modification may be required for perennial weeds which propagate from both vegetative parts and seeds.

9.10 Conclusion

The weeds are continuous threat for cotton production despite the adaptation of modern weed management tool. In general, weeds may reduce the yield up to 90% in case of severe weed infestation. The extent of losses varies from moderate to very high depending on weed competitive ability which differs among weed species. The time and effectiveness of weed control can be improved through considering the concept of critical weed-crop competition. Significant amount is invested for weed control which represents about 2.5–14.8% of the total cost of production. The weed control with herbicide is common in cotton; however, it resulted in the evolution of herbicide-resistant biotypes. The resistance phenomenon was explored which led to develop glyphosate-resistant genotypes of crops. The genetic approaches are also being utilized in the form of glyphosate-tolerant varieties to address the issue of weed management in cotton. It resulted in flexibility of in-season nondirected application of nonselective herbicides. The evolution of glyphosate-resistant biotypes is the serious challenge for many cotton-growing countries which would require integrated approaches of weed management. The key aims of integrated weed management are to reduce the residual seed bank, weed survival, and the risk of weed seed production.

References

Abbas Q, Ahmad S (2018) Effect of different sowing times and cultivars on cotton fiber quality under stable cotton-wheat cropping system in southern Punjab, Pakistan. Pak J Life Soc Sci 16:77–84

Abdullahi AE (2004) Weed survey in cotton (*Gossypium hirsutum* L.) and sunflower (*Helianthus annuus* L.) fields in the Pandamatenga plains of northeastern Botswana. S Afr J Plant Soil 21:21–24

Ahanchede A (1996) Agronomic impacts of the dispersal of two morphological types of *Commelina benghalensis* L. in Benin. In: Proceedings of 10th international symposium on biological control of weeds, Dijon, France, pp 75–82

Ahmad S, Raza I (2014) Optimization of management practices to improve cotton fiber quality under irrigated arid environment. J Food Agri Environ 2(2):609–613

Ahmad S, Raza I, Ali H, Shahzad AN, Atiq-ur-Rehman, Sarwar N (2014) Response of cotton crop to exogenous application of glycinebetaine under sufficient and scarce water conditions. Braz J Bot 37(4):407–415

Ahmad S, Abbas Q, Abbas G, Fatima Z, Atique-ur-Rehman, Naz S, Younis H, Khan RJ, Nasim W, Habib urRehman M, Ahmad A, Rasul G, Khan MA, Hasanuzzaman M (2017) Quantification of climate warming and crop management impacts on cotton phenology. Plan Theory 6(7):1–16

Ahmad S, Iqbal M, Muhammad T, Mehmood A, Ahmad S, Hasanuzzaman M (2018) Cotton productivity enhanced through transplanting and early sowing. Acta Sci Biol Sci 40:e34610

Ali H, Afzal MN, Ahmad F, Ahmad S, Akhtar M, Atif R (2011) Effect of sowing dates, plant spacing and nitrogen application on growth and productivity on cotton crop. Int J Sci Eng Res 2(9):1–6

Ali H, Abid SA, Ahmad S, Sarwar N, Arooj M, Mahmood A, Shahzad AN (2013a) Integrated weed management in cotton cultivated in the alternate-furrow planting system. J Food Agri Environ 11(3&4):1664–1669

Ali H, Abid SA, Ahmad S, Sarwar N, Arooj M, Mahmood A, Shahzad AN (2013b) Impact of integrated weed management on flat-sown cotton (*Gossypium hirsutum* L.). J Anim Plant Sci 23 (4):1185–1192

Ali H, Hameed RA, Ahmad S, Shahzad AN, Sarwar N (2014a) Efficacy of different techniques of nitrogen application on American cotton under semi-arid conditions. J Food Agri Environ 12 (1):157–160

Ali H, Hussain GS, Hussain S, Shahzad AN, Ahmad S, Javeed HMR, Sarwar N (2014b) Early sowing reduces cotton leaf curl virus occurrence and improves cotton productivity. Cercetări Agronomice în Moldova XLVII(4):71–81

Amin A, Nasim W, Mubeen M, Nadeem M, Ali L, Hammad HM, Sultana SR, Jabran K, Habib urRehman M, Ahmad S, Awais M, Rasool A, Fahad S, Saud S, Shah AN, Ihsan Z, Ali S, Bajwa AA, Hakeem KR, Ameen A, Amanullah, Hafeez Ur Rehman, Alghabar F, Jatoi GH, Akram M, Khan A, Islam F, Ata-Ul-Karim ST, Rehmani MIA, Hussain S, Razaq M, Fathi A (2017) Optimizing the phosphorus use in cotton by using CSM-CROPGRO-cotton model for semi-arid climate of Vehari-Punjab, Pakistan. Environ Sci Pollut Res 24(6):5811–5823

Amin A, Nasim W, Mubeen M, Ahmad A, Nadeem M, Urich P, Fahad S, Ahmad S, Wajid A, Tabassum F, Hammad HM, Sultana SR, Anwar S, Baloch SK, Wahid A, Wilkerson CJ, Hoogenboom G (2018) Simulated CSM-CROPGRO-cotton yield under projected future climate by SimCLIM for southern Punjab, Pakistan. Agric Syst 167:213–222

Ara A, Akram A, Ajmal M, Akhund S, Nayyar BG (2015) Pharmacological, nutritional and allelopathic attributes of noxious weed, *Trianthema portulacastrum* L. (Horse purslane). Pure Appl Biol 4(3):340–352

Bailey WA, Wilcut JW, Askew SD (1999) Velvetleaf (*Abutilon theophrasti*) interference and seed-rain dynamics in cotton. Proc South Weed Sci Soc 52:163

Barnett KA, Steckel LE (2013) Giant ragweed (*Ambrosia trifida* L.) competition in cotton. Weed Sci 61(4):543–548

Bharathi K, Muthukrishnan N (2017) Survey and records of mealy bugs species on cotton and alternate host of key mealy bug *Phenacoccus solenopsis* Tinsley and its natural enemies. Int J Curr Microbiol App Sci 6(12):1047–1054

Bishnoi LK, Panwar RS, Malik RK, Rathi SS (1993) Effect of varieties and weed free maintenance period on weed competition in cotton. In: Integrated weed management for sustainable agriculture. Proceedings of an Indian Society of Weed Science, International symposium, Hisar, India, 18–20 November 1993, III, pp 182–183

Buchanan GA, Crowley RH, McLaughlin RD (1977) Competition of prickly sida with cotton. Weed Sci 25:106–110

Buhler DD (2002) Challenges and opportunities for integrated weed management. Weed Sci 50:273–280

Bukun B (2004) Critical periods for weed control in cotton in Turkey. Weed Res 44:404–412

Bükün B (2005) Weed flora changes in cotton growing areas during the last decade after irrigation of harran plain in Şanliurfa, Turkey. Pak J Bot 37(3):667–672

Chandler JM, Meredith WR (1983) Yields of three cotton (*Gossypium hirsutum*) cultivars as influenced by spurred anoda (*Anoda cristata*) competition. Weed Sci 31:303–307

Chandler JM, Oliver LR (1979) Spurredanoda: A potential weed in southern crops. US Dep Agric ARM-S-2, 19 pp

Charles G, Roberts G (2013) Managing weeds in cotton. WEEDpak, October 2013, pp 1–30

Charudattan R, Walker HL, Boyette CD, Ridings WH, TeBeest DO, Van Dyke CG, Worsham AD (1986) Evaluation of *Alternaria cassia* as mycoherbicide for sickle pod (*Cassia obtusifolia)* in regional field tests. Southern Cooperative Series Bulletin 317, Alabama Agricultural Experiment Station. Auburn University, Auburn, AL, pp 18

Culpepper AS (2006) Glyphosate-induced weed shifts. Weed Technol 20:277–282

da Silva MP, Mariana CP, Fernanda NB, Alves D-A, Luis P (2016) Periods of weed interference on transgenic cotton 'imacd 6001ll'. Revista Caatinga 29(2):375–383

Douti PY (1997) Cotton crops versus weeds: when is the competition period? Agriculture et Développement (SI), pp 11–16

Duke SO, Powles SB (2008) Glyphosate: a once-in-a-century herbicide. Pest Manag Sci 64 (4):319–325

Economou G, Bilalis D, Avgoulas C (2005) Weed flora distribution in Greek cotton fields and its possible influence by herbicides. Phytoparasitica 33(4):406–419

García-Ruiz E, Loureiro Í, Farinós GP, Gómez P, Gutiérrez E, Saínchez FJ, Escorial MC, Ortego F, Chueca MC, Castañera P (2018) Weeds and ground-dwelling predator's response to two different weed management systems in glyphosate-tolerant cotton: A farm-scale study. PLoS One 13(1):e0191408

Giri AN, Bhosle RH (1997) Economics of integrated weed management and nitrogen fertilization in cotton (*Gossypium hirsutum* L.) under rainfed conditions. Indian J Agron 42(4):705–708

Gwathmey CO, Steckel LE, Larson JA (2008) Solid and skip-row spacings for irrigated and non-irrigated upland cotton. Agron J 100:672–680

Hall MR, Swanton CJ, Anderson GW (1992) The critical period of weed control in grain corn (*Zea mays*). Weed Sci 40:441–447

Heap I (2015) The international survey of herbicide resistant weeds. Herbicide resistant by site of action [online]. http://www.weedscience.org/summary/SOASummary.aspx

ICAC (2016) Cost of production of raw cotton. International Cotton Advisory Committee, October 2016, Washington DC, USA, pp 25–30

Iqbal J, Cheema ZA (2008) Purple nutsedge (*Cyperus rotundus* L.) management in cotton with combined application of sorgaab and s-metolachlor. Pak J Bot 40(6):2383–2391

Jadhav JK, Degaonkar AM, Narkhede MN (1995) Integrated weed management in cotton. Haryana J Agron 11(1):62–65

Jasieniuk M, Brule-Babel AL, Morrison IN (1996) The evolution and genetics of herbicide resistance in weeds. Weed Sci 44:176–193

Keeley PE, Thullen RJ (1991) Growth and interaction of barnyard grass (Echinochloa crus-galli) with cotton (*Gossypium hirsutum*). Weed Sci 39:369–375

Khan NU, Khan SU (2003) Integrated weed management in upland cotton. Pak J Weed Sci Res 9 (3–4):185–192

Khan MB, Khaliq A, Ahmad S (2004) Performance of mashbean intercropped in cotton planted in different planting patterns. J Res (Sci) 15(2):191–197

Latif A, Rao AQ, Khan MAU, Shahid N, Bajwa KS, Ashraf MA, Abbas MA, Azam M, Shahid AA, Nasir IA, Husnain T (2015) Herbicide-resistant cotton (*Gossypium hirsutum*) plants: an alternative way of manual weed removal. BMC Res Notes 8:453

Lee LJ, Ngim J (2000) A first report of glyphosate-resistant goose grass [*Eleusine indica* (L) Gaertn] in Malaysia. Pest Manag Sci 56:336–339

Luqman M, Musa M, Iqbal MS, Arshad M, Rehman HS (2017) Horse persulane management in cotton by allelopathic crop water extracts along with reduced doses of herbicides. SAARC J Agri 15(1):23–31

Ma X, Wu H, Jiang W, Ma Y, Ma Y (2015) Interference between redroot Pigweed (*Amaranthus retroflexus* L.) and cotton (*Gossypium hirsutum* L.): Growth analysis. PLoS One 10(6): e0130475

McCloskey WB (2012) Innovations in cotton weed management. California Weed SciSoc, 64. East Cabrillo Boulevard Santa Barbara, California, 23–25 Jan 2012, pp 51–55

Munger PH, Abernathy JR, Jipson JR (1984) The influence of selected plant residues on cotton and sorghum establishment. Proc South Weed Sci Soc 37:320

Nalini K, Murhukrishnan P, Chinnusamzy C, Vennila C (2015) Weeds of cotton – A Review. Agri Review 36(2):140–146

Nieto HJ, Brondo MA, Gonzales JT (1968) Critical periods of the crop growth cycle for competition from weeds. PANS(C) 14:159–166

Norsworthy JK, Smith KL, Scott RC, Gbur EE (2007) Consultant perspectives on weed management needs in Arkansas cotton. Weed Technol 21:825–831

Norsworthy JK, Schwartz LM, Barber LT (2016) The incidence and ramification of glyphosate resistance in cotton. Outlook Pest Manage 27(1):31–35

Oad FC, Siddiqui MH, Buriro UA, Solangi GS (2007) Weed management practices in cotton. Asian J Plant Sci 6(2):344–348

Oerke E-C, Dehne H-W (2004) Safeguarding production-losses in major crops and the role of crop protection. Crop Prot 23:275–285

Orr CC, Abernathy JR, Hudspeth EB (1975) *Nothanguinaphyllobia*, a parasitic nematode of silverleaf nightshade (*Solanum elaeagnifolium*). Proc South Weed Sci Soc 28:111

Özaslan C, Akin S, Gürsoy S (2015) Weed control and crop production practices in cotton production in diyarbakir province of Turkey. YYU J Agr Sci 25(1):41–47

Pala F, Mennan H (2017) Weed control methods in cotton fields – example Diyarbakir. In: International conference on agriculture, forest, food, sciences and technologies 15–17 May 2017, Cappadocia, Turkey

Panwar RS, Balyan RS, Malik RS, Kumar S (1998) Studies on weed control in cotton. Haryana J Agron 14(2):260–262

Papamichail D, Eleftherohorinos I, Frouct-Williams R, Gravanis E (2002) Critical periods of weed competition in cotton in Greece. Phytoparasitica 30(1):105–111

Powles SB, Lorraine-Colwill DF, Dellow JJ, Preston C (1998) Evolved resistance to glyphosate in rigid rye grass (*Lolium rigidum*) in Australia. Weed Sci 46:604–607

Rahman MH, Ahmad A, Wang X, Wajid A, Nasim W, Hussain M, Ahmad B, Ahmad I, Ali Z, Ishaque W, Awais M, Shelia V, Ahmad S, Fahad S, Alam M, Ullah H, Hoogenboom G (2018) Multi-model projections of future climate and climate change impacts uncertainty assessment for cotton production in Pakistan. Agric For Meteorol 253-254:94–113

Reddy KN, Nandula VK (2012) Herbicide resistant crops: history, development and current technologies. Indian J Agron 57(1):1–7

Rezakhanlou A, Mirshekari B, Zand E, Farahvash F, Baghestani MA (2013) Evaluation of competitiveness of cotton varieties to cocklebur (*Xanthium strumarium* L.). J Food Agric Environ 11:308–311

Riffle MS, Waller GR, Murray DS (1987) Plant growth inhibitors extracted and steam distilled from uni-corn plant (*Proboscidea louisianica*). Proc Okla Acad Sci 71:35–42

Rowland MW, Murray DS, Verhalen LM (1999) Full-season Palmer amaranth (*Amaranthus palmeri*) interference with cotton (*Gossypium hirsutum*). Weed Sci 47:305–309

Scott GH, Askew SD, Wilcut JW, Brownie C (2000) *Datura stramonium* interference and seed rain in *Gossypium hirsutum*. Weed Sci 48:613–617

Shabbir A (2014) Ragweed emerging a major host for the cotton mealy bug in Pakistan. Indian J Weed Sci 46(3):291–293

Silva PSL, Silva JCV, Carvalho LP, Silva KMB, Freitas FCL (2009) Weed control via intercropping with gliricidia. I. cotton crop. Planta Daninha 27:97–104

Singh M, Pathania M (2017) Yield losses due to weeds in different cotton genotypes. Biennial ConfIndSoc Weed Sci "Doubling farmers' income by 2022: The role of weed science", MPUA&T, Udaipur, India during 1–3 Mar 2017

Singh K, Singh HP, Singh K (2016) Weed management in cotton with pre- and post-emergence herbicides. Indian J Weed Sci 48(3):348–350

Stanton RA, Pratley JE, Hudson D, Dill GM (2008) A risk calculator for Glyphosate resistant in Lolium rigidum (Gaud.). Pest Manag Sci 64:402–408

Tariq M, Yasmeen A, Ahmad S, Hussain N, Afzal MN, Hasanuzzaman M (2017) Shedding of fruiting structures in cotton: factors, compensation and prevention. Trop Subtrop Agroecosyst 20(2):251–262

Tariq M, Afzal MN, Muhammad D, Ahmad S, Shahzad AN, Kiran A, Wakeel A (2018a) Relationship of tissue potassium content with yield and fiber quality components of Bt cotton as influenced by potassium application methods. Field Crops Res 229:37–43

Tariq M, Afzal MN, Ahmad M, Qayyum A, Khan MA (2018b) Performance of pre and post-emergence herbicides for weed control in cotton (*Gossypium hirsutum* L.). Pak J Weed Sci Res 24(2):147–154

Tauseef M, Ihsan F, Nazir W, Farooq J (2012) Weed flora and importance value index (ivi) of the weeds in cotton crop fields in the region of Khanewal, Pakistan. Pak J Weed Sci Res 18 (3):319–330

Tingle CH, Steele GL, Chandler JM (2003) Competition and control of smellmelon (*Cucumismelo* var. dudaimNaud.) in cotton. Weed Sci 51:586–591

Usman M, Ahmad A, Ahmad S, Irshad M, Khaliq T, Wajid A, Hussain K, Nasim W, Chattha TM, Trethowan R, Hoogenboom G (2009) Development and application of crop water stress index for scheduling irrigation in cotton (*Gossypium hirsutum* L.) under semiarid environment. J Food Agri Environ 7(3&4):386–391

Veeraputhiran R, Srinivasan G (2015) Post-emergence herbicides effect on weeds, yield and economics of Bt cotton. Indian J Weed Sci 47(4):379–382

Ward SM, Webster TM, Steckel LE (2013) Palmer amaranth (*Amaranthus palmeri*): a review. Weed Technol 27:12–28

Webster TM, Burton MG, Culpepper AS, York AC, Prostko EP (2005) Tropical spiderwort (*Commelina benghalensis*): a tropical invader threatens agroecosystems of the southern United States. Weed Technol 19:501–508

Werth J, Thornby D, Walker S (2011) Assessing weeds at risk of evolving glyphosate resistance in Australian sub-tropical glyphosate-resistant cotton systems. Crop Pasture Sci 62:1002–1009

Werth J, Boucher L, Thornby D, Walker S, Charles G (2013) Changes in weed species since the introduction of glyphosate-resistant cotton. Crop Pasture Sci 64:791–798

Wood ML, Murray DS, Banks JC, Verhalen LM, Westerman RB, Anderson KB (2002) Johnson grass (*Sorghum halepense*) density effects on cotton (*Gossypium hirsutum*) harvest and economic value. Weed Technol 16:495–502

Xiao-yan MA, Han-wen WU, Wei-li J, Ya-jie MA, Yan MA (2015) Goosegrass (*Eleusine indica*) density effects on cotton (*Gossypium hirsutum*). J Integ Agric 14(9):1778–1785

Zhang ZP (2003) Development of chemical weed control and integrated weed management in China. Weed Biol Manag 3:197–203

Zhang XM, Yang NW, Wan FH, Lövei GL (2014) Density and seasonal dynamics of *Bemisia tabaci* (Gennadius) mediterranean on common crops and weeds around cotton fields in Northern China. J Integ Agric 13(10):2211–2220

Chapter 10
Pollination Behavior of Cotton Crop and Its Management

Wali Muhammad, Munir Ahmad, and Ijaz Ahmad

Abstract Cotton is the most important non-food crop in the world producing natural fiber, edible oil, and seed cake. Approximately 249 million bales of cotton were produced worldwide in 2017. The United States, China, and India are the major cotton-producing countries in the world. Brazil is emphasizing on organic farming of cotton utilizing the natural services of insect pollinators for pollination and natural enemies for cotton pest management. Cotton products are obtained from its fruit (bolls) for lint and seed which takes place after pollination. Although cotton is commonly known as a self-pollinating crop, it is often cross-pollinated with 5–50% cross-pollination by natural mechanism of out-cross pollination. Insects are important natural pollinators, foraging cotton crop due to the presence of floral and extrafloral nectaries. Three main pollinating insect species, namely, *Allograpta exotica* (Diptera), *Apis mellifera*, and *Melissodes tepaneca* (Hymenoptera), have been reported as effective pollinators in cotton agroecosystems with 12% increase in fiber weight and 17% increase in number of seeds in insect-pollinated cotton crops than conventional cotton. Climate change, pesticide use, and agronomic practices are the main reasons for the decline of pollinators in cotton-growing areas. Pollination efficacy and pollinator diversity can be increased by manipulating the habitat for conservation of pollinators, reducing pesticide use, and maintaining natural flora near cotton-growing areas for alternate pollen source to pollinators for better cotton yield and quality. Small farmers can earn extra income if they get engaged in land use management to improve pollination services.

Keywords Cotton · Pollination · Cotton pollinators · Cotton hybridization · Organic farming

W. Muhammad (✉) · I. Ahmad
Agriculture Pest Warning & Quality Control of Pesticides, Government of Punjab, Layyah, Pakistan

M. Ahmad
Department of Entomology, Pir Mehr Ali Shah, Arid Agriculture University Rawalpindi, Rawalpindi, Pakistan

S. Ahmad, M. Hasanuzzaman (eds.), *Cotton Production and Uses*,
https://doi.org/10.1007/978-981-15-1472-2_10

Abbreviations

A.m	*Apis mellifera*
BB	Bumble bees
BE	Bees excluded
BP	Bee pollination
C	Cross-pollination
CE	Cross-pollination essential
H	Hand pollination
IE	All insects excluded
INM	Integrated nutrient management
IPM	Integrated pest management
IWM	Integrated weed management
OP	Open pollination
S	Self-pollination
TI	Tiny insects permitted
USDA	United States Department of Agriculture

10.1 Introduction

Cotton is an important fiber crop known as the most used natural fiber in the world (Cusser et al. 2016). It is known as the major component of textile industry with total supply of approximately 248.63 million bales all over the world in 2017 (Johnson et al. 2018). The USDA annual report showed the United States, China, and India as major producers of cotton. The last 40 years of agricultural development shows that the yield and production per unit area is increasing but not in a sustainable manner. It has been observed that the genetic diversity is capable of playing an important role in crop yield and quality improvements which occurs through the Mendelian variation from the wild crop species. Genetic diversity is more in cross-pollinated crops (apples, grapes, plums, pears, raspberries, blackberries, strawberries, runner beans, pumpkins, daffodils, tulips, lavender) than self-pollinated crops (peas, wheat, barley, oats, rice, potatoes, apricots, and peaches). Scientists are dealing with new technologies for improving the production of crops, especially oilseed and fiber crops, but Mendelian variation is not so far concerned by most. Often, cross-pollinated crops like tobacco, maize, and cotton have a brighter scope for higher production yields (Dastagiri et al. 2013) than that of self-pollinated plots. Cotton flower anatomy supports cross-pollination in addition to the self-pollination mechanism as it has many stamens around a five-lobed pistil in the center which releases the pollens.

Agriculture and natural ecosystems have great concern for declining pollinators in the developed world like North America and Europe. The economic and ecosystem losses due to declining pollinators are detrimental for these regions but are considered less detrimental for developing areas like South Africa and Pakistan.

Fig. 10.1 A bee visiting a cotton flower dusted with pollen

Pollinators decline in developing countries due to reduction of crop yield, increase in food prices, reduction of food quality, decrease in micronutrient availability through natural resources, and serious threat to honey production industry (Gallai et al. 2009; Potts et al. 2010).

Brazil is focusing on organic cotton production and emphasizing the cotton pollination through insect pollinators for increase in yield, lint quality, and oil contents (Araujo et al. 2012). Cotton crop has shown positive output in organic production system especially in the presence of more pollinators as compared with no or less pollinators in conventional cotton production programs. There was 14% and 17% weight increase of fiber and number of seeds, respectively, in cotton fields with natural cross-pollination as compared with fields that were lacking bees (Dutra et al. 2012). Farmers must favor the pollinators by managing their conservation and increasing bee flora and natural vegetation. Dutra et al. (2012) revealed that by increasing the nearby vegetation of cotton field, bee visitation on cotton flowers may increase from 0.45 bees/h which is inadequate to increase production (Fig. 10.1). A maximum of 27% production increase has been observed near vegetation field increasing the bee diversity and frequency. Four bee species were the major visiting pollinators of cotton crop with greater output in diverse pollinator visitation than a single bee species (Araujo et al. 2012). Small farm holders can adopt organic farming of cotton crop and increase the pollinators by reducing the use of pesticides and chemicals. However, only a few varieties were studied for efficacy of pollinators, while a lot of work is desired to attain sustainable high yields. Better economic achievements are possible under pollinator-friendly cotton production technology.

10.2 Mode of Cotton Plant Pollination

Cotton belongs to angiosperms, and the transfer or placement of pollens from anther (male flower part) to stigma (female flower part) of the same flower or the other flower is called pollination. After this transfer, pollen germinates through the pollen tube and fertilizes the egg. Artificial breeding programs are common for artificial selection to produce new high yielding cultivars. Cotton is commonly known as a self-pollinated crop, but pollinators have also been observed on cotton flowers if it is near cross-pollinated crop boundary and included in often cross-pollinated crop (McGregor 1976). The details are given below.

10.2.1 Self-Pollination

The transfer of pollen from anther to stigma of the same flower is called self-pollination. This process is a basic feature of bisexual flowers, e.g., cotton which is considered as a self-pollinated crop by many ecologists (Free 1970; Rhodes 2002). However, cross-pollination is also possible in such self-pollinated crops. In cotton, self-pollination is responsible for less genetic diversity due to which artificial breeding programs are adopted. In the past, self-pollination for cotton breeding was adopted just due to the myth that natural crossing is responsible for hybridization of seed stock.

10.2.2 Cross-Pollination

Cross-pollination is the transfer of pollen from the flower of one plant to the stigma of the flower of other plant. There exist different modes of cross-pollination like wind, water, mammals, birds, arthropods, etc. Cotton crop is pollinated by insect pollinators mainly wasp, flies, butterflies, moths, and bees which provides them nectar and pollen as a reward. Self-pollination also occurs in cross-pollinated crops from 5% to 10%. Mostly cross-pollinated crops include maize, millets, sunflower, sesame, alfalfa, and castor. In earlier days, cotton was not considered to be cross-pollinated.

10.2.3 Often Cross-Pollination

Often cross-pollinated crops include millets, safflower, cotton, pigeon pea, etc. In these crops, self-pollination is the basic phenomenon, but cross-pollination also occurs from 5% to 50%, so they are named as often cross-pollinated crops. These crops are intermediate between cross- and self-pollinated crops due to their genetic architecture. Cross-pollination is adopted by plants to maintain genetic diversity.

Insects are considered as potential cotton crop pollinators among the natural crossing agents. Honeybee, *Apis mellifera*, was identified to be the most effective pollinator of cotton crop, and bumblebees (*Bombus* species) were identified as frequently visiting pollinators with pollen loads (Meade 1918). Birds are also sometimes attracted to flowers and cause pollen transfer when collecting the nectar as their food requirement (Dellinger et al. 2014; Armbruster 2011; Proctor et al. 1996). Food reward is not always on active sexual parts of the flower as it can occur on non-sexual parts of plants like in cotton plant where nectaries are present on the back side of the leaves (Simpson and Neff 1981).

10.3 Biodiversity of Cotton Crop Pollinators

Cotton belongs to Malvaceae family which has almost thirty-two *Gossypium* species like Deccan hemp, okra, cotton, etc. These species have large size flowers and nectaries other than flowers to attract pollinators. Pollinator biodiversity is important for cotton crop in an area as most visiting flower insects are not good pollinators, e.g., European honeybee (Vaissiere et al. 1984). Many pollinators have been reported from different communities, but all are not effective cotton pollinators. The area, climate, surrounding environment, pollen, and nectar flow of that area may alter the situation. Mostly cotton pollinators are reported from the class Insecta. Hymenoptera and Diptera are the orders having most of the cotton pollinators. Three main species *Allograpta exotica* (Diptera), *Apis mellifera*, and *Melissodes tepaneca* (Hymenoptera) were reported as effective pollinators in cotton agroecosystems. They have a great variation in their life history which makes them adjustable to agriculture land use change.

FAO (2018) issued a list of cotton crop pollinators effective in cotton seed production programs. Some of the main pollinators visiting cotton flower are *Megachile monticola*, *Apis dorsata*, *Xylocopa* species, *Anthophora confusa*, *A. quadrifasciata*, and scoliid wasps. *Apis florea* is also reported to visit the cotton flowers but seems to be a less effective pollinator due to its small size. *Halictus* species, *Andrena ilerda*, *Nomia* species, *Coelioxys* species, *Nomada* species, *Xylocopa* species, and *Pithitis smaragdula* have also been reported in Pakistan and India as cotton pollinators. *Scolia* and *Xylocopa pubescens* have been reported in Egypt pollinating the cotton flowers. In North America, the bodies of *Melissodes* (solitary bees) and *Elis plumipes* (scoliid wasps) were found dusted with cotton pollens when constantly visiting cotton flowers. Carpenter bees and bumblebees are the most effective cotton pollinators in the United States because of their floral nectary visiting behavior and touching stigma and stamens while foraging. Foraging frequency of bumblebee was 1.2 flowers/plant, and there were 2750 maximum flowers potentially visited by the bumblebees.

Pollinators may have different habitats and can adjust according to their life history in different agroecosystems. *Melissodes tepaneca*, a solitary bee, lives in underground soil nest. In contrast, *A.m* is living in complex social nests above the ground (Michener 2007). Both of these bees are distance specific in their foraging and are the best central flower foragers. Some foraging insects are not central flower visitors like a native hoverfly (*A. exotica*), while some may forage through landscapes to fulfill their nectar and pollen requirements like *Allograpta* adults, which help in controlling aphid on crops through their developing larvae (Bugg 2008). Many factors can be in force for affection of pollinator's abundance and diversity. Landscape nature and complexity have a strong relationship with the diversity of foraging insects (Gaines-Day and Gratton 2016; Steffan-Dewenter et al. 2002; Petersen et al. 2013). The reason may be the different food (pollen and nectaries), nesting space, environment, specific interaction with flowering host needs, etc., for different foraging insects (Brown and Paxton 2009).

Diversity of pollinator in cotton crop resulted in a total of 841 foraging insects belonging to 52 species of four insect orders, Hymenoptera, Diptera, Lepidoptera, and Coleoptera with 37, 5, 7, and 3 species, respectively (Roubik 1995). The maximum number of specimens was of solitary bee (*M. tepaneca*) sharing 21% of the total specimens and the European honey bee (*A. mellifera*) having 19% share followed by the syrphid fly (*Allograpta exotica*) with 16% share. The remaining 15% of the pollinator specimens were from different nine species of insect order Hymenoptera. Birds have also been reported on cotton crop visiting for nectar. Passerines and hummingbirds are nectar lovers and mostly visit the flowers for sugar preference (Cronk and Ojeda 2008).

10.4 Importance and Scope of Insect Pollination in Cotton Cultivation

Cotton breeding and cotton seed production are greatly affected by natural cross-pollination. Earlier studies showed that insects are the main source of cross-pollination in cotton playing their role for enhancement of cotton yield as shown in Table 10.1. Some of the important factors have been discussed below.

10.4.1 Seed Production Technology

Seeds are produced when ovaries from female plants and pollens from male plants meet. There is a natural mechanism of self-pollination in cotton, but for cross-pollination, insects play their role in cotton crop. Insect-pollinated cotton has shown 17% increase of seeds/boll, while comparing different pollination means including BE, BP, C, CE, H, IE, TI, OP, and S (Pires et al. 2014). In different

Table 10.1 Pollinating species and their benefits to cotton crop under different ecological zones

Ecological zone/ country	Pollinating species	Insect order	Economic benefit	Ecological benefit	Reference/ source
India	Honeybees (*Apis mellifera*)	Hymenoptera	Increase the boll setting per plant	Maintenance of bee fauna	Santhy et al. (2008)
South Texas	*Apis mellifera* *Allograpta exotica*	Hymenoptera Diptera	Increased seed-cotton weight by 18%	Conservation of pollinators	Cusser et al. (2016)
Asia	*Apis dorsata*, *Xylocopa*, *Anthophora confusa*	Hymenoptera	27–30% extra boll retention (fruit set)	Conservation of pollinators	FAO (2018)
Brazil	Honeybees (*Apis mellifera*)	Hymenoptera	Increase cotton quality	0.25% dependence on pollination	Giannini et al. (2015)
West Africa	Honeybee (*Apis mellifera*), twenty-six wild bee species and four species of wasps	Hymenoptera	33–43% more cotton weight	Enhanced germination rate	Stein et al. (2017)
China	*Bombus ignites* and *Apis mellifera*	Hymenoptera	Increased cotton production	Risk of gene flow	Yan et al. (2015)

studies, it is proved that abundance and richness of bees near organic farms (with nearby natural vegetation and without pesticides) play an important role in seed production as compared with conventional farms (less vegetative diversity and use of pesticides). Pollinators are helpful to increase the number of seeds and fiber weight.

10.4.2 Yield Improvement

Cotton is capable of producing good yield in conventional agriculture depending on self-pollination, but the yield can be improved by adopting pollinator management practices by cross-pollination. As a result of insect pollination, large boll size was produced resulting in high cotton yield and creating more profits for the farmers (Cusser et al. 2016). Cotton yield increased up to 12% by maintaining pollinator communities on cotton farms (Pires et al. 2014).

10.4.3 Oil and Lint Quality

Insect pollination increased the quality and yield by improving oil contents and lint quality in fiber crops. If there is loss in number of seed per boll due to lack of pollination, reduction in cake production and oil occurred (Silva 2007). Cotton pollinated by foraging insects showed at least 10% fiber production increase along with betterment in oil contents (Klein et al. 2007).

10.4.4 Survival of Beehives

The purpose of honeybee visitation in cotton field is not always to maximize the cotton crop production but the important pollen source for managed *A.m* colonies near cotton fields. One hectare of cotton can support two to three hives for honey production along with increase of cotton crop yield (Cusser et al. 2016). In summer, exact season of cotton flowering in Indo-Pak region can support honey bee farmers. There is natural attraction for honey bees to visit cotton flowers from at least 8.5 miles away to get their pollen (Minkov 1956).

10.5 Challenges of Cotton Crop-Pollination

Different pollinators have quite different life history with variable efficacy in different regions depending on land management cultural practices and chemical used for pest control and so on. Some devastating factors affecting the pollinator's diversity in cotton are discussed below.

10.5.1 Climate Change

Plant-pollinator mismatch is resulting due to the phenological changes in the environment. Changing climate is affecting a lot of plant and pollinator species potentially damaging their interactive benefits. Timing of flowering plants hosting bumblebees has shifted earlier, but their activities are not. In this way, overall pollen chain and flowering plant seasoning is disturbing these pollinators. They are dying due to starvation as there are no pollens, and crop yields have decreased due to a decline in pollinator's diversity established for long period of evolutionary process (Hughes 2000; Parmesan and Yohe 2003).

This important phenomenon of climate change has affected the foraging activity of general pollinators by altering their body physiology. Different sized pollinating insects have shown different responses to increased stress. Pollinators bigger in size can better perform through their thermoregulatory system as compared to smaller ones (Bishop and Armbruster 1999). Mostly pollinators decrease their activity in summer with increasing temperature beyond the limits, e.g., *A.m* stops its activity when the temperature is above 40 °C (Cooper et al. 1985). These studies showed that cotton, a summer crop, will greatly face decline in pollinator's activity.

10.5.2 Farm Management and Pesticide Use

Ground nesting bees (*Melissodes*) can be destroyed through deep plowing which is common practice in cotton cultivation areas. *Apis* and *Allograpta* species, however, are not disturbed through such plowing practices (Shuler et al. 2005). Application of poisonous pesticides for pests and disease management in crops is limiting the activity of pollinators in cotton crop. Pesticides on cotton crop may affect larval development of *Allograpta* flies in addition to killing the honey bees and flying insects (Moens et al. 2011). Intensity of cultural practices used at farms negatively affects their diversity (Adger et al. 2002; Kehoe et al. 2015; Roulston and Goodell 2011). Intensively managed cotton farms are deficient in beneficial insects (pollinators, predators, and parasitoids) resulting in increased pest management cost in addition to decreased pollination activity. Ecosystem services reduce, and input cost of production is increased. Yield and its quality also reduced due to reduction of these natural farmer friends (Kevan et al. 1990).

10.6 Future Scope of Cotton Crop Pollination Management

Natural and human-managed terrestrial ecosystems are basically based on pollination which is critical to fulfill the human requirements of food and clothing. The process of pollination is connecting wild ecosystem with human-managed agroecosystems. Pollinators and pollinated crops have a strong symbiotic relationship with each other. As the coming era is focusing on organic farming which should be chemical-free for mankind, there is intense need to overcome low yield problems of our cotton crop utilizing natural services. Some important steps are discussed below which can extend the relationship between cotton and pollinators in favor of mankind.

10.6.1 Conservation of Pollinators

Pollinator communities have greatly been influenced by the human activities in cotton-producing regions. Provision of diversified natural habitat on local and regional levels will increase their diversity and the pollination activity (Garibaldi et al. 2013). It is already discussed that the increasing distance between natural habitat and cotton fields will reduce the foraging activity of pollinators. Natural habitats will play their role to conserve them during harsh environmental condition, and plant-pollinator relationship will be stronger, and richness of species will be increased in that area (Sih and Baltus 1987).

10.6.2 Target-Oriented Pesticides

Techniques for pest management in cotton crop have changed over time. In earlier days, the use of broad-spectrum pesticides was common which has been replaced with selective ones with comparative less toxicity to beneficial insects including pollinators. However, it is still dangerous to move a beehive near cotton crop because of a large number of pesticide applications in a short period of time (Goodwin 2012). The following steps can save the pollinators while increasing the cotton yield:

- Delay of the first foliar pesticide application as much possible to benefit from pollinators and other beneficial insects.
- Minimum pesticide uses on cotton crop in flowering season.
- Awareness program for pollinator safety and nonchemical pest management techniques through seminars and electronic, print, and social media.
- Use of selective safer pesticides with plant- and animal-based rather than broad-spectrum chemicals.
- Molecular insecticides safe for non-target animals.
- In case of spray, proper and early warning to beekeepers in nearby areas.

10.6.3 Ecological Intensification

Agricultural approaches should be modified to promote ecological pest management and the pollinators. These approaches include integrated pest management (IPM), integrated nutrient management (INM), integrated weed management (IWM), precision farming, and plant breeding that are good for the humanity and least disturbing to the surrounding environment. Research should focus on improving the crop and pollinator interaction while minimizing the loss caused by other pests. These are some points which may be helpful for ecologically promoting the pollinator in cotton crop.

- Maximizing the natural resources utilization like wild flowers and native bees to improve crop yield and generate income.
- Maintenance of natural habitat like oak wood lands and shrub lands around the cotton farms to boost cotton crop by strengthening pollinator communities and increasing diversity.

- Plantation of wild flower rows between the cotton crops to provide food and shelter to the pollinators. Farmers can also utilize economic crops and vegetables like okra, sunflower, and melons to increase diversity.
- Reduction in tillage practices especially nearby area of cotton crop can enhance the soil-nesting pollinators, and ultimately better pollination service will be provided.
- Farming of wild flowers and flowering plants other than cotton fields by all citizens will increase natural flora and pollinator's diversity.

These small ecological developments can change our environment and increase our cotton production with minimum use of chemicals. The coming generations may compete for their pesticide-free food and clothing by our adoption to these small steps.

10.7 Conclusion

Cotton is often a cross-pollinated crop with potential to increase fiber yield, number of seeds, oil contents, and lint quality in the presence of wild bees and foraging insects. Insects belonging to order Hymenoptera are potential cotton pollinators, e.g., *Apis mellifera* and *Xylocopa* species that visit the cotton flowers are found abundant. The out-cross cotton crop has improved its yield factor as compared with the self-pollinated crop. The decline of pollinators is observed due to injudicious pesticide use and climate change including many other factors like loss of habitat and deep plowing of land. Farmers can achieve 18% extra seed production by adopting some small steps. Pollinator diversity can be enhanced by managing land use practices, reduced use of pesticides, and intercropping of wild flowers or other pollen sources, e.g., okra, sunflower, and sesame in our cotton crop. Cotton farmers should be trained regarding the benefits of pollinators and their importance for mankind. Convincing conventional farmers to utilize natural pollination services of nature to increase cotton yield under pollinator-friendly conditions can work when they will observe their benefits.

References

Adger WN, Kelly PM, Winkels A, Huy LQ, Locke C (2002) Migration, remittances, livelihood trajectories, and social resilience. AMBIO 31:358–366

Araujo ER, Albuquerque FA, Santiago FS, Jalfim FT, Arriel NH Blackburn RM (2012) Agro ecological intercropping with cotton and food: an alternative to food production and income generation for rural families in the semi-arid region of northeast Brazil. http://www.projetodomhelder.gov.br/site/images/PDHC/Artigos_e_Publicacoes/Algodao_em_consorcio_de_base_agroecologica/SociologiaLisboa_Artigo_ConsorcioAgroecologico_English.pdf. Accessed 30 Jan 2019

Armbruster WS (2011) Evolution and implications of "specialized" pollinator-rewards. In: Patiny S (ed) Evolution of plant-pollinator relationships. Cambridge University Press, Cambridge, pp 44–67

Bishop JA, Armbruster WS (1999) Thermoregulatory abilities of Alaskan bees: effects of size, phylogeny and ecology. Funct Ecol 13:711–724

Brown MJ, Paxton RJ (2009) The conservation of bees: a global perspective. Apidologie 40 (3):410–416

Bugg RL (2008) Flower flies syrphidae and other biological control agents for aphids in vegetable crops. University of California, Division of Agriculture and Natural Resources. ANR Publication 8285: p. 1–25. http://anrcatalgo.ucdavis.edu/

Cooper PD, Schaffer WM, Buchmann SL (1985) Temperature regulation of honey bees Apis mellifera foraging in the Sonoran Desert. J Exp Biol 114:1–15

Cronk Q, Ojeda I (2008) Bird-pollinated flowers in an evolutionary and molecular context. J Exp Bot 59:715–727

Cusser S, Neff JL, Jha S (2016) Natural land cover drives pollinator abundance and richness, leading to reductions in pollen limitation in cotton agroecosystems. Agr Ecosyst Environ 226:33–42

Dastagiri MB, Immanuelraj TK, Patil GI (2013) India's self, cross & often cross-pollinated crops in the global crop production: Mendelian variation perspectives. Sci Disc 3:16–27. https://doi.org/10.11648/j.sd.20130103.11

Dellinger AS, Penneys DS, Staedler YM, Fragner L, Weckworth W, Schönenberger J (2014) A specialized bird pollination system with a bellows mechanism for pollen transfer and staminal food body rewards. Curr Biol 24:1615–1619

Dutra CC, Meoti C, Fernandes MG Raizer J (2012) Riqueza e composicao de especies de insetosvisitantesfloraisde algodoeiroBt e nao-Bt. Arq Inst Biol São Paulo 79:353–361

FAO (2018) The pollination of cultivated plants: A compendium for practitioners, vol 1. Food and Agriculture Organization of the United Nations, Italy, Rome, pp 1–62. http://www.fao.org/pollination/resources/en/

Free JB (1970) Insect pollination of crops. Academic Press, London

Gaines-Day HR, Gratton C (2016) Crop yield is correlated with honey bee hive density but not in high-woodland landscapes. Agric Ecosyst Environ 218:53–57

Gallai N, Salles JM, Settele J, Vaissiere BE (2009) Economic valuation of the vulnerability of world agriculture confronted with pollinator decline. Ecol Econ 68:810–821

Garibaldi LA, Steffan-Dewenter I, Winfree R, Aizen MA, Bommarco R, Cunningham SA, Kremen C, Carvalheiro LG, Harder LD, Afik O, Bartomeus I, Benjamin F, Boreux V, Cariveau D, Chacoff NP, Dudenhöffer JH, Freitas BM, Ghazoul J, Greenleaf S, Hipólito J, Holzschuh A, Howlett B, Isaacs R, Javorek SK, Kennedy CM, Krewenka KM, Krishnan S, Mandelik Y, Mayfield MM, Motzke I, Munyuli T, Nault BA, Otieno M, Petersen J, Pisanty G, Potts SG, Rader R, Ricketts TH, Rundlöf M, Seymour CL, Schüepp C, Szentgyörgyi H, Taki H, Tscharntke T, Vergara CH, Viana BF, Wanger TC, Westphal C, Williams N, Klein AM (2013) Wild pollinators enhance fruit set of crops regardless of honey bee abundance. Science 339:1608–1611

Giannini TC, Boff S, Cordeiro GD, Cartolano EA, Veiga AK, Imperatriz-Fonseca VL, Saraiva AM (2015) Crop pollinators in Brazil, a review of reported interactions. Apidologie 46:209–223

Goodwin M (2012) Pollination of crops in Australia and New Zealand. RIRDC. Publication No. 12/059 Project No. HG09058

Hughes L (2000) Biological consequences of global warming: is the signal already apparent? Trends Ecol Evol 15:56–61

Johnson J, MacDonald S, Meyer L, Stone L (2018) Cotton Outlook. USDA - United States Department of Agriculture. https://www.usda.gov/oce/forum/2018/commodities/Cotton.pdf

Kehoe L, Kuemmerle T, Meyer C, Levers C, Václavík T, Kreft H (2015) Global patterns of agricultural land-use intensity and vertebrate diversity. Divers Distrib 21:1308–1318

Kevan PG, Clark EA, Thomas VG (1990) Insect pollinators and sustainable agriculture. Am J Altern Agric 5:13–22

Klein AM, Vaissiere BE, Cane JH, Steffan-Dewenter I, Cunningham SA, Kremen C, Tscharntke T (2007) Importance of pollinators in changing landscapes for world crops. Proc R Soc B 274:303–313

McGregor SE (1976) Insect pollination of cultivated crop plants. US Department of Agriculture, Agriculture Handbook, pp 496, US Gov. Print Office, Washington DC

Meade RM (1918) Bee keeping may increase the cotton crop. J Hered 9:282–285

Michener CD (2007) The bees of the world. John Hopkins University Press, Baltimore

Minkov SG (1956) Nectar productivity of cotton and the role of bees in cross pollination. Trudy Kazakh Opyt Sta Pchelov 1:109–150

Moens J, De Clercq P, Tirry L (2011) Side effects of pesticides on the larvae of the hoverfly Episyrphusbalteatus in the laboratory. Phytoparasitica 39:1–9

Parmesan C, Yohe G (2003) A globally coherent fingerprint of climate change impacts across natural systems. Nature 421:37–42

Petersen JD, Reiners S, Nault BA (2013) Pollination services provided by bees in pumpkin fields supplemented with either Apis mellifera or Bombus impatiens or not supplemented. PLoS One 8 (7):e69819

Pires VC, Silveira FA, Sujii ER, Torezani KRS, Rodrigues WA, Albuquerque FA, Rodrigues SMM, Salomão AN, Pires CSS (2014) Importance of bee pollination for cotton production in conventional and organic farms in Brazil. J Pollination Ecol 13(16):151–160

Potts SG, Biesmeijer JC, Kremen C, Neumann P, Schweiger O, Kunin WE (2010) Global pollinator declines, trends, impacts and drivers. Trends Ecol Evol 25:345–353

Proctor MCF, Yeo P, Lack A (1996) The natural history of pollination. Timber Press, Portland

Rhodes J (2002) Cotton pollination by honeybees. Anim Prod Sci 42:513–518

Roubik DW (1995) Pollination of cultivated plants in the Tropics. FAO Agricultural Services Bulletin, Bulletin No. 118, Rome, FAO

Roulston TH, Goodell K (2011) The role of resources and risks in regulating wild bee populations. Annu Rev Entomol 56:293–312

Santhy V, Khadi BM, Singh P, Kumari PRV, Deshmukh RK, Vishwanathan A (2008) Hybrid seed production in cotton. CICR Technical Bulletin No. 35, Central Institute for Cotton Research Nagpur, pp 25. www.cicr.org.in

Shuler RE, Roulston TH, Farris GE (2005) Farming practices influence wild pollinator populations on squash and pumpkin. J Econ Entomol 98:790–795

Sih A, Baltus MS (1987) Patch size, pollinator behavior, and pollinator limitation in catnip. Ecology 68(6):1679–1690

Silva EMS (2007) Abelhasvisitantesflorais do algodoeiro (*Gossypium hirsutum*) emQuixeramobim e Quixeré, Estado do Ceará, e seusefeitosnaqualidade da fibra e da semente. Tese de doutorado. http://www.zootecnia.ufc.br/wa_files/tese2007_eva_20monica_20sarmento_20da_20silva.pdf

Simpson BB, Neff JL (1981) Floral rewards: alternatives to pollen and nectar. Ann Mo Bot Gard 68:301–322

Steffan-Dewenter I, Münzenberg U, Bürger C, Thies C, Tscharntke T (2002) Scale-dependent effects of landscape context on three pollinator guilds. Ecology 83:1421–1432

Stein K, Coulibaly D, Stenchly K, Goetze D, Porembski S, Lindner A, Konate S, Linsenmair EK (2017) Bee pollination increases yield quantity and quality of cash crops in Burkina Faso, West Africa. Sci Rep 7(1):17691

Vaissiere BE, Moffett JO, Loper GM (1984) Honey bees as pollinators for hybrid cotton seed production on the Texas high plains. Argon J 76:1005–1010

Yan S, Zhu J, Zhu W, Li Z, Shelton AM, Luo J, Cui L, Zhang Q, Liu X (2015) Pollen-mediated gene flow from transgenic cotton under greenhouse conditions is dependent on different pollinators. Sci Rep 5:15917

Chapter 11
Insect Pests of Cotton and Their Management

Muhammad Anees and Sarfraz Ali Shad

Abstract Cotton is a cash and industrial crop in many parts of the world. Nowadays, production of cotton is very low, and issues of yield stagnation exist in many developing countries. Major causes of lower yields are climate change soil salinity, lack of adaptation of agronomic practices, pathological, and entomological problems in nature. Insect pests of cotton crop, are categorized in sucking and chewing in nature. In order to harvest high yield, it is recommended to control all types of insect pests of cotton through integrated pest management techniques.

Keywords American bollworm · Armyworm · Spotted bollworm · Pink bollworm · Jassid · Whitefly · Cotton mealybug

11.1 Insect Pests of Cotton

Insect pests of cotton are categorized into two categories.

1. Chewing insect pests.
2. Sucking insect pests.

11.1.1 Chewing Insect Pests

11.1.2 American Bollworm/Fruit Borer

1. **Scientific Name:** *Helicoverpa armigera* Hubner.

M. Anees (✉) · S. A. Shad (✉)
Department of Entomology, Faculty of Agricultural Sciences and Technology, Bahauddin Zakariya University, Multan, Punjab, Pakistan
e-mail: sarfrazshad@bzu.edu.pk

S. Ahmad, M. Hasanuzzaman (eds.), *Cotton Production and Uses*,
https://doi.org/10.1007/978-981-15-1472-2_11

Fig. 11.1 Egg of American bollworm

2. **Common Names:** American bollworm, African bollworm, cotton bollworm, tomato bollworm, corn earworm, tobacco budworm, common bollworm, bean pod borer, etc. (Broadley 1977).
3. **Family:** Noctuidae **Order:** Lepidoptera.
4. **Identification.**

   (a) **Eggs:** Spherical in shape and creamy white in color. Females lay eggs singly (Fig. 11.1).
   (b) **Larvae:** Variety of colors from greenish to brown. The body of larvae has dark brown-gray line with precise white lines. The body has dark and pale bands. On first abdominal segment of dorsal surface, saddle markings are present and legs have dark pigmentations (Stanley 1978). White color cervical shield hair is present in sixth instar larvae on the anterior dorsal side (Zalucki et al. 1986) (Fig. 11.2).
   (c) **Pupae:** Pupate in crop debris, soil, leaf, and pods. Mostly brown in color. The distance between the outer edges of the cremaster spines at the junction is >0.22 mm (Cantrell 1980; Kirkpatrick 1961) (Fig. 11.3).
   (d) **Adult:** Brownish yellow light pale with stout mouth. Forewings are olive green to pale brown in color with dark brown circular spot in the center. Hind wings are pale smoky white with blackish outer margin. Pale patch between M3 and Cu2 present in hind wings. Genital plate is cup shaped (Common 1953) (Fig. 11.4).

5. **Biology.**

   (a) **Adult:** Mostly male moths prefer mating during night time. Female moth starts egg laying from second to seventh day of their life span (Karim 2000) (Fig. 11.4).
   (b) **Eggs:** A female moth lays almost 150–1500 eggs during its life span, but average egg laying is 450 eggs (Karim 2000) (Fig. 11.1).

Fig. 11.2 Larvae of American bollworm

Fig. 11.3 Pupa of American bollworm

(c) **Larvae:** The life cycle of larvae consists of six instars. The size of first instar larvae is1.75 mm, second instar is 3.5–4 mm, third instar is 9–10 mm, and 35–42 mm. Larvae of fourth, fifth, and sixth instars change their color according to food and weather conditions, and larval stage is completed in 15–30 days (Karim 2000) (Fig. 11.2).

(d) **Pupae:** Pupae take 5 to 8 days for their emergence from pupae to adult in summer, but in case of any type of diapause, it takes several months for their emergence (Karim 2000) (Fig. 11.3).

Fig. 11.4 Adult of American bollworm

6. **Number of Generations:** 4–8 per year.
7. **Hosts**

   American bollworm is a polyphagous pest which attacks more than 200 plant species. Cotton, chickpea, pigeon pea, sunflower, sorghum, tobacco, peanut, maize, many fruits, vegetables, and tree plants are important hosts of American bollworm in Asia (Fitt 1989).
8. **Damage**

   Mostly American bollworm larvae feed on reproductive parts (flowers and fruits) of the host plants. Larvae also attacks on foliage. Larvae bore the reproductive parts and feed within the plants. Pathogens may attack the plant due to wounding of plant parts. In cotton, bore holes can be seen at the base of flower buds. Larvae consume the leaves and shoots. Full-grown larvae bore into mature green bolls, and due to the damage caused by larvae, green bolls start to fall. Adults lay eggs on the vertical side of the leaves (CABI 2007) (Fig. 11.5).
9. **ETL:** 3–5 eggs or larvae per 25 plants.
10. **Control.**

    (a) **Chemical Control:** (1) Profenofos + cypermethrin (Polytrin C, 440EC, Syngenta; Basel, Switzerland) 600 ml/acre for American bollworm, eggs.

    (2) Profenofos (Curacron, 500EC, Syngenta; Basel, Switzerland) 1000 ml/acre.

    (3) Emamectin benzoate (Proclaim, 019EC, Syngenta; Basel, Switzerland) 200 ml/acre

    (b) **Nonchemical Control:** Bt cotton has a great potential to resist bollworms, especially American bollworm, so Bt cotton is considered the most effective nonchemical control against *H. armigera* (Fitt 2000). But now a days there are few reports of failure of Bt cotton to control chewing pests due to pest resistance against Bt.

Fig. 11.5 Damage of American bollworm

11. **Natural Enemies:** *Menochilus sexmaculatus*, *Syrphus corolla*, and *Chrysoperla carnea* considered as effective biocontrol agents against American bollworm (Ghosh et al. 2010).

11.1.3 Armyworm/Tobacco Cutworm

1. **Scientific Name:** *Spodoptera litura* Fabricius.
2. **Common Names:** Cotton cutworm, rice cutworm, tobacco budworm, cotton leaf worm, luster caterpillar, Egyptian cotton worm, tobacco caterpillar, tobacco cutworm, and tobacco leaf caterpillar.
3. **Family:** Noctuidae **Order:** Lepidoptera.
4. **Identification.**

   (a) **Eggs:** Spherical, flattened in shape, 0.4–0.7 mm in diameter. Pearly green color that turns into black with changing of stage. Female lays in the form of batches. Eggs are covered with pale orange-brown or pink hair-like scales from the female body (Pearson 1958) (Fig. 11.6).

Fig. 11.6 Eggs of armyworm

Fig. 11.7 (**a–b**) Larvae of armyworm

(b) **Larvae:** First instar larvae are tiny blackish having black band on first abdominal segment. Fully grown larvae are firm and flat with speckled short setae. Head has long hair on each segment with black shiny appearance. The color of full-grown larvae is not constant; it changes dark gray to dark brown or black. In size, mature larvae are 40–50 mm. Two large black spots can be seen on first abdominal segment besides eight abdominal segments (Hill 1975) (Fig. 11.7a, b).

(c) **Pupae:** Pupae are reddish brown in color and with rough earthen cases. It pupates in soil.

Pupae are 18–22 mm long. Two hooks are present on the last abdominal segment (USDA 1982).

(d) **Adults:** Body color is whitish to yellowish, immersed with pale red. Forewings have shaded lines and strips with dark brown color. Hind wings are

Fig. 11.8 Adult of armyworm

whitish with violet sheen; margins are dark brown besides venation brown. Thorax and abdomen have hair-like tufts on the dorsal side with orange to light brown color. Head is covered with tufts of light and dark brown scales. The size of wing span is 28–38 mm, and body length is 14–18 mm (Hill 1975) (Fig. 11.8).

5. **Biology.**
   (a) **Adult:** Females lay 50–300 eggs in masses on the lower surface of the leaves after 5 days of emergence.
   (b) **Eggs:** After 3–4 days, eggs hatch. The capacity of single female to lay eggs is 1500–2500 eggs in about 6–8 days (Chari and Patel 1983).
   (c) **Larvae:** Larval stage of *S. litura* has six instars. It takes 23–24 days for emergence of adults from first instar larvae (Etman and Hooper 1980).
   (d) **Pupae:** Larvae of armyworm pupate in the soil in the form of cocoon, and pupal duration varies from 7 to 11 days (Etman and Hooper 1980).
6. **Number of Generations:** Several.
7. **Hosts**
   Tobacco caterpillar considered is a polyphagous pest (Srivastava et al., 2015). The plant species are considered major host plants for *S. litura* in Pakistan are: cotton, castor, jantar, and cauliflower (Ahmad et al. 2013).
8. **Damage:** Extensive feeding of larvae causes damage in most host plants. Due to feeding of larvae, complete skeletonization of leaves of whole plants can be observed. Usually, larvae are leaf feeders but sometimes also attack seedlings. Heavy feeding of larvae causes stunted growth of young seedlings and small fruits. Larvae of *S. litura* feeds on undersides of the leaves and cause scars and skeletonization of leaves. Damage symptoms of *S. litura* are holes and bare section on leaves, fresh stalks, bolls, and buds on later stages of plant. Mines also can be observed on young shoots (Hill 1975) (Fig. 11.9).
9. **ETL:** On appearance.

Fig. 11.9 Damage of armyworm

10. **Control.**
    (a) **Chemical Control:** (1) Emamectin benzoate 200 ml/acre. (2) Lufenuron 200 ml/acre.
11. **Natural Enemies.**
    (a) **Egg Parasites:** (1) *Chelonus helipae* Gupta, (2) *Trichogramma australicum* Girault, (3) *T. chilonis* Ishii (Rao et al. 1993).
    (b) **Larval Parasites:** (1) *C. carhonator* Marshall, (2) *C. formosanus* (Sonan) (Rao et al. 1993).
    (c) **Predators:** (1) *Liposcelis* sp., (2) *C. carnea* (Rao et al. 1993).

11.1.4 Spotted Bollworm/Spiny Bollworm

1. **Scientific Name:** *Earias vittella* (Fabricius).
2. **Common Names:** Spiny bollworm, okra shoot and fruit borer.
3. **Family:** Noctuidae **Order:** Lepidoptera.
4. **Identification.**
    (a) **Eggs:** Eggs of *E. vittella* are light bluish green in color with longitudinal ridges (Alam 1969).
    (b) **Larvae:** The color of full-grown larvae is green and black besides orange, but in some studies, brownish color of larvae with white streaks dorsally

Fig. 11.10 Larva of spotted bollworm

besides pale yellow ventrally is also reported. The length of full-grown larvae is about 1.64 cm (Alam 1969; Mazed et al. 2016) (Fig. 11.10).

(c) **Pupae:** Larvae pupate in the form of boat-shaped gray cocoon on fruits or in soil, but in some studies, it is described that pupae were chocolate brown and bluntly rounded and enclosed in gray-colored inverted boat-shaped cocoons (Alam 1962; Butani and Jotwani 1984).

(d) **Adult:** The color of head and thorax of spotted bollworm adults is ochreous white. The length of adult is about 1.25 cm across the forewings. Forewings are pale whitish with a wedge-shaped horizontal green patch in middle. Hind wings are silvery white in color (Alam 1962; Butani and Jotwani 1984). In some studies, it is reported that narrow light green bands are present longitudinally on the middle of the forewings (Atwal and Dhaliwal 1997; Mazed et al. 2016). Female is larger in size than male. Female is V-shaped at the end of the anal part, while male has thick hairs on the anal part (Mazed et al. 2016) (Fig. 11.11a–c).

5. **Biology.**

   (a) **Eggs:** A single female can lay 63–697 eggs during oviposition period. The incubation period varies from 3 to 7 days (Kranz et al. 1978).
   (b) **Larvae:** Larvae take 10–19 days for full growth in various seasons (Pant 1960).
   (c) **Pupae:** Pupation period is 8–14 days (Mihra 1935).
   (d) **Adults:** Life of adults is 2–4 weeks as reported in previous studies (Cherian and Kylasam 1947).

6. **Hosts:** Spiny bollworm is considered a polyphagous pest all over the world. The plant species which are considered main hosts of spiny bollworm: cotton, okra, hollyhock, and shoe flower (Syed et al. 2011).
7. **Damage:** In cotton, larvae of spotted bollworm attacks the tender top growing shoots and bore the shoots leading to the tunnel down the stem. As a result, attacked shoots become dry. Larvae of spotted bollworm also attack bolls; enter

Fig. 11.11 (**a–c**) Adults of spotted bollworm

into the bolls and feed internally; after feeding on one boll, they move to another (Ahmed et al. 2012) (Fig. 11.12a–e).

8. **ETL:** 3–5 larvae/25 plants or 10% infested shoots or and squares or and bolls.
9. **Control.**

   (a) **Chemical Control:** (1) Profenofos + cypermethrin 600 ml/acre, (2) Spinetoram 50–60 ml/acre, (3) Gamma-cyhalothrin 100 ml/acre, (4) Cypermethrin 330 ml/acre, (5) Deltamethrin 80 ml/acre.
   (b) **Nonchemical Control:** Control of spotted bollworm through *Bacillus thuringiensis* (Bt) varieties are considered the most effective nonchemical control (Entwistle 1993).

10. **Natural Enemies.**

   (a) **Egg Parasitoids:** (1) *T. achaeae*, (2) *T. chilonis* Ichii.
   (b) **Larval Parasitoids:** (1) *Activa aegypti* Vill and *Activa* sp., (2) *A. hyalinata* Mall, (3) *Agathus aciculatus* (Brues), (4) *Agathus* sp., (5) *Apanteles* spp.
   (c) **Predators:** (1) *Brumoides suturalis* (F.), (2) *Geocoris* sp., (3) *Cartheconidea furcellata* (Woeff), (4) *Eumenes petiolata*, (5) *M. sexmaculatus*, (6) *Phanerotoma hendecasisella* (Kashyap and Verma 1987).

Fig. 11.12 (**a–e**) Damages caused by spotted bollworm

11.1.5 Pink Bollworm

1. **Scientific Name:** *Pectinophora gossypiella* (Saunders).
2. **Family:** Gelechiidae **Order:** Lepidoptera.
3. **Identification.**

   (a) **Eggs:** Eggs of *P. gossypiella* are elongate, oval, flattened, and about 1 mm long; 0.5 mm broad. The color of new laid eggs has a slightly greenish shade and becomes reddish on maturity (Busck 1917).

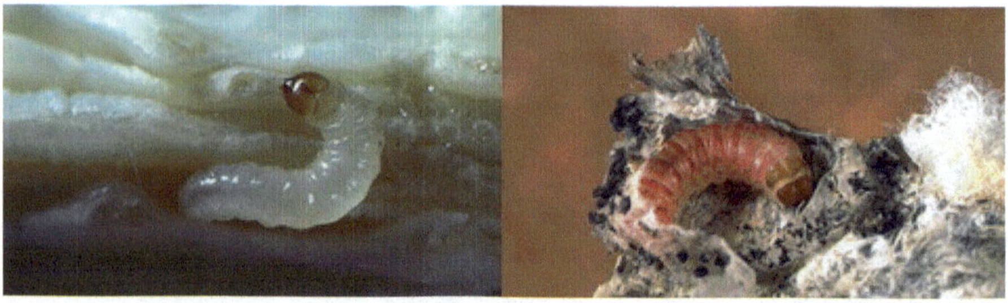

Fig. 11.13 Larvae of pink bollworm

(b) **Larvae:** The first instar larvae are white in color and have black heads. The length of mature larva is 10–12 mm. Full-grown larvae are white in color having red band on the upper side of each segment (Mukuka et al. 2002) (Fig. 11.13).

(c) **Pupae:** The color of pupae is plump, reddish brown and becomes darker on maturity, and the length is 8–10 mm. The posterior end of pupae is pointed and terminating in a short stout looks hooklike; setae and spines are present on the last segment. Imago's eyes can be seen clearly under the gena of pupal skin (Busck 1917).

(d) **Adult:** Adults of pink bollworm are 12–20 mm long having brown and dark brown color. The color of head is reddish brown having pale, rainbow-like scales. Basal segment having brown color antennae bears a pectin of 5 or 6 hair-like scales. Adult forewings are elongated, oval, and pointed at the tips with a wide fringe. In inner side, forewings are in brown color with fine dark scales on medial cells and wing base. Hind wings are broader than forewings, trapezoidal in form, and silvery gray with a darker, iridescent hind margin. Wing fringe is ochreous and darker at the base and apex. Legs are brownish black having transverse, ochreous bands in ring form. Abdomen is ochreous toward the upper side and dark brown laterally and covered with ochreous-brown scales on the underside.

In genitalia, male uncus is broad at base, tapering to a point, and aedeagus has a hooked tip. Female ovipositor is weakly sclerotized (Mukuka et al. 2002) (Fig. 11.14a, b).

4. **Biology.**

(a) **Eggs:** Singly laid eggs are found on green cotton bolls and flowers in the form of batches. Eggs hatch in 3–6 days, and a sole female can lay 300 eggs (Busck 1917; Mukuka et al. 2002; Noble 1969).

(b) **Larvae:** Larval duration is completed in 20–30 days and has four larval instars, and the last instar of larvae is of pink color. That is why it is known as pink bollworm (Busck 1917; Noble 1969).

(c) **Pupae:** Larvae pupate in seed or in soil in the form of cocoon, and pupation duration is 10–20 days (Busck 1917; Noble 1969).

Fig. 11.14 (**a–b**) Adults of pink bollworm

(d) **Adult:** Life period of adults is 1.5–2 weeks, and whole life cycle is completed in 35 days in favorable conditions (Busck 1917; Noble 1969).

5. **Number of Generations:** 5–6 generations in a cotton season (Kaçkavalci and Öncüer n.d.).
6. **Hosts:** Cotton is considered a major host of *P. gossypiella*, while okra is also reported as a host of that particular pest (Busck 1917).
7. **Damage:** Larvae make tunnels on the bolls after hatching from eggs; usually they feed on the soft portion of the bolls (Busck 1917). It is very difficult to identify the entry point of the larvae. Due to the damage of the larvae, complete opening of bolls failed (Mukuka et al. 2002). Shedding of fruits, lint damage, rotted bolls, seed loss, flower damage, and discolored seed or lint are also caused by larvae feeding (Ingram 1994; Leigh et al. 1996) (Fig. 11.15a–e).
8. **ETL:** 10% boll damage (Mohan et al. 2016).
9. **Control.**

(a) **Chemical Control:** (1) Lambda cyhalothrin 400 ml/acre, (2) Spinetoram 100–120 ml/acre, (3) Deltamethrin 80 ml/acre, (4) Triazofos 1000 ml/acre.

(b) **Nonchemical Control:** (1) Mating disruption through the usage of gossyplure in high dose is considered the most successful nonchemical control (Flint et al. 1985; Staten et al. 1987).

(2) Bt cotton is considered a long-term nonchemical control against pink bollworm (Carriere et al. 2003).

10. **Natural Enemies:** All common predators which are found in cotton field are capable of controlling the pink bollworm, but *Labidura riparia* (Pallas) attack on all stages of pink bollworm (Orphanides et al. 1971).

Fig. 11.15 (**a–e**) Damages caused by pink bollworm

11.2 Sucking Insect Pests of Cotton

11.2.1 Cotton Jassid

1. **Scientific Name:** *Amrasca biguttula* (Ishida), *Amrasca devastans.*
2. **Common Names:** Cotton jassid, cotton leafhopper, brinjal leafhopper.
3. **Family:** Cicadellidae **Order:** Homoptera.
4. **Identification.**

Fig. 11.16 Nymphs of jassid

(a) **Eggs:** Female lays eggs singly. The color of eggs is pale whitish. A hook is present at the anterior end, and the other end is broad pointed. The length of eggs is 0.50–0.63 mm (Jayarao et al. 2015).
(b) **Nymph:** Nymphal period consists of five instars. First nymphal instar is in transparent and yellowish color. The color of second instar is dark reddish with white color eyes, and rudimentary wing pads can be seen on meso- and metathorax. The third instar nymph is in yellowish green color having small wing pads. In fourth nymphal instar, the size of wing pads increases, and fourth instar is in yellowish green color. Fifth and fully developed nymphal instar has greenish yellow color with fully developed eyes, and white color wing pads can be seen up to ninth abdominal segment, while two black dots are present on the base of wing pads (Jayarao et al. 2015) (Fig. 11.16).
(c) **Adult:** Green color adults have conspicuous black spots on the apical part of the wing and top of the head. Females can be identified with the help of genitalia present at the end of the abdomen, and females are larger than males in size. The average length of a female is 2.76 mm (Jayarao et al. 2015) (Fig. 11.17).

5. **Biology.**

(a) **Eggs:** The incubation period is 6–7 days, and a single female can lay 17–18 eggs (Jayarao et al. 2015).
(b) **Nymphal Period:** Cotton jassid completes its nymphal period in 6–11 days, but that period varies host to host (Jayarao et al. 2015).

Fig. 11.17 Adult of jassid

Fig. 11.18 **(a–b)** Damages caused by jassid

(c) **Adult:** Longevity of male and female varies according to their food, so longevity of female is 16–18 days and of male is 14–17 days reported in previous studies. Cotton jassid completes its whole life cycle in 19–35 days (Jayarao et al. 2015; Shivanna et al. 2009).

6. **Number of Generations:** 7 generations per year.
7. **Hosts:** Cotton, sunflower, okra, turnip, sarson, radish, cucumber, watermelon, gourd, sponge gourd, and peas are the common hosts of cotton jassid (Jayarao et al. 2015; Saeed et al. 2015).
8. **Damage:** Cotton jassid is considered a polyphagous pest in Pakistan (Akram et al. 2011; Khan and Khaliq 2004). Cotton jassid caused heavy damage in plants after sucking the sap from leaves and injecting toxic saliva. Due to the heavy damage of cotton jassid, the following symptoms can be observed: stunted plant growth and curling of leaves downward; first, leaves become yellow and then change into brown and become dry, and there is shedding of fruiting bodies (Narayan and Singh 1994; Rehman 1940). In cotton, 37–67% losses are reported in previous studies due to the damage of cotton jassid (Ahmad et al. 1986; Bhat et al. 1984; Manzoor 1982) (Fig. 11.18a, b).

9. **ETL:** 1 jassid per leaf.
10. **Control.**

    (a) **Chemical Control:** (1) Acetamiprid 125 g/acre, (2) dinotefuran 100 g/acre, (3) imidacloprid 80–100 ml/acre, (4) buprofezin 600 g/acre, (5) diafenthiuron 200 ml/acre, (6) nitenpyram 130 g/acre, (7) thiamethoxam 24 g/acre.
    (b) **Non-chemical Control:** Spray of *Bacillus thuringiensis* 500 g/acre can be effective as a nonchemical control against cotton jassid (Mehmood et al. 2001).

11. **Natural Enemies:** *Chrysoperla carnea* is considered an effective predator of cotton leafhopper (Balasubramani and Swamiappan 1994).

11.2.2 Cotton Whitefly

1. **Scientific Name:** *Bemisia tabaci* (Gennadius).
2. **Common Names. Whitefly**
3. **Family:** Aleyrodidae **Order:** Homoptera.
4. **Identification.**

    (a) **Eggs:** Female lays eggs in the form of clusters or scattered on smooth leaves. Eggs are in yellowish white color and become pale brown, semi-transparent, and dark brown after hatching (Morr 2004).
    (b) **Nymph:** Nymphs are yellow or cream in color with oval or elongated shape. Well-developed 1–7 setae are present on the dorsal side (Morr 2004) (Fig. 11.19).

Fig. 11.19 Eggs and nymphs of whitefly

Fig. 11.20 Adult of whitefly

(c) **Adult:** Adults are dark yellow in color, and forewings are straight with anterior margin (Morr 2004). Wings are covered with waxy powder. The color of compound eyes is red (Fig. 11.20).

5. **Biology.**

   (a) **Eggs:** Whitefly complete its life cycle from egg to adult in 15–70 days depending on temperature and host. A single female can lay 28–300 eggs; it also depends on temperature and host (Butler Jr et al. 1983). Eggs hatch in 5–22 days on various temperatures (El-Helaly et al. 1971).
   (b) **Nymph:** The first stage of nymph is known as crawler, and last nymphal instar is named as pupa. Nymphal period is 2–7 weeks.
   (c) **Adults:** Female can lay eggs in 1–8 days post-mating. Total life span is 6–55 days depending upon temperature (Anonymous 2018).

6. **Number of Generations:** 10–12.
7. **Hosts:** More than 500 plant species are the hosts of that particular pest throughout the world (Greathead 1986). Cotton, sweet potatoes, cassava, tobacco, and tomato are major hosts (Li et al. 2011).
8. **Damage:** Damage caused by whitefly is categorized in three ways: (1) direct damage, (2) indirect damage, and (3) virus transmission (Berlinger 1986).

   Direct damage caused by whitefly nymphs and adults through piercing and sucking of the sap from foliage of the plants causes weakness, early wilting, and reduced plant growth and yield (Berlinger 1986).

   Whiteflies excrete the honeydew on the leaves of a plant which is responsible of black sooty mold to grow on it. Due to that sooty mold, photosynthesis slows down and plant yield is low. This known as indirect damage (Berlinger 1986) (Fig. 11.21).
9. **ETL:** 10–15 nymphs or adults per leaf

   More than 40 diseases in plants are caused by the transmission of plant viruses through whitefly (Cohen and Berlinger 1986).

Fig. 11.21 Attack and damage of whitefly

10. **Control.**

(a) **Chemical Control:** (1) Acetamiprid 125 gm/acre, (2) pyriproxyfen 400 ml/acre, (3) imidacloprid 250/acre, (4) spirotetramat 240 ml/acre, (5) buprofezin 600 gm/acre, (6) diafenthiuron 200 ml/acre (7) flonicamid 60 gm/acre.

(b) **Nonchemical Control.**

- **Cultural Control:** Row covers, repellent mulches, and soil mulching with yellow polyethylene sheet could be helpful to lower the population of whitefly (Cohen and Melamed-Madjar 1978).
- **Host Plant Resistance:** It has a lot of potential to suppress the population of whitefly and is considered a very important component of integrated pest management against whitefly (Berlinger 1986).

(c) **Biological Control.**

- **Parasites:** (1) Species belonging to genus *Encarsia* (Polaszek et al. 1992).
- **Predators:** The following predators attack whiteflies:

  – *Coccinella undecimpunctata* (Cock 1993).
  – *Coccinella septempunctata* (Cock 1993).
  – *Orius albidipennis* (Cock 1993).
  – *Cardiastethus assimilis* (Dean 1996).
  – *Orius insidiosus* (Dean 1996).
  – *Chrysoperla carnea* (Cock 1993).

11.2.3 Cotton Aphid

1. **Scientific Name:** *Aphis gossypii* Glover.
2. **Common Names:** Melon aphid, cotton aphid (Takalloozadeh, 2010).
3. **Family:** Aphididae **Order:** Homoptera.
4. **Identification.**

   (a) **Eggs:** Eggs are laid singly in the form of loose masses. Each egg is covered with wax secreted by female. Newly laid eggs are elongated in shape with a flat surface and are bright; these eggs are yellow and change into black with the passage of time.

   (b) **Nymph:** Nymphal stage consists of four instars.

   - First instar nymph is yellowish in color and translucent and changes into grayish with elongated shape after 1 day. On that stage, red noticeable compound eyes can be seen, but legs are not well developed. Two clear antennae can also be observed in first instar with four antennal segments. In first instar, abdomen is small and nine segmented.
   - Second instar nymph is similar to first instar, but second instar has five antennal segments; through this characteristic, we can differentiate first instar and second instar.
   - Third instar nymph is yellow in color, but the ventral side is pale. Clear red eyes and antennae can be observed easily. Prominent wing pads are also present in third nymphal instar.
   - Fourth nymphal instar is dark in color, and red dark eyes are present. Abdomen is more elongated than thorax. Third instar has no setae on genital plate, while fourth instar has setae on genital plate. That is the distinguishing characteristic of third and fourth instars (Begum et al. 2018; Wool and Hales 1996).

   (c) **Adult:** Adults are 1–1.5 mm long, and the color is yellow to dark green depending on temperature and winter and summer season. Cornicles are present at the end of the abdomen (Bethke and Paine 1991; Setokuchi 1981; Wall 1933). We can separate *Aphis gossypii* through these cornicles which are equally sclerotized from tip to base and darkly pigmented. They are longer than cauda which gradually tapers toward the apex with a small opening (Blackman and Eastop 2000).

5. **Biology.**

   (a) **Eggs:** On cotton plant, 1–4 nymphs/day are reported in previous studies, and incubation period is 15–19 h as reported in the laboratory (Begum et al. 2018; Saha et al. 2016).

   (b) **Nymph:** Nymphal duration is reported 5–6 days on cotton plant in earlier studies (Saha et al. 2016).

   (c) **Adult:** Adult longevity of 9–11 days is reported in earlier studies (Saha et al. 2016).

6. **Number of Generations:** 15.
7. **Hosts:** *Aphis gossypii* attacks 912 plant species belonging to 166 families. Cotton, okra, potato, chili pepper, sweet pepper, eggplant, melon, and watermelon are the major plant hosts (Ebert and Cartwright 1997; Inaizumi 1980).
8. **Damage:** Aphids attack their host plants through direct and indirect ways. Aphids are phloem feeders; they consume all nutrients which are essential for plant growth. When aphids feed on different parts of the host plant, they inject saliva which is highly toxic for plants (Dedryver et al. 2010; Nault 1997). Aphids transmit almost 275 plant diseases as a vector in their host, while *Aphis gossypii* vectored 60 diseases in their hosts (Dedryver et al. 2010; Kersting et al. 1999). Aphids secrete honeydew which helps black sooty mold to grow on it, and this inhibits photosynthesis in host plants (Michaud and Sloderbeck 2005). Formation of sticky fiber in open bolls of cotton is due to honeydew secretion by cotton aphid (Dedryver et al. 2010). Direct damage caused by cotton aphid in cotton induced plant deformation (Kersting et al. 1999). Curling of leaves downward, wrinkled leaves, stunted growth of young plants, and death of young plants are the major damage symptoms of cotton aphid (Ebert and Cartwright 1997; Matthews 1989).
9. **ETL:** 10–15 nymphs or adults per leaf.
10. **Control.**

   (a) **Chemical Control:** (1) Acetamiprid 50g/l of water, (2) carbosulfan 330 ml/acre, (3) pymetrozine 80 g/l of water, (4) imidacloprid 250 ml/acre.
   (b) **Nonchemical Control.**

   - **Cultural Control:** Topping (remove the top portion or leaves) of plants is considered a cultural control in cotton against cotton aphid (Deguine et al. 2000).
   - **Host Plant Resistance:** Host plant resistance is a very vital component of integrated pest management against aphids (Dedryver et al. 2010). In previous studies, some resistant cotton cultivars are very effective against *Aphis gossypii* (Du et al. 2004).

   (c) **Biological Control.**

   - **Parasitoids:** (1) *Lysiphlebus testaceipes*, (2) *Aphidius colemani*, (3) *Aphidius matricariae* (van Steenis 1995; Zamani et al. 2007).
   - **Predators:** Predators attack all stages except the egg of *Aphis gossypii*, and these predators belong to the following families: Coccinellidae, Syrphidae, Chrysopidae, and Hemerobiidae (Ebert and Cartwright 1997).

     Some important predators of cotton aphid are (1) *Adalia bipunctata*, (2) *Coleomegilla maculata*, (3) *Cycloneda sanguinea*, and (4) *Chrysoperla carnea* (van Steenis 1992).

11.2.4 Onion Thrips

1. **Scientific Name:** *Thrips tabaci* Lindeman.
2. **Common Names:** Onion thrips, thrips.
3. **Family:** Thripidae **Order:** Thysanoptera.
4. **Identification and Biology.**
   (a) **Eggs:** Female lays white- or yellow-color kidney-shaped eggs and puts them on the lower or upper surface of the leaves (Abdel-Gawaad and El-Shazli 1971). Eggs are approximately 0.26 mm in length (Ghabn 1948).
   (b) **Nymph:** Nymphs are white to pale yellow in color and have short micro setae with three segmented antennae. Abdomen has ten segments with a pair of spiracles on second and eighth abdominal segments (Vierbergen et al. 2010).
   (c) **Pre-pupa and Pupa:** Pre-pupal and pupal stages have wing sheath. We differentiate these stages through development and orientation of antennae, length of antennae, and wing sheath (Ghabn 1948).
   (d) **Adult:** Adults are pale yellow to dark brown in color; however, they change their color as the temperature changes (Morison 1957; Murai et al. 2002; Nakahara 1991; Sakimura 1937; Tripplehorn and Johnson 2005). Adults have seven segmented antennae and four setae on front wing (Diaz-Montano et al. 2011; Mound and Walker 1982). Males are smaller than females; the length of male is approximately 0.7 mm (Pergande 1895), and of female is 1 to 1.3 mm (Morison 1957) (Fig. 11.22).
5. **Number of Generations:** 5–8 per year.
6. **Hosts:** There are 141 plant species belonging to 41 families that are reported as hosts of *tabaci* in previous studies (Ghabn 1948). Onion, beans, cabbage, beets, cauliflower, cotton, cucumber, tomato, melon, chilies, tobacco, potato, clover,

Fig. 11.22 Adult of thrips

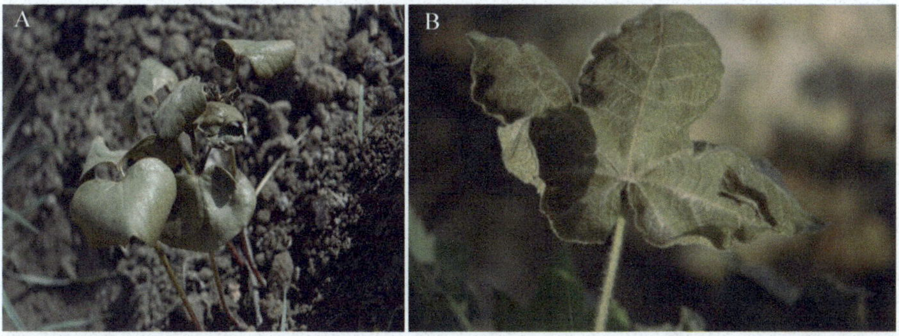

Fig. 11.23 (**a**) Damages caused by thrips

and squash are the major hosts of *tabaci* (Attique and Ahmad 1990; Doederlein and Sites 1993).

7. **Damage:** Due to sap sucking from the leaves of cotton, the following symptoms can be observed in cotton: stunted plant growth; delay in fruiting, maturity, and harvesting; faded leaf tips; and damage of the lower parts which leads to plant death (Ananthakrishnan 1993; Hawkins et al. 1966) (Fig. 11.23a, b).
8. **ETL:** 8–10 adults/leaf.
9. **Control.**

   (a) **Chemical Control.**
   (1) Acetamiprid 125 g/acre, (2) spinetoram 50–60 ml/acre, (3) chlorfenapyr 100 ml/acre, (4) acephate 300–400 g/acre, (5) profenofos 800 ml/acre

   (b) **Biological Control:** (1) *Coccinella septempunctata* is the best control agent in controlled environment (Deligeorgidis et al. 2005).

11.2.5 Red Cotton Bug

1. **Scientific Name:** *Dysdercus koenigii* (Fabricius).
2. **Common Names:** Cotton stainer.
3. **Family:** Pyrrhocoridae **Order:** Hemiptera.
4. **Identification and Biology.**

   (a) **Eggs:** Eggs are creamy white and change into yellowish orange with the passage of time. Incubation period varies from 4 to 9 days at different temperature (Fig. 11.24).

   (b) **Nymphs:** Nymphal period of red cotton bug consists of five instars.

   (c) **First Nymphal Instar: First** instar nymphs are light orange and change into blood red within one day. The duration of first instar is 2–3 days in previous studies. The average body length is 1.58 mm, and width is 0.94 mm. In first

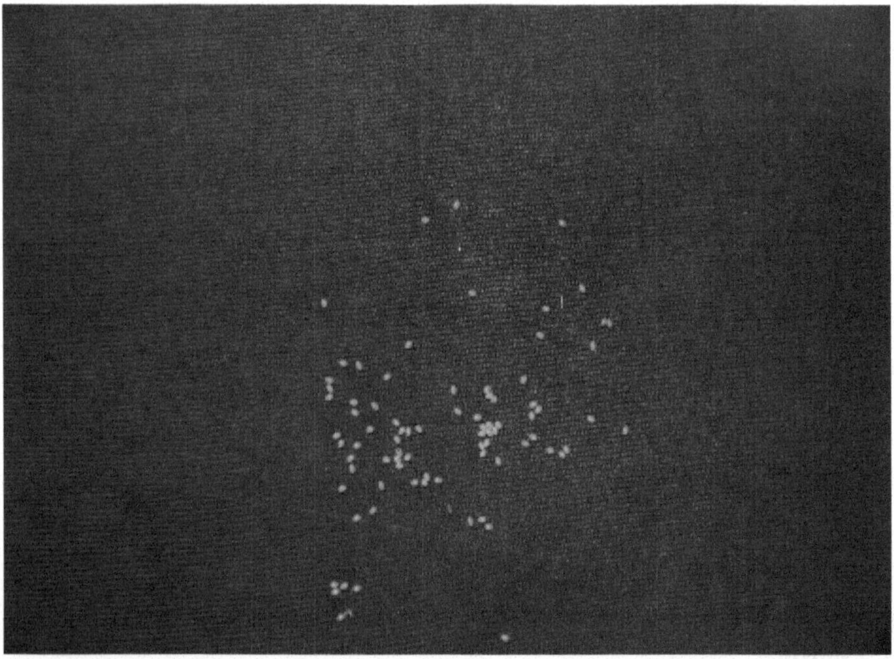

Fig. 11.24 Eggs of red cotton bug

instar, antennae are shorter than the body length of red cotton bug. No wing pads are present in first instar.

(d) **Second Nymphal Instar:** First and second instars are similar in appearance, but second instar is larger in size than first instar. The average duration of second instar is 3–4 days. The average length of second instar is 3.02 mm, and width is 1.25 mm.

(e) **Third Nymphal Instar:** Third instar nymphs are red in color and change into reddish within 1 day. Wing pads are clear on thorax in third instar. The duration of third instar is 4–5 days. The average length is 5.52 mm, and width is 2.36 mm.

(f) **Fourth Nymphal Instar:** Fourth instar is tubular in shape and crimson red in color. Wing pads are developed from metathorax. White transverse bands can be seen on third to seventh abdominal segments. The duration of fourth instar is 5–6 days. The length of fourth instar is 9–9.5 mm, while average width is 3.64 mm.

(g) **Fifth Nymphal Instar:** Fifth nymphal is crimson red in color and cylindrical in shape. Wing pads are well developed. Legs and antennae are black in color, while antennae are five segmented. The average duration of fifth instar is 6.12 days. The average length of fifth nymphal instar is 12.22 mm, and width is 4.98 mm (Fig. 11.25a–c).

Fig. 11.25 (a–c) Nymphs of red cotton bug

(h) **Adult:** Adult of red cotton bug is also crimson red in color. Hind wings are membranous and broader than forewings. Black spots can be observed in the center of forewings. The average duration of male adult is 16.18 days, while for female adult, it is 20.85 days. Female is larger (15.30 mm length and 8.44 mm width) than male (14 mm length and 7.32 mm width) (Fig. 11.26).

5. **Hosts:** The following are the major host plants of red cotton bug: cotton, okra, and hollyhock.
6. **Damage:** Red cotton bug feeds on emerging bolls and mature seed and transmits the fungi *Nematospora gossypii* which develop on immature lint and seed (Jaleel et al. 2013) (Fig. 11.27a, b).
7. **Control.**

(a) **Chemical Control:** In Pakistan, red cotton bug is considered a minor pest. That is why no research-based chemical recommendations are available for red cotton.

Fig. 11.26 Adults of red cotton bug

Fig. 11.27 (a) Damages caused by red cotton bug

11.2.6 Dusky Cotton Bug

1. **Scientific Name:** *Oxycarenus hyalinipennis* Costa.
2. **Common Names**: Cotton seed bug.
3. **Family:** Lygaeidae **Order:** Hemiptera.
4. **Biology and Identification.**
   (a) **Eggs:** Eggs laid by females are single or in patches form, and the color of eggs is yellow and becomes red near hatching.
   (b) **Nymph:** The nymphs are red in color, and abdomen of nymphs is orange in color (Smith and Brambila 2008).
   (c) **Adults:** Adults of dusky cotton bug are black and white translucent in color; thorax of adults is black, and wings are in white translucent color. Life span of

Fig. 11.28 Adults of dusky cotton bug

males is 33–49 days, and for females, it is 35–51 days. Single female can lay up to 110 eggs (Fig. 11.28).

5. **Number of Generations**: 6–7 per year (Srinivas and Patil 2010).
6. **Hosts:** The following plants are considered major host plants of dusky cotton bug: cotton, lemon, chilies, guava, mango, moringa, and okra (Shah et al. 2016).
7. **Damage:** Nymph and adults of dusky cotton bug suck the sap from tender parts and oil from seed which reduced the quality and quantity of cotton (Ijaz and Shad 2018). Due to sap and oil sucking, up to 32% yield losses and 88% germination losses were recorded in previous studies (Ahmed et al. 2015) (Figs. 11.29 and 11.30).
8. **Control.**

   (a) **Chemical Control:** No official recommendations against dusky cotton bug are available, but according to my personal observations, fipronil and matrine are very effective against dusky cotton bug.

11.2.7 Cotton Mealybug

1. **Scientific Name:** *Phenacoccus solenopsis* Tinsley.
2. **Family:** Pseudococcidae **Order:** Hemiptera.
3. **Biology and Identification.**

Fig. 11.29 Damage of dusky cotton bug

Fig. 11.30 Damage of cotton stainers

(a) **Eggs:** Females of cotton mealybug are ovoviviparous, and a single female lays 150–600 eggs which are covered in waxy ovisac (Lu et al. 2008).

(b) **Nymphs:** Nymphs of cotton mealybug are known as crawlers. Nymphal period of cotton mealybug consists of three nymphal instars for female and four instars for male. First instar nymph has six segmented antennae and quinquelocular openings on head, thorax, and abdomen. The second instar

Fig. 11.31 Nymphs of cotton mealybug

nymphs are distinguished by having 18 pairs of distinct cerarii around the margin of the body, lack of quinquelocular pores on the body, and claw having a distinct denticle. The third instar nymph differs by having seven segmented antennae and a circulus. Nymphal period consists of 25–30 days depending on temperature, and cotton mealybug can survive from 0 °C to 45 °C (Hodgson et al. 2008; Sharma 2007) (Fig. 11.31).

(c) **Adults:** Cotton mealybug can be distinguished through the morphology of adult female. Adult females of cotton mealybug are covered with waxy and powdery secretion, and six pairs of dark transverse bands are present on pro- to meta-thoracic segments. The length of adult females is 2–5 mm, and width is 2–4 mm (Kosztarab 1996; McKenzie 1967) (Fig. 11.32).

4. **Hosts:** According to a survey report, cotton mealybug is considered a serious pest, and 202 plants of different categories are reported as hosts of cotton mealybug. Some major host plants are cotton, brinjal, okra, sesame, China rose, tomato, and sunflower (Arif et al. 2009; Sharma 2007).
5. **Number of Generations:** Several.
6. **Damage:** Nymphs of cotton mealybug suck cell sap from plants and their leaves. Due to sucking of the sap, leaves become yellow and whitish and drop from the plant. As a result, the following symptoms can be observed: small fruits, crinkled

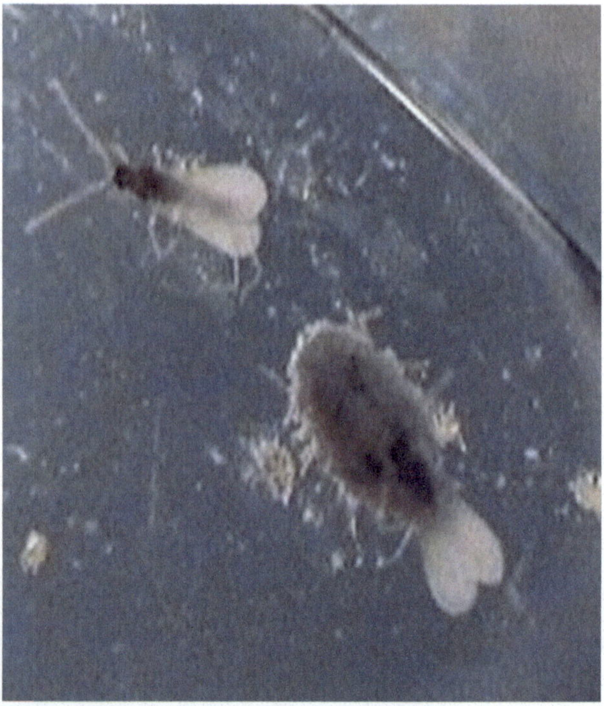

Fig. 11.32 Adults of cotton mealybug

Fig. 11.33 Damage of cotton mealybug

fruits, malformed fruits, loss of plant vigor, damaged foliage, fruit drop, and potential plant death (Culik and Gullan 2005; Dhawan and Saini 2009) (Fig. 11.33).

Cotton mealybug also feeds on phloem which leads to stunted growth of plants (Culik and Gullan 2005; Dhawan and Saini 2009). Cotton mealybug produces honeydew which creates problem in photosynthesis and also vectors some plant diseases like cocoa swollen shoot virus, cocoa mottle leaf virus, and cotton leaf curl virus (Cudjoe et al. 1993; Culik and Gullan 2005; Saeed et al. 2007).

7. **ETL:** On appearance.
8. **Control.**

   (a) **Chemical Control:** (1) Profenofos 500 EC 800 ml/acre.

   (b) **Biological Control:** Biological control which includes predators and parasitoids is considered the most effective control against cotton mealybug. Some important biocontrol agents are (1) *Aprostocetus minutus*, (2) *Cheiloneurus* sp., (3) *Chalcaspis arizonensis*, (4) *Aenasius* sp., and (5) *Aenasius bambawalei* (Fuchs et al. 1991; Hayat 2009; Sharma 2007).

References

Abdel-Gawaad AA-W, El-Shazli AY (1971) Studies on *Thrips tabaci* Lindman: VII. Effect of food on the life cycle. J Appl Entomol 67:27–30

Ahmad Z, Attique MR, Rashid A (1986) An estimate of the loss in cotton yield in Pakistan attributable to the jassid *Amrasca devastans* Dist. Crop Protect 5:105–108

Ahmad M, Ghaffar A, Rafiq M (2013) Host plants of leaf worm, *Spodoptera litura* (Fabricius) (Lepidoptera: Noctuidae) in Pakistan. Asian J Agric Biol 1:23–28

Ahmed S, Iqbal MK, Shahid M, Khan RR (2012) Comparison of antibiosis of spotted bollworm, *Earias vittella* (Fab.) on two Bt- and one non Bt- cotton varieties. Pak J Zool 44:463–468

Ahmed R, Nadeem I, Yousaf MJ, Niaz T, Ali A, Ullah Z (2015) Impact of dusky cotton bug (*Oxycarenus laetus* Kirby) on seed germination, lint color and seed weight in cotton crop. J Entomol Zool Stud 3:335–338

Akram W, Naz I, Ali S (2011) An empirical analysis of household income in rural Pakistan: evidences from tehsil Samundri. Pak Econ Soc Rev 49:231–249

Alam MZ (1962) A list of insects and mites of East Pakistan. East Pakistan Agricultural Research Institute, Tejgaon, Dacca, pp 44–48

Alam MZ (1969) Insect pests of vegetables and their control in East Pakistan. Agricultural Information Service, in collaboration with East Pakistan Agricultural Research Institute, pp 146

Ananthakrishnan TN (1993) Bionomics of thrips. Annu Rev Entomol 38:71–92

Anonymous (2018). http://www.extento.hawaii.edu/kbase/crop/type/b_tabaci.htm

Arif MI, Rafiq MW, Ghaffar A (2009) Host plants of cotton mealybug (*Phenacoccus solenopsis*): a new menace to cotton agroecosystem of Punjab. Pakistan Int J Agric Biol 11:163–167

Attique MR, Ahmad Z (1990) Investigation of *Thrips tabaci* Lind. As a cotton pest and the development of strategies for its control in Punjab. Crop Protect 9:469–473

Atwal A, Dhaliwal G (1997) Agricultural pests of South Asia and their management. J Entomol Res 21:389–390

Balasubramani V, Swamiappan M (1994) Development and feeding potential of the green lacewing *Chrysoperla carnea* Steph.(Neur. Chrysopidae) on different insect pests of cotton. Anzeiger für Schädlingskunde, Pflanzenschutz, Umweltschutz 67:165–167

Begum M, Mandal MK, Islam MA, Howlader MA (2018) Biology, nature of infestation and control of the aphid, *Aphis gossypii* (Glover, 1877) (Hemiptera: Aphididae) on arum plant, *Colocasia esculenta*. Bangladesh J Zool 46:63–70

Berlinger MJ (1986) Host plant resistance to *Bemisia tabaci*. Agric Ecosyst Environ 17:69–82

Bethke JA, Paine TD (1991) Screen hole size and barriers for exclusion of insect pests of glasshouse crops. J Entomol Sci 26:169–177

Bhat MG, Joshi AB, Singh M (1984) Relative loss of seed cotton yield by jassid and bollworms in some cotton genotypes (*Gossypium hirsutum* L.). Indian J Entomol 46:169–173

Blackman RL, Eastop VF (2000) Aphids on the world's crops: an identification and information guide. Wiley, Chichester

Broadley R (1977) Heliothis: serious agricultural pests in Queensland. Queensland Agr J 103:536–545

Busck A (1917) The pink bollworm, *Pectinophora gossypiella*. J Agric Res 9:343–370

Butani DK, Jotwani M (1984) Insects in vegetables. Periodical Expert Book Agency, Delhi

Butler GD Jr, Henneberry TJ, Clayton TE (1983) *Bemisia tabaci* (Homoptera: Aleyrodidae): development, oviposition, and longevity in relation to temperature. Ann Entomol Soc Am 76:310–313

CABI (2007) CABI. 2007 Crop Protection Compendium. Commonweath Agricultural Bureau, International. http://www.cabicompendium.org/

Cantrell B (1980) Further studies on the immature stages of the common species of armyworms, cutworms and budworms found in Queensland. Queensland J Agr Anim Sci 37:167–175

Carriere Y, Ellers-Kirk C, Sisterson M, Antilla L, Whitlow M, Dennehy TJ, Tabashnik BE (2003) Long-term regional suppression of pink bollworm by *Bacillus thuringiensis* cotton. PNAS 100:1519–1523

Chari MS, Patel SN (1983) Cotton leaf worm *Spodoptera litura* Fabricius: its biology and integrated control measures. Cotton Dev 13:465–482

Cherian MC, Kylasam MS (1947) Studies on the spotted bollworms of cotton *Earias fabia* S., and *E. insulana* B. J Bombay Nat Hist Soc 46:658–667

Cock MJW (1993) Bemisia tabaci an UPDATE 1986–1992: on the cotton whitefly with an annotated bibliography. CAB International, Ascot

Cohen S, Berlinger MJ (1986) Transmission and cultural control of whitefly-borne viruses. Agric Ecosyst Environ 17:89–97

Cohen S, Melamed-Madjar V (1978) Prevention by soil mulching of the spread of tomato yellow leaf curl virus transmitted by *Bemisia tabaci* (Gennadius) (Hemiptera: Aleyrodidae) in Israel. Bull Entomol Res 68:465–470

Common IFB (1953) The Australian species of Heliothis (Lepidoptera: Noctuidae) and their pest status. Aust J Zool 1:319–344

Cudjoe AR, Neuenschwander P, Copland MJW (1993) Interference by ants in biological control of the cassava mealybug *Phenacoccus manihoti* (Hemiptera: Pseudococcidae) in Ghana. Bull Entomol Res 83:15–22

Culik MP, Gullan PJ (2005) A new pest of tomato and other records of mealybugs (Hemiptera: Pseudococcidae) from Espírito Santo, Brazil. Zootaxa 964:1–8

Dean DE (1996) Predaceous arthopods of the sweetpotato whitefly, *Bemisia tabaci* (Gennadius), on tomatoes in Florida

Dedryver CA, Le Ralec A, Fabre F (2010) The conflicting relationships between aphids and men: a review of aphid damage and control strategies. C R Biol 333:539–553

Deguine J-P, Goze E, Leclant F (2000) The consequences of late outbreaks of the aphid *Aphis gossypii* in cotton growing in central Africa: towards a possible method for the prevention of cotton stickiness. Int J Pest Manag 46:85–89

Deligeorgidis PN, Ipsilandis CG, Vaiopoulou M, Kaltsoudas G, Sidiropoulos G (2005) Predatory effect of *Coccinella septempunctata* on *Thrips tabaci* and *Trialeurodes vaporariorum*. J Appl Entomol 129:246–249

Dhawan AK, Saini K (2009) First record of *Phenacoccus solenopsis* Tinsley (Homoptera: Pseudococcidae) on cotton in Punjab. J Insect Sci 22:309–310

Diaz-Montano J, Fuchs M, Nault BA, Fail J, Shelton AM (2011) Onion thrips (Thysanoptera: Thripidae): a global pest of increasing concern in onion. J Econ Entomol 104:1–13

Doederlein TA, Sites RW (1993) Host plant preferences of *Frankliniella occidentalis* and *Thrips tabaci* (Thysanoptera: Thripidae) for onions and associated weeds on the southern High Plains. J Econ Entomol 86:1706–1713

Du L, Ge F, Zhu S, Parajulee MN (2004) Effect of cotton cultivar on development and reproduction of *Aphis gossypii* (Homoptera: Aphididae) and its predator *Propylaea japonica* (Coleoptera: Coccinellidae). J Econ Entomol 97:1278–1283

Ebert TA, Cartwright B (1997) Biology and ecology of *Aphis gossypii* Glover (Homoptera: aphididae). Southwest Entomol 22:116–153

El-Helaly MS, El-Shazli AY, El-Gayar FH (1971) Biological studies on *Bemisia tabaci* Genn. (Homopt., Aleyrodidae) in Egypt. J Appl Ecol 69:48–55

Entwistle PF (1993) *Bacillus thuringiensis*: an environmental biopesticide: theory and practice. Wiley, Chichester

Etman AAM, Hooper GHS (1980) Developmental and reproductive biology of *Spodoptera litura* (F.) (Lepidoptera: Noctuidae). Austral Entomol 18:363–372

Fitt GP (1989) The ecology of Heliothis species in relation to agroecosystems. Annu Rev Entomol 34:17–53

Fitt GP (2000) An Australian approach to IPM in cotton: integrating new technologies to minimise insecticide dependence. Crop Protect 19:793–800

Flint HM, Merkle JR, Yamamoto A (1985) Pink bollworm (Lepidoptera: Gelechiidae): field testing a new polyethylene tube dispenser for gossyplure. J Econ Entomol 78:1431–1436

Fuchs TW, Stewart JW, Minzenmayer R, Rose M (1991) First record of *Phenacoccus solenopsis* Tinsley in cultivated cotton in the United States. Southwestern Entomol 16:215–221

Ghabn AAAE (1948) Contribution to the knowledge of the biology of *Thrips tabaci* Lind. in Egypt. Bulletin de la Societe Fouad 1er d'entomologie, vol 32, pp 123–174

Ghosh A, Chatterjee M, Roy A (2010) Bio-efficacy of spinosad against tomato fruit borer (*Helicoverpa armigera* Hub.) (Lepidoptera: Noctuidae) and its natural enemies. J Hort For 2:108–111

Greathead A (1986) Host plants. *Bemisia tabaci*: a literature survey on the cotton whitefly with an annotated bibliography, pp 17–25

Hawkins BS, Peacock HA, Steele TE (1966) Thrips injury to upland cotton (*Gossypium hirsutum* L.) varieties. Crop Sci 6:256–258

Hayat M (2009) Description of a new species of *Aenasius* Walker (Hymenoptera: Encyrtidae), parasitoid of the mealybug, *Phenacoccus solenopsis* Tinsley (Homoptera: Pseudococcidae) in India. Biosystematica 3:21–26

Hill DS (1975) Agricultural insect pests of the tropics and their control

Hodgson CJ, Abbas G, Arif MJ, Saeed S, Karar H (2008) *Phenacoccus solenopsis* Tinsley (Sternorrhyncha: Coccoidea: Pseudococcidae), an invasive mealybug damaging cotton in Pakistan and India, with a discussion on seasonal morphological variation. Zootaxa 1913:1–35

Ijaz M, Shad SA (2018) Inheritance mode and realized heritability of resistance to imidacloprid in *Oxycarenus hyalinipennis* Costa (Hemiptera: Lygaeidae). Crop Protect 112:90–95

Inaizumi M (1980) Studies on the life-cycle and polymorphisms of Aphis gossypii Glover (Homoptera: Aphididae). Spec Bull Coll Agric Utsunomiya Univ 37:1–132

Ingram W (1994) Pectinophora (Lepidoptera: Gelechiidae). In: Matthews GA, Tunstall JP (eds) Insect pests of cotton. CAB International, Wallingford, pp 107–149

Jaleel W, Saeed S, Naqqash MN (2013) Biology and bionomics of *Dysdercus koenigii* F. (Hemiptera: Pyrrhocoridae) under laboratory conditions. Pak J Agric Sci 50:373–378

Jayarao B, Abulkhader SB, Naik LK, Vinaykumar MM (2015) Assessment of biology and morphometric characteristics of different stages of leafhopper, *Amrasca biguttula biguttula* (Ishida) on okra. Bioscan 10:671–674

Kaçkavalci AA, Öncüer C (n.d.) Soke (Aydm) ovasmda pamuklarda zararh *Pectinophora gossypiella* (Saunders) (Lepidoptera: Gelechiidae)'nin populasyon degnimi ve zarar oraninin saptanmasi

Karim S (2000) Management of *Helicoverpa armigera*: a review and prospectus for Pakistan. Pak J Biol Sci 3:1213–1222

Kashyap RK, Verma AN (1987) Management of spotted bollworms (*Earias* spp.) in cotton-a review. Int J Trop Agric 5:1–27

Kersting U, Satar S, Uygun N (1999) Effect of temperature on development rate and fecundity of apterous *Aphis gossypii* Glover (Hom., Aphididae) reared on *Gossypium hirsutum* L. J Appl Entomol 123:23–27

Khan MB, Khaliq A (2004) Production of soybean (Glycine max L.) as cotton based intercrop. J Res Sci 15:79–84

Kirkpatrick TH (1961) Comparative morphological studies of Heliothis species (Lepidoptera: Noctuidae) in Queensland. Queensland J Agric Sci 18:179–194

Kosztarab M (1996) Scale insects of northeastern North America: identification, biology, and distribution. Virginia Museum of Natural History, Martinsville, p 650

Kranz J, Schmutterer H, Koch W (eds) (1978) Diseases, pest and weeds in tropical crops. Wiley, Chichester, p 663

Leigh T, Roach S, Watson T (1996) Biology and ecology of important insect and mite pests of cotton. Cotton insects and mites. The Cotton Foundation Publisher, Memphis, TN, pp 17–69

Li SJ, Xue X, Ahmed MZ, Ren SX, Du YZ, Wu JH, Cuthbertson AGS, Qiu BL (2011) Host plants and natural enemies of *Bemisia tabaci* (Hemiptera: Aleyrodidae) in China. Insect Sci 18:101–120

Lu YY, Zeng L, Wang L, Xu YJ, Chen KW (2008) Precaution of solenopsis mealybug *Phenacoccus solenopsis* Tinsley. J Environ Entomol 30:386–387

Manzoor A (1982) Evaluation of yield losses in brinjal (*Solanum melongena*) by *Amrasca devastans*. Pak J Agric Res 3:277–280

Matthews GA (1989) Cotton insect pests and their management. Longman Scientific & Technical, Harlow, Essex

Mazed M, Alam M, Miah M, Hossain M, Mian M (2016) Identification of okra shoot and fruit borer infesting okra and their distribution in Bangladesh. Bangladesh J Agric Res 41:657–665

McKenzie HL (1967) Mealybugs of California: with taxonomy, biology, and control of North American species (Homoptera, Coccoidea, Pseudococcidae). University of California Press, Berkeley, CA

Mehmood K, Afzal M, Amjad M (2001) Non-traditional insecticides: a new approach for the control of okra jassid. J Biol Sci 1:36–37

Michaud JP, Sloderbeck PE (2005) Russian Wheat Aphid: an introduced pest of small grains in the High Plains. Kansas State University Agricultural Experiment Station and Cooperative Extension Service. MF-2666

Mihra R (1935) Spotted boll-worms in South Gujarat. Spotted boll-worms in South Gujarat, Bombay, Indian Cent Cott Comm 24

Mohan KS, Ravi KC, Suresh PJ, Sumerford D, Head GP (2016) Field resistance to the Bacillus thuringiensis protein Cry1Ac expressed in Bollgard® hybrid cotton in pink bollworm, *Pectinophora gossypiella* (Saunders), populations in India. Pest Manag Sci 72:738–746

Morison GD (1957) A review of British glasshouse Thysanoptera. Ecol Entomol 109:467–520

Morr I (2004) Diagnostic protocols for regulated pests protocoles de diagnostic pourles organismes reglementes. OEPP/EPPO Bull 34:271–279

Mound LA, Walker AK (1982) Terebrantia (Insecta: Thysanoptera). Fauna of New Zealand, vol 1, 120pp

Mukuka J, Sumani A, Chalabesa A (2002) Agricultural field insect pests of Zambia and their management. Mount Makulu, Chilanga, Zambia: Plant Protection and Quaratntine Division, Soils and Crops Research Branch

Murai T, Toda S, Marullo R, Mound L (2002) Variation of Thrips tabaci in colour and size, Thrips and Tospoviruses. In: Proceedings of the 7th international symposium on Thysanoptera. Australian National Insect Collection, Canberra, Australia, pp 377–378
Nakahara S (1991) Systematics of Thysanoptera, pear thrips and other economic species
Narayan S, Singh P (1994) Resistance to Heliothis and other serious insect pests in *Gossypium* spp. A Rev J Indian Soc Cotton Improvement 19:10–24
Nault LR (1997) Arthropod transmission of plant viruses: a new synthesis. Ann Entomol Soc Am 90:521–541
Noble LW (1969) Fifty years of research on the pink bollworm in the United States. United States Department of Agriculture, USDA, Washington DC, p 62
Orphanides GM, Gonzalez D, Bartlett BR (1971) Identification and evaluation of pink bollworm predators in southern California. J Econ Entomol 64:421–424
Pant C (1960) Some aspects of the bionomics of Earias spp. at Kanpur. Agra Univ J Res (Science) 9:31–40
Pearson EO (1958) Insect pests of cotton in tropical Africa. Commonwealth Institute of Entomology, London
Pergande T (1895) Observations on certain Thripidae. Insect Life 7:390–395
Polaszek A, Evans GA, Bennett FD (1992) Encarsia parasitoids of *Bemisia tabaci* (Hymenoptera: Aphelinidae, Homoptera: Aleyrodidae): a preliminary guide to identification. Bull Entomol Res 82:375–392
Rao GVR, Wightman JA, Rao DVR (1993) World review of the natural enemies and diseases of *Spodoptera litura* (F.) (Lepidoptera: Noctuidae). Int J Trop Insect Sci 14:273–284
Rehman K (1940) Insect pest number. Punjab Agric Coll Mag 7:1–82
Saeed S, Ahmad M, Ahmad M, Kwon YJ (2007) Insecticidal control of the mealybug *Phenacoccus gossypiphilous* (Hemiptera: Pseudococcidae), a new pest of cotton in Pakistan. Entomol Res 37:76–80
Saeed R, Razaq M, Hardy ICW (2015) The importance of alternative host plants as reservoirs of the cotton leaf hopper, *Amrasca devastans*, and its natural enemies. J Pest Sci 88:517–531
Saha J, Chakraborty K, Chatterjee T (2016) Biology of cotton Aphid *Aphis gossypii* Glover. J Glob Biosci 5:4467–4473
Sakimura K (1937) The life cycle and seasonal histories of Thrips tabaci Lind. in the vicinity of Tokyo Japan. Oyodobutu Zasshi 9:1–24
Setokuchi O (1981) Occurrence and fecundity of two color forms in *Aphis gossypii* Glover (Homoptera: Aphidiae) on Dasheen leaves. Appl Entomol Zool 16:50–52
Shah ZU, Ali A, Ibrar-Ul-Haq, Hafeez F (2016) Seasonal history of dusky cotton bug (*Oxycarenus hyalinipennis* Costa). J Entomol Zool Stud 4:228–233
Sharma SS (2007) Aenasius sp. nov. effective parasitoid of mealy bug (*Phenacoccus solenopsis*) on okra. Haryana J Hort Sci 36:412
Shivanna BK, Nagaraja DN, Manjunatha M, Gayathridevi S, Pradeep S, Girijesh GK (2009) Bionomics of leafhopper, *Amrasca biguttula biguttula* (Ishida) on transgenic Bt cotton. Karnataka J Agric Sci 22:538–540
Smith TR, Brambila J (2008) A major pest of cotton, *Oxycarenus hyalinipennis* (Heteroptera: Oxycarenidae) in the Bahamas. Fla Entomol 91:479–483
Srinivas M, Patil BV (2010) Biology of dusky cotton bug, *Oxycarenus laetus* Kirby (Hemiptera: Lygaeidae) on cotton. Karnataka J Agric Sci 17:341–344
Srivastava K, Sharma D, Anal AKD, Sharma S (2015) Integrated Management of *Spodoptera litura*: a review. Int J Life Sci Scienti Res 4(1):1536–1538
Stanley SM (1978) Competitive interactions between the larvae of *Heliothis armigera* (Hübner) and *Heliothis punctigera* Wallengren (Lepidoptera: Noctuidae). PhD thesis, Australian National University. http://hdl.handle.net/1885/110252
Staten RT, Flint HM, Weddle RC, Quintero E, Zarate RE, Finnell CM, Hernandes M, Yamamoto A (1987) Pink bollworm (Lepidoptera: Gelechiidae): large-scale field trials with a high-rate gossyplure formulation. J Econ Entomol 80:1267–1271

Syed TS, Abro GH, Khanum A, Sattar M (2011) Effect of host plants on the biology of *Earias vittella* (Fab) (Noctuidae: Lepidoptera) under laboratory conditions. Pak J Zool 43:127–132

Takalloozadeh HM (2010) Effects of host plants and various temperatures on population growth parameters of *Aphis gossypii* Glover (Hom.: Aphididae). Middle East. J Sci Res 6:25–30

Tripplehorn C, Johnson N (2005) Borror and DeLong's introduction to the study of insects. Thomson Brooks/Cole, Belmont, California

USDA (1982) Pests not known to occur in the United States or of limited distribution: Rice cutworm. USDA-APHIS-PPQ

van Steenis MJ (1992) Biological control of the cotton aphid, *Aphis gossypii* Glover (Hom., Aphididae): pre-introduction evaluation of natural enemies. J Appl Entomol 114:362–380

van Steenis MJ (1995) Evaluation of four aphidiine parasitoids for biological control of *Aphis gossypii*. Entomol Exp Appl 75:151–157

Vierbergen G, Kucharczyk H, Kirk WDJ (2010) A key to the second instar larvae of the Thripidae of the Western Palaearctic region (Thysanoptera). Tijdschr Entomol 153:99–160

Wall RE (1933) A study of color and color-variation in *Aphis Gossypii* Glover. Ann Entomol Soc Am 26:425–463

Wool D, Hales D (1996) Components of variation of morphological characters in Australian *Aphis gossypii*: host-plant effects predominate. In: Proceedings of the 9th international symposium on insect-plant relationships. Springer, pp 166–168

Zalucki MP, Daglish G, Firempong S, Twine P (1986) The biology and ecology of *Heliothis-armigera* (Hubner) and *Heliothis-punctigera* Wallengren (Lepidoptera, Noctuidae) in Australia-what do we know. Aust J Zool 34:779–814

Zamani AA, Talebi A, Fathipour Y, Baniameri V (2007) Effect of temperature on life history of *Aphidius colemani* and *Aphidius matricariae* (Hymenoptera: Braconidae), two parasitoids of *Aphis gossypii* and *Myzus persicae* (Homoptera: Aphididae). Environ Entomol 36:263–271

Chapter 12
Ecological Management of Cotton Insect Pests

Munir Ahmad, Wali Muhammad, and Asif Sajjad

Abstract Ecological pest management (EPM) shares many fundamentals of integrated pest management (IPM) considering the stability of natural ecosystems as the most important goal along with good yielding crop cultivation. Cotton is grown in less than 10% of cultivable area but it is the only crop which share more than 45% usage of insecticides. Considering the harmful impacts of pesticides on agroecosystem, stability of natural environment and higher biodiversity in cotton zones is suggested through adopting whole system management approaches. In this approach, combination of many crops, lands, natural vegetation, and cultural practices can suppress the pest populations in cotton areas. Weed management through biocontrol agents is also supported in this system which increases food for predators and parasitoid development ultimately helping to suppress cotton pests. The IPM tactics like classical biological control, clean cropping, crop rotation, etc., can also be easily adopted in ecological based pest management. Soil and natural resources conservation-based biological control will be the vital component of EPM replacing the IPM in the future. It is a fundamental shift to total system approach with "built-in" preventive control measures addressing the increased demand for organic agriculture. Sustainable organic agriculture is interlinked with precise knowledge of multitrophic nutritional levels and their efficient management for pest management strategies. The pesticide or other chemical option will be used at last by adopting precision agriculture equipment, e.g., specific site application through sensor-based applicators (drone sprayer mounted with GPS and WSN).

M. Ahmad (✉)
Department of Entomology, Pir Mehr Ali Shah, Arid Agriculture University, Rawalpindi, Pakistan
e-mail: munirahmad@uaar.edu.pk

W. Muhammad
Agriculture Pest Warning & Quality Control of Pesticides, Government of Punjab, Layyah, Pakistan

A. Sajjad
Department of Entomology, University College of Agriculture and Environmental Sciences, The Islamia University of Bahawalpur, Bahawalpur, Pakistan

S. Ahmad, M. Hasanuzzaman (eds.), *Cotton Production and Uses*,
https://doi.org/10.1007/978-981-15-1472-2_12

Keywords Ecological pest management · Cotton pest control · IPM · EPM · Organic agriculture

Abbreviations

Bt	*Bacillus thuringiensis*
CIE	Copper inducible elicitor
DDT	Dichlorodiphenyltrichloroethane
EFN	Extrafloral nectar
EPM	Ecological pest management
ETL	Economic threshold level
GM	Genetically modified
GMOs	Genetically modified organisms
GPS	Global positioning system
HPR	Host-plant resistance
INM	Integrated nutrient management
IPM	Integrated pest management
IWM	Integrated weed management
PA	Precision agriculture
VRT	Variable rate technology
WSN	Wireless sensor network

12.1 Introduction

World population is increasing day by day having no proportion with food availability and food production. This increasing human population will demand high quantity of food and shelter for their safety and survival in coming days (Thomas 1999). The world human population has crossed the 7 billion limit (Smith 2011) and is expected to increase to 8 billion by 2025 (Hinrichsen and Robey 2000). There will be a need to enhance our food production for feeding this increasing world population (Nwilene et al. 2008; Dhaliwal and Arora 2001). Agriculture was started about 10,000 years ago for fulfilling the need of food, feed, and fiber through cultivating more and more area and adopting new crops as well. The emphasis was on increasing cultivable area in the start of the agricultural era, but during recent development in the world, the focus is shifted in increasing per unit area production instead of cultivable areas during the last 100 years. Cultivable land is also decreasing by increasing residential and industrial units each year to fulfill the needs of increasing population (FAO 2001). There are many factors limiting per unit area production including fertilizer application, water availability, genetic setup of crop, soil fertility, and some living organisms like arthropods, viruses, fungi, nematodes, birds, rodents, vertebrates, and sometimes human itself competing for food, feed, and fiber. In the beginning of commercial agriculture, humans and insects were living and surviving on the same resources with no effort to control the insect pests. Simply put, the insects were not categorized as pest, but with the increase in human population and ultimately the need of more food supply, humans feel that insects are competing for the same sources, so they named them as pest. These pests were suppressed by the use of different tactics like biological, cultural, and physical control. In the mid-twentieth century, the invention of DDT opened new ways to synthetic organic insecticides for controlling the insect pest of our crops (Dhaliwal et al. 1998; Metcalf 1980). Insecticide resistance was the major problem developed from the excessive use of pesticides, and almost 500 insects are reported showing resistance against insecticide along with pest resurgence, polluting the environment, and imposing health threats to human's health (Georghiou 1990). The concept of IPM was given for the first time in 1962 through the book of Rachel Carson to reduce the hazards and harmful effects of pesticides being used for crop protection.

Abrol and Shankar (2012) have suggested the adaptation of ecological approach for managing the insect pest according to the local agroecosystems by using actions and some interactions of fully understood technologies. It can also be called as truly integrated pest management. The nature is complicated, and the solution for pests and diseases we get from pesticides or other short-term sources is not sustainable. Focus on long-term solutions depending on a thirty-year experience to just control the insects apparently is needed. However, it is possible that initially it may not be economical, but the greater turn toward saving the nature and our crops too (Conway 1997). The actual understanding and implementation of ecological research is the basic challenge to effectively use all components and their interactions.

After discussing the basic theme of ecological pest management (EPM), we can understand the need of this approach to be used in cotton crop pest management program. Ecological pest management (EPM) is an approach to increasing the strengths of natural systems to reinforce the natural processes of pest regulation and improve agricultural production. The basis of this natural method of controlling pests is the biodiversity of the agroecological system. A lot of pests are attracted to cotton crop in addition to disease incidence and weed infestations. There are many techniques used to control these pests including biocontrol, hormones and pheromones, resistant cultivars, cultural and physical control, and application of both synthetic and natural insecticides. Recently, the concept of genetically modified crop plants has attracted the researchers and plant breeders, but unfortunately, small farmers and laggards are mostly dependent on chemical use for pest management, causing severe threats to human health, non-target organisms, water sources, and food crops. Cotton crop has attraction for more than 1300 herbivorous insect species (Hargreaves 1948). All these are not of economic importance, but few have a devastating role in production losses of cotton crop all over the world. Pesticide use is being reduced by the advancement in pest handling technologies (Brookes and Barfoot 2006; Wilson et al. 2004; Naranjo and Luttrell 2008). The biotechnological manipulation for protein production hazardous to some most severe insect pests of cotton is the main factor in the above-said technologies. The ecological pest management of cotton pests is a long-term program which can be adopted to decrease the reliance on insecticides and save the money needed for pest outbreaks as a result of short-term and quick control methods. It is obvious that GMOs are not proving a long-term solution; insect resistance varieties are not adaptable due to poor performance for many insect pests of cotton, so there is a need to use the biology of insect pests and their environment to achieve sustainable control. The criteria for successful EPM is based on the following points:

- Management of resistance.
- Regular monitoring of crop health and natural fauna.
- Prevention of pests and disease spread.
- Preservation and development of beneficial organism's population.
- Maintaining the crop health.

12.2 Crop Management for Pest Management

Pest management is for better crop production or, in terms of economy, for better economic benefits in the shape of food, feed, and fiber. When pests are managed below ETL by adopting a few crop management practices or altering a little in prevailing cultural practice, the net results have extra economic and ecological benefits minimizing uninvited environmental effects and increasing net benefits for the farmers. Short-term and long-term changes in management practices of crop production are carried out for timely and permanent pest management issues. These include INM, IPM, IWM, and those related to site selection, cover cropping, and specific crop production approaches. Collectively, these are referred as ecological modification for better crop production managed with less hazards to the environment. Figure 12.1 presented by Magdoff and Van (2000) is a diagrammatic concept of all system managements including crop, soil and pest management for best crop yields.

Ecological pest management will include growing of pest-resistant cultivar and physical, cultural, and biological control tactics. EPM always recommends only safe limits of pesticides as compared with IPM ideology. Crop management can provide eco-friendly pest management without lowering the crop quality and economic yield. This chapter provides information to cotton crop stakeholders to understand the basic theme of farming with no negative effects on the environment and reduce expenses and pesticide use in cotton crop, which is acceptable to farmers, workers, and consumers. Better understanding of pests (insects, weeds, diseases, mites, and vertebrates), their biology, and factors affecting pest populations (macro- and micro-climate, natural enemies) will provide applicable solutions to grower's problems. Some of these manageable factors are discussed in this section.

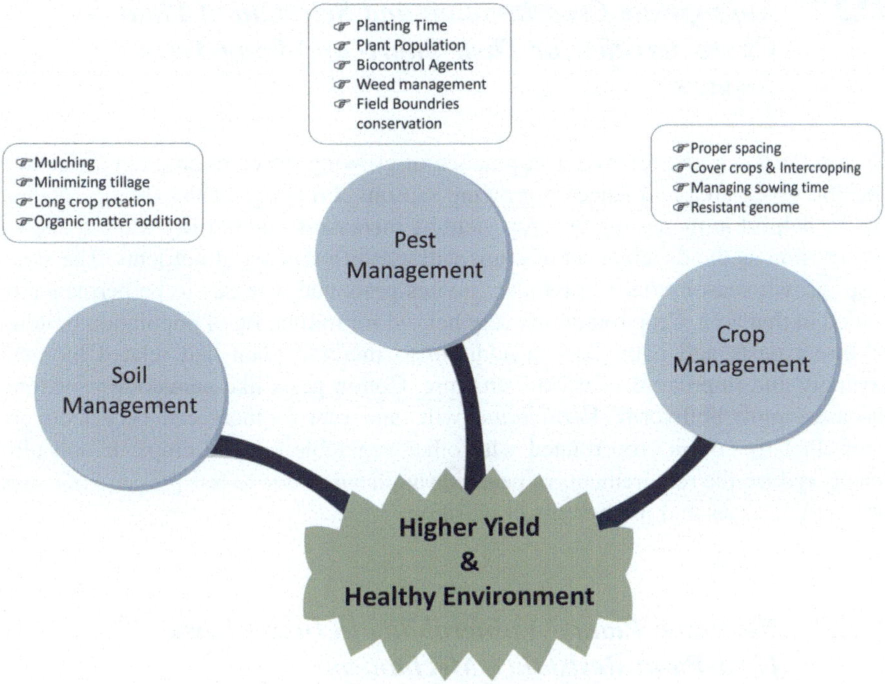

Fig. 12.1 Concept of all-system management for suppression of pest population on cotton

12.2.1 Crop Physiological Characteristics of Indigenous Varieties and Their Exploitation Under Ecological Zones

Physiological plant characteristics are important to influence the insect's population involving the metabolism based plant chemicals. These may be primary or secondary plant metabolites. Primary metabolites of plant include proteins, lipids, enzymes, carbohydrates, etc., for building the plant tissues and supply of energy. Some of these primary plant metabolites are toxicants, nutrients, and antifeedants. Secondary metabolites are not essential for plant growth but act as defense partner of plants. These are usually stored in outer layers of plants and occasionally exudates from the plants. These are detected by the different insects and act as token stimuli (Pedigo and Rice 2014). This physiological mechanism of interaction between crop plants and insects can be used to manage the insect pest of our valuable crops. The local environment provides a higher level of interaction between indigenous plants and insect pests. The need to explore the local plant material is the primary need of ecological pest management.

12.2.2 Appropriate Crop Rotation and Selection of Plant Characteristics for Their Short- and Long-Term Impacts

Crop rotation can be referred as a practice of growing different crops in series in a specific given area and adjacent growing seasons. Properly managed crop rotation can be helpful in increasing the crop yield by increasing soil fertility while reducing soil erosion as the different set of crops will use different set of nutrients. The same crop, grown season after season also, invites pests and diseases to be permanently settled in that area. Crop rotation is very helpful for mitigation of population buildup of insect pests and pathogens in addition to increase plant and related biomass diversity and improvement of soil structure. Cotton pests like nematodes, seedling diseases, pink bollworm, *Verticillium* wilt, and many others can be effectively controlled by cotton crop rotated with other vegetable or field crops. It will ultimately reduce the requirements of pesticide application due to less pressure of insect pest and diseases and improved soil nutrition.

12.2.3 Reducing Plants' Vulnerability to Insect Pests: Host-Plant Resistance Mechanisms

Plants show different response to the induced resistance through environmental factors through changing their morphology or physiology. These modifications may occur in phytochemicals or in the physical defense system of plants. Mostly plants emit volatile compounds in response to induced resistance that may attract different beneficial organisms (parasitoids or predators of pests), and the nutritional quality of host plant is also changed to the unfavorable for the attacking pest species. Induced resistance is found in almost all the plants including cotton. Cotton plants have shown resistance to herbivore insects through three volatile compounds, i.e., tannin, flavonoids, and terpenoid, referred to as secondary metabolites. Secondary metabolite's concentration in cotton plant is changed after insect attack. These changes may be used for pest management and improvement of defense mechanism through induced resistance. The use of CIE is one example to be used in IPM and EPM (Ti and Zhang 2009). Big-eyed bugs feeding on cotton thrips can survive on extrafloral nectar of cotton plants. Manipulation of the extrafloral nectar of cotton plant can be very helpful for managing insect pests ecologically inviting the predators and parasitoids (Yokoyama 1978).

The plant volatiles play a role in attracting or repelling the herbivores and pathogens. The studies regarding volatile compounds of cotton proved that high concentration and species-specific volatiles are released by cotton plants for foraging predators and parasitoids when under attack of specific types of pests. The clarity of odors is found clear when a single species is attacking the plants as compared with clarity when a pest complex is present on plant. These signals are produced during the daytime to efficiently attract the foraging beneficial arthropods (Turlings et al. 1995). This mechanism is stopped when plants are free from attack of pests. The use of this mechanism will ultimately reduce the pest management expenses and efforts of farmer community.

12.2.4 Maintaining Arthropod Biodiversity with Respect to Crop Phases Under Different Agroecological Conditions

In the world, cotton is a unique crop supporting a maximum number of arthropod's community including a diverse natural enemy's community (Dippenaar-Schoeman et al. 1999). It is obvious that these natural enemies can play a pivotal role in cotton pest management. The importance of these has been proved to control the insect pest of cotton crop (Sharma et al. 2007). The natural enemies were greatly affected by the large-scale use of pesticide in cotton crop. In recent years, the use of *Bt* cotton cultivars has played a vital role in the reduction of insecticide usage in addition to adaptation of suitable management tactics of IPM programs. This factor is supporting the option of biological control in cotton crop which requires a large number of natural enemies' maintenance in cotton fields according to the crop with respect to pest attack. The target can be attained by many conducts. Introductory and classical biological control measures have been experienced on many exotic pests of cotton but show minimal impacts (King et al. 1996). China has many active projects for the mass rearing of natural enemies for the control of cotton bollworms through releasing egg and larval parasitoids (Wu and Guo 2003). India is also working on increasing arthropod density in cotton fields for insect pest management but only on a limited scale mostly using *Trichogramma* spp. and *Chrysoperla carnea* (Sithanantham et al. 2005). Augmentation strategy also showed limited results as compared with the conservation of biological control agents through their preservation and manipulation of the local environment to enhance efficacy of local natural enemies. This approach is widely performed in the world against cotton pests through very little efforts of the farmers only in selection of suitable insecticides and not caring about other conservation options. The population of natural enemies has shown an increasing trend all over the world after adaptation of *Bt* cotton (Wilson et al. 2004). Improved conservation strategy for biological control agents through the use of *Bt* cotton has shown sustainable results in many systems. Insecticide resistance in aphid was successfully controlled in *Bt* cotton crop through the conserved natural enemies where sprays were delayed for controlling the cotton bollworms. The aphid attack was epidemic in non-*Bt* cotton fields where the natural enemies were in less numbers due to continuous use of insecticides for controlling the cotton bollworms (Wu and Guo 2003). In the USA and Australia, *Bt* cotton has also provided best options to control whitefly by reduction in repetitive sprays against bollworms (Naranjo 2001).

Many of the natural enemies are found to feed on EFN of cotton plants including *Geocoris pallens* (Hemiptera: Lygaeidae), *F. occidentalis* (Thysanoptera: Thripidae), *Collops vittatus* (Coleoptera: Melyridae), *Orius tristicolor* (Hemiptera: Anthocoridae), and *Notoxus calcaratus* (Coleoptera: Anthicidae). The maximum EFN is produced in late July on cotton plants. The concentration of proteins and soluble solids in nectar is high in premature plants and decreases by maturity which shifts the natural enemies to thrips feeding for fulfilling the protein requirements. These observations have realized the importance of cotton EFN for preservation of natural enemies and their consumption in EPM (Yokoyama 1978). EFN can support many minute predators and parasitoids in the absence of cotton pests in field when floral nectar is also not available to fulfill the moisture and food requirement of these arthropods. Longevity and reproduction of many parasitoids increased when feeding on EFN as compared with parasitoids feeding on nectar-less cotton plants. Predators and parasitoids are frequently visiting the EFN-bearing cotton plants, resultantly providing better biological control options. EFN is useful for retention of parasitoid activity for longer period due to attraction on odor from shorter distances (Rose et al. 2006).

These techniques that help to maintain the natural enemy's population in cotton field can help the cotton farmers to adopt eco-friendly management practices to reduce the expenses and pesticide use.

12.2.5 Incorporation of Genetic Plant Resources in Combination with Ecological Crop Needs

Genetically modified crops are the modern source technology needed for integrated pest management in high valued crops like cotton. GM crops are controlling the pests by posing or altering detrimental biological activity against insect pests, so these are ecologically safe as compared with insecticide spraying on cotton crop. *Bt* cotton is a selective form of host-plant resistance producing Cry toxins from the incorporated *Bacillus thuringiensis*, virulent for lepidopteran pests meeting the requirements of present challenges to cotton crop. Biological control can enhance due to reduction of insecticide sprays only used for lepidopteran pests on cotton crop (Naranjo 2001). Cry toxins are specific against a specific group of insects. In the beginning, only lepidopteran pests were the target of GM cotton producing toxins. Cry1 and Cry2 toxins controlled by single gene were specific against cotton bollworms, i.e., *Helicoverpa armigera* and *Pectinophora gossypiella*, in addition to 32 target species. Nowadays, leafhoppers, leaf worms, and semiloopers are also targeted by these modifications in cotton genetics (Benedict and Ring 2004). The recent advancement in GM cotton having double gene (Bollgard-II, Widestrike) is capable of wide range targeting of lepidopteran insect pests of cotton (Adamczyk and Gore 2004; Adamczyk Jr et al. 2001; Fitt and Wilson 2000; Chitkowski et al. 2003). Double gene *Bt* cotton is grown on wide areas in the USA, Australia, Mexico, and China with constant increase in area from 1994 onward. The success of GM cotton is maintained by constant monitoring of resistance development in insect pests and sowing of non-GM cotton in area along with GM cotton for dilution of resistant alleles in target pest populations. The new developments are based on targeting other than lepidopteran cotton pests. The scope and safety of GM cotton and other GMOs are debatable and under consideration by many countries (Naranjo et al. 2008). Wei-Di et al. (2003) have described the positive impact of transgenic cotton sowing as compared with the conventional farming regarding the composition and structure of pest and predator's community. The observations showed that nineteen pest species from 15 different families were found on transgenic cotton while seventeen pest species from 14 different families were found on conventional cotton treated with pesticides. The farmer-friendly arthropod community was more diverse on transgenic cotton, i.e., 39.3 from 24.3 families, as compared with insecticide-treated conventional cotton, i.e., 38 species from 25 families.

12.2.6 Use of Proper Sanitation System, Agroforestry, and Intercropping Approaches

Cotton crop is always facing many of arthropods (beneficial and harmful), diseases, and weeds at the same stage of growth which makes creation of decisions very complex. Many cultural practices like sanitation, tillage, and crop rotation have wide range impacts on cropping system. The measures taken to suppress a pest can cause the disturbance in other related crop factors. The pests can be managed by understanding and manipulating the interaction of different organisms within agroecosystem without disturbing the balance of system. However, it will need interdisciplinary and multidisciplinary extensive research for implementation of EPM in cotton crop (Abrol and Shankar 2012).

Smith and Reynolds (1972) have recommended the manipulation of cotton agroecosystem based on both conventional cotton cultivation practices and novel management practices. Intercropping of lucerne with cotton crop and maintenance of the pest population below ETL were reported to allow the survival of predators and parasitoids. Clean culture through weed management, maintenance of vegetative borders around the cotton fields, and spatiotemporal structure management of cotton field were also proved best options for the EPM (Cook et al. 2007). Tropical developing countries are already using the multiple cropping system for increasing the economic yield as well as food items. It is an important factor that crop diversification mostly results in pest population reduction. The term "trap crop" was used to describe this phenomenon of pest management. The positive role of trap crops, alley crops, and mixed crops can be enhanced using best crop geometries ideal for pest reduction (Shelton and Badenez-Perez 2006; Altieri and Nicholls 2004).

Intercropping is practiced mostly in China and diversified on large scale. Winter wheat and cotton are intercropped for the management of early season cotton aphid.

Fig. 12.2 Intercropping of maize and cotton for yielding higher along with pest management strategy

Medicago sativa is also grown around the cotton field for development of ladybird beetles (*Propylea quatuordecimpunctata*, *Coccinella septempunctata*, and *Hippodamia variegata*) and other natural enemies of bollworms. Cutting of lucerne every time results in movement of natural enemies feeding on non-cotton aphid (*Therioaphis maculata*) on lucerne to the cotton crop where they feed on cotton aphids (*A. gossypii*) (Lin et al. 2003). Figure 12.2 shows the concept of corn intercropping with high benefits on the same area and cropping season as compared with cotton or corn single sowing. Comparisons between the treatments showed significantly higher average incomes for the cotton. De-Sousa (2007) has supported this approach revealing the less number of pesticide application on cotton intercropped with corn as compared with pure cotton cultivation.

Alley cropping of forest trees in cotton fields, e.g., *Paulownia* and elm, in most regions of China attracts the *H. armigera* female for egg laying (Yin and He 1997). The young ones of *H. armigera* cannot survive on these plants later on. The utilization of forest trees as wind breaker, fuel, and furniture wood is additional and off-season income for the farmers. Phytosanitary should be maintained for better results of cotton and these agroforest trees. These are termed as ecological infrastructures providing physical linkage and survival space to indigenous fauna, i.e., hedgerows, corridors, and other generalist predators (Altieri and Nicholls 2004). EMP program must have the biological control, cultural control like crop rotation, clean culture, and agroforestry-based cultivation as it was practiced before World War II (Mason 1928).

12.3 Soil Management for Pest Management

A vigorous and healthy soil is a vital source of economical agricultural output using minimum inputs. Cotton is an exhaustive crop depleting the soil resources which are of ultimate importance. If soil is not properly managed, the ultimate reduction in yield occurs along with off-site contamination. Crop management should be ensured to be safe for the soil health ensuring the good crop yields. Biodiversity of an area is always based on the habitat quality and structure. Removal of natural habitats and vegetation for crop production is a negative factor of land use causing the biodiversity loss. Biodiversity of a specific area is linked to survival of many communities in that area. The disturbance in land use system can disturb the breeding sites, migratory routes, and nesting places of many species including natural enemies of cotton insect pests. The single crop cultivation on a large area for many years can promote the population development of specific organisms which might attain the status of pest. A healthy soil should be maintained to support maximum biodiversity along with cultivated cotton for management of potential pests (Fig. 12.1). To lessen the negative impact on biodiversity, cotton farmers should preserve the soil health through adopting proper soil management practices. Some of the prominent soil features needing attention are discussed below.

12.3.1 Maintaining Soil Characteristics Under Different Resource Input Systems

Growing of healthy crops needs right soil, right time, and proper ways in maintaining the soil health by crop rotation, farmyard manures, green manures, and fallowing the land which is lacking in modern farming systems. The diversity of soil and climate of cotton-producing areas all over the world have been explored through experimentation and revealed that genetically modified cotton varieties are insect and weedicide tolerant allowing the phytosanitary conditions but depleting soil nutrition. For sustainable cotton production, soil health and environment preservation are of utmost importance. The plant cover and mulches are the solution of soil fertility loss in tropical areas due to soil erosion and weeds (AFD 2006). In many areas of Brazil, cotton crop is followed by cover crop producing maximum biomass for maintaining the soil health (Seguy et al. 2004). Care should be taken as some cover crops may lead to the development of certain pests like *Spodoptera frugiperda* (Ratnadass et al. 2006). It is important to replenish the soil nutrients through time-to-time balance fertilization. For sustainable cotton production, proper dose of fertilizer at right time according to the soil condition and cotton varieties should be adopted. Other than cover crops and crop rotation with legumes, cotton stubbles are a rich source of nutrition containing nitrogen, phosphorus, and potash with a percentage of 1.11, 0.1, and 3.98, respectively. Cotton stubbles can be incorporated into the soil after chopping into pieces or powder. These can also be used for vermicomposting and then application into the cotton fields for increased nutrition and soil health. Microbial cultures are available commercially to hasten the process of decomposition. The role of healthy soil for ecological pest management is well described by Hodson and Lewis (2016) in words "Managing for soil health could become an integral component of pest management, resulting in more resilient and productive cropping systems that provide multiple agronomic and environmental benefits."

12.3.2 Least Tillage and Soil Compaction, Practices for Soil Management from Crop Sowing to Harvest

In traditional agriculture system, tillage and plowing are primary practices for crop production. Cotton cultivation is mainly considered to base on heavy tillage practices. The EPM requires minimum level of soil disturbance, and when disturbance is required from sowing to harvesting, intensity and frequency should be managed to a minimum for increasing the density and diversity of beneficial organisms (Peter-Duelli et al. 1999). The studies have revealed that minimum or zero tillage fields provide heathy habitat to earthworms and many other farmer-friendly invertebrates having capacity to maintain the minimum constant pest pressure. The pest management cost is not applied in these habitats or is at minimum level (Stinner and House 1990). More stable habitats are reported from non-till agricultural systems having an increased number of predators and parasitoids controlling increasing pest populations throughout the year. The pesticide application in these areas can reduce the predator's diversity and density opposing the positive impacts of zero tillage or reduced tillage areas resultantly increasing pest populations or resurgence of specific crop pests (Blumberg and Crossley Jr 1983). Furlan and Kreutzweiser (2015) have cleared the situation that benefits of non-till system can be wasted by adopting other counter-action pest management tactics, i.e., pesticide use. To avoid these conditions, reduced tillage system is more helpful in modifying the physicochemical system for sustainable cotton production using other cultural pest management techniques as well. Many invertebrates and vertebrates are helpful to increase the fertility and porosity of soil in reduced tillage system (Stinner and House 1990; Fawcett and Towery 2002). The studies conducted in the US Cotton Belt proved that all pests are not reduced in this system, but cutworms, grasshoppers, and aphids are controlled below ETL after adopting reduced tillage system (McCutcheon 2000).

12.3.3 Nutrient Cycling, Energy Flow, and Impact of Crop Residues on Soil Improvement

Natural systems are based on efficient energy flow systems. One of these energy flow systems is agroecosystem where energy is captured by chlorophyll of green plants and consumed by many organisms (mainly animals). Bacteria and fungi are facultative organisms that decompose the dead matters of some organisms to consumable compounds of other organisms of food web. Natural ecosystem is also efficient in capturing the rainfall and preserving it for nutrient cycling. The loss of nutrients by leaching and contamination of ground water is managed by natural recycling of nutrients and efficient energy flow from the low level to the uppermost trophic level. These nutrients and successful energy flow provide healthy plants and help us in suppressing pest pressure and diseases (Magdoff and Van 2000). Farmers can get benefits from natural nutrient cycling at their farms for sustainable cotton production adopting the ecological strategies. On farm, cycling of nutrients should be promoted, decreasing long-distance energy (nutrients) in cotton farmers. Increasing organism's diversity through minimum disturbance of natural ecosystem will show healthier impacts on soil and its related organisms. Arthropods from different orders and families play an important role in ecosystem functions including energy flow, pollination, nutrient cycling, and maintenance of soil structure (Tscharntke et al. 2005; Petchey and Gaston 2002).

12.3.4 Crop Rotation to Improve Soil Microbiota and Reduction of Pest Stresses

Soil nutrition levels are affected due to different crops grown on the soil for a specific period, and this has an impact on the yield of the next coming crops. Crop rotation, based on previous cropping history, is proved as a key factor in recommendation of production technology of the next crop. Nutrition uptake and consumption are different for all different crops; understanding of this factor will lead to adaptation of the best strategy for maintaining the soil health when deciding for next crop sowing (Karlen et al. 1994). Nitrogen-deficit soils can be chosen for the cultivation of nitrogen-fixing leguminous crops, e.g., mung bean and alfalfa or lucerne (Russelle et al. 1987). In California, fertilizer plan for cotton is approved on the base of fertilizer applied to previously grown crop (Forbes and Rosenheim 2011). Many crops are causing the abundance of many insect pests and plant diseases (Benson 1985). *Spodoptera litura* is a potential pest of cotton attracted to *Sesbania sesban*, if grown nearby cotton fields. The susceptibility and resistance of different crops to insects vary according to the physico-morphic and biochemical composition of these crops and season. The identification of peak pest population period and susceptible crop can help the farmers to rotate their crop accordingly. Continuous planting of susceptible crop may lead to the pest outbreak in this area. The crop rotation with resistant or non-preferred crop of specific pests will reduce pest population in an area (Rothamsted 2006). Weed management is also based on the crop rotational history, ultimately affecting pest population (Liebman and Dyck 1993).

Meisner and Rosenheim (2014) have described the option of crop rotation for sustainable cotton yield. Bayesian models were used to analyze the eco-informatics data for the cotton-growing areas of California. The studies revealed that many crops grown in a field before cotton were influencing pest positively and some other crops grown before cotton were negatively associated with certain pests, e.g., corn cultivation was negatively associated with *Lygus Hesperus* population. The further area-wise research on different crops compatible with cotton cultivation may lead to liable cropping pattern for EPM of cotton insect pests.

12.3.5 Developing Sound Agroecological System for Soil Manipulation and Improvement

Ecosystem is very important as well as the human beings. All approaches based on human gains will ultimately destroy the ecosystem, and as a result, humans will suffer. Human needs are not only the priority. Agriculture is based on the human-regulated system as a subsystem of natural ecosystem. The subsystem products should not be harmful to the natural ecosystem; otherwise, agriculture and human-regulated systems will collapse (Ohlander et al. 1999). As discussed earlier, soil health and condition are important for pest incidence and ecology of crops. In case of bad soil health, plants are more susceptible to pests and diseases. Safe soil management and improvement in its health including increased organic matter will help to better cotton production. Biodiversity of plants and arthropods on earth is a parameter to understand soil quality and capacity to support life. Healthy soils are in strong association with ecosystem function like water storage, nutrient cycling, energy flow, and supporting living organisms. Land management decisions are a guarantee for ultimate productivity of crops and health of animals including humans (Doran and Zeiss 2000).

12.4 Mechanisms and Tools for Pest Management

Pest management itself is different from the pest killing tactics. Management also differs from the pest control philosophically. Management refers to the judicious use of different tactics and strategies in reducing the damage to tolerable levels. EPM is one of the oldest and less expensive pest management strategies. This approach utilizes the weaknesses in seasonal cycles of pests and limitation of resources for pests. The tools of EPM are to narrow the host range, limit their dispersal, and stop their seasonal movements. Preventing pest early is preferred over controlling on later stages (Pedigo and Rice 2014).

12.4.1 Population Dynamics of Pests and Their Interaction at Trophic Levels Under Multi-Pest Situations

The best way to control a living organism is to manage the weaker stage or to disturb it at the most susceptible stage. Trophic levels also play an important role in the survival of insect pests in multitrophic food webs. Plants are considered as the most crucial component of multitrophic interaction between insects, crops, and other organisms. Plants are necessary for the pests and natural enemies (Lewis et al. 1997). There may be many pests on cotton crop categorized as primary, secondary, and tertiary pests. Some living organisms and arthropods are also part of food web and may play a role of potential pest. Tri-trophic interactions are reported between plants, pests, and beneficial arthropods. These multilevel interactions are interwoven tightly and have the capacity to play a significant role in EPM on sustainable bases. Dicke and Sabelis (1987) have reported the impact of plants releasing token stimuli on pest population. These token stimuli attract the predators on crop plants which feed on pest population, and plants are benefitted manipulating three trophic levels of food web. Some cotton varieties have more potential to attract the predators and parasitoids as compared with traditional cultivars, resultantly attacking herbivores. Pest population and natural enemy's growth rates are also linked affecting the overall efficacy of EPM. The missing trophic level can disturb overall population dynamics, so maintenance of each level in sufficient quantity is important in EPM.

12.4.2 Preventive Strategies Through Natural Defenses for Better Ecological Pest Management

EPM is mainly dealing with the before the time management and not waiting to attain ETL by pest population. Preventive control measures are briefed by Morales (2002) based on the agroecological theory. Preventive measures include repellents, trap crops, site selection and soil management, intercropping, and post-harvest management. For efficient results of preventive strategy, the knowledge of crop and pest's ecology, experience of implementation, and strong social cooperation and links are required. Other preventive measures as crop rotation, avoidance from mono-cropping provision of agro forest trees on boundaries of cotton field, leaving unsprayed strips, and conservation of natural enemies are helpful for pest outbreaks in cotton crop (Altieri 1994). The role of crop rotation and soil management for EPM is described in the beginning of this chapter.

12.4.3 Insect Behavioral (Push-Pull Strategy), Physiological, and Molecular Mechanisms for Exploitation of Supplementary Pest Management

Pest management through push-pull strategy is basically based on the behavioral manipulation of natural enemies and insects to be controlled using the stimuli. Stimuli are used for pushing of the pests from the protected crops making them unsuitable or unacceptable while pulling these pests to another attractive source and finally removing them from the targeted crop. Push-pull strategy is based on the non-toxic components and safe for the environment. This strategy is one of the potential components of IPM and EPM of cotton pests. Cook et al. (2007) described the components of push-pull strategy and correlated with management of many agricultural pests. There are two parts of this strategy. First one is to push away the undesirable organisms (pests) from protected crops using avoidance behavior or making the crops undetectable due to non-acceptable plant chemicals and morphology. Long-range and short-range stimuli are found deterring the pests from crops. Long-range stimulus acts as first line of defense, while the short-range stimulus is mostly pest specific. Second is to pull the pests and divert from protected crops. It is achieved mostly through lures, trap crops, or traps. This is useful to detract the pests from the long distances. According to Cook et al. (2007), visual cues, synthetic repellents, non-host volatiles, host-derived semichemicals, anti-aggregating pheromones, alarm pheromones, oviposition deterrents, and oviposition deterring pheromones and antifeedants are the stimuli for the push component of push-pull strategy, while visual stimulants, host volatiles, sex and aggregating pheromones, and gustatory and oviposition stimulants are stimuli for the pull component of push-pull strategy.

In cotton crop, *Helicoverpa armigera* (a polyphagous lepidopteran) can be controlled through using neem-based spray as a push component and an effective trap crop, i.e., pigeon pea, as a pull component of the strategy. Management of *Helicoverpa punctigera* and *H. armigera* has been reported in Australia by Pyke et al. (1987). The egg deposition was up to 92% on cotton crop in the presence of pigeon pea as a trap crop. The efficacy of trap crops may be increased using the sugar spray. Neem-based spray on cotton and okra as a trap crop has also managed the *H. armigera* successfully (Duraimurugan and Regupathy 2005). Trap crops with increased concentration of sucrose and the molecular level change in cotton plant physiology will open new ways for effective pest management without the use of insecticides. Figure 12.3 is concept clearing for understanding the mechanism of push-pull strategy. The use of repellent plant extracts or plants with repellant volatile compounds can be substituted according to dominant pest species of the area. The okra planting on cotton boundaries will act as a pulling component for *H. armigera.*

12.4.4 Relationship Between Insects, Their Biological Control Organisms, and Surrounding Crops and Wild Vegetation

Agroecosystem management is important in EPM of cotton herbivores. The efficacy of biological control agents in increased through stability of ecological conditions. The relationship understanding between insects and their surroundings and its manipulation for forecasting of pest population development is important to keep them at equilibrium level. Increase in crop production at the expense of natural vegetation destruction has led to the development of monoculture. Monoculture failed to support natural biodiversity of biological control agents and the outbreak of single or few pest species in cotton fields. Destruction of natural and wild vegetation seriously disturbs the density and function of natural enemies due to their dependence on natural vegetation for alternate food sources, shelter, nesting, pollen, and

Fig. 12.3 Theoretical demonstration of push-pull strategy for ecological based pest management

hibernation sites. Diverse cropping system should be designed for the equilibrium of insects, biocontrol agents, and vegetation. Altieri and Letourneau (1982) have suggested manipulating the vegetation on field banks, road sides, and water channel bunds for maintenance of arthropod composition in an area. It will also work as windbreaks and shelter belts of birds and generalist predators and pollinators. Within cotton field, diversification can be attained through multi-cropping, trap crop strips, and even planting non-variety plants susceptible to pests. This will affect the population dynamics of insects and pests. The populations dynamics can also be manipulated through cover crops, weed management, and mulching practices.

12.4.5 Host-Plant Resistance Mechanisms: Antixenosis, Antibiosis, and Tolerance for Insect Pests and Their Interaction with Natural Biocontrol Systems (Insects, Pathogens, Other Arthropods, Vertebrates, Etc.)

Host-plant resistance (HPR) is a successfully used tactic of IPM and has a wide scope in EPM also. HPR utilizes the interspecific plant variation for pest management options. The development of resistant varieties through plant breeding for lowering the pest population and increasing the plant tolerance level of integration of both is a basic mechanism of HPR. Tumlinson et al. (1993) have described plant volatile compounds as part of HPR. Plants have many toxins and other chemicals deterring the herbivore insects. The recent studies are supporting HPR for its utilization in IPM and EPM at interactive and multitrophic situations.

The resistant varieties have increased level of already existing allomones or newly induced allomones. Decreased concentrations of kairomones also act as mechanism of resistance. Physical properties are also part of resistance mechanism and used in cotton pest management. Pedigo and Rice (2014) stated that the plant characteristics involved in resistance mechanism can be mediated by the environmental conditions. The genetic resistance is categorized by many scientists, but the widely used classification is that of R.H. Painter. According to this classification, non-preference (antixenosis), tolerance, and antibiosis are the major modes of resistance mechanism.

Non-preference utilizes the plant characteristics deterring pests away from the plants which need to be protected. This mechanism uses behavior of plants and insect pests as well. Morphological non-preference and allelochemical non-preference are two major types of this mechanism. First one results from the structural plant parts that disrupt the insect's normal behavior, e.g., feeding and oviposition. Second one is common in plants causing their rejection by pests due to the presence of unacceptable physiological and chemical composition. *Antibiosis* is the presence of plant metabolites which impairs the insect metabolism after feeding on these plants, and an insect is not able to complete its life cycle normally. Allelochemicals are mostly working as antibiosis type of resistance. Gossypol compounds in cotton plant are examples of antibiosis causing agents. The ability of a plant to give satisfactory yield even after the attack of insect pests up to injury level is called *tolerance*. General vigor of crop plants is considered to tolerate the attack by many scientists, but the background of this type of resistance is not much clear. Wound-healing ability of plants, mechanical tissue support, and photosynthesis partitioning are major components for tolerance.

12.4.6 Utilizing Environmental Conditions for Effective Utility of Parasites, Pathogens, Parasitoids, and Predators

Environmental conditions play a vital role in regularization of pest population. The augmentation of biological control agents through the manipulation of environmental conditions is the best way to increase their population and efficacy. The surrounding crops and plants of a cotton field may play their role as insectaries, naturally conserving the natural enemies and helping the existing numbers. The provision of nesting and breeding place, overwintering sites, and alternative food sources will ultimately increase the reproduction of natural enemies (Pedigo and Rice 2014). Extrafloral nectaries are best examples of alternation nutrition provision to the foraging predators and parasitoids. These EFN increase the efficacy and retention of potential parasitoids through attracting them from the nearby fields. Armyworms and budworms in cotton fields are controlled by adopting this strategy through manipulating the population of *Cardiochiles nigriceps*, *C. marginiventris*, and *Microplitis croceipes* (Stapel et al. 1997; Lewis and Takasu 1990).

12.4.7 Identification and Conservation of Natural Resources Needed for Ecologically Based IPM

Natural resources are God's gift for the farmers, if successfully manipulated. Identification and conservation of natural resources are important part of any sustainable strategy to control pests. The need of conservation includes manipulating their environment for increasing beneficial impact on natural ecosystems. Natural resources conservation refers to reduction of the human disturbance in lieu of mechanization and urbanization. Natural areas and resource around the cotton fields are providing biodiversity and natural control of insect, pests, and diseases. These can be referred as ecosystem services for farmers. Pest suppression by natural enemies is flourishing on natural vegetation, and pollinators increasing the oil content of cotton seed are also habitants of natural vegetation. Resistance management in cotton pests through providing them non-spray zone in the shape of natural vegetation and increasing their susceptibility to pesticides justifies the need of natural resources conservation. Natural resources like soil fertility, nutrient recycling, and carbon sequestration are basic needs of a healthy agroecosystem. For natural control of cotton pests, healthy landscape is required with standing vegetation providing shelter and food to birds, predators, and parasitoids. Natural vegetation may be of varying types, e.g., land covering shrubs, trees, grasses, etc. These are managing weeds, protecting soils from erosion, saving water from evaporation, and habituating a lot of natural fauna.

The effective EPM is only possible when the farmers are identifying and conserving the natural resources near their field crops such as natural vegetation, bushlands, clean and healthy rivers with protected aquatic habitat, and healthy soils with ground covers improving the soil biology. To improve the natural environment, the farmer should adopt the following conservation tactics:

- Identification of natural flora and fauna and their impact on cotton production.
- Improvement of soil health through cover crops and mulching.
- Water management for farm irrigation through efficient irrigation systems.
- Proper draining of rain and flood water.
- Maintenance of natural vegetation and its protection from livestock.
- Plantation of year-round flowering plants to provide shelter for the natural enemies of cotton pests and pollinators.
- In case of *Bt* cotton varieties, sowing of non-*Bt* cotton in strips for managing resistance development in pests.

12.4.8 Grower-Friendly Diagnostic and Monitoring Methods of Precision Agriculture and Their Stable Delivery for Effective Ecological Pest Management

Precision agriculture has taken its place through the advances in centralized and Big Data. Traditional agriculture is being occupied by precision agriculture in developing countries. After launching the GMOs and high yielding varieties, yield optimization is in way through advances in data collection analyzing, decision making, and management of issues on molecular level rather than traditional approaches of present agriculture system. There is need of precision agriculture for lowering down the environmental hazards, input costs and wasting and deterioration of natural resources and addressing human health problems caused by injudicious use of chemical in agriculture industry. The use of yield mapping, GPS, and VRT technology for the optimization of yield and profitability on sustainable bases reducing the environmental hazards has opened a new era of site-specific crop management. Cotton cultivation all over the world is utilizing fertile soils, fit for irrigation water, and a lot of chemical applications in shape of fertilizers and pesticides. The precision agriculture will reduce input cost while considering the producer, consumer, and environmental health problems. Precision agriculture techniques help the farmers through recognizing complex situations and their handling on bases of efficient monitoring and detection methods (Kaushik and Mishra 2013).

The notion of precision agriculture was started with the idea of grid sampling, fertilizer plan mapping, soil pH calculations, and impact of these factors on yield factors. Utilization of GPS in agriculture for accurate positioning and site-specific applications through drone technology on sensor and remote-based decisions is the base of precision agriculture. The materials used for grower-friendly diagnostic and monitoring methods include GPS mounted drones, sensors, and remote equipment for the data collection and decision making by real-time adjustment equipment and geo-specific mapping and air mounted artificial intelligence-based applicators. New technologies like automatic steering systems, yield monitors, precision seedling systems, cell phone applications, farm advisory software, centralized weather data forecasting online system, and variable field applicators are also part of precision agriculture helping the farmers. Nowadays, GPS has an efficient system to locate position within centimeters, and farmers are able to grow crops accurately without depleting the natural resources and minimizing the input costs.

One of the grower-friendly diagnostic systems was reported by Ranjitha et al. (2014) using the hyperspectral radiometry for monitoring of pest and disease attack on crops. Satellite remote sensing was used to interfere the physical structure of crop plants, and diagnosis of thrips damage was done based on spectral indices. Azfar et al. (2015) reported wireless sensor network (WSN) as an efficient tool for grower-friendly diagnostic services and farm monitoring. WSN has efficient real time data monitoring and ability of specific application especially for pest management parameters. Pest status, identification, and disease detection are main problems of cotton growers which are covered by the WSN including proper fertilizer use indication, irrigation management, sowing and harvesting season optimization, and evaluation of crop condition. WSN will use many data stations and remote sensing systems for monitoring of pest status and controlling through trapping applications. Battery-operated low-cost system with wireless image sensing is also in progress to capture and analyze the pictures of crop damaging factors (Lopez et al. 2012). During these all processes, no human intervention will be required. According to the authors, the cost and efforts for monitoring of pest problems will be reduced to much lower level. This approach will help in EM of cotton pests without any hazards to the environment and agroecosystems.

12.4.9 Implementation and Evaluation of Tested and New Ecologically Based IPM Technologies Under Multidisciplinary Cooperation

The pest management technologies, developed by the researchers and institutions, are beneficial only when these enable the farmers to manage their problems efficiently. So, joint ventures are required to implement the EPM successfully on field level equally beneficial for huge farming land to a few acres of culturable land systems. Improved linkage between the universities, research institutes, and private and public stakeholders of cotton-growing community can make this concept of pest management viable and efficient. Monitoring and implementation by government authorities will make it possible to be adopted by farmers. Strong and sustainable efforts are required to dissipate this approach among the common farming community of developing countries. Proper communication with farming stakeholders emphasizing benefits of this approach is of great importance. Comparison of different pest control systems and their impacts on environment and human health in front of farmers will create easiness for EPM adaptation. Government bodies, ministers, researchers, and public-sector agencies are counselled to adopt first these initiatives.

12.5 Concerns Related to EPM

12.5.1 Advantages of EPM

- Consumption of pesticides and fossil fuels will be reduced.
- Cost of equipment and labor will be minimized.
- Energy saving up to 30% (Pimentel et al. 2005).
- Safety of natural flora and fauna including pollinators, predators, and parasitoids.
- Due to conservational strategy, reduction in soil erosion, run-off water, etc.
- Restoring ground water quality with minimized used of chemicals.

12.5.2 Disadvantages of EPM

- Strong pests which cannot be controlled without pesticides.
- Adoptable only by educated farmers.
- Biological control is not quick fix.
- Need long-term policies.
- Biocontrol agents will be different for different environments, e.g., one biocontrol agent will not perform equally in all soil types.
- Slow incorporation among society.
- Natural vegetation can also exploit some pest species.

12.5.3 Obstacles to Adopt EPM

- Need multidisciplinary holistic research.
- The lack of practical agriculture scientist able to integrate theory and on-farm application.
- Reluctance of private firms and stakeholders.
- Difference of opinions between pest management experts.
- Ecological based strategy is not business oriented as pesticide control.
- Monoculture developed and hard to reverse.

12.6 Conclusion

Ecological based management for cotton pest is better as preventive strategy rather than reactive. On farm management for site selection, cropping system, fertilizer plan, tillage, and water management plan will shape our farm's natural defense. The rigorous research will ultimately address the arthropod management in organic cotton cultivation programs. The concept of applied ecology seems prevailing in EPM approach including the crop and crop stage specific agroecological factors. Biochemical and agroforest ecology will have a broad scope in cotton industry for sustainable arthropod management plans. The understanding of silent war between arthropods and our grown crops is itself a weapon to deal with it. Insect-plant interaction, soil-plant interaction, and other organisms included in food webs related to human communities are continuously in process of evolution to compete on each trophic level. Number of plant traits including defense mechanism, biochemical interaction with insects, and morphological structure development for anti-adaptation of pests are developed over millions of years. The maximum cotton production achievement by adopting all these plant traits and interactions will decrease reliance on pesticides and synthetic chemicals. Cotton breeders should focus on plant induced resistance mechanisms against insects through gene manipulation. GM cotton is safe for biodiversity and ecosystem without open chemical use for controlling the pest population. Conservation of natural resources including natural enemies is a major component of EPM. The natural resource conservation is dealing with minimum or no tillage, efficient use of irrigation water, maintenance of biodiversity of all local flora and fauna, crop rotational plans, use of semiochemicals as natural defense, host-plant resistance, acceleration of biological control agents, and all manipulations safe for agroecological system. Spatiotemporal crop management through crop rotation, intercropping, and sowing time management will reduce the pest adaptation rate on crops and increase the biocontrol agents. The only option of pesticide and chemical applications should be retained only to precisely operating systems after monitoring based on sensors and GPS. The use of GIS technology, air mounted monitoring, and site-specific application devices is the future of educated and research-oriented cotton farmers. The change of pest management strategy from conventional to EPM needs field-to-field, farm-to-farm, and ecosystem-to-ecosystem change in cropping system and socio-economic relations. The EPM is actually stability of cropping system involving multidisciplinary fields and social and political interactions. The benefits of EPM in cotton crop are tangible for human and ecosystem. However, a chance of unwanted pest proliferation remains a threat in chemical-free cotton cultivation, which may cause a short-term loss, but long-term policies will reduce these threats and enable the harvesting of other benefits.

References

Abrol DP, Shankar U (2012) History, overview and principles of ecologically-based pest management. In: Abrol DP, Shankar U (eds) Integrated pest management: principles and practice. CAB International, Cambridge, MA, pp 1–26

Adamczyk JJ Jr, Adams LC, Hardee DD (2001) Field efficacy and seasonal expression profiles for terminal leaves or single and double *Bacillus thuringiensis* toxin cotton genotypes. J Econ Entomol 94:1589–1593

Adamczyk JJ, Gore J (2004) Laboratory and field performance of cotton containing Cry1ac, Cry1f, and both Cry1ac and Cry1f (Widestrike™) against beet armyworm and fall armyworm larvae (Lepidoptera: Noctuidae). Fla Entomol 87:427–432

AFD (2006) Le semis direct sur couverture végétalepermanente (SCV). AFD, Paris, p 68

Altieri MA (1994) Biodiversity and pest management in agroecosystems. Food Products Press, New York

Altieri MA, Letourneau DK (1982) Vegetation management and biological control in agroecosystems. Crop Prot 1(4):405–430

Altieri MA, Nicholls CI (2004) Biodiversity and pest management in agroecosystems. Food Product Press, New York

Azfar S, Nadeem A, Shaikh AB (2015) Pest detection and control techniques using wireless sensor network: a review. J Entomol Zool Stud 3(2):92–99

Benedict JH, Ring DR (2004) Transgenic crops expressing Bt proteins: current status, challenges and outlook. In: Koul O, Dhaliwal GS (eds) Transgenic crop protection: concepts and strategies. Science Publishers, Enfield, NH, pp 15–84

Benson GO (1985) Why the reduced yields when corn follows corn and possible management responses. Proc 40th Corn Sorghum Res Conf. pp 161–174

Blumberg AY, Crossley DA Jr (1983) Comparison of soil surface arthropod populations in conventional tillage, no-tillage and old field systems. Agro-Ecosystems 8:247–253

Brookes G, Barfoot P (2006) Global impact of biotech crops: socio-economic and environmental effects in the first ten years of commercial use. Ag Bio Forum 9:139–151

Chitkowski RL, Turnipseed SG, Sullivan MJ, WCJr B (2003) Field and laboratory evaluations of transgenic cottons expressing two proteins compared with one of *Bacillus thuringiensis* var. *kurstaki* Berliner for management of noctuid (Lepidoptera) pests. J Econ Entomol 96:755–762

Conway G (1997) The doubly green revolution: food for all in the twenty-first century. Penguin, London, UK

Cook SM, Khan ZR, Pickett JA (2007) Theuse of push-pull strategies in integrated pest management. Annu Rev Entomol 52:375–400

De-Sousa HFA (2007) Effect of strip intercropping of cotton and maize on pest's incidence and yield in Morrumbala District, Mozambique. In: 8th African Crop Science Society Conference, El-Minia, Egypt, 27–31 October, 2007. African Crop Science Society. pp. 1053–1055

Dhaliwal GS, Arora R (2001) Integrated pest management: concepts and approaches. Kalyani Publishers, New Delhi

Dhaliwal BK, Walia US, Brar LS (1998) Response of *Phalaris minor* Retz. Biotypes to various herbicides. Indian J Weed Sci 30:116–120

Dicke M, Sabelis MW (1987) How plants obtain predatory mites as bodyguards. Netherlands J Zool 38(2):148–165

Dippenaar-Schoeman AS, van den Berg AM, van den Berg A (1999) Spiders in south African cotton fields: species diversity and abundance (Arachnida: Araneae). Afr Plant Prot 5:93–103

Doran JW, Zeiss MR (2000) Soil health and sustainability: managing the biotic component of soil quality. Appl Soil Ecol 15(1):3–11

Duraimurugan P, Regupathy A (2005) Push-pull strategy with trap crops, neem and nuclear polyhedrosis virus for insecticide resistance management in *Helicoverpa armigera* (Hubner) in cotton. Am J Appl Sci 2:1042–1048

FAO (2001) World review of the state of food and agriculture. Food and Agriculture Organization of the United Nations, Rome

Fawcett R, Towery D (2002) Conservation tillage and plant biotechnology. How new technologies can improve the environment by reducing the need to plow, CTIC, Purdue University, pp. 20

Fitt GP, Wilson LJ (2000) Genetic engineering in IPM: Bt cotton. In: Kennedy GG, Sutton TB (eds) Emerging technologies for integrated pest management: concepts, research and implementation, proceedings of a conference, Raleigh, NC, USA, 8–10 March, 1999, pp. 108–125

Forbes AA, Rosenheim JA (2011) Plant responses to insect herbivore damage are modulated by phosphorus nutrition. Entomol Exp Appl 139:242–249

Furlan L, Kreutzweiser D (2015) Alternatives to neonicotinoid insecticides for pest control: case studies in agriculture and forestry. Environ Sci Pollut Res 22:135–147

Georghiou GP (1990) Overview of insecticide resistance. In: Green MB, LeBaron HM, Moberg WK (eds) Managing resistance to agrochemicals. American Chemical Society, Washington, DC, pp 18–41

Hargreaves H (1948) List of the recorded cotton insects of the world. Commonwealth Institute of Entomology, London, UK, p 50

Hinrichsen D, Robey B (2000) Population and the environment: the global challenge. Population Reports, Series M, No. 5. Johns Hopkins University School of Public Health, Baltimore, MD

Hodson A, Lewis EE (2016) Managing for soil health can suppress pests. Calif Agr 70(3):137–141

Karlen DL, Varvel GE, Bullock D, Cruse RM (1994) Crop rotations for the 21st century. Adv Agron 53:1–45

Kaushik L, Mishra A (2013) Role of precision farming in sustainable development of hill agriculture. In: ISAE 2013. Proceedings of the international symposium on agriculture and environment 2013, 28 November 2013. University of Ruhuna, Sri Lanka, pp 83–86

King EG, Coleman RJ, Morales-Ramos JA, Summy KR, Bell MR, Snodgras GL (1996) Biological control. In: King EG, Phillips JR, Coleman RJ (eds) Cotton insects and mites: characterization and management. The Cotton Foundation Publisher, Memphis, TN, pp 511–538

Lewis WJ, Takasu K (1990) Use of learned odours by a parasitic wasp in accordance with host and food needs. Nature 348:635–636

Lewis WJ, van Lenteren JC, Phatak SC, Tumlinson JH (1997) A total system approach to sustainable pest management. Proc Natl Acad Sci 94(23):12243–12248

Liebman M, Dyck E (1993) Crop rotation and intercropping strategies for weed management. Ecol Appl 3:92–122

Lin R, Liang H, Zhang R, Tian C, Ma Y (2003) Impact of alfalfa/cotton intercropping and management on some aphid predators in China. J Appl Entomol 127:33–36

Lopez O, Rach MM, Migallon H, Malaumbres MP, Bonastre A, Serrano JJ (2012) Monitoring pest insect traps by means of low-power image sensor technologies. Sensors 12(11):15801–15819

Magdoff F, Van EH (2000) Building soils for better crops. Sustainable Agriculture Network, Beltsville, pp 80–82

Mason A (1928) Spraying, dusting and *fumigating* of plants. New York. In: Macmillan CD, Shrestha A (eds) (2004) new dimensions in Agroecology. The Haworth Press, Binghampton, NY

McCutcheon GS (2000) Beneficial arthropods in conservation tillage cotton – a three year study. In: Dugger CP, Richter DA (eds) Proceedings of 2000 Beltwide Cotton Conference, San Antonio, National Cotton Council, Memphis, pp. 1302–1306

Meisner MH, Rosenheim JA (2014) Ecoinformatics reveals effects of crop rotational histories on cotton yield. PLoS One 9(1):e85710

Metcalf RL (1980) Changing role of insecticides in crop protection. Annu Rev Entomol 125:219–256

Morales H (2002) Pest management in traditional tropical agroecosystems: lessons for pest prevention research and extension. Integrated Pest Manag Rev 7(3):145–163

Naranjo SE (2001) Conservation and evaluation of natural enemies in IPM systems for *Bemisia tabaci*. Crop Prot 20:835–852

Naranjo SE, Luttrell RG (2008) Cotton arthropod IPM. In: Radcliff EB, Hutchison WD (eds) Integrated pest management. Cambridge University Press, Cambridge

Naranjo SE, Ruberson JR, Sharma HC, Wilson L, Wu K (2008) The present and future role of insect-resistant genetically modified cotton in IPM. In: Romeis J, Shelton AM, Kennedy GG (eds) Integration of insect-resistant genetically modified crops within IPM programs. Springer, Dordrecht, pp 159–194

Nwilene FE, Nwanze KF, Youdeowei A (2008) Impact of integrated pest management on food and horticultural crops in Africa. Entomol Exp Appl 128:355–363

Ohlander L, Lagerberg C, Gertsson U (1999) Visions for ecologically sound agricultural systems. J Sustain Agr 14(1):73–79

Pedigo LP, Rice ME (2014) Managing insects by resistance plants. In: Entomology and pest management. Waveland Press, Illionis, USA, pp 441–474

Petchey OL, Gaston KJ (2002) Functional diversity (FD), species richness and community composition. Ecol Lett 5:402–411

Peter-Duelli MKO, Duelli P, Obrist MK, Schmatz DR (1999) Biodiversity evaluation in agricultural landscapes: above-ground insects. Agric Ecosyst Environ 74:33–64

Pimentel D, Hepperly P, Hanson J, Douds D, Seidel R (2005) Environmental, energetic, and economic comparisons of organic and conventional farming systems. Bioscience 55 (7):573–582

Pyke B, Rice M, Sabine B, Zalucki MP (1987) The push-pull strategy—behavioural control of *Heliothis*. Aust Cotton Grow May–July: 7–9

Ranjitha G, Srinivasan MR, Rajesh A (2014) Detection and estimation of damage caused by thrips *Thrips tabaci* (Lind) of cotton using hyperspectral radiometer. Agrotechnology 3(1):123

Ratnadass A, Michellon R, Randriamanantsoa R, Seguy L (2006) Effects of soil and plant management on crop pests and diseases. In: Uphoff N, Ball AS, Fernandes E, Herren H, Husson O, Laing M, Palm C, Pretty J, Sanchez P, Sanginga N, Thies J (eds) Biological approaches to sustainable soil systems. CRC Press, Boca Raton, pp 589–602

Rose USR, Lewis J, Tumlinson JH (2006) Extra floral nectar from cotton (*Gossypium hirsutum*) as a food source for parasitic wasps. Funct Ecol 20(1):67–74

Rothamsted Research Center (2006) Guide to the classical and other long-term experiments, datasets and sample archive. Technical report, Rothamsted Research Center, pp. 1–60

Russelle MP, Hesterman OB, Shaeffer CC, Heichel GH (1987) Estimating nitrogen and rotation effects in legume-corn rotations. In: Power JF (ed) The role of legumes in conservation tillage systems. Soil Conserv Soc Am, Ankeny, IA, pp. 41–42

Seguy L, Bouzinac S, Belot JL, Martin J (2004) Sustainable cotton production systems for the humid savannas of Central Brazil. In: Swanepoel A (ed) Proceedings world cotton Conf.–3, cotton production for the new millennium. Cape Town, 2003. ARC, Institute for Industrial Crops, Pretoria, pp 481–496

Sharma HC, Arora R, Pampapathy G (2007) Influence of transgenic cottons with *Bacillus thuringiensis cry1Ac* gene on the natural enemies of *Helicoverpa armigera*. BioControl 52:469–489

Shelton AM, Badenez-Perez FR (2006) Concepts and applications of trap cropping in pest management. Annu Rev Entomol 51:285–308

Sithanantham S, Singh SP, Romeis J (2005) Biological control of *Helicoverpa*: research status, constraints and opportunities. In: Sharma HC (ed) *Heliothis/Helicoverpa* management: emerging trends and strategies for future research. Oxford/IBH, New Delhi, India, pp 329–369

Smith K (2011) We are seven billion. Nat Clim Chang 1:331–335

Smith RF, Reynolds HT (1972) Effects of manipulation of cotton agro-ecosystems on insect populations. In: Farvar MT, Milton JP (eds) The careless technology: ecology and international development. The Natural History Press, Garden City, NY, pp 373–406

Stapel JO, Cortesero AM, De Moraes CM, Tumlinson JH, Lewis WJ (1997) Extra floral nectar, honeydew, and sucrose effects on searching behavior and efficiency of *Microplitis croceipes* (hymenoptera: Braconidae) in cotton. Environ Entomol 26(3):617–623

Stinner BR, House GJ (1990) Arthropods and other invertebrates in conservation-tillage agriculture. Annu Rev Entomol 35:299–318

Thomas MB (1999) Ecological approaches and the development of 'truly integrated' pest management. Proc Natl Acad Sci U S A 96:5944–5951

Ti X, Zhang Q (2009) Advances in research of induced resistance to insects in cotton. Front Biol China 4(3):289–297

Tscharntke T, Klein AM, Kruess A, Steffan-Dewenter I, Thies C (2005) Landscape perspectives on agricultural intensification and biodiversity – ecosystem service management. Ecol Lett 8:857–874

Tumlinson J, Lewis WJ, Vet LEM (1993) How parasitic wasps find their hosts. Sci Am 268 (3):100–106

Turlings TC, Loughrin JH, McCall PJ, Röse US, Lewis WJ, Tumlinson JH (1995) How caterpillar-damaged plants protect themselves by attracting parasitic wasps. Proc Natl Acad Sci U S A 92 (10):4169–4174

Wei-Di L, Wu KM, Chen XX (2003) Effects of transgenic cottons carrying Cry1A+ CpTI and Cry1A genes on the structures and composition of pest and beneficial arthropod communities in cotton field in North China. J Agr Biotechnol 11(5):494–499

Wilson LJ, Mensah RK, Fitt GP (2004) Implementing integrated pest management in Australian cotton. In: Horowitz AR, Ishaaya I (eds) Insect pest management. Springer, Berlin, pp 97–118

Wu K, Guo Y (2003) Influences of *Bacillus thuringiensis* Berliner cotton planting on population dynamics of the cotton aphid, *Aphis gossypii* glover, in northern China. Environ Entomol 32:312–318

Yin R, He Q (1997) The spatial and temporal effects of paulownia intercropping: the case of northern China. Agrofor Syst 37:91–109

Yokoyama VY (1978) Relation of seasonal changes in extra floral nectar and foliar protein and arthropod populations in cotton. Environ Entomol 7(6):799–802

Chapter 13
Cotton Diseases and Their Management

Sobia Chohan, Rashida Perveen, Muhammad Abid, Muhammad Nouman Tahir, and Muhammad Sajid

Abstract Cotton industry is an important sector of the economy in all agriculture-based countries. Nevertheless, cotton production is constantly endangered by pathogens that cause considerable economic losses. Worldwide, numerous different diseases have been identified in cotton. *Fusarium* and *Verticillium* wilt, *Alternaria* leaf spot and seedling diseases, boll rot, leaf curl disease, and bacterial blight are the major constraints to the cotton fiber production. Maintaining the disease incidence at low level is the ultimate preference of the researchers. Understanding the etiology is the main factor to estimate the economic impact of diseases, which eventually helps to develop the management strategies. Presently, cotton leaf curl has emerged as main risk to all cotton-growing areas because of the changes in viral disease complex. In this chapter, brief history of the major diseases, the host-pathogen interactions, the taxonomy of the recognized causal agents, and different control strategies applicable to each disease including some rising techniques such as genome modification for enhanced resistance are discussed.

Keywords Cotton viral diseases · Fungal diseases · Bacterial diseases · Wilt diseases

13.1 Introduction

Cotton belongs to Malvaceae family and is the fabric of our lives as it is the prime source of fiber worldwide. Cotton is a leading cash crop as it plays a key role in the lives of millions of people in Asia, Africa, Australia, South America, and North

S. Chohan (✉) · R. Perveen · M. N. Tahir · M. Sajid
Department of Plant Pathology, Bahauddin Zakariya University, Multan, Pakistan
e-mail: sobiachohan@bzu.edu.pk

M. Abid
Department of Plant Pathology, Bahauddin Zakariya University, Multan, Pakistan

Institute of Plant Protection and Agro-products Safety, Anhui Academy of Agricultural Sciences, Hefei, China

S. Ahmad, M. Hasanuzzaman (eds.), *Cotton Production and Uses*,
https://doi.org/10.1007/978-981-15-1472-2_13

America, and its fiber is a valuable product that provides income to farmers and industrialists (Ahmad et al. 2014, 2017, 2018; Abbas and Ahmad 2018; Ahmad and Raza 2014; Ali et al. 2011, 2013a, b, 2014a, b). Cotton has been cultivated for thousands of years across the world to make fabrics. It is a valuable component of farming systems in approximately 60 countries worldwide (Amin et al. 2017, 2018; Khan et al. 2004; Rahman et al. 2018; Tariq et al. 2017, 2018; Usman et al. 2009). The top three leading countries in cotton production are China, the United States, and India. Pakistan is the fourth largest cotton-producing country with ~2.17-million-ton annual production. Cotton grows well in tropical and subtropical regions with warm and humid climatic conditions. There are many biotic factors that are responsible for crop yield losses, plant vigor, and situations that eventually result in poor fiber quality. The reduced cotton production by biotic stresses is caused by plant fungal, bacterial, and viral pathogens.

To date, the struggle against cotton diseases continues for its proper management under sustainable agriculture. It has been revealed that cotton was affected by more than sixty diseases, which brought heavy yield losses. The work carried out so far shows that seedling rots, boll rot, different types of leaf spots, stunting, reduction in the size of leaves, premature opening of bolls, attack of nematodes, blights by bacteria, and leaf curling due to virus are found responsible for causing huge damage to cotton crop. The extent of disease damage depends on the environment and cultivar genotype.

Wilts, boll rots, and root rot are more important among other fungal diseases. The loss caused by certain fungal diseases is obvious, as in the case of root-rot plants that died, while in others, losses caused are not evident besides hard to quantify at numerous stages of development. The most significant bacterial disease is bacterial blight incited by *Xanthomonas campestris* pv. *malvacearum*, which occurs mainly in South Asia. The bacterium destroys the protoplasm of leaves that ultimately causes disruption in the process of photosynthesis. In case of severe attack incidence, the lesions appear on stem and bolls.

The crop record revealed that root and boll rots were considered as the most severe and destructive diseases of cotton, and since the last decade, cotton leaf curl disease (CLCuD) has been found to be a vital cotton disease. The CLCuD caused by whitefly-transmitted begomoviruses (family *Geminiviridae*) is predominant in South Asia and Africa and is a key constraint to productivity besides fiber quality.

This chapter includes some important pathogens of cotton crop and describes their control measures. Generally, single control strategy is not wholly effective to control diseases; thus, integrated approaches are more useful. The strategies which contribute to control one disease usually help in controlling others too. Besides chemical control, sowing disease-free seed and resistant varieties, employing crop rotation, and removing infected plant debris along with suitable practices should be part of integrated disease management strategy. Strict quarantine regulation should also be followed, as the exchange of diseased planting material is the key factor in the disease dispersion.

13.2 Fungal Diseases

13.2.1 Seedling Diseases

Seedling diseases are a worldwide problem in cotton-growing areas causing up to 5% estimated average yield loss annually. There are different fungal species involved in cotton seed rots and seedling diseases. The primary fungal genera associated to seed deterioration are *Fusarium* (Klich 1986), *Rhizoctonia* (Brown and McCarter 1976), *Pythium* (Devay et al. 1982), and *Thielaviopsis* (King and Barker 1934). These pathogens cause damping-off disease and colonize the weak cotton seedling plants.

13.2.1.1 Symptoms

The disease-causing organisms attack seeds and seedlings at pre-emergence and post-emergence stages. Symptoms include decay of seeds and young seedlings, partial or complete stem girdling, stunted growth, and seedling rot. The fungal pathogens invade the seedlings at soil level producing water-soaked, reddish-brown sunken lesions and girdling the hypocotyl, and the seedling may collapse. On examination of infected seedlings, dark lesions may expose on stem and roots. Seedling diseases not only kill the entire seedling population but also result in uneven stands with skips in rows. Surviving seedlings become pale and stunted and die soon. This condition of hypocotyl damage has been known as "sore shin" in the United States (Atkinson 1892).

13.2.1.2 Disease Cycle

Under favorable conditions, hyphae of *Rhizoctonia solani* grow rapidly in soil and convert into mycelia. The fungus may survive for years in soil/plant debris as sclerotia. *Pythium* spp. overwinter as oospores and infect host tissues through germ tube produced from an encysted zoospore. *Fusarium* spp. and *Thielaviopsis basicola* overwinter as chlamydospores in soil and plant residues for years.

13.2.1.3 Predisposing Factors

Seedling diseases are more prevalent under cool and wet climate. Sowing of cotton seeds in sandy soils with low organic matter increases the susceptibility of cotton to fungal pathogens. Some factors like deep planting, poor seed bed conditions, and compacted soil besides nematode or insect infestations may increase the problem.

13.2.1.4 Management

Cotton seedling disease control is based on preventive rather than curative treatments. Rotation with non-host monocotyledonous crops like wheat, corn, and sorghum can be useful in reducing the inoculum rate. Cultural practices like planting in raised beds can help in controlling seedling diseases by improving soil drainage. Cultivation of disease-resistant varieties and planting of good quality seeds are recommended. Eradication of the debris of the infected plant parts may help to control seedling diseases of cotton. Seed treatment with suitable fungicides like thiram, azoxystrobin, and metalaxyl is effective against these diseases. Nemli and Sayar (2002), while examining the effects of different fungicides, found that combinations of carboxin+thiram+metalaxyl besides fludioxonil+metalaxyl are more effective against seedling root rot. The use of bioagents to reduce fungal population and inoculum level in soil is also practiced by many researchers. Certain biofungicides are commercially available such as Kodiak, Subtilex, and Deny; these are suggested against seedling diseases (McSpadden Gardener and Fravel 2002). Moreover, Erdoğan et al. (2016), Wang et al. (2004), and Pleban et al. (1995) determined the effects of fluorescent *Pseudomonas* (FP) and *Bacillus subtilis* bacteria against *Rhizoctonia solani*, *Colletotrichum gossypii*, and *Fusarium* spp.

13.2.2 Foliar Diseases

13.2.2.1 *Alternaria* Leaf Spot

Leaf blight is a common foliar disease found almost in every cotton-growing area around the world (Fig. 13.1). This disease was firstly stated in the United States (Atkinson 1892; Paulwetter 1918). Later, similar leaf blight and spot were also found on upland cotton in Nigeria (Jones 1928), in Zimbabwe (Hopkins 1932), and in India (Rane and Patel 1956).

13.2.2.2 Symptoms

Initial symptoms appear as small, circular brown, and gray brown to tan lesions with purple margins on green leaves. These spots vary in size from 1 to 10 mm in diameter exhibiting concentric zonation on older leaves (Fig. 13.1). As the disease progresses, mature lesions may coalesce and become irregular and necrotic. The affected leaves become blighted and brittle and often crack exhibiting shot hole appearance. The disease is more severe on lower leaves than upper leaves, except leaves are affected by premature defoliation. Under humid weather conditions, prolific sporulation of the fungus may result in black sooty masses on necrotic lesions. Lesions may also appear on stem, bracts, and bolls. *Alternaria* leaf spot

Fig. 13.1 *Alternaria* leaf blight of cotton

may be mixed with angular spots of bacterial blight that is described later in this chapter.

13.2.2.3 Causal Organism

Taxonomy

Leaf spot is caused by *Alternaria macrospora* Zimm. It is an ascomycetous fungus belonging to class Dothideomycetes, order Pleosporales, and family Pleosporaceae. Earlier, it was recognized as *A. alternata* (Fr.) Keissler in Egypt and Russia (Kamel et al. 1971; Dzhamalov 1973), but recent reports from Zimbabwe described it as *A. macrospora* (Hillocks 1991).

Morphology

Alternaria macrospora comprises cylindrical to slightly tapering conidiophores which are formed solitarily or in clusters. They are septate, erect, flexuous, and pale brown in color. The conidia are produced singly or sometimes in chains of two, light to dark brown in color with 4–9 transverse septa and several longitudinal septa, ellipsoidal, melanized, and obclavate to obpyriform with narrow beak (Ellis 1971). There is substantial variation in conidial size, but most descriptions suggest as 70–180 × 15–22 μm (Ellis 1971; Sangeetha and Ashtaputre 2015; Venkatesh and Darvin 2016; Waghunde et al. 2018).

13.2.2.4 Disease Cycle

Crop residues and infected seeds are the main cause of inoculum and give rise to infected seedlings that support early stages of an epidemic. The conidia spread through air currents and water splashes onto healthy plants. Prolonged wet and humid weather conditions and temperature (about 27 °C) favor disease development initially from cotyledons to the lower leaves. Under favorable conditions, pathogen kills surrounding leaf tissues and produces abundant spores on the surface of the lesions within few days. Defoliation of leaves of susceptible varieties encourages maximum sporulation of *A. macrospora* (Bashi et al. 1983), and damaged bolls are responsible for the seed infection (Bashan 1984). Symptom development may have favored by nutritional or physiological stress to plants like premature senescence and heavy fruit loads. The disease cycle is completed with the shedding of infected leaves or the planting of infected seed.

13.2.2.5 Predisposing Factors

Extended periods of wet weather and high humidity favor the disease development. Minimum temperature for the disease to occur is 10 °C and maximum is 30 °C, whereas optimal temperature for disease development ranges between 20 and 30 °C. Soils deficient in potassium favor the disease development (Hillocks 1991).

13.2.2.6 Management

As fungus can survive on infested crop residues in soil, thus, residue management through tillage may reduce inoculum production in the field. Rotation of cotton with cereals helps in reducing seedling infection to a sufficient level. Application of fertilizers especially potassium is adequate to maintain soil fertility level (Hillocks 1991).

Planting healthy seeds and cultivation of resistant varieties are recommended. Bashan (1986) considered that higher phenol contents might be responsible for resistance to the disease.

Controlling the disease with foliar fungicides is usually considered economically useful. Foliar spray with copper fungicides, such as mancozeb at 2.5 g/L and difenoconazole at 1 ml/L, helps in reducing the primary inoculum. Similarly, fungicidal seed treatment with broad spectrum fungicides like strobilurins (trifloxystrobin) and sterol biosynthesis inhibitors (ipconazole) is found effective in protecting the cotyledons of emerging seedlings from the fungus.

Seed treatment with bioagent *Pseudomonas fluorescens* at 10 g/kg seeds after every 10 days of interval may reduce the disease intensity.

13.2.3 Grey Mildew Disease

The disease known as grey mildew or dahiya in India, false mildew/areolate mildew in the United States, and white mildew in South America (Hillocks 1991) was first reported in the United States (Atkinson 1891). The disease is of little importance in the United States; however, it is quite common in India, East Africa, and South America.

13.2.3.1 Symptoms

Initial symptoms appeared firstly on lower leaves as irregular-angular lesions measuring 1–10 mm in diameter after first boll set. They are light green to yellow green translucent spots bounded by veinlets (called areolate) on upper surface, but on under surface of the leaves, white mildew-like growth is observed as a result of abundant sporulation. Under high humidity, lesions may also become white on upper surface of the leaves. This is conidial stage of causal fungus. Later, the lesions turn dark brown in color and become necrotic. Symptoms also appear on cotyledons as circular water-soaked patches which turn reddish brown and chlorotic. Severe infection leads to defoliation and premature boll opening.

13.2.3.2 Causal Organism

Taxonomy

The causal agent of grey mildew is *Ramularia areola* (Atk.) (synonym: *Ramularia gossypii* Speg.). Its anamorph is *Cercospora gossypina* Cooke (Ehrlich and Wolf 1983). The teleomorph stage of fungus is known as *Mycosphaerella areola* Earle (Ehrlich and Wolf 1983; Gouws et al. 2001).

Morphology

Conidiophores of *R. areola* bear hyaline septate conidia measuring 14–30 × 4–5 μm. On lower surface of fallen leaves, spermogonia appear as black dots on the lesions in addition to conidial stage. They are 28–75 μm in diameter and on maturity liberate rod-shaped spermatia (2–4 × 0.4–2 μm) in a matrix through the ostiolum. Later, these conidia and spermogonia are replaced by dark brown-colored perithecia (70–80 μm diameter) with a slight papilla. These perithecia produce fusiform asci measuring 35–40 × 6–8 μm having eight elongated, biseriate ascospores which are 12.4–15.6 × 3.2–3.8 μm (Ehrlich and Wolf 1983).

13.2.3.3 Disease Cycle

The fungus passes through three separate phases during its whole life cycle. One stage is conidial stage that appears on the underside surface of the leaves. Next is spermagonial stage that develops on fallen leaves followed by the third stage, i.e., ascogenous, which produces on partially decayed leaves. Conidia and ascospores are the primary sources of inoculum. It is disseminated by wind and irrigation water. For germination of conidia and ascospores, free moisture is required with temperature range of 16–34 °C. The optimum temperature for growth is 25–30 °C.

13.2.3.4 Predisposing Factors

Humid conditions with sporadic rains are favorable for development of the disease.

13.2.3.5 Management

Destruction of crop residues, deep plowing and crop rotation are such cultural practices which reduce the multiplication and spread of primary inoculum.

Foliar application of benomyl at 200–300 g ha^{-1} is effective in controlling grey mildew of cotton.

Using resistant cultivars is the best approach to control the disease.

13.2.4 Boll Rot Disease

Boll rot occurs to some extent in most of cotton-growing regions in the world. However, yield losses occur only in areas of high humidity during late summer and fall. As a result of this disease, poor quality of lint is produced. Numerous microorganisms are associated with boll rots. Some of these organisms directly invade the cotton bolls, while others enter through insect wounds or as secondary invaders. Nearly hundred microorganisms have been isolated from rotted bolls (Hillocks 1991). Most commonly isolated fungi from rotted bolls of cotton is *Fusarium* spp. throughout the cotton-growing countries. In America, *F. oxysporum*, *F. roseum*, *F. solani*, and *F. moniliforme* are mainly isolated from rotted bolls (McCarter et al. 1970). *F. moniliforme*, *F. roseum*, *F. solani*, and *Colletotrichum* spp. are involved in boll rot in Africa (Follen and Goebel 1973), whereas in India, *F. equiseti* has been reported as a cause of boll rot (Sharma and Sandhu 1985). Recently, a species complex of *F. incarnatum-equiseti* is reported to cause boll rot in cotton (Chohan and Abid 2018).

Infection of boll first starts as appearance of water-soaked necrotic lesions on the margins of the bracts. During moist conditions, these lesions enlarge, and white to gray or salmon-colored/pale pinkish fungal growth covers the infected bolls. Severely infected bolls may drop from plants (Figs. 13.2 and 13.3).

Fig. 13.2 Cotton bolls infected with *Fusarium* spp

Fig. 13.3 Infected cotton bolls with *Fusarium* spp

To prevent boll rots, farmers should adopt agronomic practices such as to avoid excessive application of nitrogen, maintain low humidity by minimizing the size of crop canopy and burning of diseased crop residues. Acid delinting helps in eliminating seed-borne infections. Chemical approach may also be useful in controlling boll rot.

13.2.5 Wilt Diseases

Cotton plants are attacked by two destructive wilts; one is named as *Fusarium* wilt and the other as *Verticillium* wilt. In case of both wilts, the vascular system of plant is colonized by the pathogen. The microorganism invades the root of cotton plant, penetrates and proliferates within xylem tissues, and eventually spreads throughout the plant. Vascular wilt fungi are soil-inhabiting pathogens, but these may grow on crop residues in the absence of host for long periods in the form of thick-walled resting structures. On favorable conditions, the fungi sporulate and cause infection by blocking the vascular system of cotton plant.

13.2.5.1 Fusarium Wilt

Fusarium wilt of cotton was first reported by Atkinson (1892) from America. It is generally found wherever cotton is grown around the world. It has been originated from Mexico and spread to South America, the United States, Egypt, West Indies, Italy, Africa, Greece, Zimbabwe, China, France, Russia, and India (Menlikiev 1962; Cook 1981; Hillocks 1992).

Symptoms

Wilting symptoms may appear any time of plant growth, i.e., from seedling to maturity. If seedlings are infected, first symptoms appear as vein darkening, yellowing, and shriveling of young cotyledons. Later, the cotyledons become necrotic and shed, and eventually seedlings may wilt and die. In older plants, initial symptoms appear as marginal yellowing of lower leaves. The leaves become flaccid resulting in drooping and wilting of the whole plant. Sometimes, symptoms appeared on one side of the plant (Figs. 13.4 and 13.5). When conditions are conducive for disease expansion, the symptoms appear within 2 months after sowing resulting into wilting and death of plants with a few boll settings.

Fig. 13.4 Cotton plants showing symptoms of *Fusarium* wilt disease

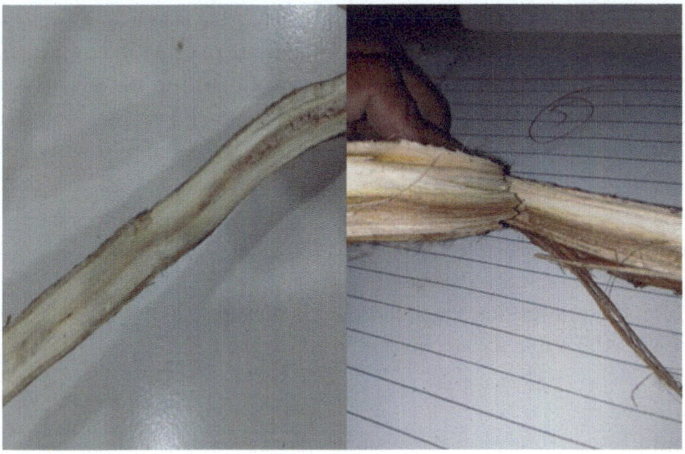

Fig. 13.5 Cross section of cotton stem showing symptoms of *Fusarium* wilt disease

Causal Organism

Taxonomy

The causal microorganism is an asexual ascomycetous fungus belonging to the class Hyphomycetes. It is recognized as *Fusarium oxysporum* Schlecht f. sp. *vasinfectum* Atk. Syn. & Hans. Snyder and Hansen (1940) grouped the parasitic forms into *formae speciales* based on host specificity of the strains. There are more than 120 *formae speciales*.

Morphology

Mycelium of *F. oxysporum* f. sp. *vasinfectum* is initially white to grayish white or bluish purple in color and produces two types of conidia: microconidia and macroconidia. Microconidia are small, one to two-celled, and elliptical in shape measuring 5–20 × 2.2–3.5 μm. Macroconidia are multinucleate, usually three to five septate, fusiform, sickle-shaped, and light buff to salmon colored measuring 27–48 × 2.5–4.5 μm. Resting spores are called chlamydospores which are mostly spherical, single or in chains, terminal or intercalary measuring 7–13 μm in diameter. The distinguishing feature of *F. oxysporum* from other *Fusarium* species is formation of chlamydospores with short conidiophores.

Disease Cycle

Pathogens are soil borne and survive as chlamydospores in the absence of host for several years. Fungus also survives as saprophyte on plant debris. Chlamydospores are sources of primary infection. Infection starts as conidia germinate, 6 h after inoculation on root surface, and form a mycelial mat covering the root surface. Later, the penetrating hyphae become systemic and proliferate in the xylem vessels, and conidia are transported upward in transpiration stream and produce more mycelium. Wilt symptoms usually appear when cotton plants are about 5–6 weeks old.

Predisposing Factors

Optimum temperature for disease to develop is between 20 and 27 °C, and moisture contents of 80–90% capacity are favorable for disease.

Management

Wilt disease can be managed through using resistant varieties, implementation of cultural practices like mixed cropping, field sanitation, proper use of fertilizers and micronutrients, and crop rotation.

Biological control with non-pathogenic bacteria like *Pseudomonas fluorescens* is found effective in reducing incidence of wilt disease of cotton.

13.2.5.2 Verticillium Wilt

Verticillium wilt is the second vascular wilt disease affecting the cotton crop. Vascular pathogens have the ability to grow inside the vascular system, but these fungi grow outside the vascular tissues on advance stage of infection. *Verticillium* wilt may sporulate on plant residues after the death of host and specialized soil-invading pathogen with narrow host ranges. *Verticillium* wilt has the ability to survive for a long time in the absence of host plant in soil and is susceptible to toxic fungal combinations.

Verticillium wilt of cotton was first reported in 1974 after *Verticillium dahliae* was isolated from a few diseased plant of upland cotton growing in Arlington, Virginia. In 1932, Miles and Persons reported that *Verticillium* wilt of cotton occurred near Mississippi River heavy loam soil. In 1936, Barker and Sherbakoff confirmed the causal agent after the thorough survey of US cotton wilts and found that *Verticillium* wilt was common in Texas, Arizona, New Mexico, and Missouri as well as in California. After in the United States, it was also reported in Central Asia, Brazil, China, Turkey, and South Africa. Outbreak of cotton wilt occurred in Australia when cotton first grows on land (Evans 1967).

Symptoms

Initially, the young plants infected show yellowing, epinasty, and defoliation of the leaves, and in warm weather, they recover quickly and show stunted growth. Under high humid conditions, early infection causes little loss of yield and quality. The plant leaves show mosaic pattern with yellowing of tissues along margins besides between veins. Leaves and bolls of the plant remain defoliated and ultimately killed.

Taxonomy

The causal organism of *Verticillium* wilt of cotton was often given as *V. albo-atrum*. Five species of *Verticillium* are *V. albo-atrum* Reinke & Berthold 1879, *V. dahliae* Klebahn 1913, *V. nigrescens* Pethybridge 1919, *V. nubilum Pethybridge* 1919, and *V. tricorpus* Isaac 1953.

Morphology

Initially white or light cream, but later become black with formation of microsclerotia colonies on PDA of *Verticillium dahliae* grow moderately fast (2.0–3.5 mm at 20–25 °C). Microsclerotia are dark brown to black and then torulose. Conidia are ellipsoidal to short cylindrical with erect and prostrate conidiophores.

Disease Cycle

Verticillium wilt survives as dormant microsclerotia present in the soil debris and soil depths down to 40 cm. Microsclerotia produced colonies on cotton root surface in response to exudates. Hyphae from colonies penetrate the xylem vessels through wounds. Wounds are not required but increase wilt infection. Hyphae surrounding throughout the necrotic tissue of death leaves, stems, and roots, where microsclerotia are formed after several weeks and month and depending on moisture amount, which again dispersed in soil (Schnathorst 1981; Huisman and Gerik 1989; Bell 1992).

Predisposing Factors

Verticillium wilt grows well on simple sugars and amino acid that are normally found in root exudates and xylem sap. The optimal temperature for growth is 22–27 °C.

Management

Several methods are used for the control of wilt disease. A combination of cultural, chemical, and biological methods is also used to minimize the losses caused by wilt. Selection of resistant cultivars is another way to control the wilt disease because the cultivars have the capacity to resist against disease infection.

In cultural control, using different methods and resistant varieties is practiced to prevent the introduction and spread of the disease in soil and reduce the inoculum rate. The methods are rotation, use of fertilizer, control of soil moisture, planting time and tillage method, reduction of seed transmission, planting density, removal of weeds and crop residues, and solarization.

In chemical control, fungicides such as carbendazim and ethylene thiosulphonate can control seedling pathogens as well as prevent seed transmission of pathogen (Shen 1985). Benzimidazole fungicides are systematic and complete control of the *Verticillium* wilt in glasshouse and field with different concentration, for example, 100 ppm of Benlate in water drenchers and 10–20 kg of Benlate in field.

In biological control, more than 20 years *Trichoderma viride* (*T. lignorum*) has been used to control wilt in Russia (Fedorinchik 1964). In low organic matter soil, *Gliocladium roseum* may be better than *T. viride* as an antagonist of *V. dahliae* (Globus and Muromtsev 1990).

13.3 Viral Diseases

13.3.1 Cotton Leaf Curl Disease

CLCuD is the most noteworthy restraining factor in cotton productivity in Pakistan (Briddon and Markham 2000; Sattar et al. 2013, 2017). Whitefly-transmitted

CLCuD, prevalent from South Asia to China, is favored by high temperature conditions. The disease was initially stated in Nigeria in 1912 (Kirkpatrick 1931) and then also reported from other African countries such as Sudan, Tanzania, Egypt, and Malawi along with South Africa. In Pakistan, the first disease incidence was reported as a minor attack in 1967 near Multan. The disease came into noticed when cotton production of Pakistan suffered due to the epidemic of CLCuD in 1988. Later on, CLCuD was spread out across Pakistan as well as into northwest India. In India, the first incidence of CLCuD was reported in 1989 near Sri Ganganagar. In Pakistan, introduction of highly susceptible varieties in the 1990s resulted in high yield loss causing financial loss of about US$5 billion (Briddon and Markham 2000). During the late 1990s, cultivating locally developed tolerant varieties in Pakistan restored the cotton production. However, in 2001–2002, characteristic disease symptoms were observed on resistant cotton cultivars in Burewala region in Punjab, Pakistan, suggesting the second epidemic of CLCuD. Currently, this dominant resistance-breaking recombinant strain of CLCuD is spread throughout Punjab, Pakistan, and into northwest India.

The CLCuD-affected plants show distinctive symptoms like vein swelling, upward/downward leaf curling, and stunted plant growth along with the formation of cup-shaped leaf-like-outgrowth undersides of leaves known as enations (Fig. 13.6). Early infection results in severe stunting of plants with high yield loss, while late infection causes mild symptoms (Sattar et al. 2013).

Fig. 13.6 Cotton plant showing leaf curl symptoms with enations

13.3.1.1 Etiology of CLCuD

Causative agent of CLCuD from Africa besides Asia has been determined. In these continents, the disease is associated with begomovirus complexes, which consist of a monopartite begomovirus and a symptom-modulating satellite molecule called betasatellite, previously known as DNA β. CLCuD infection is also shown to be linked with a satellite-like molecule called alphasatellite, previously known as DNA 1.

13.3.1.2 Begomoviruses Associated with CLCuD

The most destructive whitefly-transmitted genus *Begomovirus* (family *Geminiviridae)* consists of small, single-stranded, circular DNA genomes encapsidated in twinned quasi-icosahedral particles. Excluding begomoviruses, all geminiviruses are monopartite, having a single genomic component that is capable of replication, systemic movement, and infections. The begomoviruses are either bipartite or monopartite (consisting of a component that is homologue of DNA-A). Begomoviruses from the New World have bipartite genome DNA A and DNA B, both of which are essential for successful infection (Stanley 1983). Although in the Old World there are bipartite begomoviruses causing disease in field crops, however, a large number of diseases are caused by monopartite begomoviruses.

In spite of early identification of CLCuD in Africa, causative agent was recognized too later. Single begomovirus *Cotton leaf curl Gezira virus* (CLCuGeV) has been identified with CLCuD in Africa (Idris and Brown 2002). CLCuGeV has been identified from numerous plant species as well as cotton, hollyhock, okra, and *Sida* spp. (Tahir et al. 2011). CLCuGeV is a geographically widespread begomovirus as was identified from different host plants including cotton from diverse areas in Asia (Tahir et al. 2011; Khan et al. 2012; Idris et al. 2014).

In Asia, the scenario regarding CLCuD is much more complex. Several begomoviruses have been associated with CLCuD. In Pakistan, during the first epidemic of CLCuD, three diverse species of begomoviruses were identified: *Cotton leaf curl Alabad virus* (CLCuAlV), *Cotton leaf curl Kokhran virus* (CLCuKoV), and *Cotton leaf curl Multan virus* (CLCuMuV) (Zhou et al. 1998). Later, analysis identified other species of begomoviruses: *Cotton leaf curl Rajasthan virus* (CLCuRaV), *Papaya leaf curl virus* (PaLCuV), and *Tomato leaf curl Bangalore virus* (ToLCBaV) (Mansoor et al. 2003; Kirthi et al. 2004).

However, during 2001, the change in the genetic makeup of begomoviruses in Pakistan results in the appearance of distinct resistance-breaking recombinant strain, named CLCuKoV-Burewala strain (CLCuKoV-Bu) (previously known as Cotton leaf curl Burewala virus) (Amrao et al. 2010a). Recombinant strain genome is derived from two begomovirus species linked with first CLCuD epidemic, CLCuMuV along with CLCuKoV. Following resistance breakdown, only CLCuKoV-Bu was found to be linked with CLCuD across Punjab, Pakistan.

Although during the 1990s, southern parts in Pakistan remained unaffected. However, disease appeared in Sindh province since 2004 (Mansoor et al. 2006) and extended to northwest Indian states during 2005 (Rajagopalan et al. 2012). Additionally, another species of begomovirus associated with CLCuD was identified in southern India, named *Cotton leaf curl Banglore virus* (CLCuBaV) (Reddy et al. 2005). A new recombinant species was characterized which has been shown to be associated with CLCuD in Sindh province, named CLCuKoV-Shahdadpur strain (CLCuKoV-Sha), formerly known as Cotton leaf curl Shahdadpur virus (Amrao et al. 2010b). In Sindh province, in addition to CLCuKoV-Sha and CLCuKoV-Bu, CLCuD-associated African begomovirus, CLCuGeV has been also reported in cotton (Tahir et al. 2011).

During both epidemics, from Indo-Pak subcontinent region, only monopartite begomoviruses with DNA satellites are associated with CLCuD until 2013–2014. However, in recent times, bipartite begomoviruses *Tomato leaf curl Gujarat virus* (Zaidi et al. 2015) and *Tomato leaf curl New Delhi virus* (ToLCNDV) (Zaidi et al. 2016) have been found from infected cotton plants in Pakistan. Furthermore, *Okra enation leaf curl virus* and a *Mastrevirus Chickpea chlorotic dwarf virus* have also been identified from cotton plants. Recent studies show that ToLCNDV has not only established in Asia but also spread out to Africa and Europe (Mnari-Hattab et al. 2015; Zammouri et al. 2017; Ruiz et al. 2015; Panno et al. 2016; Juárez et al. 2014). Under experimental conditions, DNA A of ToLCNDV trans-replicates CLCuD-associated betasatellite resulting in accumulation of symptom-modulating satellite components (Saeed et al. 2007). The presence of bipartite ToLCNDV along with monopartite begomoviruses suggests a possible future epidemic in areas under cotton cultivation by CLCuD complexes in South Asia (Sattar et al. 2017).

Similarly, in Africa, like Asia, a bipartite begomovirus *Cotton yellow mosaic virus* (CYMV) has been reported first time in cultivated cotton (Leke et al. 2016), suggesting accumulation of distinct begomoviruses from their hosts to susceptible cotton plants in the OW.

13.3.1.3 DNA Satellites Associated with CLCuD

Satellites are termed as viruses/nucleic acids which depend on a helper virus for their replication but lack extensive nucleotide-sequence homology to the helper virus and are dispensable for proliferation. In OW, majority of the monopartite begomoviruses are associated with more frequently found ssDNA satellite components recognized as betasatellites (Briddon and Stanley 2006), satellite-like components alphasatellites, and newly characterized deltasatellites (Zhou 2013; Fiallo-Olivé et al. 2012).

Betasatellites are around half of the size of helper begomoviruses (~1350 nt) having no sequence homology with helper begomoviruses, excluding non-nucleotide sequence TAATATTAC (Briddon et al. 2003). Betasatellites genome consists of single coding sequence in complimentary sense known as βC1. Pathogenicity determinant gene βC1 has been involved in upregulating viral titer in host and overcoming the host defenses (Qazi et al. 2007; Briddon et al. 2001; Amin et al. 2011). All betasatellites have sequence rich in adenine (A-rich region) along with highly conserved region recognized as satellite conserved region (SCR).

In OW, both Asian and African CLCuD complexes require their cognate symptom-modulating betasatellites. In Africa, a distinct betasatellite, Cotton leaf curl Gezira betasatellite (CLCuGeB), is found to be linked with CLCuGeV (Idris and Brown 2002). In Asia, different strains of the most important betasatellite CLCuMuB are linked with CLCuD (Akhtar et al. 2014).

In the 1990s, during the first epidemic of CLCuD, although six distinct monopartite begomoviruses were identified, only a single betasatellite Cotton leaf curl Multan betasatellite-Multan strain (CLCuMuB^{Mul}) was found to be linked with CLCuD (Mansoor et al. 2003). In the second epidemic of CLCuD, with resistance-breaking begomovirus CLCuKoV-Bur, a distinct "Burewala strain" of CLCuMuB (CLCuMuB^{Bur}) became prominent. CLCuMuB^{Bur}was also found to be a recombinant, containing some sequence within the SCR from Tomato leaf curl betasatellite (ToLCB) (Amin et al. 2006). Similarly, CLCuD in Sindh, which was found to be linked with CLCuKoV-Sha, a new recombinant betasatellite of CLCuMuB called "Shahdadpur" strain (CLCuMuB^{Sha}), was characterized (Amrao et al. 2010b). As compared to CLCuMuB^{Bur}, CLCuMuB^{Sha} contains a smaller fragment from ToLCB in SCR. Along with malvaceous betasatellites, CLCuMuB^{Bur}and CLCuMuB^{Sha}, a non-malvaceous betasatellite Chili leaf curl betasatellite has also been reported from cotton in Sindh (Tahir et al. 2011). The presence of a distinct biotype of *B. tabaci* may be the reason for the association of different begomoviruses with distinct strains of betasatellites in Sindh, Pakistan.

Alphasatellite is the third component of CLCuD complex. Alphasatellites are able to replicate autonomously in host; thus, they are known as satellite-like molecules. These molecules depend on their helper virus for transmission along with encapsidation (Mansoor et al. 1999; Briddon et al. 2004).

Alphasatellites are circular ssDNA molecules (~1400 nt) having highly conserved genome with a hairpin structure with a non-nucleotide sequence (TAGTATTAC), an A-rich region, and a single large virion-sense gene, encoding a rolling-circle replication initiator protein (the replication-associated protein [Rep]) (Briddon et al. 2004). Alphasatellite Rep shows high level of sequence homology with the Rep encoded by another family of DNA viruses, *Nanoviridae* (Saunders and Stanley 1999). Thus, the origin of alphasatellites is supposed to be nanoviruses (Briddon et al. 2004).

In Africa, Cotton leaf curl Gezira alphasatellite (CLCuGeA) has been involved in CLCuD complexes in association with CLCuGeV and CLCuGeB. In Asia, Cotton leaf curl Multan alphasatellite (CLCuMuA) is found to be linked with CLCuD complexes.

Although alphasatellites were first identified in CLCuD-affected cotton in 1999 in Pakistan (Mansoor et al. 1999), however, following resistance-breaking epidemic of CLCuD, alphasatellites were not detected with CLCuD until 2009. In recent studies, alphasatellites are not only identified with CLCuD complexes but also found to be associated with other begomovirus and betasatellite complexes. Another alphasatellite

Gossypium darwinii symptomless alphasatellite has been reported from wild cotton in Pakistan (Nawaz-ul-Rehman et al. 2012).

From OW, alphasatellites have only been described in relationship with monopartite begomoviruses and betasatellite complexes. However, recently alphasatellites are reported in NW in association with bipartite begomoviruses but in the absence of betasatellites (Paprotka et al. 2010; Romay et al. 2010; Ferro et al. 2017). Due to the rising number of alphasatellites, scientists create a new system of nomenclature and classification of alphasatellites. Alphasatellites are assigned to a new family *Alphasatellitidae*, which is divided into two subfamilies: *Geminialphasatellitinae* that includes geminivirus-associated alphasatellites and *Nanoalphasatellitinae* consisting of nanovirus-associated alphasatellites (Briddon et al. 2018). According to the current classification, subfamily *Geminialphasatellitinae* consists of four genera, and CLCuD-associated alphasatellites are grouped in the genus *Colecusatellite* (Briddon et al. 2018).

The precise reason for the presence of alphasatellites with begomovirus-betasatellite complexes yet needs to be cleared. However, evidence regarding the function of Rep encoded by alphasatellites in suppression of posttranscriptional gene silencing shows their involvement in overcoming host defense mechanism (Nawaz-ul-Rehman et al. 2010).

13.3.2 Cotton Leaf Crumple Disease

A bipartite begomovirus *Cotton leaf crumple virus* (CLCrV), transmitted by whitefly in persistent manner, is found to be associated with cotton leaf crumple disease (CLCrD) in the NW. CLCrV infection produced characteristic foliar discoloration and veinal hypertrophy resulting in puckering or crumpling, downward leaf curling, and shortening of internodes along with stunting of host (Brown et al. 1987). CLCrD losses depend on plant age at infection time. Disease severity is more pronounced if plants become infected during early stage (Brown et al. 1987).

Cook made the first report of CLCrD in 1924 and named it as crazy top at that time (Cook 1924). The disease became prominent due to an outbreak in the 1950s when it was reported in California (Dickson et al. 1954) and in Arizona later (Allen et al. 1960). CLCrD mainly occurs in California, Arizona, Texas, and Mexico. In California in 1958, 71–85% yield losses due to CLCrD were reported (Van Schaik et al. 1962). In 1983, CLCrD resulted in 23–55% yield reduction in Arizona (United States). Although the seed index was reduced, however, the lint weight or lint index was not affected (Allen et al. 1960; Van Schaik et al. 1962; Brown et al. 1987). Ratooning cotton with early increase in whitefly populations was the contributing factor of CLCrD. Elimination of stub cotton in southwestern United States helps in controlling CLCrD since 1982. However, CLCrD tolerant lines have also been designated for developing resistant cotton cultivars Idris and Brown 2002.

13.3.3 Perspectives for the Viral Disease Management

Begomoviral disease can be managed by controlling the insect vectors using pesticides, as a single viruliferous whitefly is capable of transmitting virus. On the other hand, removal of viral inoculum sources (alternate hosts) is also efficient in reducing the diseases spread by whitefly feeding on infected plants. However, in cotton-vegetable cropping systems when there is no host-free period, it is difficult to attain significant control of the whitefly populations. In Pakistan, such conditions prevail due to prevalence of distinct viruses with wide host range. Thus, cultural control of whiteflies is not very effective because of broad host range of whitefly.

Resistance to CLCuD is the effective approach to control the disease, when disease incidence occurs regularly in the production season. The genetic makeup of the host and whitefly populations with the concentration of inoculum at a particular area influences the natural incidence of CLCuD.

In the late 1990s, using conventional breeding approaches, a number of CLCuD-resistant varieties were developed. Cultivation of these varieties resulted in alleviating losses due to CLCuD in Pakistan till the appearance of resistance-breaking CLCuKoV-Bur in 2001–2002. There is no single variety that is genetically resistant to the prevalent CLCuD complex. Breeding efforts are again underway in South Asia for the identification of resistant germplasm.

Currently, a number of efforts have been deployed to develop CLCuD-resistant plants by using genetic engineering approach. Such strategies are mainly based on interference with the begomovirus replication or movement within the cotton plants to reduce symptom expression in the host, which is the prime objective for plant protection without using pesticides.

Anti-sense RNA strategy has been conducted against CLCuD complex to inhibit the target mRNA expression in transgenic cotton (Amudha et al. 2011; Hashmi et al. 2011). Diversity between CLCuD-associated begomoviruses urges scientists for broad-spectrum resistance against all the prevailing viruses in the field. Successful results using RNA-interference (RNAi) technique against *African cassava mosaic virus* (Chellappan et al. 2004) and *Mungbean yellow mosaic virus* (Pooggin et al. 2003) have been reported. In recent times, RNAi-based construct targeting V2 gene of CLCuKoV-Bur limits virus replication in cotton plants (Yasmeen et al. 2016).

Site-specific genome editing using the clustered regularly interspaced short palindromic repeats/CRISPR-associated protein 9 (CRISPR/Cas9) system has recently emerged as a groundbreaking tool to introduce required traits in eukaryotic species, including plants. Scientists have used CRISPR/Cas9 system to engineer resistance in plants against geminiviruses. Recent studies demonstrated the efficient control of geminiviruses in plants using CRISPR/Cas9 system (Baltes et al. 2015; Ji et al. 2015; Ali et al. 2015, 2016). Targeting the conserved non-nucleotide sequence of CLCuKoV-Bur via CRISPR/Cas9 can be used for broad-spectrum geminivirus resistance (Ali et al. 2016).

Moreover, CRISPR/Cas9-based multiplexing technique to target CLCuD begomoviruses and associated satellites has been proposed (Iqbal et al. 2016).

Recently, the expression of insecticidal proteins and RNAi against whitefly has been demonstrated (Shukla et al. 2016; Raza et al. 2016; Javaid et al. 2016). Thus, the ability of begomoviruses and associated satellites to cause mixed infections could be obstructed by engineering dual *Begomovirus-Bemisia tabaci* resistance in plants (Zaidi et al. 2017).

13.4 Bacterial Diseases

13.4.1 Bacterial Blight of Cotton

13.4.1.1 Introduction

Bacterial blight of cotton is the most destructive disease that causes significant losses in yield in the rainy season (Delannoy et al. 2005). It was first reported during 1891 in Alabama in the United States (Atkinson 1891). In Pakistan, this disease appeared in Burewala near Multan. The world area cultivated with cotton is 33.4 M ha with a production of 121.4 M bales, while in Pakistan, it is cultivated on an area of ≈3.0 M ha during 2013–2014 with a production of 9.5 M bales. This disease causes 30% yield losses in diverse cotton-growing areas of the world (Ramapandu et al. 1979; Chidambaram and Kannan 1989). It was assumed that the yield damages generally reached between 10% and 30% in Asian countries, and up to 50% were registered in African countries (Bayles and Verhalen 2007). Approximately 37–40% of yield losses were observed in the Faisalabad area (Bhutta and Bhatti 1983; Khan et al. 1999).

13.4.1.2 Pathogen and Disease Spread

The bacterium can survive in the field on debris from previously harvested crops, and its initial inoculum was seed born (Mohan 1983). The bacterium sticks to the leaf surface, enters the leaf via the stomata or open wounds, and then produces symptoms in susceptible plants. It causes defoliation, swelling of the stem, black arm, breakage of the weakened trunk, and detachment of the bolls. The enhancement of disease reduces the quality of the cotton from lint staining with the consequent loss of yield (Verma 1986). Damage to the tissues of the stem and bolls has occurred, which gradually produce necrotic, angular, waxy, and marshy lesions on the leaf surface; these lesions are called bacterial blight, angular leaf spot, lesion of the black arm, and boll rot (Hillocks 1992). The first symptom appeared as tiny lesions impregnated with water (dark green flaccid lesions) and then spread to the bottom of young leaves (Verma 1986). As the disease progresses, diseased leaves defoliate early (Ridgway et al. 1984), and the disease spreads along the veins of the host plant known as the bacterial veins (Verma 1986). Bacterial ooze stained cotton fiber in diseased bolls (Brown and Ware 1958). Fruit positions become vulnerable to lesions of the black arm due to the delicate nature of the infected stems (Innes 1983; Akello and Hillocks 2002). The bacterium has overwintered on infected seeds and on plant residues and could survive at least 22 months upon seed (Kirkpatrick and Rothrock 2001). Wind-driven rain and running water along with rain were main ways of spreading this bacterium (Brown and Ware 1958). The blowing sand was a general source of spreading pathogen. Dust along with storms first produced wounds in the plant tissues and then later caused infections in plants. Furthermore, seeds, machines, insects, and animals are also responsible for the transmission of this pathogen (Thaxton and Zik 2001).

13.4.1.3 Epidemiology

This disease was more severe in subhumid than semiarid regions with wind, rainfall ranging from 25.4 to 76.2 mm, and dust events in growing season (Kirkpatrick and Rothrock 2001). Disease infestation was higher in high-humidity areas that favored growth besides the spread of pathogen (Voloudakis et al. 2006). In natural environmental conditions, black cotton arm infections damage 35% of bolls.

13.4.1.4 Symptoms

Bacterial blight begins as small water soaked lesions (spots) on the leaves and seedlings and mature plants. The lesions progress into characteristic angular shapes when leaf veins limit bacterial movement (Fig. 13.7a). Unlike many other lesions on more or less circular cotton leaves, those associated with bacterial blight are more triangular or rectangular, although the shape may be more difficult to distinguish with leaf aging. Bacterial lesions may appear on the upper surface of the leaf; however, the wet or "greasy" appearance of the lesions is often observed more clearly on the underside of the leaf (Fig. 13.7b).

Fig. 13.7 Bacterial blight of cotton (**a**) and bottom of an infected leaf (**b**)

13.4.1.5 Management

- Resistant varieties against pathogen should be sown.
- Remove the infected stems and cotton debris present in the fields as soon as possible.
- Avoid applying agronomic practices in the field during wet conditions.
- Use growth regulators to activate the defense mechanism of the plant against pathogen.
- Use an alternate crop in the fields and apply irrigation timely.
- Manage seed sanitation to overcome the diseases by using acids, copper compounds, or chlorine derivatives and heat treatments.
- Disease forecasting models are very important to the farmers for disease prediction.

13.4.2 Bacterial Seed Rot and Boll Rot of Cotton

13.4.2.1 Introduction

The bacterial seed and boll rot of cotton represent a new threat for cotton growers. This disease was first discovered in South Carolina, United States (Hudson 2000), with a yield loss of 10–15%. In 2006, it was reported in the Chinese province of Xinjiang (Ren et al. 2008) causing 20% of stimulated performance losses. In Pakistan, firstly, it was reported in Faisalabad District (Ehetisham-ul-Haq et al. 2014). Usually, the disease is observed when the bolls open, and diseased locules are recognized as "hard lock" having dense lint with rotted seeds (Fig. 13.8a).

At the time of collection, partially open infected bolls fall sooner or are not harvested efficiently (Hudson 2000). Despite the reduction in cotton, even rot of

Fig. 13.8 Symptoms of boll rot of cotton (**a**) and hard lock (**b**)

seeds deteriorates the quality of the cotton fiber. The infected seeds do not support the development of cotton fibers. The quality of the cotton became poor when immature fiber was mixed with the healthy one, which eventually decreases the yield. The affected bolls do not open normally and remain attached to plants or fall to ground. Yield besides quality of cotton decreases when infected/immature fiber is mixed with the healthy one (Hudson 2000).

13.4.2.2 Pathogen and Disease Cycle

According to Stewart (2007), who made a comprehensive note on boll rot in cotton, the pathogen enters the plant by penetrating sucking insects. If the bolls are attacked in the first 10 days of their formation by the insect, they are unable to develop further due to wound, hormonal imbalance, or digestive juice of the insect. After flowering, the pathogen invades the wounds produced by insect feeding. Bolls remain susceptible to invasion of pathogens till the initial 3 weeks. After that, no damages occur when the bolls go even under secondary deposition, and insects cannot pierce into their stilettos. Symptoms did not appear on the external side of the carpel. Immature fiber and discoloration of brown seeds on diseased bolls are important symptoms.

13.4.2.3 Epidemiology

Boll rot of cotton is favored by wet and humid conditions and especially by high humidity. Rainfall that splashes soil up onto lower bolls enables infection. Immature bolls due to disease attack and lodged plants are major sources, and there is a high risk of disease spread in rainy season. In wet weather conditions, the chances of boll rot and hard lock development are high at boll opening time; high humid conditions delay boll opening and provide enough time for pathogen development.

13.4.2.4 Symptoms

Hudson (2000) describes the symptoms of the disease. According to him, discolored and semi-dead seed is observed when the infected immature boll is open or dissected. In particular, the immature locule has been attributed to the development of "hard locks" (Fig. 13.8b). In contrast to typically white and downy fiber, lint from infected locules (normally 4 locules/boll) is brown and dense. Simply, the disease pattern for seed besides rotting of SC cotton bolls includes (1) the asymptomatic external carpels of the infected bolls, (2) incomplete maturity of fibers and seeds, and (3) necrotic brown color of fibers besides the seed tissues. Mauney et al. (2003) describe structural differences and development between embryos dissected from typical cotton seeds and SC capsules that exhibit "semi-empty," an idiom adopted while awaiting identification of one or more random agents. They discovered that normal fertilization took place in both situations; however, a discrepancy in successive growth was evident (Mauney et al. 2004).

13.4.2.5 Management

- Sowing of resistant or tolerant varieties, crop rotation, and seed treatment with chemicals and insecticides (for pests and disease vectors).
- Avoid very low plant populations that give rise to expose the soil that can be splashed on low bolls at the end of the season.
- Avoid dense vegetation.
- Evaluate the incidence before and after defoliation by counting all bolls on 10 plants of 10 designated sites in the field.
- Completely remove the crop residues.

13.5 Conclusion

In order to increase cotton yield, it is very much important for cotton growers to have knowledge about cotton diseases, causal organisms, and disease attack time along with their control measures through integrated disease management options including cultural, biological, and chemical means.

References

Abbas Q, Ahmad S (2018) Effect of different sowing times and cultivars on cotton fiber quality under stable cotton-wheat cropping system in southern Punjab, Pakistan. Pak J Life Soc Sci 16:77–84

Ahmad S, Abbas Q, Abbas G, Fatima Z, Atique-ur-Rehman NS, Younis H, Khan RJ, Nasim W, Habib ur Rehman M, Ahmad A, Rasul G, Khan MA, Hasanuzzaman M (2017) Quantification of climate warming and crop management impacts on cotton phenology. Plants 6(7):1–16

Ahmad S, Iqbal M, Muhammad T, Mehmood A, Ahmad S, Hasanuzzaman M (2018) Cotton productivity enhanced through transplanting and early sowing. Acta Sci Biol Sci 40:e34610

Ahmad S, Raza I (2014) Optimization of management practices to improve cotton fiber quality under irrigated arid environment. J Food Agric Environ 2(2):609–613

Ahmad S, Raza I, Ali H, Shahzad AN, Atiq-ur-Rehman SN (2014) Response of cotton crop to exogenous application of glycinebetaine under sufficient and scarce water conditions. Braz J Bot 37(4):407–415

Akello B, Hillocks RJ (2002) Distribution and races of Xanthomonas axonopodis pv. Malvacearum on cotton (Gossypium hirsutum) in Uganda. J Phytopathol 150:65–69

Akhtar S, Tahir MN, Baloch GR, Javaid S, Khan AQ, Amin I, Briddon RW, Mansoor S (2014) Regional changes in the sequence of cotton leaf curl Multan betasatellite. Viruses 6:2186–2203

Ali H, Abid SA, Ahmad S, Sarwar N, Arooj M, Mahmood A, Shahzad AN (2013a) Integrated weed management in cotton cultivated in the alternate-furrow planting system. J Food Agric Environ 11(3&4):1664–1669

Ali H, Abid SA, Ahmad S, Sarwar N, Arooj M, Mahmood A, Shahzad AN (2013b) Impact of integrated weed management on flat-sown cotton (*Gossypium hirsutum* L.). J Anim Plant Sci 23 (4):1185–1192

Ali H, Afzal MN, Ahmad F, Ahmad S, Akhtar M, Atif R (2011) Effect of sowing dates, plant spacing and nitrogen application on growth and productivity on cotton crop. Int J Sci Eng Res 2 (9):1–6

Ali H, Hameed RA, Ahmad S, Shahzad AN, Sarwar N (2014a) Efficacy of different techniques of nitrogen application on American cotton under semi-arid conditions. J Food Agric Environ 12 (1):157–160

Ali H, Hussain GS, Hussain S, Shahzad AN, Ahmad S, Javeed HMR, Sarwar N (2014b) Early sowing reduces cotton leaf curl virus occurrence and improves cotton productivity. Cercetări Agronomice în Moldova XLVII(4):71–81

Ali Z, Abulfaraj A, Idris A, Ali S, Tashkandi M, Mahfouz MM (2015) CRISPR/Cas9-mediated viral interference in plants. Genome Biol 16:238

Ali Z, Ali S, Tashkandi M, Zaidi SS, Mahfouz MM (2016) CRISPR/Cas9-mediated immunity to geminiviruses: differential interference and evasion. Sci Rep 6:26912

Allen RM, Tucker H, Nelson RA (1960) Leaf crumple disease of cotton in Arizona. Plant Dis Rep 44:246–250

Amin A, Nasim W, Mubeen M, Ahmad A, Nadeem M, Urich P, Fahad S, Ahmad S, Wajid A, Tabassum F, Hammad HM, Sultana SR, Anwar S, Baloch SK, Wahid A, Wilkerson CJ, Hoogenboom G (2018) Simulated CSM-CROPGRO-cotton yield under projected future climate by SimCLIM for southern Punjab, Pakistan. Agric Syst 167:213–222

Amin A, Nasim W, Mubeen M, Nadeem M, Ali L, Hammad HM, Sultana SR, Jabran K, Habib ur Rehman M, Ahmad S, Awais M, Rasool A, Fahad S, Saud S, Shah AN, Ihsan Z, Ali S, Bajwa AA, Hakeem KR, Ameen A, Amanullah, Rehman HU, Alghabar F, Jatoi GH, Akram M, Khan A, Islam F, Ata-Ul-Karim ST, Rehmani MIA, Hussain S, Razaq M, Fathi A (2017) Optimizing the phosphorus use in cotton by using CSM-CROPGRO-cotton model for semi-arid climate of Vehari-Punjab, Pakistan. Environ Sci Pollut Res 24(6):5811–5823

Amin I, Hussain K, Akbergenov R, Yadav JS, Qazi J, Mansoor S, Hohn T, Fauquet CM, Briddon RW (2011) Suppressors of RNA silencing encoded by the components of the cotton leaf curl begomovirus-betasatellite complex. Mol Plant-Microbe Interact 24:973–983

Amin I, Mansoor S, Amrao L, Hussain M, Irum S, Zafar Y, Bull SE, Briddon RW (2006) Mobilisation into cotton and spread of a recombinant cotton leaf curl disease satellite. Arch Virol 151:2055–2065

Amrao L, Akhter S, Tahir MN, Amin I, Briddon RW, Mansoor S (2010b) Cotton leaf curl disease in Sindh province of Pakistan is associated with recombinant begomovirus components. Virus Res 153:161–165

Amrao L, Amin I, Shahid SM, Briddon RW, Mansoor S (2010a) Cotton leaf curl disease in resistant cotton is associated with a single begomovirus that lacks an intact transcriptional activator protein. Virus Res 152:153–163

Amudha J, Balasubramani G, Malathi VG, Monga D, Kranthi KR (2011) Cotton leaf curl virus resistance transgenics with antisense coat protein gene (AV1). Curr Sci 101:300–307

Atkinson CF (1892) Some disease of cotton: 3. Frenching. Bull Alabama Agric Exp Station 41:19–29

Atkinson GF (1891) Black rust of cotton: a preliminary note. Bot Gaz 16:61–65

Baltes NJ, Hummel AW, Konecna E, Cegan R, Bruns AN, Bisaro DM, Voytas DF (2015) Conferring resistance to geminiviruses with the CRISPR-Cas prokaryotic immune system. Nat Plants 1:15145

Bashan Y (1984) Transmission of *Alternaria macrospora* in the cotton seeds. J Phytopathol 110:110–118

Bashan Y (1986) Phenols in cotton resistant seedling and susceptible to *Alternaria macrospora*. J Phytopathol 116:1–10

Bashi E, Rotem J, Hans P, Kranz J (1983) Influence of controlled environment and age on development of *Alternaria macrospora* and on shedding of leaves in cotton. Phytopathology 73:1145–1147

Bayles MB, Verhalen LM (2007) Bacterial blight reactions of sixty-one upland cotton cultivars. J Cotton Sci 11:40–51

Bell AA (1992) Biology and ecology of *Verticillium dahliae*. In: Lyda SD (ed) Comparative pathology of Sclerotial-forming plant pathogens: a Phymatotrichum Omnivorum symposium. Texas A & M University Press, College Station

Bhutta AR, Bhatti MAR (1983) Incidence of bacterial blight of cotton and reaction of different cultivars to Xanthomonas pv. Malvacearum. Pak Cottons 27(I):75–78

Briddon RW, Bull SE, Amin I, Idris AM, Mansoor S, Bedford ID, Dhawan P, Rishi N, Siwatch SS, Abdel-Salam AM, Brown JK, Zafar Y, Markham PG (2003) Diversity of DNA ß, a satellite molecule associated with some monopartite begomoviruses. Virology 312:106–121

Briddon RW, Bull SE, Amin I, Mansoor S, Bedford ID, Rishi N, Siwatch SS, Zafar Y, Abdel-Salam AM, Markham PG (2004) Diversity of DNA 1: a satellite-like molecule associated with monopartite begomovirus-DNA ß complexes. Virology 324:462–474

Briddon RW, Mansoor S, Bedford ID, Pinner MS, Saunders K, Stanley J, Zafar Y, Malik KA, Markham PG (2001) Identification of DNA components required for induction of cotton leaf curl disease. Virology 285:234–243

Briddon RW, Markham PG (2000) Cotton leaf curl virus disease. Virus Res 71:151–159

Briddon RW, Martin DP, Roumagnac P, Navas-Castillo J, Fiallo-Olivé E, Moriones E, Lett JM, Zerbini FM, Varsani A (2018) Alphasatellitidae: a new family with two subfamilies for the classification of geminivirus- and nanovirus-associated alphasatellites. Arch Virol 163:2587–2600

Briddon RW, Stanley J (2006) Subviral agents associated with plant single-stranded DNA viruses. Virology 344:198–210

Brown EA, McCarter SM (1976) Effect of seedling disease caused by the *Rhizoctonia solani* on subsequent growth and yield of cotton. Phytopathology 66:111–115

Brown HB, Ware JO (1958) Cotton, 3rd edn. McGraw-Hill Book Company, Inc, New York, p 411

Brown JK, Mihail JD, Nelson MR (1987) Effects of cotton leaf crumple virus on cotton inoculated at different growth stages. Plant Dis 71:699–703

Chellappan P, Masona MV, Vanitharani R, Taylor NJ, Fauquet CM (2004) Broad spectrum resistance to ssDNA viruses associated with transgene- induced gene silencing in cassava. Plant Mol Biol 56:601–611

Chidambaram P, Kannan A (1989) Grey mildew of cotton. Tech Bull Central Inst Cotton Res Regional Station Coimbatore, India

Chohan S, Abid M (2018) First report of *Fusarium incarnatum-equiseti* species complex associated with boll rot of cotton in Pakistan. Plant Dis 103:151

Cook OF (1924) Acromania or 'crazy top', a growth disorder of cotton. J Agric Res 28:803–828

Cook RJ (1981) Fusarium disease in the People's republic of China. In: Nelson PE, Toussoun TA, Cook RJ (eds) Fusarium diseases: biology and control. Pennsylvania State University Press, University Park and London, pp 53–55

Delannoy E, Lyon BR, Marmey P, Jalloul A, Daniel JF, Montillet JL, Essenberg M, Nicole M (2005) Resistance of cotton towards Xanthomonas axonopodis pv. Malvacearum. Annu Rev Phytopathol 43:63–82

Devay JE, Garber RH, Matherson D (1982) Role of Pythium species in the seedling disease complex of cotton in California. Plant Dis 66:151–154

Dickson RC, Johnson MM, Laird EF (1954) Leaf crumple, a virus disease of cotton. Phytopathology 44:479–480

Dzhamalov A (1973) Irrigation and Alternaria leaf spot of cotton. Zaschita Rastenii 12:48

Ehetisham-ul-Haq M, Khan MA, Javed MT, Atiq M, Rashid A (2014) Management of newly emerging bacterial seed and boll rot of cotton with antibiotics, homeopathic products and plant extracts. J Tropic Agric 92:196–206

Ehrlich J, Wolf FA (1983) Areolate mildew of cotton. Phytopathology 22:229–240
Ellis MB (1971) Dematiaceous Hyphomycetes. CAB International, Wallingford, p 608
Erdoğan O, Bölek Y, Göre ME (2016) Biological control of cotton seedling diseases by fluorescent *Pseudomonas spp*. Tar Bil Der 22(3):398–407
Evans G (1967) Verticillium wilt of cotton the situation in the Namoi Valley. Agric Gaz N S W 78:581–583
Fedorinchik NS (1964) Biological method of controlling plant diseases. Vses Naucb -Issled Inst Zashch Rast Tr 23:201–210
Ferro CG, Silva JP, Xavier CAD, Godinho MT, Lima ATM, Mar TB, Lau D, Zerbini FM (2017) The ever increasing diversity of begomoviruses infecting non-cultivated hosts: new species from Sida spp. and Leonurus sibiricus, plus two New World alphasatellites. Ann Appl Biol 170:204–218
Fiallo-Olivé E, Martínez-Zubiaur Y, Moriones E, Navas-Castillo J (2012) A novel class of DNA satellites associated with New World begomoviruses. Virology 426:1–6
Follen JC, Goebel S (1973) Rots of cotton capsules in irrigated crops in Côte d'Ivoire. Relationship with varietal characteristics, irrigation method and date of sowing. Cott Trop Fibers 28 (3):401–407
Globus GA, Muromtsev GS (1990) The use of *Gliocladium roseum* as antagonist for defense of cotton from phytopathogenic fungi. In: Proceedings of the fifth international verticillium symposium, Leningrad, USSR, p. 90
Gouws MA, Prinsloo GC, Van der Linde EJ (2001) First report of *Mycosphaerella areola*, teleomorph of *Ramulariopsis gossypii*, on cotton in South Africa. Afr Plant Prot 7(2):115–116
Hashmi JA, Zafar Y, Arshad M, Mansoor S, Asad S (2011) Engineering cotton (*Gossypium hirsutum* L.) for resistance to cotton leaf curl disease using viral truncated AC1 DNA sequences. Virus Genes 42:286–296
Hillocks RJ (1991) *Alternaria* leaf spot of cotton with special reference to Zimbabwe. Trop Pest Manag 37(2):124–128
Hillocks RJ (1992) Cotton diseases. Melksham UK, pp. 39–86
Hopkins JCE (1932) Some disease of cotton in southern Rhodesia. Empire Cotton Growing Rev 9:109–118
Hudson J (2000) Seed rot hits South Carolina cotton. Southeast Farm Press. Available via http://southeastfarmpress.com/mag/farming_seed_rot_hits/. Accessed June 2018
Huisman OC, Gerik JS (1989) Dynamic of colonization of plant roots by *Verticillium dahlia* and other fungi. In: Tjamos EC, Beckman CH (eds) Vascular wilt disease of plants, NATO ASI (series H: cell biology), vol 28. Springer, Berlin, Heidelberg, pp 1–17
Idris A, Al-Saleh M, Amer M, Abdalla O, Brown JK (2014) Introduction of cotton leaf curl Gezira virus into the United Arab Emirates. Plant Dis 98:1593
Idris AM, Brown JK (2002) Molecular analysis of cotton leaf curl virus-Sudan reveals an evolutionary history of recombination. Virus Genes 24:249–256
Innes NL (1983) Bacterial blight of cotton. Biol Rev 58:157–176
Iqbal Z, Sattar MN, Shafiq M (2016) CRISPR/Cas9: a tool to circumscribe cotton leaf curl disease. Front Plant Sci 7:475
Javaid S, Amin I, Jander G, Mukhtar Z, Saeed NA, Mansoor S (2016) A transgenic approach to control hemipteran insects by expressing insecticidal genes under phloem-specific promoters. Sci Rep 6:34706
Ji X, Zhang H, Zhang Y, Wang Y, Gao C (2015) Establishing a CRISPR-Cas-like immune system conferring DNA virus resistance in plants. Nat Plants 1:15144
Jones GH (1928) An Alternaria disease of the cotton plant. Ann Bot 42:935–947
Juárez M, Tovar R, Fiallo-Olivé E, Aranda MA, Gosálvez B, Castillo P, Moriones E, Navas-Castillo J (2014) First detection of tomato leaf curl New Delhi virus infecting zucchini in Spain. Plant Dis 98:857
Kamel M, Ibrahim AN, Kamal SA, El-fahl AM (1971) Spore germination of Alternaria leaf spot disease. U A R J Bot 14:245–254

Khan AJ, Akhtar S, Al-Shihi AA, Al-Hinai FM, Briddon RW (2012) Identification of cotton leaf curl Gezira virus in papaya in Oman. Plant Dis 96:1704

Khan MA, Khan HA, Ilyas MB, Rashid A (1999) Correlation of environmental conditions with bacterial blight disease on six commercially grown cotton cultivars in five districts of the Punjab. Pak J Agric Sci 36(1–2):1–5

Khan MB, Khaliq A, Ahmad S (2004) Performance of mashbean intercropped in cotton planted in different planting patterns. J Res (Sci) 15(2):191–197

King CJ, Barker HD (1934) An interval collar rot on cotton. Phytopathology 29:75

Kirkpatrick TL, Rothrock CS (2001) Compendium of cotton diseases, 2nd edn. APS Press, St. Paul, MN, p 77

Kirkpatrick TW (1931) Further studies on leaf-curl of cotton in the Sudan. Bull Entomol Res 22:323–363

Kirthi N, Priyadarshini CG, Sharma P, Maiya SP, Hemalatha V, Sivaraman P, Dhawan P, Rishi N, Savithri HS (2004) Genetic variability of begomoviruses associated with cotton leaf curl disease originating from India. Arch Virol 149:2047–2057

Klich MA (1986) Mycoflora of cotton seeds from southern USA. A three-year study of distribution and frequency. Mycologia 78:706–712

Leke WN, Khatabi B, Mignouna DB, Brown JK, Fondong VN (2016) Complete genome sequence of a new bipartite begomovirus infecting cotton in the Republic of Benin in West Africa. Arch Virol 161:2329–2333

Mansoor S, Amrao L, Amin I, Briddon RW, Malik KA, Zafar Y (2006) First report of cotton leaf curl disease in central and southern Sindh province in Pakistan. Plant Dis 90:826

Mansoor S, Briddon RW, Bull SE, Bedford ID, Bashir A, Hussain M, Saeed M, Zafar Y, Malik KA, Fauquet C, Markham PG (2003) Cotton leaf curl disease is associated with multiple monopartite begomoviruses supported by single DNA β. Arch Virol 148:1969–1986

Mansoor S, Khan SH, Bashir A, Saeed M, Zafar Y, Malik KA, Briddon RW, Stanley J, Markham PG (1999) Identification of a novel circular single-stranded DNA associated with cotton leaf curl disease in Pakistan. Virology 259:190–199

Mauney JR, Scottsdale AZ, Stewart JMcD (2003) Embryo development in bolls exhibiting the "Hollow Seed" syndrome in South Carolina. In: 2003 Beltwide Cotton Conference, Nashville, TN January 6–10, pp. 1653–1655

Mauney JR, Scottsdale AZ, Stewart JMcD, Fayetteville AR, Jones M (2004) Onset and progression of the "Hollow Seed" (seed rot) Malady of South Carolina. In: 2004 Beltwide Cotton Conference, San Antonio, TX, January 5–9, pp. 1967–1969

Mccarter SM, Roncardi RW, Crawford JL (1970) Microorganisms associated with *Aspergillus flavus* boll rot in Georgia. Plant Dis Resporter 54:586–590

McSpadden Gardener BB, Fravel DR (2002) Biological control of plant pathogens: research, commercialization, and application in the USA. Plant Health Prog. https://doi.org/10.1094/PHP-2002-0510-01-RV

Menlikiev NY (1962) Fusarium wilt of fine-staple cotton and a study of *Fusarium oxysporum* f. sp. *vasinfectum* strains as the causal agent of the disease in conditions of the Vakash Valley. Rev Appl Mycol 43:3381

Mnari-Hattab M, Zammouri S, Belkadhi MS, Doña DB, ben Nahia E, Hajlaoui MR (2015) First report of tomato leaf curl New Delhi virus infecting cucurbits in Tunisia. New Dis Rep 31:21

Mohan SK (1983) Seed transmission and epidemiology of Xanthamonas campestris pv. Malvacearum. Seed Sci Technol 11:569–571

Nawaz-ul-Rehman MS, Briddon RW, Fauquet CM (2012) A melting pot of Old World begomoviruses and their satellites infecting a collection of Gossypium species in Pakistan. PLoS One 7:e40050

Nawaz-ul-Rehman MS, Nahid N, Mansoor S, Briddon RW, Fauquet CM (2010) Post-transcriptional gene silencing suppressor activity of two non-pathogenic alphasatellites associated with a begomovirus. Virology 405:300–308

Nemli T, Sayar I (2002) Aydin Söke region the prevalence of cotton-depleting disease the factors and prevention of the possibility of investigation. Scientific and Technical Research Council of Turkey the TARP-2535 Ankara, p. 57

Panno S, Iacono G, Davino M, Marchione S, Zappardo V, Bella P, Tomassoli L, Accotto GP, Davino S (2016) First report of tomato leaf curl New Delhi virus affecting zucchini squash in an important horticultural area of southern Italy. New Dis Rep 33:6

Paprotka T, Metzler V, Jeske H (2010) The first DNA 1-like alpha satellites in association with New World begomoviruses in natural infections. Virology 404:148–157

Paulwetter RC (1918) The Alternaria leaf spot of cotton. Phytopathology 8:98–115

Pleban S, Ingel F, Chet I (1995) Control of *R. solani* and *S. rolfsii* in the greenhouse using Endophytic bacillus spp. Eur J Plant Pathol 101(6):665–672

Pooggin M, Shivaprasad PV, Veluthambi K, Hohn T (2003) RNAi targeting of DNA virus in plants. Nat Biotechnol 21:131–132

Qazi J, Amin I, Mansoor S, Iqbal MJ, Briddon RW (2007) Contribution of the satellite encoded gene betaC1 to cotton leaf curl disease symptoms. Virus Res 128:135–139

Rahman MH, Ahmad A, Wang X, Wajid A, Nasim W, Hussain M, Ahmad B, Ahmad I, Ali Z, Ishaque W, Awais M, Shelia V, Ahmad S, Fahad S, Alam M, Ullah H, Hoogenboom G (2018) Multi-model projections of future climate and climate change impacts uncertainty assessment for cotton production in Pakistan. Agric For Meteorol 253-254:94–113

Rajagopalan PA, Naik A, Katturi P, Kurulekar M, Kankanallu RS, Anandalakshmi R (2012) Dominance of resistance-breaking cotton leaf curl Burewala virus (CLCuBuV) in northwestern India. Arch Virol 157:855–868

Ramapandu S, Sitaramaiah K, Subbarao K, Prasada Rao MP (1979) Screening of cotton germplasm against bacterial blight caused by Xanthomonas axonopodis pv. Malvacearum. Indian Phytopathol 32:486–487

Rane MS, Patel MK (1956) Diseases of cotton in Bombay 1. Alternaria leaf spot. Indian Phytopathol 9:106–113

Raza A, Malik HJ, Shafiq M, Amin I, Scheffler JA, Scheffler BE, Mansoor S (2016) RNA interference based approach to down regulate osmoregulators of whitefly (*Bemisia tabaci*): potential technology for the control of whitefly. PLoS One 11:e0153883

Reddy RVC, Muniyappa V, Colvin J, Seal S (2005) A new begomovirus isolated from *Gossypium barbadense* in southern India. Plant Pathol 54:570–570

Ren YZ, Liu YQ, Ding SL, Li GY, Zhang H (2008) First report of boll rot of cotton caused by *Pantoea agglomerans* in China. Plant Dis 92:1364–1364

Ridgway RL, Bell AA, Veech JA, Chandler JM (1984) Cotton protection practices in the USA and the world, pp. 265–365. In: Kohel RJ, Lewis CF (eds) Cotton. American Society of Agronomy/ Crop Science Society of America/Soil Science Society of America, Madison, pp 226–365

Romay G, Chirinos D, Geraud-Pouey F, Desbiez C (2010) Association of an atypical alphasatellite with a bipartite New World begomovirus. Arch Virol 155:1843–1847

Ruiz ML, Simón A, Velasco L, García MC, Janssen D (2015) First report of tomato leaf curl New Delhi virus infecting tomato in Spain. Plant Dis 99:894

Saeed M, Zafar Y, Randles JW, Rezaian MA (2007) A monopartite begomovirus-associated DNA ß satellite substitutes for the DNA B of a bipartite begomovirus to permit systemic infection. J Gen Virol 88:2881–2889

Sangeetha KD, Ashtaputre SA (2015) Morphological and cultural variability in isolates of Alternaria sp. causing leaf blight of cotton. Karnataka J Agric Sci 28(2):214–219

Sattar MN, Iqbal Z, Tahir MN, Ullah S (2017) The prediction of a new CLCuD epidemic in the Old World. Front Microbiol 8:631

Sattar MN, Kvarnheden A, Saeed M, Briddon RW (2013) Cotton leaf curl disease–an emerging threat to cotton production worldwide. J Gen Virol 94:695–710

Saunders K, Stanley J (1999) A nanovirus-like component associated with yellow vein disease of Ageratum conyzoides: evidence for interfamilial recombination between plant DNA viruses. Virology 264:142–152

Schnathorst WC (1981) Life cycle and epidemiology of verticillium. In: Mace ME, Bell AA, Beckman CH (eds) Fungal wilt disease of plants. Academic Press, New York, pp 81–111

Sharma YR, Sandhu BS (1985) A new fungus associated with boll rot of *G. arboreum* cotton. Curr Sci 54:936

Shen CY (1985) Integrated management of *Fusarium* and *Verticillium* wilt of cotton in China. Crop Prot 4:337–345

Shukla AK, Upadhyay SK, Mishra M, Saurabh S, Singh R, Singh H, Thakur N, Rai P, Pandey P, Hans AL, Srivastava S, Rajapure V, Yadav SK, Singh MK, Kumar J, Chandrashekar K, Verma PC, Singh AP, Nair KN, Bhadauria S, Wahajuddin M, Singh S, Sharma S, Omkar URS, Ranade SA, Tuli PK, Singh PK (2016) Expression of an insecticidal fern protein in cotton protects against whitefly. Nat Biotechnol 34:1046–1051

Snyder WC, Hansen HN (1940) The species concept in Fusarium. Am J Bot 27:64–67

Stanley J (1983) Infectivity of the cloned geminivirus genome requires sequences from both DNAs. Nature 305:643–645

Stewart J (2007) Boll development and seed rot. In: World Cotton Research Conference-4, Lubbock, TX, USA, 10–14 September, 2007. International Cotton Advisory Committee (ICAC)

Tahir MN, Amin I, Briddon RW, Mansoor S (2011) The merging of two dynasties - identification of an African cotton leaf curl disease-associated begomovirus with cotton in Pakistan. PLoS One 6: e20366

Tariq M, Afzal MN, Muhammad D, Ahmad S, Shahzad AN, Kiran A, Wakeel A (2018) Relationship of tissue potassium content with yield and fiber quality components of Bt cotton as influenced by potassium application methods. Field Crop Res 229:37–43

Tariq M, Yasmeen A, Ahmad S, Hussain N, Afzal MN, Hasanuzzaman M (2017) Shedding of fruiting structures in cotton: factors, compensation and prevention. Trop Subtrop Agroecosyst 20(2):251–262

Thaxton PM, Zik KME (2001) Bacterial blight. In: Kirkpatrick TL, Rothrock CS (eds) Compendium of cotton diseases, 2nd edn. American Phytopathological Society, St. Paul, MN, pp 34–35

Usman M, Ahmad A, Ahmad S, Irshad M, Khaliq T, Wajid A, Hussain K, Nasim W, Chattha TM, Trethowan R, Hoogenboom G (2009) Development and application of crop water stress index for scheduling irrigation in cotton (*Gossypium hirsutum* L.) under semiarid environment. J Food Agric Environ 7(3&4):386–391

van Schaik PH, Erwin DC, Garber MJ (1962) Effects of time on symptom expression of the leaf-crumple virus on yield and quality of fiber cotton. Crop Sci 2:275–277

Venkatesh I, Darvin G (2016) An overview on cotton Alternaria leaf spot and its management. Int J Appl Bio Pharm Tech 7(2):135

Verma JP (1986) Bacterial blight of cotton. CRC Press, Boca Raton, FL, pp 278–279

Voloudakis AE, Marmey P, Delannoy E, Jalloul A, Martinez C, Nicole M (2006) Molecular cloning and characterization of (Gossypium hirsutum) superoxide dismutase genes during cotton Xanthomonas axonopodis pv. Malvacearum interaction. Physiol Mol Plant Pathol 68:119–127

Waghunde RR, Patel UT, Vahunia B (2018) Morphological and cultural variability of *Alternaria Macrospora* causing leaf blight in cotton. J Pharmacogn Phytochem 7(3):3096–3099

Wang C, Wang D, Zhou Q (2004) Colonization and persistence of a plant growth-promoting bacterium *Pseudomonas fluorescens* strain CS85 on roots of cotton seedlings. Can J Microbiol 50(7):475–481

Yasmeen A, Kiani S, Butt A, Rao AQ, Akram F, Ahmad A, Nasir IA, Husnain T, Mansoor S, Amin I, Aftab S, Zubair M, Tahir MN, Akhtar S, Scheffler J, Scheffler B (2016) Amplicon-based RNA interference targeting V2 gene of cotton leaf curl Kokhran virus-Burewala strain can provide resistance in transgenic cotton plants. Mol Biotechnol 58:807–820

Zaidi SS, Briddon RW, Mansoor S (2017) Engineering dual begomovirus-Bemisia tabaci resistance in plants. Trends Plant Sci 22:6–8

Zaidi SSA, Iqbal Z, Amin I, Mansoor S (2015) First report of tomato leaf curl Gujarat virus, a bipartite begomovirus on cotton showing leaf curl symptoms in Pakistan. Plant Dis 99:1655

Zaidi SSA, Shafiq M, Amin I, Scheffler BE, Scheffler JA, Briddon RW, Mansoor S (2016) Frequent occurrence of tomato leaf curl New Delhi virus in cotton leaf curl disease affected cotton in Pakistan. PLoS One 11:e0155520

Zammouri S, Zaagueri T, Eddouzi J, Belkhadhi MS, Hajlaoui MR, Mnari-Hattab M (2017) First report of tomato leaf curl New Delhi virus on tomato crop in Tunisia. J Plant Pathol 99:799–818

Zhou X (2013) Advances in understanding begomovirus satellites. Annu Rev Phytopathol 51:357–381

Zhou X, Liu Y, Robinson DJ, Harrison BD (1998) Four DNA-A variants among Pakistani isolates of cotton leaf curl virus and their affinities to DNA-A of geminivirus isolates from okra. J Gen Virol 79:915–923

Chapter 14
Cotton Diseases and Disorders Under Changing Climate

Ateeq-ur-Rehman, Muhammad Mohsin Alam Bhatti, Ummad-ud din Umar, and Syed Atif Hasan Naqvi

Abstract Cotton is the most important fibre crop and second oil seed crop of the world. The crop is facing many abiotic and biotic factors which are disturbing its physiology and productivity. Among abiotic factors the climate of the world is changing every year and this is posing a great threat to the world food security. This factor has already disturbed the resources and the human health in the African countries and poses the same threat to many other underdeveloped as well as developed nations of the world. With advances in studies on changing climate, their influences on the agricultural systems have become a major concern. It is imperative to note that climate variation is one of the most important factors disturbing annual crop production, even in those areas which are known to us for their high yields by employing advanced agricultural technologies. It is planned to provide useful information for the agricultural scientists and other stakeholders who are interested in the changing environmental conditions and their impacts on insect pests and plant diseases. The negative effects of the climate will be the increased attacks of pests and diseases. Agriculture is tremendously susceptible to climate transformation. Extreme temperatures will in due course of time lessen growers produces of cotton and upsurge incidence of economically harmful insects including sucking as well as chewing insects and increase disease causing pathogens. Rainfall variations and floods are expected to understand the prospects towards the crop failures and yield loss. Though there might be few advantages liable on different crop cultivating regions, however the global influence on cotton cultivation is anticipated to be adverse due to upsurge of sucking insect vectors and disease causing pathogens. This chapter focuses on the impacts of climatic changes on few economically important insect pests and cotton diseases and disorders and tackles the problems as well.

Keywords Climate change · Temperature · Rainfall · Cotton diseases

Ateeq-ur-Rehman (✉) · M. M. A. Bhatti · U.-u. d. Umar · S. A. H. Naqvi
Department of Plant Pathology, Bahauddin Zakariya University, Multan, Pakistan
e-mail: ateequrrehman@bzu.edu.pk

S. Ahmad, M. Hasanuzzaman (eds.), *Cotton Production and Uses*,
https://doi.org/10.1007/978-981-15-1472-2_14

14.1 Introduction

Cotton is the most important fibre crop and second one oil seed crop of the world (Cetin and Basbag 2010). The crop is facing many abiotic and biotic factors which are disturbing its physiology and also productivity. Among abiotic factors the climate of the world is changing every year and this is posing a great threat to the world food security. This factor has already disturbed the resources and the human health in the African countries and poses the same threat to many other developed as well as underdeveloped nations of the world (Ahmad et al. 2014, 2018; Abbas and Ahmad 2018; Ahmad and Raza 2014; Ali et al. 2011, 2013a, b, 2014a, b; Khan et al. 2004; Usman et al. 2009).

As a result, there is a boundless apprehension relating to the influence of changing climate on production of economically imperative crops in agriculture particularly in relationship to the rainfall (Hulme 1996).

It is worth stating that changing climate is the most significant issue globally (Ahmad et al. 2017; Amin et al. 2017, 2018; Rahman et al. 2018; Tariq et al. 2017, 2018). The changes in conditions will ultimately alter and disturb the physiology of the crop plants. As number and severity of days will be amplified with temperatures that are expected to be very high throughout the cotton spell, the effects of this might be negative on cotton. This will affect the shape of decline in routine photosynthesis and occasionally increased frequency of respiration at night and consumption of deposited assimilates which cause intensifications in shedding of cotton squares and bolls with increased dropping quantity of seeds for each boll (Bang and Constable 2009). In arid and semi-arid regions of cotton sown areas, if temperature increases at high during summer months, fruit shedding is increased. High air temperature in the form of heat stress is responsible for the less fruit retention and affects cotton production (Brown 2008). However, the impact of heat stress on cotton could be dissimilar conferring to cotton cultivar and environmental conditions (Shah et al. 2001).

14.2 Temperature

Climate of the world is changing and going towards the extreme of the conditions like high and low temperatures. Young cotton plants are sensitive at above 65 °C temperature. The cotton areas with sandy soils may reach these temperatures of surface sands during the hot sunny days. If this hot sand is thrown to plants during early time of cultivation, then the affected plants are girdled, wilted, and toppled down. If high temperature hits the emerging plants, then they are burned or prevented from standing in the field.

Symptoms of heat on the properly irrigated cotton plants during summer are; more vegetative growth, premature leaf senescence, early maturity of plants, reduction of boll size, less number of seeds per boll, and falling of bolls.

The cool temperatures during the cotton sowing result in delayed plant emergence and uneven crop stand. The frozen temperatures during boll development may kill the bolls (Oosterhuis and Hake 2001).

Cotton grows well at 15 °C which is minimum atmospheric temperature requirement for germination. For vegetative growth optimum temperature requirement is 21–27 °C, and for healthy fruiting, it requires temperature of 27–32 °C. Commercial cotton varieties cultivated today mostly require more than 15 °C to produce and become indolent at temperatures below 15 °C and are probably killed due to freezing (Waddle 1984). More incidences of stress caused by heat can also harm cotton bolls and cause parrot beaked bolls, and boll freezing as well as cavity formation also reduces yield. In contrast, Ali et al. in 2004 observed the relationship between average air temperature after September, pest population, and fruit retention. The scientists observed that when average air temperature decreased after the month of September, it increased the pest populations, whereas the fruit retention on cotton was reduced when the temperature was increased. Sarvella in 1966 noticed that pollen sterility at high temperatures of day and night above 28 and 20 °C, respectively, in the field was responsible for the poor boll setting. Mauney et al. in 1978 and Guinn and Mauney in 1984 detected that at high temperatures, availability of water and nutrition impaired, and they triggered fruit loss.

At high temperatures the harm of fruit may cause crop to grow in vegetative forms. Heat stress not only reduces the availability of the assimilates for plant growth but also damages the cotton plant tissues.

According to Bang (2007), the consequences of severe heat stress on tissue exposure are:

- Bolls resembling parrot beak: This is the first effect of the high temperature that reduces the capability of the pollen during blossoming phase. The phenomenon decreases the size of bolls and can decrease the cotton production. This causes the insignificant boll development which has not evenly designed and seed numbers sandwiched between locks.
- Boll freeze or cavitation or boll dangle. This occurs before the formations of abscission layer and the young bolls die. The process will result in the increased vegetative growth of the crop.

Changing climate is a great hurdle for agriculture in many areas of sub-Saharan Africa (Burke et al. 2009). The growing temperature and unstable rainfall together will impact destructively on agriculture (IPCC 2000; Muller 2009).

While on global level, the temperature has augmented >0.5 °C in the previous century. Similar degree of warming of 0.05 °C each decade has been seen in Southern African regions (Hulme 1996).

Different cotton cultivars are influenced by the changing climatic and environmental factors. These responses may vary depending on the developing stage of the cotton plant and climatic conditions. Important developing phases in cotton plant are:

1. Climatic circumstances during sowing.
2. Early season sowing.
3. Climatic condition during flowering.
4. Environmental conditions during boll formation.
5. Weather situations near crop termination.

Advancing degree of temperature leads to increase in climate transformation, changes in water cycle besides rainfall patterns, and an upsurge of certain dangerous climatic events. Liable on area, high temperature may, for instance, guides towards extended growth season besides added rainfall or lesser rainfall resulting into smaller growing season. The extreme meteorological conditions may disturb plants at any time of growth period and are tough in the direction of forecast. Elevated temperatures could affect multiple regions in multiple ways for each particular region. Soil temperature if low at the sowing time may hamper timely planting of cotton in many countries. Increasing temperatures will advantage many countries and regions that might be able to sow cotton much prior than they do now. However, in contrary, elevated temperatures in cotton cultivating parts and countries even now woe from increased temperatures could have adverse influence because of intensified shedding of floret buds. Escalation in temperature could have encouraging impacts on produces in those hubs of cotton where actual fruiting period is reduced between two phases of lesser temperatures: one primary in time to start active flowering besides boll development and the other when plant is fully mature that outcomes in expiry of fruit development. Boll retention is more subtle to increased temperatures than any other condition, except for nutritional unavailability and imbalance. The heat traumas, yield is still feasible in warmer milieus (ICAC 2009). *Gossypium* is efficaciously grown at 82.7 °F in China, 99.6 °F in India, 98.2 °F in Pakistan, and 107.2 °F in Sudan. There is no confirmation that 107.2 °F is the highest limit, but understanding of some countries, predominantly in India, Pakistan, and Syrian Arab Republic, revealed that heat trauma is a huge restriction to maximize cotton productivity. The above mentioned countries effectively established heat-resistant cultivars throughout the period of decade from the 1970s to 1980s. These nations provide almost 75% of the world's cotton-growing land; besides most of them will be able to grow at somewhat escalated temperatures in comparison with prevailing averages. On the other hand, there are cotton cultivating zones that have low yield because too much temperature at 104 °F would seem to be a hindrance. They have extended cultivating seasons and any further escalation in temperature might encourage sterility which prevents boll development. Cotton breeders and seed sector in the countries will have to concentrate on heat-tolerant varieties (ICAC 2009). Escalating temperatures causes a complex outcome on plant growing stages besides yield, liable to location. This will definitely impact harmfully on fibre features. Reported scientific literature discloses that higher temperatures may result in higher micronaire, stronger fibres, as well as mature fibres.

Global warming has certain impacts on pests as they can adjust their body temperatures according to surroundings. Some studies have shown that global warming impacts metabolism and increases their population rate, dispersal to cooler topographies in the North and South, and subsequent in incidence of diverse plant differences besides different kind. Therefore, with temperature fluctuations, an upsurge in pest problem is anticipated (Karl et al. 2009).

Insect/pest management has certainly developed further perilous in accomplishing best growth besides yield. Atmospheric CO_2 levels besides elevated temperatures have influence over efficacy of certain pest managing tools presently in use, some seed cultivars and pesticides. Wu et al. (2007) revealed that genetically improved *Bacillus thuringiensis* (Bt) cotton displays a lesser amount of Bt toxin after exposure of raised CO_2, which distressed plant-bollworm interactions. Karl et al. (2009) stated that escalated temperature lessens efficacy of some specific classes of insecticides which include different chemicals that are currently of much use including pyrethroids and spinosad.

This phenomenon will also be responsible for the increase in the cotton insect pests besides diseases.

Insects are familiar menace to cotton yield all around the globe. Most insects could become accustomed, so they adjust their body temperature to temperature of surrounding atmosphere. Consequence of global warming on existing creatures is sluggish, sufficient for cotton pests to regulate towards intensifying temperatures besides other variations. Hence, the insects at present troubling cotton are anticipated to endure their survival and probably flourish better in changing environments (ICAC 2007).

Several researchers distress over the increasing level of global warming will impact pest's metabolism, letting insects increase their growth and reproduction proportion. Increasing temperatures drive to open new extents for annexation by insects, and most might be able to settle at new developing areas. Advancement in the populations of presently imperative insects, like bollworms, might also arise at higher reproduction rates, with the abolition of diapauses throughout winter to evade cooler seasons. Impacts may have additional well-augmented conditions wherever alternative host plants already exist for wintering (ICAC 2007). Global warming possibly will influence disease control in three ways: (1) its impact over pathogens, (2) by generating disease pacifying situations, and (3) by disturbing tissues of host. It is dreaded that escalation in temperature disturbs some disease management approaches because of variations in the pathogen appearance time. It impacts on the chemical control method which becomes non-effective as there is likelihood of faster disintegration of chemicals under higher temperatures. Elevated CO_2 levels drive upsurge severity of infection caused by diseases and encourage fungal development and spore formation. It kills further tissues of the host plants. Overall, the disease problem will turn out to be more significant (ICAC 2007).

The regions of the world where the cotton is already grown in the high temperature zones will be prone to more insect pest attacks, and this will make the crop to be attacked by the diseases. The dry weather if it prevails will give increase in the populations of the white fly which will not only directly damage the crop but also transmit the *Cotton leaf curl virus* in the crop. The dry areas will also favour the development of thrips and mites on the cotton crop.

14.3 Rainfall

Under the changing climatic conditions, there are chances of low rainfall and drought conditions which may be harmful for the cotton crop. Cotton plants prone to water deficiency are easily recognized by dark green colour of leaves and leaves wilting. The stems of affected plants show considerable reddening and growth of the plants stops. The intermodal distance of plants is also shortened and severe water stress causes the abortions of the squares (Oosterhuis and Hake 2001).

Cotton seedlings are sensitive to hail because the top portion of the plant is open to damage. The hail storm may cause the ragged cotyledons and complete removal of leaf tissues. The plant becomes bushy as sprouting of many vegetative branches due to the damage of the top portion of the plant (Oosterhuis and Hake 2001).

Bibi et al. (2008) stated that during environmental changes, there would be continuous rains during flowering and boll opening in few areas of the world that would decrease pollination and ultimately decrease fibre quality. Heavy rains during flowering and fruit setting of cotton cause falling of flower buds and bolls.

In the changing climate, smog is also becoming a problem in the cotton-growing areas of the world. The smog affects the growth of the plants by damaging the photosynthetic materials in leaves. The leaves show senescence, and they become green to speckled yellow and ultimately red (Oosterhuis and Hake 2001).

The excessive water-logging particularly with younger plants is responsible for root damage due to lack of oxygen, the foliage becomes yellow due to ethylene production or poor nutrient uptake and wilting of plants, square abscission and shorter internodes may occur (Oosterhuis and Hake 2001).

Symptoms of soil saturation are plant wilting and low nitrogen contents and growth of plants stops. If these conditions prevail longtime, then the affected plants die.

In view of cotton diseases, black rot of roots, *Verticillium* wilt, and *Fusarium* wilt in the crop have high incidence when the cold wet conditions prevail. When the climatic conditions are warm aiding with moist environment, *Alternaria* leaf spot attacks the cotton plant (Bang and Constable 2009).

Rains during morning reduce the pollination as the flowers are open. Other effects of these conditions include fewer seeds in bolls and many motes resulting in unfilled, malformed or misshapen bolls (Oosterhuis and Hake 2001).

14.4 Effects of Changing Climate on Cotton Diseases and Disorders in Cotton-Growing Countries

Effects of changing macro- as well as microclimate with regard to cotton diseases and productivity will be harmful because they relay to the compact accessibility of water for irrigation, especially in countries like China, western United States, Australia and Pakistan. Stress caused by high temperature and heat will decrease yields especially in countries like Pakistan which lack abundant availability of water and other resources. The influences of climate transformation on precipitation will likely be encouraging in the Yellow River area (China), India, southeastern United States and southeastern Anatolia (Turkey). Effects of precipitation in Brazil and West and Central Africa are uncertain. More than 70 different countries cultivate cotton on large scale with commercial prospects, common in the longitudinal band between 37°N and 32°S. *Gossypium* production is exclusively improved to semi-arid and arid atmospheres, where it is either cultivated rain-fed or through irrigation. Approximately 53% of cultivated lands on earth are *Gossypium* growing zones and 73% fibre development zones use from full or additional irrigation (ITC 2011).

The historical studies reflect that among many abiotic stresses, heat stress is the most imperative problem in the production of quality cotton from various countries including Pakistan, India and Syria, and additional increase in temperature can abolish cotton economy.

The impact of changing climate on the diseases and disorders of cotton in various cotton-growing countries is discussed as below:

14.4.1 China

China is known to be the largest cotton producer and consumer. The temperature during the cotton season remains favourable for the growth and development of cotton (Fig. 14.1).

The overall temperature and the rainfall in the areas of the cotton will increase. Elevated precipitation later in summer and beginning of fall often disturbs the cotton quality by providing suitable conditions for cotton pests and diseases (ITC 2011). There is an escalating trend towards extreme temperatures causing heat waves and more increase in intensity of rainfall. This may cause harm to cotton crop by the attack of diseases like bacterial blight of cotton.

With changing climate and global warming, there are chances to face unpredictable disasters like higher temperature and freezing, and escalated drought and flooding may occur in an unusual manner (China 2004). These climatic changes may affect the cotton crop through biotic and abiotic factors.

14.4.2 India

Cotton crop may take advantage from increase in production because of the establishment of new cotton-growing areas which were not known for cotton cultivation in the past. The complete impact of climate variation on cotton production, trade, economy and our dependency on cotton are very difficult to foresee, although firm observations can be made like temperatures are foreseeable to rise in India. These rising temperatures during the cotton season are elaborated in Fig. 14.1.

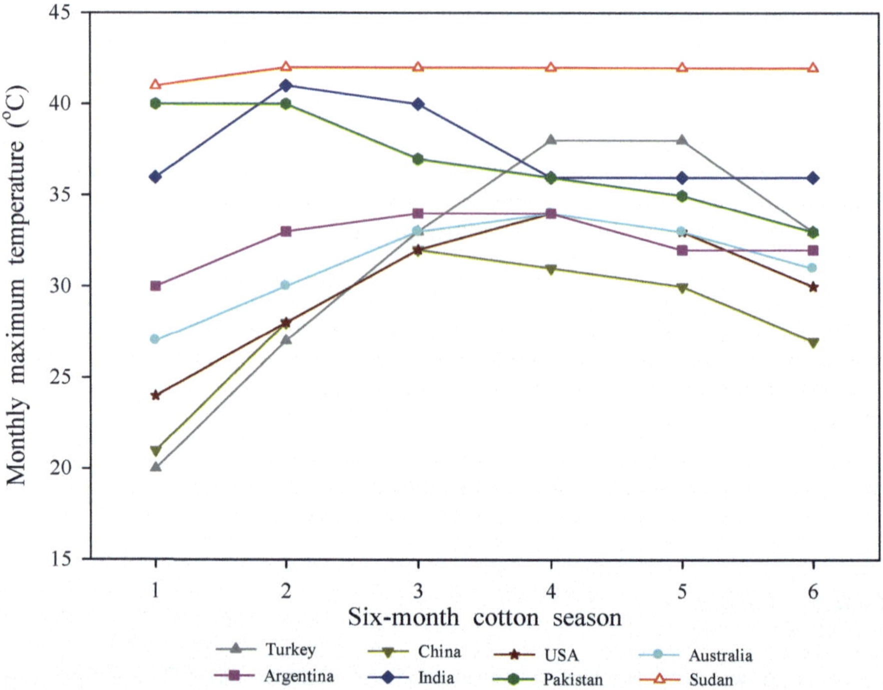

Fig. 14.1 Monthly average maximum temperature (in °C) for a 6-month cotton season (*Source*: Modified and adapted from ICAC 2009; Global warming and cotton production—Part 2)

High temperature in most cotton-growing areas is one of the reasons in poor and retarded development of cotton plant and ultimately fruit formation. High rainfall is expected to be another rising problem during monsoon season. Cotton yield may suffer from higher climate unpredictability in areas which are highly rain-fed which leads to different periods of flooding and drought to the cotton crop. Flooded cotton, mostly in north India, may grieve from water scarcity which is because of reduction of snow from Himalayan besides Tibetan Plateau glaciers (ITC 2011).

14.4.3 Turkey

In Turkey, about 50% of cotton is produced in southeastern Anatolia, which is reported to have semi-arid climate. The summer mean temperatures are recoded more than 30 °C during the hottest months of July and August. The summer dryness is severe and for long time (Turkey 2007). These situations may damage the cotton growth during the growing periods.

14.4.4 Pakistan

Cotton crop sowing is popular in Pakistan. Pakistan is sensitive to heat stress and water scarcity during flowering and seed formation. Rise in temperature is responsible for the pest attack and forfeiture of vegetative and fruiting parts (ITC 2011).

At elevated temperature and increased humidity, the shedding of small squares is accompanied with leaves falling and increased insect pests attack. However, the problems of sucking pest attack like whitefly, might serve as a vector for *Cotton leaf curl virus* have become very serious.

Many areas of Pakistan are facing heat stress due to high temperature (Fig. 14.1) accompanied with water shortage during the cotton-growing season and increased pest population and pathogens. This imposes a threat on the well-being of people associated with cotton industry and those districts of Pakistan which directly contribute to country GDP through cotton production in days to come.

In Pakistan during cotton season, the temperature generally shows increasing trends. However, the rainfall reflects the decreasing trends during the months of March, April and May in all districts with few variations. During the month of May, temperature shows a significant increase during the last 30 years, while these variations during other months of the cotton growth are non-significant in many other districts.

The remedies to lessen effect of changing climate on cotton include:

- The cotton-growing countries should focus on the breeding, screening and approval of the cotton varieties which may perform better under the changing climatic conditions of temperature and rainfall.
- Reducing the overdoses of nitrogen which if increased makes the cotton attractive for the pests and the diseases.
- The use of biotechnology for the improvement resistance of cotton varieties against the insect pests, diseases and adverse environmental conditions may combat the problems faced under the changing climate. The cotton industry should use an integrated disease management tactic, and breeding efforts are heavily focused on emerging germplasm with genetically developed disease resistance (Bang and Constable 2009).
- Fluctuating time of planting offers a 'systematic solution' to deliver assistances both in maintaining fibre quality as well as yield and dropping the jeopardy of hostile effects of high temperatures and low humidity (high evaporative demand) and dropping the incidence of seedling diseases early in the season (Bang and Constable 2009).

14.5 Conclusion

The world's climatic conditions are changing and becoming harsh, and the cotton diseases and disorders are becoming serious globally in cotton-growing countries. The problem may be managed through integrated crop management strategies which include cotton breeding for resistance, low-dose application of nitrogen, early or late sowing of cotton, and integrated disease and pest management tactics.

References

Abbas Q, Ahmad S (2018) Effect of different sowing times and cultivars on cotton fiber quality under stable cotton-wheat cropping system in southern Punjab, Pakistan. Pak J Life Soc Sci 16:77–84

Ahmad S, Abbas Q, Abbas G, Fatima Z, Atique-ur-Rehman NS, Younis H, Khan RJ, Nasim W, Habib ur Rehman M, Ahmad A, Rasul G, Khan MA, Hasanuzzaman M (2017) Quantification of climate warming and crop management impacts on cotton phenology. Plan Theory 6(7):1–16

Ahmad S, Iqbal M, Muhammad T, Mehmood A, Ahmad S, Hasanuzzaman M (2018) Cotton productivity enhanced through transplanting and early sowing. Acta Sci Biol Sci 40:e34610

Ahmad S, Raza I (2014) Optimization of management practices to improve cotton fiber quality under irrigated arid environment. J Food Agric Environ 2(2):609–613

Ahmad S, Raza I, Ali H, Shahzad AN, Atiq-ur-Rehman SN (2014) Response of cotton crop to exogenous application of glycinebetaine under sufficient and scarce water conditions. Braz J Bot 37(4):407–415

Ali H, Abid SA, Ahmad S, Sarwar N, Arooj M, Mahmood A, Shahzad AN (2013a) Integrated weed management in cotton cultivated in the alternate-furrow planting system. J Food Agric Environ 11(3&4):1664–1669

Ali H, Abid SA, Ahmad S, Sarwar N, Arooj M, Mahmood A, Shahzad AN (2013b) Impact of integrated weed management on flat-sown cotton (*Gossypium hirsutum* L.). J Anim Plant Sci 23 (4):1185–1192

Ali H, Afzal MN, Ahmad F, Ahmad S, Akhtar M, Atif R (2011) Effect of sowing dates, plant spacing and nitrogen application on growth and productivity on cotton crop. Int J Sci Eng Res 2 (9):1–6

Ali H, Hameed RA, Ahmad S, Shahzad AN, Sarwar N (2014a) Efficacy of different techniques of nitrogen application on American cotton under semi-arid conditions. J Food Agric Environ 12 (1):157–160

Ali H, Hussain GS, Hussain S, Shahzad AN, Ahmad S, Javeed HMR, Sarwar N (2014b) Early sowing reduces cotton leaf curl virus occurrence and improves cotton productivity. Cercetări Agronomice în Moldova XLVII(4):71–81

Ali M, Mohy-Ud-Din Q, Ali MA, Sabir S, Ali L (2004) Cotton yield as influenced by different sowing dates under the climatic conditions of Vehari, Pakistan. Int J Agric Biol 6:644–646

Amin A, Nasim W, Mubeen M, Ahmad A, Nadeem M, Urich P, Fahad S, Ahmad S, Wajid A, Tabassum F, Hammad HM, Sultana SR, Anwar S, Baloch SK, Wahid A, Wilkerson CJ, Hoogenboom G (2018) Simulated CSM-CROPGRO-cotton yield under projected future climate by SimCLIM for southern Punjab, Pakistan. Agric Syst 167:213–222

Amin A, Nasim W, Mubeen M, Nadeem M, Ali L, Hammad HM, Sultana SR, Jabran K, Habib ur Rehman M, Ahmad S, Awais M, Rasool A, Fahad S, Saud S, Shah AN, Ihsan Z, Ali S, Bajwa AA, Hakeem KR, Ameen A, Amanullah HUR, Alghabar F, Jatoi GH, Akram M, Khan A, Islam F, Ata-Ul-Karim ST, Rehmani MIA, Hussain S, Razaq M, Fathi A (2017) Optimizing the

phosphorus use in cotton by using CSM-CROPGRO-cotton model for semi-arid climate of Vehari-Punjab, Pakistan. Environ Sci Pollut Res 24(6):5811–5823
Bang M (2007) Effects of climate change on cotton growth and development. The Austral Cotton Grow, June–July. pp. 41–45
Bang M, Constable G (2009) Cotton farming system for a changing climate. australiancottonconference.com.au/app/tinyfckupload/File
Bibi A, Oosterhuis DM, Gonias ED (2008) Photosynthesis, quantum yield of photosystem II and membrane leakage as affected by high temperature in cotton genotypes. J Cotton Sci 12:150–159
Brown PW (2008) Cotton heat stress. Arizona cooperative extension. University of Arizona. AZ1448
Burke MB, Lobell DB, Guarino L (2009) Shifts in African crop climates by 2050 and the implications for crop improvement and genetic resources conservation. Glob Environ Chang 3:317–325
Cetin O, Basbag S (2010) Effect of climatic factors on cotton production in semi-arid regions – a review. Res Crops 11(3):785–791
China (2004) The peoples' republic of China's initial national communication under the United Nations framework convention on climate change. UNFCCC, Switzerland, p 156
Guinn G, Mauney JR (1984) Fruiting of cotton. II. Effects of plant moisture status and active load on boll retention. Agron J 76:94–98
Hulme M. ed. (1996) Climate change in South Africa: An exploration of some potential impacts and implications in the SADC Region. Climate Research Unit, University of East Anglia
ICAC (2007) Global warming and cotton production- part 1. In: ICAC Recorder, Vol. 25, No. 4 (December 2007). pp. 12–16. International Cotton Advisory Committee (ICAC). United States of America
ICAC (2009) Global warming and cotton production- part 2. In: ICAC Recorder, Vol. 27, No. 1 (March 2009). pp. 9–13. International Cotton Advisory Committee (ICAC). United States of America
International Trade Centre (2011) Cotton and climate change: Impacts and options to mitigate and adopt. Geneva: ITC, 2011. xii, 32 p. (Technical paper) Doc. No. MAR-11-200.E
IPCC (2000) Emission scenario, summary for policy makers. Online at http://www.grida.no/climate/ipcc/spmpdf/srese.pdf
Karl TR, Melillo JM, Peterson TC (2009) Global climate change impacts in the United States. Cabmridge University Press, United States of America, p 196
Khan MB, Khaliq A, Ahmad S (2004) Performance of mashbean intercropped in cotton planted in different planting patterns. J Res (Sci) 15(2):191–197
Mauney JR, Fry KE, Guinn G (1978) Relationship of photosynthetic rate to growth and fruiting of cotton, soybean, sorghum and sunflower. Crop Sci 18:259–263
Muller C (2009) Climate change impact on Sub-Saharan Africa: an overview and analysis of scenarios and models. Discussion paper
Oosterhuis DM, Hake KD (2001) Environmental disorders of the cotton crop. In: Krickpatrick TL, Rothrock CS (eds) Compendium of cotton diseases, 2nd edn. American Phytopathological Society, St. Paul, Minnesota, United States of America, pp 61–63
Rahman MH, Ahmad A, Wang X, Wajid A, Nasim W, Hussain M, Ahmad B, Ahmad I, Ali Z, Ishaque W, Awais M, Shelia V, Ahmad S, Fahad S, Alam M, Ullah H, Hoogenboom G (2018) Multi-model projections of future climate and climate change impacts uncertainty assessment for cotton production in Pakistan. Agric For Meteorol 253-254:94–113
Sarvella P (1966) Environmental influences on sterility in cytoplasmic male sterile cotton. Crop Sci 6:361–364
Shah S, Jabeen S, Zamir SI (2001) Yield and quality of two cotton cultivars to pre-sowing heat stress. Int J Agric Biol 3:389–390

Tariq M, Afzal MN, Muhammad D, Ahmad S, Shahzad AN, Kiran A, Wakeel A (2018) Relationship of tissue potassium content with yield and fiber quality components of Bt cotton as influenced by potassium application methods. Field Crop Res 229:37–43

Tariq M, Yasmeen A, Ahmad S, Hussain N, Afzal MN, Hasanuzzaman M (2017) Shedding of fruiting structures in cotton: factors, compensation and prevention. Trop Subtrop Agroecosyst 20(2):251–262

Turkey (2007) First National Communication of the republic of Turkey under the United Nations framework convention on climate change. UNFCCC, Switzerland, p 276

Usman M, Ahmad A, Ahmad S, Irshad M, Khaliq T, Wajid A, Hussain K, Nasim W, Chattha TM, Trethowan R, Hoogenboom G (2009) Development and application of crop water stress index for scheduling irrigation in cotton (*Gossypium hirsutum* L.) under semiarid environment. J Food Agric Environ 7(3&4):386–391

Waddle BA (1984) Crop growing practices. Cotton Agronomy Monograph No. 24 ASACSSASSSA, Madison, WI

Wu G, Chen FJ, Ge F, Sun YC (2007) Effects of elevated carbon dioxide on the growth and foliar chemistry of transgenic BT cotton. J Integr Plant Biol 49(9):1361–1369

Chapter 15
Cotton-Based Cropping Systems and Their Impacts on Production

Amar Matloob, Farhena Aslam, Haseeb Ur Rehman, Abdul Khaliq, Shakeel Ahmad, Azra Yasmeen, and Nazim Hussain

Abstract Cotton farming symbolizes single largest use of arable land for fiber production on earth, and cotton-based cropping systems are practiced under diverse agro-climatic environments in more than 100 countries. World cotton production has escalated in recent past and has undergone numerous technological transformations and socioeconomic interventions in quest of productivity and sustainability. Cotton-based cropping systems range from low-input rainfed systems in Australia and Africa to highly mechanized intensive farming systems in the United States, Brazil, and China. In India and Pakistan, multiplicity of cotton varieties, weather extremes, uncertainty of climatic optima, spurious seeds, non-remunerative markets, and low quality plus adulterated chemicals or pesticides are key problems leading to low yields besides net profits in otherwise high productivity cotton-based cropping systems. Resource conserving, eco-efficient, climate smart, and economically viable cropping systems that rotate/intercrop cotton with cereals, oilseeds, and legumes are required. Relay or intercropping and crop rotations will lead to the ecological intensification of cotton-based cropping systems. An ideal cotton-based cropping system should aim at higher yields and net profits per unit area, bring stability into the production system, ensure optimal utilization of the available resources, be able to meet domestic requirements of farmer, and avoid ecological uncertainty in the form of shifts in insect pests or weed populations or evolution of pesticide resistance in the long run. Another area requiring significant improvement is integrating current curative pest management options with other cultural methods to avoid insecticide/herbicide resistance development in an era of transgenics. The transgenics have their own pros and cons, and due deliberations in the best interest of agro-ecosystem

A. Matloob (✉)
Department of Agronomy, Faculty of Agriculture and Environmental Science, MNS-University of Agriculture, Multan, Pakistan
e-mail: amar.matloob@mnsuam.edu.pk

F. Aslam · H. U. Rehman · S. Ahmad · A. Yasmeen · N. Hussain
Department of Agronomy, Faculty of Agricultural Sciences and Technology, Bahauddin Zakariya University, Multan, Pakistan

A. Khaliq
Department of Agronomy, University of Agriculture, Faisalabad, Pakistan

S. Ahmad, M. Hasanuzzaman (eds.), *Cotton Production and Uses*,
https://doi.org/10.1007/978-981-15-1472-2_15

sustainability and small landholders be made with involvement of all stakeholders. Biotech seed industry should plan safe mechanisms for herbicide-tolerant crop development to evade resistance development or gene introgression in weeds. Productivity and profitability of cotton-based cropping systems needs to be explored with greater ecological orientation under conventional and organic management systems. This chapter documents the productivity and resource use efficiency of cotton-based cropping systems based on existing agronomic and experimental evidences. Crop growth and development, productivity, quality, resource use efficiencies, and profitability of various systems have been discussed at the plant, field, and system levels.

Keywords Climate smart cotton production · Insect pests · Lint yield and quality · Productivity and profitability · Resource use efficiency · Sustainable cropping systems

Abbreviations

g ha^{-1}	Gram per hectare
K	Potassium
kg ha^{-1}	Kilogram per hectare
t ha^{-1}	Tons per hectare
ATER	Area time equivalent ratio
LER	Land equivalent ratio
LUE	Light use efficiency
N	Nitrogen
NUE	Nitrogen use efficiency
P	Phosphorus
PAR	Photosynthetically active radiation
Zn	Zinc

15.1 Introduction

The cotton plant includes 40 species in genus *Gossypium* that is native to tropical plus subtropical areas (Lee and Fang 2015). Cotton cultivation is an integral component of crop production since centuries, and more than hundred countries produce cotton. Worldwide it is grown on an area of 35 M ha primarily in longitudinal band between 37°N and 32°S with a total production of 26 M tons of lint (ICAC 2015). Besides production, its trade involves more than 150 countries which either export or import cotton (Ahmad et al. 2014, 2017, 2018; Abbas and Ahmad 2018; Ahmad and Raza 2014; Ali et al. 2011, 2013a, b, 2014a, b). At farm level, it provides source of income to more than 100 million farm families. About 2/3 of the world's cotton is produced in four countries (China, the USA, India, Pakistan) (Khan

Table 15.1 Country-wise area and production of seed cotton in the world during year 2017

Sr. No.	Country	Area (000 ha)	Production (000 tons)
1	India	12200.00	18530.00
2	China	3625.39	17148.46
3	USA	4492.22	12000.00
4	Pakistan	2699.00	5700.30
5	Brazil	927.99	3842.87
6	Uzbekistan	1201.18	2900.18
7	Turkey	501.48	2450.00
8	Australia	518.59	2150.96
9	Mexico	211.92	1009.10
10	Burkina Faso	844.90	844.34
11	Greece	262.62	792.00
12	Argentina	253.31	616.16
13	Mali	630.81	591.64
14	Turkmenistan	531.67	483.19
15	Syria	119.77	441.44

Source: FAO (http://www.fao.org/faostat/en/)

Fig. 15.1 Distribution of cotton production worldwide (2017)

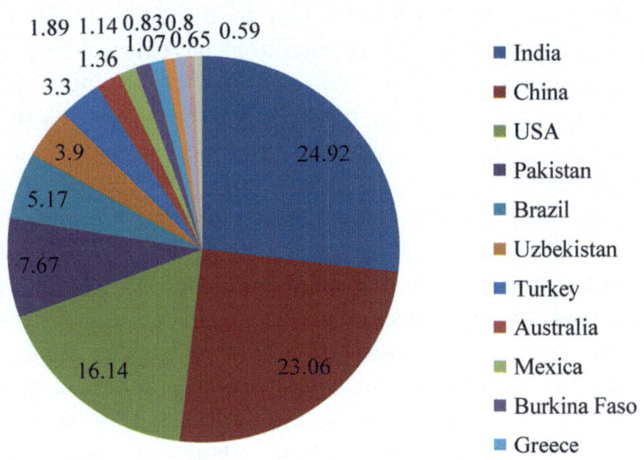

et al. 2004; Usman et al. 2009; Amin et al. 2017, 2018; Rahman et al. 2018; Tariq et al. 2017, 2018). India and China are largest producers having share of 25 and 23%, respectively (Table 15.1; Fig. 15.1), followed by the USA (16%), Pakistan (7.67%), and Brazil (5.17%). China, India, and Pakistan consume approximately 2/3 of the world's cotton having shares of 35, 15, and 10%, respectively. About 1/3 of worldwide cotton production is being traded internationally. The USA is the largest cotton exporter having a share of 41%, and then China is the largest importer having a share of 19% in global imports.

India is the only country where all four cultivated cotton species are being sown. Area wise, India ranks first globally (about 20%), but with regard to production, it is ranked second, after China (Gopalakrishnan et al. 2007). Three cotton-growing zones include northern *G. hirsutum* and *G. arboreum* zone (Punjab, Haryana, and Rajasthan), central *G. hirsutum*, *G. arboreum*, and *G. herbaceum* zone (Gujrat, Madhya Pradesh, and Maharashtra), and composite southern *G. hirsutum*, *G. arboreum*, *G. herbaceum*, and *G. barbadense* zone (Tamil Nadu, Andhra Pradesh, Karnataka) (Blaise 2017).

The three core cotton-growing zones in China are Yangtze River basin, Yellow River basin, and northwestern inland (Dai and Dong 2014). These zones include 13 provinces, autonomous regions, and direct-controlled municipalities: Tianjin, Hebei, Shanxi, Jiangsu, Anhui, Jiangxi, Henan, Shandong, Hubei, Hunan, Shaanxi, Gansu, and Xinjiang. In 2007, these three zones represented 99.85% of total cotton-growing parts in China (Yang and Cui 2010).

In the USA, the Cotton Belt includes 17 states in 4 geographical regions, viz., the Southeast (22% of the total; Alabama, Arkansas, Georgia, Florida), mid-South (34% of the total; Arkansas, Louisiana, Mississippi, Missouri besides Tennessee), Southwest (35% of the total; Texas, Kansas plus Oklahoma), and the West (9% of the total; Arizona, California and New Mexico) with 70% of the production concentrated in the mid-South, Southwest regions (ITC 2011). Cotton production is highly mechanized and capital-intensive precision agriculture.

In Brazil, cotton production occurs primarily in states of Mato Grosso, MatoGrosso do Sul besides Goi as of the Centre-West region. Balanced fertilizer use and favorable weather conditions have remarkably increased the cotton productivity during last one and half decade transforming Brazil from a net importer to an exporter. Subsequently, moving to Centre-West, its production has become most efficient, and growers are now amid most technologically advanced in world (Graham 2009).

Pakistan is world's fourth largest raw cotton producer, with almost 8.0% share in worldwide cotton productivity (Fig. 15.1). Cotton is mostly cultivated in Punjab and Sindh province which contribute about 79% of the total cotton production of Pakistan (Anonymous 2018). Pakistan's cotton belt extends above 1200 km beside River Indus from latitudes of 27°N–33°N besides altitudes from 27 m to 153 m. Soils vary from sandy to clay loam with clay dominant towards south (Gillham et al. 1995). In Punjab, it is mainly cultivated in Bahawalpur, Rahim Yar Khan, Bahawalnagar, Multan, Vehari, Khanewal, Dera Ghazi Khan, Muzafar Garh, Rajanpur, besides Lodhran (Salma et al. 2012), while, in Sindh, it is grown in Nawabshah, Ghotki, Sanghar, and Nosheroferoz districts (Khushk et al. 1990).

Ranking sixth regarding global cotton production, the Uzbek cotton is grown in periphery of Aydar Lake and also to some extent in Tashkent along Syr Darya and along Amu Darya in border areas with Turkmenistan.

Around 50% of Turkish cotton produces are in southeastern Anatolia, having semi-arid climate. Summers are very hot along with mean temperatures higher than 30 °C in July and August. Mean temperature in January is between 2 and 5 °C. Annual rainfall varies amid 350 and 800 mm. Other major production areas are the Aegean and Cukurova regions (Karademir 2006).

Australia is the world's largest exporters of raw cotton with over 90% of domestic produce exported to China, Indonesia, and Thailand. Cotton production in Australia is located in New South Wales (NSW) besides Queensland, particularly in Murray-Darling river basin (ITC 2011). Major production zone in NSW stretches south from Macintyre River on Queensland border and covers Gwydir, Namoi, and Macquarie valleys. In NSW, cotton is also grown along Barwon besides Darling Rivers in west and Lachlan and Murrumbidgee Rivers in south and spreading into new areas like Forbes. In Queensland, cotton is grown mostly in south in Darling Downs, St George, Dirranbandi, and Macintyre Valley regions and also near Emerald, Theodore, and Biloela in Central Queensland (Farrell 2017). In Africa, Burkina Faso and Mali are the important cotton-producing countries.

15.2 Cotton-Based Cropping Systems

Cotton production is carried out in five continents under diverse agro-climatic conditions with contrasting productivity levels and production constraints. Developing eco-efficient cropping systems that are productive and sustainable is a key to food/agricultural security (Mao et al. 2015). Often rotation, relay or intercropping of major crops with minor/special crops is exploited to bring diversity and increase productivity in agro-ecosystems. In cotton-based cropping systems, intercropping has been reported to cause a yield reduction to the tune of 8–31% than sole stand of cotton. Nevertheless, total productivity and net returns greater in intercropping compared to sole cropping (Mohammad et al. 1991) and cotton-based intercropping system reported to more revenue up to 30–40% (Saeed et al. 1999). Various cotton-based cropping systems are discussed in the following sections. Moreover, issues and possible strategies to address these are given in Table 15.2.

15.2.1 Wheat-Cotton Relay Cropping Systems

China is a major country regarding cotton production and 200 million Chinese farmers are engaged in cotton production. Major cotton production in China is focused in three distinct areas, namely, Yellow River Valley, Yangtze River Valley, and northwestern regions, representing 47, 26 and 21% of the total area under cotton, respectively (Zhang 2007). Nevertheless, average farm size is quite less corresponding to 0.10–0.13 ha per person. The highest lint yield of 1393 kg ha^{-1} is harvested in Gansu and Xinjiang provinces of northwestern region. Lint yield is low in Yellow River Valley (810 kg ha^{-1}) compared to Yangtze River Valley (947 kg ha^{-1}). In the northwestern provinces of China, cotton is sown as monocrop at higher densities. Contrarily, in other two regions, cotton is either intercropped with wheat (Yellow River Valley) or double cropped with winter wheat or rapeseed (Yangtze River Valley). Intercropping of cotton in wheat ensures production of cash as well as food crop simultaneously. Area under cotton-wheat intercropping systems in three provinces (Hebei, Henan, and Shandong) was 0.12 million ha in 1980 that

Table 15.2 Issues of various cotton-based cropping systems and possible strategies

Cotton-based cropping systems	Issues	Strategy	References
Intercropping	Greater N requirements of Bt cotton than other cotton cultivars	Intercropping with groundnut	Rochester et al. (2001); Ghosh et al. (2008)
	Nutrient management	Cotton intercrop with leguminous crops	Singh et al. (2009)
		Integrated nutrient management	Singh et al. (2013); Singh and Ahlawat (2014)
	Weed infestation	Cotton intercrop with sesbania/green gram along with bioinoculants such as *Azospirillium* and *Pseudomonas* not only suppresses the weed dry biomass but also increases seed cotton yield	Sivakumar (2004); Vaiyapuri et al. (2010); Marimuthu and Subbian (2013)
	Aphid attack	Cotton-cowpea intercropping increases the population of natural enemies/predators of aphid	Fernandes et al. (2018)
		Cotton-fennel intercropping increases the population of predators	Ramalho et al. (2012); Fernandes et al. (2015)
		Cotton intercropping or field boundary with alfalfa increases the population of predators	Lin et al. (2003); Zhang et al. (2004)
	Insect pest management	Cotton-basil intercropping significantly reduces overall pest population and results in about 50% reduction in pink bollworm infestation	Schader et al. (2005)
Wheat-cotton relay cropping system	Late sowing of wheat	Relay cropping with zero till	Singh et al. (2016)
	Time conflict between Bt cotton sowing and wheat harvest	Adjusting sowing time of Bt cotton by transplanting one and half month old cotton nursery	Shah et al. (2017)
	Low yield	Relay cropping of wheat increases the productivity of wheat-cotton system	Buttar et al. (2017); Singh et al. (2016); Sajjad et al. (2018)
	Nutrient management	Integrated nutrient management (inorganic+ FYM+ bioinoculants)	Ramprakash and Prasad (2000); Das et al. (2004)
	Insect pest management	Wheat-cotton relay intercropping controlling cotton aphid	Parajulee et al. (1997); Li (2001)

(continued)

Table 15.2 (continued)

Cotton-based cropping systems	Issues	Strategy	References
Wheat-cotton double cropping system	Soil potassium deficiency	Incorporation of wheat straw	Sui et al. (2015); Yu et al. (2016)
	Soil nutrient management	Crop residue management	Liebman and Gallandt (2002); Takahashi et al. (2003); Wang et al. (2007); Rafique et al. (2012); Northup and Rao (2015); Suriyagoda et al. (2014)
		Inclusion of legumes in crop rotation	Rocheser and Peoples (2005)
	Low yield	Wheat-cotton double cropping system improves the productivity per land area	Dai and Dong (2014); Feng et al. (2017)
	Nematode management	Crop rotation (cotton-peanut, maize, grain sorghum), deep tillage to bury the nematode and weed seeds, nematicide application at appropriate rates	Mueller et al. (2012)
Rainfed cotton system	Nutrient management	Integrated nutrient management (inorganic + organic sources)	Mannikar (1993); Katyal et al. (1997); Venugopalan and Pundarikakshudu (1999)
	Water management under limited supply	Skip-row planting/row width and orientation	Milroy et al. (2004); Bange et al. (2005)
	Low yield	Intercropping with oilseeds and legumes improves cotton yield and net returns from cropping system	Reddy et al. (1985); Sarkar et al. (1995); Aladakatti et al. (2011)

increased to about 1.2 million ha in 1988 and witnessed an increase of 1.6 million ha in 1990 and has remain stabilized since then (Zhang and Li 1997; Zhang 2007). A relay strip intercropping approach is being followed for wheat and cotton, in which at the onset of spring, cotton is intersown in wheat strips that were planted in fall. The whole field is occupied by cotton after the wheat harvest in early summer. Once cotton is harvested in fall, two crops have successfully completed their life cycle in 1 year on a single piece of land. There is overlapping of seedling phase of cotton with maturation phase of wheat from April to June in this system (Zhang et al. 2007, 2008a, b). However, cotton harvest is early enough to permit timely sowing of wheat. Various variations based on a number of alternating rows of wheat and cotton are in practice, i.e., 3:1. 3:2, 4:1, and 6:2. The 3:1 pattern means 3 rows of wheat and

1 of cotton in an alternating fashion. Although crop production per unit of land area is significantly increased compared to the monocrops, relay intercropping can have some tradeoffs manifested as shading effect of wheat on young cotton seedlings and competition for resources. These limitations might hinder cotton growth and development hampering lint yield and quality.

Wheat yield in intercropping stretched from 4.6–5.2 t ha^{-1} and was 70–79% of the yield recorded for monoculture of wheat (6.55 t ha^{-1}). Upper limit of wheat yield (79%) under intercropping was realized for 3:1 system followed by 6:2 (73%), 4:2 (70%), and 3:2 (70%) systems, respectively. Likewise, cotton lint yield averaged over 3 years under relay intercropping with wheat range from 0.59 to 0.74 t ha^{-1}. These yield levels corresponded to 54–69% of the cotton grown in monoculture, and 3:2 and 4:2 systems produced 69 and 68% of cotton lint yield realized in monoculture. Lowest (1.28) LER was observed for 6:2 system as compared to 1.39 recorded for the remaining three systems. Moreover, estimated growth rate of cotton observed for 3:2 (8.4 g m^{-2} d^{-1}) and 4:2 (7.7 g m^{-2} d^{-1}) systems was not significantly different from that (8.9 g m^{-2} d^{-1}) observed for monoculture (Zhang et al. 2007). A developmental delay in cotton growth compared to its monoculture was also noticed. Adding further, Zhang et al. (2008a) proposed that such developmental delay in cotton under intercropping was 10–15 calendar days which corresponded to 4.7 physiological days and 115 growing degree days (thermal time calculated from sowing to first square). This delay resulted in reduced fruiting branches, nodes, and fruits. Fruit formation was also delayed resulting in less number of fruit with low average fruit age having negative implications for sink capacity of cotton plant and its harvest index and lint yield. Such findings were attributed to the competitive effects of wheat on the young cotton seedlings and suboptimal thermal climate that prevails in wheat-cotton relay intercrops. Such effects were quite pronounced in 3:1 system than other systems.

Zhang et al. (2008b) quantified the light interception and its utilization efficiency by cotton and wheat grown as monocultures or intercrops (3:1. 3:2, 4:1, and 6:2) under varying row width besides rows $strip^{-1}$. These authors concluded that LUE of wheat plus cotton remained unaffected by intercropping systems and intercrops intercepted more light than monocultures of cotton. The wheat strips in intercropping harnessed 20% more light per unit strip area than sole stand of wheat. Three years average revealed that wheat under aforementioned intercropping systems intercepted 83, 71, 73, and 75% as much light as intercepted by pure stand of wheat (471 MJ m^{-2} PAR). The corresponding values for cotton were 73, 93, 86, and 67%, respectively, as against 3 year's average 471 MJ m^{-2} PAR intercepted by sole crop. The increased productivity of intercrops relative to monoculture was attributed to greater accumulation of intercepted light per unit land area suggesting complementary light interception over time and space. The water productivity of wheat (0.95–1.28 kg m^{-3}) and cotton (0.11–0.22 kg m^{-3}) under these relay intercropping systems was 27 and 40% less as compared to the monocultures of these crops, respectively (Zhang 2007). The total N uptake by wheat did not vary between mono and relay intercropping systems. The N-yield per unit area for wheat and cotton ranged from 203–288 kg ha^{-1} and 110–127 kg ha^{-1} and was greater in monoculture than intercrops. Cotton N uptake was diminished through intercropping phase which recovered later on. Intercrops utilized more N per unit production as revealed by relative N yield total (1.4–1.7) and relative yield total (1.3–1.4) values. Yearly N surplus under intercropping (400 kg ha^{-1}) far exceeded the N surplus under monoculture of wheat (220 kg ha^{-1}) besides cotton (140 kg ha^{-1}) suggesting a potential environmental risk necessitating the need of demand-based rate besides N timing application under such intercropping systems (Zhang et al. 2008c).

In India and Pakistan, about 80–90% of area under cotton is devoted to cotton-wheat rotation corresponding to 1.40 and 2.62 million ha, respectively (Mayee et al. 2008, 2009). Wheat sowing is often delayed past its optimum planting time due to late picking of cotton and time spent in land preparation (Singh et al. 2016; Shah et al. 2016, 2017). To resolve time conflict in this system, relay sowing of wheat seems a promising option that ensures timely planting of wheat with concurrent increase in wheat productivity and overall system profitability (Sajjad et al. 2018). Relay cropping of wheat allowed one bonus picking compared to conventional tillage wheat, and this improved cotton yield by 12%. Likewise, sowing of wheat was advanced by 31 days which also enhanced wheat yield up to 19% over conventional practice. The relay sowing of wheat in cotton using strip rotor or zero-till double-disc openers was better for wheat emergence besides final productivity than zero-till tine openers. The practice of relay cropping was also advantageous economically and net benefits were increased to the tune of 311 to 425 US$ ha^{-1} (Singh et al. 2016). With recent introduction of Bt-cotton hybrids in this rotation (usually grown at a wider row spacing of 90–120 cm), possibility exists of successfully growing a short period intercrop such as groundnut (Singh et al. 2009). This notion was tested by Singh and Ahlawat (2014) who appraised the two-tiered intercropping of cotton besides peanut when 50% of the recommended N dose for wheat was substituted through farm yard manure. Wheat grain yields were 5% more numerous under this management scenario than grown solely after cotton. Soil physicochemical properties were also positively influenced under cotton + groundnut-wheat rotation. In Pakistan, wheat is usually harvested in late April, while optimal time of planting of Bt cotton is first fortnight of March (Shah et al. 2016, 2017). Cotton planting on a fallow land in March resulted in greater cotton yields compared with planting in April after wheat (Shah et al. 2017). Productivity of Bt cotton is constrained by 30–35-day time conflict which can be overcome by transplanting of one and half month old cotton nursery transplants. This study concluded that yield of wheat sown on beds was significantly higher than wheat sown on a flat bed or ridges. Moreover, bed-transplanted cotton had highest yield at both locations, i.e., Vehari and Multan. Upper limits of net benefits, benefit-cost ratio and marginal rate of return were realized when 45-day-old cotton seedlings were transplanted on beds after bed sown wheat.

Under relay cropping of wheat in cotton, Zohry - (2005) reported seed cotton yield similar to its sole planting under Egyptian environments. Wheat yield was 80% of the sole planting under this system. Moreover, irrigation given to wheat (harvested in April) was also used by cotton (sown in March) (Zohry and Ouda 2015). The timing of the last two irrigations given to the wheat crop affected the LER and ATER in this system (Sultan et al. 2012a). Sultan et al. (2012b) studied the response of sole and relay intercropping of cotton and wheat to sowing date and row width. These authors found better performance of both sole and intercropped cotton in terms of fruiting branches, bolls, boll weight and seed cotton yield at wider ridge width of 100 cm compared to 80 and 90 cm. However, seed and lint yield of sole crop was highest in case of 15 March sowing, while intercrops had highest yield when sown on 1 April.

15.2.2 Cotton-Wheat Double Cropping System

Cotton-wheat double cropping in sequence integrates potential higher yield and income using mechanization (Dai and Dong 2014). Du et al. (2015) suggested that cotton productivity under double cropping systems is limited by lower interception of PAR rather than radiation use efficiency. Interception of PAR by cotton cultivars (Siza 30 and CCRI 50) was significantly less compared to their monoculture when these were sown under wheat/intercropped cotton, wheat/transplanted cotton, and wheat/direct-seeded cotton systems. Nitrogen uptake, besides use efficiency, was quantified by Du et al. (2016) under aforementioned cropping systems. These authors reported that decline in N accumulation rate and shortening of period during which faster accumulation of N occurred were conducive to lower N uptake by cotton under wheat-cotton system compared to cotton monoculture. The N uptake was less for both cultivars than their monocultures, although early-maturing cultivar CCRI 50 had high NUE than mid-late maturing Siza 30. A perusal crop N balance revealed significant N surpluses for preceding wheat in wheat-cotton rotations than cotton monoculture. The study by Hussain et al. (2014) involving omission plot technique in Bt cotton-wheat system revealed that omitting N, P, and K reduced cotton yield up to 28, 6.5, and 14.5% during the first year of study with N being the most limiting factor. The corresponding reduction during the second year of study amounted to 26.5, 15.5, and 12.4%, respectively. The greater yield reduction due to P omission during the second year of study led to the conclusion that P supply is depleted at a faster rate than K in cotton-wheat rotation. To overcome such issues, Rochester et al. (2001) had advocated the use of legumes to improve N fertility besides soil properties in cotton-based systems. Recently, Feng et al. (2017) appraised the effects of sowing patterns (monoculture cotton, cotton intercropped in wheat, cotton transplanted after wheat, besides cotton direct-seeded after wheat) on growth and yield plus economic benefits in wheat-cotton cropping system. Regardless of the cropping systems and fields, cotton yields were lower in double cropping relative to monoculture. However, such decrements were quite small for cotton sown in field with high fertility status. Yield differences among cropping patterns were attributed to difference in number of bolls besides weight. Yield variation between two fields with contrasting soil fertility was due to differences in boll numbers. Over the 2-year study period, net revenue varied from 11 to 35% in low fertility field and 32 to 74% in high fertility field compared to cotton monoculture. Luo et al. (2018) proposed that N application rate can be reduced to the tune of 20–30% without sacrificing yield as N applied at 264 kg ha^{-1} via fertigation in high density crop stand of cotton (19.5 plants m^{-2}) was effective in improving agronomic NUE and N recovery efficiency as compared to higher N application rates (319 and 375 kg ha^{-1}) under conventional application method or low plant density (12.0 plants m^{-2}). Abid et al. (2013) founded that single application of 7.5 kg Zn ha^{-1}was adequate for two cycles of cotton-wheat system, while Zn total uptake by wheat (134.9–289.6 g ha^{-1}) was more superior compared to cotton (92.3–192.5 g ha^{-1}). Two-year constant application of 5.0–7.5 kg Zn ha^{-1} did not depress yields under the rotation.

The K deficiency has been acknowledged as a major factor limiting cotton yield (Mullins and Burmester 2010; Yu et al. 2016). A substantial amount of wheat residues (estimate at 6 hundred million tons per annum) are produced in this cropping system. Out of these, only 16% of wheat straw is recycled, while 23% is burned causing wastage of resources and environmental pollution (Zeng et al. 2007; Cao et al. 2008). To cope with K demand of cotton, inorganic salts of K (rich in chlorides or sulfates) are used which have potential to cause soil degradation through crust formation, acidification, and groundwater contamination (Pernes-Debuyser and Tessier 2004; Paradelo et al. 2013; Udeigwe et al. 2015). Wheat straw incorporation has been suggested as an alternative of inorganic K sources (Sui et al. 2015) which significantly improved the availability of macronutrients NPK and also enhanced fraction of water-soluble organic carbon. Residue recycling in cotton-wheat system resulted in 2–7% higher yields of cotton, and 2–10% improvements in wheat yields in 5-year crop residue management study undertaken at two alkaline calcareous aridisols soils of Pakistan. The balance nutrient management and integrated nutrient management systems were found superior to farmer fertilizer use, and all apparent N balances were positive with substantial improvements in nitrate-N and organic matter (Rafique et al. 2012). Under the influence of wheat straw incorporation at 4.5 and 9.0 t ha^{-1} in cotton-wheat double cropping systems of Nanjing and Defang, China, the K concentration of cotton and seed cotton yield was similar to that achieved with inorganic K application at 150 kg ha^{-1} of K_2O (Yu et al. 2016). These authors postulated that wheat straw incorporation at 9.0 t ha^{-1} can substitute inorganic K application at 150 kg ha^{-1} of K in cotton-wheat double cropping systems within less than 3 years.

Application of saline-sodic water for cotton-wheat rotation decreased cotton yield by 21% compared with good-quality canal water over a 7-year study period. Reduction in wheat yield was not observed during the course of study. Farm yard manure and gypsum application improved seed cotton yield, and reduction under these amendments was only 10 and 9% compared to good-quality water (Buttar et al. 2017). Besides agronomic limitations, the wheat-cotton relay intercropping system is also constrained by demand for manual labor and a steady decline has been observed in the use of this system (Feike et al. 2012).

15.2.3 Rainfed Cotton Systems

Modern cotton cultivars like their wild ancestors possess the ability to adapt and survive periods of extremely dry weather and intermittent water supply that occurs under rainfed conditions (Hearn 1990). In Australia, rainfed cotton production has expanded and accounts for 17% acreage of the total cotton area in Australia (Dowling 2002; Bange et al. 2005). The rainfed cotton production system uses "skip-row orientation" in order to maximize the use of limited water supply, reduce the production risk, improve lint quality, and lower the input costs (Milroy et al. 2004; Bange et al. 2005). Gwathmey et al. (2008) suggested that cotton growers interested in skip-row sowing should consider rows spacing of 76 cm or less to minimize weed problems besides productivity loss since skip-row cotton matured later than solid sowing, but the effect was diminished using narrower rows. Moreover, weed growth in narrowly spaced skipped rows was less compared to wider skip-row spacing of 102 cm. In India, the central cotton zone comprising of states of Gujarat, Madhya Pradesh, and Maharashtra is primarily rainfed accounting for 68% of the cotton area. Maharashtra alone occupies one-third of the country's cotton area; however, only less than 10% of the area is irrigated (Blaise 2017). Cotton is grown on an area of 7.10 M ha on vertisols in this rainfed region. Opening of furrows of alternative rows during the month of August facilitates in situ rainwater conservation. Historically cotton was rotated with grain sorghum in the central India, but low prices of sorghum favored continuous cropping of cotton in rainfed area. The rainfed cotton productivity (230 kg ha^{-1}) is lower than India's average cotton productivity of 478 kg lint ha^{-1} (Singh et al. 2012). A system of high density plantation of straight cotton varieties with 25% more NPK is becoming popular in the wake of early canopy closure and limits evaporation losses with additional benefits of optimizing bolls $plant^{-1}$besides boll weight. Response of Bt cotton hybrids to K application in vertisols low in exchangeable K has been reported (Blaise 2012). Such responses were not observed in the past with old cotton varieties (Mannikar 1993). In Benin (Africa), cotton cultivation is also rainfed and compact and early genotypes could be used in crossbreeding to produce varieties adapted to these environments (Sekloka et al. 2018). In the USA, low input rainfed cotton production systems exist in Texas contrary to intensive cotton farming in California and Arizona (James and Choudhary 2010).

15.2.4 Systems Involving Intercropping of Cotton with Other Crops

Agroforestry as an opportunity cropping for cotton was recently evaluated in China by Wang et al. (2016). The authors tested different densities of cotton, viz., 13.5, 18.0, and 22.5 plants m^{-2} sown within 6 m path between jujube plantations. Leaf area and dry matter accumulation by cotton was reduced in rows adjacent to the jujube trees, and biomass allocation to the cotton fruiting parts was also restricted. Shading effect was two rows deep in case of 6-year-old stand of jujube, while it was extended to three rows deep in a 7-year-old jujube stand. These authors inferred that shading effect on cotton yield can be compensated by having more number of bolls per unit land area by increasing cotton density.

In Egypt, Lamlom et al. (2018) studied the combinations of four systems in winter (double cropping systems of Egyptian clover besides cotton, relay intercropping cotton with faba bean, onion or wheat) besides three systems during summer (sole-cotton, intercropping cowpea-or-sesame with cotton). They found highest LER and ATER under Egyptian clover/cotton + cowpea followed by onion + cotton/cotton + cowpea. The later combination recorded maximum monetary benefits and improved fiber quality traits compared to conventional cropping system consisting of Egyptian clover/cotton.

Pigeon pea/cotton intercropping system has been reported as a cash cropping system of black cotton soil of Deccan Plateau, India. Interactive effect of land configuration systems (flat, ridge and furrow and broad bed and furrow) with strip widths (1.5:4.5, 3.0:3.0, 4.5:1.5, and 6:0 m) was nonsignificant for yield of both pigeon pea and cotton crops besides total biomass production. A reduction in strip width was accompanied with decrease in yield of both cotton and pigeon pea. However, greater yield reduction was recorded in case of cotton compared to pigeon pea. Strip combination of 4.5 m pigeon pea and 1.5 m cotton gave highest land equivalent and monetary value equivalent ratios (Potdar et al. 1996). In cotton + black gram intercropping system, cotton is mainly irrigated one time in 15–20 days. Intercrop of black gram is often affected by surplus water and resulted in reduced productivity. In such circumstances, skip furrow method of irrigation is advocated.

15.3 Productivity and Cost-Effectiveness of Cotton-Based Cropping Systems

Studying productivity besides cost-effectiveness of cotton-based cropping systems under conventional and conservation agriculture systems is important (Daujanov et al. 2016). In the first crop cycle (2007–2008), the cotton and wheat yields were lowered by 29 and 27%, respectively, in organic production system compared to conventional system. However, such differences for yield were not observed in the second crop cycle (2009–2010), and yields were similar under all systems. During the first crop cycle, higher gross margins (+29%) were recorded for conventional systems. However, in the second crop cycle, these were higher by 25% in organic system due to lower variation in cost of production (Forster et al. 2014). Across 4 harvest years, gross margins associated with soybean under organic systems were always higher. These authors suggested that soybean production is a viable option for small landholders under cotton-based cropping systems of India. This notion of maximizing the profitability of cotton-based cropping system through legume incorporation has been explored in Australia and India (Williams et al. 2011; Turkhede et al. 2017). Inclusion of vetch in continuous cotton or cotton-wheat system increased gross margins per ha by 23 and 12%, respectively, in Australia. The gross margins increased because of higher yields and reduction in input cost of N since the same was contributed by vetch. The cost of N produced by vetch more than paid for its cost of production. In India, the sowing of cotton on one side of the permanent raised broad bed and intercropped with mung bean resulted in 37 and 10% higher productivity over conventional tilled flat cotton-wheat systems (CT), and permanent raised beds with cotton sown in the center without mung bean. This treatment combination improved system irrigation water productivity by 131%, energy productivity by 54% and net returns by 69% as compared to CT (Choudhary et al. 2016). Conservation agriculture practices in cotton-wheat cropping systems of Indo-Gangetic Plains were evaluated in terms of crop and water productivity and cost-effectiveness (Das et al. 2014). These authors concluded that seed cotton yield under zero-tilled broad bed with residue retention was 24 and 51% more numerous as compared to zero-tilled narrow bed without residue retention and sowing on flat beds after conventional tillage without residue recycling. Higher water productivity (12–48%), productivity of system in terms of wheat-equivalent yield (13–15%), and net returns (13–36%) were also associated with this treatment combination.

Braunack (2013) appraised factors that influence cotton yield besides quality in Australian cotton-based farming systems and how these have changed over time (2004–2011) since the adoption of transgenic cotton Bollgard II® varieties in irrigated and dryland systems. The study revealed that over the study period, the use of N fertilizers was increased, while P use remained steady in the irrigated cotton system. Nevertheless, use of both these nutrients was decreased in dryland cotton. The number of insecticide sprays was decreased in both systems. Lint yield and fiber length showed an increasing trend in irrigated cotton with a decrease observed for fiber strength and micronaire. The yield of dryland cotton was decreased with increase in fiber strength, length, miconaire, and trash level.

Additional income and monetary benefits can be achieved by sowing cotton in wider rows and planting a relay crop in between these rows (Rao 1991). Wheat and barley sown as intercrops in cotton recorded 69 and 23% higher yields, respectively, compared with sole planting of these crops, although cotton yield remains unaffected in relay cropping (Khan and Khaliq 2005). Wheat relayed in cotton recorded yield of 2964 kg ha^{-1} as well as 1750 kg ha^{-1} recorded for wheat that was sown after cotton. Depending upon the sowing time of cotton, relay cropping increased land use efficiency by 81–213% (Hussein 2005). Bio-economic efficiency of cotton in 80 cm rows and 120/40 cm spaced rows besides cowpea and sorghum as intercrops was evaluated in Pakistan (Aasim et al. 2008). Advantage of 32–46% for LER over sole cropping was realized in both planting designs. Sorghum intercropping in cotton resulted in greater reduction in cotton yield than cowpea.

Dhaliwal and Sandhu (2015) evaluated the productivity and economics of seven cotton-based cropping systems (S_1, Bt cotton (*Gossypiumhirsutum*)-wheat (*Triticumaestivum*L.); S_2, American cotton-wheat; S_3, desi cotton (*Gossypiumarboreum*)-wheat; S_4, alternatively Bt cotton and desi cotton-wheat;

S_5, Bt cotton-barley (*Hordeum vulgare*); S_6, Bt cotton-onion (*Allium cepa* L.); and S_7, Bt cotton-transplanted gobhi sarson (*Brassica napus*) in Indian Punjab. Greater rice-equivalent productivity as compared to all-other systems were realized under Bt-cotton-onion (235.5 q ha^{-1}) followed by desi cotton-wheat (123.3 q ha^{-1}) systems. Greater rice-equivalent yield under Bt cotton-onion system was attributed to greater bulb yield besides market price of onion crop only; however, *Kharif* crop under this system failed to produce any difference amid cotton yield from other cotton-based system. In cotton-based cropping systems of Mahrashtra, India, intercropping with pigeon pea and mix cropping with green gram, maize, sesame, and pearl millet resulted in greater net profit than cotton as sole crop (Gahukar 2017).

Compared to conventional tillage system and rotation (cotton-wheat-maize), cotton yields were consistently higher over a 3-year study period in Uzbekistan under conservation tillage system and rotation involving soybean as legume (cotton-wheat-soybean) suggesting improvement in soil health and crop productivity (Khaitov and Allanov 2014). Cong et al. (2014) suggested that intercrops have greater belowground productivity and sequester more C and N as compared to sole crops. This notion was supported by the results of their 7-year field study in which soil organic C and N contents and C and N sequestration rates were higher in case of intercrops over sole crops.

15.4 Insect Pests

Lepidoptera insects like *Helicoverpa armigera* (American bollworm), *Earias vittella* (spotted bollworm), *Pectinophora gossypiella* (pink bollworm), *Earias* spp. (spiny bollworm), *Argotis* spp. (cutworms), *Helicoverpa zea* (bollworm), and *Heliothis virescens* (tobacco budworm) are the major pests infesting 88% of the global cotton area (James 2002). Of the total pesticide used worldwide, cotton crop receives over 18% of all insecticides used and substantial amounts of other plant protection chemicals (Deguine et al. 2008). Till the first half of the 1980s, chlorinated hydrocarbons were the major pesticides used to control bollworms in cotton, and later organophosphates and pyrethroids were used for this purpose (Zhao et al. 2011). These insecticide groups were rendered ineffective due to evolution of resistance in bollworms against insecticide molecules owing to their indiscriminate use in ever-rising polluting quantities (Deguine et al. 2008). These insecticides also exerted negative influence on the beneficial insects or natural enemies of bollworms with a concurrent increase in the damage caused by bollworms as well as other secondary pests. Transgenic Bt cotton was developed with an aim to reduce insecticide input and lower down the production costs. Studies document multidimensional benefits for the farmers with the use of Bt cotton such as reduced production costs, high yield and net profits, and biodiversity conservation (Qaim 2003; Pray et al. 2001, 2002; Vitale et al. 2008; Abedullah and Qaim 2015). The advent of Bt cotton is regarded as a major breakthrough that has revolutionized the pest management in a pest-intensive crop like cotton (Alvi et al. 2012). The Bt cotton is now grown on more than 25 M ha in 2 developed (the USA and Australia) and 11 developing (China, India, Pakistan, Argentina, Brazil, Mexico, Colombia, South Africa, Costa Rica, Myanmar, and Burkina Faso) countries (James 2012).

However, acclaimed benefits of Bt cotton are often challenged due to methodological issues and effectiveness of pest control achieved herein due to low or variable level of Bt gene expression and Cry endotoxin protein level in plant tissue (Xu et al. 2008; Shantharam et al. 2008; Jost et al. 2008; Zhao et al. 2011). Theoretically, for effective pest control, whole Bt cotton plant should produce enough toxin that is lethal to arthropod pests especially in period of attack of pest (Naik et al. 2018; Ahmad et al. 2019). Nevertheless, Bt toxin production varies among cotton genotypes (Cheema et al. 2015) and plant parts/tissue (Khan et al. 2018) and has been reported to decline during course of plant growth (Zaman et al. 2015). Unfortunately, the reports of insect resistance to Cry1Ac proteins from different geographical regions of world (Tabashnik et al. 2013; Tabashnik and Carriere 2017; Naik et al. 2018) have challenged the efficacy, prospect, and sustainability of the Bt technology. Additionally, uncertainty about its anticipated ecological impacts such as evolution of resistance in bollworms, negative effect on non-target species/natural enemies of bollworms/predators, soil biota and nutrient dynamics, and increased insurgence of secondary pests might take years or so to manifest (Zhao et al. 2011). Furthermore, Bt cotton is not effective against secondary pests which are otherwise killed by insecticides in conventional cotton. Hence, in the long run, these secondary pests might become the primary pests themselves or will need greater insecticide application, thereby counteracting the benefits of Bt cotton. Displacement of vegetative or boll feeding caterpillars by sucking insects as major pests has been suggested (Deguine et al. 2008). Men et al. (2004a) reported that total pesticide application was similar between Bt and non-Bt cotton due to additional spraying to control secondary insect pests. Likewise, Zhao et al. (2011) concluded that reduction in pesticide use due to adoption of Bt cotton was far less than that reported in literature. Contrarily, Wang et al. (2009) advocated that increased fraction of pesticides for controlling secondary insects was quite small when compared to the overall reduction in pesticide input owing to Bt cotton cultivation. This scenario suggests the need of in-depth studies at plant, field, and agro-ecosystem level while considering the spatiotemporal variations in natural factors regulating pest population, agronomic practices, and socioeconomic factors. In doing so, we would be better able to portray a wider picture of the benefits of Bt cotton and overcome pest-related apprehensions.

Pyke et al. (1987) proposed the concept of "Push-Pull" strategy which consisted of insect diversion via *stimulo-deterrent* techniques. The idea was further advanced and refined by Miller and Cowles (1990). The "Push" includes the use of repellent plants that repel or deter away the pests from the resource (main crop), thereby discouraging the settlement of pests on main crops by masking the host appearance (Cook et al. 2007). The "Pull" on the other hand involves the simultaneous attraction toward the neighboring trap plants with attractive highly apparent stimuli (Cook et al. 2007; Ratnadass et al. 2012). The stimuli may be trophic, visual, or chemical (Gaba et al. 2015). Introduction of a new crop either as push or pull modifies the predator population via creation of new habitat. For Australian cotton production systems, Mensah (1999) recommended the growing of 9–12 wide single median strip of lucerne between two cotton fields 300 m wide. Growing lucerne as a nursery crop around cotton field margins has been reported to act as nursery for ladybird beetles (*Coccinella septempunctata*, *Propylea quatuordecimpunctata*, and *Hypodama variagata*), chrysopids, and other beneficial arthropods. Once lucerne is cut, these beneficial insects move into the cotton and help control cotton aphid-significant pest of cotton in Xinjiang province of China (Lin et al. 2003). Likewise, *Helicoverpa* spp. infestation in cotton was significantly less using field peas and sorghum in Australia and the USA, respectively (Grundy et al. 2004; Tillman and Mullinix Jr. 2004). Another study from India found that neem as a push coupled with pigeon pea or okra as trap crop (pull) was effective against *Helicoverpa* spp. (Duraimurugan and Regupathy 2005).

Trap and intercropping have been reported to offset the pest populations by maintaining favorable beneficial insect complexes (Deguine et al. 2008). In cotton grown in Texas, USA, relay intercropped cotton had lower average abundance of aphid than sole stand of cotton crop. Predators appeared earlier in summer and in higher numbers in relay intercropped cotton (Parajulee et al. 1997). In another study, sowing wheat as winter relay crop in cotton caused a significant reduction in the population of cotton aphid and increased the population of predators of cotton pests (Men et al. 2004b). Highest mass of natural enemies besides low populations of cotton aphids were found in cotton relay-intercropped in a wheat variety (Lovrin10) that was susceptible to wheat aphid as compared to aphid-resistant wheat variety (KOK1679). These authors concluded that background of wheat variety (resistance or susceptibility to wheat aphid) could affect natural enemies in cotton and wheat varieties with susceptibility or moderate resistance to wheat aphid might decrease cotton aphid more efficiently by enhancing predators of cotton at seedling stage (Ma et al. 2006).

15.5 Weeds

Cotton owing to its C_3 metabolism competes poorly with weeds especially during early vegetative phase (Bryson et al. 1999; Deguine et al. 2008). Cotton is extremely infested with grasses, sedges, and broadleaf which reduces up to 30% cotton yield (Jabran 2016). Losses as high as 90% have also been reported (Manalil et al. 2017). Major weeds of cotton crop are *Amaranthus* spp. *Convolvulus* spp., *Cyperus rotundus, Cynodon dactylon* (L.) Pers., *Echinochloa* spp., *Dicanthium annulatum* (Forssk.) Stapf., *Ipomoea* spp., *Lolium perenne* L., *Leptochloa chinensis* L. Nees., *Polygonum aviculare* L., *Sporobolus clandestinus* (Biehler) A.S. Hitchc., *T. portulacastrum*, and *Xanthium* spp. (Zhang 2003; Economou et al. 2005; Rajput et al. 2008; Kruger et al. 2009; Hiremath et al. 2013; Manalil et al. 2017). Being a wide-rowed crop, growers control weeds intensively through chemical besides mechanical means.

Torres et al. (2009) described four herbicide-tolerant cottons covering the global cotton market, viz., glyphosate-tolerant Roundup Ready® and Roundup Ready Flex® and bromoxynil-tolerant BXN® varieties and glufosinate ammonium-tolerant LibertyLink®. Research work to develop 2,4-D-tolerant varieties to overcome drift effects and for controlling broadleaf weeds is under way. Overreliance on glyphosate as sole weed control in herbicide-resistant cotton has resulted in 43 resistant weeds across the globe with 16 species stated from Australia alone (Iqbal et al. 2019). The widespread adoption of glyphosate-tolerant cotton in Australia from 2000/2001 growing season till now has caused evolution of numerous glyphosate-resistant weed biotypes (Werth et al. 2012, 2013; Heap and Duke 2018). Weeds like *Amaranthus tuberculatus*, palmer amaranth, annual ragweed, giant ragweed, fleabane, horseweed (*Conyza canadensis* L.), liverseed grass, windmill grass, annual ryegrass, and Johnson grass have been reported to develop glyphosate resistance (Guest et al. 2014; Sosnoskie and Culpepper 2014; Green 2018).

In an era of herbicide-resistant weeds, due consideration should be given to improve weed management since available evidences have led to the conclusion that weed management system merely relying on herbicides loses its sustainability in the long term (Charles and Taylor 2004; Lamichhane et al. 2017). The weed-suppressive effects of narrow row spacing (Gwathmey et al. 2008; Stephenson and Brecke 2010) in cotton are well documented. Manalil et al. (2017) suggested the use of competitive cotton cultivars with high seedling vigor, manipulation of row spacing, and its orientation with respect to sunlight and high seeding densities as tools that can impart cotton crop a competitive edge over weeds. Furthermore, certain weeds act as host of disease-causing microorganisms and refuge for insect pest; yet many also harbor and sustain population of natural enemies/predators (Showler and Greenberg 2003). In this backdrop, weed management should take into account the phytosanitary context of the whole cropping system (Deguine et al. 2008). Weed density and dry biomass were lowest in cotton intercropped in sesbania in addition to application of *Azospirillum* besides *Pseudomonas*. This system suppressed weed growth by 54.5 and 44% compared to pure-crop-cotton through 2007 and 2008. Higher cotton equivalent productivity of 2052 and 1895 kg ha^{-1} was documented in cotton + onion rotation that was at par with cotton + sesbania rotation with cotton equivalent productivity of 2010 and 1894 kg ha^{-1} in both years (Marimuthu and Subbian 2013).

Cotton unit yield and total output have increased due to intensive farming technologies like increased use of nutrients, advent of Bt cotton varieties, seedling transplanting, double cropping, manipulation of plant architecture, plant training, use of growth regulators, high seeding densities, use of plastic mulch, mechanized picking, etc. However, these have made cotton cultivation labor and input intensive, and endemic ecological effects associated with cotton productivity are increasing in spotlight and are often questioned. Moreover, data regarding sustainability of cotton-based cropping systems are either scanty or not readily available. It is unequivocal that water and soil management, pests and pesticide management, biodiversity and land use patterns, and productivity and profitability of cotton-based cropping systems need to be revisited with greater ecological orientation in pursuit of sustainability. Resource conservation practices like intercropping with legumes, permanently raised beds, zero tillage, drip irrigation, fertigation, residue retention/recycling, mulching, and green manuring can be opted to bring in the much needed diversity and stability while enhancing the productivity and resource use efficiency of cropping systems. Research efforts and extension recommendations should consider the spatiotemporal variations and cropping system-specific issues and be able to differentiate between various cotton-based cropping systems. Cotton cultivation needs to switch from input-based to cropping system-based growth, and in this paradigm shift, the knowledge generation and its dissemination could play a vital role. The problems faced especially in the backdrop of dwindling natural resource base and climate change are multifarious, and to cope with these challenges, it is imperative to make adjustments and/or fine-tune technology, management practices, and legislation.

References

Abbas Q, Ahmad S (2018) Effect of different sowing times and cultivars on cotton fiber quality under stable cotton-wheat cropping system in southern Punjab, Pakistan. Pak J Life Soc Sci 16:77–84

Ahmad S, Abbas Q, Abbas G, Fatima Z, Atique-ur-Rehman NS, Younis H, Khan RJ, Nasim W, Habib ur Rehman M, Ahmad A, Rasul G, Khan MA, Hasanuzzaman M (2017) Quantification of climate warming and crop management impacts on cotton phenology. Plan Theory 6(7):1–16

Ahmad S, Iqbal M, Muhammad T, Mehmood A, Ahmad S, Hasanuzzaman M (2018) Cotton productivity enhanced through transplanting and early sowing. Acta Sci Biol Sci 40:e34610

Ahmad S, Raza I (2014) Optimization of management practices to improve cotton fiber quality under irrigated arid environment. J Food Agric Environ 2(2):609–613

Ahmad S, Raza I, Ali H, Shahzad AN, Atiq-ur-Rehman SN (2014) Response of cotton crop to exogenous application of glycine betaine under sufficient and scarce water conditions. Braz J Bot 37(4):407–415

Ali H, Abid SA, Ahmad S, Sarwar N, Arooj M, Mahmood A, Shahzad AN (2013a) Integrated weed management in cotton cultivated in the alternate-furrow planting system. J Food Agric Environ 11(3&4):1664–1669

Ali H, Abid SA, Ahmad S, Sarwar N, Arooj M, Mahmood A, Shahzad AN (2013b) Impact of integrated weed management on flat-sown cotton (Gossypium hirsutum L.). J Anim Plant Sci 23 (4):1185–1192

Ali H, Afzal MN, Ahmad F, Ahmad S, Akhtar M, Atif R (2011) Effect of sowing dates, plant spacing and nitrogen application on growth and productivity on cotton crop. Int J Sci Eng Res 2 (9):1–6

Ali H, Hameed RA, Ahmad S, Shahzad AN, Sarwar N (2014a) Efficacy of different techniques of nitrogen application on American cotton under semi-arid conditions. J Food Agric Environ 12 (1):157–160

Ali H, Hussain GS, Hussain S, Shahzad AN, Ahmad S, Javeed HMR, Sarwar N (2014b) Early sowing reduces cotton leaf curl virus occurrence and improves cotton productivity. Cercetări Agronomice în Moldova XLVII(4):71–81

Amin A, Nasim W, Mubeen M, Ahmad A, Nadeem M, Urich P, Fahad S, Ahmad S, Wajid A, Tabassum F, Hammad HM, Sultana SR, Anwar S, Baloch SK, Wahid A, Wilkerson CJ, Hoogenboom G (2018) Simulated CSM-CROPGRO-cotton yield under projected future climate by SimCLIM for southern Punjab, Pakistan. Agric Syst 167:213–222

Amin A, Nasim W, Mubeen M, Nadeem M, Ali L, Hammad HM, Sultana SR, Jabran K, Habib urRehman M, Ahmad S, Awais M, Rasool A, Fahad S, Saud S, Shah AN, Ihsan Z, Ali S, Bajwa AA, Hakeem KR, Ameen A, Amanullah HUR, Alghabar F, Jatoi GH, Akram M, Khan A, Islam F, Ata-Ul-Karim ST, Rehmani MIA, Hussain S, Razaq M, Fathi A (2017) Optimizing the phosphorus use in cotton by using CSM-CROPGRO-cotton model for semi-arid climate of Vehari-Punjab, Pakistan. Environ Sci Pollut Res 24(6):5811–5823

Anonymous (2018) (https://en.wikipedia.org/wiki/Cotton_production_in_Pakistan)

Aasim M, Umer EM, Karim A (2008) Yield and competition indices of intercropping cotton (Gossypium hirsutum L.) using different planting patterns. Tarim Bilim Derg 14:326–333

Abedullah KS, Qaim M (2015) Bt cotton, pesticide use and environmental efficiency in Pakistan. J Agric Econ 66(1):66–86

Abid M, Ahmed N, Qayyum MF, Shaaban M, Rashid A (2013) Residual and cumulative effect of fertilizer zinc applied in wheat-cotton production system in an irrigated aridisol. Plant Soil Environ 11:505–510

Ahmad S, Cheema HMN, Khan AA, Khan RSA, Ahmad JN (2019) Resistance status of Helicoverpa armigera against Bt cotton in Pakistan. Transgenic Res 28(2):199–212

Aladakatti YR, Hallikeri SS, Nandagavi RA, Hugar AY, Naveen NE (2011) Effect of intercropping of oilseed crops on growth, yield and economics of cotton (Gossypium hirsutum) under rainfed conditions. Karnataka J Agric Sci 24:280–282

Alvi AHK, Sayyed AH, Naeem M, Ali M (2012) Field evolved resistance in Helicoverpa armigera (Lepidoptera: Noctuidae) to Bacillus thuringiensis toxin Cry1Ac in Pakistan. PLoS One 7: e47309

Bange MP, Carberry PS, Marshall J, Milroy SP (2005) Row configuration as a tool for managing rain-fed cotton systems: review and analysis. Aust J Exp Agric 45:65–77

Blaise D (2012) Fertilizer-K recommendation for cotton grown on Vertisols: is there a need for revision? In: IPI-FAI-IPNI Roundtable on Refinement of K recommendations in Vertisols, 20 March 2012, New Delhi. http://www.ipipotash.org/udocs/presentation_dr_blaise.pdf

Blaise D (2017) Cotton based cropping systems. In: Ramesh K, Biswas AK, Lakaria B, Srivastava S, Patra AK (eds) Enhancing nutrient use efficiency. New India Publishing Agency, New Delhi, pp 369–384

Braunack MV (2013) Cotton farming systems in Australia: factors contributing to changed yield and fibre quality. Crop Pasture Sci 64:834–844

Bryson CT, Salisbury C, McCloskey WB (1999) Weeds and their control. In: Cothren JT, Smith CW (eds) Cotton: origin, technology, and production. John Wiley & Sons, New York, pp 617–658

Buttar GS, Thind HS, Sekhon KS, Kaur A, Gill RS, Sidhu BS, Aujla MS (2017) Management of saline-sodic water in cotton-wheat cropping system. J Agric Sci Technol 19:465–474

Cao GL, Zhang XY, Wang YQ, Zheng FC (2008) Estimation of emissions from field burning of crop straw in China. Chin Sci Bull 53:784–790

Charles GW, Taylor IN (2004) Herbicide resistance and species shift in cotton: The need for an Integrated Weed Management (IWM) approach. In: Swanepoel A (ed) Proc. World Cotton Research Conf.–3, Cotton production for the new millenium, Cape Town, 2003, ARC, Institute for Industrial Crops, Pretoria, pp. 817–828

Cheema HMN, Khan AA, Khan MI, Aslam U, Rana IA, Khan IA (2015) Assessment of Bt cotton genotypes for the Cry1Ac transgene and its expression. J Agric Sci 154:109–117

Choudhary R, Singh P, Sidhu HS, Nandala DP, Jat HS, Singh Y, Jat ML (2016) Evaluation of tillage and crop establishment methods integrated with relay seeding of wheat and mungbean for sustainable intensification of cotton-wheat system in South Asia. Field Crops Res 199:31–41

Cong WF, Hoffland E, Li L, Six J, Sun JH, Bao XG, Zhang FS, Van Der Werf W (2014) Intercropping enhances soil carbon and nitrogen. Glob Chang Biol 21:1715–1726

Cook SM, Khan ZR, Pickett JA (2007) The use of push–pull strategies in integrated pest management. Annu Rev Entomol 52:375–400

Dai JL, Dong HZ (2014) Intensive cotton farming technologies in China: achievements, challenges and countermeasures. Field Crops Res 155:99–110

Das A, Prasad M, Shivay YS, Subha KM (2004) Productivity and sustainability of cotton (Gossypium hirsutum L.)–wheat (Triticuma estivum L.) cropping system as influenced by prilled urea, farmyard manure and azotobacter. J Agron Crop Sci 190:298–304

Das TK, Bhattacharyya R, Sudhishri S, Sharma AR, Saharawat YS, Bandyopadhyay KK, Sepat S, Bana RS, Aggarwal P, Sharma RK, Bhatia A, Singh G, Datta SP, Kar A, Singh B, Singh P, Pathak H, Vyas AK, Jat ML (2014) Conservation agriculture in an irrigated cotton–wheat system of the western Indo-Gangetic Plains: Crop and water productivity and economic profitability. Field Crops Res 158:24–33

Daujanov A, Groeneveld R, Pulatov A, Heijman WJM (2016) Cost-benefit analysis of conservation agriculture implementation in Syrdarya Province of Uzbekistan. Visegrad J Bioecon Sustain Dev 5:48–52

Deguine JP, Ferron P, Russell D (2008) Sustainable pest management for cotton production. A review. Agron Sustain Dev 28:113–137

Dhaliwal NS, Sandhu BS (2015) Yield production and economics of different cropping system in south-western part of Punjab. Int Res J Econ Stat 6:414–418

Dowling D (2002) Cotton year book. The Australian Cotton Grower, Toowoomba

Du X, Chen B, Shen T, Zhang Y, Zhou Z (2015) Effect of cropping system on radiation use efficiency in double-cropped wheat–cotton. Field Crops Res 170:21–31

Du X, Chen B, Zhang Y, Zhao W, Shen T, Zhou Z, Meng Y (2016) Nitrogen use efficiency of cotton (Gossypium hirsutum L.) as influenced by wheat–cotton cropping systems. Eur J Agron 75:72–79

Duraimurugan P, Regupathy A (2005) Push-pull strategy with trap crops, neem and nuclear polyhedrosis virus for insecticide resistance management in Helicoverpa armigera (Hubner) in cotton. Am J Appl Sci 2:1042–1048

Economou G, Bilalis D, Avgoulas C (2005) Weed flora distribution in Greek cotton fields and its possible influence by herbicides. Phytoparasitica 33:406–419

Farrell R (2017) Australia: cotton and products annual (April 2017). USDA Foreign agricultural service, global agriculture information network. Report No. As1705. p. 7

Feike T, Doluschitz R, Chen Q, Graeff-Hönninger S, Claupein W (2012) How to overcome the slow death of intercropping in the North China Plain. Sustainability 4:2550–2565

Feng L, Wang G, Han Y, Li Y, Zhu Y, Zhou Z, Cao W (2017) Effects of planting pattern on growth and yield and economic benefits of cotton in a wheat-cotton double cropping system versus monoculture cotton. Field Crops Res 213:100–108

Fernandes FS, Godoy WAC, Ramalho FS, Garcia AG, Santos BDB, Malaquias JB (2018) Population dynamics of Aphis gossypii Glover and Aphis craccivora Koch (Hemiptera: Heteroptera: Aphididae) in sole and intercropping systems of cotton and cowpea. Annals Braz Acad Sci 90:311–323

Fernandes FS, Ramalho FS, Malaquias JB, Godoy WAC, Santos BDB (2015) Interspecific associations between Cycloneda sanguinea and two aphid species (Aphis gossypii and Hyadaphis foeniculi) in sole-crop and fennel-cotton intercropping systems. PLoS One 10(8):e0131449

Forster D, Andres C, Verma R, Zundel C, Messmer MM, Mäder P (2014) Productivity and profitability of cotton-based production systems under organic and conventional management in India. In: Rahmann G, Aksoy U (eds) Proceedings of the 4th ISOFAR Scientific Conference. 'Building Organic Bridges', at the Organic World Congress 2014, 13–15 Oct, Istanbul, Turkey (eprint ID 23660) pp. 647–650

Gaba S, Lescourret F, Boudsocq S, Enjalbert J, Hinsinger P, Journet EP, Navas ML, Wery J, Louarn G, Malézieux E, Pelzer E, Prudent M, Ozier-Lafontaine H (2015) Multiple cropping systems as drivers for providing multiple ecosystem services: from concepts to design. Agron Sustain Dev 35:607–623

Gahukar RT (2017) Cotton based cropping systems in Maharashtra: economic analysis and future needs. J Cotton Res Dev 31:152–156

Ghosh PK, Bandypadhyay KK, Wanjari RH, Manna MC, Mishra AK, Mohanty M (2008) Legume effect for enhancing productivity and nutrient-use efficiency in major cropping systems—an Indian perspective: a review. J Sustain Agric 30:59–86

Gillham FEM, Bell TM, Arin T, Matthews GA, Le Rumeur C, Hearn AB (1995) Cotton production prospects in the next decade. World Bank, USA, 277

Gopalakrishnan N, Manickam S, Prakash AH (2007) Problems and prospectus of cotton in different zones of India. Central Institute for Cotton Research, Regional Station, Coimbatore. –641 003 (December 15–22, 2007)

Graham P (2009) Brazil. In: Cotton Outlook (2009) Cotton trading relationships with China. Cotton Outlook, Special Feature. June 2009. pp. 22–26

Green JM (2018) The rise and future of glyphosate and glyphosate-resistant crops. Pest Manag Sci 74:1035–1039

Grundy PR, Sequeira RV, Short KS (2004) Evaluating legume species as alternative trap crops to chickpea for management of Helicoverpa spp. (Lepidoptera: Noctuidae) in central Queensland cotton cropping systems. Bull Entomol Res 94:481–486

Guest A, Mass S, Taylor I, Werth J, Thornby D, Charles G (2014) Herbicide resistance in Australian cotton farming systems. Cotton Pest Management Guide 2014–15. pp. 85–91. http://www.insidecotton.com/xmlui/bitstream/handle/1/1079/CPMG1415_08_herbicide_%20resistance_in_aus_cotton_FS.pdf?sequence=11&isAllowed=y

Gwathmey CO, Steckel LE, Larson JA (2008) Solid and skip-row spacings for irrigated and nonirrigated upland cotton. Agron J 100:672–680

Heap I, Duke SO (2018) Overview of glyphosate-resistant weeds worldwide. Pest Manag Sci 74:1040–1049

Hearn AB (1990) Prospects for rain-fed cotton. In: Proceedings of the 5th Australian Cotton Conference, 8–9 August. Australian Cotton Growers Research Association, Broadbeach, Queensland, pp 135–144

Hiremath R, Yadahalli GS, Chittapur BM, Siddapur AD, YadahalliVG KBRG (2013) Integrated weed management in Bt cotton (Gossypium hirsutum L.) under UKP command area of Karnataka. Acta Biol Indica 2:400–405

Hussain A, Kumar D, Dwivedi BS, Rana DS, Gangaiah B (2014) Relative response of Bt cotton (Gossypium hirsutum) to balanced fertilization in irrigated cotton-wheat cropping system. Afr J Agric Res 9:21–33

Hussein SMA (2005) Planting date, pattern and fertilizers levels for cotton grown in relay intercropping with wheat. Zagazig J Agric Res 32:1403–1425

ICAC (2015) International Cotton Advisory Committee. Available at: http://www.ic.ac.org. Accessed 28 Nov 2015

Iqbal N, Manalil S, Chauhan BS, Adkins SW (2019) Glyphosate-tolerant cotton in Australia: successes and failures. Arch Agron Soil Sci. https://doi.org/10.1080/03650340.2019.1566720

ITC (International Trade Centre) (2011) Cotton and climate change: impacts and options to mitigate and adapt. ITC, Geneva, p 32. (Technical paper). Doc. No. MAR-11-200.E

Jabran K (2016) Weed flora, yield losses and weed control in cotton crop. Julius-Kühn-Archiv 452:177–182

James C (2012) Global status of commercialized biotech/GM crops: 2011. International Service for the Acquisition of Agri-biotech Application (ISAAA)

James C, Choudhary B (2010) Global adoption of biotech cotton, 1996 to 2007. In: Zehr UB (ed) Cotton, biotechnology in agriculture and forestry, vol 65. Springer-Verlag, Berlin Heidelberg, pp 177–196

James C (2002) Global review of commercialized transgenic crops: 2001. Feature: Bt Cotton. ISAAA Briefs No. 26. ISAAA: Ithaca, NY

Jost P, Shurley D, Culpepper S, Roberts P, Nichols R, Reeves J, Anthony S (2008) Economic comparison of transgenic and non-transgenic cotton production systems in Georgia. Agron J 100:42–51

Karademir C (2006) Cotton situation in Turkey. Presentation (.ppt) at ICAC Research Associate Programme, Washington, D.C., USA. April 2006, p. 22

Katyal JC, Sharma KL, Srinivas K, Reddy MN (1997) Balanced fertilizer use in semi-arid soils. Fert News 42:59–69

Khaitov B, Allanov K (2014) Crop rotation with no-till methods in cotton production of Uzbekistan. Eur J Soil Sci 3:28–32

Khan B, Khaliq A (2005) Production of winter cereals as relay crops by surface seeding in cotton based cropping system. J Res (Sci) 16:79–86

Khan MB, Khaliq A, Ahmad S (2004) Performance of mashbean intercropped in cotton planted in different planting patterns. J Res (Sci) 15(2):191–197

Khan MI, Khan AA, Cheema HMN, Khan RSA (2018) Spatiotemporal and intra-plant expression variability of insecticidal gene (Cry1Ac) in upland cotton. Int J Agric Biol 20:715–722

Khushk AM, Mernon MY, Lashari MI, Longmire J (1990) Wheat in the cotton-based cropping systems of the irrigated Sindh. PARC/CIMMYT Paper 90–4. p. 38

Kruger GR, Johnson WG, Weller SC, Owen MDK, Shaw DR, Wilcut JW, Jordan DL, Wilson RG, Bernards ML, Young BG (2009) US grower views on problematic weeds and changes in weed pressure in glyphosate-resistant corn, cotton, and soybean cropping systems. Weed Technol 23:162–166

Lamichhane JR, Devos Y, Beckie HJ, Owen MD, Tillie P, Messéan A, Kudsk P (2017) Integrated weed management systems with herbicide-tolerant crops in the European Union: lessons learnt from home and abroad. Crit Rev Biotechnol 37:459–475

Lamlom MM, Abdel-Wahab SI, Abdel-Wahab TI, Ibrahim MAA (2018) Crop interference effects of some winter and summer field crops on Egyptian cotton characters. Adv Crop Sci Tech 6 (5):394. https://doi.org/10.4172/2329-8863.1000394

Lee JA, Fang DD (2015) Cotton as a world crop: origin, history, and current status. In: Fang DD, Percy RG (eds) Cotton, 2nd edn. ASA-CSSA-SSSA, Madison, WI, pp 1–23

Li W (2001) Agro-ecological farming systems in China: man and the biosphere. Taylor & Francis Ltd., New York

Liebman M, Gallandt ER (2002) Differential responses to red clover residue and ammonium nitrate by common bean and wild mustard. Weed Sci 50:521–529

Lin R, Liang H, Zhang R, Tian C, Ma Y (2003) Impact of alfalfa/cotton intercropping and management on some aphid predators in China. J Appl Entomol 127:33–36

Luo Z, Liu H, Li W, Zhao Q, Dai J, Tian L, Dong H (2018) Effects of reduced nitrogen rate on cotton yield and nitrogen use efficiency as mediated by application mode or plant density. Field Crops Res 218:150–157

Ma XM, Liu XX, Zhang QW, Zhao JZ, Cai QN, Ma YA, Chen DM (2006) Assessment of cotton aphids, Aphis gossypii, and their natural enemies on aphid-resistant and aphid-susceptible wheat varieties in a wheat–cotton relay intercropping system. Ento Eperiet Appl 121:235–241

Manalil S, Coast O, Werth J, Chauhan BS (2017) Weed management in cotton (Gossypium hirsutum L.) through weed-crop competition: A review. Crop Prot 95:53–59

Mannikar ND (1993) Fertilizer management in cotton. In: Tandon HLS (ed) Fertilizer management in commercial crops. Fertilizer Development and Consultation Organization (FDCO), New Delhi, pp 26–46

Mao L, Zhang L, Zhang S, Evers JB, van der Werf W, WANG J, Sun H, Su Z, Huub S (2015) Resource use efficiency, ecological intensification and sustainability of intercropping systems. J Integr Agric 14:1542–1550

Marimuthu S, Subbian P (2013) Integrated nutrient management on weed dynamics of cotton based cropping systems in South India. SAARC J Agric 11:7–22

Mayee CD, Monga D, Dhillon SS, Nehra PL, Pundhir P (2008) Cotton–wheat production system in South Asia: A success story. Asia-Pacific Association of Agricultural Research Institutions, Bangkok, pp 1–48

Mayee CD, Singh P, Dongre AB, Rao MRK, Raj S (2009) Transgenic Bt cotton. Central Institute of Cotton Research, Nagpur, Maharashtra

Men X, Ge F, Edwards CA, Yardim EN (2004a) Influence of pesticide applications on pest and predatory arthropods associated with transgenic Bt cotton and non-transgenic cotton plants in China. Phytoparasitica 32:246–254

Men X, Ge F, Yardim EN, Parajulee MN (2004b) Evaluation of winter wheat as a potential relay crop for enhancing biological control of cotton aphids in seedling cotton. BioControl 49:701–714

Mensah RK (1999) Habitat diversity: implications for the conservation and the use of predatory insects of Helicoverpa spp. in cotton systems in Australia. Int J Pest Manag 45:91–100

Miller JR, Cowles RS (1990) Stimulo-deterrent diversion: a concept and its possible application to onion maggot control. J Chem Ecol 16:3197–3212

Milroy SP, Bange MP, Hearn AB (2004) Row configuration in rainfed cotton systems: modification of the OZCOT simulation model. Agric Syst 82:1–16

Mohammad MK, El-din GMS, Hosny AA (1991) Evaluating three patterns of intercropping cotton and forage cowpeas. Ann Agric Sci Moshotor 29:1269–1284

Mueller J, Kirkpatrick T, Overstreet C, Koenning S, Kemerait B, Nichols B (2012) Managing nematodes in cotton-based cropping systems. https://www.cottoninc.com/wpcontent/.../2015/12/2012-Managing-Nematodes.pdf

Mullins GL, Burmester CH (2010) Relation of growth and development to mineral nutrition. In: Stewart JM, Oosterhuis DM, Heitholt JJ, Mauney JR (eds) Physiological of cotton. Springer Publishing Co., New York, pp 97–105

Naik VC, Kumbhare S, Kranthi S, Satija U, Kranthi KR (2018) Field evolved-resistance of pink bollworm, Pectinophora gossypiella (Saunders) (Lepidoptera: Gelechiidae) to transgenic Bt-cotton expressing Cry1Ac and Cry2Ab in India. Pest Manag Sci 74:2544–2554

Northup BK, Rao SC (2015) Green manures in continuous wheat systems affect grain yield and nitrogen content. Agron J 107:1666–1672

Paradelo R, van Oort F, Chenu C (2013) Water-dispersible clay in bare fallow soils after 80 years of continuous fertilizer addition. Geoderma 200-201:40–44

Parajulee MN, Montandon R, Slosser JE (1997) Relay intercropping to enhance abundance of insect predators of cotton aphid (Aphis gossypii Glover) in Texas cotton. Int J Pest Manag 43:227–232

Pernes-Debuyser A, Tessier D (2004) Soil physical properties affected by long-term fertilization. Eur J Soil Sci 55:505–512

Potdar MV, Anders M, Sharma MM (1996) Yield advantages and economic returns from pigeon pea/cotton strip intercropping rotations on a vertisol in the Indian semi-arid tropics. In: Ito O, Johansen C, Adu-Gyamfi JJ, Katayama K, JVDKK R, Rego TJ (eds) Dynamics of roots and nitrogen in cropping systems of the semi-arid tropics. Japan International Research Center for Agricultural Sciences, Tsukuba, pp 59–72

Pray CE, Huang J, Hu R, Rozelle S (2002) Five years of Bt cotton in China: The benefits continue. Plant J 31:423–430

Pray C, Ma D, Huang J, Qiao F (2001) Impact of Bt cotton in China. World Dev 29:813–825

Pyke B, Rice R, Sabine B, Zalucki MP (1987) The push–pull strategy–behavioural control of Heliothis. Aust. Cotton Grow. May–July, 7–9

Qaim M (2003) Bt cotton in India: Field trial results and economic projections. World Dev 31:2115–2127

Rafique E, Mahmood-ul-Hassan M, Rashid A, Chaudhary MF (2012) Nutrient balances as affected by integrated nutrient and crop residue management in cotton-wheat system in aridisols. I. Nitrogen. J Plant Nutr 35:591–616

Rahman MH, Ahmad A, Wang X, Wajid A, Nasim W, Hussain M, Ahmad B, Ahmad I, Ali Z, Ishaque W, Awais M, Shelia V, Ahmad S, Fahad S, Alam M, Ullah H, Hoogenboom G (2018) Multi-model projections of future climate and climate change impacts uncertainty assessment for cotton production in Pakistan. Agric For Meteorol 253-254:94–113

Rajput MT, Tahir SS, Ahmed B, Arain MA (2008) Check list of the weeds found in cotton crops, cultivated in Taluka Ubauro, district Sukkur, Pakistan. Pak J Bot 40:65–70

Ramalho FS, Fernandes FS, Nascimento AR, Nascimento Junior JL, Malaquias JB, Silva CA (2012) Assessment of fennel aphids (Hemiptera: Aphididae) and their predators in fennel intercropped with cotton with colored fibers. J Econ Entomol 105:113–119

Ramprakash, Prasad M (2000) Effect of nitrogen, chlormequat chloride and farmyard manure applied to cotton (Gossypium hirsutum) and their residual effect on succeeding wheat (Triticum aestivum) crop. Indian J Agron 45:263–268

Rao VP (1991) A study on intercropping of cotton with grain legumes under rainfed conditions. J Res APAU 19:73–74

Ratnadass A, Fernandes P, Avelino J, Habib R (2012) Plant species diversity for sustainable management of crop pests and diseases in agroecosystems: a review. Agron Sustain Dev 32:273–303

Reddy MS, Natarajan M, Rao MR, Willy RW, Ong CK (1985) Cropping system for rainfed situations with particular reference to ICRISAT an experience. Paper presented in: national symposium on cropping system at Karnal, p. 13

Rocheser IJ, Peoples M (2005) Growing vetches in irrigated cotton systems inputs of fixed N, N fertilizer savings, and cotton productivity. Plant Soil 271:251–264

Rochester IJ, Peoples MB, Hullugalle NR, Gault RR, Constable GA (2001) Using legumes to enhance nitrogen fertility and improve soil condition in cotton cropping systems. Field Crops Res 70:27–41

Saeed M, Shahid MRM, Jabar A, Ullah E, Khan MB (1999) Agroeconomic assessment of different cotton-based inter-relay cropping systems in two geometrical patterns. Int J Agric Biol 4:234–237

Sajjad A, Anjum SA, Ahmad R, Waraich EA (2018) Relay cropping of wheat (Triticum aestivum L.) in cotton (Gossypium hirsutum L.) improves the profitability of cotton-wheat cropping system in Punjab, Pakistan. Environ Sci Pollut Res 25:782–789

Salma S, Rehman S, Shah MA (2012) Rainfall trends in different climate zones of Pakistan. Pak J Meterol 9:37–47

Sarkar RK, Chakraborty A, Mazumdar RC (1995) Effect of intercropping of oilseed and pulse crops in upland cotton for productivity and monetary advantage in the system. Indian J Agric Sci 65:246–249

Schader C, Zaller JG, Köpke U (2005) Cotton-Basil intercropping: Effects on pests, yields and economical parameters in an organic field in Fayoum, Egypt. Biol Agric Hort 23:59–72

Sekloka E, Sabi AK, Zinsou VA, Aboudou A, Ndogbe CK, Afouda L, Baba-Moussa L (2018) Phenological, morphological and agronomic characterization of sixteen genotypes of cotton plant (Gossypium hirsutum L.) in rainfed condition in Benin. J Plant Breeding Crop Sci 10:33–40

Shah MA, Farooq M, Hussain M (2016) Productivity and profitability of cotton–wheat system as influenced by relay intercropping of insect resistant transgenic cotton in bed planted wheat. Eur J Agron 75:33–41

Shah MA, Farooq M, Hussain M (2017) Evaluation of transplanting Bt cotton in a cotton–wheat cropping system. Exp Agric 53:227–241

Shantharam S, Sullia SB, Shivakumara GS (2008) Peer review contestations in the era of transgenic crops. Curr Sci 95:167–168

Showler AT, Greenberg SM (2003) Effects of weeds on selected arthropod herbivore and natural enemy populations, and on cotton growth and yield. Environ Entomol 32:39–50

Singh J, Babar S, Abraham S, Venugopalan MV, Majumdar G (2012) Fertilization of high density, rainfed cotton grown on vertisols of India. Better Crops 96:26–28

Singh M, Sidhu HS, Mahal JS, Manes GS, Jat ML, Mahal AK, Singh P, Singh Y (2016) Relay sowing of wheat in the cotton–wheat cropping system in North-West India: technical and economic aspects. Exp Agric 53:539–552

Singh RJ, Ahlawat IPS (2014) Effects of transgenic cotton-based cropping systems and their fertility levels on succeeding wheat crop. Commun Soil Sci Plant Anal 45:2385–2396

Singh RJ, Ahlawat IPS, Singh S (2013) Effects of transgenic Bt cotton on soil fertility and biology under field conditions in subtropical Inceptisol. Environ Monit Assess 185:485–495
Singh RJ, Ahlawat IPS, Gangaiah B (2009) Direct and residual effects of nitrogen requirement in Bt cotton–wheat cropping system. Indian J Agron 54:401–408
Sivakumar SD (2004) Performance of vegetable intercropping systems and sources of nutrients supply on sustainable yield of rainfed cotton. Ph.D. Dissertation, Agricultural College and Research Institute, Madurai. Tamil Nadu Agric. Univ., Coimbatore, India
Sosnoskie LM, Culpepper AS (2014) Glyphosate-resistant palmer amaranth (Amaranthus palmeri) increases herbicide use, tillage, and hand-weeding in Georgia cotton. Weed Sci 62:393–402
Stephenson DO, Brecke BJ (2010) Weed management in single- vs. twin-row cotton (Gossypium hirsutum). Weed Technol 24:275–280
Sui N, Zhou Z, Yu C, Liu R, Yang C, Zhang F, Song G, Meng Y (2015) Yield and potassium use efficiency of cotton with wheat straw incorporation and potassium fertilization on soils with various conditions in the wheat-cotton rotation system. Field Crops Res 172:132–144
Sultan MS, El-Kassaby AT, Ghonema MH, Ageez AA, Abd-Allah AMM (2012a) Relay intercropping wheat and cotton studies: i- effect of times of two last irrigations and ridge width on growth and yield of wheat. J Plant Prod Mansoura Univ 3:679–689
Sultan MS, El-Kassaby AT, Ghonema MH, Ogeaz AA, Abd-Allah AMM (2012b) Relay intercropping wheat and cotton studies: II-Effect of sowing date and ridge width on cotton. J Biol Sci 12:349–354
Suriyagoda L, De Costa WAJM, Lambers H (2014) Growth and phosphorus nutrition of rice when inorganic fertiliser application is partly replaced by straw under varying moisture availability in sandy and clay soils. Plant Soil 384:53–68
Tabashnik BE, Bre'vault T, Carrie're Y (2013) Insect resistance to Bt crops: lessons from the first billion acres. Nat Biotechnol 31:510–521
Tabashnik BE, Carriere Y (2017) Surge in insect resistance to transgenic crops and prospects for sustainability. Nat Biotechnol 35:926–935
Takahashi S, Uenosono S, Ono S (2003) Short- and long-term effects of rice straw application on nitrogen uptake by crops and nitrogen mineralization under flooded and upland conditions. Plant Soil 251:291–301
Tariq M, Afzal MN, Muhammad D, Ahmad S, Shahzad AN, Kiran A, Wakeel A (2018) Relationship of tissue potassium content with yield and fiber quality components of Bt cotton as influenced by potassium application methods. Field Crops Res 229:37–43
Tariq M, Yasmeen A, Ahmad S, Hussain N, Afzal MN, Hasanuzzaman M (2017) Shedding of fruiting structures in cotton: factors, compensation and prevention. Trop Subtrop Agroecosyst 20(2):251–262
Tillman PG, Mullinix BG Jr (2004) Grain sorghum as a trap crop for corn earworm (Lepidoptera: Noctuidae) in cotton. Environ Entomol 33:1371–1380
Torres JB, Ruberson JR, Whitehouse M (2009) Transgenic cotton for sustainable pest management: a review. In: Lichtfouse E (ed) Sustainable agriculture reviews: organic farming, pest control and remediation of soil pollutants, 1st edn. Springer, New York, pp 45–82
Turkhede AB, Nagdeve MB, Karunakar AP, Gabhane VV, Mohod VD, Mali RS (2017) Diversification in cotton based cropping system under mechanization in rainfed condition of vidarbha of Maharashtra, India. Int J Curr Microbiol App Sci 6:2189–2206
Udeigwe TK, Teboh JM, Eze PN, Stietiya MH, Kumar V, Hendrix J, Mascagni HJ Jr, Ying T, Kandakji T (2015) Implications of leading crop production practices on environmental quality and human health. J Environ Manag 151:267–279
Usman M, Ahmad A, Ahmad S, Irshad M, Khaliq T, Wajid A, Hussain K, Nasim W, Chattha TM, Trethowan R, Hoogenboom G (2009) Development and application of crop water stress index for scheduling irrigation in cotton (Gossypium hirsutum L.) under semiarid environment. J Food Agric Environ 7(3&4):386–391

Vaiyapuri K, Amanullah MM, Rajendran K, Sathyamoorthi K (2010) Intercropping unconventional green manures in cotton: An organic approach for multiple benefits – A review. Asian J Plant Sci 9:223–226

Venugopalan MV, Pundarikakshudu R (1999) Long-term effect of nutrient management and cropping system on cotton yield and soil fertility in rainfed vertisols. Nutr Cycl Agroecosyst 55:159–164

Vitale J, Glick H, Greenplate JT, Abdeennadher M, Traoré O (2008) Second-generation Bt cotton field trials in Burkina Faso: Analyzing the potential benefits to West African farmers. Crop Sci 48:1958–1966

Wang Q, Han S, Zhang L, Zhang D, van der Werf W, Evers JB, Sun H, Su Z, Zhang S (2016) Density responses and spatial distribution of cotton yield and yield components in jujube (Zizyphus jujube)/cotton (Gossypium hirsutum) agroforestry. Eur J Agron 79:58–65

Wang XB, Cai DX, Hoogmoed WB, Perdok UD, Oenema O (2007) Crop residue, manure and fertilizer in dryland maize under reduced tillage in northern China: I grain yields and nutrient use efficiencies. Nutr Cycl Agroecosyst 79:1–16

Wang ZJ, Lin H, Huang JK, Hu RF, Rozelle S, Pray C (2009) Bt Cotton in China: are secondary insect infestations offsetting the benefits in farmer fields? Agric Sci China 8:83–90

Werth J, Boucher L, Thornby D, Walker S, Charles G (2013) Changes in weed species since the introduction of glyphosate-resistant cotton. Crop Pasture Sci 64:791–798

Werth J, Thornby D, Walker S (2012) Assessing weeds at risk of evolving glyphosate resistance in Australian sub-tropical glyphosate-resistant cotton systems. Crop Pasture Sci 62:1002–1009

Williams EJ, Rochester I, Constable G (2011) Maximizing the profitability of cotton cropping systems with legumes. Available at: http://www.insidecotton.com/jspui/bitstream/1/418/1/5100601_Emma_Williams.pdf

Xu N, Fok M, Bai L, Zhou Z (2008) Effectiveness and chemical pest control of Bt-cotton in the Yangtze River, Valley, China. Crop Prot 27:1269–1276

Yang HQ, Cui WG (2010) Cotton industry in China, status and development strategies. Crops 5:13–17. (in Chinese)

Yu C, Wang X, Hu B, Yang C, Sui N, Liu R, Meng Y, Zhou Z (2016) Effects of wheat straw incorporation in cotton-wheat double cropping system on nutrient status and growth in cotton. Field Crops Res 197:39–51

Zaman M, Mirza MS, Irem S, Zafar Y, Mehmoob-ur-Rehman (2015) A temporal expression of Cry1Ac protein in cotton plant and its impact on soil health. Int J Agric Biol 17:280–288

Zeng XY, Ma YT, Ma LR (2007) Utilization of straw in biomass energy in China. Renew Sust Energ Rev 11:976–987

Zhang L (2007) Productivity and resource use in cotton and wheat relay intercropping. Ph.D. thesis Wageningen University. ISBN: 978-90-8504-759-9

Zhang L, Li Y (1997) The technical approach on wheat-cotton double cropping system sustainable development in Huanghuaihai Plain. Proceedings of International Symposium of Sustainable Agricultural Technologies (ISSAT'97), Beijing

Zhang L, Spiertz JHJ, Zhang S, Li B, van der Werf W (2008c) Nitrogen economy and use efficiency in cotton and wheat relay intercropping system. Plant Soil 303:55–68

Zhang L, van der Werf W, Bastiaans L, Zhang S, Li B, Spiertz JHJ (2008b) Light interception and utilization in relay intercrops of wheat and cotton. Field Crops Res 107:29–42

Zhang L, van der Werf W, Zhang S, Li B, Spiertz JHJ (2008a) Temperature-mediated developmental delay may limit yield of cotton in relay intercrops with wheat. Field Crops Res 106:258–268

Zhang L, van der Werf W, Zhang S, Li B, Spiertz JHJ (2007) Growth, yield and quality of wheat and cotton in relay strip intercropping systems. Field Crops Res 103:178–188

Zhang R, Ren L, Wang C, Lin R, Tian C (2004) Cotton aphid predators on alfalfa and their impact on cotton aphid abundance. Appl Entomol Zool 39:235–241

Zhang ZP (2003) Development of chemical weed control and integrated weed management in China. Weed Biol Manag 3:197–203

Zhao JH, Ho P, Azadi H (2011) Benefits of Bt cotton counterbalanced by secondary pests? Perceptions of ecological change in China. Environ Monit Assess 173:985–994

Zohry AA (2005) Effect of relaying cotton on some crops under bio-mineral N fertilization rates on yield and yield components. Ann Agric Sci 431:89–103

Zohry AE, Ouda S (2015) Management of crops intensification in Egypt to overcome water scarcity. Global J Adv Res 2:1824–1831

Chapter 16
Cotton Relay Intercropping Under Continuous Cotton-Wheat Cropping System

Khawar Jabran, Ahmad Nawaz, Ahmet Uludag, Shakeel Ahmad, and Mubshar Hussain

Abstract Cotton-wheat is an important cropping system of the world in which cotton is sown after harvest of wheat in the start of summer season. In the recent decades, Bt cotton cultivars have been introduced to better combat the bollworms. However, the Bt cotton cultivars have a growth period longer than the conventional cotton cultivars. This situation pressured the farmers to opt to grow either wheat or cotton in a single year. This not only could result in economic loss to farmers but also could threaten the food security of the cropping region. Relay cropping of cotton in wheat was suggested by an innovative solution for maintaining the productivity and sustainability of cotton-wheat cropping system. Relay cropping of cotton in wheat could be done either by inter-seeding the seeds of cotton in free space between the wheat strips (while wheat is at reproductive phase) or by transplanting the 5–7-week-old cotton seedlings between the wheat strips. Subsequent research work indicated that relay cropping could improve the resource use efficiency and overall productivity of the cotton-wheat cropping system. In a 2-year study in Punjab, Pakistan, conducted at two locations, intercropping cotton in bed–/ridge-sown wheat in early March improved the overall system productivity and cotton fiber quality as

K. Jabran
Department of Plant Production and Technologies, Faculty of Agricultural Sciences and Technologies, Niğde Ömer Halisdemir University, Niğde, Turkey

A. Nawaz
College of Agriculture, BZU, Layyah, Pakistan

A. Uludag
Plant Protection Department, Faculty of Agriculture, Canakkale Onsekiz Mart University, Canakkale, Turkey

S. Ahmad
Department of Agronomy, Faculty of Agricultural Sciences and Technology, Bahauddin Zakariya University, Multan, Pakistan

M. Hussain (✉)
Department of Agronomy, Faculty of Agricultural Sciences and Technology, Bahauddin Zakariya University, Multan, Pakistan

School of Veterinary and Life Sciences, Murdoch University, Murdoch, WA, Australia

S. Ahmad, M. Hasanuzzaman (eds.), *Cotton Production and Uses*,
https://doi.org/10.1007/978-981-15-1472-2_16

compared with conventionally tilled cotton sown after harvest of flat-sown wheat in late April. Future research may investigate the weed control and incorporation of conservation agricultural practices in the cotton-wheat relay intercropping systems.

Keywords Cotton · Wheat · Cropping system · Relay cropping

Abbreviations

BSW	Bed-sown wheat
Bt	*Bacillus thuringiensis*
BtCWCS	Bt cotton-wheat cropping system
CTC	Conventionally tilled cotton
RSW	Ridge-sown wheat

16.1 Introduction

Cotton (*Gossypium hirsutum* L.) is a cash crop in many countries of world, plays a role to provide livelihood to poor farming, and is a source for raw material for textile industry (Ahmad et al. 2014, 2017, 2018; Abbas and Ahmad 2018; Ahmad and Raza 2014; Ali et al. 2011, 2013a, b, 2014a, b). The leading cotton growers in the world are China, India, USA, Pakistan, and Turkey (Jabran and Chauhan 2019; Jabran et al. 2019). Cotton has unlimited uses and products and also possesses an inevitable role in the global economy (Jabran et al. 2019).

Globally cotton is grown as cotton-fallow as well as in-rotation with winter crops such as wheat (*Triticum aestivum* L.), etc. (Hulugalle et al. 2010; Ali et al. 2019). Major biotic stresses to cotton include insect pests, weeds, fungal pathogens, and viruses (Dogan et al. 2014; Jabran 2016; Saeed et al. 2016; Perge and Tolessa 2019; Razaq et al. 2019). Cotton-wheat is a popular cropping system and followed in several areas of world (Amin et al. 2017, 2018; Khan et al. 2004; Rahman et al. 2018; Tariq et al. 2017, 2018; Usman et al. 2009). In the recent decades, Bt cotton has been introduced in many parts of world (including the areas wherever cotton-wheat system is followed) in order to control bollworms. Although resistant or tolerant to bollworms, the Bt cotton has a crop duration longer than the conventional cotton cultivars. Usually Bt cotton cultivars are required to be sown at the time wheat is witnessing its reproductive phase. There is option to sow the Bt cotton after harvest of wheat; however, this results in a decreased growth and productivity of Bt cotton cultivars. Relay intercropping cotton in the wheat (while it is at its reproductive stage) is another option to sustain optimal production of together the wheat besides Bt cotton in the cotton-wheat cropping system. Intercropping has multiple benefits including the timely sowing of crops grown in rotation; management of weeds, insect pests, and diseases; efficient utilization of water/nutrient resources; improvement in soil quality; and system productivity (Willey 1985; Paltridge et al. 2014; Wu and Wu 2014;

Shahzad et al. 2016; Shah et al. 2016). Sometimes, intercropping may result in crop competition between both crops due to increased competition for water and nutrients. In this chapter we have discussed relay intercropping of Bt cotton for purpose of maintaining productivity of cotton-wheat system.

16.2 Cotton Plants and Environment

Cotton takes place among the sensitive crop plants and responds strongly to the climatic fluctuations. Being a tropical or subtropical crop, it requires a reasonably warm soil for seed germination and healthy stand establishment. On the other hand, a chilling stress during or immediately after sowing may not only result in poor crop but may also cause some disease development (Zhao et al. 2012). Low temperature also decreases the biomass production in cotton in addition to causing a poor crop stand (Reddy et al. 1999). Additionally, the optimum temperature also facilitates the proper morphological development, boll setting, and boll development (Reddy et al. 1999). However, a high rise in temperature (particularly in the wake of climate change) may damage the cotton plants through heat stress (DeRidder and Salvucci 2007; Jabran and Dogan 2018). Hence, the sensitivity of cotton crop plants is considered while selecting a sowing date. Selection of an optimum sowing date will give sufficient time for optimum germination, crop stand establishment, development of canopy and leaf area, and boll formation and boll development.

16.3 Relay Cropping of Cotton in Wheat (Inter-Seeding or Transplanting)

As described in the "Introduction" section, the introduction of Bt cotton in the cotton-wheat system compromised wheat cultivation. This is because the crop cultivation of Bt cotton expands to a part of the wheat crop growth cycle. The process of intercropping of seeds and transplanting of seedlings of Bt cotton is initiated when the wheat has completed its vegetative cycle and is continuing its reproductive phase. This intercropping of cotton seeds or transplanting of cotton seedlings in the wheat crop provides an opportunity to evade land preparation process and hence improve economic efficiency of cotton-wheat system. Upon maturity of wheat crop, the cotton plants gain a height of 30–50 cm and do not restrict the harvesting of wheat either manually or through a combine harvester. Combine harvester machine can easily harvest the wheat leaving lower 45–55 cm part, and in this way the cotton plants also remain safe from any damage (Fig. 16.1). Intercropping of cotton seeds or transplanting of seedlings also does restrict the process of interculture if required for weed control or fertilizer mixing, etc. (Fig. 16.2). Intercropping of cotton and wheat had higher light interception than

Fig. 16.1 Description of mechanical wheat harvesting with combined harvester with relay intercropped cotton (Source: CCRI, Multan)

Fig. 16.2 Description of mechanical interculturing to incorporate wheat stubbles in relay intercropped cotton

the individually sown cotton and wheat (Zhang et al. 2008). Similar research work was conducted in China where cotton was intercropped in the standing wheat (Zhang et al. 2007). In addition to sole wheat and cotton, the other treatments were one or two rows of cotton intercropped in three rows of wheat and two rows of cotton intercropped in three, four, or six rows of cotton. In general, although the yields of intercrops were decreased compared to sole crops, system productivity and land equivalent ratio were increased for most of the intercrops (Zhang et al. 2007).

Another research from Pakistan evaluated relay cropping of cotton in wheat (Nasrullah et al. 2017). Cotton relay cropping in wheat had higher income and benefit-cost ratio than the conventional cotton-wheat system. The research showed that wheat could be sown within the ridges that were spaced 150 cm apart, while later in the growing season (mid-March), cotton could be seeded on top of the ridges (Nasrullah et al. 2017).

One option is to sow the wheat on raised beds or ridges and, later in the season, transplant or sow the seeds of cotton on the free space between the wheat beds or ridges. Sowing (intercropping) of cotton may be a difficult task than transplanting of the 4–6-week-old seedlings. Either way will let the cotton to has full of its growing period in the field while also allowing the cultivation of wheat. Transplanting of cotton seedlings in the wheat may be advantageous over the sowing (intercropping) of seeds. For example, a better cotton crop establishment can be witnessed in case of transplanting because direct sowing may result in poor germination and may require a replanting. Also, transplanting can evade some of the laborious steps such as thinning (Jahromi and Mahboubi 2012). Another advantage of transplanting over seeding comprises the higher tolerance of 5–7-week-old seedlings against the harshness than the seedlings emerging from seeds. Ultimately the transplanting of 5–7-week-old seedlings provides an adequate and uniform crop stand, optimum crop growth, boll formation, and boll development. A higher yield in case of transplanting compared to direct seed sowing has also been reported (Salama et al. 1995).

The yield of intercrops is often low intercropping/relay cropping system against the sole stands of crops owing to competition for growth factors such as nutrients and water. However the yield gains in case of intercropping/relay cropping are frequently credited to complementation amid intercrops in mixture, leading to overall better use efficiency of available resources and higher system productivity (Shah et al. 2016).

16.4 Case Study on Relay Intercropping of Cotton in Wheat

Shah et al. (2016) conducted a 2-year study at two locations (Vehari and Multan) in Punjab, Pakistan, to examine production besides profitability of relay intercropping of cotton in wheat in Bt cotton-wheat system (BtCWS). The BtCWS included in research were (1) conventionally tilled cotton in fallow soil (sown in early/late March), (2) conventionally tilled cotton sown in late April after flat-sown wheat, (3) intercropping cotton on bed-sown wheat in early and late March, and (4) intercropping of cotton in ridge-sown wheat in early and late March. In bed sowing treatment, the Bt cotton was relay-intercropped on both sides of beds in bed-sown wheat or in the furrows amid 75 cm ridges in ridge-sown wheat. The layout of all the experimental treatment used in this study is given in Fig. 16.3. The results of the study indicated that the sowing of cotton in fallow soil in early March had the highest cotton yield, while the conventionally tilled cotton sown after

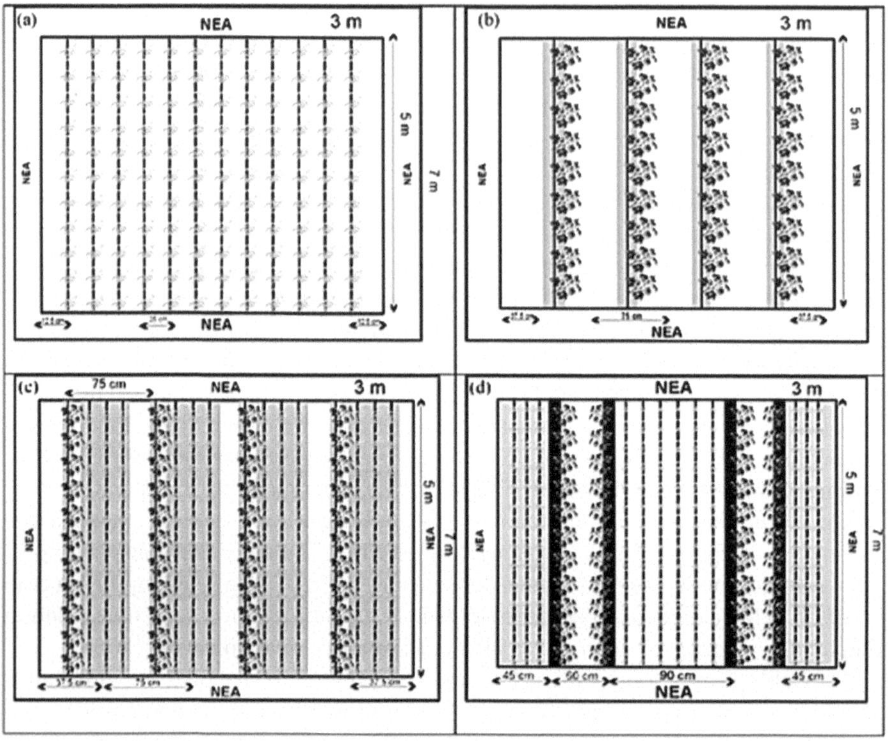

Fig. 16.3 Explanation of flat-sown wheat (**a**), conventionally tilled cotton (**b**), ridge-sown wheat with cotton intercropped (**c**), bed-sown wheat with cotton intercropped (**d**). *NEA* nonexperimental area. Net plot size = 3 m × 5 m (12 wheat rows besides 4 Bt cotton rows), gross plot size = 3 m × 7 m. (Adopted from Shah et al. (2016))

harvesting flat-sown wheat in late April had the lowest yield. However, the system production regarding benefit-cost ratio, net income, and marginal rate of returns was the highest when Bt cotton was intercropped in bed-sown wheat during early March in both years at both locations (Table 16.1; Shah et al. 2016). The plant height, biomass production, and fiber quality for fiber strength, fiber length, fiber uniformity, as well as fiber firmness were also highest in the relay intercropping Bt cotton in bed-sown wheat in early March as compared with sowing of conventionally tilled cotton after wheat harvest in late April (unpublished data; Table 16.2).

16.5 Conclusions

Wheat is planted in winter and harvested at the start of summer season and a vice versa pattern is followed for the cotton. In an effort to introduce Bt cotton cultivars in cotton-wheat systems, an overlap between wheat and cotton growth cycles was

Table 16.1 Impact of relay intercropping of Bt cotton in wheat on system productivity of Bt cotton-wheat system

	Net income[a] (US$ ha^{-1})		System productivity[b]	
Experimental treatments	Multan	Vehari	Multan	Vehari
CTC on fallow soil during early-March	1057	1186	1.68	1.77
CTC on fallow soil during late-March	836	926	1.54	1.6
CTC during late April after flat-sown wheat	2655	2891	2.05	2.13
Cotton intercropping in BSW during early March	3344	3415	2.37	2.39
Cotton intercropping in BSW during late March	3132	3168	2.28	2.3
Cotton intercropping in RSW during early March	3170	3241	2.3	2.33
Cotton intercropping in RSW during late March	3102	3120	2.27	2.28

All values are mean of 2 years
[a]Net income was acquired by subtracting total expenses ($ ha^{-1}) from gross income ($ ha^{-1})
[b]System productivity was appraised as ratio of output harvested ($ ha^{-1}) to input applied ($ ha^{-1}) on yearly basis

Table 16.2 Impact of cotton relay cropping in wheat on fiber quality of Bt cotton in Bt cotton-wheat system

	Fiber length (mm)		Fiber strength (g tex^{-1})		Fiber fineness (μg $inch^{-1}$)	
Experimental treatments	Multan	Vehari	Multan	Vehari	Multan	Vehari
CTC on fallow-land during early-March	28.72	28.80	28.80	28.95	5.04	5.27
CTC on fallow-land during late -March	28.51	28.61	28.71	28.75	4.91	5.15
CTC during late April after flat-sown wheat	27.70	27.80	27.79	27.89	4.46	4.55
Cotton intercropping in BSW during early March	28.34	28.42	28.61	28.67	4.78	4.98
Cotton intercropping in BSW during late March	28.04	28.10	28.37	28.44	4.66	4.70
Cotton intercropping in RSW during early March	28.24	28.25	28.50	28.57	4.74	4.89
Cotton intercropping in RSW during late March	27.90	27.96	28.32	28.40	4.61	4.75

All values are mean of 2 years; (Shah et al. 2019)

witnessed. This was because Bt cotton cultivars had a longer growth period than the conventional cotton cultivars. Under the situation, farmers had to opt one from the two crops (cotton or wheat). Relay cropping of cotton in wheat (inter-seeding of seeds or transplanting of cotton seedlings) was investigated and suggested as an innovative solution to maintain the productivity of cotton-wheat cropping system. Recent literature indicated that cotton relay cropping in wheat results in efficient use of resources and improved system productivity. Such relay cropping may also allow the use of conservation agriculture practices (such as zero tillage, soil cover). Ridge and bed sowing of cotton and wheat may also be useful to facilitate the relay cropping of cotton in wheat. We suggest that conservation agriculture practices, ridge sowing and bed sowing, should be focused as future research for the sustainable development of cotton-wheat relay cropping system. Another important aspect is to investigate the comparative weed biology and management in the conventional and relay cotton-wheat systems.

References

Abbas Q, Ahmad S (2018) Effect of different sowing times and cultivars on cotton fiber quality under stable cotton-wheat cropping system in southern Punjab, Pakistan. Pak J Life Soc Sci 16:77–84

Ahmad S, Abbas Q, Abbas G, Fatima Z, Atique-ur-Rehman NS, Younis H, Khan RJ, Nasim W, Habib urRehman M, Ahmad A, Rasul G, Khan MA, Hasanuzzaman M (2017) Quantification of climate warming and crop management impacts on cotton phenology. Plan Theory 6(7):1–16

Ahmad S, Iqbal M, Muhammad T, Mehmood A, Ahmad S, Hasanuzzaman M (2018) Cotton productivity enhanced through transplanting and early sowing. Acta Sci Biol Sci 40:e34610

Ahmad S, Raza I (2014) Optimization of management practices to improve cotton fiber quality under irrigated arid environment. J Food Agric Environ 2(2):609–613

Ahmad S, Raza I, Ali H, Shahzad AN, Atiq-ur-Rehman SN (2014) Response of cotton crop to exogenous application of glycinebetaine under sufficient and scarce water conditions. Braz J Bot 37(4):407–415

Ali H, Abid SA, Ahmad S, Sarwar N, Arooj M, Mahmood A, Shahzad AN (2013a) Integrated weed management in cotton cultivated in the alternate-furrow planting system. J Food Agric Environ 11(3&4):1664–1669

Ali H, Abid SA, Ahmad S, Sarwar N, Arooj M, Mahmood A, Shahzad AN (2013b) Impact of integrated weed management on flat-sown cotton (*Gossypium hirsutum* L.). J Anim Plant Sci 23 (4):1185–1192

Ali H, Afzal MN, Ahmad F, Ahmad S, Akhtar M, Atif R (2011) Effect of sowing dates, plant spacing and nitrogen application on growth and productivity on cotton crop. Int J Sci Eng Res 2 (9):1–6

Ali H, Hameed RA, Ahmad S, Shahzad AN, Sarwar N (2014a) Efficacy of different techniques of nitrogen application on American cotton under semi-arid conditions. J Food Agric Environ 12 (1):157–160

Ali H, Hussain GS, Hussain S, Shahzad AN, Ahmad S, Javeed HMR, Sarwar N (2014b) Early sowing reduces cotton leaf curl virus occurrence and improves cotton productivity. Cercetări Agronomice în Moldova XLVII(4):71–81

Ali MA, Farooq J, Batool A, Zahoor A, Azeem F, Mahmood A, Jabran K (2019) Cotton production in Pakistan. In: Jabran K, Chauhan BS (eds) Cotton production. John Wiley & Sons, Inc., USA

Amin A, Nasim W, Mubeen M, Ahmad A, Nadeem M, Urich P, Fahad S, Ahmad S, Wajid A, Tabassum F, Hammad HM, Sultana SR, Anwar S, Baloch SK, Wahid A, Wilkerson CJ, Hoogenboom G (2018) Simulated CSM-CROPGRO-cotton yield under projected future climate by SimCLIM for southern Punjab, Pakistan. Agric Syst 167:213–222

Amin A, Nasim W, Mubeen M, Nadeem M, Ali L, Hammad HM, Sultana SR, Jabran K, Habib urRehman M, Ahmad S, Awais M, Rasool A, Fahad S, Saud S, Shah AN, Ihsan Z, Ali S, Bajwa AA, Hakeem KR, Ameen A, Amanullah HUR, Alghabar F, Jatoi GH, Akram M, Khan A, Islam F, Ata-Ul-Karim ST, Rehmani MIA, Hussain S, Razaq M, Fathi A (2017) Optimizing the phosphorus use in cotton by using CSM-CROPGRO-cotton model for semi-arid climate of Vehari-Punjab, Pakistan. Environ Sci Pollut Res 24(6):5811–5823

DeRidder BP, Salvucci ME (2007) Modulation of Rubisco activase gene expression during heat stress in cotton (*Gossypium hirsutum* L.) involves post-transcriptional mechanisms. Plant Sci 172:246–254

Dogan MN, Jabran K, Unay A (2014) Integrated weed management in cotton. In: Chauhan BS, Mahajan G (eds) Recent advances in weed management. Springer, The Netherlands, pp 197–222. https://doi.org/10.1007/978-1-4939-1019-9-9

Hulugalle NR, Weaver TB, Finlay LA (2010) Soil water storage and drainage under cotton-based cropping systems in a furrow-irrigated vertisol. Agr Water Manag 97:1703–1710

Jabran K (2016) Weed flora, yield losses and weed control in cotton crop. Julius-Kühn-Archiv 452:177–182

Jabran K, Chauhan BS (2019) Cotton production. John Wiley & Sons, Inc., USA

Jabran K, Dogan MN (2018) High carbon dioxide concentration and elevated temperature impact the growth of weeds but do not change the efficacy of glyphosate. Pest Manag Sci 74:766–771

Jabran K, Ul-Allah S, Chauhan BS, Bakhsh A (2019) An introduction to global production trends and uses, history and evolution, and, genetic and biotechnological improvements in cotton. In: Jabran K, Chauhan BS (eds) Cotton production. John Wiley & Sons, Inc., USA

Jahromi AM, Mahboubi M (2012) Evaluate the cotton transplantation and its impact on yield performance components in saline lands. Adv Environ Biol 6:1304–1306

Khan MB, Khaliq A, Ahmad S (2004) Performance of mashbean intercropped in cotton planted in different planting patterns. J Res (Sci) 15(2):191–197

Nasrullah HM, Aslam M, Akhtar M, Ali B, Majid A, Akram M, Farooq U (2017) Relay cropping of cotton in standing wheat: an innovative approach for enhancing the productivity and income of small farm. Romamian Agric Res 34:87–195

Paltridge NG, Coventry DR, Tao J, Heath TJ, Tashi N (2014) Intensifying grain and fodder production in Tibet by using cereal-forage intercrops. Agron J 106:337–342

Perge LL, Tolessa TT (2019) Integrated management of major fungal, bacterial, viral and nematode diseases of cotton. In: Jabran K, Chauhan BS (eds) Cotton production. John Wiley & Sons, Inc., USA

Rahman MH, Ahmad A, Wang X, Wajid A, Nasim W, Hussain M, Ahmad B, Ahmad I, Ali Z, Ishaque W, Awais M, Shelia V, Ahmad S, Fahad S, Alam M, Ullah H, Hoogenboom G (2018) Multi-model projections of future climate and climate change impacts uncertainty assessment for cotton production in Pakistan. Agric For Meteorol 253-254:94–113

Razaq M, Mensha R, Athar HR (2019) Insect pest management in cotton. In: Jabran K, Chauhan BS (eds) Cotton production. John Wiley & Sons, Inc., USA

Reddy KR, Davidonis GH, Johnson AS, Vinyard BT (1999) Temperature regime and carbon dioxide enrichment alter cotton boll development and fiber properties. Agron J 91:851–858

Saeed R, Razaq M, Hardy IC (2016) Impact of neonicotinoid seed treatment of cotton on the cotton leafhopper, *Amrasca devastans* (Hemiptera: Cicadellidae), and its natural enemies. Pest Manag Sci 72:1260–1267

Salama GM, Youssef AM, Farge SSA (1995) Tomato plant growth and productivity as affected by method transplanting. Vegetable Res Sect Hort Res Inst Agric Res Cairo Egypt 22:109–115

Shah MA, Farooq M, Hussain M (2016) Productivity and profitability of cotton–wheat system as influenced by relay intercropping of insect resistant transgenic cotton in bed planted wheat. Eur J Agron 75:33–41

Shah MA, Hussain M, Nawaz A, Jabran K, Farooq S, Farooq M (2019) Relay intercropping improves growth and fiber quality of cotton. Int J Agric Biol 22:1539–1546

Shahzad M, Farooq M, Jabran K, Hussain M (2016) Impact of different crop rotations and tillage systems on weed infestation and productivity of bread wheat. Crop Prot 89:161–169

Tariq M, Afzal MN, Muhammad D, Ahmad S, Shahzad AN, Kiran A, Wakeel A (2018) Relationship of tissue potassium content with yield and fiber quality components of Bt cotton as influenced by potassium application methods. Field Crops Res 229:37–43

Tariq M, Yasmeen A, Ahmad S, Hussain N, Afzal MN, Hasanuzzaman M (2017) Shedding of fruiting structures in cotton: factors, compensation and prevention. Trop Subtrop Agroecosyst 20(2):251–262

Usman M, Ahmad A, Ahmad S, Irshad M, Khaliq T, Wajid A, Hussain K, Nasim W, Chattha TM, Trethowan R, Hoogenboom G (2009) Development and application of crop water stress index for scheduling irrigation in cotton (*Gossypium hirsutum* L.) under semiarid environment. J Food Agric Environ 7(3&4):386–391

Willey R (1985) Evaluation and presentation of intercropping advantages. Exp Agric 21:119–133

Wu K, Wu B (2014) Potential environmental benefits of intercropping annual with leguminous perennial crops in Chinese agriculture. Agric Ecosyst Environ 188:147–149

Zhang L, van der Werf W, Zhang S, Li B, Spiertz JHJ (2007) Growth, yield and quality of wheat and cotton in relay strip intercropping systems. Field Crops Res 103:178–188

Zhang L, van der Werf W, Zhang S, Li B, Spiertz JHJ (2008) Light interception and use efficiency in cotton and wheat relay strip intercropping systems. Field Crops Res 107:29–42n 8n

Zhao J, Li S, Jiang T, Liu Z, Zhang W, Jian G, Qi F (2012) Chilling stress—the key predisposing factor for causing *Alternaria alternata* infection and leading to cotton (*Gossypium hirsutum* L.) leaf senescence. PLoS One 7:e36126

Chapter 17
Cotton-Based Intercropping Systems

Atique-ur-Rehman, Hakoomat Ali, Naeem Sarwar, Shakeel Ahmad, Omer Farooq, Kamrun Nahar, and Mirza Hasanuzzaman

Abstract Coping with the threat of climate change and food security, crop productivity needs improvement. Low-price situation for cotton in some developing countries is discouraging cotton cultivation. Cotton-based intercropping offers a better solution to cope with the situation. The system of companion plantation of one crop with another during the same growing season on the same piece of land offers an option. The basic objective is to augment yields by multiplying the available growing area. Across other merits of intercropping, the biodiversity of growing crops attracts a variety of predatory insects making possible an integrated pest control of cotton. Intercropping may be classified into three types, viz., mixed cropping, which includes planting a variety of harmonious plants together; alley or row cropping, in which diverse crops are grown with each other in rows; or temporal intercropping, that is a system of the fast-growing crop with a slow growth. Due to a relatively longer duration with slow-growing habit during initial stages, cotton-based cropping system is ideally suitable for intercropping. The cotton-based intercropping aims to maximize the yield of cotton along with extra profits from intercrops. There is a wide range of crops including cereals, legumes, and vegetables, which are possible to grow with or in standing cotton crop. Each group offers versatile advantages in terms of yields of both companion crops and monetary returns of the system. However, there is a crop-wise variation in monetary advantages of diverse intercropping systems.

Atique-ur-Rehman (✉) · H. Ali · N. Sarwar · O. Farooq
Department of Agronomy, Bahauddin Zakariya University, Multan, Pakistan
e-mail: dr.atique@bzu.edu.pk

S. Ahmad
Department of Agronomy, Faculty of Agricultural Sciences and Technology, Bahauddin Zakariya University, Multan, Pakistan

K. Nahar
Department of Agricultural Botany, Sher-e-Bangla Agricultural University, Dhaka, Bangladesh

M. Hasanuzzaman
Department of Agronomy, Faculty of Agriculture, Sher-e-Bangla Agricultural University, Dhaka, Bangladesh

S. Ahmad, M. Hasanuzzaman (eds.), *Cotton Production and Uses*,
https://doi.org/10.1007/978-981-15-1472-2_17

Keywords Cotton · Seed cotton yield · Net returns · Cereals · Legumes

17.1 Introduction

In the scenario of terrifyingly increasing populace of the world, improving crop productivity becomes a common challenge for agriculturists (Ahmad et al. 2014, 2017, 2018; Abbas and Ahmad 2018; Ahmad and Raza 2014; Ali et al. 2011, 2013a, b, 2014a, b). Increasing output per unit land area is strategic for sustainable agriculture, along with the expansion of cultivable land. Nevertheless, as the degree of proper agricultural land is becoming static or fading, multiple cropping on the same piece of land is advocated vigorously as a vital means to augment agricultural produce. Among solutions, mechanized rotational monoculture and traditional polyculture production systems, commonly practiced in the developing countries, are useful (Tsubo et al. 2003). However, in case of small landholdings, multiple cropping systems are the most viable option due to the integration of different crops by space and other inputs more proficiently (Baldy and Stigter 1997). Moreover, it has superior yield stability in inconstant environmental conditions besides soil conservation practices.

In developing countries, the majority of farmers have small landholdings, usually less than 5 ha (Tariq et al. 2017, 2018; Usman et al. 2009). Our small farms are overloaded with surplus family labor, and yields on these farms are far low. Cotton is the most important cash crop of Pakistan; however, its yield is low (Amin et al. 2017, 2018a, b; Khan et al. 2004; Rahman et al. 2018). Cotton has a position of foremost fiber besides cash crop, playing a crucial role to uphold the national economy. Intercropping is a companion planting technique of growing one crop alongside another on the same piece of land, aiming to increase yields from available growing space. The biodiversity of growing crops offers a variety of benefits besides predatory insects, which is not likely with monoculture. It may be mixed cropping, by planting a variety of well-suited plants collectively; alley or row cropping, in which different crops are planted with each other in rows; or temporal intercropping, which is a system of the fast-growing crop with a slow-growing crop. Cotton-based cropping systems have been reported as a promising strategy, particularly for small landholdings. Although a reduction in cotton yield by 8–31% is reported, however, more crop productivity from the single field with better net returns was recorded in intercropping than sole crop (Khan et al. 2012; Mohammad et al. 1991). An increase of 30–40% in farm income from different intercropping systems in cotton has been reported (Saeed et al. 1999). However, degree of such agroeconomic advantages mainly depends on the growth habit and nature of intercrop (Rao 1991). In traditional closely spaced single-row method of cotton planting, growing intercrops does not remain convenient. It is particular that cotton must be planted in widely spaced multi-rows to ensure enough space for the facilitation of intercropping.

Cotton-based cropping systems are idyllically suitable for intercropping since a relatively longer duration and a slow initial growth of cotton. Generally, cotton cultivation is done as mixing in interior space with pulses. Although this practice is much feasible, however, the pecuniary benefit of cotton-pulse intercropping is reported inadequate. Wide spacing between rows and slow growth at initial stages of cotton are advantageous. Vacant space between rows of cotton crop gives enough time of 2–3 months for growing a short-duration intercrop. Legumes, growing quickly and completing their life cycle within a short duration, including grams, beans, and cowpea, are reported more attuned for intercropping in cotton (Surendran et al. 2016). Legume intercrops in cotton-based cropping system have been advocated as successful component crops with diverse nutrient and moisture requirements with wide-ranging feeding zones in soil. These features with varied growth duration enable legumes to utilize different available resources optimally (Sankaranarayanan et al. 2011).

Furthermore, slow-growing habit with a lot of unfilled interspaces during initial stages of cotton offers an immense possibility for growing intercrops (Nehra et al. 1990a, b). Likewise, early-maturing legumes such as gram and cluster bean can be grown advantageously as intercropping by exploiting empty spaces in between cotton rows. Thus the fertility status of soil can be improved in this way. Moreover, growing legumes in cotton could utilize the environment in a better way in addition to providing insure against the inclement weather situation and consequent crops (Balasubramanian 1987). Weed infestation remained a big challenge in cotton-based systems; however, control of these is an important aspect of intercropping of legumes. Further, growing legumes in cotton can reduce the leaching losses of nutrients, provide help in pest control, and reduce soil erosion losses.

Relay cropping is a system of intercropping in which the second crop is planted during growth of the first crop. After harvest of the first crop, crop plants fade to make space for the second crop. Row cropping is widespread because it also uses space that would otherwise remain unexploited. Space between rows of first crop has no use except for providing space for weeds and other pests. The crops that fill space between rows and do not compete with main crop for nutrients are excellent for intercropping. Cotton-based intercropping ensures a better yield as compared to that obtained from any of the sole crops. However, efficiency of the intercropping system can be considerably improved through appropriate choice of crops and their apposite cultivars, keeping in view the difference in life span, morphology, and growing habit from the cotton. It will ensure proper utilization of available resources like moisture, nutrients, solar radiation, etc. during the growing season.

In addition to that existing low-price situation for cotton, less produce from single field is discouraging cotton cultivation. To overcome above situation, intercropping of high-value vegetable crops is one of the viable options. In cotton-vegetable intercropping, there is a wide range of this system, for example, radish + beetroot + coriander between the cotton rows, radish + cluster bean + beetroot system, cotton + coriander + vegetable cowpea + cluster bean, etc. These systems significantly improved the yield of cotton and its quality and more net returns from limited land area.

Owing to this significant improvement in yield of cotton and intercrops and net benefits, cotton-based intercropping systems are advocated particularly for the farmers with limited landholdings. The present chapter ascertains different aspects of cotton-based intercropping systems.

17.2 Advantages of Intercropping

17.2.1 Resource Use Efficiency

Intercropping is reported to net returns as compared to sole crop, which is greatly attributed to improved resource use. In intercropping system, an absolute and efficient use of available resources like solar radiation, soil nutrients, and water is advantage of intercropping over sole crop. A variation in growth patterns of both crops enables different growth resources availability at diverse times in the field (Fukai and Trenbath 1993; Marimuthua et al. 2014). Generally, a minimum difference of 25% in the duration of both crops is recommended for maximum yield advantages of crops in an intercropping system (Anitha et al. 2001). Early-maturing crops should be grown with little obstruction from the late-maturing crop to improve resource use efficiency of both crops. Although there may be some problem for later intercrop from the associated crop, however, after the harvest of first crop, there would be long time period for further growth, which ensures good recovery and complete use of resources. It permits efficient consumption of different resources by crop intensification in dimensions of space besides time. Intercropping makes possible to intensify crop production in time and space; however, in space dimension, intensification of land besides resource use is a more imperative feature of intercropping. For instance, crops with different growth patterns and dissimilar foliage display will improve efficient use of light from the same field during growing season. Likewise, different types of root systems will explore soil from various depths, making an efficient use of soil nutrients and moisture (Francis 1986). Difference in nutrient demand and time of maturity among different species in intercropping will lead to efficient consumption of applied inputs and natural resources through decreasing competition between component intercrops. In intercropping, more spread and faster canopy covers the field and increases the interception of solar radiation, leading to an efficient utilization of solar light as compared to monocrop (Keating and Carberry 1993). This system improves radiation use efficiency resulting in a greater productivity per unit land area and minimizes the fraction of radiation energy reaching at ground (Ramakrishna and Ong 1994).

Improving nutrient use efficiency is vital for good production and net benefits from single piece of land, and it is particular for farmers with limited resources. It is, therefore, important to improve fertilizer or nutrient use efficiency, and individual crops have mostly lower nutrient use efficiency. In intercropping system, however, the communal nutrient use efficiency is higher than sole crops (Chowdhury and

Rosario 1994). During early growth stage of main crops like cotton, different natural and applied resources like solar radiation, irrigation water, and fertilizer may be wasted; however, growing an associated crop between the rows of main crop will improve their utilization (Midmore 1993). Intercropping systems have higher land equivalent ratio as compared to sole crop which makes a higher total absorption of nutrients. Moreover, intercrops have improved efficiency for converting absorbed nutrients to seeds or grains as compared to sole crops, which contribute toward higher yield (Chowdhury and Rosario 1994). According to Morris and Garrity (1993), intercrops absorbed 43% and 35% more phosphorus and potassium as compared to sole crop, respectively. One possible reason for more uptakes of nutrients by intercrops is bigger and long-lasting functional root system than either of the sole crop. Expanded and enlarged root system provides more soil profile from which nutrients can be taken up (Morris and Garrity 1993), thus improving and maintaining fertility of soil. This is particularly important when an exhaustive crop (such as cotton) and restorative crop (such as legumes) are grown in association (Geno and Geno 2001). Pulses or legumes have deep rooting system, by which these can take up nutrients from lower soil profile; thus can recycle leached nutrients from the surface. Another important aspect of intercropping is water use, which may be affected by intercropping system. For example, in wheat-lucerne rotation, lucerne water uptakes from a root depth was twofold than wheat (Dunin et al. 1999).

17.2.2 Modification of Microclimate

In intercropping system, reduced light intensity, temperature over the field, desiccating wind, and change in other climatic components, microclimate is modified. In a sorghum-tomato intercropping system, shaded tomato crop yielded more than pure crop, and this may be at the cost of little yield loss of sorghum; hence in this system, land equivalent ratio ranged from 2.58 to 2.99 (Kamel et al. 2004). It was found that the microclimate over the field especially for tomato was changed due to reduced air temperatures (less 5–7 °C than sole tomato) by shade of sorghum. In sole tomato crop, poor development and consequently reduced yield were observed due to meager foliage development, reducing of blossom, deprived fruit set, rupture of leaves besides branches, and lodging of plants besides high dust coverage on leaves (Lemma 2002). On the other side, strip intercropped tomato in maize or sorghum was protected from wind and has higher yield (Lemma 2002). In this regard, maize is particularly selected for intercropping to keep cover on the understory crop because of its sowing in lines and wide adaptation in a wide variation of climates. Great modifications in temperatures of both air and soil occur in an intercropping system. Microclimate within canopy of intercrops regulates temperature extremes, and in summer season, a widespread canopy of main crop like cotton lowers the temperature and reduces air movement, leading to less evaporation loss and augmented relative humidity (Farrell and Altieri 1995). In the cropping system, where capital is a main constraint for the modification of microclimate like shading by synthetic materials, this feature of intercropping is of significant worth. It is the reason that intercropping is advocated for temperate crops in tropics, which change soil microclimate classification (Jaya et al. 2001). Modification in microclimate intercropping is acceptable as well as affordable.

17.2.3 Light Interception and Radiation Use Efficiency

Light interception by the crop canopy may vary according to planting geometry and population density (Jaya et al. 2001). Furthermore, orientation and density of crop may change the temperature over the cropped area. In a medium density field of maize (7.1 plant m^{-2}) with orientation of N-S, reduced the canopy temperatures by 1.2 °C. This decreased temperature was found to be related with decline in irradiance up to 70% and was greatly subtle to row orientation and density (Jaya et al. 2001). For estimating efficiency of intercropping system, light interception and its use efficiency are used as potent models. Moreover, the role of intercepted radiation and light use efficiency in improving productivity is greatly postulated in any cropping system (Willey 1990).

The extent of the radiation interception by the canopies of component crop of intercropping system is primarily depending upon the canopy structure and leaf area index (Bastiaans et al. 2000). A widespread canopy cover over the entire field or increased duration of soil coverage of intercrops is the main reason for a high light interception and consequent light use efficiency (Keating and Carberry 1993). In an intercropping system where both intercrops have non-synchronous canopy patterns with variation in maturation times, a greater light interception occurs due to more and extended leaf area during the whole growing season, may be 30–40% more as compared to monocrops (Carandang 1980).

17.2.4 Soil Fertility Improvement

Sustaining fertility status of soil is very crucial in modern agriculture. Soil fertility status in intercropping system is greatly improved as a result of rotation in each season on the same piece of land. This is particular when a legume crop is a part of intercropping system. Symbiotic relationship of rhizobium bacteria with plants, especially legumes, can fix atmospheric nitrogen for plant uptake. There are a lot of reports for the increase in the nitrogen content in accompanying non-legume intercrop (Fujita et al. 1992; Anil et al. 1998; Eskandari et al. 2009). Legumes are able to grow well in low phosphate soils (Geno and Geno 2001), and after harvesting of intercrop, their roots besides fallen leaves after decaying deliver nutrients to other crops. However, harvesting of legumes crops with more biomass, the addition of nitrogen is decreased (Geno and Geno 2001). Moreover, nitrogen fixed by intercrop legume is not sufficient to provide soil fertility for complete crop period. It is, therefore, to start a more successful nitrogen fixation; a basal nitrogenous fertilizer is required for both intercrops. Intercropping system improves the fertilizer use efficiency as a result of augmented humus and the difference in rooting systems and nutrient uptake by both crops.

Plant interactions in growing media are generally more extreme as compared to aboveground, and a limited uptake of nutrients occurs owing to this competition. There is a great variation in specific nutrient zones of soil, which is mainly due to specific environmental conditions, management practices, or nutrient solubility (Snaydon and Harris 1981). Variation in rooting patterns of intercrops enables nutrient uptake from different zones of soil. Furthermore, roots use soil nutrients contrarily to fulfill nutrient requirement of crops, e.g., legumes use nitrogen from atmosphere in gas form, while non-legumes utilize available nitrogen as NO^{-3} or NH^{+4}, greatly differing in their nutrient requirement.

17.2.5 Pest Management

An important benefit of intercropping is the significant reduction by damage from pest and disease. Intercrops reduce pest infestation by breakdown of pest's ability to attach to its host and attraction of intercrops towards their pest avoiding main species (Danso et al. 1987). To avoid pest infestation, intercropping in cotton with different crops is found useful. There are a number of crops used for a purpose like groundnut (Waterworth 1994), with basil (Schader et al. 2005). Intercropping in cotton with different spring and winter crops (relay cropping) is commonly practiced to conserve and improve cotton from aphid in different regions of the world (Parajulee et al. 1997; Xia et al. 1998).

Although intercrops reduce the extent of pest infestation, however, complete abolition of pest damage does not occur (Fujita et al. 1992). Greater species variety of agricultural ecosystems acts as barrier for plant pathogens, which increased by sowing more than one crop in the same field (Anil et al. 1998). In an intercropping system of marigold and tomato, it was noted that there was an adverse effect on the germination and density of *Alternaria solani* conidia; as a result leaf damage of tomato was significantly reduced (Gomez-Rodriguez et al. 2007). Furthermore, significant discount in tomato blight produced by *A. solani* was recorded which was induced by marigold. The mechanisms of survival were mainly allelopathic effect of marigold on *A. solani* and altered microclimate around the canopy, providing a physical barrier against spreading of conidia, and owing to these reasons, tomato intercropping is usual in different countries.

Likewise, suppression of weed infestation is effectively done by intercropping patterns as compared to monocropping; however, their effectiveness varies greatly according to intercrop species (Girjesh and Patil 1991). Intercropping is proposed as an important option for integrated weed management by a number of researchers (Rana and Pal 1999; Schoofs and Entz 2000). Intercrops suppress weed by obtaining more crop yield than monocrop and less weed growth as a result of seizing resources from weeds or through allelochemicals. In monocropping systems, weeds use natural resources such as light, moisture, and nutrients, which must be used by crop. However, in the intercropping, use of these resources is complementary by both crops, consequently resource use efficiency is improved in case of intercropping system, and this act actually fills empty niche, resulting in a better and effective weed control monocropping system (Saudy and El-Metwally 2009).

Liebman and Dyck (1993) concluded that weed density besides biomass production may be noticeably decreased using intercropping. Intercrops may validate weed control benefits over sole crops in two means. First, higher crop productivity besides lesser weed growth may be achieved if intercrops are more effective than sole crops through alleopathy. Alleopathic interactions may occur among weeds, among weeds and crops, or even among crops. Allelopathic potential of crops for weed control greatly varies among species, for example, spring-planted rye reduced weed biomass by 94%. Similarly, sunflower is known as an alleopathic crop and can control weeds equivalent to herbicide (Geno and Geno 2001). Likewise, in an intercropping system of lucerne with oat, oat helped lucerne to fight against weeds and consequently augmented forage yield of lucerne (Lanini et al. 1992). It is thus ascertained as an economical and efficient substitute to herbicide use for weed control. Likewise, alleopathic water extract of mature sorghum can be used as spray on wheat to control weeds, where 35–50% and 10–21% higher weed control and yield were recorded, respectively (Cheema and Khaliq 2000).

Otherwise, intercrops may deliver yield benefits without suppressing weeds if intercrops unitized resources which are not exploitable by weeds or convert resources to harvestable material more efficiently than sole crops (Geno and Geno 2001). According to different researchers, the main aim of intercropping must be to improve use of resources like light, moisture, nutrients, and space (Ndakidemi 2006), along with increase in quality and quantity of the crop (Mpairwe et al. 2002). Moreover, water quality can be controlled by minimum use of synthetic fertilizers, which can pollute the environment, causing a big hazard for humans (Crew and Peoples 2004).

17.3 Cotton-Based Intercropping

The yield and productivity of cotton in some developing countries is not as good as in most of cotton-growing countries of the world. Farmers are trying their best effort to attain maximum potential yield with their available resources. However, with limited landholdings, their efforts are going in vain and still there is a huge difference. It is, therefore, need of the time to adopt some latest appropriate management practices like intercropping along with prudent use of inputs to gain maximum yield potential. Cotton-based intercropping appears as an auspicious strategy for this purpose. Although there may be some penalty for cotton yield with intercrops by 8–31%, however, total productivity of both crops and net return per unit area are the advantages of intercropping that can be advocated (Mohammad et al. 1991). Yield advantages are reported from different cotton-based intercropping (Table 17.1). Total increase in farm income from different cotton-based intercropping systems is reported by 30–40% (Saeed et al. 1999). However, degree of such agroeconomic benefits differs according to the type of intercrop (Rao 1991). Yield advantage of cotton in intercropping is often credited to improve weed control, which may be due to allelopathic impact of component crops (Narwal 2004) and exploitation of various farm resources (Iqbal et al. 2007). Moreover, suitability or need of intercrop may be different according to region, available resources, and most importantly the adoptability of intercrop with main crop. There are different intercropping systems commonly practiced in different areas with variable planting geometry (Fig. 17.1). In conventional system of cotton production, cotton is sown in narrow-spaced single line in which intercropping is not so convenient. In the latest method, cotton is planted in widely spaced multi-rows with seed cotton yields similar to that obtained in conventional single-row system and option of intercropping (Deshpande et al. 1989).

17.3.1 Cotton-Wheat Intercropping

Sowing of wheat in cotton is an important intercropping system consisting of the two main cash and staple crops. Intercropping system of cotton and wheat may improve productivity and profitability of this important system (Shah et al. 2016). Sowing of cotton in standing wheat as relay intercropping is a good option particularly when there is risk to exclude cotton from wheat-cotton system due to time conflict between these crops (Shah et al. 2016). From the results of experiments on cotton-wheat system, it is clear that both yield and net returns were either at par or improved (Table 17.1). However, the form of this system may be different in various parts of the world. For instance, in some areas of China, wheat is grown in the end of autumn season and harvested before or in the start of summer of the succeeding year (Zhang et al. 2008). Cotton is sown prior to the harvest of wheat in space left over in the field, and this forms a strip-based cotton-wheat system. Sowing of both crops done in rows or on ridges and the system prevails from April, the time of sowing of cotton, to June, the time of wheat harvesting (Zhang et al. 2008).

After wheat harvesting, cottonseeds are initially sown in strips since the younger plants require some time to develop the canopy and fill the gap and become empty after the wheat harvesting. In this system, wheat crop covers the younger cotton seedlings, which need little space to grow. After a few weeks, harvesting of wheat is done and the entire field is free for cotton crop. Soon after harvesting, the gaps due to wheat harvest are bridged by the developing leaf canopy of cotton. The entire cropping season comprises of three stages: (1) only wheat crop prior to cotton sowing, (2) an intercropping stage with both wheat and cotton in the field, and (3) only cotton crop after harvest of wheat. Different intercropping strategies are practiced with some spatial arrangement as the number of wheat lines in each strip and the number of cotton lines in each strip: 3:1, 3:2, 4:2, and 6:2 (Zhang et al. 2007).

Table 17.1 Cotton yield and net returns from different cotton based intercropping systems

Intercropping Systems	Seed Cotton (t ha^{-1})	Net Returns ($)	Reference
Cotton alone	1.97	485.24	Iqbal et al. (2007)
Cotton + single-row sorghum	1.21	534.16	Iqbal et al. (2007)
Cotton double-row sorghum	1.15	600.49	Iqbal et al. (2007)
Cotton + single-row soybean	1.36	397.48	Iqbal et al. (2007)
Cotton + double-row soybean	1.21	371.18	Iqbal et al. (2007)
Cotton + single-row sesame	1.19	486.51	Iqbal et al. (2007)
Cotton + double-row sesame	1.14	579.24	Iqbal et al. (2007)
Cotton flat sown (early march)	3.57	979.00	Shah et al. (2016)
Cotton flat sown (late march)	3.37	830.00	Shah et al. (2016)
Bed sowing wheat + cotton (early march)	3.21	3364.00	Shah et al. (2016)
Bed sowing wheat + cotton (late march)	3.05	3129.00	Shah et al. (2016)
Ridge sowing wheat + cotton (early march)	3.33	3137.00	Shah et al. (2016)
Ridge sowing wheat + cotton (late march)	3.09	3077.00	Shah et al. (2016)
Cotton alone	2.62	442.69	Sankaranarayanan et al. (2012)
Cotton + radish + v. cowpea + beet root	2.74	492.96	Sankaranarayanan et al. (2012)
Cotton + radish + cluster bean + beet root	2.55	424.82	Sankaranarayanan et al. (2012)
Cotton + radish + dolichos + beet root	2.68	472.64	Sankaranarayanan et al. (2012)
Cotton + coriander + v. cowpea + cluster bean	2.78	505.17	Sankaranarayanan et al. (2012)
Cotton + coriander + dolichos + cluster bean	2.83	511.35	Sankaranarayanan et al. (2012)
Cotton + beet root + v. cowpea + cluster bean	2.91	546.53	Sankaranarayanan et al. (2012)
Cotton + cluster bean + v. cowpea + dolichos	2.88	530.83	Sankaranarayanan et al. (2012)
Cotton + radish + beet root + coriander	2.58	448.46	Sankaranarayanan et al. (2012)
Cotton alone	1.78	506.6	Marimuthua et al. (2014)
Cotton + dhaincha	1.95	577.2	Marimuthua et al. (2014)
Cotton + green gram	1.59	524.5	Marimuthua et al. (2014)
Cotton + onion	1.53	502.6	Marimuthua et al. (2014)

(continued)

Table 17.1 (continued)

Intercropping Systems	Seed Cotton (t ha^{-1})	Net Returns ($)	Reference
Cotton alone	1.97	–	Jayakumar and Surendran (2017)
Cotton + onion	2.39	–	Jayakumar and Surendran (2017)
Cotton + black gram	2.24	–	Jayakumar and Surendran (2017)
Cotton + green gram	1.99	–	Jayakumar and Surendran (2017)
Cotton + lucerne	1.91	–	Jayakumar and Surendran (2017)

In cotton-wheat system, timely sowing of wheat is a challenge. The problem arises from late cultivation of wheat and thus is a main reason for decrease in the area under wheat cultivation. This delay in the time of wheat sowing and reduction in area under wheat cultivation are due to delayed cotton picking. Among different factors, certain are unrestrained; however, some can be controlled by farmer, for example, punctual sowing of wheat. Completing the last picking from cotton field before sowing of wheat can do this. Moreover, to cope with the problem, wheat sowing can be done in standing cotton, a new technology for cotton-based cultivation systems. To cultivate wheat in the standing cotton field, firstly the picking of the mature cotton bolls, ready to harvest, must be completed. After that, field is irrigated between the rows of cotton. Meanwhile, wheat seed is dipped in water for a certain period of 8–10 h at the rate of 60 kg ha^{-1}. After that, this seed is broadcasted into the thin layer of irrigation water in standing cotton field. To ensure a good emergence rate of wheat seed, the field must remain wet for the next few days after sowing. The fertilizers required as basal dose can be applied with irrigation water. Phosphatic fertilizers are usually not easily soluble; therefore, a solution of these can be good. During irrigation operation, this solubilized fertilizer can be added in irrigation water at the head of field water course. In cotton-based cropping systems, this technique is a good option for completion of timely wheat sowing in standing cotton. Yield analyses of this system validated a high land equivalent ratio of 1.28 in 6:2 rows system, and 1.39 in other systems of 3:1, 3:2, and 4:2 row system of wheat and cotton intercropping (Zhang et al. 2008). However, the extent of yield improvement of this intercropping system by different resource use such as light use efficiency, nutrient, and water uptake is unknown.

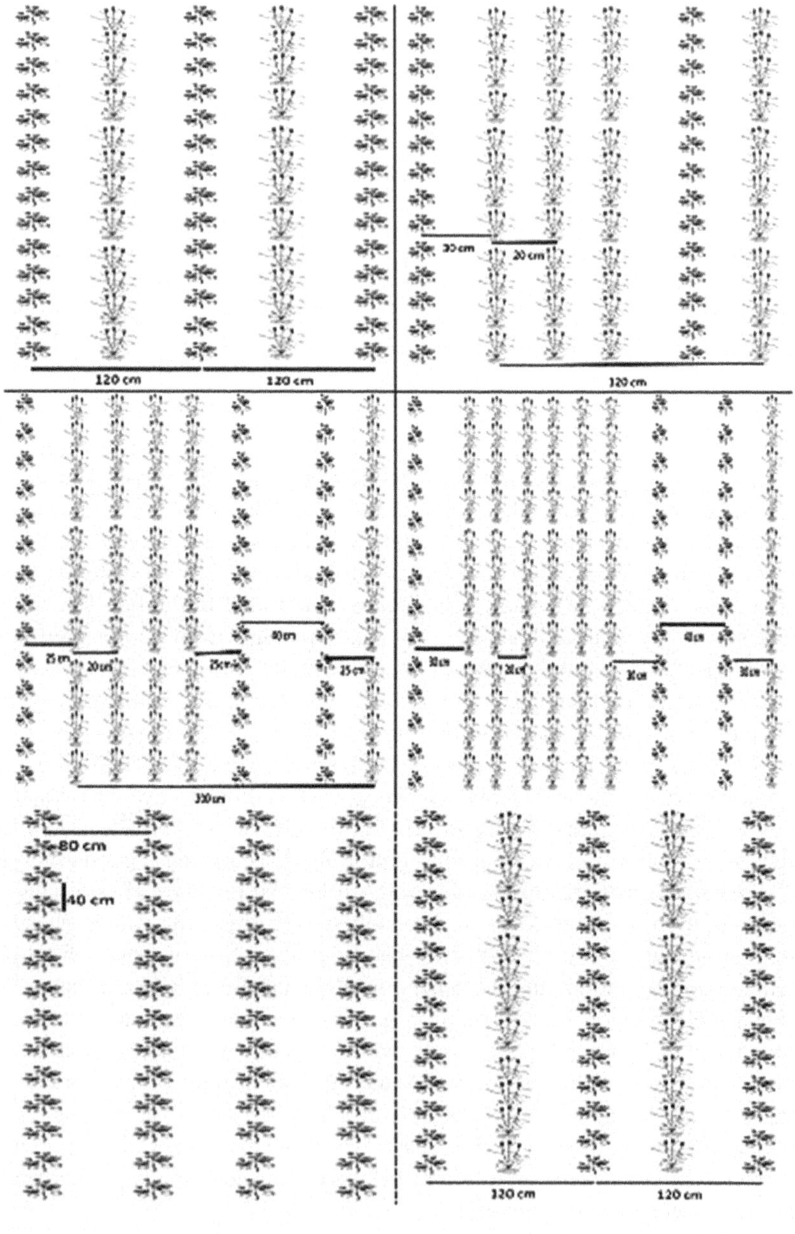

Fig. 17.1 Different cotton-based intercropping systems with variable planting geometry

Fig. 17.2 Cotton-rice intercropping in Bangladesh

17.3.2 Cotton-Rice Intercropping

Intercropping cotton in rice is promising under particular environments of erratic monsoon seasons. Although limited reports are available, however, rice cotton intercropping improved the performance of this system. Considering rice water requirements, rice is grown in furrows while cotton on ridges. In this ridge furrow system, the width of ridges is maintained at 15 cm high and 40 cm at base to provide optimum space for cotton. Rice transplanting is done in non-puddled but soft moist soil in furrows without standing water. The distance between rice seedlings is kept at 30 cm to keep them exposed for light. Planting geometry is maintained as one row of cotton after two rows of rice. Overall performance of the rice cotton system was improved by attaining 1.5, 2.5 and 3.9 t ha^{-1} yield of cotton, rice grain and rice straw, respectively (Kaddar and Ray 1985). A picture of cotton-rice intercropping is shown in Fig. 17.2.

17.3.3 Cotton-Legume Intercropping

Productivity of cotton may be affected by different intercrops; therefore, selection of suitable intercrop for growing it with cotton is vital to attain maximum advantages from intercropping. While selecting a companion crop, the time of utilization of different available resources by both crops must be considered. Utilization of resources by intercrops will define the extent of competition between both crops. Difference in root system and canopy structure are important features in this regard. Legumes and some oilseeds offer such characteristics and are commonly used in present farming systems due to their less water requirement and resistance to different environmental stresses. Pulses and some oilseed crops have advantage over cereals due to their deep rooting system. Moreover, these have better ability to withstand various abiotic stresses with minimum competition with cotton and complete their life cycle in small time. Owing to these characteristics, legumes are well recognized in intercropping. Different legumes such as chickpea, mung bean, cowpea, and groundnut are reported most appropriate in intercropping with cotton because of their short period and habit of drought tolerance. Various legumes such as chickpea, groundnut, mung bean, mash bean, and cowpea have been reported for improved yields of cotton when intercropped with cotton (Khan et al. 2001; Khan and Khaliq 2004). Likewise different oilseed crops like linseed and sesame also have yield advantages when grown with cotton, particularly in rainfed conditions.

Growing legumes with cotton as intercrops is very fruitful in terms of use efficiency of different nutrients and soil moisture in various zones of soil profile (Sankaranarayanan et al. 2011). Moreover, intercropping of legumes in cotton is an advantageous feature for weed and pest control in addition to decreasing the leaching losses of nutrients and soil erosion reduction (Kumar and Kumar 2006). Pest management effect of legumes is particular in cotton-based system where pest infestation is a big problem. Furthermore, legumes' ability to improve nitrogen status of soil by atmospheric nitrogen fixation is also expedient. In general, cotton is exhaustive crop, which removes more quantities of soil N, P, and K when it was intercropped with different legumes like cowpea and chickpea as compared to groundnut, soybean, or sesame. Cotton-based intercropping with short-season legumes such as cowpea is found to be a more productive system (Rusinamhodzi et al. 2006). Fixation capacity of legumes for atmospheric nitrogen is improved when grown as intercrops than sole crops (Rusinamhodzi et al. 2006). In an experiment, significantly higher nitrogen (38.97 kg ha^{-1}), phosphorus (13.39 kg ha^{-1}), and potassium (34.85 kg ha^{-1}) were recorded by cotton, when grown as sole crop. Conversely, the uptake of N and P by cotton was as much as in sole crop when it was grown with green gram, black gram, cluster bean, and cowpea as intercropping. Cotton produced more dry matter in this system, and a high produce by cotton showed a greater availability of these nutrients, which consequently increase the uptake of nitrogen, phosphorus and potassium by sole cotton (Katkar et al. 2002; Yakudu et al. 2010; Jayakumar and Surendran 2017). Improved N balance by legume intercrops is not advantageous only for companion cotton; however, it may be opportune for a subsequent high-yielding cereal crop with excessive nutrient demands (Rusinamhodzi et al. 2006; Adu-Gyamfi et al. 2007).

Leguminous crops can be used as bio-mulches to reduce evaporation losses, serving as obstacle for vapor diffusion and protecting soil surface from direct incidence of radiation. These mulches contain organic carbon and other nutrients. The carbon and other nutrient concentrations vary among different legumes, for example, maximum carbon is reported in chickpea mulching, while cowpea and sun hemp contain higher available P_2O_5 and K_2O as compared to no mulch conditions. In crux, legumes intercropping are favorable for soil health and fertility (Reddy 2005; Wankhade et al. 2001; Sankaranarayanan et al. 2010, 2011). Likewise, legumes intercropping with cotton are also advantageous with respect to monetary benefits. More yield of both crops and reduced expenses for different inputs are the main reasons for these rewards. Superior net returns were obtained from systems of cotton green gram and lucerne intercropping as compared to sole cotton (Jayakumar and Surendran 2017). Similarly, greater seed cotton yield with more monetary returns was obtained when cotton was grown as paired rows and with three rows of gram (Sanandachari 1980), onions, cluster bean (Shanmugham and Basu 1989), cowpea (Sheshadri and Natarajan 1989), and mung bean (Amin et al. 2018a, b). However, the extent of returns varies according to crop species or planting geometry of intercropping. For example, cotton intercropping in paired rows is reported to be more successful and profitable as compared to when grown in alternate rows. Likewise, among different legume intercrops, beetroot, chickpea, and cluster bean have shown to be more lucrative in cotton-based intercropping system.

17.3.4 Cotton-Vegetable Intercropping

Intercropping of high-value vegetable crops in cotton is also one of the viable options. While considering the inconsistency in performance and price fluctuation of vegetable crops, intercropping of multiple vegetables with different growth habits are aimed. In cotton-vegetable intercropping, there is a wide range of this system, for example, radish + beetroot + coriander between the cotton rows, radish + cluster bean + beetroot system, cotton + coriander + vegetable cowpea + cluster bean, etc. These systems significantly may improve the yield of cotton and its quality and more net returns from limited land area. Rajpoot et al. (2018) planted okra in cotton and recorded more seed cotton yield and net returns as compared to sole cotton. Earlier, more seed yield of cotton and net returns are reported from cotton-onion intercropping system (Shanmugham and Basu 1989). In another intercropping of cotton with cantaloupe, yield of cantaloupe was significantly improved as compared to sole cantaloupe. Although seed yield of companion cotton crop was reduced than sole cotton crop, however, the net returns in intercropping systems surpassed those of monoculture crops either cantaloupe or cotton (Eure et al. 2015). Surprisingly, in a comparison of different cotton-based intercropping systems, cotton-onion system recorded maximum cottonseed yield as compared to other cropping systems (Marimuthua et al. 2014). Owing to noteworthy improvement in yields and net returns in cotton-vegetable intercropping, the system may be recommended for small farmers who completely rely on cropping for daily food and monetary benefits.

17.4 Conclusion

Cotton-based intercropping systems are important for farmers predominantly with small landholdings. These systems are feasible and involve intercrops intensively to use resources more effectively, resulting in improved productivity with a simultaneous improvement in agricultural area. Improved resource use efficiency, light interception by cotton and intercrops, pest and weed management are promising features of the systems. These characteristics lead toward augmented biological diversity and greater production constancy and restore soil fertility, consequently a better cropping system.

References

Abbas Q, Ahmad S (2018) Effect of different sowing times and cultivars on cotton fiber quality under stable cotton-wheat cropping system in southern Punjab, Pakistan. Pak J Life Soc Sci 16:77–84

Adu-Gyamfi JJ, Myaka FA, Sakala WD, Odgaard R, Vesterager JM, Hogh-Jensen H (2007) Biological nitro- gen fixation and nitrogen and phosphorus budgets in farmer-managed intercrops of maize-pigeonpea in semiarid southern and eastern Africa. Plant Soil 295:127–136

Ahmad S, Abbas Q, Abbas G, Fatima Z, Atique-ur-Rehman NS, Younis H, Khan RJ, Nasim W, Habib urRehman M, Ahmad A, Rasul G, Khan MA, Hasanuzzaman M (2017) Quantification of climate warming and crop management impacts on cotton phenology. Plan Theory 6(7):1–16

Ahmad S, Iqbal M, Muhammad T, Mehmood A, Ahmad S, Hasanuzzaman M (2018) Cotton productivity enhanced through transplanting and early sowing. Acta Sci Biol Sci 40:e34610

Ahmad S, Raza I (2014) Optimization of management practices to improve cotton fiber quality under irrigated arid environment. J Food Agric Environ 2(2):609–613

Ahmad S, Raza I, Ali H, Shahzad AN, Atiq-ur-Rehman SN (2014) Response of cotton crop to exogenous application of glycinebetaine under sufficient and scarce water conditions. Braz J Bot 37(4):407–415

Ali H, Abid SA, Ahmad S, Sarwar N, Arooj M, Mahmood A, Shahzad AN (2013a) Integrated weed management in cotton cultivated in the alternate-furrow planting system. J Food Agric Environ 11(3&4):1664–1669

Ali H, Abid SA, Ahmad S, Sarwar N, Arooj M, Mahmood A, Shahzad AN (2013b) Impact of integrated weed management on flat-sown cotton (*Gossypium hirsutum* L.). J Anim Plant Sci 23 (4):1185–1192

Ali H, Afzal MN, Ahmad F, Ahmad S, Akhtar M, Atif R (2011) Effect of sowing dates, plant spacing and nitrogen application on growth and productivity on cotton crop. Int J Sci Eng Res 2 (9):1–6

Ali H, Hameed RA, Ahmad S, Shahzad AN, Sarwar N (2014a) Efficacy of different techniques of nitrogen application on American cotton under semi-arid conditions. J Food Agric Environ 12 (1):157–160

Ali H, Hussain GS, Hussain S, Shahzad AN, Ahmad S, Javeed HMR, Sarwar N (2014b) Early sowing reduces cotton leaf curl virus occurrence and improves cotton productivity. Cercetări Agronomice în Moldova XLVII(4):71–81

Amin A, Nasim W, Mubeen M, Ahmad A, Nadeem M, Urich P, Fahad S, Ahmad S, Wajid A, Tabassum F, Hammad HM, Sultana SR, Anwar S, Baloch SK, Wahid A, Wilkerson CJ, Hoogenboom G (2018a) Simulated CSM-CROPGRO-cotton yield under projected future climate by SimCLIM for southern Punjab, Pakistan. Agric Syst 167:213–222

Amin A, Nasim W, Mubeen M, Nadeem M, Ali L, Hammad HM, Sultana SR, Jabran K, Habib urRehman M, Ahmad S, Awais M, Rasool A, Fahad S, Saud S, Shah AN, Ihsan Z, Ali S, Bajwa AA, Hakeem KR, Ameen A, Amanullah HUR, Alghabar F, Jatoi GH, Akram M, Khan A, Islam F, Ata-Ul-Karim ST, Rehmani MIA, Hussain S, Razaq M, Fathi A (2017) Optimizing the phosphorus use in cotton by using CSM-CROPGRO-cotton model for semi-arid climate of Vehari-Punjab, Pakistan. Environ Sci Pollut Res 24(6):5811–5823

Amin MR, Islam MS, Hasan MK, Sabagh AE (2018b) Improvement of production and net economic return through intercropping of upland cotton with mungbean. Azarian J Agric 5:67–75

Anil L, Park J, Phipps RH, Miller FA (1998) Temperate intercropping of cereals for forage: a review of potential for growth and utilization with particular reference to the UK. Grass Forage Sci 53:301–317

Anitha S, Geethakumari VL, Filial GR (2001) Effect of intercrops on nutrient uptake and productivity of chilli-based cropping system. J Trop Agric 39:60–61

Balasubramanian TN (1987) Performance of arboretum and hirsutum cotton under intercropping blackgram and land management practices. Ph.D. Thesis, the Tamil Nadu Agricultural University, Coimbatore, India

Baldy C, Stigter CJ (1997) Agrometeorology of multiple cropping in warm climates. INRA, Paris

Bastiaans L, Kropff MJ, Goudriaan J, van Laar HH (2000) Design of weed management systems with a reduced reliance on herbicides poses new challenges and prerequisites for modelling crop-weed interactions. Field Crops Res 67:161–179

Carandang DA (1980) Resource utilization in integrated farming system with crops as the major enterprise. In: Telangco MH (ed) Integrated crop-livestock-fish farming. Laguna, Philippines, pp 8–21

Cheema ZA, Khaliq A (2000) Use of sorghum alleopathic properties to control weeds in irrigated wheat in a semiarid region of Punjab. Agric Ecosyst Environ 79:105–112

Chowdhury MK, Rosario EL (1994) Comparison of nitrogen, phosphorous and potassium utilization efficiency in maize/mung bean intercropping. J Agric Sci 122(2):193–199

Crew TE, Peoples MB (2004) Legume versus fertilizer source of nitrogen: ecological tradeoffs and human needs. Agric Ecosyst Environ 102:279–297

Danso SKA, Zapata F, Hardarson G, Fried M (1987) Nitrogen fixation in faba beans as affected by plant population density in sole or intercropped systems with barley. Soil Biol Biochem 19:411–415

Deshpande RM, Kharche SG, Rawankar HN (1989) Studies on intercropping with legumes in relation to planting pattern of hybrid cotton. PKV Res J 13:100–104

Dunin FX, Williams J, Verberg K, Keating BA (1999) Can agricultural catchment emulate natural ecosystems in recharge control in southeastern Australia? Agric Syst 45:343–364

Eskandari H, Ghanbari A, Javanmard A (2009) Intercropping of cereals and legumes for forage production. Not Sci Biol 1:07–13

Eure PM, Culpepper AS, Merchant RM, Roberts PM, Collins GC (2015) Weed control, crop response, and profitability when intercropping cantaloupe and cotton. Weed Technol 29:217–225

Farrell JG, Altieri MA (1995) Agroforestry systems in agroecology. In: Altieri MA (ed) The science of sustainable agriculture. Intermediate Technology Publications, London, pp 219–263

Francis CA (1986) Future perspectives of multiple cropping. In: Francis CA (ed) Multiple cropping systems. Macmillan, New York, pp 351–370

Fujita K, Ofosu KG, Ogata S (1992) Biological nitrogen fixation in mixed legume-cereal cropping system. Plant Soil 144:155–175

Fukai S, Trenbath BR (1993) Processes determining intercrop productivity and yields of component crops. Field Crops Res 34:247–271

Geno L, Geno B (2001) Polyculture Production: Principle, benefits and risk of multiple cropping. A report for the Rural Industry Research and Development Corporation (RIRDC), Publication, No. 01134

Girjesh GK, Patil VC (1991) Weed management studies in groundnut and sunflower intercropping system. J Oilseeds Res 8:7–13

Gomez-Rodriguez O, Zavaleta-Mejia E, Gonzalez-Hernandez AV, Livera-Munoz M, Cardenas-Soriano E (2007) Physiological and morphological adaptations in tomato intercropped with Tagetes erecta and Amaranthus hypochondriacus. Rev Fitotec Mex 30(4):421–428

Iqbal J, Cheema ZA, An M (2007) Intercropping of field crops in cotton for the management of purple nutsedge (*Cyperus rotundus* L.). Plant Soil 300:163–171

Jaya IKD, Bell CJ, Sale PW (2001) Modification of within- microclimate in maize for intercropping in the lowland tropics. Proceedings of the Australian Agronomy Conference. Australia Society of Agronomy, Australia, pp 123–133

Jayakumar M, Surendran U (2017) Intercropping and balanced nutrient management for sustainable cotton production. J Plant Nutr 40:632–644

Kaddar VK, Ray N (1985) Rice-based cropping systems. Int Rice Res Notes 10(4)

Kamel AS, Sahar AS, Ibrahim ST (2004) New agro-techniques in intensive crop rotations under marginal conditions of Upper Egypt. New directions for a diverse planet: Proceedings of the 4th International Crop Science Congress, Brisbane, Australia, 26 Sep–1 Oct 2004

Katkar RN, Turkhede AB, Solanke VM, Wankhade ST, Patil MR (2002) Effect of integrated management of organic manures and fertilizers on soil properties and yield of cotton. J Cotton Res Dev 16:89–92

Keating BA, Carberry PS (1993) Resource capture and use in intercropping: solar radiation. Field Crops Res 34:273–301

Khan MB, Akhtar M, Khaliq A (2001) Effect of planting patterns and different intercropping systems on the productivity of cotton (*Gossypium hirsutum* L.) under irrigated conditions of Faisalabad. Int J Agric Biol 3:432–435

Khan MB, Khaliq A (2004) Study of mungbean intercropping in cotton planted with different techniques. J Res (Sci) 15:23–31

Khan MB, Khaliq A, Ahmad S (2004) Performance of mashbean intercropped in cotton planted in different planting patterns. J Res (Sci) 15(2):191–197

Khan MB, Khan M, Hussain M, Farooq M, Jabran K, Lee DJ (2012) Bio-economic assessment of different wheat-canola intercropping systems. Int J Agric Biol 14:769–774

Kumar P, Kumar BU (2006) Use of organics for crops for crop production under rainfed situation-a review. Agric Rev 27:208–215

Lanini WT, Orloff SB, Vargas RN, Orr JP (1992) Using oats as companion crop in establishing alfalfa. Calif Agric 46(4):25–27

Lemma D (2002) Tomatoes: research experiences and production prospects. Ethiopian Agricultural Research Organization (EARO), Report (43):1–47. Addis Ababa, Ethiopia

Liebman M, Dyck E (1993) Crop rotation and intercropping strategies for weed management. Ecol Appl 3(1):92–122

Marimuthua S, Surendran U, Subbian P (2014) Productivity, nutrient uptake and post-harvest soil fertility as influenced by cotton-based cropping system with integrated nutrient management practices in semi-arid tropics. Arch Agron Soil Sci 60:87–101

Midmore DJ (1993) Agronomic modification of resource use and intercrop productivity. Field Crops Res 34:357–380

Mohammad MK, El-din GMS, Hosny AA (1991) Evaluating three patterns of intercropping cotton and forage cowpeas. Ann Agric Sci Moshtohor 29:1269–1284

Morris RA, Garrity DP (1993) Resource capture and utilization in intercropping: non nitrogen nutrients. Field Crops Res 34:319–334

Mpairwe DR, Sabiiti EN, Umunna NN, Tegegne A, Osuji PO (2002) Effect of intercropping cereal crops with forage legumes and source of nutrients on cereal grain yield and fodder dry matter yields. Afr Crop Sci J 10:81–97

Narwal SS (2004) Allelopathy in crop production. Scientific Publishers, Jodhpur, India

Ndakidemi P (2006) Manipulating legume/cereal mixtures to optimize the above and below ground interactions in the traditional African cropping systems. Afr J Biotechnol 525:2526–2533

Nehra DS, Bishnoi LK, Kairon MS (1990a) Effect of nitrogen and azotobacter on yield of seed cotton (*Gossypium hirsutum* L.). J Cotton Res Dev 4:61–64

Nehra DS, Kairon MS, Bishnoi LK (1990b) Effect of rate and methods of nitrogen application on hirsutum cotton genotypes. J Cotton Res Dev 4:209–212

Parajulee MN, Montandon R, Slosser JE (1997) Relay intercropping to enhance abundance of insect predators of cotton aphid (*Aphis gossypii* glover) in Texas cotton. Int J Pest Manag 43:227–232

Rahman MH, Ahmad A, Wang X, Wajid A, Nasim W, Hussain M, Ahmad B, Ahmad I, Ali Z, Ishaque W, Awais M, Shelia V, Ahmad S, Fahad S, Alam M, Ullah H, Hoogenboom G (2018) Multi-model projections of future climate and climate change impacts uncertainty assessment for cotton production in Pakistan. Agric For Meteorol 253-254:94–113

Rajpoot SK, Rana DS, Choudhary AK (2018) *Bt*-cotton–vegetable-based intercropping systems as influenced by crop establishment method and planting geometry of *Bt*-cotton in indo-Gangetic plains region. Curr Sci 115:516–522

Ramakrishna A, Ong CK (1994) Productivity and light interception in upland rice /legume intercropping systems. Trop Agric 71(1):5–11

Rana KS, Pal M (1999) Effect of intercropping systems and weed control on crop weed competition and grain yield of pigeonpea. Crop Res 17:179–182

Rao VP (1991) A study on intercropping of cotton with grain legumes under rainfed conditions. The J Res APAU 19(20):73–74

Reddy RRP (2005) Planting techniques for intercropping studies in cotton. Ph.D. thesis submitted to Acharya N.G. Ranga Agricultural University

Rusinamhodzi L, Murwira HK, Nyamangara J (2006) Cotton–cowpea intercropping and its N2 fixation capacity improves yield of a subsequent maize crop under Zimbabwean rain-fed conditions. Plant Soil 287:327–336

Saeed M, Shahid MRM, Jabbar A, Ullah E, Bismillah M (1999) Agro-economic assessment of different cotton-based inter/relay cropping systems in two geometrical patterns. Int J Agric Biol 4(1):234–237

Sanandachari A (1980) Inter cropping of rice fallow cotton with pulses. Cotton Dev 10(1&2):5–6

Sankaranarayanan K, Nalayini P, Praharaj CS (2012) Multi-tier cropping system to enhance resource utilization, profitability and sustainability of *Bt* cotton *(Gossypium hirsutum)* production system. Indian J Agric Sci 82:1044–1050

Sankaranarayanan K, Nalayini P, Sabesh M, Rajendran K, Nachane RP, Gopalakrishnan N (2011) Multi-tier cropping system for profitability and stability in *Bt. cotton* production. Central Institute for Cotton Research, Technical Bulletin No. 2, pp. 1–19

Sankaranarayanan K, Prahara CS, Nalayini P, Bandyopadhyay KK, Gopalakrishnan N (2010) Legume as companion crop for cotton. J Cotton Res Dev 24:115–126

Saudy HS, El-Metwally IM (2009) Weed management under different patterns of sunflower-soybean intercropping. J Cent Eur Agric 10(1):41–52

Schader C, Zaller JG, Kopke U (2005) Cotton-basil intercropping, effects on pests, yields and economical parameters in an organic field in Fayoum, Egypt. Biol Agric Hortic 23:59–72

Schoofs A, Entz MH (2000) Influence of annual forages on weed dynamics in a cropping system. Can J Plant Sci 80:187–198

Shah MA, Farooq M, Hussain M (2016) Productivity and profitability of cotton–wheat system as influenced by relay intercropping of insect resistant transgenic cotton in bed planted wheat. Eur J Agron 75:33–41

Shanmugham K, Basu AK (1989) Intercropping in cotton increases production and profit. Indian Farming 41(3):21–22 & 30

Sheshadri V, Natarajan K (1989) Effect of different intercrops on the growth and yield of cotton (*Gossypium hirsutum* L.) and their economics. Indian J Agric Sci 59(4):227–230

Snaydon RW, Harris PM (1981) Interactions below ground- the use of nutrients and water. In: Willey R (ed) Proceedings of the international workshop on intercropping. ICRISAT, Patanancheru, India, pp 188–201

Surendran U, Rama Subramoniam S, Raja P, Kumar V, Murugappan V (2016) Budgeting of major nutrients and the mitigation options for nutrient mining in semi-arid tropical agro-ecosystem of Tamil Nadu, India using NUT-MON model. Environ Monit Assess 188(4):250

Tariq M, Afzal MN, Muhammad D, Ahmad S, Shahzad AN, Kiran A, Wakeel A (2018) Relationship of tissue potassium content with yield and fiber quality components of Bt cotton as influenced by potassium application methods. Field Crops Res 229:37–43

Tariq M, Yasmeen A, Ahmad S, Hussain N, Afzal MN, Hasanuzzaman M (2017) Shedding of fruiting structures in cotton: factors, compensation and prevention. Trop Subtrop Agroecosyst 20(2):251–262

Tsubo M, Mukhala E, Ogindo HO, Walker S (2003) Productivity of maize-bean intercropping in a semi- arid region of South Africa. Water SA 29(4):381–388

Usman M, Ahmad A, Ahmad S, Irshad M, Khaliq T, Wajid A, Hussain K, Nasim W, Chattha TM, Trethowan R, Hoogenboom G (2009) Development and application of crop water stress index for scheduling irrigation in cotton (*Gossypium hirsutum* L.) under semiarid environment. J Food Agric Environ 7(3&4):386–391

Wankhade ST, Turkhede AB, Katkar RN, Solanke VM, Sakhare BA (2001) Effect of bio-mulching and fertilizer on growth and yield of cotton. Crop Res 21:34–37

Waterworth JV (1994) Intercropping cotton and groundnut in low and high rainfall areas in eastern Zambia. Exp Agric 30:461–465

Willey RW (1990) Resource use in intercropping systems. Agric Water Manag 17:215–231

Xia J, Cui J, Ma L, Dong S, Cui X (1998) The role of transgenic BT cotton in integrated insect pest management. Acta Gossypii Sinica 11:57–64

Yakudu H, Kwari JD, Sandade MK (2010) Effect of phosphorus on nitrogen fixation by some grain legume varieties in Sudan Asahelian zone of north. Nig J Basic Appl Sci 189:19–26

Zhang L, van der Werf W, Bastiaans L, Zhang S, Li B, Spiertz JHJ (2008) Light interception and utilization in relay intercrops of wheat and cotton. Field Crops Res 107:29–42

Zhang L, van der Werf W, Zhang S, Li B, Spiertz JHJ (2007) Growth, yield and quality of wheat and cotton in relay strip intercropping systems. Field Crops Res 103:178–188

Chapter 18
Abiotic Stresses Mediated Changes in Morphophysiology of Cotton Plant

Sibgha Noreen, Shakeel Ahmad, Zartash Fatima, Iqra Zakir, Pakeeza Iqbal, Kamrun Nahar, and Mirza Hasanuzzaman

Abstract Cotton plant is a warm-weather-loving perennial shrub and now has been domesticated to an annual crop cycle for commercial purposes. It belongs to genus *Gossypium* (*G. hirsutum* L., *G. barbadense* L., *G. herbaceum* L., *G. arboreum* L.), widely grown in arid, semiarid, and tropical climates. Globally, of these, *G. hirsutum* L. (the upland cottons) occupies about 95% of total 33–35 million hectares (2.5% of arable land) of land under cotton cultivation. Cotton crop is not only a natural fiber resource but also a food and feed for billions of humans and livestock. The projected increase in population is 9.0 billion by 2030, which would require an additional quantum of fiber and cotton seed production by more than 70% over the current level of productivity. Cotton plant having an indeterminate growth habit is highly vulnerable to occurrence of persistent and/or intermittent changes in the environments. The footprints of abiotic stresses are more visible on growth and development than those of biotic stresses. In the days to come, under the aegis of climate change, the sustainability of cotton productivity from productive and marginal lands rests by maintaining balance between vegetative and reproductive development from seedlings through maturity. The prevalence of imbalance state (either short or long

S. Noreen (✉)
Institute of Pure and Applied Biology, Bahauddin Zakariya University, Multan, Pakistan
e-mail: sibgha.noreen@bzu.edu.pk

S. Ahmad (✉)
Department of Agronomy, Faculty of Agricultural Sciences and Technology, Bahauddin Zakariya University, Multan, Pakistan
e-mail: shakeelahmad@bzu.edu.pk

Z. Fatima · I. Zakir
Department of Agronomy, Bahauddin Zakariya University, Multan, Pakistan

P. Iqbal
Department of Botany, University of Agriculture, Faisalabad, Pakistan

K. Nahar
Department of Agricultural Botany, Sher-e-Bangla Agricultural University, Dhaka, Bangladesh

M. Hasanuzzaman
Department of Agronomy, Faculty of Agriculture, Sher-e-Bangla Agricultural University, Dhaka, Bangladesh

S. Ahmad, M. Hasanuzzaman (eds.), *Cotton Production and Uses*,
https://doi.org/10.1007/978-981-15-1472-2_18

duration) could lead to loss in farm income. The potential yield could be harvested by transitioning cotton plant from "green cotton" to "white cotton." This is an effort to manipulate the plant for transporting its greater photo-assimilates from source to sink organs. Farm manager is ought to be proactive and skillful in adopting certain management tools, monitoring crop development, selection of tolerant/resistant cultivars, nutrient management, and phytosanitary measures to reinforce cotton plant for abreasting the external vagaries.

Keywords Morphophysiological attributes · Abiotic stresses · Cotton · Plant mapping · Source-sink relationship

Abbreviations

AP	Ascorbate peroxidase
CAT	Catalase
CO_2	Carbon dioxide
DD	Degree-days
DPA	Days post-anthesis
EDU	Ethylene diurea
ET	Evapotranspiration
GHG	Greenhouse gas
GR	Glutathione reductase
HSP	Heat shock proteins
LEA	Late embryogenesis abundant
O_3	Ozone
Pn	Net photosynthesis
PPFD	Photosynthetic photon flux density
ROS	Reactive oxygen species
SOD	Superoxide dismutase

18.1 Introduction

Cotton plant is a perennial shrub or tree, which has been domesticated to an annual growth habit and highly prone to climatic conditions. It belongs to genus *Gossypium* (*G. hirsutum* L., *G. barbadense* L., *G. herbaceum* L., *G. arboreum* L.) and is comprised of 50 species, of which 45 are diploid and the remaining 5 being tetraploid. The major cultivated cotton types are "the Upland cottons" (*G. hirsutum* L.) and "Pima/Sea Island/Egyptian cotton/extra long staple" (*G. barbadense* L.). Cotton plant is a unique one and follows sigmoidal curve for its growth and development. Having an indeterminate growth habit has a periodic and well-defined predictable growth patterns. It is mainly grown in subtropical and tropical climates between 37 °N latitude and 32 °S latitude, covering an area of 33–37 million hectares across 100 countries.

Cotton is a rapidly renewable resource, a leading natural fiber and second largest oilseed crop in agricultural production (Ahmad et al. 2014, 2017, 2018; Abbas and Ahmad 2018; Ahmad and Raza 2014; Ali et al. 2011, 2013a, b, 2014a, b). It produces spin able fibers for commercial use in textile industry. It is also a major source of edible oil for humans and meal as a vegetable protein for livestock production (www.icac.org) (Giband et al. 2010). The best cotton produces now achieve more than two bales of cotton per mega liter (ML) of water. It is a vital agricultural commodity and multibillion US Dollar industries that underpin both developed and developing economies (Amin et al. 2017, 2018; Khan et al. 2004; Rahman et al. 2018; Tariq et al. 2017, 2018; Usman et al. 2009). More than 400 million people derive their livelihood by engaging in farming, processing, textile, and garment industries.

Currently, the primary challenge is to enhance tolerance to drought and salt stresses to maintain its productivity on marginal lands (Rasapula et al. 2011). Thereby, there is a greater need to understand the basics of certain physiological processes in response to changing climatic conditions (Boyles et al. 2005). The productivity of cotton will have to be increased by 70% over the current level of production, with simultaneous reduction in greenhouse gas (GHG) emissions without any loss in biodiversity and habitats under the expected vagaries of climate changes (Burney et al. 2010).

To make cotton crop to be sustainable, commercial, and cost-effective, it is necessary to know regarding plant growth along with its response to ecological stresses. A comprehensive understanding of physiological processes along with their response toward stress is necessary to design strategies for managing stresses for maximizing productivity and profitability.

18.2 Growth and Development

Cotton being a perennial one exhibits as an indeterminate growth habit. This means that fruiting bodies are developed over a longer period of times, and compensates its loss of fruiting bodies following the footprints of certain stresses, e.g., drought, salinity, pollutants, extreme temperature, insect-pests and disease attack, and other physiological disorders.

Cotton seedlings emerge 5–10 days after planting, and cotyledons are arranged directly opposite on main stem. Leaves on main stem along with branches are spirally arranged on the stem in a three-eighths polyllotoxy about the cotyledonary node (Fig. 18.1). The development of reproductive organs starts about 4–5 weeks after planting. The bolls develop rapidly after fertilization and reach at their full size within 3 weeks. During the growth period, plant produces two types of branches, monopodial (vegetative) and sympodial (reproductive) ones. The first sympodial branch generally arises from sixth or seventh node on main stem. The monopodial

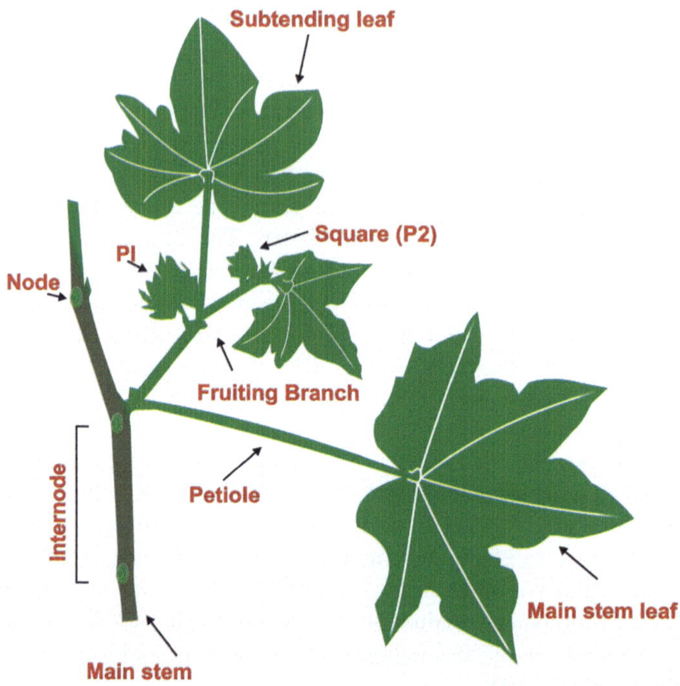

Fig. 18.1 Developing fruiting branch and other structures. (Source: Adapted after William and Bange 2018)

branch grows in zigzag manner along with arise manifold fruiting positions. With progressive growth, new sub-standing leaves are produced and supply carbohydrates to the developing fruits. As the plant enters into the reproductive phase, the vegetative growth is slowed down and the greater proportion of photo assimilates are diverted toward developing bolls. This is the pivotal stage, where balance between vegetative and reproductive growth has to be maintained progressive as long as weather favors for turning into cut out stage.

18.3 Source-Sink Relationship

The environmental stresses impact largely on the production of metabolites and its redistribution between source and sink organs. The development of imbalance between source and sink quantum and occurrence of osmotic adjustment results in reduction of photosynthetic process, leaf and root growth, and abscission of fruiting bodies (Kirkham et al. 1972; Terry et al. 1971; Baker and Boker 2010). The productivity of assimilates is dependent upon genetic make-up of species, architectural characteristics of cultivars degree of lobing, and eco-adophic factors (Baker and

Boker 2010). The rate of net photosynthesis (Pn) is maximal during leaf expansion (16–20 days) period and starts declining toward ontogeny (Constable and Rawson 1980; Reddy et al. 1991; Wullschleger and Oosterhuis 1990).

Photosynthetic rate and production of fruiting bodies is maximal at optimum temperature of 30/20 °C and declines at 40/30 °C (Reddy et al. 1990), while seedling growth is stopped below 16 °C (Warner and Burke 1993). The photosynthetic rate is declined drastically at water potential (ψ_i) of −1.2 MPa drought stress condition (Marani et al. 1985) and results in irreversible process (Bielorai and Hopmans 1975; Hsiao et al. 1982). The decrease in leaf expansion, vegetative growth, and leaf abscission leads to lowering in amount of capturing light (Sinclair and Ludlow 1985). Among the nutrients, nitrogen deficiency syndrome results in reduction in stem extension, leaf area index, and photosynthetic efficiency (Fernandez et al. 1993) and increases stomatal closure due to production of higher quantity of cytokinins and abscisic acid (Radin et al. 1987).

18.3.1 Sources of Assimilates

The synthesis of assimilates is affected by atmospheric CO_2, O_3, and agrochemicals, and exogenous application applied mineral nutrients and osmoprotectants (Baker and Boker 2010). These enrichment of CO_2 causes increase in photosynthesis and biological yield to the proportion of 37% and 40%, respectively (Reddy et al. 1997), whereas, its rate is reduced by 5.9% due to elevated O_3 concentration from 0.03 to 0.107 $\mu L\ L^{-1}$ (Reddy et al. 1989). However, application of plant growth regulators, viz., mepiquat chloride, gibberellic acid, and indolebutyric acid, causes increase in proliferation of roots and photosynthetic rate (Oosterhuis and Zhao 1993).

18.3.2 Sinks of Assimilates

Cotton plant accumulates starch in phloem and chloroplasts through photosynthetic process in the presence of higher quantum of CO_2 and radiance (photosynthetic photon flux density, PPFD) during day time and is translocated to growing points at night (Baker and Boker 2010). The supply of assimilates is shared equally between shoots and roots (Kimball and Mauney 1993); however, roots maintain sinks strength by more than six times than that of fruiting bodies (Fye et al. 1984). The growth rate is reduced at −1.2 MPa (leaf water potential (ψ_i)) during midday, which affects in balancing share between resource and sink (Marani et al. 1985). The growth and development of reproductive organs is enhanced at 27 °C and declines at 33 °C temperature (Baker et al. 1983), while greater proportion bolls are shed at 35/25 °C and eventually stopped at 40/30 °C temperature. The dray matter yield capsule of bolls is increased linearly with quantum of sink up to 20 DPA (days post-anthesis) and maximal at 30 DPA (Stewart 1986).

The occurrence of various physiological events is interacted by genetic makeup of genotypes, interception of light, photosynthetic rate, eco-edaphic factors, and farm management practices (Stewart 1986). The plastochron at main stem is 2–3 days $leaf^{-1}$, and 6–7 days $leaf^{-1}$ at sympodial branches occurs at 27 °C temperature (Reddy et al. 1993). The source-sink relationship is imbalanced on the occurrence of reproductive organs (Baker et al. 1983). The shortfall of assimilates delays in initiation of nodal positions on main stem as well as on sympodia (Reddy et al. 1993). The quantum of production and redistribution pattern of assimilates leverage the loss in bolls and productivity (Constable et al. 1991).

18.4 Mapping of Cotton Plant

Plant diagrams are utilized to map bolls positions along with stages of development of all flowering on the day of sampling and to evaluate success of productivity input resources (Oosterhuis 1990). Opening of successive flowers on sympodia appears at interval of 6–7 days (horizontal interval), and first follower on successive branches opens at 3 days interval (vertical interval), besides second, third, and later flowers are likewise separated. This order is thus spirally outward and upward. The number of fruiting positions along sympodia varies greatly due to genetic constitution of the genotype, nutritional management, controlling insect-pest attack, and eco-edaphic and farm management factors. The retention of bolls on the first, second, and third sympodial positions contributes around 60%, 30%, and 10% toward total cotton productivity (Pervez et al. 2005a, b). Among these contributing factors, the meeting of nutritional needs cause substantial retention of harvestable bolls on the sympodia (Pervez et al. 2005a, b). Furthermore, the lint quality gathered from bolls on sympodia also tends to reduce away from main stem (Jenkins et al. 1990) (Figs. 18.2, 18.3, and 18.4).

Growth and maturation of cotton are temperature dependent, thereby occurrence of various growth stages can be predicted by calculating degree-days (DD) at threshold temperature of 15.6 °C (60 °F) and no upper threshold (Fry 1983). DD means the accumulation of heat units associated with everyday maximum besides minimum temperature during each day (El-Zik et al. 1980). The accumulated heat units for a particular growth stage help in maintaining the events during growing season (Landivar and Benedict 1996) (Table 18.1).

Under favorable growing environments, the plenty of irrigation water and excessive nitrogenous fertilization would result in rank growth, which causes heavy economic loss. The excessive vegetative growth may be checked by foliar spray of mepiquat chloride, for diverting its photo-assimilates from vegetative to reproductive organs. Thereby, developing bolls become great sinks for carbohydrates, H_2O, besides nutrients, and leaving a little food for vegetative development.

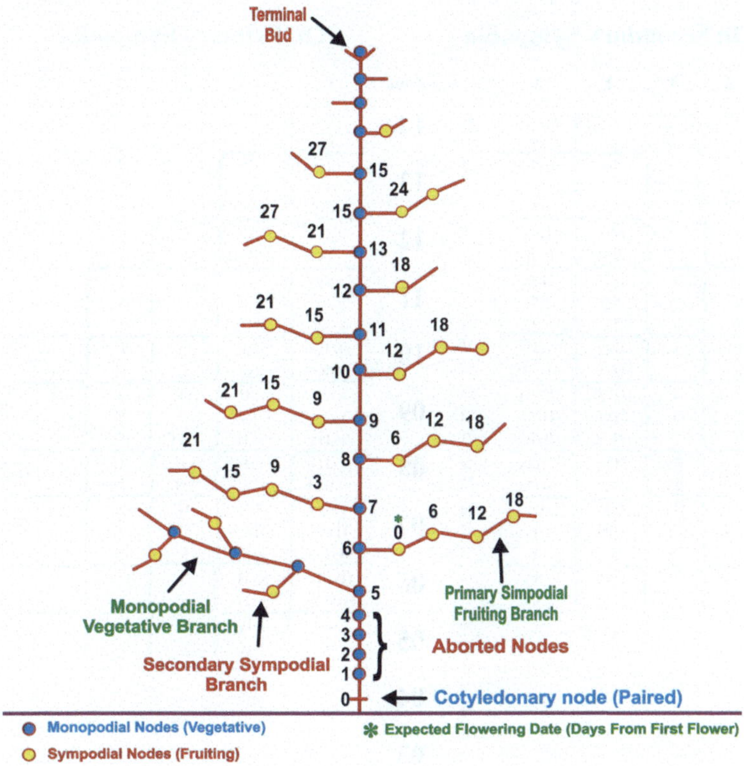

Fig. 18.2 Schematic cotton flowering pattern. (Redrawn from Oosterhuis 1990 and Pervez et al. 2005a, b)

18.5 Fiber Development

Cotton fibers begin development as single cell that begin to form on unfertilized seeds called ovules (William and Bange 2018). On the physiological basis, quality of fiber is determined by interaction of genetic potential of genotype and environmental fluctuations experienced during growth period by cotton plant (Bradow and Davidonis 2010) (Table 18.2).

18.6 Abiotic Stresses

18.6.1 Extreme Temperature Stress

The temperature is the predominant among cardinal ecological aspects which influences crop growth and productivity since the beginning of germination/emergence

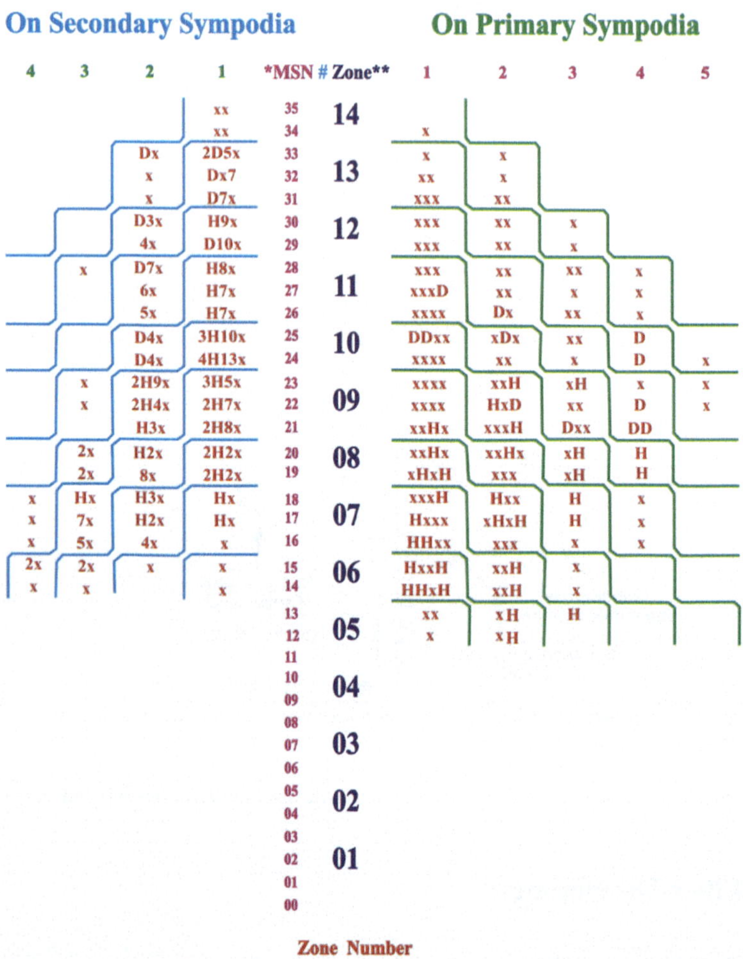

On Secondary Sympodia											On Primary Sympodia										
Σ	14	13	12	11	10	9	8	7	6	Sym	5	6	7	8	9	10	11	12	13	14	Σ
214	7	24	37	32	46	46	16	4	2	x	3	5	12	14	18	17	19	15	17	11	131
10	1	5	2	2						D						3	3	4	2		12
33			1	3	12	9	6	2		H		7	6	6	4	5	3				31
257	8	29	40	37	58	55	22	6	2	Σ	3	12	18	22	22	25	25	19	19	11	176
												2	6	22	55	58	37	40	29	8	257
										Σ	3	14	24	44	77	83	62	55	48	19	433

* Main stem node number

** Bold numerals denote zone number

Fig. 18.3 Fruit production efficiency without fertilizer application. (Modified and adapted after Pervez et al. 2005a, b)

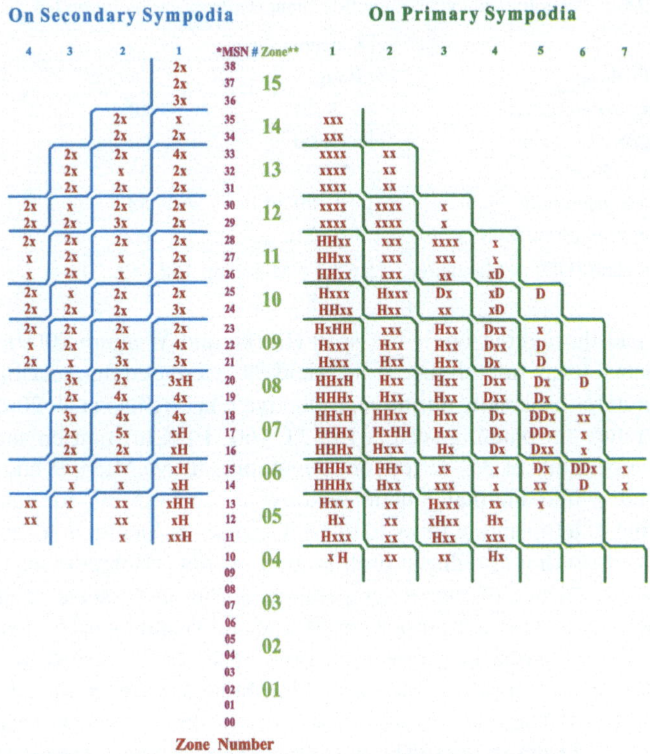

On Secondary Sympodia												On Primary Sympodia													
Σ	15	14	13	12	11	10	9	8	7	6	5	Sym	4	5	6	7	8	9	10	11	12	13	14	15	Σ
172	21	15	22	16	15	17	20	25	10	7	4	x	1	10	10	15	19	19	21	26	33	33	17		131
												D							4	10	6	5	2		12
8								1	2	2	3	H	1	3	6	16	13	10	8	11					31
180	21	15	22	16	15	17	20	26	12	9	7	Σ	2	13	16	31	32	29	33	47	39	38	19		176
														7	9	12	26	20	17	15	16	22	15	21	180
												Σ	2	20	25	43	58	49	50	62	55	60	34	21	479

* Main stem node number

** Bold numerals denote zone number

Fig. 18.4 Fruit production efficiency at 125 kg K ha^{-1}. (Modified and adapted after Pervez et al. 2005a, b)

Table 18.1 Phenological stages

Phenological event	Temp (°C)	Days after planting	References
First square	27	27	Reddy et al. (1993)
Square to bloom	27	20	Hesketh and Low (1968)
		24	Reddy et al. (1993)
		26	Hesketh et al. (1972)
Bloom to open boll	27	48	Reddy et al. (1992)
		61	Hesketh and Low (1968)

Table 18.2 Phenological events of cotton plant (*Gossypium hirsutum* L.)

Growth stages	Calendar days		Degree days[a] (Base 15.6 °C)
	Range	Average	
Sowing to emergence	5–20	10	50
Emergence to square	40–60	50	450
Square to bloom	20–27	23	330
Bloom to open boll	40–80	58	950
Normal crop production	190–210	200	>2800

[a]Degree days (DD) = Max temp −15.6 °C + Min temp 15.6 °C/2

stage and throughout one's life span (Burke and Wanjura 2009). Although cotton originates from hot climates, its yield is decreased appreciably due to higher temperatures especially during reproductive phase (Zhao et al. 2005). The minimum temperature for planting seed is 15.5 °C (60 °F) (Christiansen and Rowland 1986) and temperature of 35 °C for root development (McMichael and Burke 1994) for irrigated, while thermal kinetic window is 23.5–25 °C for rainfed cotton. The lowering of temperature from 30 to 18 °C causes reduction in hydraulic conductivity of roots, resulting in reduced proliferation of roots (Bolger et al. 1992).

The prevalence of higher temperature during early stage of growth affects the productivity to a greater proportion (Burke and Wanjura 2010). During reproductive phase, the increased in temperature from 18 to 28 °C resorts to increased fruiting branches from 5 to 16, while no fruiting branches are produced beyond 36 °C in Pima cottons. The phenomenon of sterility in flower occurs at temperature greater than 38 °C (Taha et al. 1981) and progression of fruiting structures (Reddy et al. 1995). The higher temperature causes increase in oxidative stress, lowers photosynthesis, and depletes ATP and carbohydrates (Oosterhuis and Snider 2010). The efficiency of metabolic activity in Upland cotton is the highest at its optimal thermal window of 23–32 °C (Snider et al. 2009). However, metabolic activity and membrane functions are diminished at 20–23 °C and below 15 °C, respectively.

18.6.2 Mineral Nutrients Stress

The success of cotton cultivation depends upon an adequate availability of macro- and micronutrients and eco-edaphic factors during the season (Mullins and Burmester 2010). Cotton having an indeterminate growth habit, contrarily to determinate ones, requires greater quantity of nutrients during its reproductive phase compared to vegetative phase (Pervez et al. 2005a, b). However, availability of nutrients during each growth period is pertinent to avoid any deficiency syndrome (Hodges and Constable 2010; Rochester 2012) (Table 18.3). The proportionate amount of translocation and/or relative redistribution of nutrients from vegetative to reproductive organs causes causal effects on cotton productivity. Nitrogen nutrient is more mobile than those of K, S, Ca, Mg, Fe, Mn, and B for their translocation

Table 18.3 Physiological basis of fiber development

Fiber development	Description
Initiation	It occurs just pre-flowering and at flowering. It is initiation of fiber cells on seed coat which can take up to 3 days. Thereafter, second set of fiber cells are initiated and develop into the fuzz
Elongation	This is rapid expansion besides growth of fiber cell's primary wall. Finally, fiber length is affected by length of besides rate of fiber elongation
Secondary wall thickening or fiber thickening	The secondary wall is formed where cellulose is laid down in layers inside fiber cell's primary wall. Deposition is affected due to fluctuations and formation of fiber growth rings. Due to fluctuations in photosynthesis on an everyday basis and formation of fiber growth rings. Cellulose layers are composed of two layers. The thicker is formed during day besides porous layer is formed at night
Maturation	Fiber cells dry out and fiber becomes a twisted ribbon-like structure. Mature fiber is easily detached from fuzzy seed

from leaves, stems, and capsule wall (Rosolem and Mikkelsen 1991). A greater quantum of N, P, and Zn is accumulated in the bolls and thereafter removed to seedcotton produce (Rochester 2007). The in-season nutritional status could be assessed by leaf tissue and petiole-nitrate analysis (Constable et al. 1991). The chlorophyll meters (Makhdum et al. 2002) and leaf/canopy reflectance sensors are also becoming more commonplace (Constable et al. 1991). The deficiency syndrome could be corrected by foliar and/or side dressing of nutrients (Makhdum et al. 2002).

Cotton plant follows sigmoidal curve for growth, dry matter production, and nutrient uptake after greatly during course of development (Oosterhuis 1990). The accumulation of nutrient is maximal at peak flowering for utilization of assimilates between vegetative and reproductive organs (Schwab et al. 2000). It accumulates 29%, 22%, and 21% of nitrogen, phosphorus, and potassium nutrient, at full flowering stage under irrigated condition (Halevy et al. 1987) (Tables 18.4, 18.5, 18.6, 18.7, and 18.8).

Cultivars vary in their nutrient uptake due to difference in demand between upper ground parts and root system (Kerby and Adams 1985), and also amount of externally application of fertilizers; e.g., N uptake is increased from 110 to 322 kg N ha^{-1} by adding nitrogenous fertilizer from 0 to 180 kg N ha^{-1} (Pervez et al. 2005a, b).

The appearance of deficiency syndrome is an outcome of inhibition of chlorophyll formation and/or occurrence of oxidative stress due to limited utilization of photo assimilates (Hodges and Constable 2010). During the discourse of growth, concentration of N, P, K, Fe, Cu, and Zn drops, while Ca, Mg, Na, Mn, S, and B increased in leaf tissues with advancement in age (Boquet and Breitenback 2000). The fruiting bodies especially bolls are the major sinks of nutrients, and their accumulation vary appreciably due to eco-edaphic factors, genetic makeup of

Table 18.4 Uptake of nutrients at maturity

Nutrients	Maximum uptake (kg ha^{-1}), (g ha^{-1})	Maximum uptake rate (kg ha^{-1} day^{-1})	Time of maximum uptake (days from sowing)	Percentage taken up during flowering
Nitrogen	332	2.1	102	55
Phosphorus	49	0.7	110	75
Potassium	312	3.2	115	61
Sulfur	71	0.8	101	63
Calcium	289	2.6	112	55
Magnesium	72	0.7	108	61
Iron	2592	24.0	130	46
Manganese	829	6.5	123	49
Boron	652	6.5	118	60
Copper	77	0.9	119	61
Zinc	272	3.7	109	73

Source: Hodges (1992)

cultivars, temperature and drought stresses, amount of nutrients in the rhizosphere, and other agronomic practices under different ecologies (Constable et al. 1988).

The removal of nutrients by harvestable portion determines the amount of nutrients required for gathering the targeted yield. Thereby replenishment of nutrients in consonance with nutrient(s) removal is pertinent to maintain soil fertility (Rochester 2007).

18.6.2.1 Boron Stress

Boron is most vital micronutrient for cotton (Rosolem and Costa 1999). The deficiency syndrome appears on younger growing parts due to limited translocation in plant system (Rosolem and Bogiani 2011). There is a very narrow range between sufficiency and toxicity levels of boron. Its toxicity causes negative effects on photosynthesis, chlorophyll constituents, cell division, and lignin development (Reid 2007). The reproductive phase is highly prone to boron deficiency (Zhao and Oosterhuis 2002). The requirement is about 340 g B ha^{-1}, of which 12% is retained in seed cotton and remaining is stored in other plant parts (Zhao and Oosterhuis 2003).

18.6.3 Drought Stress

Among the abiotic stresses, drought is the most limiting aspect for growth besides development of cotton. Cotton is being grown extensively is arid and semiarid regions, where irrigation supplies are limited most of times. Cotton is "xerophyte,"

Table 18.5 Pattern of nutrients uptake at full bloom stage

Nutrient	Total uptake (kg ha^{-1}), *(g ha^{-1})	Uptake index (kg/100 kg lint), *(g/100 kg seed cotton)	Peak uptake rate (kg ha^{-1} day^{-1}), *(g/ha^{-1} day^{-1})	% of total uptake at full bloom stage	Reference
Nitrogen	51–301	8–51	2.54–3.87	23–39	Mullins and Burmester (1991)
Phosphorus	8.2–72.3	1.3–3.3	0.31–0.48	21.36	Mullins and Burmester (1991)
Potassium	53–393	12.1–27.0	2.2–3.5	35	Mullins and Burmester (1991)
Magnesium	35–104	3.0–6.4	0.3–0.8	30–52	Mullins and Burmester (1992)
Calcium	60–70	6–77	1.5–3.1	46–49	Mullins and Burmester (1992)
Sulfur	15.6–25.1	1.0–6.8	0.34–0.49	30	Mullins and Burmester (1993)
Zinc	25–38	5–7	1.9–4.1	25–45	Mullins and Burmester (1993)
Manganese	451 ± 175	30.0	8.2–14.4	35–47	Mullins and Burmester (1993)
Copper	28 ± 14	4.0	0.34–1.33	29–58	Mullins and Burmester (1993)
Iron	600–814	242	23–27	41–60	Mullins and Burmester (1993)
Boron	66–17	9.3	–	–	Alimov and Ibragimov (1976)

(continued)

Table 18.5 (continued)

Nutrient	Total uptake (kg ha^{-1}), *(g ha^{-1})	Uptake index (kg/100 kg lint), *(g/100 kg seed cotton)	Peak uptake rate (kg ha^{-1} day^{-1}), *(g/ha^{-1} day^{-1})	% of total uptake at full bloom stage	Reference
Molydenum	1.97–4.03	–	–	–	Alimov and Ibragimov (1976)
Cobolt	2.44–4.35	–	–	–	Alimov and Ibragimov (1976)
Sodium	4.3–17.1	–	–	–	Bassett et al. (1970)

Table 18.6 Uptake of nutrients at various growth stages

Uptake (kg ha^{-1})				
Stage of growth	Plant organ	Nitrogen	Phosphorus	Potassium
First flower bud	Leaves	7.3	0.47	6.0
	Stems	1.6	0.14	8.9
	Total	8.9	0.61	14.9
First flower	Leaves	21.4	1.23	19.7
	Stems	9.1	0.64	19.7
	Capsule	5.7	0.66	4.9
	Total	36.2	2.53	44.3
Peak flowering	Leaves	28.0	1.32	28.3
	Stems	11.2	0.97	26.4
	Capsule	3.7	0.77	7.3
	Seed	13.5	1.47	10.1
	Lint	1.1	0.27	4.3
	Total	57.5	4.80	76.4
First boll split	Leaves	29.8	1.29	35.1
	Stems	12.6	1.15	31.4
	Capsule	3.2	1.09	15.0
	Seed	28.5	3.97	24.3
	Lint	1.2	0.71	9.4
	Total	75.3	8.21	115.2
Maturity	Leaves	10.6	0.54	11.2
	Stems	8.9	0.75	16.9
	Capsule	3.2	0.72	18.3
	Seed	46.5	7.61	22.8
	Lint	1.0	0.40	7.6
	Total	70.2	10.02	76.8

Table 18.7 Pattern of accumulation and timings of nutrients by cotton boll

Nutrient	Maximum uptake per boll (mg/boll), (μg/boll)	Maximum uptake (per day)	Time of maximum uptake (days from anthesis)
Nitrogen	111	36	19
Phosphorus	21.4	0.71	19
Potassium	10.3	3.2	19
Sulfur	17.5	0.37	26
Calcium	31.0	0.82	27
Magnesium	17.2	0.45	21
Iron	221	5.6	24
Manganese	111	2.5	22
Boron	118	3.8	18
Copper	30	0.91	19
Zinc	104	3.0	18

Adapted after Constable et al. (1988)

Table 18.8 Proportional nutrients uptake towards differential yield level

	Nutrients uptake (kg ha^{-1}), [a](g ha^{-1})			% Exported		
Lint yield (kg ha^{-1})	1000	1800	2400	1000	1800	2400
Nitrogen	63	175	290	66	52	46
Phosphorus	13	27	41	82	69	60
Potassium	77	167	250	21	17	15
Sulfur	10	39	62	42	21	18
Calcium	71	94	155	3	3	2
Magnesium	16	36	63	45	34	25
Iron[a]	227	820	1620	40	17	11
Manganese[a]	152	355	655	5	3	2
Boron[a]	75	320	560	22	13	11
Copper[a]	25	52	81	51	38	31
Zinc[a]	58	119	203	99	73	61

Adapted after Rochester (2007)
[a]Nutrients uptake (g ha^{-1})

a plant which requires less water, and tolerant to heat besides drought. Cotton plants avoid adverse weather possibly due to deep well-distributed root system along with indeterminate growth pattern. Evapotranspiration (ETc)-based requirement of water is to be 2.0 mm day^{-1} (20,000 L ha^{-1}) through vegetative stage besides 6 and 8 mm day^{-1} during flowering and early-bulling period (termed as critical window). It requires around 80–85% of total water during this “critical window”; however, moisture stress during this stage caused severe yield losses. Contrarily, excessive moisture coupled with higher amount of nitrogenous fertilizers during vegetative and boll opening stages results in reduced growth and lowering of yields.

Plants undergo a series of integrated events, varying from signaling stress and transduction to the gene expression as an effort to acclimatize under the stressful conditions. The drought stress is evidenced at whole plant, cellular and molecular levels (Chaves et al. 2009). Water stress caused reduction in photosynthesis and growth, because of stomatal closure along with lowered activity of photosynthetic enzymes (Chaves et al. 2009) and efficiency of chloroplast to fix carbon dioxide (Bota et al. 2004). At the cellular level, the oxidative stress is occurred due to generation of reactive oxygen species (ROS) with concurrent working in stress signal transduction pathway (Foyer and Noctor 2009). The appreciable changes occur in protein synthesis and biological functions at molecular level in response to drought stress. The gas exchange parameter, viz., net photosynthesis, stomatal conductance, and transpiration rate, while fluorescence parameters, i.e., effective quantum yield of PSII (Φ psII), and electron transport rates are declined in response to drought. However, quantum of hydrogen peroxide, malondialdehyde (MDA), and anthocyanin levels are enhanced under drought stress (Deeba et al. 2012).

The plant having intrinsic self-defense system accumulates heat shock proteins (HSP) besides late embryogenesis abundant (LEA) proteins and also accumulates compatible solutes and potassium to maintain water potential gradient (Loka et al. 2011; Oosterhuis and Wullschleger 1987). Under the stressful conditions, the production of abscisic acid and ethylene causes abscission of bolls and other fruiting bodies (Dumka et al. 2004), thereby, results in retention of lower amount of fruits, boll weight, and loss in yield (Saini 1997; Ritchie et al. 2009).

The water use efficiency is reduced because of reduction in photosynthesis and transpiration rate (Loka et al. 2011). It varies greatly due to fruiting habits of varieties, load of fruiting bodies, and stage of growth (Parida et al. 2007). Water under water stress ROS, like peroxide radicals, hydrogen peroxide besides hydroxyl radicals are generated (Faria et al. 1997) and cause oxidative stress. Under this condition, antioxidant defense comes into play its role as scavenger. The major role is played by antioxidant species, viz., superoxide dismutase (SOD), catalase (CAT), ascorbate peroxidase (AP), glutathione reductase (GR), and carotenoids along with α-tocopherol come in to action to work as scavenger (Gaspar et al. 2002).

Deleterious effects of draught stress can be ameliorated by foliar spray of some asmoprotectants, viz., glycine betaine, salicylic acid, RGR-IV (containing gibberellic acid and indolebutyric acid), and 1-methylcyclopropene. These chemicals enhance production and accumulation of osmolytes in the plant system to protect enzymatic system, lipid peroxidation, and photosynthetic apparatus (Allakhverdiev et al. 2003; Gorham et al. 2000; Waseem et al. 2006; Zhao and Oosterhuis 1997; Loka and Oosterhuis 2011).

On an average, irrigation water usage is 1214 L to produce 1.0 kg lint plus 2.0 kg seed. Globally, 87% of total production is harvested by using 644 L irrigation water kg^{-1} lint. The water productivity can be enhanced by rainwater harvesting, irrigation with precision timing based on ET, irrigation through alternate furrows, sprinklers, or subsurface drip irrigation. The soil and moisture conservation methods like minimum tillage, mulching, cover crops or intercrop, and efficient pest besides weed control could result in improving water use efficiency and water productivity.

18.6.4 Salinity Stress

Salinity in topsoil and subsoil is one of key abiotic ecological stresses to cotton production. Globally, one-third of total agricultural land is salt affected, which lies in arid and semiarid environments. Although, cotton is categorized as one of the most tolerant crops (Maas 1990), however, its growth and development and economic yield are affected to a greater proportion (Higbie et al. 2010; Khorsandi and Anagholi 2009). The salt stress reflects differential response in cotton, due to quality of irrigation water, amount of rainfall, and proportionate amount of salts in soil. Among the cotton species, viz., varieties of *Gossypium barbadense* L. has greater tolerance capacity compared to *G. hirsutum* L., besides *G. arboretum* L. (Abul-Naas and Omran 1974).

At onset of growth and development, germination besides seedling stages is highly prone to salt stress (Oliveira et al. 1998; Malik and Makhdum 1987; Gorham et al. 2010). The presence of salinity at >282 mol m^{-3} (NaCl) causes damaging effects on root growth (Silberbush and Ben-Asher 1987). The shoot growth is inhibited due to reduction in soil water potential and vapor pressure deficit (Shalhevet and Hsiao 1986; Gorham et al. 2010). The toxicity of salts can be ameliorated through adequate nutrition (Brugnoli and Björkman 1992). During the reproductive development, the photo-assimilates from source to sink are restricted, causing burden on reduced number of fruiting bodies, heavy fruit drop, and lowering in retention of bolls, thereby leading to economic loss in yield and fiber quality (Moreno et al. 2000; Ahmad 1994), and reducing the values of fiber length, fiber strength, fiber fineness, and amount of oil content in seed (Muhammed and Makhdum 1973; Ahmad 1994).

The salt stress causes stomatal closure and increased resistance to CO_2 diffusion rapid senescence (Gorham et al. 2009), chlorophyll constituents ‘a’ and ‘b’ (Ahmad and Abdullah 1980; Jafri and Ahmad 1995), excessive buildup of Na^+ along with Cl^- in leaf tissues leading to osmotic stress (Zhang et al. 2012) and reduced movement of osmolytes from source to sink (Jafri and Ahmad 1995).

The salt-tolerant varieties maintain lower K^+/Na^+ ratio than salt-sensitive ones (Läuchli and Stelter 1982; Nawaz et al. 1986). The higher K^+/Na^+ ratios occur due to restricted movement of K^+ and Cl^- in the phloem (Abdullah and Ahmad 1986). The assimilation of Ca^{2+} and Na^+ ions from root cells causes greater efflux of K^+ (Cramer et al. 1985) and maintenance of selectivity of K^+ over Na^+ in plasma membrane (Gorham et al. 2010). The plant accumulates greater proportion of protein (Brugnoli and Björkman 1992), while assimilation of N, P, and K is decreased (Subbarao et al. 1995) enhancing leaf phosphorylase activity in leaf (Rathert 1983). Cotton plants develop salt tolerance and water stress by greater production besides accumulation of K^+, sucrose, glucose, amino acids, proline, and glycine betaine (Lin et al. 1995; Gorham 1996). The presence of antioxidant defense system (catalase, ascorbate peroxidase, superoxide dismutase, glutathione reductase) improves salt tolerance in cotton (Banks et al. 2000; Li et al. 1998).

The tolerance capacity can be enhanced by soaking cotton seed with $CaSO_4$ and foliar spray of kinetin solution, MC BU TTB, and polystimuline K (10–20 ppm) (Gorham et al. 2000; Stark and Schmidt 1991). The exogenous application of gibberellic acid (GA3) mitigates the burden of salinity by enhancing growth, greater uptake of K^+ with simultaneous reduction in Na^+ ion (Ibrahim 1984; Gossett et al. 2000). Cotton may be successfully cultivated by adopting certain agronomic measures, e.g., cultivation on furrows with plastic mulching (Dong et al. 2010), maintaining plant density at 4–5 plants m^{-2} (Zhang et al. 2012), alternate irrigation with saline and non-saline water (Moreno et al. 1998), furrow irrigation and/or drip irrigation (Ghani et al. 2007), sprinkling system (Meiri et al. 1992), application of nitrate nitrogen rather than ammonium nitrogen (Leidi et al. 1991), addition of soil amendments (sand, gypsum (calcium sulphate)), growing sesbania alone, or intercropping (Tiwari et al. 1993; Tiwari 1994). Apart from these, cultivation of salt-tolerant cultivars would be more valuable and cost-effective (Malik and Makhdum 1987; Iqbal et al. 1991; Gorham et al. 2010).

18.6.5 Air Pollution Stress

The footprints of air pollution comprised of primary (N_2, O_2, CO_2, methane, and anthropogenic compounds) and secondary (O_3, peroxyacetyl nitrate, H_2O_2, and oxygenated compounds) pollutants affect growth and development (Temple and Grants 2010). Among these, ozone (O_3) causes deterioration in cell membrane and partial and/or complete loss of turgor pressure at 0.25 ppm on higher (Runeckles and Chevone 1992; Heagle et al. 1986). The net photosynthesis (Pn) in appreciably reduced at >0.20 ppm (Grantz and Farrar 2000).

In response to O_3 exposure, the attributes of photosynthetic efficiency, abscission of leaves, and yield are reduced due to decreased efficiency in CO_2 assimilation (Miller et al. 1988). However, cotton plant has an in-built compensatory mechanism to tolerate the adverse effects of O_3 (Temple 1990). The root organ is highly prone to O_3 stress compared to shoot organ, because of reduced root hydraulic capacity (Grantz and Yang 1996). Cotton varieties vary greatly in their relative tolerance to O_3 and other pollutants (Runeckles and Chevone 1992), and yield is reduced from 15% to 20% (Heck et al. 1988).

Cotton is quite responsive to increased concentration of CO_2 from 550 to 650 ppm and causes enhancement I growth by 65% and yield by 50% (Kimball and Mauney 1993; Mauney 2010), net photo synthesis from 65% to 70% (Inoue et al. 1990); total biomass by 37%, increase in LAI of 4 (Mauney et al. 1994), C/N in ratio leaves and stems (Hendrix 1992), boll loading period and yield (Mauney et al. 1994), and water use efficiency by 19–28% (Dugas et al. 1994). However, under changing climatic condition, the cotton yield may decrease due to interactive effect of increased temperature and carbon dioxide (Reddy et al. 1996). The adverse effect of O_3 could be scavenged by exogenous application of some antioxidants, viz., citrate, ascorbate, and ethylene di-urea (EDU) ([*N*-2-(2-oxo-1-imidozolidimyl) ethyl]-*N*′-phenyl urea) (Manning and Krupa 1992), overhead sprinkler (Grantz et al. 1997) and breeding of varieties resistant to O_3 by employing conventional and molecular engineering technology (Grantz and McCool 1992).

18.7 Future Perspective

A number of advances have been made in revealing basics of physiology of cotton in consonance with rapid development of highly productive varieties and be resilient to external environment. In the present times, primary challenge is to enhance the tolerance level to drought and salinity for maintaining the productivity on the marginal lands. Presently much progress has been made in the development of biotech cotton varieties, which has accounted for more than 50% of world cotton production. The development of drought tolerant and/or water-efficient varieties would be required in the wake of declining freshwater supplies for irrigation purpose.

References

Abbas Q, Ahmad S (2018) Effect of different sowing times and cultivars on cotton fiber quality under stable cotton-wheat cropping system in southern Punjab, Pakistan. Pak J Life Soc Sci 16:77–84

Abdullah Z, Ahmad R (1986) Salinity induced changes in the reproductive physiology of cotton plants. In: Ahmad R, San Pietro A (eds) Prospects for biosaline research. Pakistan, Department of Botany, University of Karachi, pp 125–137

Abul-Naas AA, Omran MS (1974) Salt tolerance of seventeen cotton cultivars during germination and early seedling development. Z Ack Pflanzenbau 140:229–236

Ahmad FM (1994) Effect of saline water irrigation at different stages of growth on cotton plant. Assiut J Agric Sci 25:63–74

Ahmad R, Abdullah Z (1980) Biomass production of food and fiber crops using highly saline water under desert conditions. In: San Peitro A (ed) Biosaline research. Plenum Press, New York, pp 149–163

Ahmad S, Raza I (2014) Optimization of management practices to improve cotton fiber quality under irrigated arid environment. J Food Agric Environ 2(2):609–613

Ahmad S, Raza I, Ali H, Shahzad AN, Atiq-ur-Rehman, Sarwar N (2014) Response of cotton crop to exogenous application of glycinebetaine under sufficient and scarce water conditions. Braz J Bot 37(4):407–415

Ahmad S, Abbas Q, Abbas G, Fatima Z, Atique-ur-Rehman, Naz S, Younis H, Khan RJ, Nasim W, Habib ur Rehman M, Ahmad A, Rasul G, Khan MA, Hasanuzzaman M (2017) Quantification of climate warming and crop management impacts on cotton phenology. Plants 6(7):1–16

Ahmad S, Iqbal M, Muhammad T, Mehmood A, Ahmad S, Hasanuzzaman M (2018) Cotton productivity enhanced through transplanting and early sowing. Acta Sci Biol Sci 40:e34610

Ali H, Afzal MN, Ahmad F, Ahmad S, Akhtar M, Atif R (2011) Effect of sowing dates, plant spacing and nitrogen application on growth and productivity on cotton crop. Int J Sci Eng Res 2 (9):1–6

Ali H, Abid SA, Ahmad S, Sarwar N, Arooj M, Mahmood A, Shahzad AN (2013a) Integrated weed management in cotton cultivated in the alternate-furrow planting system. J Food Agric Environ 11(3&4):1664–1669

Ali H, Abid SA, Ahmad S, Sarwar N, Arooj M, Mahmood A, Shahzad AN (2013b) Impact of integrated weed management on flat-sown cotton (*Gossypium hirsutum* L.). J Anim Plant Sci 23 (4):1185–1192

Ali H, Hameed RA, Ahmad S, Shahzad AN, Sarwar N (2014a) Efficacy of different techniques of nitrogen application on American cotton under semi-arid conditions. J Food Agric Environ 12 (1):157–160

Ali H, Hussain GS, Hussain S, Shahzad AN, Ahmad S, Javeed HMR, Sarwar N (2014b) Early sowing reduces cotton leaf curl virus occurrence and improves cotton productivity. Cer Agron Moldova XLVII(4):71–81

Alimov KH, Ibragimov S (1976) Trace elements in different cotton cultivars. Field Crop Abstr 29 (9):7404

Allakhverdiev SI, Hayashi H, Nishiyama Y, Ivanov AG, Aliev JA, Klimov VV, Murata N, Carpentier R (2003) Glycine betaine protects the D1/D2/Cytb559 complex of PS II against the photo-induced and heat-induced inactivation. J Plant Physiol 160:41–49

Amin A, Nasim W, Mubeen M, Nadeem M, Ali L, Hammad HM, Sultana SR, Jabran K, Habib urRehman M, Ahmad S, Awais M, Rasool A, Fahad S, Saud S, Shah AN, Ihsan Z, Ali S, Bajwa AA, Hakeem KR, Ameen A, Amanullah, Rehman HU, Alghabar F, Jatoi GH, Akram M, Khan A, Islam F, Ata-Ul-Karim ST, Rehmani MIA, Hussain S, Razaq M, Fathi A (2017) Optimizing the phosphorus use in cotton by using CSM-CROPGRO-cotton model for semi-arid climate of Vehari-Punjab, Pakistan. Environ Sci Pollut Res 24(6):5811–5823

Amin A, Nasim W, Mubeen M, Ahmad A, Nadeem M, Urich P, Fahad S, Ahmad S, Wajid A, Tabassum F, Hammad HM, Sultana SR, Anwar S, Baloch SK, Wahid A, Wilkerson CJ, Hoogenboom G (2018) Simulated CSM-CROPGRO-cotton yield under projected future climate by SimCLIM for southern Punjab, Pakistan. Agric Syst 167:213–222

Baker DN, Boker JT (2010) Cotton source/sink relationships. In: Stewart JM, Oosterhuis D, Heitholt JJ, Manuney J (eds) Physiology of cotton. The National Cotton Council, Springer Science + Business Media, New York, pp 80–96

Baker DN, Lambert JR, McKinion JM (1983) GOSSYM: a simulator of cotton crop growth and yield. South Carolina Agric Exp Stn Bull 1089:134p

Banks SW, Gossett DR, Manchandia A, Bellaire B, Lucas MC, Millhollon EP, Dugger P, Richter D (2000) The influence of alpha-amanitin on the induction of antioxidant enzymes during salt stress. In: 1998 Proc. Beltwide Cotton Conf., San Diego, California, USA, 5–9 January 1998, pp 1393–1395

Bassett DM, Anderson WD, Werkhoven CHE (1970) Dry matter production and nutrient uptake in irrigated cotton (*Gossypium hirsutum*). Agron J 62:299–303

Bielorai H, Hopmans PAM (1975) Recovery of leaf water potential, transpiration, and photosynthesis of cotton during irrigation cycles. Agron J 67:629–632

Bolger TP, Upchurch DP, McMichael BL (1992) Temperature effects on cotton root hydraulic conductance. Environ Exp Bot 32:49–54

Boquet DJ, Breitenback GA (2000) Nitrogen rate effects on partitioning of nitrogen and dry matters by cotton. Crop Sci 40:1685–1693

Bota J, Medrano H, Flexas J (2004) Is photosynthesis limited by decreased Rubisco activity and RuBP content under progressive water stress? New Phytol 162:671–681

Boyles MB, Verhalin LM, Johnson WM, Barnes BR (2005) Trends over time among cotton cultivation released by the Oklahoma Agricultural Experimental Station. Crop Sci 45:966–980

Bradow JM, Davidonis GH (2010) Effect of environment on fiber quality. In: Stewart JM, Oosterhuis D, Heitholt JJ, Manuney J (eds) Physiology of cotton. The National Cotton Council, Springer Science + Business Media, New York, pp 229–245

Brugnoli E, Björkman O (1992) Growth of cotton under continuous salinity stress - influence on allocation pattern, stomatal and nonstomatal components of photosynthesis and dissipation of excess light energy. Planta 187:335–347

Burke JJ, Wanjura DF (2009) Plant responses to temperature extremes. In: Stewart JM, Oosterhuis DM, Heitholt JJ, Mauney JR (eds) Physiology of cotton. Springer, New York, pp 123–128

Burke JJ, Wanjura DF (2010) Plant responses to temperature extremes. In: Stewart JM, Oosterhuis D, Heitholt JJ, Manuney J (eds) Physiology of cotton. The Cotton Foundation, Springer Science + Business Media, New York, pp 123–128

Burney JA, Davis SJ, Lobell DB (2010) Greenhouse gas mitigation by agricultural intensification. Proc Natl Acad Sci U S A 197:12052–12057

Chaves MM, Flexas J, Pinheiro C (2009) Photosynthesis under drought and salt stress: regulation mechanisms from whole plant to cell. Ann Bot 103(2009):551–560

Christiansen MN, Rowland RA (1986) Germination and stand establishment. In: Mauney JR, Stewart JMD (eds) Cotton physiology. The Cotton Foundation, Memphis, TN, pp 535–541

Constable GA, Rawson HM (1980) Photosynthesis, respiration and transpiration of cotton fruit. Photosynthetica 14:557–563

Constable GA, Rochester IJ, Cook JB (1988) Zinc, copper, iron, manganese, and boron uptake by cotton on cracking clay soils of high pH. Aust J Exp Agric 28:351–356

Constable GA, Rochester IJ, Betts JH, Herridge DF (1991) Prediction of nitrogen fertilizer requirement in cotton using petiole and sap nitrate. Commun Soil Sci Plant Anal 22:1315–1324

Cramer GR, Läuchli A, Polito VS (1985) Displacement of Ca^{2+} by Na^{+} from the plasmalemma of root cells. A primary response to salt stress. Plant Physiol 79:207

Deeba F, Pandey AK, Ranjan S, Mishra A, Singh R, Sharma YK, Shirke PA, Pandey V (2012) Physiological and potassium responses of cotton (*Gossypium hirsutum* L.) to drought stress. Plant Physiol Biochem 53:6–18

Dong HZ, Kong XQ, Luo Z, Li WJ, Xin CS (2010) Unequal salt distribution in the root zone increases growth and yield of cotton. Eur J Agron 33:285–292

Dugas WA, Heuer ML, Hunsaker D, Kimball BA, Lewin KF, Nagy J, Johnson M (1994) Sap flow measurements of transpiration from cotton grown under ambient and enriched CO_2 concentrations. Agric For Meterol 70:231–246

Dumka D, Bednarz CW, Maw BW (2004) Delayed initiation of fruiting as a mechanism of improved drought avoidance in cotton. Crop Sci 44:528–544

El-Zik KM, Yamada H, Walhood VT (1980) The effects of management on blooming boll-retention, and productivity of Upland cotton (*Gossypium hirsutum* L.). In: Proc. Beltwide Cotton Prod. Res. Conf. National Cotton Council, Memphis, TN

Faria T, Vaz M, Schwanz P, Polle A, Pereira JS, Chaves MM (1997) Responses of photosynthetic and defense systems to high temperature stress in Quercussuber seedlings grown under elevated CO_2. Plant Biol 1:365–371

Fernandez CJ, Cothren JT, McInnes KJ (1993) Whole-plant photosynthetic rates of cotton under nitrogen stress. In: 1993 Proc. Beltwide Cotton Confs. National Cotton Council of America, Memphis, TN, pp 1256–1258

Foyer CH, Noctor G (2009) Redox regulation in photosynthetic organisms: signaling, acclimation, and practical implications. Antioxid Redox Signal 11:861–905

Fry KE (1983) Heat unit calculation in cotton crop and inset models. USDA, ARS, Adv Agric Tech AAT-W-23, p 23

Fye RE, Reddy VR, Baker DN (1984) The validation of GOSSYM: Part I. Arizona conditions. Agric Syst 14:85–105

Gaspar T, Franck T, Bisbis B, Kevers C, Jouve L, Hausman JF, Dommes J (2002) Concepts in plant stress physiology. Application to plant tissue cultures. Plant Growth Regul 37:263–285

Ghani Akbar G, Hussain Z, Yasin M (2007) Problems and potentials of permanent raised bed cropping systems in Pakistan. Pak J Water Res 11(1):11

Giband M, Dessauuw D, Barroso PAV (2010) Cotton taxonomy, origin and domestication. In: Wakelyn PJ, Chaudhry MR (eds) Cotton: technology for 21st century. The International Cotton Advisory Committee, Washington, DC, pp 15–17
Gorham J (1996) Mechanisms of salt tolerance of halophytes. In: Choukr-Allah R, Malcolm CV, Hamdy A (eds) Halophytes and biosaline agriculture. Marcel Dekker, New York, pp 31–53
Gorham J, Jokinen J, Malik NA, Khan IA (2000) Glycine betaine treatment improves cotton yields in field trials in Pakistan. In: Proc. World Cotton Res. Conf. II. Athens, Greece, pp 624–627
Gorham J, Lauchli A, Leidi EO (2009) Plant responses to salinity. In: Stewart JM, Oosterhuis DM, Heitholt JJ, Mauney JR (eds) Physiology of cotton. National Cotton Council of America, Memphis, TN. Springer, London, pp 130–142
Gorham J, Läuchli A, Leidi EO (2010) Plant responses to salinity. In: Stewart JM, Oosterhuis DM, Heitholt JJ, Mauney JR (eds) Physiology of cotton. Springer, New York, pp 129–141
Gossett DR, Bellaire B, Banks SW, Lucas MC, Manchandia A, Millhollon EP, Dugger P, Richter D (2000) The influence of abscisic acid on the induction of antioxidant enzymes during salt stress. In: Proc. Beltwide Cotton Conf., San Diego, California, USA, 5–9 January 1998, pp 1396–1399
Grantz DA, Farrar JR (2000) Ozone inhibits phloem loading from a transport pool: compartmental efflux analysis in Pima cotton. Aust J Plant Physiol 27:859–868
Grantz DA, McCool PH (1992) Effect of ozone on Pima and Acala cottons in the San Joaquin Valley. In: Proc. 1992 Beltwide Cotton Conferences, vol 3. National Cotton Council of America, Memphis, TN, pp 1082–1084
Grantz DA, Yang S (1996) Effect of O_3 on hydraulic architecture in Pima cotton. Plant Physiol 112:1649–1657
Grantz DA, Zhang XJ, Massmann WJ, Delany A, Pederson JR (1997) Ozone deposition to a cotton (*Gossypium hirsutum* L.) field: stomatal and surface wetness effects during the California Ozone Deposition Experiment. Agric For Meteorol 85:19–31
Halevy J, Marani A, Markovitz T (1987) Growth and N P K uptake of high-yielding cotton grown at different nitrogen levels in a permanent-plot experiment. Plant Soil 103:39–44
Heagle AS, Heck WW, Lesser VM, Rawlings JO, Mowry FL (1986) Injury and yield response of cotton to chronic doses of ozone and sulfur dioxide. J Environ Qual 15:375–382
Heck WW, Taylor OC, Tingey DT (eds) (1988) Assessment of crop loss from air pollutants. Elsevier Appl. Sci, London
Hendrix DL (1992) Influence of elevated CO_2 on leaf starch of field-grown cotton. Crit Rev Plant Sci 11:223–226
Hesketh JD, Low A (1968) The effect of temperature on components of yield and fiber quality of cotton varieties of diverse origin. Cotton Grow Rev 45:243–257
Hesketh JD, Baker DN, Duncan WG (1972) Simulation of growth and yield in cotton: II. Environmental control of morphogenesis. Crop Sci 12:436–439
Higbie SM, Wang F, Stewart JM, Sterling TM, Lindemann WC, Hughs E, Zhang J (2010) Physiological response to salt (NaCl) stress in selected cultivated tetraploid cottons. Int J Agron 1:1–12
Hodges SC (1992) Nutrient deficiency disorders. In: Hillocks R (ed) Cotton diseases. CAB International, Wallingford, UK, pp 355–403
Hodges SC, Constable G (2010) Plant responses to mineral deficiencies and toxicities. In: Stewart JM, Oosterhuis D, Heitholt JJ, Mauney J (eds) Physiology of cotton. The National Cotton Council, Springer Science + Business Media, New York, pp 142–173
Hsiao TC, Oliveira EC, Radulovich R (1982) Physiology and productivity of cotton under water stress. In: Proc. 1982 Cotton Prod. Res. Confs. Las Vegas, NV. National Cotton Council of America, Memphis, TN, p 60
Ibrahim AA (1984) Effect of GA3 and boron on growth, yield and accumulation of Na, K, and Cl in cotton grown under saline conditions. Ann Agric Sci Moshtohor 21:519–531
Inoue Y, Kimball BA, Mauney JR, Jackson RD, Pinter PJ Jr, Reginato RJ (1990) Stomatal behavior and relationship between photosynthesis and transpiration in field-grown cotton as affected by CO_2 enrichment. Jap J Crop Sci 59:510–517

Iqbal RMS, Chaudhry MB, Aslam M, Bandesha AA (1991) Economic and agricultural impact of mutation breeding in cotton in Pakistan. A review. In: Kitto PH (ed) Plant mutation for crop improvement. IAEA, Vienna, pp 187–201

Jafri AZ, Ahmad R (1995) Effect of soil salinity on leaf development, stomatal size, and its distribution in cotton (*Gossypium hirsutum* L.). Pak J Bot 27:297–303

Jenkins JN, McCarty LC, Parrot WL Jr (1990) Effectiveness of fruiting sites in cotton: yield. Crop Sci 30:365–369

Kerby TA, Adams F (1985) Potassium nutrition of cotton. In: Munson RD (ed) Potassium in agriculture. ASA, CSSA, and SSSA, Madison, WI, pp 843–860

Khan MB, Khaliq A, Ahmad S (2004) Performance of mashbean intercropped in cotton planted in different planting patterns. J Res (Sci) 15(2):191–197

Khorsandi F, Anagholi A (2009) Reproductive compensation of cotton after salt stress relief at different growth stages. J Agron Crop Sci 195:278–283

Kimball BA, Mauney JR (1993) Response of cotton to varying CO_2, irrigation, and nitrogen: yield and growth. Agron J 85:706–712

Kirkham MB, Gardner WR, Gerloff GC (1972) Regulation of cell division and cell enlargement by turgor pressure. Plant Physiol 49:961–962

Landivar JA, Benedict JH (1996) Monitoring system for the management of cotton growth and fruiting. Bulletin B-2. Texas A&M University Agricultural Research and Extension Center, Corpus Christi, TX

Läuchli A, Stelter W (1982) Salt tolerance of cotton genotypes in relation to K/Na-selectivity. In: San Pietro A (ed) Biosaline research. Plenum Press, New York, pp 511–514

Leidi EO, Nogales R, Lips SH (1991) Effect of salinity on cotton plants grown under nitrate or ammonium nutrition at different calcium levels. Field Crops Res 26:35–44

Li WJ, Dong HZ, Guo QZ, Pang JQ, Zhang J (1998) Physiological response of a good upland hybrid and its parent to PEG and NaCl stresses. China Cottons 25:7–10

Lin J, Zhu Z, Fan B, Lin JD, Zhu ZY, Fan BX (1995) Physiological reaction of cotton varieties under different levels of salt stress. China Cottons 22:16–17

Loka DA, Oosterhuis DM (2011) Effect of 1-MCP on the cotton flower under water-deficit stress. In: Oosterhuis DM (ed) Summaries of Arkansas Cotton Research, Ark Agric Exp Station. University of Arkansas System, Fayetteville

Loka DA, Oosterhuis DM, Ritchie GL (2011) Water deficit stress in cotton. In: Oosterhuis DM (ed) Stress physiology in cotton. Cotton Foundation, Memphis, TN, pp 33–72

Maas EV (1990) Crop salt tolerance. In: Tanji KJ (ed) Agricultural salinity assessment and management. American Society of Civil Engineers, New York, pp 262–304

Makhdum MI, Shababuddin, Ahmad F, Chaudhry FI (2002) Using a chlorophyll meter to improve nitrogen management in cotton (*Gossypium hirsutum* L.). Pak J Soil Sci 21:60–67

Malik MN, Makhdum MI (1987) Salinity tolerance of cotton cultivars (*G. hirsutum* L.) at germination. Pak Cottons 31:171–174

Manning WJ, Krupa SV (1992) Experimental methodology for studying the effects of ozone on crops and trees. In: Lefohn AS (ed) Surface level ozone exposures and their effects on vegetation. Lewis Publ. Inc., Chelsea, MI, pp 93–156

Marani A, Baker DN, Reddy VR, McKinion JM (1985) Effect of water stress on canopy senescence and carbon exchange rates in cotton. Crop Sci 25:798–802

Mauney JR (2010) Responses of cotton to CO_2 enrichment. In: Stewart JM, Oosterhuis DM, Heitholt JJ, Mauney JR (eds) Physiology of cotton. Springer, New York, pp 174–178

Mauney JR, Kimball BA, Pinter PJ Jr, LaMorte RL, Lewin KF, Nagy J, Hendrey GR (1994) Growth and yield in cotton in response to a free-air carbon dioxide enrichment (FACE) environment. Agric For Meterol 70:49–67

McMichael BL, Burke JJ (1994) Metabolic activity of cotton roots in response to temperature. Environ Exp Bot 34:201–206

Meiri A, Frenkel H, Mantell A (1992) Cotton response to water and salinity under sprinkler and drip irrigation. Agron J 84(1):44

Miller JE, Patterson RP, Heagle AS, Pursley WA, Heck WW (1988) Growth of cotton under chronic ozone stress at two levels of soil moisture. J Environ Qual 17:635–643
Moreno F, Fernandez-Boy E, Cabrera F, Fernandez JE, Palomo MJ, Giron IF, Bellido B (1998) Irrigation with saline water in the reclaimed marsh soils of Southwest Spain: impact on soil properties and cotton crop. In: Ragab R (ed) The use of saline and brackish water for irrigation. Implications for the management of irrigation, drainage and crops. Proc. international workshop tenth ICID Afro-Asian Regional Conf. on Irrigation and Drainage, Denpasas, Bali, Indonesia, 19–26 July, 1998. Pearce-G, Indonesian National Committee on Irrigation and Drainage (INACID), Directorate General of Water Resources Development, Ministry of Public Works, Jakarta, Indonesia, pp 51–58
Moreno F, Fernandez BE, Cabrera F, Fernandez JE, Palomo MJ, Giron IF, Bellido B (2000) Irrigation with saline water in the reclaimed marsh soils of southwest Spain: impact on soil properties and cotton crop. The use of saline and brackish water for irrigation drainage and crops. In: Proc. international workshop at the tenth ICID Afro-Asian Regional Conference on Irrigation and Drainage, Denpasas, Bali, Indonesia, 19–26 July, 1998, pp 51–58; 4 ref., Denpasas, Bali, Indonesia
Muhammed S, Makhdum MI (1973) Effect of soil salinity on the composition of oil and amino acids and on the oil content of sunflower seed. Pak J Agric Sci 10:71–76
Mullins GL, Burmester CH (1991) Dry matter, nitrogen, phosphorus, and potassium accumulation by four cotton varieties. Agron J 82:729–736
Mullins GL, Burmester CH (1992) Uptake of calcium and magnesium by cotton grown under dryland conditions. Agron J 84:564–569
Mullins GL, Burmester CH (1993) Accumulation of copper, iron, manganese and zinc by four cotton cultivars. Field Crops Res 32:129–140
Mullins GL, Burmester CH (2010) Relation of growth and development to mineral nutrition. In: Stewart JM, Oosterhuis D, Heitholt JJ, Mauney J (eds) Physiology of cotton. The National Cotton Council, Springer Science + Business Media, New York, pp 97–105
Nawaz A, Ahmad N, Qureshi RH (1986) Salt tolerance of cotton. In: Ahmad R, Pietro AS (eds) Prospects for biosaline research. Karachi University, Pakistan, pp 285–291
Oliveira FA, de TGS C, da Oliveira BC (1998) Effect of saline substrate on germination, vigor, and growth of herbaceous cotton. Engenharia Agricola 18:1–10
Oosterhuis DM (1990) Growth and development of a cotton plant. In: Miley WN, Oosterhuis DM (eds) Nitrogen nutrition of cotton: practical issues. ASA, CSSA, and SSSA, Madison, WI, pp 1–24
Oosterhuis DM, Snider JL (2010) High temperature stress on floral development and yield of cotton. In: Oosterhuis DM (ed) Stress physiology in cotton, vol 7. The Cotton Foundation, Cordova, TN, pp 1–24
Oosterhuis DM, Wullschleger SD (1987) Osmotic adjustment in cotton (*Gossypium hirsutum* L.) leaves and roots in response to water stress. Plant Physiol 84:1154–1157
Oosterhuis DM, Zhao D (1993) Physiological effects of PGR-IV on the growth and yield of cotton. In: Proc. Beltwide Cotton Prod. Res. Confs. National Cotton Council of America, Memphis, TN, p 1270
Parida AK, Dagaonkar VS, Phalak MS, Umalkar GV, Aurangabadkar LP (2007) Alterations in photosynthetic pigments, protein and osmotic components in cotton genotypes subjected to short-term drought stress followed by recovery. Plant Biotechnol Rep 1(1):37–48
Pervez H, Ashraf M, Makhdum MI (2005a) Response of cotton to potassium fertilizer on effectiveness of fruiting sites in Aridisoles. J Plant Nutr 28:1023–1039
Pervez H, Makhdum MI, Ashraf M (2005b) The interactive effects of potassium nutrition on the uptake of other nutrients in cotton (*Gossypium hirsutum* L.) under an arid environment. J Chem Soc Pak 27:1–17
Radin JW, Kimball BA, Hendrix DL, Mauney JR (1987) Photosynthesis of cotton plants exposed to elevated levels of carbon dioxide in the field. Photosynth Res 12:191–203

Rahman MH, Ahmad A, Wang X, Wajid A, Nasim W, Hussain M, Ahmad B, Ahmad I, Ali Z, Ishaque W, Awais M, Shelia V, Ahmad S, Fahad S, Alam M, Ullah H, Hoogenboom G (2018) Multi-model projections of future climate and climate change impacts uncertainty assessment for cotton production in Pakistan. Agric For Meteorol 253–254:94–113

Rasapula V, Shen G, Kuppu S, Paez-Valencia J, Mendoza M, Hou P, Chen J, Qiu X, Zhu L, Zhang X, Auld D, Blumwald E, Zhang H, Gaxiola R, Payton P (2011) Expression of an Arabidopsis vacuolar H^+ - pyrophosphatase gene (AVP 1) in cotton improves drought and salt tolerance and increases fiber yield in the field conditions. Plant Biotechnol J 9:88–99

Rathert G (1983) Effects of high salinity stress on mineral and carbohydrate metabolism of two cotton varieties. Plant Soil 73:247–256

Reddy VR, Acock B, Baker DN, Acock M (1989) Seasonal leaf area-leaf weight relationships in the cotton canopy. Agron J 81:1–4

Reddy VR, Baker DN, Hodges HF (1990) Temperature and mepiquat chloride effects on cotton canopy architecture. Agron J 82:190–195

Reddy VR, Reddy KR, Baker DN (1991) Temperature effect on growth and development of cotton during the fruiting period. Agron J 83:211–217

Reddy KR, Hodges HF, Reddy VR (1992) Temperature effects on cotton fruit retention. Agron J 84:26–30

Reddy KR, Hodges HF, McKinion JM (1993) A temperature model for cotton phenology. Biotronics 22:47–59

Reddy VR, Reddy KR, Acock B (1995) Carbon dioxide and temperature interactions on stem extension, node initiation, and fruiting in cotton. Agric Ecosyst Environ 55:17–28

Reddy KR, Hodges HF, McKinion JM (1996) Weather and cotton growth: present and future. Bulletin 1061. Mississippi Agricultural and Forestry Experiment Station, Mississippi State University, Mississippi, p 23

Reddy KR, Hodges HF, McKinion JM (1997) A comparison of scenarios for the effect of global climate change on cotton growth and yield. Aust J Plant Physiol 24:707–713

Reid R (2007) Update on boron toxicity and tolerance in plants. In: Xu F, Goldbach HE, Brown PH, Bell RW, Fujiwara T, Hunt CD, Goldberg S, Shi J (eds) Advances in plant and animal boron nutrition. Springer, Dordrecht, The Netherlands, pp 83–90

Ritchie GL, Whitaker JR, Bednarz CW, Hook JE (2009) Subsurface drip and over-head irrigation of cotton: a comparison of plant boll distribution in upland cotton. Agron J 101:1336–1344

Rochester IJ (2007) Nutrient uptake and export from an Australian cotton field. Nutr Cycl Agroecosyst 77:213–223

Rochester IJ (2012) Using seed nitrogen concentration to estimate crop N use efficiency in high-yielding irrigated cotton. Field Crop Res 127:140–145

Rosolem CA, Bogiani JC (2011) Physiology of boron stress in cotton. In: Oosterhuis DM (ed) Stress physiology in cotton, vol 7. The Cotton Foundation, Cordova, TN, pp 113–124

Rosolem CA, Costa A (1999) Boron nutrition and growth of cotton as a function of temporary boron deficiency. Anain II Congresso Brasileiro de Algodao: O algodao no seculo XX, perspectivas para a seculo XXI, Ribeirao Preto, SP, Brasil, 5–10 Setembro 1999, pp 403–406

Rosolem CA, Mikkelsen DS (1991) Potassium absorption and partitioning in cotton as affected by periods of potassium deficiency. J Plant Nutr 14:1001–1016

Runeckles VC, Chevone BI (1992) Crop responses to ozone. In: Lefohn AS (ed) Surface level ozone exposures and their effects on vegetation. Lewis Publ. Inc., Chelsea, MI, pp 189–270

Saini HS (1997) Injuries to reproductive development under water stress, and their consequences for crop productivity. J Crop Prod 1:137–144

Schwab GJ, Mullins GL, Burmester CH (2000) Growth and nutrient uptake by cotton roots under field conditions. Commun Soil Sci Plant Anal 31(1&2):149–164

Shalhevet J, Hsiao TC (1986) Salinity and drought. A comparison of their effects on osmotic adjustment, assimilation, transpiration and growth. Irrig Sci 7:249–264

Silberbush M, Ben-Asher J (1987) The effect of salinity on parameters of potassium and nitrate uptake of cotton. Commun Soil Sci Plant Anal 18:65–81

Sinclair TR, Ludlow MM (1985) Who taught plants thermodynamics? The unfulfilled potential of plant water potential. Aust J Plant Physiol 12:213–217

Snider JL, Oosterhuis DM, Skulman BW, Kawakami EM (2009) Heat stress-induced limitations to reproductive success in *Gossypium hirsutum*. Physiol Plant 137:125–138

Stark C, Schmidt R (1991) Behaviour of 22Na in salt-stressed crops as affected by the growth regulator MCBuTTB. Beitragezur Tropischen Landwirtschaft und Veterinarmedizin 29:435–443

Stewart JMD (1986) Integrated events in the flower and fruit. In: Mauney J Jr, Stewart JMD (eds) Cotton physiology. Cotton Foundation, Memphis, TN, pp 261–297

Subbarao KV, Chassot A, Gordon TR, Hubbard JC, Bonello P, Mullin R, Okamoto D, Davis RM, Koike ST (1995) Genetic relationships and cross pathogenicities of *Verticillium dahliae* isolates from cauliflower and other crops. Phytopathology 85:1105–1112

Taha MA, Malik MNA, Chaudhry FI, Makhdum I (1981) Heat induced sterility in cotton sown during early April in West Punjab. Exp Agric 17:189–194

Tariq M, Yasmeen A, Ahmad S, Hussain N, Afzal MN, Hasanuzzaman M (2017) Shedding of fruiting structures in cotton: factors, compensation and prevention. Trop Subtrop Agroecosyst 20(2):251–262

Tariq M, Afzal MN, Muhammad D, Ahmad S, Shahzad AN, Kiran A, Wakeel A (2018) Relationship of tissue potassium content with yield and fiber quality components of Bt cotton as influenced by potassium application methods. Field Crops Res 229:37–43

Temple PJ (1990) Growth form and yield responses of four cotton cultivars to ozone. Agron J 82:1045–1050

Temple PJ, Grants DA (2010) Air pollution stress. In: Stewart JM, Oosterhuis DM, Heitholt JJ, Mauney JR (eds) Physiology of cotton. Springer, New York, pp 162–178

Terry N, Waldron LJ, Ulrich A (1971) Effects of moisture stress on the multiplication and expansion of cells in leaves of sugar beet. Planta (Berl) 97:281–289

Tiwari RJ (1994) Response of gypsum on morpho-physiochemical properties of cotton cultivars under salt affected vertisols of Madhya Pradesh. Crop Res Hisar 7:197–200

Tiwari RJ, Dwivedi K, Verma SK (1993) Effect of gypsum on leaf-water potential of cottons (*Gossypium hirsutum*, *G. herbaceum*, and *G. arboreum*) varieties grown in salt-affected vertisol of Madhya Pradesh. Indian J Agric Sci 63:734–736

Usman M, Ahmad A, Ahmad S, Irshad M, Khaliq T, Wajid A, Hussain K, Nasim W, Chattha TM, Trethowan R, Hoogenboom G (2009) Development and application of crop water stress index for scheduling irrigation in cotton (*Gossypium hirsutum* L.) under semiarid environment. J Food Agric Environ 7(3&4):386–391

Warner DA, Burke JJ (1993) Cool night temperatures alter leaf starch and photosystem II: chlorophyll fluorescence in cotton. Agron J 85:836–840

Waseem M, Athar HU, Ashraf M (2006) Effect of salicylic acid applied through rooting medium on drought tolerance of wheat. Pak J Bot 38:1127–1136

William S, Bange M (2018) The cotton plant. In: Australian cotton production manual. Cotton Research and Development Corporation, Australia, p 2018

Wullschleger SD, Oosterhuis DM (1990) Photosynthesis of individual field-grown cotton leaves during ontogeny. Photosynth Res 23:163–170

Zhang HJ, Dong HZ, Li WJ, Zhang DM (2012) Effects of soil salinity and plant density on yield leaf senescence of field grown cotton. J Agron Crop Sci 198:27–37

Zhao D, Oosterhuis DM (1997) Physiological response of growth chamber-grown cotton plants to the plant growth regulator PGR-IV under water-deficit stress. Environ Exp Bot 38:7–14

Zhao D, Oosterhuis DM (2002) Cotton carbon exchange, nonstructural carbohydrates, and boron distribution in tissues during development of boron deficiency. Field Crops Res 78:75–87

Zhao D, Oosterhuis DM (2003) Cotton: growth and physiological responses of cotton to boron deficiency. J Plant Nutr 26:855–867

Zhao D, Reddy KR, Kakani VG, Koti S, Gao W (2005) Physiological causes of cotton fruit abscission under conditions of high temperature and enhanced ultraviolet-B radiation. Physiol Plant 124:189–199

Chapter 19
Salinity Tolerance in Cotton

Niaz Ahmed, Usman Khalid Chaudhry, Muhammad Arif Ali, Fiaz Ahmad, Muhammad Sarfraz, and Sajjad Hussain

Abstract Cotton is the chief crop and main pillar of textile industry. Its fiber and seed have significant economic importance. However, salinity interferes with the normal growth functioning and results in halted growth and declined yield of fiber and seed. Salinity effects are more obvious at early growth stages of cotton, limiting final yield. Salt decreases boll formation per plant which ultimately gives decreased fiber yield and poor lint quality. Salinity is a global issue increasing every year due to uncontrolled measures and improper land management. Application of saline irrigation water is adding increments to already existing salts and deteriorating the productive soil. Arid regions are totally dependent upon rain for growth of cotton. Salt problem is more in arid regions due least availability of moisture and water for flushing salts from cotton root zone. Moreover, higher temperature favors excessive evaporation under arid conditions and leaving salt on the upper surface of soil. Salts at the surface soil impede cotton seed germination. In this chapter, we discussed formation of saline soils and their sources which deter cotton growth. Physiological changes, oxidative stress caused due to salinity, role of molecular transporters involved in detoxification and specific gene expression is also illuminated.

Keywords Cotton · Salinity · Growth of cotton · Agronomic approaches · Physiology · Molecular techniques

---

N. Ahmed (✉) · M. A. Ali
Department of Soil Science, Bahauddin Zakariya University, Multan, Pakistan
e-mail: niaz.ahmad@bzu.edu.pk

U. K. Chaudhry
Department of Agricultural Genetic Engineering, Ayhan Sahenk Faculty of Agricultural Sciences and Technologies, Niğde Ömer Halisdemir University, Niğde, Turkey

F. Ahmad
Central Cotton Research Institute, Multan, Pakistan

M. Sarfraz
Soil Salinity Research Institute, Pindi Bhattian, Punjab, Pakistan

S. Hussain (✉)
Department of Horticulture, Bahauddin Zakariya University, Multan, Pakistan

S. Ahmad, M. Hasanuzzaman (eds.), *Cotton Production and Uses*,
https://doi.org/10.1007/978-981-15-1472-2_19

Abbreviations

ABA	Abscisic acid
AMF	Arbuscular mycorrhizal fungi
APX	Ascorbate peroxidase
CAT	Catalase
H_2O_2	Hydrogen peroxide
IPT	Isopentenyl transferase
$^1O_2$	Singlet oxygen
$O_2^{\bullet -}$	Superoxide anions
$^{\bullet}OH$	Hydroxyl radicals
POD	Peroxidases
ROS	Reactive oxygen species
SOD	Superoxide dismutase

19.1 Introduction

Soil salinity is foremost burden on agricultural lands, becoming hurdles for productive exploitation of agricultural lands for vigorous crop growth (Haque 2006; Lobell et al. 2007). Globally utilization of natural resources is increasing day by day and burgeoning population which severely influencing agriculture and different factors contributing toward worse soil conditions creating saline environment (Shahbaz and Ashraf 2013). Generally saline soils exhibit the characteristics of 4 dS m^{-1} and 15% exchangeable sodium percentage that inhibits the functioning of crops and ultimately result in retard growth and loss in yield (Munns 2005; Shahzad et al. 2017). Worldwide salt-affected soils cover the area of 20% of cultivable lands, and the problem on agricultural lands is 33%; moreover, it is increasing on annual basis due to change in climate. Several factors are currently under investigation which includes decreased supply of water and higher evaporation rate, weathering of rocks, saline water irrigation, and mismanagement of cultural practices (Jamil et al. 2011). No continent is free from saline problem (Table 19.1); it is estimated that 800 million lands are salt affected (FAO 2019), and salt affected land area would be increased by the year 2050 unless no precautionary measures cannot be taken with proper amendments (Ashraf 2009). It mostly occurs in arid and semiarid regions of the world which exist in all continents except Antarctica (Rengasamy 2006). Cotton (*Gossypium* sp.) is a cash crop named as white gold and king of fiber according its economic importance (Moseley 2001; Ahmad et al. 2014, 2017, 2018; Abbas and Ahmad 2018; Ahmad and Raza 2014; Ali et al. 2011, 2013a, b, 2014a, b). In addition to its fiber, its seed contains 15–20% oil contents, and seed cake is a rich source of protein for animal feed (Kothari et al. 2016). Cotton seed cake is used as manure having 6.5%, 3.0%, and 2.3% NPK. Cotton has significant importance for

Table 19.1 Salt affected soils in different continents of the world

Regions	Area (million hectares)		
	Saline	Sodic	Total area
North America	6.2	9.6	15.8
Central America	2.0	–	2.0
South America	69.4	59.6	129.0
Africa	53.5	27.0	80.5
South Asia	83.3	1.8	85.1
North and Central Asia	91.6	120.1	211.7
Southeast Asia	20.0	–	20.0
Europe	7.8	22.9	30.7
Australia	17.4	340.0	357.4
Total	351.5	581.0	932.2

Source: Szabolcs (1989)

textile industry, and it is the oldest known crop employed for fiber (Amin et al. 2017, 2018; Bakhsh et al. 2012; Rahman et al. 2018; Tariq et al. 2017, 2018; Usman et al. 2009). Salinity obstructs its growth and quality traits (Ashraf 2010). Cotton is one of the salt-tolerant crops, but it is sensitive at early stages to salt stress, i.e., germination and seedling emergence as compared to later growth stages (Ahmad et al. 2002). Cotton has the ability of retaining sodium contents more than 95% in its tissues (Gouia et al. 1994). It is also reported to retain Na^+ accumulation in leaves and roots of salt-tolerant cultivars (Sun and Liu 2001). Accumulation of salt in soils deters seed germination and plant growth and creates osmotic imbalance and toxicity leading to poor stand establishment (Ahmad et al. 2002; Bednarz et al. 2002). Under high salinity stress, it halts the physiological functioning of cotton by limiting photosynthesis and respiration which result in lower boll formation and poor fiber quality (Brugnoli and Lauteri 1991). Cotton is considered as moderately salt tolerant; however, it is depicted that its tolerance level varies from cultivars to cultivars (Leidi and Saiz 1997). Its tolerance level is also up to 7.7 dS m^{-1} (Maas and Hoffman 1977). Cotton salt tolerance level is dependent upon its ability of regulating sodium ion homeostasis in its tissues to minimize detrimental effects of cytotoxicity and by adjusting osmotic balance (Munns and Tester 2008). Salt stress disrupts photosynthetic machinery by firstly closing stomata of plants (Brugnoli and Lauteri 1991). Moreover, several reactive oxygen species (ROS) are formed under stress that further aggravates the oxidative stress to cotton (Meloni et al. 2003). On the other hand, contemporary modern approaches are available to cope with salt stress conditions to avail higher yield returns (Qadir et al. 2000; Gao et al. 2009; Shahzad et al. 2017). Saline soils can be reclaimed with plenty of good quality irrigation water (Murtaza et al. 2006). Chemical approach includes application of gypsum that improves the physical structure of soil and efficiently removes the soil from root zone (Murtaza et al. 2013). Arbuscular mycorrhizal fungi (AMF) are widely used to improve salinity tolerance in crops (Wu et al. 2010; Hajiboland et al. 2010; He et al. 2007). Improvements in cotton against salt stress tolerance is also conferred through

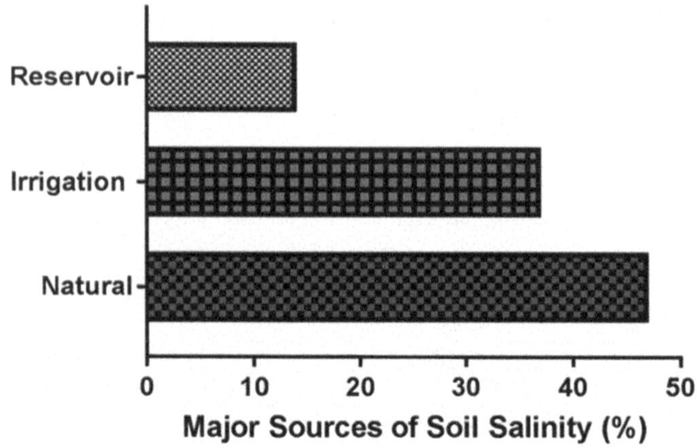

Fig. 19.1 Different sources contribution in forming soil salinity

biotechnological techniques (Zhang et al. 2011a, b, 2007; Ashraf et al. 2018a, b). It has broadened our horizon to understand the genes responsible for contributing stress tolerance (Liang et al. 2018).

19.2 Saline Soil Genesis and Distribution

Salinization originate from number of variety sources, and primary sources include weathering of rocks and minerals in the earth crust which contributes soluble salt concentration to soil and sea (Van Breemen and Buurman 2002). Weathering of earth crusts is also distributing salts to ocean and other water bodies in its surrounding. Weathering of rocks is the primary origin of salinization, while secondary origins are irrigation with saline water (Bui 2013). Saline water irrigation deposits salts in soils under harch climatic conditions. Water evaporation increases resulting in accumulation of salts on the surface of the soil which give rise to saline soil (Fig. 19.1). Underground water utilized for irrigating crops is a principal source of soil sodication because it also contains sodium fairly in high concentration (Bauder et al. 2011). The other important process of saline soil genesis is the rise in ground table in some parts around the globe (Fan et al. 2013). Groundwater table is increasing at the rate of 1–2 m annually. It is mineralized with salts, and due to capillary action, water is moving upward and enhancing salts in the root zone of the plant and leave behind on the surface after evaporation (Xie et al. 2011). Fossil salts often forms in arid region due to earlier depositions of salts in the form of marine deposits (Thomas 2011). Seepage of salts from upper soil parts to the lower soil layer which ultimately finish their journey in underground water contaminates it with accumulation of excessive salt. Agricultural lands in the vicinity of ocean and seas absorb salt either via wind or groundwater movement (Ondrasek et al. 2011). It is chiefly responsible for movement of salts up to many kilometers in the surrounding area. It was observed that rate of deposition by this mean is 20–100 kg ha^{-1} per year in coastal areas (Rengasamy 2010). Anthropogenic activities disturb the natural environment and intentionally increasing salinity of soil. It comprises construction of roads, dams, canals, and other irrigation system on saline strata (Metternicht and Zink 2003; Oldeman et al. 1991). Likely all these human practices block the movement of water which causes severe increase to rise in water table creates water logging and finally salinization of land (Day Jr et al. 2013). Moreover, mostly farmers are unaware of the water requirement of the crops, and they tend to over-irrigate the crops which result in increase in water table (Carreira et al. 2014). Fertilization is another practice which works as a double-edged sword to increase yield of crop while on the other hand is harmful for soil health and properties leading toward contribution of salt to soil (Savci 2012). Poor waste management practices and dumping of animal waste on farm increase salt content for agricultural land. Salt stress declined agricultural productivity in many parts of world (Rozema and Flowers 2008). It was observed salinity covers more than 397 million hectares of land globally (FAO 2005). Salinity problem is exacerbated especially in arid regions because of less availability of moisture, and these regions also receive less rainfall (Pasternak and De Malach 1994; Villa-Castorena et al. 2003). Irrigated lands receive salts in the form of saline irrigated water with least leaching and poor drainage practices. Approximately 20% of the world's land is irrigated with saline water (Sumner 1999). Almost 75 countries have been marked in the red zone of salinity stress with a moderate to severe salinity problems covering a total area of 831 million hectare, which was productive for crops in the past (Martinez-Beltran and LiconaManzur 2005; Qadir et al. 2000). Land coverage of salinity in across different continents is illustrated in Fig. 19.2 (Hoang et al. 2016).

19.3 Production of Cotton on Saline Soil

Cotton production halts due to presence of excessive salt concentration in soil, disturbing uptake of other essential nutrients for the growth and yield of crop (Jafri and Rafiq 1994). The main mineral ion that causes salinity stress is sodium chloride and sodium sulfate (Reich et al. 2017). Salinity influences soil nutrients for cotton by osmotic effect by increasing its concentration in the vicinity of cotton roots for uptake by plant and likely in the roots (Wang et al. 2015). On another hand NaCl deposition in roots creates dehydration by pulling out water from the roots (Younis et al. 2014). Moreover, it also disrupts the ion solution inside the plant cells that results in hindrance of physiological processes of cotton (Ashraf et al. 2017). Under saline environment, the nutrients that becomes inhibited are nitrogen, potassium, phosphorus, and zinc, so it is imperative to consider the application of these nutrients on such soils for ensuring better supply of these nutrients to combat with toxic effects of sodium and providing maintenance strategy to keep the other function of cotton to normal limits (Hu and Schmidhalter 2005; Dong et al. 2010c). Cotton is

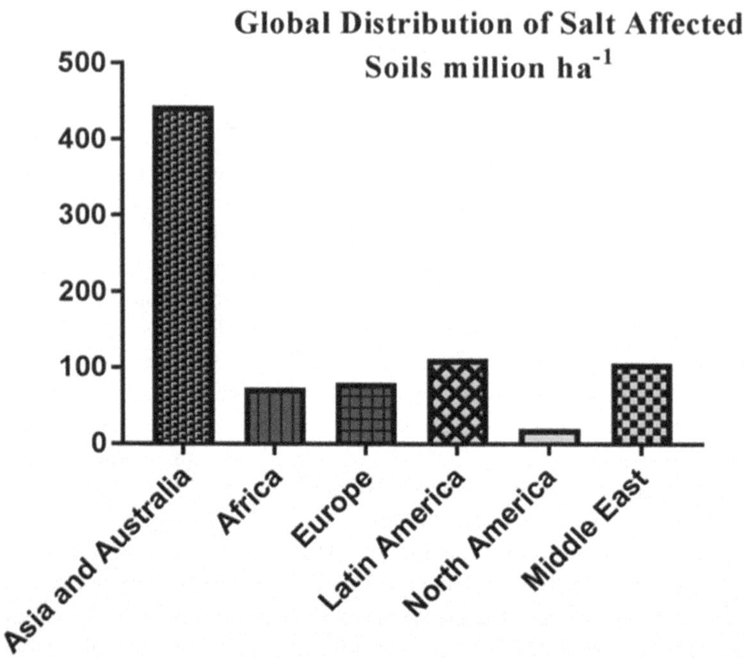

Fig. 19.2 Distribution of salt affected lands in different continents

sensitive to salt at early growth stages such as during seed germination, emergence, and establishment in comparison to mature plant (Chen et al. 2010). It was suggested to take extensive measures during early growth stages to ensure robust growth for final fruitful yield (Ashraf 2010). Delayed and non-uniformity in emergence rate are the first gesture of salinity after sowing cotton seeds in saline soil (Dong et al. 2009). Decrease in germination of seed is obvious with the proportional concentration of salt concentration in soil (Ashraf et al. 2002). Complete failure of germination was observed at 16 dS m^{-1} (Kahlown and Azam 2002). *Gossypium barbadense* genotypes exhibited higher tolerance toward salinity as compared to *Gossypium hirsutum* and *Gossypium arboreum* cotton (Ahmad et al. 2002). Salinity has also mild effect on root length at low concentration, while higher concentration of sodium affects its root length (Chen et al. 2018a, b). Decreased root length and delayed secondary root growth have been reported (Cramer et al. 1987). Sodium is also a competitor of calcium to limit its uptake by cotton roots (Byrt et al. 2018). Cotton is salt tolerant, but its vegetative growth is severely affected on saline soil. Shoot is more sensitive to salt than roots. Decrease in leaf area per plant, stem thickness, and shoot and root weight reduction are important morphology traits significantly influenced by uptake of salt and higher accumulation in plants via roots (Anjum et al. 2005). On the other hand, application of calcium is beneficial for limiting Na^{+} uptake by plants (Reid and Smith 2000). Biomass production of cotton reduces with adverse salt stress. Boll formation also reduces with less number of fruits per plant (Gandahi et al. 2017). Salinization results in boll shedding and senescence of leaves (Rathert 1982; Brugnoli and Björkman 1992). Cotton fiber yield and length are important traits for textile industry; however, it results in poor lint quality due to interference of salts (Soares et al. 2018). Its seed also holds significant position for feed and oil sector which also becomes affected with saline conditions (Ashraf 2010). Cotton seed is a good source of oil. Decrease of cotton seed yield causes decrease in seed oil contents. It was reported that 50% decrease in seed yield was with a salinity level of 16.75 dS m^{-1} (Ali et al. 1986). Morphologically decrease in cotton was also reported with salinity conditions as enhanced sodicity hampered growth with a significant decrease in root length, fruit number, and ultimately lint yield (Dodd et al. 2013). Cotton is deemed to tolerate salt moderately within the range of 7.7 dS m^{-1} (Maas 1986), thereby it is efficient candidate crop against salt-affected soils for its growth (Ahmad et al. 2002). However, reduction in its growth and yield traits has been reported with the increase in its salinity threshold level (Khan et al. 2001; Dong 2012b; Higbie et al. 2010). Moreover, tolerance level varies among different genotypes of cotton (Ashraf 2010; Hanif et al. 2008). Cotton grown on saline soil for years exhibited elevation of Na^+ concentration and decrease in phosphorus and potassium concentration in plant tissues (Rochester 2010). Similar negative correlation was observed from young and mature leaves at different growth stages of cotton (Dodd 2007). Its tolerance mechanism is dependent on genotypes for different growth stages. In order to make it tolerant, it is essential to have basic knowledge of cotton regarding tolerance at varying growth stages (Ashraf and Ahmad 2000). General perception is that halophyte plant accumulates enormous quantity of NaCl ions in tissues to adapt themselves to saline conditions, while mesophytes restrict the entry of these ions (Flowers and Colmer 2015). Higher Na^+ concentration perturbs other nutrient in plants as mentioned earlier that it disturbs osmotic balance, thereby disturbed K^+/Na^+ uptake and its interference with each other for uptake is important mechanism for considering tolerance among cultivars (Leidi and Saiz 1997). Negative correlation of these ions uptake confers positive correlation toward salinity tolerance such as higher K^+ allows lower Na^+ ions uptake (Cramer et al. 1987). Moreover Na^+ exclusion also resulted in tolerance, and it was also observed in tolerant cultivars that K^+ concentration was measured to be higher in leaves (Ahmad et al. 2002). Oxidative stress also triggered with the entry of Na^+ ion and disturbing other ions, so ROS needs to be eliminated or their effects needs to be suppressed; in this regard ROS scavenger comes to play their role which is mentioned later in this chapter. Another important strategy that is being adapted by breeders is development of salt-tolerant cotton cultivars to improve the characteristics of their local high-yielding cultivar (Harshavardhan et al. 2018). Breeding for salinity tolerance has reported potential results for improving yield of crops (Blum 2018). Cotton conventional breeding improved tolerance against salinity with a 7.4% increase in yield (Ledbetter 1987; Ashraf and Akram 2009). Later selection method was exploited which was also potential method for developing salt-tolerant cultivars (Da Silva et al. 1992).

19.4 Physiological Changes and Role of Antioxidant Enzymes

Salt stress is an environmental factor that interrupts physiological and biochemical changes in cotton (Meloni et al. 2003; Zhang et al. 2014). Salinity stress affects physiological processes by increasing respiration rate and disturbing mineral ion distribution especially displacement of calcium and potassium with Na^+, sometimes leading to ion toxicity (Kinraide 1999). Salinity limits the photosynthesis process as well as cell growth (Munns et al. 2006). It directly inhibits CO_2 availability due to limited diffusion via stomata, mesophyll, and disrupting metabolism of photosynthesis (Lawlor and Cornic 2002; Flexas et al. 2007). Salt interfere the photosynthetic machinery of cotton by decreasing its photosynthetic rate (Meloni et al. 2003). It results from the decline in chlorophyll contents of plants under adverse conditions (Jaleel et al. 2008). Foremost mechanism during salinity stress is the adjustment of stomata to limits transpiration and maintains cell turgor (Miller et al. 2010). Stomatal closure is the response against dehydration with the impaired supply of CO_2 to shifting plants to water-saving strategy (Brodribb and Holbrook 2003). In this regard abscisic acid (ABA) plays a crucial role as a signaling molecule from its production site (roots) to the leaves for closure of stomata (Wilkinson and Davies 2002). Moreover, under saline conditions Na^+ uptake is increased which halts the uptake of other essential nutrients such as potassium, calcium, and manganese (Hasegawa et al. 2000). Salt uptake in tissues of plant brutally influences older leaves with higher accumulation of salts (Munns 2002; Chaves et al. 2003). Decrease in nutrient uptake and impaired photosynthetic process was observed in cotton in response to salinity (Liu et al. 2014a, b). Salinity stress causes excessive production of ROS (You and Chan 2015). ROS are produced in different cell compartments of plant, generally the sites of its production are chloroplast and mitochondria (Jubany-Marí et al. 2009). It includes oxygen free radicals, i.e., singlet oxygen ($^1O_2$), superoxide anions ($O_2^{\bullet -}$), hydrogen peroxide (H_2O_2), and hydroxyl radicals ($^{\bullet}OH$) (Zheng et al. 2009). These ROS species have detrimental effects on plant functioning by causing damage to DNA and protein (Foyer and Noctor 2005). Oxidative damages are alleviated by ROS-scavenging enzymes (Hussain et al. 2018). Antioxidant enzymes are superoxide dismutase (SOD), catalase (CAT), peroxidases (POD), and ascorbate peroxidase (APX) (Hossain and Dietz 2016). These enzymes work in a sequence to alleviate oxidative stress of cotton under salinity stress (Garratt et al. 2002). SOD is the first ROS-scavenger enzyme to start its function for alleviating ROS species; it dismutases O_2^- into H_2O_2 and O_2 (Azarabadi et al. 2017). Immediately after that POD starts catalyzing the H_2O_2 to H_2O and O_2 (Waszczak et al. 2018). Later CAT and APX capture H_2O_2 and convert in to H_2O (Mittler 2017). Antioxidant enzymes work as indicator of salt-tolerant and susceptible cultivars (Ashraf and Harris 2004). It was observed that enzymes activities increased during salinity stress to cope with ROS species and aids in tolerance to stress (Koca et al. 2007). It was reported from studies on cotton that tolerant cultivar exhibited a higher level of antioxidant enzymes (Zhang et al. 2014; Liu et al. 2014a, b; Ibrahim et al. 2017). Increase in

SOD activity in cotton alleviated oxidative stress caused by salinity (Xie et al. 2008). Lipid peroxidation results from ROS species where malondialdehyde is the product formed due to lipid peroxidation, and it represents how much damage has occurred to the plant in stress condition (Sharma et al. 2012). MDA is a marker for observing oxidative damage to the plants (Davey et al. 2005). Cotton exhibited increment in its amount due to salinity stress (Tang et al. 2007; Meloni et al. 2003). It also assists in differentiating the tolerant and susceptible cultivars (Liu et al. 2014a, b). Proline also accumulates during salinity stress to work in osmotic adjustment, alleviates free radicals, and maintains cellular redox potential (Ashraf and Foolad 2007). It accumulates normally in cytosol for osmotic regulations to cytoplasm (Ashraf et al. 2018a, b). Higher concentration of proline is correlated with tolerance of cultivar as tolerant cultivar depicts increase level of proline during salinity stress (Hayat et al. 2012). Furthermore, exogenous application also conferred significant results against salinity tolerance (Heuer 2003). Influence of proline for osmotic adjustment was observed in cotton (Meloni et al. 2001). Cotton was subjected to salt stress conditions, and marked increase proline concentration was observed to tolerate the stress (Golan-Goldhirsh et al. 1990). Proline is also associated for improving fiber quality of cotton (Xu et al. 2013).

19.5 Genetic Engineering and Molecular Biological Tool

Salts in high amount in soil result in higher uptake by plants and ultimately causing disruption of membrane functions and inhibit cell division, photosynthesis, and development of plants (Flowers 1999; Horie and Schroeder 2004). Plants as being immobile to move to favorable growth conditions thus have to survive under existing environmental conditions. First plant organs that come in contact with saline environment are root hairs from where it is taken up and transported to epidermis and cortex of plants (Cao et al. 2017). Sodium is transported to the shoots via transpiration stream in xylem, and rarely it is returned to the roots through phloem (Wu et al. 2018). Therefore, it is observed that its movement is unidirectional and results in higher accumulation of sodium in shoots (Ishikawa and Shabala 2019). Sodium accumulates in higher concentration in shoots as compared to roots, and from shoots it is transported to the leaves (Farooq et al. 2015). Potassium is a crucial nutrient for plant especially under salt stress because it is the competitor of sodium; if K^+ concentration is higher, it inhibits the entry of Na^+ in the plant cell and protects it from detrimental effects of Na^+ (Adams and Shin 2014). Potassium is also essential for photosynthetic apparatus to aid in its functioning (Lu et al. 2016). Concentration of Na^+ in high quantity disrupts the membrane functions causing disturbance in ion homeostasis which result in stunted growth and sometimes lead to death of cell (Flowers et al. 2014). Plants both halophytes and glycophytes utilize an identical strategy of regulating and maintaining Na^+ ion homeostasis by coordinated functions of ion transporters for controlling the flow into the plants (Wang et al. 2017). Moreover, there are numerous selective pumps that favor the uptake of

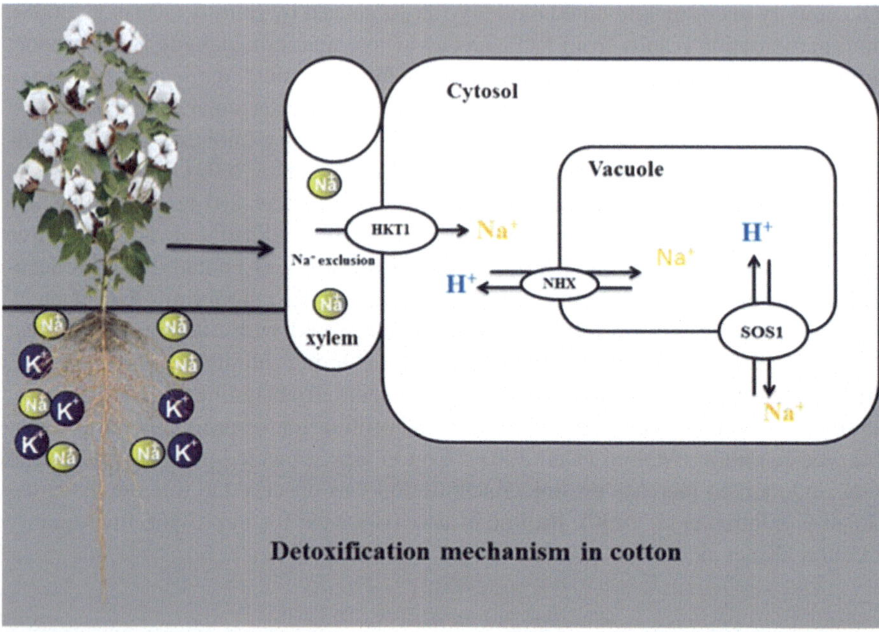

Fig. 19.3 Illustration of Na^+ transport mechanisms against salt stress response and role of molecular transporters. Sodium concentration increases under saline conditions for uptake via plant roots. High-affinity potassium transporter (HKT1) unloads Na^+ from the xylem. Na^+/H^+ exchanger (NHX) and salt overly sensitive (SOS1) assist in detoxification mechanism by decreasing the concentration of Na^+ with H^+ ion concentration

K^+ compared to Na^+ (Zhang et al. 2018). High-affinity potassium transporter (HKT) proteins are reported in numerous plants for playing selective role of K^+ uptake against Na^+ (Fairbairn et al. 2000; Golldack et al. 2002; Horie et al. 2001; Sunarpi et al. 2005). HKT was the first potassium-selective transporter found in plants (Schachtman and Schroeder 1994). It also plays function in Na^+ exclusion from leaves and maintains K^+ homeostasis in leaves (Horie et al. 2005). Currently Na^+/H^+ transporter located in tonoplast was identified that play role in outward movement of Na^+ from cytosol to apoplast or vacuole (Zeng et al. 2018). However, it is an energy-consuming process for cell, and proton pumps give force for transporting Na^+ contrary to electrochemical gradient (Blumwald et al. 2000). It was reported in cotton roots that Na^+ concentration was lower due to higher influx of H^+ via potential role of Na^+/H^+ antiporter (Kong et al. 2011). Another transporter is salt overly sensitive pathway (SOS); it also works as an exchanger in plasma membrane (Qiu et al. 2002). It becomes activated and plays crucial role in Na^+ exclusion mechanism to render plant salt tolerance (Zhu 2000). Mechanism of Na^+ transport by different pumps is depicted in Fig. 19.3. Engineered transgenic cotton is also worth mentioning for their contribution toward development of salt-tolerant cotton cultivars (Liu et al. 2014a, b). Transgenic plants contain any foreign DNA that plant does not contain naturally to improve traits and quality of plants (Rao et al. 2009). Transgenic cotton exhibited higher root development and minimized the transpiration rate; in that way it became tolerant to survive on saline soil conditions (Liu et al. 2014a, b). Cotton plants were transformed with H^+-PPase gene exhibited tolerance in transgenic cotton lines by improving vegetative growth and higher photosynthetic rate resultantly lower ion leakage from the plants under salt stress as compared with non-transgenic plants (Bock 2010). Overexpression of Arabidopsis vacuolar pyrophosphatase gene (AVP1) in cotton contributed to 20% increase in fiber due to a number of boll formations under salt stress (Zhang et al. 2011a, b). Senescence results from decline in cytokinin contents in plant under stress conditions thereby isopentenyl transferase (IPT) gene has potential role in supplementing cytokinin and enhancing chlorophyll contents and delayed in senescence was reported in transgenic cotton to survive on salt affected soils (Liu et al. 2012a, b). Numerous studies have been accomplished successfully in manipulating cotton for salinity stress tolerance (Shen et al. 2015; Zhang et al. 2007; Yu et al. 2016; Cheng et al. 2018; Song et al. 2018). Plants sense stress environment and generate signals to try to adapt and adjust themselves to the salinity stress (Zhu 2016). Principal factor is regulation of genes that response under salinity stresses (Egea et al. 2018). However, it is a complex set of many genes that regulate to make the plant tolerant, while some genes are down regulated (Albaladejo et al. 2018). Salt stress induces gene expression; it was reported that the expression of salt overly sensitive (SOS_2) gene and plasma membrane H^+ ATPase (PMA_2) gene was observed to be higher in cotton (Peng et al. 2016). In another study conducted on cotton gene expression profiles were documented for salt stress by exploiting microarray technique, and all the observed genes conferred cotton tolerant to salinity stress, furthermore some transporters also exhibited their role in rendering tolerance (Zhu et al. 2013). Genetic transporter expression is efficient for tolerance, and they also improve quality traits of cotton under adverse saline conditions (He et al. 2005). Some of the genes and their potential role under salt stress in cotton are given in Table 19.2.

19.6 Agronomic Practices to Circumvent Salinity for Cotton

Proper sowing method and good cultivation practices circumvent the effect of salt stress to cotton (Anjum et al. 2005). Cultivation methods such as mulching maintain moisture and protect from evaporation (Dong 2012b). Mulching practice to protect cotton from salinity effect is not new (Sandoval and Benz 1966). Straw mulching for 3 years reduced soil salinity (Benz et al. 1967). Mulching of cotton burr at the rate of 90 t ha^{-1} with intermittent sprinkling assisted in removing salts from root zone (Carter and Fanning 1964). This mulching technique is adapted for last many decades for growing cotton (Liu et al. 2012a, b). Plastic mulching with polyethylene is a general practice among cotton growers to protect seeds from complete failure of seed germination and safeguard early emergence on salt affected soils (Dong et al.

Table 19.2 Genes for salinity tolerance in cotton and their functions

Gene name	Function	Reference
Arabidopsis vacuolar H^+-Pyrophosphatase (AVP)	Stimulate auxin transport and improves root system	Pasapula et al. (2011)
Thellungiella halophila vacuolar H^+-pyrophosphatase (TsVP)	Improves seed yield/fiber quality	Zhang et al. (2016)
Sodium/hydrogen exchanger (NHX1)	Tonoplast Na^+/H^+ antiporter	Wu et al. (2004)
Salt overly sensitive (SOS)	Ion homeostasis/exclusion of Na^+	Wei et al. (2017)
Calcium dependent protein kinases (CPKs)	Stress signaling	Gao et al. (2018)
Cotton Bax inhibitor-1 (GhBI-1)	Suppressed stress-induced cell death	Zhang et al. (2018)
Glycine sarcosine methyltransferase (GSMT)	Improved glycine betaine accumulation and intracellular osmoregulation	Song et al. (2018)
High affinity potassium transporter (HAK)	Improved uptake of potassium	Liu et al. (2015)
Choline monooxygenase (CMO)	Glycine betaine synthesis and salinity tolerance	Zhang et al. (2009)

2009). Plastic mulching improved growth and lint yield of cotton by raising temperature of the soil (Dong et al. 2007). Benefits of conserving moisture are vital after its adoption that aided in control of saline environment in root zone (Bezborodov et al. 2010). Moreover, mulching is documented for unequal dispersion of salts which imparted suitable growth of roots and reduced damage of salt (Bezborodov et al. 2010). Sowing method has positive effect to overcome salinity stress (Sarangi et al. 2017). Cotton crops sown on ridges showed better growth and development as compared to crop sown on flat beds (Dong et al. 2010b). It was inferred that by exploiting this ridge sowing method there was non-uniform distribution of salts across the field as well as reduced deposition of Na^+ in the root zone (Dong et al. 2008). Time of sowing is also important factor as cotton sown during normal growing season in temperate regions exhibited weak stand establishment and resulted in late maturity (Dong et al. 2007). However, late sowing efficiently enhanced seedling emergence and vigorous stand establishment due to rise in temperature with a declined Na^+ contents in tissues of cotton (Dong et al. 2010a). Planting density also played a role in mitigating saline conditions. Cotton yield increase was observed by increasing planting density because enormous amount of salts might reduce plant size (Feinerman 1983). Increasing population density under severe salinity improved cotton seed yield (Dong 2012a). Plant density increased vegetative growth production, and it had positive effect for seed yield of cotton (Zhang et al. 2012a, b).

19.7 Reclamation Options and Fertilizer Management of Cotton for Saline Soils

Saline soils with excessive accumulated salts exacerbate soil physic-chemical properties which ultimately create harmful growth conditions for plants (Chabra 1996). Salinity deteriorates the soil physical structure by creating dispersion of particles, soil erosion, and sometimes water logged conditions (Warrence et al. 2002). In order to amend these soils, different reclamation strategies are in practice nowadays (Qadir et al. 2000). Reclamation can be done by considering different features appropriate in the selection of site, soil depth, and presence of hard pan and finally the most important presence of salt (Murtaza et al. 2006; Ghafoor et al. 2004). Freshwater availability free from salt can be used for removing salts from root zone and upper surface of soil with considerable leaching characteristics of soil (Bezborodov et al. 2010). It should also be considered that leaching should not be too much in order to protect the groundwater table form salts and also the amount of water applied for removal of salts. Proper irrigation management practices also contribute their role in minimizing salinity both for cotton crop and reclaiming soil to some extent (Wang et al. 2012). Sprinkler and drip irrigation are the preferred ways to remove the excessive salt from root zone of plant as well as protecting the groundwater table by limiting excessive leaching of salts as compared to flood irrigation (Karlberg and de Vries 2004). Leaching of soluble salts is an effective way of protecting rhizosphere from its toxic effects. It was documented from one study that drip irrigations conferred fruitful results for sustaining cotton productivity with the alleviation of sodium ion from roots of cotton at different growth stages (Kang et al. 2012). Chemical amendments are also beneficial of removing exchangeable sodium from cation exchange sites (Sahin et al. 2002). Gypsum is the most commonly used chemical amendment because it is cheap and easy availability (Ilyas et al. 1997). Organic amendments are also deployed for remediation of saline soils. Different organic amendments manure, mulch, and compost proved to have reclaiming characteristics of saline soil (Diacono and Montemurro 2015; Suzuki 1999). Emergence of cotton seedling increased on saline soil with organic manure application (da Costa et al. 2016). These amendments have multifaceted role in soil by enhancing aggregation of particles, improving water holding capacity to protect soil from drying via evaporation which causes accumulation of more Na^+ in soil pockets (Lu et al. 2015). Moreover, they also provide nutrient that are essential to combat with Na^+ for uptake by plant (Zhang et al. 2015). Therefore, fertilizer selection is crucial for saline soils. Potassium is normally applied as potassium chloride (muriate of potash) which is not suitable for saline soil. Potassium magnesium sulfate fertilizer is effective to cope with NaCl stress (Khare et al. 2015). Furthermore, nitrate has also potential to alleviate influence of higher chloride concentration in soil (Shrivastava and Kumar 2015). Management of fertilizers is a good option for cotton growing on salinized soils. Application of nitrogen to soil enhanced uptake by plant and confined Na^+ uptake in cotton (Kawakami et al. 2010). Potassium and nitrogen foliar sprays are direct source of absorption by cotton leaves and bolls for sturdy growth of vegetative parts (Jabeen and Ahmad 2009). Interference of salt with nitrogen was illustrated by varying their levels and concluded that higher salt contents had profound effects on cotton plants (Chen et al. 2010). Good supply of nitrogen to the bolls either by foliar or basal resulted in higher lint yield of cotton on saline soils (Zhang et al. 2012a, b). Fertigation is another method for simultaneous application of water and fertilizers to crops (Castellanos et al. 2012). It is a modern agriculture approach for reduction of environmental pollution and feasible for cumulative chemical fertilizer usage efficiency (Hagin et al. 2002). It was employed for cotton growth to eradicate salt and give salt-free environment to the roots of cotton (Min et al. 2016). The benefits of fertigation for cotton crop was proved in another study that irrigation with suitable nitrogen fertilizer eradicated the devastating effecting of Na^+ and provided nitrogen nutrient to the cotton for higher growth and yield (Min et al. 2017).

19.8 Conclusion

Salinity is an emerging global issue; cotton is being grown in more than 80 countries and plays a key role in the economy of various countries. Salinity is causing soil degradation at an alarming rate. Cotton productivity is severely hindered by degradation of soil. Cotton needs sustainable development under this situation; in order to cope with salinity, there are various mitigation tactics which can lead to sustainable productivity of cotton. Development of adaptable cotton cultivars is a basic step to face this challenge in the long run. Various scientists are keen to develop those techniques to enable crop to battle with various hazards faced during their life cycle. Adaptation in plants can be developed using various biotechnological tools. Development of appropriate surrounding situations for optimum development of crop is also considered; reclamation of soil, the use of appropriate fertilizer application, and good quality irrigation water are essential to minimize saline conditions and provide better growth of cotton.

Saline soils are formed; some decades now, the interference of human activities is worsening the already existing salt-affected soils. Moreover, adding increment to the salinized land area, proper control measures should be taken to control such activities polluting our natural environment and water resources. Salinity issues should be addressed, and awareness needs to create among farming communities directly involved for agriculture on such soils. Proper reclamation strategies potentially improved salt-affected soils and eradicating salts from root zone for improving growth of cotton. Irrigation water analysis should be done for irrigation saline soils. High EC water should not be applied to salt-affected soil. Sowing method for cotton is an efficient way for overcoming salinity problem. Likely density and plant methods should be considered. Fertilizer management is also important to supplement essential nutrients that become limited due to presence of sodium. High salt-tolerant cotton genotypes are suitable for growth on saline soil. Proper selection of cultivar also confers higher yield returns. If you don't have any local salt-tolerant cultivar, breeding approaches should be exploited for improving traits of your local cultivar. The conventional breeding assist in fixing traits in susceptible but high yielding cotton genotypes it renders susceptible cotton genotype a tolerant one after several crossing for breeding. Physiological changes also occur in cotton due to the intake of Na^+. Molecular approaches paved the path for improving antioxidant enzyme activities to combat salinity problem in cotton tissues. Development of transgenic also exhibited tolerance to salt stress which can be used growing cotton on salt-affected soil.

References

Abbas Q, Ahmad S (2018) Effect of different sowing times and cultivars on cotton fiber quality under stable cotton-wheat cropping system in southern Punjab, Pakistan. Pak J Life Soc Sci 16:77–84

Adams E, Shin R (2014) Transport, signaling, and homeostasis of potassium and sodium in plants. J Integr Plant Biol 56(3):231–249

Ahmad S, Raza I (2014) Optimization of management practices to improve cotton fiber quality under irrigated arid environment. J Food Agric Environ 2(2):609–613

Ahmad S, Khan N, Iqbal MZ, Hussain A, Hassan M (2002) Salt tolerance of cotton (*Gossypium hirsutum* L.). Asian J Plant Sci 1(6):715–719

Ahmad S, Raza I, Ali H, Shahzad AN, Atiq-ur-Rehman, Sarwar N (2014) Response of cotton crop to exogenous application of glycinebetaine under sufficient and scarce water conditions. Braz J Bot 37(4):407–415

Ahmad S, Abbas Q, Abbas G, Fatima Z, Atique-ur-Rehman, Naz S, Younis H, Khan RJ, Nasim W, Habib urRehman M, Ahmad A, Rasul G, Khan MA, Hasanuzzaman M (2017) Quantification of climate warming and crop management impacts on cotton phenology. Plants 6(7):1–16

Ahmad S, Iqbal M, Muhammad T, Mehmood A, Ahmad S, Hasanuzzaman M (2018) Cotton productivity enhanced through transplanting and early sowing. Acta Sci Biol Sci 40:e34610

Albaladejo I, Egea I, Morales B, Flores FB, Capel C, Lozano R, Bolarin MC (2018) Identification of key genes involved in the phenotypic alterations of res (*restored cell structure by salinity*) tomato mutant and its recovery induced by salt stress through transcriptomic analysis. BMC Plant Biol 18(1):213

Ali A, Ahmad N, Makhdum MI, Gill KH (1986) Effect of soil salinity on cotton (*Gossypium hirsutum*) on early stage of growth. Pak J Sci 1:37–40

Ali H, Afzal MN, Ahmad F, Ahmad S, Akhtar M, Atif R (2011) Effect of sowing dates, plant spacing and nitrogen application on growth and productivity on cotton crop. Int J Sci Eng Res 2 (9):1–6

Ali H, Abid SA, Ahmad S, Sarwar N, Arooj M, Mahmood A, Shahzad AN (2013a) Integrated weed management in cotton cultivated in the alternate-furrow planting system. J Food Agric Environ 11(3&4):1664–1669

Ali H, Abid SA, Ahmad S, Sarwar N, Arooj M, Mahmood A, Shahzad AN (2013b) Impact of integrated weed management on flat-sown cotton (*Gossypium hirsutum* L.). J Anim Plant Sci 23 (4):1185–1192

Ali H, Hameed RA, Ahmad S, Shahzad AN, Sarwar N (2014a) Efficacy of different techniques of nitrogen application on American cotton under semi-arid conditions. J Food Agric Environ 12 (1):157–160

Ali H, Hussain GS, Hussain S, Shahzad AN, Ahmad S, Javeed HMR, Sarwar N (2014b) Early sowing reduces cotton leaf curl virus occurrence and improves cotton productivity. Cer Agron Moldova XLVII(4):71–81

Amin A, Nasim W, Mubeen M, Nadeem M, Ali L, Hammad HM, Sultana SR, Jabran K, Habib urRehman M, Ahmad S, Awais M, Rasool A, Fahad S, Saud S, Shah AN, Ihsan Z, Ali S, Bajwa AA, Hakeem KR, Ameen A, Amanullah, Rehman HU, Alghabar F, Jatoi GH, Akram M, Khan A, Islam F, Ata-Ul-Karim ST, Rehmani MIA, Hussain S, Razaq M, Fathi A (2017) Optimizing the phosphorus use in cotton by using CSM-CROPGRO-cotton model for semi-arid climate of Vehari-Punjab, Pakistan. Environ Sci Pollut Res 24(6):5811–5823

Amin A, Nasim W, Mubeen M, Ahmad A, Nadeem M, Urich P, Fahad S, Ahmad S, Wajid A, Tabassum F, Hammad HM, Sultana SR, Anwar S, Baloch SK, Wahid A, Wilkerson CJ, Hoogenboom G (2018) Simulated CSM-CROPGRO-cotton yield under projected future climate by SimCLIM for southern Punjab, Pakistan. Agric Syst 167:213–222

Anjum R, Ahmed A, Ullah R, Jahangir M, Yousaf M (2005) Effect of soil salinity/sodicity on the growth and yield of different varieties of cotton. Int J Agric Biol 7(4):606–608

Ashraf M, Ahmad S (2000) Influence of sodium chloride on ion accumulation, yield components and fibre characteristics in salt-tolerant and salt-sensitive lines of cotton (Gossypium hirsutum L.). Field Crop Res 66:115–127

Ashraf M (2009) Biotechnological approach of improving plant salt tolerance using antioxidants as markers. Biotechnol Adv 27:84–93

Ashraf M, Akram NA (2009) Improving salinity tolerance of plants through conventional breeding and genetic engineering: an analytical comparison. Biotechnol Adv 27(6):744–752

Ashraf M (2010) Salt tolerance of cotton: some new advances. Crit Rev Plant Sci 21(1):1–30

Ashraf M, Foolad MR (2007) Roles of glycine betaine and proline in improving plant abiotic stress resistance. Environ Exp Bot 59(2):206–216

Ashraf M, Harris PJC (2004) Potential biochemical indicators of salinity tolerance in plants. Plant Sci 166(1):3–16

Ashraf MY, Sarwar G, Ashraf M, Afaf R, Sattar A (2002) Salinity induced changes in α-amylase activity during germination and early cotton seedling growth. Biol Plantarum 45(4):589–591

Ashraf M, Shahzad SM, Imtiaz M, Rizwan MS, Iqbal MM (2017) Ameliorative effects of potassium nutrition on yield and fiber quality characteristics of cotton (*Gossypium hirsutum* L.) under NaCl stress. Soil Environ 36(1):51–58

Ashraf J, Zuo D, Wang Q, Malik W, Zhang Y, Abid MA, Cheng H, Yang Q, Song G (2018a) Recent insights into cotton functional genomics: progress and future perspectives. Plant Biotechnol J 16(3):699–713

Ashraf MA, Iqbal M, Rasheed R, Hussain I, Perveen S, Mahmood S (2018b) Dynamic proline metabolism: importance and regulation in water-limited environments. In: Plant metabolites and regulation under environmental stress. Academic, Cambridge, MA, pp 323–336

Azarabadi S, Abdollahi H, Torabi M, Salehi Z, Nasiri J (2017) ROS generation, oxidative burst and dynamic expression profiles of ROS-scavenging enzymes of superoxide dismutase (*SOD*), catalase (*CAT*) and ascorbate peroxidase (*APX*) in response to *Erwinia amylovora* in pear (*Pyrus communis* L.). Eur J Plant Pathol 147(2):279–294

Bakhsh A, Rao AQ, Shahid AA, Husnain T (2012) Spatio temporal expression pattern of an insecticidal gene (cry2A) in transgenic cotton lines. Not Sci Biol 4:115–119.

Bauder TA, Waskom RM, Sutherland PL, Davis JG, Follett RH, Soltanpour PN (2011) Irrigation water quality criteria. Service in Action; No. 0.506

Bednarz CW, Shurley WD, Anthony WS (2002) Losses in yield, quality, and profitability of cotton from improper harvest timing. Agron J 94(5):1004–1011

Benz LC, Sandoval FM, Willis WO (1967) Soil salinity changes with fallow and a straw mulch on fallow. Soil Sci 104:63–68

Bezborodov GA, Shadmanov DK, Mirhashimov RT, Yuldashev T, Qureshi AS, Noble AD, Qadir M (2010) Mulching and water quality effects on soil salinity and sodicity dynamics and cotton productivity in Central Asia. Agric Ecosyst Environ 138(1–2):95–102

Blum A (2018) Plant breeding for stress environments. CRC Press, Taylor & Francis Group, Boca Raton, FL, p 231

Blumwald E, Aharon GS, Apse MP (2000) Sodium transport in plant cells. Biochim Biophys Acta 1465:140–151

Bock R (2010) The give-and-take of DNA: horizontal gene transfer in plants. Trends Plant Sci 15:11–22

Brodribb TJ, Holbrook NM (2003) Stomatal closure during leaf dehydration, correlation with other leaf physiological traits. Plant Physiol 132(4):2166–2173

Brugnoli E, Björkman O (1992) Growth of cotton under continuous salinity stress-influence on allocation pattern, stomatal and nonstomatal components of photosynthesis and dissipation of excess light energy. Planta 187:335–347

Brugnoli E, Lauteri M (1991) Effects of salinity on stomatal conductance, photosynthetic capacity, and carbon isotope discrimination of salt-tolerant (*Gossypium hirsutum* L.) and salt-sensitive (*Phaseolus vulgaris* L.) C_3 non-halophytes. Plant Physiol 95(2):628–635

Bui EN (2013) Soil salinity: a neglected factor in plant ecology and biogeography. J Arid Environ 92:14–25

Byrt CS, Munns R, Burton RA, Gilliham M, Wege S (2018) Root cell wall solutions for crop plants in saline soils. Plant Sci 269:47–55

Cao D, Lutz A, Hill CB, Callahan DL, Roessner U (2017) A quantitative profiling method of phytohormones and other metabolites applied to barley roots subjected to salinity stress. Front Plant Sci 7:2070

Carreira PM, Marques JM, Nunes D (2014) Source of groundwater salinity in coastline aquifers based on environmental isotopes (Portugal): natural vs. human interference. A review and reinterpretation. Appl Geochem 41:163–175

Carter DL, Fanning CD (1964) Combining surface mulches and periodic water applications for reclaiming saline soils. Soil Sci Soc Am J 28:564–567

Castellanos MT, Cabello MJ, Cartagena MC, Tarquis AM, Arce A, Ribas F (2012) Nitrogen uptake dynamics, yield and quality as influenced by nitrogen fertilization in 'Piel de sapo' melon. Span J Agric Res 10(3):756–767

Chabra R (1996) Soil salinity and water quality. CRC Press, Taylor & Francis Group, Balkema Publishers, Brookfield, VT

Chaves MM, Maroco JP, Pereira JS (2003) Understanding plant responses to drought—from genes to the whole plant. Funct Plant Biol 30(3):239–264

Chen W, Hou Z, Wu L, Liang Y, Wei C (2010) Effects of salinity and nitrogen on cotton growth in arid environment. Plant Soil 326(1–2):61–73

Chen W, Jin M, Ferré TPA, Liu Y, Xian Y, Shan T, Ping X (2018a) Spatial distribution of soil moisture, soil salinity, and root density beneath a cotton field under mulched drip irrigation with brackish and fresh water. Field Crop Res 215:207–221

Chen W, Wang Z, Jin M, Ferré TPA, Wang J, Huang J, Wang X (2018b) Effect of sodium chloride and manganese in irrigation water on cotton growth. Agron J 110:900–909

Cheng C, Zhang Y, Chen X, Song J, Guo Z, Li K, Zhang K (2018) Co-expression of *AtNHX1* and *TsVP* improves the salt tolerance of transgenic cotton and increases seed cotton yield in a saline field. Mol Breed 38(2):19

Cramer GR, Lynch J, Läuchli A, Epstein E (1987) Influx of Na^+, K^+, and Ca^{2+} into roots of salt-stressed cotton seedlings: effects of supplemental Ca^{2+}. Plant Physiol 83(3):510–516

da Costa MML, Gomes Nobre R, Soares de Lima G, Raj Gheyi H, Wesley Alves Pinheiro F, Sudário Dias A, Almeida dos AnjosSoares L (2016) Saline-sodic soil and organic matter addition in the cultivation of the colored cotton 'BRS Topázio'. Semina: Ciências Agrárias 37 (2):701–704

Da Silva MJ, De Souza JG, Neto MB, Da Silva JV (1992) Selection on 3 cotton cultivars for tolerance to germination under saline conditions. Pesquisa Agropec Brasileira 27:655–659

Davey MW, Stals E, Panis B, Keulemans J, Swennen RL (2005) High-throughput determination of malondialdehyde in plant tissues. Anal Biochem 347(2):201–207

Day JW Jr, Yanez-Arancibia A, Kemp WM, Crump BC (2013) Introduction to estuarine ecology. In: Day JW, Crump BC, Kemp WM, Yáñez-Arancibia A (eds) Estuarine ecology, 2nd edn. Wiley-Blackwell, Hoboken NJ, p 550

Diacono M, Montemurro F (2015) Effectiveness of organic wastes as fertilizers and amendments in salt-affected soils. Agriculture 5(2):221–230

Dodd K (2007) Characterising the soil and plant interactions that affect the growth and nutrition of cotton in sodic Vertosols. PhD Thesis, The University of New England, Armidale, NSW, Australia

Dodd K, Guppy CN, Lockwood PV, Rochester IJ (2013) The effect of sodicity on cotton: does soil chemistry or soil physical condition have the greater role? Crop Pasture Sci 64(8):806–815

Dong H (2012a) Technology and field management for controlling soil salinity effects on cotton. Aus J Crop Sci 6(2):333

Dong HZ (2012b) Underlying mechanisms and related techniques of stand establishment of cotton on coastal saline-alkali soil. China J Appl Ecol 23(2):566–572

Dong H, Li W, Tang W, Li Z, Zhang D (2007) Enhanced plant growth, development and fiber yield of Bt transgenic cotton by an integration of plastic mulching and seedling transplanting. Ind Crop Prod 26:298–306

Dong H, Li W, Tang W, Zhang D (2008) Furrow seeding with plastic mulching increases stand establishment and lint yield of cotton in a saline field. Agron J 100:1640–1646

Dong H, Li W, Tang W, Zhang D (2009) Early plastic mulching increases stand establishment and lint yield of cotton in saline fields. Field Crop Res 111:269–275

Dong H, Li W, Xin C, Tang W, Zhang D (2010a) Late planting of short-season cotton in saline fields of the Yellow River Delta. Crop Sci 50:292–300

Dong H, Kong X, Luo Z, Li W, Xin C (2010b) Unequal salt distribution in the root zone increases growth and yield of cotton. Eur J Agron 33:285–292

Dong H, Kong X, Li W, Tang W, Zhang D (2010c) Effects of plant density and nitrogen and potassium fertilization on cotton yield and uptake of major nutrients in two fields with varying fertility. Field Crop Res 119(1):106–113

Egea I, Pineda B, Ortíz-Atienza A, Plasencia FA, Drevensek S, García-Sogo B, Yuste-Lisbona FJ, Barrero-Gil J, Atarés A, Flores FB, Barneche F, Angosto T, Capel C, Salinas J, Vriezen W, Esch E, Bowler C, Bolarín MC, Moreno V, Lozano R (2018) The SlCBL10 calcineurin B-like protein ensures plant growth under salt stress by regulating Na^+ and Ca^{2+} homeostasis. Plant Physiol 176(2):1676–1693

Fairbairn DJ, Liu W, Schachtman DP, Gomez-Gallego S, Day SR, Teasdale RD (2000) Characterisation of two distinct HKT1-like potassium transporters from Eucalyptus camaldulensis. Plant Mol Biol 43:515–525

Fan Y, Li H, Miguez-Macho G (2013) Global patterns of groundwater table depth. Science 339 (6122):940–943

FAO (2005) Global network on integrated soil management for sustainable use of salt effected soils. http://www.fao.org/ag/AGL/agll/spush/intro.htm

FAO (2019) Management of salt affected soils. Soil Sci Soc Am J. http://www.fao.org/soils-portal/soil-management/management-of-some-problem-soils/salt-affected-soils/more-information-on-salt-affected-soils/en/. Accessed 21 June 2019

Farooq M, Hussain M, Wakeel A, Siddique KHM (2015) Salt stress in maize: effects, resistance mechanisms, and management. A review. Agron Sustain Dev 35(2):461–481

Feinerman E (1983) Crop density and irrigation with saline water. West J Agric Econ 8:134–140

Flexas J, Diaz-Espejo A, Galmés J, Kaldenhoff R, Medrano H, Ribas-Carbo M (2007) Rapid variations of mesophyll conductance in response to changes in CO_2 concentration around leaves. Plant Cell Environ 30(10):1284–1298

Flowers TJ (1999) Salinisation and horticultural production. Sci Hortic 78:1–4

Flowers TJ, Colmer TD (2015) Plant salt tolerance: adaptations in halophytes. Ann Bot 115 (3):327–331

Flowers TJ, Munns R, Colmer TD (2014) Sodium chloride toxicity and the cellular basis of salt tolerance in halophytes. Ann Bot 115(3):419–431

Foyer CH, Noctor G (2005) Redox homeostasis and antioxidant signaling: a metabolic interface between stress perception and physiological responses. Plant Cell 17:1866–1875

Gandahi AW, Kubar A, Sarki MS, Talpur N, Gandahi M (2017) Response of conjunctive use of fresh and saline water on growth and biomass of cotton genotypes. J Basic Appl Sci 13:326–334

Gao SQ, Chen M, Xia LQ, Xiu HJ, Xu ZS, Li LC, Zhao CP, Cheng XG, Ma YZ (2009) A cotton (*Gossypium hirsutum*) DRE-binding transcription factor gene, GhDREB, confers enhanced tolerance to drought, high salt, and freezing stresses in transgenic wheat. Plant Cell Rep 28 (2):301–311

Gao W, Xu FC, Guo DD, Zhao JR, Liu J, Guo YW, Singh PK, Ma XN, Long L, Botella JR, Song CP (2018) Calcium-dependent protein kinases in cotton: insights into early plant responses to salt stress. BMC Plant Biol 18(1):15

Garratt LC, Janagoudar BS, Lowe KC, Anthony P, Power JB, Davey MR (2002) Salinity tolerance and antioxidant status in cotton cultures. Free Radic Biol Med 33(4):502–511

Ghafoor A, Qadir M, Murtaza G (2004) Salt-affected soils: principles of management, vol xxv. Allied Book Centre, Lahore, Pakistan. 304p

Golan-Goldhirsh A, Hankamer B, Lips SH (1990) Hydroxyproline and proline content of cell walls of sunflower, peanut and cotton grown under salt stress. Plant Sci 69(1):27–32

Golldack D, Su H, Quigley F, Kamasani UR, Munoz-Garay C, Balderas E, Popova OV, Bennett J, Bohnert HJ, Pantoja O (2002) Characterization of a HKT-type transporter in rice as a general alkali cation transporter. Plant J 31:529–542

Gouia H, Ghorbal MH, Touraine B (1994) Effects of NaCl on flows of N and mineral ions and on NO_3-reduction rate within whole plants of salt-sensitive bean and salt-tolerant cotton. Plant Physiol 105(4):1409–1418

Hagin J, Sneh M, Lowengart-Aycicegi A (2002) Fertigation – fertilization through irrigation. IPI Research Topics No. 23. Ed. by Johnston AE. International Potash Institute, Basel, Switzerland

Hajiboland R, Aliasgharzadeh N, Laiegh SF, Poschenrieder C (2010) Colonization with arbuscular mycorrhizal fungi improves salinity tolerance of tomato (*Solanum lycopersicum* L.) plants. Plant Soil 331(1–2):313–327

Hanif M, Noor E, Murtaza N, Qayyum A, Malik W (2008) Assessment of variability for salt tolerance at seedling stage in *Gossypium hirsutum* L. J Food Agric Environ 6:134–138

Haque SA (2006) Salinity problems and crop production in coastal regions of Bangladesh. Pak J Bot 38(5):1359–1365

Harshavardhan VT, Govind G, Kalladan R, Sreenivasulu N, Hong CY (2018) Cross-protection by oxidative stress: improving tolerance to abiotic stresses including salinity. In: Kumar V, Wani SH, Suprasanna P, Tran LP (eds) Salinity responses and tolerance in plants, vol 1. Springer, Berlin, pp 283–305

Hasegawa PM, Bressan RA, Zhu JK, Bohnert HJ (2000) Plant cellular and molecular response to high salinity. Annu Rev Plant Physiol Plant Mol Biol 51:463–499

Hayat S, Hayat Q, Alyemeni MN, Wani AS, Pichtel J, Ahmad A (2012) Role of proline under changing environments: a review. Plant Signal Behav 7(11):1456–1466

He C, Yan J, Shen G, Fu L, Holaday AS, Auld D, Blumwald E, Zhang H (2005) Expression of an Arabidopsis vacuolar sodium/proton antiporter gene in cotton improves photosynthetic performance under salt conditions and increases fiber yield in the field. Plant Cell Physiol 46 (11):1848–1854

He Z, He C, Zhang Z, Zou Z, Wang H (2007) Changes of antioxidative enzymes and cell membrane osmosis in tomato colonized by arbuscular mycorrhizae under NaCl stress. Colloids Surf B: Biointerfaces 59(2):128–133

Heuer B (2003) Influence of exogenous application of proline and glycinebetaine on growth of salt-stressed tomato plants. Plant Sci 165(4):693–699

Higbie SM, Wang F, Stewart JM, Sterling TM, Lindemann WC, Hughs E, Zhang J (2010) Physiological response to salt (NaCl) stress in selected cultivated Tetraploid cottons. Int J Agron 2010:1–12

Hoang TML, Tran TN, Nguyen TKT, Williams B, Wurm P, Bellairs S, Mundree S (2016) Improvement of salinity stress tolerance in rice: challenges and opportunities. Agronomy 6 (4):54

Horie T, Schroeder JI (2004) Sodium transporters in plants. Diverse genes and physiological functions. Plant Physiol 136:2457–2462

Horie T, Yoshida K, Nakayama H, Yamada K, Oiki S, Shinmyo A (2001) Two types of HKT transporters with different properties of Na^+ and K^+ transport in Oryza sativa. Plant J 27:129–138

Horie ST, Motoda J, Kubo M, Yang H, Yoda K, Horie R, Chan WY, Leung HY, Hattori K, Konomi M, Osumi M (2005) Enhanced salt tolerance mediated by AtHKT1 transporter-induced Na+ unloading from xylem vessels to xylem parenchyma cells. Plant J 44(6):928–938

Hossain MS, Dietz KJ (2016) Tuning of redox regulatory mechanisms, reactive oxygen species and redox homeostasis under salinity stress. Front Plant Sci 7:548

Hu Y, Schmidhalter U (2005) Drought and salinity: a comparison of their effects on mineral nutrition of plants. J Plant Nutr Soil Sci 168(4):541–549

Hussain S, Khalid MF, Saqib M, Ahmad S, Zafar W, Rao MJ, Morillon R, Anjum MA (2018) Drought tolerance in citrus rootstocks is associated with better antioxidant defense mechanism. Acta Physiol Plant 40(8):135

Ibrahim W, Ahmed IM, Chen X, Wu F (2017) Genotype-dependent alleviation effects of exogenous GSH on salinity stress in cotton is related to improvement in chlorophyll content, photosynthetic performance, and leaf/root ultrastructure. Environ Sci Pollut Res 24(10):9417–9427

Ilyas M, Qureshi RH, Qadir MA (1997) Chemical changes in a saline-sodic soil after gypsum application and cropping. Soil Technol 10(3):247–260

Ishikawa T, Shabala S (2019) Control of xylem Na^+ loading and transport to the shoot in rice and barley as a determinant of differential salinity stress tolerance. Physiol Plant 165(3):619–631

Jabeen R, Ahmad R (2009) Alleviation of the adverse effects of salt stress by foliar application of sodium antagonistic essential minerals of cotton (*Gossypium hirsutum* L.). Pak J Bot 41 (5):2199–2208

Jafri AZ, Rafiq A (1994) Plant growth and ionic distribution in cotton (*Gossypium hirsutum* L.) under saline environment. Pak J Bot 26:105

Jaleel CA, Sankar B, Sridharan R, Panneerselvam R (2008) Soil salinity alters growth, chlorophyll content, and secondary metabolite accumulation in Catharanthusroseus. Turk J Biol 32 (2):79–83

Jamil A, Riaz S, Ashraf M, Foolad MR (2011) Gene expression profiling of plants under salt stress. Crit Rev Plant Sci 30(5):435–458

Jubany-Marí T, Munné-Bosch S, López-Carbonell M, Alegre L (2009) Hydrogen peroxide is involved in the acclimation of the Mediterranean shrub, Cistus albidus L., to summer drought. J Exp Bot 60:107–120

Kahlown MA, Azam M (2002) Individual and combined effect of waterlogging and salinity on crop yields in the Indus basin. Irrig Drain 51(4):329–338

Kang Y, Wang R, Wan S, Hu W, Jiang S, Liu S (2012) Effects of different water levels on cotton growth and water use through drip irrigation in an arid region with saline ground water of Northwest China. Agric Water Manage 109:117–126

Karlberg L, de Vries FWTP (2004) Exploring potentials and constraints of low-cost drip irrigation with saline water in sub-Saharan Africa. Phys Chem Earth 29:1035–1042

Kawakami EM, Oosterhuis DM, Snider JL (2010) Effect of salinity on cotton nitrogen uptake and assimilation of urea applied with N-(n-Butyl) thiophosphorctriamide and dicyandiaminde. Summaries Arkansas Cotton Research. AAES Res Ser 589:40–45

Khan TM, Saeed M, Mukhtar MS, Khan AM (2001) Assesment of variation for salinity tolerance in some hybrids of cotton (*Gossypium hirsutum* L.). Int J Agric Biol 3:167–170

Khare T, Kumar V, Kishor PB (2015) Na^+ and Cl^- ions show additive effects under NaCl stress on induction of oxidative stress and the responsive antioxidative defense in rice. Protoplasma 252 (4):1149–1165

Kinraide TB (1999) Interactions among Ca^{2+}, Na^+ and K^+ in salinity toxicity: quantitative resolution of multiple toxic and ameliorative effects. J Exp Bot 50(338):1495–1505

Koca H, Bor M, Özdemir F, Türkan I (2007) The effect of salt stress on lipid peroxidation, antioxidative enzymes and proline content of sesame cultivars. Environ Exp Bot 60(3):344–351

Kong X, Luo Z, Dong H, Eneji AE, Li W (2011) Effects of non-uniform root zone salinity on water use, Na^+ recirculation, and Na^+ and H^+ flux in cotton. J Exp Bot 63(5):2105–2116

Kothari N, Campbell BT, Dever JK, Hinze LL (2016) Combining ability and performance of cotton germplasm with diverse seed oil content. Crop Sci 56(1):19–29

Lawlor DW, Cornic G (2002) Photosynthetic carbon assimilation and associated metabolism in relation to water deficits in higher plants. Plant Cell Environ 25(2):275–294

Ledbetter CA (1987) Heritability of salt tolerance during germination and emergence in short staple cotton. Diss Abstr Int Sci Eng 47(11):113

Leidi EO, Saiz JF (1997) Is salinity tolerance related to Na accumulation in upland cotton (*Gossypium hirsutum*) seedlings? Plant Soil 190(1):67–75

Liang W, Ma X, Wan P, Liu L (2018) Plant salt-tolerance mechanism: a review. Biochem Biophys Res Commun 495(1):286–291

Liu MX, Yang JS, Li XM, YU M, Wang J (2012a) Effects of irrigation water quality and drip tape arrangement on soil salinity, soil moisture distribution, and cotton yield (*Gossypium hirsutum* L.) under mulched drip irrigation in Xinjiang, China. J Integr Agric 11(3):502–511

Liu YD, Yin ZJ, Yu JW, Li J, Wei HL, Han XL, Shen FF (2012b) Improved salt tolerance and delayed leaf senescence in transgenic cotton expressing the Agrobacterium IPT gene. Biol Plantarum 56(2):237–246

Liu G, Li X, Jin S, Liu X, Zhu L, Nie Y, Zhang X, Zhang J (2014a) Overexpression of rice NAC gene SNAC1 improves drought and salt tolerance by enhancing root development and reducing transpiration rate in transgenic cotton. PLoS One 9(1):e86895

Liu S, Dong Y, Xu L, Kong J (2014b) Effects of foliar applications of nitric oxide and salicylic acid on salt-induced changes in photosynthesis and anti-oxidative metabolism of cotton seedlings. Plant Growth Regul 73(1):67–78

Liu JF, Zhang SL, Tang HL, Wu LZ, Dong LJ, Liu LD, Che WL (2015) Overexpression of an Aeluropuslittoralis Parl. potassium transporter gene, AlHAK1, in cotton enhances potassium uptake and salt tolerance. Euphytica 203(1):197–209

Lobell DB, Ortiz-Monasterio JI, Gurrola FC, Valenzuela L (2007) Identification of saline soils with multiyear remote sensing of crop yields. Soil Sci Soc Am J 71(3):777–783

Lu H, Lashari MS, Liu X, Ji H, Li L, Zheng J, Kibue GW, Joseph S, Pan G (2015) Changes in soil microbial community structure and enzyme activity with amendment of biochar-manure compost and pyroligneous solution in a saline soil from Central China. Eur J Soil Biol 70:67–76

Lu Z, Lu J, Pan Y, Lu P, Li X, Cong R, Ren T (2016) Anatomical variation of mesophyll conductance under potassium deficiency has a vital role in determining leaf photosynthesis. Plant Cell Environ 39(11):2428–2439

Maas EV (1986) Salt tolerance of plants. Appl Agric Res 1:12–26

Maas EV, Hoffman GJ (1977) Crop salt tolerance–current assessment. J Irr Drain Div 103:115–134

Martinez-Beltran J, LiconaManzur C (2005) Overview of salinity problems in the world and FAO strategies to address the problem. In: International salinity forum managing saline soils and water: science, technology and social issues, Riverside Convention Center, Riverside, California, USA, 25–28 April 2005, pp 311–314

Meloni DA, Oliva MA, Ruiz HA, Martinez CA (2001) Contribution of proline and inorganic solutes to osmotic adjustment in cotton under salt stress. J Plant Nutr 24(3):599–612

Meloni DA, Oliva MA, Martinez CA, Cambraia J (2003) Photosynthesis and activity of superoxide dismutase, peroxidase and glutathione reductase in cotton under salt stress. Environ Exp Bot 49 (1):69–76

Metternicht GI, Zink JA (2003) Remote sensing of soil salinity: potential and constraints. Remote Sens Environ 85:1–20

Miller G, Suzuki N, Ciftci-Yilmaz S, Mittler R (2010) Reactive oxygen species homeostasis and signalling during drought and salinity stresses. Plant Cell Environ 33(4):453–467

Min W, Guo H, Zhou G, Zhang W, Ma L, Ye J, Hou Z, Wu L (2016) Soil salinity, leaching, and cotton growth as affected by saline water drip irrigation and N fertigation. Acta Agric Scand B Soil Plant Sci 66(6):489–501

Min W, Guo H, Hu Z, Zhang H, Ye J, Hou Z (2017) Cotton growth and the fate of N fertilizer as affected by saline water irrigation and N fertigation in a drip-irrigated field. Acta Agric Scand B Soil Plant Sci 67(8):712–722

Mittler R (2017) ROS are good. Trends Plant Sci 22(1):11–19

Moseley WG (2001) Sahelian 'white gold' and rural poverty-environment interactions: the political ecology of cotton production, environmental change, and household food economy in Mali. Doctoral dissertation, University of Georgia

Munns R (2002) Comparative physiology of salt and water stress. Plant Cell Environ 25 (2):239–250

Munns R (2005) Genes and salt tolerance: bringing them together. New Phytol 167:645–663

Munns R, Tester M (2008) Mechanisms of salinity tolerance. Annu Rev Plant Biol 59:651–681

Munns R, James RA, Läuchli A (2006) Approaches to increasing the salt tolerance of wheat and other cereals. J Exp Bot 57(5):1025–1043

Murtaza G, Ghafoor A, Qadir M (2006) Irrigation and soil management strategies for using saline-sodic water in a cotton–wheat rotation. Agric Water Manage 81(1–2):98–114

Murtaza G, Murtaza B, Usman HM, Ghafoor A (2013) Amelioration of saline-sodic soil using gypsum and low quality water in following sorghum-berseem crop rotation. Int J Agric Biol 15 (4):640–648

Oldeman LR, Hakkeling RTA, Sombroek WG (1991) World map of the status of human-induced soil degradation: an explanatory note. United Nations Environment Programme - I11. Global Assessment of Soil Degradation GLASOD, October 1990. International Soil Reference and Information Centre, Nairobi

Ondrasek O, Rengel Z, Veres S (2011) Soil salinisation and salt stress in crop production. InTech, Croatia. https://doi.org/10.5772/22248

Pasapula V, Shen G, Kuppu S, Paez-Valencia J, Mendoza M, Hou P, Chen J, Qiu X, Zhu L, Zhang X, Auld D, Blumwald E, Zhang H, Gaxiola R, Paytn P (2011) Expression of an Arabidopsis vacuolar H^+ pyrophosphatase gene (AVP1) in cotton improves drought and salt tolerance and increases fibre yield in the field conditions. Plant Biotechnol J 9(1):88–99

Pasternak D, De Malach Y (1994) Crop irrigation with saline water. Handbook of plant and crop stress. Marcel Dekker, New York, pp 599–622

Peng Z, He S, Sun J, Pan Z, Gong WF, Lu Y, Du X (2016) Na^+ compartmentalization related to salinity stress tolerance in upland cotton (*Gossypium hirsutum*) seedlings. Sci Rep 6:34548

Qadir M, Ghafoor A, Murtaza G (2000) Amelioration strategies for saline soils: a review. Land Degrad Dev 11:501–521

Qiu QS, Guo Y, Dietrich MA, Schumaker KS, Zhu JK (2002) Regulation of SOS1, a plasma membrane Na^+/H^+ exchanger in Arabidopsis thaliana, by SOS2 and SOS3. Proc Natl Acad Sci U S A 99(12):8436–8441

Rahman MH, Ahmad A, Wang X, Wajid A, Nasim W, Hussain M, Ahmad B, Ahmad I, Ali Z, Ishaque W, Awais M, Shelia V, Ahmad S, Fahad S, Alam M, Ullah H, Hoogenboom G (2018) Multi-model projections of future climate and climate change impacts uncertainty assessment for cotton production in Pakistan. Agric For Meteorol 253–254:94–113

Rathert G (1982) Influence of extreme K/Na ratios and high substrate salinity on plant metabolism of crops differing in salt tolerance. VI. Mineral distribution variability among different salt-tolerant cotton varieties. J Plant Nutr 5:183–193

Rao AQ, Bakhsh A, Kiani S, Shahzad K, Shahid AL, Husnain T, Riazuddin S (2009) The myth of plant transformation. Biotechnol Adv 27(6):753–763

Reich M, Aghajanzadeh T, Helm J, Parmar S, Hawkesford MJ, De Kok LJ (2017) Chloride and sulfate salinity differently affect biomass, mineral nutrient composition and expression of sulfate transport and assimilation genes in Brassica rapa. Plant Soil 411(1–2):319–332
Reid RJ, Smith FA (2000) The limits of sodium/calcium interactions in plant growth. Funct Plant Biol 27(7):709–715
Rengasamy P (2006) World salinization with emphasis on Australia. J Exp Bot 57:1017–1023
Rengasamy P (2010) Soil processes affecting crop production in salt-affected soils. Funct Plant Biol 37(7):613–620
Rochester IJ (2010) Phosphorus and potassium nutrition of cotton: interaction with sodium. Crop Pasture Sci 61:825–834
Rozema J, Flowers T (2008) Crops for a salinized world. Science 322(5907):1478–1480
Sahin U, Anapali O, Hanay A (2002) The effect of consecutive applications of leaching water applied in equal, increasing or decreasing quantities on soil hydraulic conductivity of a saline–sodic soil in the laboratory. Soil Use Manage 18:152–154
Sandoval FM, Benz LC (1966) Effect of bare fallow, barley and grass on salinity of a soil over a saline water table. Soil Sci Soc Am J 30:392–397
Sarangi SK, Maji B, Mandal UK, Mahanta KK, Mandal S, Sharma PC (2017) Raised bed sowing – a climate change adaptive maize cultivation practice for coastal saline region. Paper presented orally at 5th National Seminar of Indian Society of Soil Salinity and Water Quality (ISSSWQ), Swami Keshwanand Rajasthan Agricultural University, Bikaner, 21–23 January 2017, pp 71–72
Savci S (2012) An agricultural pollutant: chemical fertilizer. Int J Environ Sci Dev 3(1):77–80
Schachtman DP, Schroeder JI (1994) Structure and transport mechanism of a high affinity potassium uptake transporter from higher plants. Nature 370:655–658
Shahbaz M, Ashraf M (2013) Improving salinity tolerance in cereals. Crit Rev Plant Sci 32:237–249
Shahzad S, Khan MY, Zahir ZA, Asghar HN, Chaudhry UK (2017) Comparative effectiveness of different carriers to improve the efficacy of bacterial consortium for enhancing wheat production under salt affected field conditions. Pak J Bot 49(4):1523–1530
Sharma P, Jha AB, Dubey RS, Pessarakli M (2012) Reactive oxygen species, oxidative damage, and antioxidative defense mechanism in plants under stressful conditions. J Bot 2012:217037
Shen G, Wei J, Qiu X, Hu R, Kuppu S, Auld D, Blumwald E, Gaxiola R, Payton P, Zhang H (2015) Co-overexpression of AVP1 and AtNHX1 in cotton further improves drought and salt tolerance in transgenic cotton plants. Plant Mol Biol Rep 33(2):167–177
Shrivastava P, Kumar R (2015) Soil salinity: a serious environmental issue and plant growth promoting bacteria as one of the tools for its alleviation. Saudi J Biol Sci 22(2):123–131
Soares LA, Fernandes PD, de Lima GS, Suassuna JF, Brito MEB, da Sá FV (2018) Growth and fiber quality of colored cotton under salinity management strategies. Rev Bras Eng Agríc Ambient 22(5):332–337
Song J, Zhang R, Yue D, Chen X, Guo Z, Cheng C, Hu M, Zhang J, Zhang K (2018) Co-expression of ApGSMT2g and ApDMT2g in cotton enhances salt tolerance and increases seed cotton yield in saline fields. Plant Sci 274:369–382
Sumner ME (ed) (1999) Handbook of soil science. CRC Press, Boca Raton, FL
Sun X, Liu Y (2001) Test on criteria of evaluating salt tolerance of cotton cultivars. Zuo Wu Xue Bao 27(6):794–801
Sunarpi, Horie T, Motoda J, Kubo M, Yang H, Yoda K, Horie R, Chan WY, Leung HY, Hattori K, Konomi M, Osumi M, Yamagami M, Schroeder JI, Uozumi N (2005) Enhanced salt tolerance mediated by AtHKT1 transporter-induced Na^+ unloading from xylem vessels to xylem parenchyma cells. Plant J 44(6):928–938
Suzuki K (1999) Effect of continuous compost application on water-stable soil macroaggregation in a field subjected to double cropping. Soil Sci Plant Nutr 45:1003–1007
Szabolcs I (1989) Salt-affected soils. CRC Press, Boca Raton, FL

Tang W, Luo Z, Wen SM, Dong HZ, Xin CS, Li WJ (2007) Comparison of inhibitory effects on leaf photosynthesis in cotton seedlings between drought and salinity stress. Cotton Sci 19:28–32

Tariq M, Yasmeen A, Ahmad S, Hussain N, Afzal MN, Hasanuzzaman M (2017) Shedding of fruiting structures in cotton: factors, compensation and prevention. Trop Subtrop Agroecosyst 20(2):251–262

Tariq M, Afzal MN, Muhammad D, Ahmad S, Shahzad AN, Kiran A, Wakeel A (2018) Relationship of tissue potassium content with yield and fiber quality components of Bt cotton as influenced by potassium application methods. Field Crop Res 229:37–43

Thomas DSG (ed) (2011) Arid zone geomorphology: process, form and change in drylands. Wiley, Hoboken, NJ, p 648

Usman M, Ahmad A, Ahmad S, Irshad M, Khaliq T, Wajid A, Hussain K, Nasim W, Chattha TM, Trethowan R, Hoogenboom G (2009) Development and application of crop water stress index for scheduling irrigation in cotton (*Gossypium hirsutum* L.) under semiarid environment. J Food Agric Environ 7(3&4):386–391

Van Breemen N, Buurman P (2002) Soil formation. Springer Science & Business Media, New York

Villa-Castorena M, Ulery AL, Catalán-Valencia EA, Remmenga MD (2003) Salinity and nitrogen rate effects on the growth and yield of Chile pepper plants. Soil Sci Soc Am J 67(6):1781–1789

Wang R, Kang Y, Wan S, Hu W, Liu S, Jiang S, Liu S (2012) Influence of different amounts of irrigation water on salt leaching and cotton growth under drip irrigation in an arid and saline area. Agric Water Manage 110:109–117

Wang R, Kang Y, Wan S (2015) Effects of different drip irrigation regimes on saline–sodic soil nutrients and cotton yield in an arid region of Northwest China. Agric Water Manage 153:1–8

Wang N, Qi H, Qiao W, Shi J, Xu Q, Zhou H, Yan G, Huang Q (2017) Cotton (*Gossypium hirsutum* L.) genotypes with contrasting K^+/Na^+ ion homeostasis: implications for salinity tolerance. Acta Physiol Plant 39(3):77

Warrence NJ, Bauder JW, Pearson KE (2002) Basics of salinity and sodicity effects on soil physical properties. Bozeman, MT, Department of Land Resources and Environmental Sciences, Montana State University, pp 1–29

Waszczak C, Carmody M, Kangasjärvi J (2018) Reactive oxygen species in plant signaling. Annu Rev Plant Biol 69:209–236

Wei Y, Xu Y, Lu P, Wang X, Li Z, Cai X, Zhou Z, Wang Y, Zhang Z, Lin Z, Liu F, Wang K (2017) Salt stress responsiveness of a wild cotton species (*Gossypium klotzschianum*) based on transcriptomic analysis. PLoS One 12(5):e0178313

Wilkinson S, Davies WJ (2002) ABA-based chemical signalling: the co-ordination of responses to stress in plants. Plant Cell Environ 25(2):195–210

Wu CA, Yang GD, Meng QW, Zheng CC (2004) The cotton GhNHX1 gene encoding a novel putative tonoplast Na^+/H^+ antiporter plays an important role in salt stress. Plant Cell Physiol 45 (5):600–607

Wu QS, Zou YN, He XH (2010) Contributions of arbuscular mycorrhizal fungi to growth, photosynthesis, root morphology and ionic balance of citrus seedlings under salt stress. Acta Physiol Plant 32(2):297–304

Wu H, Zhang X, Giraldo JP, Shabala S (2018) It is not all about sodium: revealing tissue specificity and signalling roles of potassium in plant responses to salt stress. Plant Soil 431(1–2):1–17

Xie Z, Duan L, Tian X, Wang B, Eneji AE, Li Z (2008) Coronatine alleviates salinity stress in cotton by improving the antioxidative defense system and radical-scavenging activity. J Plant Physiol 165(4):375–384

Xie T, Liu X, Sun T (2011) The effects of groundwater table and flood irrigation strategies on soil water and salt dynamics and reed water use in the Yellow River Delta, China. Ecol Model 222 (2):241–252

Xu WL, Zhang DJ, Wu YF, Qin LX, Huang GQ, Li J, Li L, Li XB (2013) Cotton PRP5 gene encoding a proline-rich protein is involved in fiber development. Plant Mol Biol 82 (4–5):353–365

You J, Chan Z (2015) ROS regulation during abiotic stress responses in crop plants. Front Plant Sci 6:1092

Younis A, Riaz A, Ahmed I, Siddique MI, Tariq U, Hameed M, Nadeem M (2014) Anatomical changes induced by NaCl stress in root and stem of *Gazania harlequin* L. Agric Commun 2:8–14

Yu LH, Wu SJ, Peng YS, Liu RN, Chen X, Zhao P, Xu P, Zhu JB, Jiao GL, Pei Y, Xiang CB (2016) Arabidopsis EDT 1/HDG 11 improves drought and salt tolerance in cotton and poplar and increases cotton yield in the field. Plant Biotechnol J 14(1):72–84

Zeng Y, Li Q, Wang H, Zhang J, Du J, Feng H, Blumwald E, Yu L, Xu G (2018) Two NHX-type transporters from Helianthus tuberosus improve the tolerance of rice to salinity and nutrient deficiency stress. Plant Biotechnol J 16(1):310–321

Zhang HJ, Dong H, Shi YJ, Chen SY, Zhu YH (2007) Transformation of cotton (Gossypium hirsutum) with *AhCMO* gene and the expression of salinity tolerance. Acta Agron Sin 33:1073–1078

Zhang H, Dong H, Li W, Sun Y, Chen S, Kong X (2009) Increased glycine betaine synthesis and salinity tolerance in AhCMO transgenic cotton lines. Mol Breed 23(2):289–298

Zhang H, Shen G, Kuppu S, Gaxiola R, Payton P (2011a) Creating drought-and salt tolerant cotton by overexpressing a vacuolar pyrophosphatase gene. Plant Signal Behav 6(6):861–863

Zhang X, Zhen J, Li Z, Kang D, Yang Y, Kong J, Hua J (2011b) Expression profile of early responsive genes under salt stress in upland cotton (*Gossypium hirsutum* L.). Plant Mol Biol Rep 29(3):626–637

Zhang D, Li W, Xin C, Tang W, Eneji AE, Dong H (2012a) Lint yield and nitrogen use efficiency of field-grown cotton vary with soil salinity and nitrogen application rate. Field Crop Res 138:63–70

Zhang HJ, Dong HZ, Li WJ, Zhang DM (2012b) Effects of soil salinity and plant density on yield and leaf senescence of field-grown cotton. J Agron Crop Sci 198(1):27–37

Zhang L, Ma H, Chen T, Pen J, Yu S, Zhao X (2014) Morphological and physiological responses of cotton (*Gossypium hirsutum* L.) plants to salinity. PLoS One 9(11):e112807

Zhang T, Wang T, Liu KS, Wang L, Wang K, Zhou Y (2015) Effects of different amendments for the reclamation of coastal saline soil on soil nutrient dynamics and electrical conductivity responses. Agric Water Manage 159:115–122

Zhang K, Song J, Chen X, Yin T, Liu C, Li K, Zhang J (2016) Expression of the Thellungiella halophila vacuolar H^+-pyrophosphatase gene (TsVP) in cotton improves salinity tolerance and increases seed cotton yield in a saline field. Euphytica 211(2):231–244

Zhang M, Cao Y, Wang Z, Wang ZQ, Shi J, Liang X, Song W, Chen Q, Lai J, Jiang C (2018) A retrotransposon in an HKT1 family sodium transporter causes variation of leaf Na^+ exclusion and salt tolerance in maize. New Phytol 217(3):1161–1176

Zheng C, Jiang D, Liu F, Dai T, Jing Q, Cao W (2009) Effects of salt and water logging stresses and their combination on leaf photosynthesis, chloroplast ATP synthesis, and antioxidant capacity in wheat. Plant Sci 176:575–582

Zhu JK (2000) Genetic analysis of plant salt tolerance using Arabidopsis. Plant Physiol 124 (3):941–948

Zhu JK (2016) Abiotic stress signaling and responses in plants. Cell 167(2):313–324

Zhu YN, Shi DQ, Ruan MB, Zhang LL, Meng ZH, Liu J, Yang WC (2013) Transcriptome analysis reveals crosstalk of responsive genes to multiple abiotic stresses in cotton (*Gossypium hirsutum* L.). PLoS One 8(11):e80218

Chapter 20
Heat Stress in Cotton: Responses and Adaptive Mechanisms

Fiaz Ahmad, Asia Perveen, Noor Mohammad, Muhammad Arif Ali, Muhammad Naeem Akhtar, Khurram Shahzad, Subhan Danish, and Niaz Ahmed

Abstract Cotton is vital cash besides fiber crop and plays pivotal role in economy in many countries. It thrives well under optimal temperature. Too high and too low temperatures affect badly its growth and yield. Too low temperature affects its germination and seedling establishment stages. Particularly, high temperatures influence many physiological and biochemical processes within cotton plant that result in poor seed cotton yield. Several researches in different agroecological zones employed different agronomic practices and modern breeding techniques to mitigate the heat stress for better cotton production. A bevy of literature regarding heat stress is presented here.

Keywords Global warming · Climate change · *Gossypium hirsutum* · Abiotic stresses

Abbreviations

APX	Ascorbate peroxidase
ASC	Ascorbate
B	Boron
Ca	Calcium
CAT	Catalase
CER	CO_2 exchange rate
CICR	Central Institute for Cotton Research
GDP	Gross domestic product
GHG	Greenhouse gas

F. Ahmad (✉) · A. Perveen · N. Mohammad
Physiology/Chemistry Section, Central Cotton Research Institute, Multan, Pakistan

M. A. Ali · K. Shahzad · S. Danish · N. Ahmed
Department of Soil Science, Bahauddin Zakariya University, Multan, Pakistan

M. N. Akhtar
Department of Soil and Environmental Sciences, MNS University of Agriculture, Multan, Pakistan

Pesticide Laboratory, Multan, Pakistan

S. Ahmad, M. Hasanuzzaman (eds.), *Cotton Production and Uses*,
https://doi.org/10.1007/978-981-15-1472-2_20

GPX	Glutathione peroxidase
GSH	Glutathione
HSFs	Heat shock factors
HSPs	Heat shock proteins
K	Potassium
LAI	Leaf area index
LAR	Leaf area ratio
LEA	Late embryogenesis abundant
LEL	Leaf electrolyte leakage
Mg	Magnesium
Mn	Manganese
N	Nitrogen
NAR	Net assimilation rate
POD	Peroxidases
PRK	Phosphoribulokinase
PSII	Photosystem II
ROS	Reactive oxygen species
RuBP	Ribulose-1,5-biphosphate
SA	Salicylic acid
Se	Selenium
SOD	Superoxide dismutase

20.1 Introduction: Climate Change Scenario

20.1.1 Global Warming and Its Impact on Agriculture

The worlds' agricultural growth has declined from 3.2% in 1980s to 2% in 2000, which is alarming and a threat to food security (Ahmad et al. 2014, 2018; Abbas and Ahmad 2018; Ahmad and Raza 2014; Ali et al. 2011, 2013a, b, 2014a, b; Usman et al. 2009). The change in climate conditions particularly rise in temperature is the major factor affecting growth of agricultural sector leading to food security at risk (Christensen and Christensen 2007; Ahmad et al. 2017; Amin et al. 2017, 2018; Rahman et al. 2018; Tariq et al. 2017, 2018). Temperature is anticipated to rise by 2–3 °C in the next 25–45 years. The rise in temperature will also affect rainfall pattern making it more erratic. Pakistan is anticipated to be one of the most vulnerable countries in South Asia to climate change. Projected increase in CO_2 concentration is anticipated to raise the mean temperature from 1.4 to 5.8 °C resulting rise of 20–149 cm in sea level in future (IPCC 2007).

Agriculture sector itself, although a noteworthy contributor to gross domestic product (GDP) and most vulnerable to climate change, may also harm the environment through greenhouse gas (GHG) emissions adding 20% in the form of methane, nitrous oxide, and carbon dioxide. About 37% of the total worlds' emissions from agriculture production are accumulating from Asia and the Pacific (ADB 2009).

Table 20.1 Greenhouse gasses emissions through different activities associated with agriculture

Source/activity	Emission ($Mt\ CO_2$)	Year of estimate	Reference
Emissions of GHG through direct agricultural activities	5120–6116	2005	Smith et al. (2007)
Emissions of GHG through indirect agricultural activities	2198–6567	2008	van Der Werf et al. (2009)
Fertilizer manufacturing	282–575	2007	Bellarby et al. (2008)
Utilization of energy for animal feed production	60	2005	Steinfield et al. (2006)
Pesticide industries	3–140	2007	Bellarby et al. (2008)

Earth temperature is increasing exponentially; it has been raised doubled in number as compared to 50 years ago; greenhouse gases (GHGs) being the major cause. Since 2000, the emission of these gases is increasing due to burgeoning population and industrialization, thereby the number has reached up to 6 billion metric tons of "CO_2 equivalent" worldwide, which is more than 20% increase. Greenhouse gases added equivalent to CO_2 to the environment by different components of agriculture sector are given in Table 20.1.

Now a days global warming is the main issue for rapid cause of environmental adversities that needs to be addressed. Temperature is increasing and causing devastating effects to our planet and our crops. It results in poor germination, poor seedling emergence, and aberrant vegetative and reproductive growth. High temperatures have direct influence on increasing the rate of plants reproductive growth that shortens period for photosynthesis thereby restricting ideal seed production. Plants need optimum growth conditions as both higher and lower temperature interferes with the robust growth of plant, contrarily crop species differ in their behavior towards changing temperature some are highly tolerant while some exhibit sensitivity.

Geographical location has significant importance for the change in climate for crops growth; therefore, Pakistan is more prone to climate change due to its geographical location (Janjua et al. 2010). Precipitation is also decreased with the raise in temperature. Mean temperature across the country has increased by 0.5 °C in the past 30 years, and forecasts indicate a further increase of 1.4–3.7 °C by 2060: higher than the expected global average. Schlenker et al. (2006) estimated impact of climate uncertainty on crop yields in United States and established threshold levels of temperatures for different crops such as 29 °C for corn and soybeans and 33 °C for cotton. Moreover, temperature more than optimum requirement halts growth of aforementioned crops with severe yield losses.

20.1.2 *Cotton Production in the Perspective of Global Warming*

A brief change in temperature leading to exceed plants thermal capability, or thermal capacity of a plant is considered heat stress (Gür et al. 2010). Cotton yield is determined by the surrounding environment which has dramatic effects on growth and development. The temperature fluctuations affect growth and developmental processes and thus determine up to 70% yield variations in cotton (Farooq et al. 2015; Luo et al. 2014; Nasim et al. 2016; Rahman et al. 2017). Heat stress is often associated with other ecological stresses like drought (Rehman 2006).

Heat stress decreases the potential of the crop, and it is estimated that crop exhibit only 25% of its potential due to such environmental stress (Boyer 1982). High temperature and change in rainfall pattern are major drawbacks in achieving higher and stable cotton yields (Bange and Milroy 2004; Gwimbi and Mundoga 2010; Iqbal et al. 2016). Cotton growth and development is maximum at 33 °C, and significant decline in fruit retention is observed above 36 °C (Luo 2011; Nasim et al. 2016; Singh et al. 2007). Heat stress is a severe threat to cotton productivity globally (Hall 2001).

20.2 Effects on Cotton Plant

High-temperature stress influences cotton plant in a number of ways such as by inducing morphological, physiological changes and biochemical alterations thus limiting the crop performance and lower seed cotton production. Heat stress has effects on seed germination, seedling and root growth. High temperature effects on diverse growth phases in cotton are depicted in Fig. 20.1.

The temperature range of 28–30 °C is considered optimum for seed germination besides cotton seedling development. Cotton root growth is maximum at day/night temperatures of 30/22–35/27 °C and rise in temperatures to 40/32 °C alter root distribution pattern resulting in limited downward extension of roots (Reddy et al. 1997a, b). Temperatures higher than 30 °C, but not exceeding 40 °C, increase seed germination rate leading to early seedling development. Increase in temperature beyond 40 °C has damaging effects on cotton seedlings. In such conditions, heat-tolerant genotypes withstand better due to activation of acquired thermo-tolerance until the temperature approaches 37.7–40 °C (Burke 2001). The extent of damage is much higher in heat-sensitive genotypes, and the entire fields may be wiped out due to rapid loss of water when hot winds blow across the cotton fields. Such events often take place in major cotton-growing countries like India and Pakistan.

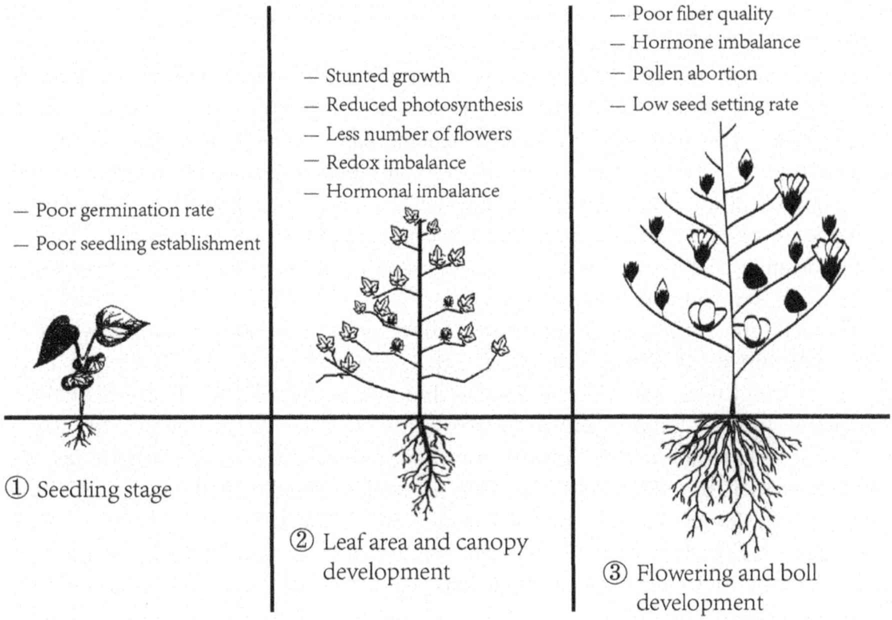

Fig. 20.1 Effect of higher temperature on agronomic besides physiological attributes of cotton at numerous stages of development (Zafar et al. 2018)

20.2.1 Effects on Plant Growth Besides Development

Although, cotton has a well-defined growth and developmental pattern, however, it is highly temperature dependent (Iqbal et al. 2016). The rise in temperature generally results in accelerated growth of plants making crop to mature early and at the same time limiting the crop to achieve its genetic potential (Reddy and Zhao 2005). The developmental processes of the plants are more rapid during increased daytime temperature (Reddy et al. 1996), while the leaf expansion rate in cotton is more under dark conditions (Krieg and Sung 1986). Continuous increase in temperature throughout the cotton-growing season shortens the crop duration up to 24 days (Reddy et al. 1996) or even 35 days earlier from germination to maturity if the average temperature at global level rises 5 °C (Reddy et al. 1992a, b, c, 1997a, b).

Leaves are more sensitive to temperature variations during early stage of development. At about 3 weeks after emergence, leaves expanded six to eight times more at 28–30 °C temperatures than those at 20–21 °C (Reddy et al. 1992a, b, c, 1997a, b). Unlike roots, shoots require higher temperature for optimum growth (Arndt 1945; Pearson et al. 1970). Cooler temperatures cause accumulation of metabolites through slowing down the plant growth and development, thus making plant to develop more vegetative branches (Reddy et al. 1992a, b, c). That is why excessive vegetative growth does not take place at higher daytime temperatures. Plant growth traits like

LAI, LAR, and NAR respond positively to an increasing temperature up to a maximum of 35 °C (Jackson 1967).

Temperatures above 30/22 °C are conducive for stem elongation, fruiting branches as well as fruiting branch nodes at early crop stage (Reddy et al. 1992a, b, c). The number of fruiting sites increases greatly with the increase in mainstem nodes (Reddy et al. 1992a, b, c). Night temperatures have relatively more effect in controlling flower initiation (Gipson and Ray 1969). For example, the night temperature of 25 °C caused delayed flowering and first fruiting branch in upland cotton (Mauney 1966), and boll maturity was delayed when the night temperatures were lower (Gipson and Joham 1969).

Temperature variations dramatically affect growth besides potential yield of cotton (Nasim et al. 2011; Luo et al. 2014; Rahman et al. 2017). Cotton plant grows efficiently at 33 °C temperature, but the effective fruit bearing declines considerably when temperature rises beyond 36 °C (Luo 2011; Singh et al. 2007). Under higher temperatures, production and assimilation of carbohydrates are inhibited which promotes boll shedding as well as smaller and malformed bolls (parrot beak), lesser lint quantity, and decreased yields (Hatfield et al. 2008, 2011; Oosterhuis 2009). However, the cotton genotypes in India and Pakistan are well adapted to the high-temperature conditions and are successfully grown at temperatures as high up to 46 °C.

Environmental stresses including high temperature at floral development stage are crucial and limit potential yields (Boyer 1982). The most sensitive growth phase for cotton to heat stress is reproductive growth stage which includes its pollen tube growth and development and fertilization (Zinn et al. 2010). Unfavorable weather variations adversely affect development of ovule, pollen fertility, and anther dehiscence or dispersal of pollens (Zinn et al. 2010; Young et al. 2004). Healthy pollen grains have key roles in fertilization process, but they are more susceptible to damage by high-temperature stress (Kakani et al. 2005). Therefore, this stress during anthesis may lead to improper fertilization resulting in lesser number of seeds and bolls (Kakani et al. 2005; Reddy et al. 1992a, b, c; Snider et al. 2009). Development of fiber takes place on seeds in the boll; therefore seed number in boll besides ovules in a locule would determine the quantity and quality of lint fiber to be produced (Stewart 1986). Variation in seed number in a boll reflects either inadequate fertilization of seed or post-fertilization growth termination of the embryo depending upon both cultivar and unfavorable ecological conditions (Karmakar et al. 2016; Stewart 1986). Heat-induced sterility has been a common issue in commercially grown varieties in Pakistan where most of the initial fruit produced is shed, due to high temperature, which often triggers extra vegetative growth (Taha et al. 1981).

Increasing temperatures accelerate crop growth and developmental processes (Rawson 1992; Ziska and Bunce 1997) but also have detrimental effects on overall crop performance if temperature exceeds than the desirable limits. While cotton is highly sensitive with raise in high temperature creating heat stress for all growth stages, fruiting phase is reported to be more sensitive among other growth phases (Snider et al. 2009, 2010, 2011). For example, 1 °C rise in temperature produced squares, flowers, and matured bolls by 1.6, 3.1, and 6.9 days earlier, respectively

(Reddy et al. 1997a, b). Disruption and shortening of fruit maturation period, as a result of high-temperature stress, limit productivity, causing subsequent yield losses in cotton (Rawson 1992; Stewart 1986; Wullschleger and Oosterhuis 1990; Ziska and Bunce 1997) owing to increased shedding of fruit forms and production of smaller bolls (Reddy et al. 1999; Yfoulis and Fasoulas 1978; Zhao et al. 2005).

Fruit retention is highly sensitive to increased temperature stress, besides duration of stress is very crucial as it determines the fruit load on a plant. The ideal temperature for healthy cotton growth is 20–32 °C (Mohamed and Abdel-Hamid 2013; Reddy et al. 1992a, b, c). It is described that maximum growth of cotton especially number of bolls which is essential for higher lint yield occurs during both day and night with a temperature of 30 °C day besides 22 °C night, respectively (Burke et al. 1988; Reddy et al. 1996). For example, Reddy et al. (1992a, b, c) found 30/32 °C temperature to be the most appropriate to gain maximum boll weight. The cotton plants retained only 50% of the squares and bolls when exposed to 33 °C average daily temperature and fruit retention declined sharply to none when daily average temperature was rose to 36 °C (Reddy et al. 1992a, b, c). Similarly, a 12 h exposure of cotton plants to 40 °C produced only 1% of their mass as bolls (Reddy et al. 1991). It has been observed that young bolls are relatively more vulnerable to temperature stress and often shed when the plant faces 32 °C or higher average daily temperatures (Reddy et al. 1996). However, the cotton crop ably tolerates to short duration stress imposed by temperatures as high up to 43/45 °C provided there is ample moisture in the soil.

Among the C_3 plants, cotton has relatively more heat-tolerance capability; however, temperatures higher than optimum stimulate shedding of squares and bolls resulting in sharp decreases in yield (Oosterhuis 1997; Schlenkera and Roberts 2009). Apart from the intensity and duration of temperature stress having a critical role in fruit retention, high night temperatures at fruiting phase of cotton are very deleterious and cause more damage to the seed cotton yield than the high day time temperatures. Investigations have revealed that high temperatures at night promote respiration rates, decrease concentrations of leaf soluble carbohydrates (Loka and Oosterhuis 2010), and increase abscission leading to significantly lower production (Gipson and Joham 1969).

20.2.2 Effects on Physiological and Biochemical Parameters

Cotton crop is usually grown in arid besides semiarid areas in different countries where temperatures are quite high during the crop growth periods. The higher temperatures have limiting effects on growth, physiological, and biochemical processes.

20.2.2.1 Effects on Water Relations

The presence of optimum leaf water content maintains cell turgidity and key cell functions like stomatal regulation, net photosynthetic rate, and translocation of assimilates to different parts of the plant. Variation in ambient temperature greatly influences plant water status. A change in the surrounding environment like increase in temperature than the desired level disturbs the plant water status. Increased transpiration rates triggered by high temperature result in water loss from the plant when soil water content is not sufficient. Although high temperatures severely disrupt the tendency to maintain cell water status under limited moisture conditions, the plants are able to maintain steady tissue water status under conditions of ample moisture content. Limited water availability under field conditions often results in high-temperature stress due to lower rate of evaporative cooling. Increased transpiration rates due to high day temperatures cause reduction in water potential that leads to perturbation of numerous physiological processes (Tsukaguchi et al. 2003).

20.2.2.2 Effects on Cell Membrane, Anther Dehiscence, and Pollen Viability

Cell membrane, that surrounds the cytoplasm, is a selectively permeable structure which separates the interior of cells from outside environment. The cell membrane should maintain its integrity for normal cell functions by allowing selective substances to move into or out of cell. Cell membrane is mainly composed of lipids (up to 80%) and proteins. Under stressful environments such as high temperature, lipid peroxidation occurs within the membrane causing increase in the fluidity of the thylakoid membranes thus negatively affecting the efficiency of the photosystem complex. Under continued heat stress, cyclic photophosphorylation is increased to disperse excess energies and preserve the more sensitive photosystem II complex (Schrader et al. 2004; Sharkey 2005). In cotton plant, the ability of membrane structure to adjust under high-temperature conditions has been identified as a physiological adaptation toward heat stress (Rahman et al. 2004). Defensive role of antioxidants cannot be understated when high-temperature stress occurs. Reactive oxidative species (ROS) increase extensively with increased levels of heat stress (Wahid et al. 2007) which also activate enzymatic pathways needed to initiate stress response (Dat et al. 1998; Foyer and Noctor 2005). Though, if stress conditions prolong, then ROS can also initiate programmed cell death (Gechev et al. 2006). During stressful periods, sufficient antioxidant pools are necessary to moderate heat-related responses for proper growth and continued development of cotton (Snider et al. 2011).

In cotton fruiting period is highly temperature sensitive, and higher temperature can disrupt the fruit setting process due to pollen abortion at temperatures of 35–39 °C (Min et al. 2014). Often growth of filaments is restricted, while stigmas elongate properly giving rise to asynchronous development (Brown 2008). Such

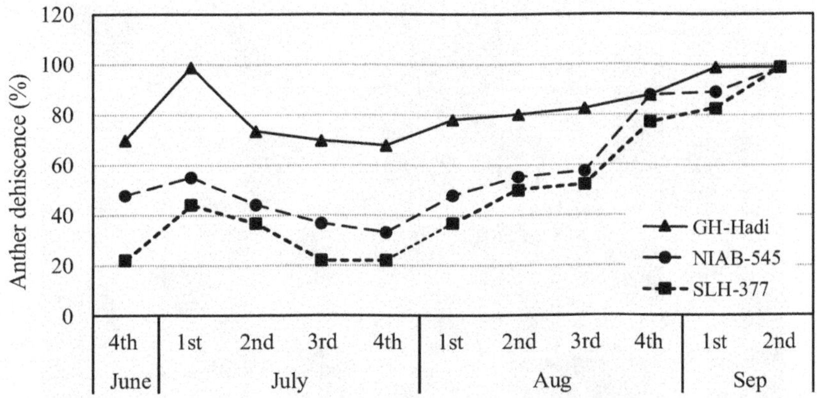

Fig. 20.2 Dehiscence of anthers in three cotton genotypes during the fruiting period (CCRI 2017)

flower abnormalities develop in cotton when high night temperatures cause rise in canopy temperatures and limit cotton reproductive performance (Zeiher et al. 1994) due to a decline of up to 84% in anther dehiscence and 78% in pollen viability when night temperature increased to 30 °C (Ahmed et al. 1992).

The dehiscence of anthers which is highly temperature sensitive is the primary step in fertilization process. Monitoring of anther dehiscence on daily basis in field evaluation trials gives handsome information about performance of cotton varieties under prevailing environmental conditions. For this purpose, up to three flowers from each varietal plot are collected between 09:30 and 10:30 h and transported to laboratory in zip-locked polythene bags. The flowers are examined under the microscope to assess the number of pollen grains that have burst out from the anthers. The anther dehiscence percentage is calculated on the basis of dehiscent anthers, i.e., 90% (fully dehiscent), 50% (partially dehiscent), and 10% (non-dehiscent).

The field studies conducted at Central Cotton Research Institute (CCRI), Multan Pakistan have revealed that dehiscence of anthers started to decline from first week of July, remained lowest during third and fourth week and then increased gradually from first week of August reaching its maximum up to 100% in September. Although heat-tolerant genotypes maintained highest anther dehiscence during the period, the trend was, however, similar in all genotypes (Fig. 20.2).

Pollen viability refers to the health of pollens, and it provides information about the ability of pollens to fertilize. Pollen viability is determined by gently tapping the inverted flower on a glass slide. The pollens collected on the slide are stained with vital dye Acetocarmine and observed under the microscope after 6 h at 200×. The viable pollens show bold red color of the dye, while nonviable pollens remain colorless (Fig. 20.3). Different physiological and other yield-contributing traits have positive correlations such as pollen viability, % boll set on first besides second positions along sympodia, and bolls per plant besides boll weight along with SCY,

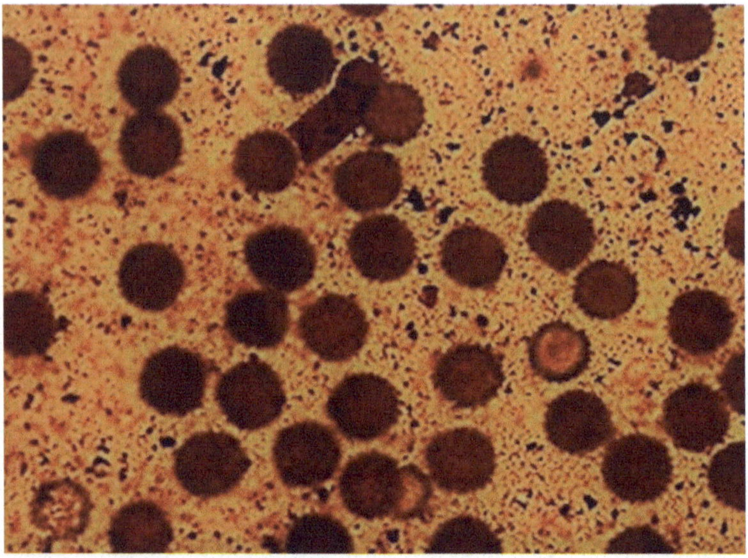

Fig. 20.3 Pollen viability in cotton: Pollens with bold red color are viable and those without color are non-viable (unpublished data of Physiology/Chemistry Section CCRI, Multan)

Table 20.2 Relationship between cotton yield and physiological traits determining heat tolerance

	AD%	PV (%)	% BSFP	% BSSP	RCIL (%)	EC	NBPP	BW (g)
PV (%)	0.97**							
% BSFP	0.90**	0.89**						
% BSSP	0.88**	0.88**	0.98**					
RCIL (%)	−0.97**	−0.97**	−0.86**	−0.84**				
EC (μS cm^{-1})	−0.90**	−0.88**	−0.90**	−0.89**	0.85**			
NBPP	0.60**	0.58**	0.58**	0.60**	−0.59**	−0.54**		
BW (g)	0.12^{ns}	0.15^{ns}	0.08^{ns}	0.10^{ns}	-0.12^{ns}	-0.08^{ns}	−0.34**	
SCY (kg ha^{-1})	0.99**	0.97**	0.91**	0.90**	−0.98**	−0.90**	0.60**	0.11^{ns}

AD anther dehiscence, *PV* pollen viability, *BSFP* boll set on first position, *BSSP* boll set on second position, *RCIL* relative ell injury level, *EC* electrical conductivity, *NBPP* number of bolls per plant, *BW* boll weight, *SCY* seed cotton yield

**Significant at $p < 0.01$ ^{ns}Non-significant

while cell injury and electrical conductivity have negative correlations with seed cotton yield (Table 20.2).

20.2.2.3 Effects on Photosynthesis and Photorespiration

The process of photosynthesis is vital for different plant functions and survival. Optimum temperature, light intensity, ample availability of water, and carbon dioxide have positive influence on photosynthesis. However, extreme temperature stress is considered the most important limiting factor for photosynthesis (Salvucci and Crafts-Brandner 2004). Restricted photosynthetic efficiency as a result of temperature stress has been stated in diverse crops counting cotton (Bibi et al. 2008; Reddy et al. 1991). Maximum net photosynthesis in cotton is noticed at optimum temperature of 28 °C, and it tended to decrease when temperature reached at 35 °C (Bibi et al. 2008; Crafts-Brandner and Salvucci 2000; Reddy and Hodges 1995; Snider et al. 2009; Wise et al. 2004) owing to significant reduction in photosynthetic pigments at higher temperature (Mohamed and Abdel-Hamid 2013). High temperatures during the vegetative stage can destroy components of leaf photosynthesis, reducing CO_2 gain rates thereby limiting the export of assimilate from leaves to developmental parts (Jiao and Benhua 1996). Higher-temperature stress affects photosynthetic efficiency of cotton through decreased chlorophyll content (Snider et al. 2009, 2010), inhibited CO_2 exchange rate through limiting activity of rubisco (Crafts-Brandner and Salvucci 2000; Law and Crafts-Brandner 1999), decreased membrane integrity (Schrader et al. 2004; Bibi et al. 2008; Rahman et al. 2004), and increased photorespiration (Perry et al. 1983). Inhibition of photosynthesis due to higher temperature stress precedes the other detectable stress symptoms (Berry and Bjorkman 1980) such as the activity of rubisco, regeneration rate of ribulose-1,5-biphosphate (RuBP), and metabolism of triose phosphate (Wise et al. 2004). Higher temperature disrupts the fixation of photosynthetic CO_2 by damaging photosystem II (PSII) electron transport mechanism in thylakoid membrane (Berry and Bjorkman 1980). Since PSII function is the most unstable component in electron transport (Havaux et al. 1996; Quinn and Williams 1985), its inhibition results in enhanced chlorophyll fluorescence (Krause and Weis 1991). That is why the magnitude of heat-induced changes in photosynthesis mechanism is quantified on the basis of chlorophyll fluorescence (Govindjee 1995; Krause and Weis 1991; Strasser 1997).

Photorespiration is a process during which plants take up oxygen instead of CO_2 when the light intensity is high. High temperatures during the daytime increase photorespiration and decrease net carbon assimilation in C_3 species and thus result in the loss of carbohydrates (Guinn 1974; Krieg and Sung 1986; Ludwig et al. 1965). The conditions of continued high temperatures adversely affect plant growth through increased photorespiration (Arevalo et al. 2004, 2008). During photorespiration, carbohydrates produced by photosynthesis get utilized for respiratory energy rather than to fulfill the need of plant developmental processes such as filling of developing bolls (Loka and Oosterhuis 2010). Moreover, under conditions of high-temperature plants are unable to accumulate and provide enough quantity of carbohydrates to match with the plants' needs (Oosterhuis 1999). Perry et al. (1983) reported that photorespiration increased linearly with the rise in temperature from 22 to 40 °C:

photorespiration representing less than 15% to almost 50%, respectively, of the net photosynthesis. This highlights that the components of photorespiration and photosynthesis are highly influenced by temperature variations.

20.2.2.4 Effects on Enzyme Activation

Enzyme activity within plants is highly temperature dependent. Elevated temperatures beyond the desirable limits significantly lower the activity of different enzymes. The enzyme Rubisco activase regulates activation of ribulose-1,5-bisphosphate carboxylase/oxegenase (Rubisco) in light (Andrews et al. 1995; Portis Jr 1992; Salvucci and Ogren 1996). Vital role for Rubisco activase in sustaining active state of Rubisco during light at levels which are sufficient for photosynthesis has been reported in numerous studies (Andrews et al. 1995; Eckardt and Portis Jr 1997; Salvucci et al. 1986). Isolated Rubisco activase is mainly sensitive to inactivation by raised temperatures (Crafts-Brandner et al. 1997; Crafts-Brandner and Salvucci 2000; Salvucci et al. 2001). Therefore, the inactivation of Rubisco activase offers a potential biochemical explanation of decreased activeness of Rubisco at raised temperatures (Kobza and Seemann 1989; Krause and Weis 1991; Weis 1981).

20.2.2.5 Effects on Reactive Oxygen Species, Antioxidants, and Heat Shock Proteins

High-temperature stress induces a number of biochemical alterations in plants as a defense mechanism including production of antioxidants and heat shock proteins. ROS are chemically active and unstable compounds which consist of singlet oxygen, superoxide radical, peroxides, hydroxyl radical, and alpha oxygen. ROS, a natural byproduct in normal metabolism of oxygen, is involved in cell signaling besides homeostasis. ROS are produced in excess, in chloroplasts, and mitochondria under environmentally stressed conditions like heat stress causing damage to cell structures; the state termed as oxidative stress (Apel and Hirt 2004). The presence of excess ROS affects normal cell functions due to oxidative damage leading to cell death if the stress conditions prevail.

As a defense mechanism, plants synthesize different antioxidants to protect cells from oxidative damage caused by production of excess ROS. For normal cell functions and growth, there needs to be maintained balance between production and breakdown of ROS by antioxidants. Antioxidants prevent the oxidation of other molecules and neutralize the free radicals making them less reactive. The defense mechanism of plant comprises of different enzymatic components like superoxide dismutase (SOD), ascorbate peroxidase (APX), ascorbate (ASC), and glutathione (GSH) (Foyer and Noctor 2005).

Proteins in plant cells are temperature sensitive and are prone to denaturation by heat stress. As a phenomenon of thermotolerance, plants synthesize proteins which are termed as HSPs. The HSPs act as chaperones and prevent denaturation of cell proteins (Moriarty et al. 2002), promote refolding of denatured proteins (Frydman 2001), and are involved in other stress response mechanisms (Wang et al. 2004). Synthesis of HSPs increases with the gradual rise in temperature (Abrol and Ingram 1996). The production and accumulation of HSPs has been evidenced in cotton under controlled temperatures of 38–41 °C (Burke et al. 1985). The HSPs have been categorized into five groups on basis of molecular weights as HSP100s, HSP90s, HSP70s, HSP60s, and sHSPs (12–40 kDa) (Wang et al. 2004). In contrast to the other parts of plants, germinating pollens upon exposure to heat do not exhibit HSPs synthesis and therefore lose viability (Hopf et al. 1992).

20.2.3 Effects on Fiber Quality

Cotton crop is grown primarily to obtain lint fiber which is composed mainly of cellulose (>85%). Cellulose is a linear chain polysaccharide made up of glucose molecules mutually linked by beta-1,4 glycosidic bonds. The ideal temperature range reported for the synthesis of cellulose is from 25 to 30 °C, and cellulose synthesis decreases if temperature drops or exceeds this range (Roberts et al. 1992). Sucrose (carbohydrate), the product of photosynthesis in plants, is the basic compound in cellulose synthesis (Tian et al. 2013); therefore, any change in the concentration of sucrose would directly affect the synthesis of cellulose. Cotton photosynthetic capacity decreases if average daily temperature rises above 32 °C (Crafts-Brandner and Salvucci 2000), thus decreasing sucrose synthesis. To produce healthy fiber, plant should be able to maintain a steady rate of photosynthesis under varied conditions. Generally, 12,000–15,000 fibers are produced by a single seed under favorable temperature conditions (Oosterhuis 1997).

Adequate supply of carbohydrates is very crucial in healthy fiber development. Unfavorable conditions such as high temperatures inhibit assimilation of carbohydrates thereby decreasing seed number, seed size, number of fibers per seed, and also the weight of fiber produced on a seed thus ultimately leading to yield reduction (Arevalo et al. 2004; Oosterhuis 1999). Rising temperatures have marked effects on cotton fiber characteristics which may be either positive or negative under different circumstances. The quality of fiber is determined on the basis of different indices like fiber length, fiber strength, and fiber micronaire (fineness) which exhibit variable degree of sensitivities to the environmental factors (Bowman and Gutiérrez 2003; Bradow and Davidonis 2000; Gokani and Thaker 2002; Gou et al. 2007; John and Keller 1996; Pettigrew 2008). Increased temperatures may lead to development of altered fiber traits such as higher micronaire value, more fiber strength, and increased fiber maturity (Ton 2011). Increased fiber maturity and strength are desirable, while fiber with higher micronaire is of lower economic value.

Temperature variations have predominant effects on fiber quality parameters (Pettigrew 2008) and particularly on fiber strength during thickening of secondary cell wall (Ruan 2007). An average daily temperature of 26 °C is considered optimal for fiber development (Rahman et al. 2007). Increase in average everyday temperature beyond 30 °C or peak day temperature above 35 °C (Pettigrew 2001; Reddy et al. 1991) inhibits development of quality fiber depending upon the duration of high-temperature stress (Oosterhuis 1999; Rahman et al. 2007; Reddy et al. 1995). Similarly, night temperatures also affect fiber quality. Fiber length was found maximum between 15 and 21 °C night temperatures, and length decreased when nigh temperature rose above 21 °C (Gipson and Joham 1969; Pettigrew 2008) or decreased below 15 °C (Gipson and Joham 1969; Zhang et al. 2012). Decrease in lint index, percentage of lint and lint produced per boll at higher and lower temperature limits have been evidenced. Fiber growth duration and rate of fiber elongation may vary among the varieties and with the change in environmental factors (Gipson and Ray 1969). Fiber elongation, however, requires lower temperature than that optimally required for boll development (Pettigrew 2001). Fiber micronaire (fineness) has been reported to deteriorate above 33/28 °C temperature regimes (Pettigrew 2008; Reddy et al. 1999).

Sensitivity of fiber to temperature varies with stage of fiber development. The early stage of fiber elongation such as up to 2 weeks after anthesis has been reported to be more to night-temperature sensitive than the later stages of fiber elongation (Gipson and Joham 1969; Xie et al. 1993; Gipson and Ray 1969). Initiation of fiber elongation starts with the flowering and continues up to 25 days after flowering, while the secondary cell wall thickening continues during 20–60 days after flowering, although these processes vary with varieties and overall temperature conditions or cumulative heat units (Bradow and Davidonis 2010). Relationships between fiber quality characters besides temperatures are mentioned in Table 20.3.

20.2.4 *Effects on Genetics and Molecular Responses*

In nature plants suffer from various abiotic stresses throughout the course of their growth, while heat stress has a unique action mode on physiology of plant cells. While mostly heat stress becomes exacerbated with the occurrence or severity of salt

Table 20.3 Relationship of fiber quality parameters with temperature conditions

Fiber trait	Correlation	Temperature condition	References
Length of fiber	Negative	Difference amid maximum and minimum temperatures	Hanson et al. (1956)
Strength of fiber	Positive	Maximum or mean maximum temperature	Hanson et al. (1956)
Strength of fiber	Positive	Heat-unit-accumulation during boll development	Snipes and Baskin (1994)
Secondary wall deposition (fiber maturity)	Positive	Temperature/heat unit accumulation	Johnson et al. (1997), Bradow et al. (1996)
Fiber fineness (micronaire increase)	Positive	Heat unit accumulation	Johnson et al. (1997), Bradow et al. (1997)

Source: Singh et al. (2007)

and drought stress, it is imperative to investigate influence of independent stress besides biological impact of heat stress in order to alleviate combined effect of these abiotic stresses. Plant susceptibility to higher temperature depends on developmental stages, high temperature negatively affecting growth stages of plants. Effects of high temperature vary within species and among genotypes (Barnabás et al. 2008; Sakata and Higashitani 2008). The exposure of plants to heat shock by an increase in 5 °C exceeding from its optimum temperature requirement significantly alters and influences metabolic and cellular machinery essential for heat stress tolerance (Guy 1999). Plant adaptation under thermal includes changes in cellular structural organization, i.e., changes in function of organelles besides cellular membrane functions (Weis and Berry 1988), inhibiting biosynthesis of essential proteins besides enhancing transcription and translation of HSPs (Bray et al. 2000; Demirel et al. 2014), and the production of phytohormones, e.g., ABA and antioxidants enzymes (Maestri et al. 2002).

Fluctuation in temperature is sensed by plants with the aid of complex group of sensors present in different cellular compartments. Fluidity of cell membrane increases to activate the lipid-based signaling pathways likely augmented Ca^{2+} influx besides cytoskeleton reorganization. Signaling between the two pathways leads to increased production of osmolytes besides osmoprotectants in reaction to heat stress. However, *Arabidopsis* CNGC2 gene encrypts a component of membrane cyclic nucleotide-gated Ca^{2+} channels which are responsible for sensing and resultantly increase the temperature in plasma membrane in order to tackle heat shock reaction (Saidi et al. 2009). Mechanism illuminates crucial role of lipid membranes against heat stress (Horváth et al. 2012). Newly, it was illustrated that signaling pathways activate specific tissues under heat stress (Mittler et al. 2011).

Heat stress triggers changes in photosynthesis besides respiration, hence leading to reduction in life cycle resulting in reduced yield of plant (Barnabás et al. 2008). Initial effect of thermal stress encompasses structural changes in chloroplast protein with decreased enzyme activity (Ahmad et al. 2010). Furthermore, it causes injuries in cellular membrane structure with the alteration in cell elongation, expansion, and differentiation (Potters et al. 2009; Rasheed 2009; Smertenko et al. 1997). Homeostasis of plants is also disturbed with heat stress including biosynthesis of, and metabolites compartmentalization in plant tissues (Maestri et al. 2002) modification in activities of starch accumulation, sucrose synthesis, carbon metabolic enzymes, and down regulation of specific genes responsible for carbohydrate metabolism (Ruan et al. 2010).

Biosynthesis of various phytohormones increases under heat stress that causes premature senescence (Larkindale et al. 2005; Larkindale and Huang 2004; Talanova et al. 2003) such as a bscisic acid synthesis increases due to heat stress causing abscission of reproductive organs (Binder and Patterson 2009).

Transcriptomic changes in plants occurr in regulating gene expression to combat with adverse effect of temperature; approximately 5% of plants genes become upregulated via heat stress, while chaperones are minor part of general heat shock reaction (Saidi et al. 2011; Aksoy et al. 2015). Most of genes are involved in primary/secondary metabolisms, transcription besides translation in response to high temperature.

Higher accumulation of ROS species works as indicator of heat stress causing oxidative/cellular damages to plants whereas peroxidation of lipids disturb consist membrane permeability besides function. Heat stress causes denaturation and misfolding of freshly synthesized protein. Heat shock factors (HSFs) activate heat shock proteins (HSP); however, their expression pattern varies between species (Banti et al. 2010).

Plant development and physiological processes are adversely effected due to heat stress. Its damage to plant varies depending upon its developmental stage with a severity during reproduction stage which handicap plants to adapt with changing environmental conditions (Hall 2001). Higher temperature affects cotton reproductive development by decrease in pollen viability (Hejnák et al. 2015).

Flowering of plants is decreased due to heat stress which ultimately limits sexual reproduction (Hedhly et al. 2009; Thakur et al. 2010; Zafar et al. 2018). Numerous researches were conducted to unravel the effect of temperature under different conditions by artificial application of high temperature under glass or subjecting plants to high light intensity in growth chambers, and it was inferred to be deleterious for bud initiation and other growth stages of plants (Hedhly et al. 2009; Nava et al. 2009). Agronomic leguminous besides cereal crops depicted higher sensitivity at flowering, whereas horticultural crops depicted decreased fruit formation (Frank et al. 2009; Saha et al. 2010), which was speculated to be due to reduced availability of water and nutrients uptake by plant organs for their normal growth and higher yield (Young et al. 2004).

In depth it was revealed that male gametophyte is sensitive to heat stress as compared to pistil and female gametophyte which exhibited tolerant behavior (Hedhly et al. 2009). Heat stress generally stimulates rather than delays the process of anthesis to hasten flower opening and abnormal reproductive development without accumulation of necessary resources (Zinn et al. 2010). As we abovementioned about the gene expression changes with heat stress, it also varies in other plant parts such as tapetum degeneration is observed with a high rate of plant sterility among a group of species (Endo et al. 2009; Oshino et al. 2007).

Heat stress is also the cause of male sterile plants especially for sensitive species with impaired pollen development, a major factor for reduced plant yield in such scenario of environmental pressure (Wassmann et al. 2009; Sakata and Higashitani 2008). For cotton the raise in temperature from 34 to 43 °C during its growth abruptly disrupts anther formation and limiting physiological processes (Zahid et al. 2016). It also alters transport of nutrients and minerals within plants due to disrupted balance of symplastic besides apoplastic phloem loading (Taiz and Zeiger 2006). Cotton plants under pressure of heat stress exhibit lower concentration of sugars (soluble) in anther walls, pollen-grains, resultantly decreased locular fluid and pollen viability (Snider et al. 2009). Heat stress promotes development of aborted tapetal cells, which causes swift progress toward meiotic prophase triggering programmed cell death and pollen sterility (Parish et al. 2012; Sakata and

Higashitani 2008). In cotton higher temperature damages developing microspores e.g., anthers (Min et al. 2013).

20.2.5 *Molecular Mechanisms of Heat Tolerance*

Plants are capable to adjust themselves according to climate drift by activating genes responsible to circumvent harsh existing conditions; in this way they protect themselves from heat stress (Sánchez-Rodríguez et al. 2011; Qi et al. 2011). This ability of plants to help them to acclimate to higher temperature by maintaining homeostasis to prevent heat injury and another unprecedented mechanism such as huge production of HSPs (Vierling 1991). The heat tolerance in plants is due to various multigenic traits directly or indirectly involved during development and maintenance of thermal tolerance, the main players are antioxidant enzymes, gene regulations, lipid membrane stability, and compatible solutes accumulations (Kaya et al. 2001).

A number of studies have highlighted the importance of HSFs which are critical for heat tolerance, while some have lesser critical part such as HSP101, HSA32, HSFA1, HSFA3, and knockout of variants showed less impact for heat tolerance (Schramm et al. 2008; Yoshida et al. 2011). It indicates that complex network is involved for conferring differential protection against heat stress. Nonetheless, HSPs have significant role for tolerance which work as molecular chaperones to circumvent denaturation of targeted proteins besides likely facilitating refolding of protein (Ahuja et al. 2010; Lohar and Peat 1998; Scharf et al. 2012).

Heat stress tolerance is dependent upon induction of heat shock proteins (HSP70 and HSP90) in cotton (Gurley et al. 2000). The inhibited expression of HSP70 and HSP90 in cotton stimulated oxidative stress and reduced the tolerance for heat stress in cotton genotypes, which exhibited that HSP70 and HSP90 are involved for heat tolerance (Sable et al. 2018). Although, HSP101 and HSP70 are not normally required for growth under normal conditions but, however, have significant importance for tolerance and protein oxidative protection in cotton (Zhang et al. 2016). It is supported from several studies that HSFs can role as molecular sensors which sense ROS species and regulate the expression of oxidative stress responsive genes in cotton (Miller and Mittler 2006; Sable et al. 2018; Sekmen et al. 2014). Moreover, in cotton HSPs are associated with membrane to form heat shock lipids which stabilize the membrane during earlier temperature stress (Cottee et al. 2014).

Plants facing any abiotic stress immediately produce ROS-scavenging enzymes to alleviate oxidative stress produced by ROS species; therefore, plants suffering from heat stress also produce antioxidant enzymes (SOD, POD, CAT, APX and GPX) (Sekmen et al. 2014). It was reported that in tolerant plant species the production of these enzymes is higher as compared to susceptible for protection from oxidative damages (Abiko et al. 2005). Antioxidant enzymes are found in almost all the cellular components of plants for detoxification and cellular survival (Asada 2006; Iba 2002; Mittler et al. 2004). It was reported from dicot model plant

(Arabidopsis) that APX gene family starts expression during heat stress which is solely dependent on HSF for heat tolerance (Panchuk et al. 2002).

The LEA proteins, i.e., ubiquitin, besides dehydrins, are key players for protecting from heat and drought stress moreover drought stress further aggravates heat stress with least availability of water for plant growth. In cotton, LEA proteins aid in preventing aggregation and protection from desiccation (Magwanga et al. 2018). The role of ubiquitin was against heat stress, and it was reported that it is important for the first 30 min for short exposure of plants to heat stress (Huang and Xu 2008).

Different approaches have been exploited to discover molecular mechanisms rendering heat tolerance during reproductive organ development especially at the stage of pollen formation which is necessary to understand and pave path for developing heat-tolerant cultivars, thereby various genome-wide approaches have been deployed for heat tolerance cotton breeding program (Min et al. 2014). It was concluded that numerous mechanisms such as several hormones, antioxidants, and HSPs are important for reproductive structures tolerance in cotton (Min et al. 2014).

Higher temperature stress has devastating effect on crop yield. In cotton, tolerant cultivars exhibit higher chlorophyll content, maintain photosynthesis machinery by adjusting stomatal conductance during heat stress (Hejnák et al. 2015), although to ensure higher fruit setting and seed cotton production, are important parameters to be considered with elevated temperatures (Reddy et al. 1992a, b, c). Genetic resources need to be exploited when there are limited options available for heat tolerance in cotton; sometimes wild cultivars are used to fix trait through breeding for achieving higher tolerance (Pradhan et al. 2012).

In heat-tolerant plants, expression of multiple proteins has been observed with an increase in concentration of phosphoribulokinase (PRK), which is the main component of calvin cycle for final RuBP production. Protective proteins (HSP70, HSP90 besides Cpn60) also accumulated with an elevated gene expression to confer protective role against heat stress (Scafaro et al. 2010).

Likewise, a proteomic study was conducted for comparing protein expression in cotton among susceptible and tolerant cultivars, and it was observed that accumulation of HSPs was higher in tolerant cultivars making them more tolerant under adverse temperature (Min et al. 2014). In cotton genetic variation occurs with the ability of each cotton cultivars to withstand heat stress with increased membrane stability and chlorophyll contents and minimum electrolyte leakage from tolerant cultivars (Asha and Ahamed 2013).

Currently global warming is the main issue as we know earth temperature is increasing every year around the globe which will have severe impact on the crops growth. On the other hand, human population is also increasing rapidly and to ensure food security heat-tolerant crops development is necessary to be sure to cope with future changing climatic conditions. That is also necessary to understand behavior of crop to changing climatic conditions that what kind of physiological adaptations take place in plants. Modern genetic approaches are also paving path for tolerant crops development within a less period of time. Figure 20.4 explains different mechanisms of heat-tolerance initiated in plants under high-temperature stress.

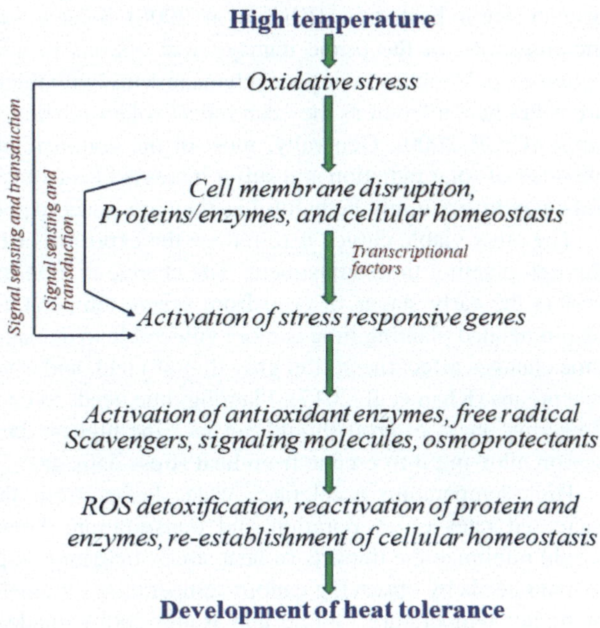

Fig. 20.4 Schematic diagram of mechanisms involved in heat tolerance of plants (Hasanuzzaman et al. 2013)

20.3 Strategies to Cope with High-Temperature Stress in Cotton

20.3.1 Variety Selection, Sowing Time Adjustment, and Irrigation Management

Cotton plant is able to respond with the severity of stress and adopt according to the harsh environmental conditions. High-temperature stress in cotton is most important due to its impact in early growth including germination, flowering besides during boll formation stages. High night temperature also increases overall mean temperature. Relative humidity has direct role to inhibit cooling phenomena during night leading to higher nighttime canopy and air temperatures. Moreover, higher humidity during daytime limits transpiration rates thus resulting in higher daytime canopy temperatures as well.

Cotton growth is influenced due to temperature stress which induces differential physiological, biochemical, and metabolic changes, by changing plant photosynthetic performance, stomatal conductance, maintaining oxidative balance, carbohydrate production, lipid peroxidation, and synthesis of protein for heat tolerance (Bibi et al. 2008; Roy and Ghosh 1996).

Heat stress is a hard to control phenomenon due to climate change. Different strategies should be applied to adapt to temperature stress. There exists genetic variability in cotton varieties and available germplasm express either susceptibility or tolerance in heat stress (Khan et al. 2008). Selection of heat-tolerant varieties is, therefore, one of the prime management options to avoid heat stress impacts on cotton crop. Varieties owing leaf characteristics with thick cuticle and waxy surfaces are better heat tolerant as they can reflect solar radiation to decrease damage of heat stress (CICR 2000). Generally, most of the commercial cotton varieties are good absorber of solar radiation and suffer increased heat stress impact. Selection of short season cultivars might be helpful as they are lesser exposed to heat stress.

The other viable choice is to reduce the exposure of fruiting phase to heat stress through planting time adjustment. The change in planting time is a good option to protect the early stages of crop from severe temperatures. Sowing of cotton after recommended planting time is more vulnerable to damages from heat stress. Sowing time changes affect the cotton growth, lint yield, and assimilate supply to reproductive organs (Khan et al. 2017). Planting time needs to be adjusted in such a way that flowering phase of crop should not face the highest day/night temperatures of the season allowing it to escape from heat stress damage.

High temperature accelerates water losses from the soil and plant through increased rates of evaporation and transpiration. Timely irrigation management would minimize the impacts of heat stress. Irrigation ought to be applied according to plant needs by observing canopy temperature for ameliorating detrimental effects of higher temperature (White and Raine 2008). Inadequate availability of water forces plant to adapt to such conditions by physiological changes through adjusting stomatal conductance, which increase the incidence of water stress in cotton. It is necessary to apply water to cope such situations which keeps the canopy cool. In arid conditions crop is totally dependent on rain, and with deficient soil moisture conditions, adjustment of row spacing helps to increase lint yield of cotton (Bange et al. 2008).

20.3.2 Screening for Heat Tolerance

In Pakistan cotton is mostly cultivated in hottest regions (Riaz et al. 2013). The genotypes commercially grown often face extremely high temperature up to 50 °C during months (May and June), which is almost 20 °C above than optimal temperature required for its normal growth, thus retarding crop's performance to a high extent. Development of heat-tolerant commercial cotton genotypes is a main challenge (Moreno and Orellana 2011; Zhang et al. 2006). Identification besides confirmation of traits that confer tolerance to high temperature remains elusive due to dynamic responses of plants subjected to heat stress (Rodríguez et al. 2005; Wahid et al. 2007). Scientists are also working to examine in what way plants could be managed in high-temperature conditions. Based on importance of high-temperature stress, physiological, biochemical, and molecular responses have been studied to screen the newly evolved cotton germplasm. Identification and development of potential genotypes which possess better tolerance to heat stress could give better yield performances in heat-prone areas.

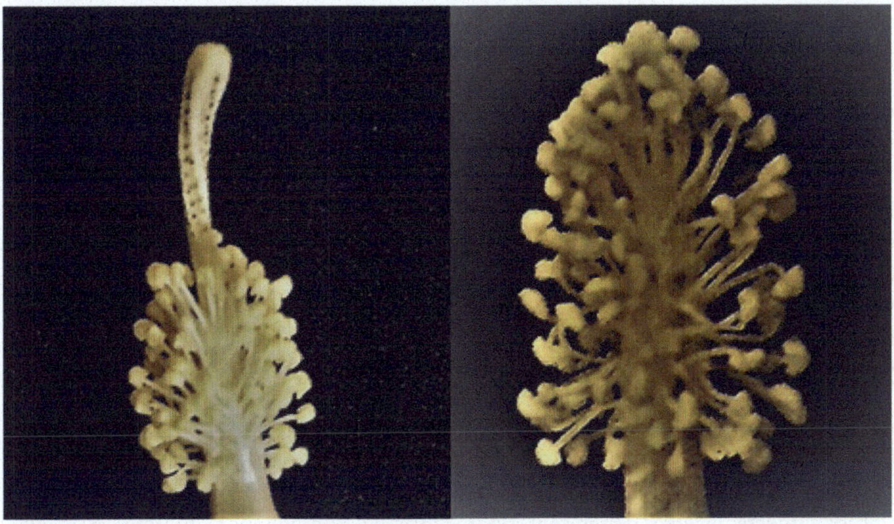

Fig. 20.5 Reproductive tissues of cotton flowers exposed to heat-stress (left) and optimum thermal conditions (right). Heat-stressed-flowers commonly have short filaments which creates illusion of an elongated stigma. (Modified and adopted from Brown 2008)

Several methods in natural field environmental conditions are generally being in practice for selection of heat tolerance. Field research is more advantageous for understanding the behavior as compared to controlled conditions; however; it has also limitations to control the environment under field conditions.

Studies have revealed that there exists a relationship between temperature besides cotton reproductive performance (Brown 2001; Zeiher et al. 1994). Heat stress damages young squares 15 days prior to flowering. Once developed into flowers most of these do not open fully owing to their smaller size. Moreover, the flowers show asynchronous development of male besides female reproductive structures, and anthers fail either to release pollen or are unable to fertilize due to incompatible elongation (Fig. 20.5) of filaments and stigmas (Brown 2008). High temperature affects anther developmental phases causing abnormalities in the structures, pollen sterility, and premature abortions. For instance, pollen germination and pollen tube growth at 82.4–86.0 °F (28–30 °C) negatively affect cotton reproductive performance. Pollen germination which was maximum at 82.4 °F/28 °C (Burke et al. 2004) showed a moderate to sharp decline when temperature increased beyond 28 °C to up to 98.6 °F/37 °C. Likewise, length of germinating pollen tubes increased maximum at temperatures between 82.4 °F/28 °C and 87.8 °F/31 °C but decreased significantly at 93.2 °F/34 °C approaching zero at 109.4 °F/43 °C. Higher temperature results in membrane modification and its composition with higher leakage of ions. Stability of cell membrane for thermal stability is evaluated with the ability of plants against hardens reaction to higher temperature besides tolerate harmful metabolic changes for heat stress (Alexandrov 1964).

Heat tolerance test was developed to understand stability of cellular membrane by measuring electrolyte leakage from leaf discs washed in deionized water immediately after exposure to heat-stress (Sullivan 1972). The EC of exudates from tissues discs is expressed as measures of CMT to stress (Blum and Ebercon 1981). The method is, however, more applicable to plants at mature stages.

Temperature variation effects on plants are multifaceted and greatly influence pollen viability, fruit setting along sympodial branches (% boll set on first and second positions), and biochemical parameters such as chlorophyll content, crude protein and proline levels. These parameters may also be considered, while testing cotton genotypes for heat tolerance as well as yield performance. Relatively new physiological techniques like CMT, LEL, and anther dehiscence have been established as reliable and efficient screening methods (Singh et al. 2007).

Stomatal conductance is worthy criteria for higher yield of irrigated crops grown under environmental adversities (Lu et al. 1998; Rahman 2005; Ulloa et al. 2000). Canopy temperature is a sensitive technique for measuring tolerance to high temperatures under filed conditions (Oosterhuis and Snider 2009; Snider et al. 2010). Photosynthesis is also dependent upon stomatal conductance, thereby it is also sensitive under heat stress conditions, nevetheless it is not practical approach for screening of germplasm against stress tolerance (Bibi et al. 2008). Important traits for selection of heat tolerance in pants are given in Table 20.4.

20.3.3 Chemical and Biochemical Interventions to Induce Heat Tolerance

Various approaches have been employed to induce thermal stress tolerance in plants such as foliar application of chemicals, seed treatment prior to sowing with some inorganic salts, oxidants (e.g., H_2O_2), and osmoprotectants (e.g., growth hormones) (Wahid et al. 2007).

Seed treatment with high temperature (42 °C) before sowing enhances tolerance ability of plant and to combat overheating and overcome dehydration; likely it also aids in higher accumulation of water soluble proteins. Exogenous application is also a good option such as Ca^{2+} applications increases plants' heat tolerance. Calcium in the form of $CaCl_2$ application just before stress treatment increases antioxidant enzyme activity leading to protection from heat stress (Kolupaev et al. 2005; Tikhomirova 1985). Glycine betaine and polyamines, low molecular weight organic compounds, which confer fruitful to induce heat tolerance in numerous plant species. Seeds pre-treated with glycine betaine produced plants having low membrane damage, improved photosynthetic rate, and enhanced leaf water potential besides more shoot growth (Wahid and Shabbir 2005).

The protective mechanisms activated within plants in response to higher-temperature stress are described schematically in Fig. 20.6. Plants tolerate to high-temperature stress by morphological, physico-biochemical changes. In physiological

Table 20.4 Important morpho-physiological selectable traits conferring heat tolerance

No.	Trait	Characteristics	References
(A) Morphological traits amenable to direct selection			
1.	Okra leaf type	High leaf N content High CO_2 exchange rate (CER) High photoelectron transport rate Reduced non photochemical quenching	Pettigrew (2004)
		Reduced individual leaf area Higher photosynthesis	Wells et al. (1986)
2.	Lower fruiting height	Greater heat tolerance	Feaster and Turcotte (1985)
3.	Thicker leaves	High N content	Hall (2001)
		High photosynthetic capacity	Wright et al. (1993)
4.	Abundant flowering and fruiting at higher temperature	Heat tolerance	Ehlig and LeMert (1973)
5.	Earliness	Reproductive heat tolerance	Ahmed et al. (1993), Ehlers and Hall (1996)
6.	Stay-green effect	Heat tolerance	Reynolds et al. (1997)
7.	Pollen selection	Pollen selection through heat treatment Reproductive stage heat tolerance	Rodriguez-Garay and Barrow (1988)
(B) Physiological traits for both direct and/or indirect selection			
1.	Cell membrane thermostability (CMT)	Measures resistance of protoplasmic proteins to denaturation	Bibi et al. (2003), Rahman et al. (2004)
		Heat and drought tolerance	Ashraf et al. (1994)
2.	Leaf electrolyte leakage (LEL)	Heat tolerance	Thiaw (2003)
3.	Anther dehiscence and pollen viability	Reproductive heat tolerance	Singh et al. (2007), Brown and Zeiher (1998)
4.	Stomatal conductance		Ulloa et al. (2000), Rahman (2005)
5.	Canopy temperature		Oosterhuis et al. (2009)
6.	Photosynthesis		Bibi et al. (2008)

Source: Singh et al. (2007)

response, plants accumulate compatible osmolytes, which aid in increasing retention of water in plants for improved stomatal regulation and photosynthetic rate. The morpho-physiological changes include decrease in cell size, stomata closure to curtail water, increased stomatal and trichomes densities besides greater xylem vessels. In biochemical alterations, accumulation of stress-related antioxidant enzymes takes place which augment activities of antioxidants enzymes in plant cells. Antioxidants alleviate ROS and reduce photo-oxidation damage with

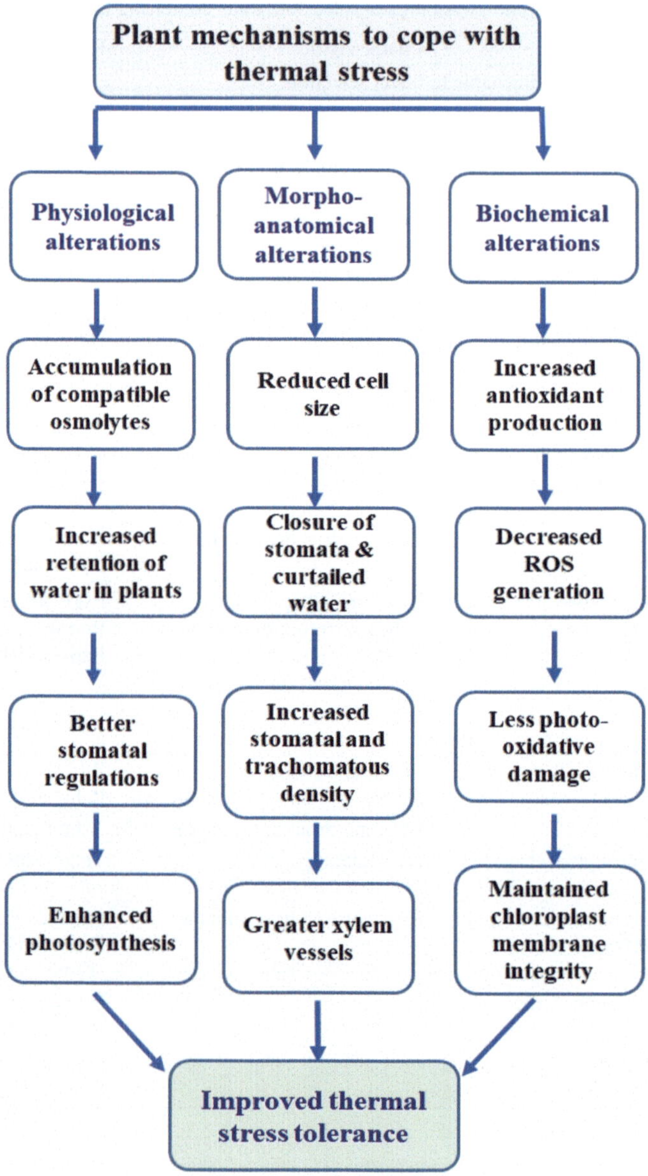

Fig. 20.6 Schematic presentation of morphological, physiological and biochemical alteration of plants to cope with high temperatures. (Modified and adopted from Waraich et al. 2012)

maintenance of chloroplast integrity. Plant growth regulator's exogenous applications are an important alternative to decrease negative impact of temperature stress by enhancing the antioxidant defense (Sarwar et al. 2017).

Foliar applications of essential macronutrients (N, K, Ca, and Mg) and micronutrients (B, Mn, Se), and Salicylic acid (SA) also help in alleviating adverse effects of temperature stress. These nutrients enhance antioxidants enzyme concentration in plant cells. Nutrients like K and Ca improve uptake of water with improved stomatal regulation that makes plant able to survive during exposure of heat stress. Moreover, K and Ca also help in osmotic balance and maintenance of higher tissue water potential under temperature stress conditions (Waraich et al. 2012).

References

Abbas Q, Ahmad S (2018) Effect of different sowing times and cultivars on cotton fiber quality under stable cotton-wheat cropping system in southern Punjab, Pakistan. Pak J Life Soc Sci 16:77–84

Abiko M, Akibayashi K, Sakata T, Kimura M, Kihara M, Itoh K, Asamizu E, Sato S, Takahashi H, Higashitani A (2005) High-temperature induction of male sterility during barley (*Hordeum vulgare* L.) anther development is mediated by transcriptional inhibition. Sex Plant Reprod 18:91–100

Abrol YP, Ingram KT (1996) Effects of higher day and night temperatures on growth and yields of some crop plants. In: Bazzaz F, Sombroek W (eds) Global climate change and agricultural production. Rome, FAO and Wiley, pp 123–140

ADB (2009) Annual Report: Asian Development Bank, p 122

Ahmad S, Raza I (2014) Optimization of management practices to improve cotton fiber quality under irrigated arid environment. J Food Agric Environ 2(2):609–613

Ahmad A, Diwan H, Abrol YP (2010) Global climate change, stress and plant productivity. In: Pareek A, Sopory SK, Bohnert HJ, Govindjee (eds) Abiotic stress adaptation in plants: physiological, molecular and genome foundation. Springer Science+Business Media B.V., New York, pp 503–531

Ahmad S, Raza I, Ali H, Shahzad AN, Atiq-ur-Rehman, Sarwar N (2014) Response of cotton crop to exogenous application of glycinebetaine under sufficient and scarce water conditions. Braz J Bot 37(4):407–415

Ahmad S, Abbas Q, Abbas G, Fatima Z, Atique-ur-Rehman, Naz S, Younis H, Khan RJ, Nasim W, Habib ur Rehman M, Ahmad A, Rasul G, Khan MA, Hasanuzzaman M (2017) Quantification of climate warming and crop management impacts on cotton phenology. Plants 6(7):1–16

Ahmad S, Iqbal M, Muhammad T, Mehmood A, Ahmad S, Hasanuzzaman M (2018) Cotton productivity enhanced through transplanting and early sowing. Acta Sci Biol Sci 40:e34610

Ahmed FE, Hall AE, DeMason DA (1992) Heat injury during floral development in cowpea (Vigna unguiculata, Fabaceae). Am J Bot 79:784–791

Ahmed FE, Mutters RG, Hall AE (1993) Interactive effects of high temperature and light quality on floral bud development in cowpea. Aust J Plant Physiol 20:661–667

Ahuja I, de Vos RC, Bones AM, Hall RD (2010) Plant molecular stress responses face climate change. Trends Plant Sci 15(12):664–674

Aksoy E, Demirel U, Öztürk ZN, Çaliskan S, Çaliskan ME (2015) Recent advances in potato genomics, transcriptomics, and transgenics under drought and heat stresses: a review. Turk J Bot 39:920–940

Alexandrov VY (1964) Cytophysiological and cytoecological investigations of heat resistance of plant cells toward the action of high and low temperature. Q Rev Biol 39:35–77

Ali H, Afzal MN, Ahmad F, Ahmad S, Akhtar M, Atif R (2011) Effect of sowing dates, plant spacing and nitrogen application on growth and productivity on cotton crop. Int J Sci Eng Res 2 (9):1–6

Ali H, Abid SA, Ahmad S, Sarwar N, Arooj M, Mahmood A, Shahzad AN (2013a) Integrated weed management in cotton cultivated in the alternate-furrow planting system. J Food Agric Environ 11(3&4):1664–1669

Ali H, Abid SA, Ahmad S, Sarwar N, Arooj M, Mahmood A, Shahzad AN (2013b) Impact of integrated weed management on flat-sown cotton (*Gossypium hirsutum* L.). J Anim Plant Sci 23 (4):1185–1192

Ali H, Hameed RA, Ahmad S, Shahzad AN, Sarwar N (2014a) Efficacy of different techniques of nitrogen application on American cotton under semi-arid conditions. J Food Agric Environ 12 (1):157–160

Ali H, Hussain GS, Hussain S, Shahzad AN, Ahmad S, Javeed HMR, Sarwar N (2014b) Early sowing reduces cotton leaf curl virus occurrence and improves cotton productivity. Cer Agron Moldova XLVII(4):71–81

Amin A, Nasim W, Mubeen M, Nadeem M, Ali L, Hammad HM, Sultana SR, Jabran K, Habib ur Rehman M, Ahmad S, Awais M, Rasool A, Fahad S, Saud S, Shah AN, Ihsan Z, Ali S, Bajwa AA, Hakeem KR, Ameen A, Amanullah, Rehman HU, Alghabar F, Jatoi GH, Akram M, Khan A, Islam F, Ata-Ul-Karim ST, Rehmani MIA, Hussain S, Razaq M, Fathi A (2017) Optimizing the phosphorus use in cotton by using CSM-CROPGRO-cotton model for semi-arid climate of Vehari-Punjab, Pakistan. Environ Sci Pollut Res 24(6):5811–5823

Amin A, Nasim W, Mubeen M, Ahmad A, Nadeem M, Urich P, Fahad S, Ahmad S, Wajid A, Tabassum F, Hammad HM, Sultana SR, Anwar S, Baloch SK, Wahid A, Wilkerson CJ, Hoogenboom G (2018) Simulated CSM-CROPGRO-cotton yield under projected future climate by SimCLIM for southern Punjab, Pakistan. Agric Syst 167:213–222

Andrews TJ, Hudson GS, Mate CJ, von Caemmerer S, Evans JR, Arvidsson YBC (1995) Rubisco: the consequences of altering its expression and activation in transgenic plants. J Exp Bot 46:1293–1300

Apel K, Hirt H (2004) Reactive oxygen species: metabolism, oxidative stress, and signal transduction. Annu Rev Plant Biol 55:373–399

Arevalo LM, Oosterhuis DM, Coker DL, Brown RS (2004) Effect of night temperatures on plant growth, boll development and yield. In: Beltwide cotton Conferences, San Antonio, Texas. National Cotton Council of America, Memphis, TN, p 1972

Arevalo LM, Oosterhuis DM, Coker DL, Brown RS (2008) Physiological response of cotton to high night temperatures. Am J Plant Sci Biotechnol 2:63–68

Arndt CH (1945) Temperature-growth relations of the roots and hypocotyls of cotton seedlings. Plant Physiol 20:200–220

Asada K (2006) Production and scavenging of reactive oxygen species in chloroplasts and their functions. Plant Physiol 141(2):391–396

Asha R, Ahamed ML (2013) Screening of cotton genotypes for heat tolerance via in vitro. Int J Bioresour Stress Manage 4:599–604

Ashraf M, Saeed MM, Qureshi MJ (1994) Tolerance to high temperature in cotton (*Gossypium hirsutum* L.) at initial growth stages. Environ and Exper Bot 34:275–283

Bange MP, Milroy SP (2004) Impact of short-term exposure to cold night temperatures on early development of cotton (*Gossypium hirsutum* L.). Aust J Agric Res 55:655–664

Bange MP, Caton SJ, Milroy SP (2008) Managing yields of high fruit retention in transgenic cotton (*Gossypium hirsutum* L.) using sowing date. Aust J Agric Res 59:733–741

Banti V, Mafessoni F, Loreti E, Alpi A, Perata P (2010) The heat-inducible transcription factor HsfA2 enhances anoxia tolerance in Arabidopsis. Plant Physiol 152:1471–1483

Barnabás B, Jäger K, Fehér A (2008) The effect of drought and heat stress on reproductive processes in cereals. Plant Cell Environ 31:11–38

Bellarby J, Foereid B, Hastings A, Smith P (2008) Cool farming: climate impacts of agriculture and mitigation potential. Greenpeace International, Amsterdam

Berry J, Bjorkman O (1980) Photosynthetic response and adaptation to temperature in higher plants. Annu Rev Plant Physiol 31:491–543

Bibi AC, Oosterhuis DM, Brown RS, Gonias ED, Bourland FM (2003) The physiological response of cotton to high temperatures for germplasm screening, pp 87–93. Summaries of Arkansas Cotton Research 2003. AAES Research Series 521

Bibi AC, Oosterhuis DM, Gonias ED (2008) Photosynthesis, quantum yield of photosystem II and membrane leakage as affected by high temperatures in cotton genotypes. J Cotton Sci 12 (12):150–159

Binder BM, Patterson SE (2009) Ethylene-dependent and -independent regulation of abscission. Stewart Postharvest Rev 5:1–10

Blum A, Ebercon A (1981) Cell membrane stability as a measure of drought and heat tolerance in wheat. Crop Sci 21:43–47

Bowman DT, Gutiérrez OA (2003) Sources of fiber strength in the U.S. upland cotton crop from 1980 to 2000. J Cotton Sci 7:164–169

Boyer JS (1982) Plant productivity and environment. Science 218:443–448

Bradow JM, Davidonis GH (2000) Quantitation of fiber quality and the cotton production-processing interface: a physiologist's perspective. J Cotton Sci 4:34–64

Bradow JM, Davidonis GH (2010) Effects of environment on fiber quality. In: Stewart JM, Oosterhuis DM, Heitholt JJ, Mauney JR (eds) Physiology of cotton. Springer, Dordrecht, pp 229–245

Bradow JM, Sassenrath-Cole GF, Hinojosa O, Wartelle LH (1996) Cotton fibre physical and physiological maturity variation in response to genotype and environment. In: Proceedings of the Beltwide cotton conference, vol 2. National Cotton Council, Memphis, TN, pp 1251–1254

Bradow JM, Wartelle LH, Bauer PJ, Sassenrath-Cole GF (1997) Small-sample cotton fiber quality quantitation. J Cotton Sci 1:48–58

Brown PW, Zeiher CA (1998) A model to estimate cotton canopy temperature in the desert southwest. In: Dugger CP, Richter DA (eds) Proceeding of the Beltwide cotton conferences. National Cotton Council of America, Memphis, TN, p 1734

Bray EA, Bailey-Serres J, Weretilnyk E (2000) Responses to abiotic stresses. In: Buchannan BB, Gruissem W, Jones RL (eds) Biochemistry and molecular biology of plants. American Society of Plant Physiologists, Rockville, pp 1158–1203

Brown PW (2001) Heat stress and cotton yields in Arizona. In: Cotton: a College of Agriculture Report. AZ1224. College of Agriculture and Life Sciences, University of Arizona, Tucson, AZ, p 7

Brown PW (2008) Cotton heat stress. The University of Arizona Cooperative Extension, AZ1448. College of Agriculture and Life Sciences, University of Arizona, Tucson, AZ, 10 p

Burke JJ (2001) Opportunities for improving cotton's tolerance to high temperature. In: Proceedings of the Beltwide Cotton Conference. National Cotton Council, Memphis, TN, pp 1453–1454

Burke JJ, Hatfield JL, Klein RR, Mullet JE (1985) Accumulation of heat shock proteins in field-grown cotton. Plant Physiol 78:394–398

Burke JJ, Mahan JR, Hatfield JL (1988) Crop-specific thermal kinetic windows in relation to wheat and cotton biomass production. Agron J 80:553–556

Burke JJ, Velten J, Oliver MJ (2004) In vitro analysis of cotton pollen germination. Agron J 96:359–368

CCRI (2017) Central Cotton Research Institute, Annual summary progress report 2017. 158 p

Christensen JH, Christensen OB (2007) A summary of the PRUDENCE model projections of changes in European climate by the end of this century. Climate Change 81:7–30

CICR (2000) Abiotic stresses in cotton: a physiological approach. CICR Tech Bull 2:1–13

Cottee NS, Wilson IW, Tan DKY, Bange MP (2014) Understanding the molecular events underpinning cultivar differences in the physiological performance and heat tolerance of cotton (*Gossypium hirsutum*). Funct Plant Biol 41:56–67

Crafts-Brandner SJ, Salvucci ME (2000) Rubisco activase constrains the photosynthetic potential of leaves at high temperature and CO_2. Proc Natl Acad Sci U S A 97:13430–13435

Crafts-Brandner SJ, Van De Loo FJ, Salvucci ME (1997) The two forms of ribulose-1,5-bisphosphate carboxylase/oxygenase activase differ in sensitivity to elevated temperature. Plant Physiol 114:439–444

Dat JF, Lopez-delgado H, Foyer CH, Scott IM (1998) Parallel changes in H_2O_2 and catalase during thermotolerance induced by salicylic acid or heat acclimation in mustard seedlings. Plant Physiol 116:1351–1357

Demirel U, Gür A, Can N, Memon AR (2014) Identification of heat responsive genes in cotton. Biol Plant 58(3):515–523

Eckardt A, Portis AR Jr (1997) Heat denaturation profiles of ribulose-1,5-bisphosphate carboxylase/oxygenase (rubisco) and rubisco activase and the inability of rubisco activase to restore activity of heat-denatured rubisco. Plant Physiol 113:243–248

Ehlers JD, Hall AE (1996) Genotypic classification of cowpea based on responses to heat and photoperiod. Crop Sci 36:673–679

Ehlig CF, LeMert RD (1973) Effects of fruit load, temperature, and relative humidity on boll retention of cotton. Crop Sci 13:168–171

Endo M, Tsuchiya T, Hamada K, Kawamura S, Yano K, Ohshima M, Higashitani A, Watanabe M, Kawagishi-Kobayashi M (2009) High temperatures cause male sterility in rice plants with transcriptional alterations during pollen development. Plant Cell Physiol 50:1911–1922

Farooq J, Mahmood K, Akram MW, Rehman AU, Javaid MI, Petrescu-Mag IV, Nawaz B (2015) High temperature stress in cotton *Gossypium hirsutum* L. Extreme Life Biospeol Astrobiol 7:34–44

Feaster CV, Turcotte EL (1985) Use of heat tolerance in cotton breeding. In: Dugger CP, Richter DA (eds) Proceeding of the Beltwide cotton conferences. National Cotton Council of America, Memphis, TN, pp 364–366

Foyer CH, Noctor G (2005) Oxidant and antioxidant signalling in plants: a re-evaluation of the concept of oxidative stress in a physiological context. Plant Cell Environ 28:1056–1071

Frank G, Pressman E, Ophir R, Althan L, Shaked R, Freedman M, Shen S, Firon N (2009) Transcriptional profiling of maturing tomato (*Solanum lycopersicum* L.) microspores reveals the involvement of heat shock proteins, ROS scavengers, hormones, and sugars in the heat stress response. J Exp Bot 60(13):3891–3908

Frydman J (2001) Folding of newly translated proteins in vivo: the role of molecular chaperones. Annu Rev Biochem 70:603–647

Gechev TS, Van Breusegem F, Stone JM, Denev I, Laloi C (2006) Reactive oxygen species as signals that modulate plant stress responses and programmed cell death. BioEssays 28:1091–1101

Gipson JR, Joham HE (1969) Influence of night temperature on growth and development of cotton (*Gossypium hirsutum* L.). I. Fruiting and boll development. Agron J 60:292–295

Gipson JR, Ray LL (1969) Fiber elongation rates in five varieties of cotton (Gossypium hirsutum L.) as influenced by night temperature. Crop Sci 9:339–341

Gokani SJ, Thaker VS (2002) Physiological and biochemical changes associated with cotton fiber development. IX. Role of IAA and PAA. Field Crop Res 77:127–136

Gou JY, Wang LJ, Chen SP, Hu WL, Chen XY (2007) Gene expression and metabolite profiles of cotton fiber during cell elongation and secondary cell wall synthesis. Cell Res 17:422–434

Govindjee (1995) Sixty-three years since Kautsky: chlorophyll a flourescence. Aust J Plant Physiol 22:131–160

Guinn G (1974) Abscission of cotton floral buds and bolls as influenced by factors affecting photosynthesis and respiration. Crop Sci 14:291–293

Gür A, Demirel U, Özden M, Kahraman A, Çopur O (2010) Diurnal gradual heat stress affects antioxidant enzymes, proline accumulation and some physiological components in cotton (Gossypium hirsutum L.). Afr J Biotechnol 9:1008–1015

Gurley WB, Caldana C, Grimbs S, Willmitzer L, Fernie AR, Nikoloski Z (2000) HSP101: a key component for the acquisition of thermotolerance in plants. Plant Cell 12(4):457–460

Guy C (1999) Molecular responses of plants to cold shock and cold. J Mol Microbiol Biotechnol 1:231–242

Gwimbi P, Mundoga T (2010) Impact of climate change on cotton production under rainfed conditions: case of Gokwe. J Sustain Dev Afr 12:59–69

Hall AE (2001) Crop response to environment. CRC Press, Boca Raton, FL

Hanson RG, Ewing EC, Ewing EC Jr (1956) Effect of environmental factors on fiber properties and yield of deltapine cottons. Agron J 48:573–581

Hasanuzzaman M, Nahar K, Alam MM, Roychowdhury R, Fujita M (2013) Physiological, biochemical, and molecular mechanisms of heat stress tolerance in plants. Int J Mol Sci 14:9643–9684

Hatfield JL, Boote KJ, Kimball BA, Wolf DW, Ort D, Izaurralde RC, Thomson AM, Morgan JA, Polley HW, Fay PA, Mader T, Hahn GL (2008) Agriculture. In: The effects of climate change on agriculture, land resources, water resources, and biodiversity in the United States. A report by the U.S. Climate Change Science Program and the Subcommittee on Global Change Research, Washington, DC, USA, p 362

Hatfield JL, Boote KJ, Kimball BA, Ziska LH, Izaurralde RC, Ort D, Thomson AM, Wolfe D (2011) Climate impacts on agriculture: implications for crop production. Agron J 103:351–370

Havaux M, Tardy F, Ravenel J, Chanu D, Parot P (1996) Thylakoid membrane stability to heat stress studied by flash spectroscopic measurements of the electrochromic shift in intact potato leaves: influence of the xanthophyll content. Plant Cell Environ 19:1359–1368

Hedhly A, Hormaza JI, Herrero M (2009) Global warming and sexual plant reproduction. Trends Plant Sci 14:30–36

Hejnák V, Tatar Ö, Atasoy GD, Martinková J, Çelen AE, Hnilička F, Skalický M (2015) Growth and photosynthesis of upland and Pima cotton: response to drought and heat stress. Plant Soil Environ 62:507–514

Hopf N, Plesofsky-Vig N, Brambl R (1992) The heat shock response of pollen and other tissues of maize. Plant Mol Biol 19:623–630

Horváth I, Glatz A, Nakamoto H, Mishkind ML, Munnik T, Saidi Y, Goloubinoff P, Harwood JL, Vigh L (2012) Heat shock response in photosynthetic organisms: membrane and lipid connections. Prog Lipid Res 51:208–220

Huang B, Xu C (2008) Identification and characterization of proteins associated with plant tolerance to heat stress. J Integr Plant Biol 50(10):1230–1237

Iba K (2002) Acclimative response to temperature stress in higher plants: approaches of gene engineering for temperature tolerance. Annu Rev Plant Biol 53:225–245

IPCC (2007) Intergovernmental Panel on Climate Change, Fourth assessment report: climate change 2007, Synthesis report, p 76

Iqbal MA, Ping Q, Abid M, Kazmi SMM, Rizwan M (2016) Assessing risk perceptions and attitude among cotton farmers: a case of Punjab province, Pakistan. Int J Disaster Risk Reduct 16:68–74

Jackson JE (1967) Studies on the sowing dates of cotton in the Sudan Gezira. J Agric Sci 69:305–315

Janjua PZ, Samad G, Khan NU (2010) Impact of climate change on wheat production: a case study of Pakistan. Pak Dev Rev 49:799–822

Jiao D, Benhua J (1996) Changes of the photosynthetic electron transport and photosyntheic enzyme activies of two rice varieties under photooxidation condition. Acta Agron Sin 22:43–49

John ME, Keller G (1996) Metabolic pathway engineering in cotton: biosynthesis of polyhydroxybutyrate in fiber cells. Proc Natl Acad Sci U S A 93:12768–12773

Johnson RM, Bradow JM, Sassenrath-Cole GF (1997) Modeling of cotton fiber quality from environmental parameters. In: Digger CP, Richter CA (eds) Proceedings of the Beltwide cotton conference, vol 2. National Cotton Council, Memphis, TN, pp 1454–1455

Kakani VG, Reddy KR, Koti S, Wallace TP, Prasad PV, Reddy VR, Zhao D (2005) Differences in in vitro pollen germination and pollen tube growth of cotton cultivars in response to high temperature. Ann Bot 96:59–67

Karmakar R, Das I, Dutta D, Rakshit A (2016) Potential effects of climate change on soil properties: a review. Sci Int 4:51–73

Kaya H, Shibahara KI, Taoka KI, Iwabuchi M, Stillman B, Araki T (2001) FASCIATA genes for chromatin assembly factor-1 in Arabidopsis maintain the cellular organization of apical meristems. Cell 104:131–142

Khan AI, Khan IA, Sadaqat HA (2008) Heat tolerance is variable in cotton (*Gossypium hirsutum* L.) and can be exploited for breeding of better yielding cultivars under high temperature regimes. Pak J Bot 40(5):2053–2058

Khan A, Najeeb U, Wang L, Tan DKY, Yang G, Munsif F, Ali S, Hafeez A (2017) Planting density and sowing date strongly influence growth and lint yield of cotton crops. Field Crop Res 209:129–135

Kobza J, Seemann JR (1989) Regulation of ribulose-1,5-bisphosphate carboxylase activity in response to diurnal changes in irradiance. Plant Physiol 89:918–924

Kolupaev YE, Akinina GE, Mokrousov AV (2005) Induction of heat tolerance in wheat coleoptiles by calcium ions and its relation to oxidative stress. Russ J Plant Physiol 52:199–204

Krause GH, Weis E (1991) Chlorophyll fluorescence and photosynthesis: the basics. Annu Rev Plant Physiol Plant Mol Biol 42:313–349

Krieg DR, Sung JFM (1986) Source-sink relations as affected by water stress. In: Mauney JR, Stewart JM (eds) Cotton physiology. The Cotton Foundation, Memphis, TN, pp 73–77

Larkindale J, Huang B (2004) Thermotolerance and antioxidant systems in Agrostis stolonifera: involvement of salicylic acid, abscisic acid, calcium, hydrogen peroxide, and ethylene. J Plant Physiol 161(4):405–413

Larkindale J, Hall JD, Knight MR, Vierling E, Larkindale J, Hall JD, Knight MR, Vierling E (2005) Heat stress phenotypes of Arabidopsis mutants implicate multiple signaling pathways in the acquisition of thermotolerance. Plant Physiol 138:882–897

Law RD, Crafts-Brandner SJ (1999) Inhibition and acclimation of photosynthesis to heat stress is closely correlated with activation of ribulose-1,5-bisphosphate carboxylase/oxygenase. Plant Physiol 120:173–182

Lohar DP, Peat WE (1998) Floral characteristics of heat-tolerant and heat-sensitive tomato (*Lycopersicon esculentum* Mill.) cultivars at high temperature. Sci Hortic 73:53–60

Loka DA, Oosterhuis DM (2010) Effect of high night temperatures on cotton respiration, ATP levels and carbohydrate content. Environ Exp Bot 68:258–263

Lu Z, Percy RG, Qualset CO, Zeiger E (1998) Stomatal conductance predicts yields in irrigated Pima cotton and bread wheat grown at high temperatures. J Exp Bot 49:453–460

Ludwig LJ, Saeki T, Evans LT (1965) Photosynthesis in artificial communities of cotton plants in relation to leaf area. I. Experiments with progressive defoliation of mature plants. Aust J Biol Sci 18:1103–1118

Luo Q (2011) Temperature thresholds and crop production: a review. Clim Chang 109:583–598

Luo Q, Bange M, Clancy L (2014) Cotton crop phenology in a new temperature regime. Ecol Model 285:22–29

Maestri E, Klueva N, Perrotta C, Gulli M, Nguyen HT, Marmiroli N (2002) Molecular genetics of heat tolerance and heat shock proteins in cereals. Plant Mol Biol 48:667–681

Magwanga RO, Lu P, Kirungu JN, Lu H, Wang X, Cai X, Zhou Z, Zhang Z, Salih H, Wang K, Liu F (2018) Characterization of the late embryogenesis abundant (LEA) proteins family and their role in drought stress tolerance in upland cotton. BMC Genet 19:6

Mauney JR (1966) Floral initiation of upland cotton Gossypium hirsutum L. in response to temperatures. J Exp Bot 17:452–459

Miller G, Mittler R (2006) Could heat shock transcription factors function as hydrogen peroxide sensors in plants? Ann Bot 98:279–288

Min L, Zhu L, Tu L, Deng F, Yuan D, Zhang X (2013) Cotton GhCKI disrupts normal male reproduction by delaying tapetum programmed cell death via inactivating starch synthase. Plant J 75:823–835

Min L, Li Y, Hu Q, Zhu L, Gao W, Wu Y, Ding Y, Liu S, Yang X, Zhang X (2014) Sugar and auxin signaling pathways respond to high-temperature stress during anther development as revealed by transcript profiling analysis in cotton. Plant Physiol 164:1293–1308

Mittler R, Vanderauwera S, Gollery M, Van Breusegem F (2004) Reactive oxygen gene network of plants. Trends Plant Sci 9:490–498

Mittler R, Vanderauwera S, Suzuki N, Miller G, Tognetti VB, Vandepoele K, Gollery M, Shulaev V, Van Breusegem F (2011) ROS signaling: the new wave? Trends Plant Sci 16:300–309

Mohamed HI, Abdel-Hamid AME (2013) Molecular and biochemical studies for heat tolerance on four cotton genotypes. Rom Biotechnol Lett 18:8823–8831

Moreno AA, Orellana A (2011) The physiological role of the unfolded protein response in plants. Biol Res 44:75–80

Moriarty T, West R, Small G, Rao D, Ristic Z (2002) Heterologous expression of maize chloroplast protein synthesis elongation factor (EF-Tu) enhances Escherichia coli viability under heat stress. Plant Sci 163:1075–1082

Nasim W, Ahmad A, Wajid A, Akhtar J (2011) Nitrogen effects on growth and development of sunflower hybrids under agro-climatic conditions of Multan. Pak J Bot 43:2083–2092

Nasim W, Belhouchette H, Tariq M, Fahad S, Hammad HM, Mubeen M, Munis MF, Chaudhary HJ, Khan I, Mahmood F, Abbas T, Rasul F, Nadeem M, Bajwa AA, Ullah N, Alghabari F, Saud S, Mubarak H, Ahmad R (2016) Correlation studies on nitrogen for sunflower crop across the agroclimatic variability. Environ Sci Pollut Res 23:3658–3670

Nava GA, Dalmago GA, Bergamaschi H, Paniz R, dos Santos RP, Marodin GAB (2009) Effect of high temperatures in the pre-blooming and blooming periods on ovule formation, pollen grains and yield of 'Granada' peach. Sci Hortic 122:37–44

Oosterhuis DM (1997) Effect of temperature extremes on cotton yields in Arkansas. In: Oosterhuis DM (ed) Proceeding of cotton research meeting and research summaries. Agricultural Experiment Station special report. University of Arkansas, Fayetteville, pp 94–98

Oosterhuis DM (1999) Yield response to environmental extremes in cotton. In: Oosterhuis DM (ed) Proceeding of the 1999 cotton research meeting and summaries of cotton research in progress, vol 193. Arkansas Agricultural Experiment Station, Fayetteville, pp 30–38

Oosterhuis DM (2009) Summaries of Arkansas cotton research. Arkansas Agricultural Experiment Station, University of Arkansas Division of Agriculture, Fayetteville. 225 p

Oosterhuis DM, Snider JL (2009) High temperature stress on floral development and yield of cotton. In: Stress physiology in cotton. The Cotton Foundation, Cordova, TN, pp 1–24

Oosterhuis DM, Bourland FM, Bibi AC, Gonias ED, Loka D, Storch D (2009) Screening for temperature tolerance in cotton. Summaries of Arkansas cotton research in 2008. AAES Research Series 573:41

Oshino T, Abiko M, Saito R, Ichiishi E, Endo M, Kawagishi-Kobayashi M, Higashitani A (2007) Premature progression of anther early developmental programs accompanied by comprehensive alterations in transcription during high-temperature injury in barley plants. Mol Gen Genomics 278(1):31–42

Panchuk II, Volkov RA, Schoffl F (2002) Heat stress- and heat shock transcription factor-dependent expression and activity of ascorbate peroxidase in Arabidopsis. Plant Physiol 129:838–853

Parish RW, Phan HA, Iacuone S, Li SF (2012) Tapetal development and abiotic stress: a centre of vulnerability. Funct Plant Biol 39:553–559

Pearson RW, Ratliff LT, Taylor HM (1970) Effect of soil temperature, strength, and PH on cotton seedling root elongation. Agron J 62:243–246

Perry SW, Krieg DR, Hutmacher RB (1983) Photosynthetic rate control in cotton: photorespiration. Plant Physiol 73:662–665

Pettigrew WT (2001) Environmental effects on cotton fiber carbohydrate concentration and quality. Crop Sci 41:1108–1113

Pettigrew WT (2004) Physiological consequences of moisture deficit stress in cotton. Crop Sci 44:1265–1272

Pettigrew WT (2008) The effect of higher temperatures on cotton lint yield production and fiber quality. Crop Sci 48:278–285

Portis AR Jr (1992) Regulation of ribulose 1,5 biphosphate carboxylase/oxygenase activity. Annu Rev Plant Physiol Plant Mol Biol 43:415–437

Potters G, Pasternak TP, Guisez Y, Jansen MA (2009) Different stresses, similar morphogenic responses: integrating a plethora of pathways. Plant Cell Environ 32:158–169

Pradhan GP, Prasad PVV, Fritz AK, Kirkham MB, Gill BS (2012) Response of Aegilops species to drought stress during reproductive stages of development. Funct Plant Biol 39:51–59

Qi Y, Wang H, Zou Y, Liu C, Liu Y, Wang Y, Zhang W (2011) Over-expression of mitochondrial heat shock protein 70 suppresses programmed cell death in rice. FEBS Lett 585(1):231–239

Quinn PJ, Williams WP (1985) Environmentally induced changes in chloroplast membranes and their effects on photosynthetic function. In: Barber J, Baker NR (eds) Photosynthetic mechanisms and the environment. Elsevier Science Publishers B.V., Amsterdam, pp 1–47

Rahman HH (2005) Genetic analysis of stomatal conductance in upland cotton (*Gossypium hirsutum* L.) under contrasting temperature regimes. J Agric Sci 143:161–168

Rahman MH, Malik SA, Saleem M (2004) Heat tolerance of upland cotton during the fruiting stage evaluated using cellular membrane thermostability. Field Crop Res 85:149–158

Rahman MH, Murtaza N, Shah MKN (2007) Study of cotton fibre traits inheritance under different temperature regimes. J Agron Crop Sci 193:45–54

Rahman MH, Ahmad A, Wajid A, Hussain M, Rasul F, Ishaque W, Islam MA, Shelia V, Awais M, Ullah A, Wahid A, Sultana SR, Saud S, Khan S, Fahad S, Hussain M, Hussain S, Nasim W (2017) Application of CSM-CROPGRO-Cotton model for cultivars and optimum planting dates: evaluation in changing semi-arid climate. Field Crop Res 238:139–152

Rahman MH, Ahmad A, Wang X, Wajid A, Nasim W, Hussain M, Ahmad B, Ahmad I, Ali Z, Ishaque W, Awais M, Shelia V, Ahmad S, Fahad S, Alam M, Ullah H, Hoogenboom G (2018) Multi-model projections of future climate and climate change impacts uncertainty assessment for cotton production in Pakistan. Agric For Meteorol 253–254:94–113

Rasheed R (2009) Salinity and extreme temperature effects on sprouting buds of sugarcane (*Saccharum officinarum* L.): some histological and biochemical studies. University of Agriculture, Faisalabad

Rawson HM (1992) Plant responses to temperature under conditions of elevated CO_2. Aust J Bot 40 (5):473–490

Reddy KR, Hodges HF (1995) Cotton crop responses to a changing environment. In: Rosenzweig C et al (eds) Climate change and agriculture: analysis of potential international impacts. American Society of Agronomy Special Publication no. 59. American Society of Agronomy, Madison, WI, pp 3–30

Reddy KR, Zhao D (2005) Interactive effects of elevated CO_2 and potassium deficiency on photosynthesis, growth, and biomass partitioning of cotton. Field Crop Res 94:201–213

Reddy VR, Baker DN, Hodges HF (1991) Temperature effects on cotton canopy growth, photosynthesis, and respiration. Agron J 83:699–704

Reddy KR, Hodges HF, McKinion JM, Wall GW (1992a) Temperature effects on Pima cotton growth and development. Agron J 84:237–243

Reddy KR, Hodges HF, Reddy VR (1992b) Temperature effects on cotton fruit retention. Agron J 84:26–30

Reddy KR, Reddy VR, Hodges HF (1992c) Temperature effects on early season cotton growth and development. Agron J 84:229–237

Reddy KR, Hodges HF, McKinion JM (1995) Carbon dioxide and temperature effects on pima cotton growth. Agric Ecosyst Environ 54:17–29

Reddy KR, Hodges HF, Mckinion JM (1996) Food and agriculture in the 21st century: a cotton example. World Resour Rev 8:80–97

Reddy KR, Hodges HF, McKinion JM (1997a) A comparison of scenarios for the effect of global climate change on cotton growth and yield. Funct Plant Biol 24:703–713

Reddy KR, Hodges HF, McKinion JM (1997b) Crop modelling and application: a cotton example. Academic, California

Reddy KR, Davidonis GH, Johnson AS, Vinyard BT (1999) Temperature regime and carbon dioxide enrichment alter cotton boll development and fiber properties. Agron J 91:851–858

Rehman H (2006) Number and weight of cotton lint fibres: variation due to high temperatures in the field. Aust J Agric Res 57:583–590

Reynolds MP, Nagarajan S, Razzaue MA, Ageeb OAA (eds) (1997) Using canopy temperature depression to select for yield potential of wheat in heat-stressed environmental. Wheat Special Report No 42:51 p CIMMYT, Mexico

Riaz M, Farooq J, Mahmood A, Sadiq MA, Yaseen M (2013) Genotypic variability for root/shoot parameters under water stress in some advanced lines of cotton (*Gossypium hirsutum* L.). Genet Mol Res 12:552–561

Roberts EM, Rao NR, Huang JY, Trolinder NL, Haigler CH (1992) Effects of cycling temperatures on fiber metabolism in cultured cotton ovules. Plant Physiol 100:979–986

Rodriguez-Garay B, Barrow JR (1988) Pollen selection for heat tolerance in cotton. Crop Sci 28:857–859

Rodríguez M, Canales E, Borrás-hidalgo O (2005) Molecular aspects of abiotic stress in plants. Biotecnología Aplicada 22:1–10

Roy M, Ghosh B (1996) Polyamines, both common and uncommon, under heat stress in rice (Oryza sativa) callus. Physiol Plant 98:196–200

Ruan YL (2007) Rapid cell expansion and cellulose synthesis regulated by plasmodesmata and sugar: insights from the single-celled cotton fibre. Funct Plant Biol 34(1):10

Ruan YL, Jin Y, Yang YJ, Li GJ, Boyer JS (2010) Sugar input, metabolism, and signaling mediated by invertase: roles in development, yield potential, and response to drought and heat. Mol Plant 3:942–955

Sable A, Rai KM, Choudhary A, Yadav VK, Agarwal SK, Sawant SV (2018) Inhibition of heat shock proteins HSP90 and HSP70 induce oxidative stress, suppressing cotton fiber development. Sci Rep 8:3620

Saha SR, Hossain MM, Rahman MM, Kuo CG, Abdullah S (2010) Effect of high temperature stress on the performance of twelve sweet pepper genotypes. Bangladesh J Agric Res 35(3):525–534

Saidi Y, Finka A, Muriset M, Bromberg Z, Weiss YG, Maathuis FJ, Goloubinoff P (2009) The heat shock response in moss plants is regulated by specific calcium-permeable channels in the plasma membrane. Plant Cell 21:2829–2843

Saidi Y, Finka A, Goloubinoff P (2011) Heat perception and signalling in plants: a tortuous path to thermotolerance. New Phytol 190:556–565

Sakata T, Higashitani A (2008) Male sterility accompanied with abnormal anther development in plants – genes and environmental stresses with special reference to high temperature injury. Int J Plant Dev Biol 2:42–51

Salvucci ME, Crafts-Brandner SJ (2004) Inhibition of photosynthesis by heat stress: the activation state of Rubisco as a limiting factor in photosynthesis. Physiol Plant 120:179–186

Salvucci ME, Ogren WL (1996) The mechanism of Rubisco activase: insights from studies of the properties and structure of the enzyme. Photosynth Res 47:1–11

Salvucci ME, Portis AR, Ogren WL (1986) Light and CO_2 response of ribulose-1,5-bisphosphate carboxylase/oxygenase activation in Arabidopsis leaves. Plant Physiol 80:655–659

Salvucci ME, Osteryoung KW, Crafts-Brandner SJ, Vierling E (2001) Exceptional sensitivity of Rubisco activase to thermal denaturation in vitro and in vivo. Plant Physiol 127:1053–1064

Sánchez-Rodríguez E, Moreno DA, Ferreres F, Rubio-Wilhelmi Mdel M, Ruiz JM (2011) Differential responses of five cherry tomato varieties to water stress: changes on phenolic metabolites and related enzymes. Phytochemistry 72:723–729

Sarwar M, Saleem MF, Najeeb U, Shakeel A, Ali S, Bilal MF (2017) Hydrogen peroxide reduces heat-induced yield losses in cotton (*Gossypium hirsutum* L.) by protecting cellular membrane damage. J Agron Crop Sci 203:429–441

Scafaro AP, Haynes PA, Atwell BJ (2010) Physiological and molecular changes in Oryza meridionalis Ng., a heat-tolerant species of wild rice. J Exp Bot 61(1):191–202
Scharf KD, Berberich T, Ebersberger I, Nover L (2012) The plant heat stress transcription factor (Hsf) family: structure, function and evolution. Biochim Biophys Acta 1819:104–119
Schlenker W, Hanemann WM, Fisher AC (2006) The impact of global warming on U.S. agriculture: an econometric analysis of optimal growing conditions. Rev Econ Stat 88:113–125
Schlenkera W, Roberts MJ (2009) Nonlinear temperature effects indicate severe damages to U.S. crop yields under climate change. Proc Natl Acad Sci U S A 106:15594–15598
Schrader SM, Wise RR, Wacholtz WF, Ort DR, Sharkey TD (2004) Thylakoid membrane responses to moderately high leaf temperature in Pima cotton. Plant Cell Environ 27:725–735
Schramm F, Larkindale J, Kiehlmann E, Ganguli A, Englich G, Vierling E, von Koskull-Döring P (2008) A cascade of transcription factor DREB2A and heat stress transcription factor HsfA3 regulates the heat stress response of Arabidopsis. Plant J 53:264–274
Sekmen AH, Ozgur R, Uzilday B, Turkan I (2014) Reactive oxygen species scavenging capacities of cotton (*Gossypium hirsutum*) cultivars under combined drought and heat induced oxidative stress. Environ Exp Bot 99:141–149
Sharkey TD (2005) Effects of moderate heat stress on photosynthesis: importance of thylakoid reactions, rubisco deactivation, reactive oxygen species, and thermotolerance provided by isoprene. Plant Cell Environ 28:269–277
Singh RP, Vara Prasad PV, Sunita K, Giri SN, Reddy KR (2007) Influence of high temperature and breeding for heat tolerance in cotton: a review. Adv Agron 93:313–385
Smertenko A, Dráber P, Viklický V, Opatrný Z (1997) Heat stress affects the organization of microtubules and cell division in Nicotiana tabacum cells. Plant Cell Environ 20:1534–1542
Smith P, Martino D, Cai Z, Gwary D, Janzen H, Kumar P, McCarl B, Ogle S, O'Mara F, Rice C, Scholes B, Sirotenko O, Howden M, McAllister T, Pan G, Romanenkov V, Schneider U, Towprayoon S, Wattenbach M, Smith J (2007) Greenhouse gas mitigation in agriculture. Philos Trans R Soc B Biol Sci 363:789–813
Snider JL, Oosterhuis DM, Skulman BW, Kawakami EM (2009) Heat stress-induced limitations to reproductive success in *Gossypium hirsutum*. Physiol Plant 137:125–138
Snider JL, Oosterhuis DM, Kawakami EM (2010) Genotypic differences in thermotolerance are dependent upon prestress capacity for antioxidant protection of the photosynthetic apparatus in Gossypium hirsutum. Physiol Plant 138:268–277
Snider JL, Oosterhuis DM, Kawakami EM (2011) Diurnal pollen tube growth rate is slowed by high temperature in field-grown *Gossypium hirsutum* pistils. J Plant Physiol 168:441–448
Snipes CE, Baskin CC (1994) Influence of early defoliation on cotton yield, seed quality, and fiber properties. Field Crops Res 37:137–143
Steinfield H, Gerber P, Wassenar T, Castel V, Rosales M, de Haan C (2006) Livestock's long shadow: environmental issues and options. FAO, Rome
Stewart JM (1986) Integrated events in the flower and fruit. In: Mauney JR, Stewart JMD (eds) Cotton physiology. The Cotton Foundation, Memphis, TN, pp 261–300
Strasser BJ (1997) Donor side capacity of photosystem II probed by chlorophyll a fluorescence transients. Photosynth Res 52:147–155
Sullivan CY (1972) Mechanisms of heat and drought resistance in grain sorghum and method of measurements. In: Rao N, House L (eds) Sorghum in the seventies. pp. 247–264, New Delhi: Oxford and IBH Publishers
Taha MA, Malik MNA, Chaudhry FL, Makhdum MI (1981) Heat-induced sterility in cotton sown during early April in West Punjab. Environ Sci 17:189–194
Taiz L, Zeiger E (2006) Plant physiology, 4th edn. Sinauer Associates, Inc., Sunderland, MA
Talanova VV, Akimova TV, Titov AF (2003) Effect of whole plant and local heating on the ABA content in cucumber seedling leaves and roots and on their heat tolerance. Russ J Plant Physiol 50:90–94

Tariq M, Yasmeen A, Ahmad S, Hussain N, Afzal MN, Hasanuzzaman M (2017) Shedding of fruiting structures in cotton: factors, compensation and prevention. Trop Subtrop Agroecosyst 20:251–262

Tariq M, Afzal MN, Muhammad D, Ahmad S, Shahzad AN, Kiran A, Wakeel A (2018) Relationship of tissue potassium content with yield and fiber quality components of Bt cotton as influenced by potassium application methods. Field Crop Res 229:37–43

Thakur P, Kumar S, Malik JA, Berger JD, Nayyar H (2010) Cold stress effects on reproductive development in grain crops: an overview. Environ Exp Bot 67:429–443

Thiaw S (2003) Association between slow leaf-electrolyte-leakage under heat stress and heat tolerance during reproductive development in cowpea. University of California, Riverside, USA

Tian J, Hu Y, Gan X, Zhang Y, Hu X, Gou L, Lou H, Zhang W (2013) Effects of increased night temperature on cellulose synthesis and the activity of sucrose metabolism enzymes in cotton fiber. J Integ Agri 12:979–988

Tikhomirova EV (1985) Changes of nitrogen metabolism in millet at elevated temperatures. Field Crop Res 11:259–264

Ton P (2011) Cotton and climate change: impacts and options to mitigate and adapt. ITC, 2011. Technical paper, Doc. No. MAR-11-200.E Technical Report, Geneva, xii, 32 p

Tsukaguchi T, Kawamitsu Y, Takeda H, Suzuki K, Egawa Y (2003) Water status of flower buds and leaves as affected by high temperature in heat-tolerant and heat-sensitive cultivars of snap bean (*Phaseolus vulgaris* L.). Plant Product Sci 6:24–27

Ulloa M, Cantrell RG, Percy RG, Zeiger E, Lu Z (2000) Breeding & genetics QTL analysis of stomatal conductance and relationship to lint yield in an interspecific cotton. J Cotton Sci 4:10–18

Usman M, Ahmad A, Ahmad S, Irshad M, Khaliq T, Wajid A, Hussain K, Nasim W, Chattha TM, Trethowan R, Hoogenboom G (2009) Development and application of crop water stress index for scheduling irrigation in cotton (*Gossypium hirsutum* L.) under semiarid environment. J Food Agric Environ 7(3&4):386–391

van der Werf GR, Morton DC, DeFries RS, Olivier JGJ, Kasibhatla PS, Jackson RB, Collatz GJ, Randerson JT (2009) CO_2 emissions from forest loss. Nat Geosci 2:737–738

Vierling E (1991) The roles of heat shock proteins in plants. Annu Rev Plant Physiol Plant Mol Biol 42(1):579–620

Wahid A, Shabbir A (2005) Induction of heat stress tolerance in barley seedlings by pre-sowing seed treatment with glycinebetaine. Plant Growth Regul 46:133–141

Wahid A, Gelani S, Ashraf M, Foolad MR (2007) Heat tolerance in plants: an overview. Environ Exp Bot 61(3):199–223

Wang W, Vinocur B, Shoseyov O, Altman A (2004) Role of plant heat-shock proteins and molecular chaperones in the abiotic stress response. Trends Plant Sci 9:244–252

Waraich EA, Ahmad R, Halim A, Aziz T (2012) Alleviation of temperature stress by nutrient management in crop plants: a review. J Soil Sci Plant Nutr 12:221–244

Wassmann R, Jagadish SVK, Sumfleth K, Pathak H, Howell G, Ismail A, Serraj R, Redona E, Singh RK, Heuer S (2009) Regional vulnerability of climate change impacts on Asian rice production and scope for adaptation. Adv Agron 102:91–133

Weis E (1981) Reversible heat-inactivation of the Calvin cycle: a possible mechanism of the temperature regulation of photosynthesis. Planta 151:33–39

Weis E, Berry JA (1988) Plants and high temperature stress. Symp Soc Exp Biol 42:329–346

Wells R, Meredith WR, Williford JR (1986) Canopy photosynthesis and its relationship to plant productivity in near-isogenic cotton lines differing in leaf morphology. Plant Physiol 82:635–640

White SC, Raine SR (2008) A grower guide to plant based sensing for irrigation scheduling. Centre for Engineering in Agriculture Publication, Toowoomba, QLD

Wise RR, Olson AJ, Schrader SM, Sharkey TD (2004) Electron transport is the functional limitation of photosynthesis in field-grown Pima cotton plants at high temperature. Plant Cell Environ 27:717–724

Wright GC, Hubick KT, Farquhar GD, Rao RNC (1993) Genetics and environmental variation in transpiration efficiency and its correlation with carbon isotope discrimination and specific leaf area in peanut. In: Ehleninger JE, Hall AE, Farquhar GD, Saugier B (eds) Stable isotopes and plant carbon- water relations. Academic, New York, pp 247–267

Wullschleger SD, Oosterhuis DM (1990) Photosynthetic carbon production and use by developing cotton leaves and bolls. Crop Sci 30:1259–1264

Xie W, Trolinder NL, Haigler CH (1993) Cool temperature effects on cotton fiber initiation and elongation clarified using in vitro cultures. Crop Sci 33:1258–1264

Yfoulis A, Fasoulas A (1978) Role of minimum and maximum environmental temperature on maturation period of the cotton boll. Agron J 70:421–425

Yoshida T, Ohama N, Nakajima J, Kidokoro S, Mizoi J, Nakashima K, Maruyama K, Kim JM, Seki M, Todaka D, Osakabe Y, Sakuma Y, Schöz F, Shinozaki F, Yamaguchi-Shinozaki K (2011) Arabidopsis HsfA1 transcription factors function as the main positive regulators in heat shock-responsive gene expression. Mol Gen Genomics 286:321–332

Young LW, Wilen RW, Bonham-Smith PC (2004) High temperature stress of Brassica napus during flowering reduces micro- and megagametophyte fertility, induces fruit abortion, and disrupts seed production. J Exp Bot 55:485–495

Zafar SA, Noor MA, Waqas MA, Wang X, Shaheen T, Raza M, Mehboob-Ur-Rahman (2018) Temperature extremes in cotton production and mitigation strategies. In: Mehmoob-Ur-Rahman, Zafar Y (eds) Past, present and future trends in cotton breeding. IntechOpen, Croatia, pp 65–91

Zahid KR, Ali F, Shah F, Younas M, Shah T, Shahwar D, Hassan W, Ahmad Z, Qi C, Lu Y, Iqbal A, Wu W (2016) Response and tolerance mechanism of cotton (*Gossypium hirsutum* L.) to elevated temperature stress: a review. Front Plant Sci 7:937

Zeiher CA, Brown PW, Silvertooth JC, Matumba N, Mitton N (1994) The effect of night temperature on cotton reproductive development. Cotton: A college of agriculture report 370096:89–96

Zhang Y, Mian MAR, Bouton JH (2006) Recent molecular and genomic studies on stress tolerance of forage and turf grasses. Crop Sci 46:497–511

Zhang K, Zhang J, Ma J, Tang S, Liu D, Teng Z, Liu D, Zhang Z (2012) Genetic mapping and quantitative trait locus analysis of fiber quality traits using a three-parent composite population in upland cotton (*Gossypium hirsutum* L.). Mol Breed 29:335–348

Zhang J, Srivastava V, Stewart JM, Underwood J (2016) Heat-tolerance in cotton is correlated with induced overexpression of heat-shock factors, heat-shock proteins, and general stress response genes. J Cotton Sci 20:253–262

Zhao D, Reddy KR, Kakani VG, Koti S, Gao W (2005) Physiological causes of cotton fruit abscission under conditions of high temperature and enhanced ultraviolet-B radiation. Physiol Plant 124:189–199

Zinn KE, Tunc-Ozdemir M, Harper JF (2010) Temperature stress and plant sexual reproduction: uncovering the weakest links. J Exp Bot 61:1959–1968

Ziska LH, Bunce JA (1997) Influence of increasing carbon dioxide concentration on the photosynthetic and growth stimulation of selected C_4 crops and weeds. Photosynth Res 54:199–208

Chapter 21
Applications of Crop Modeling in Cotton Production

Ghulam Abbas, Zartash Fatima, Muhammad Tariq, Mukhtar Ahmed, Muhammad Habib ur Rahman, Wajid Nasim, Ghulam Rasul, and Shakeel Ahmad

Abstract Cotton growth models are being generally used by cotton scientists as well as policy makers across globe as an important and effective research tool. Cotton simulation models have been applied during last and current decades for the analysis of the cotton plant responses to drought, heat, and nutrients stress as well as to test the alternating optimum sowing window under climate warming trend in cotton belt. Cotton growth models are useful research tools in worldwide. Mostly cotton models were applied for climatic changes, cotton management practices, and irrigation strategies on lint and cottonseed yield in worldwide. All cotton models were successfully used at local, regional, and national levels in worldwide, but among all cotton growth models, CROPGRO-Cotton model was mostly used by researchers and policy makers. For irrigation management strategies, mostly AquaCrop model was used by researchers.

G. Abbas (✉) · Z. Fatima · S. Ahmad (✉)
Department of Agronomy, Bahauddin Zakariya University, Multan, Pakistan
e-mail: shakeelahmad@bzu.edu.pk

M. Tariq
Central Cotton Research Institute, Multan, Pakistan

M. Ahmed
Department of Agronomy, PMAS Arid Agriculture University, Rawalpindi, Pakistan

M. H. ur. Rahman
Department of Agronomy, Muhammad Nawaz Shareef University of Agriculture, Multan, Pakistan

W. Nasim
Department of Agronomy, University College of Agriculture and Environmental Sciences, Islamia University of Bahawalpur, Bahawalpur, Pakistan

G. Rasul
International Center for Integrated Mountain Development, Kathmandu, Nepal

Pakistan Meteorological Department, Islamabad, Pakistan

S. Ahmad, M. Hasanuzzaman (eds.), *Cotton Production and Uses*,
https://doi.org/10.1007/978-981-15-1472-2_21

Keywords Model · Climate change · Irrigation · Management · CROPGRO · Phenology

Abbreviations

CRM	Coefficient residual mass
DSSAT	Decision support system for agro-technology transfer
GCM	General circulation model
LAI	Leaf area index
LFMAX	Maximum leaf area
NRMSE	Normalized root mean square error
RCP	Representative concentration pathway
RMSE	Root mean square error
RUE	Radiation use efficiency
SCY	Seed cotton yield
TDM	Total dry matter
WUE	Water use efficiency

21.1 Introduction

Cotton (*Gossypium hirsutum* L.) is a main fiber crop, which is playing an elemental role in economy of various countries, mostly for the less developing countries, and yet well developed countries retain definite states and provinces in worldwide (Usman et al. 2009; Ahmad et al. 2014, 2018; Abbas and Ahmad 2018; Ahmad and Raza 2014; Ali et al. 2011, 2013a, b, 2014a, b; Amouzou et al. 2018; Koukouli and Georgiou 2018; Rahman et al. 2018). Cotton crop is grown as a fiber crop across more than 70 countries in the world (Khan et al. 2004; Amin et al. 2017, 2018; Tariq et al. 2017, 2018). Cotton crop provides lint raw material for an ever extending textile industry for cloth making. Cottonseed oil is used for the cooking food intentions as well as edible oil and protein rich oil cake residues are used for animal feeding purpose (Adhikari et al. 2017; Ali et al. 2018; Tsakmakis et al. 2019). The cotton is a one of the most excellent fiber, as well as main agronomic crop, which is spreading for one of the global leading textiles industries having an annually economic influence of no less than $600 billion globally. Cotton crop is, hence, precisely called the life blood of financial system of several countries. China, India, America, Pakistan, Brazil, Australia, etc. are the top mainly cotton-producing countries at global level. Most cotton-producing country is China, which produced cotton of 6.51 million metric tons, and second most cotton producing country is India, which produces cotton 6.42 million metric tons, while the USA had a top third position for cotton raising country, which produces total cotton production of 3.61 million metric tons (Shuli et al. 2018).

Table 21.1 History of cotton modeling in the world

Model name	Events for model improvements	References
SIMCOT	SIMCOT-I Daily photosynthesis and simulated the distribution of the photosynthate produced among the plant parts	Duncan et al. (1967), Duncan (1972)
	SIMCOT-II Addition of a nitrogen balance subroutine	McKinion et al. (1975)
	Addition of RHIZOS model into SIMCOT-II model	Lambert and Baker (1984)
KUTUN	Mechanistic based model	Mutsaers (1984)
SIRATAC	Field management simulation	Hearn and da Roza (1985)
GOSSYM	Incorporate many of the physical properties unique to individual soils in model	Baker et al. (1983)
	Considerable work on testing and improving crop model	Brown et al. (1985)
	Addition of expert system COMAX into models	Lemmon (1986)
COMAX	Some growers are satisfied, and others are unsatisfied with the model use	Ladewig and Thomas (1992)
	Computerize the optimization of production decisions by simplifying GOSSYM	Marani et al. (1992a)
CALGOS	Simplify the GOSSYM model for site specific	Marani et al. (1992b)
OZCOT	A temperature-driven model of fruiting dynamics to the soil water balance of Ritchie (1972)	Hearn (1994)
COTCO2	An attempt to simulate changes in cotton growth caused by changes in atmospheric CO_2 concentrations	Wall et al. (1994)
COTMAN	Site-specific simple model	Zhang et al. (1994)
COTGROW	Specific purpose based model	Pan et al. (1996)
	Model testing at farmer field	Hodges et al. (1998)
COTONS	Linking physiological and architectural models enhanced the cotton model in function	Jallas et al. (2000)
CROPGRO	Use template model for cotton sensitivity analysis	Pathak et al. (2007)

Source: Modified and adapted and from Landivar et al. (2010)

Nowadays, the cotton crop growth modeling is a useful approach for obtain decision-making information about the crop growth, development, and production sufficiently well (Richardson et al. 2002; Li et al. 2009; Amin et al. 2018; Koukouli and Georgiou 2018). History of cotton growth models is given in Table 21.1. Cotton growth models like CROPGRO, CropSyst, EPIC, AquaCrop, InfoCrop etc. can concurrently integrates nonlinear interactions among soil, water, cotton plant, weather parameters, and crop husbandry practices for determination of production, environmental stresses, and water requirement besides nutrients needs the cotton industry across the globe (Ko et al. 2009; Venugopalan et al. 2014; Linker et al. 2016; Kumar et al. 2017; Amouzou et al. 2018). Cotton models are being broadly employed by cotton scientists as well as policy makers as an imperative assessment making tool for study of the influences of climatic changes, cotton management practices and irrigation strategies on lint and cottonseed yield (Fig. 21.1). Cotton

Fig. 21.1 Crop growth models used for cotton for different applications purpose. (Source: Published literature)

simulation models have been applied for last decades for the analyzing of the cotton plant responses to drought, heat, and nutrients stress as well as to test the alternating optimum sowing window under climate warming trend in cotton belt (Guerra et al. 2005; McCarthy et al. 2014; Dzotsi et al. 2015; Wang et al. 2017a, b; Li et al. 2019a, b). Cotton prediction models also quantify the impact of pest and diseases on cotton productivity at the farm level. Most of cotton growth models, nevertheless, are quite complicated, which require of advanced expertise for their calibration, evaluation, validation, and application requires abundance of inputs. Cotton Modeling is helpful for assessment of the influence of seasonal changes on cotton phenology stages and phases, to develop deficit irrigation strategies, and to predict the expected yield and fertilizer, water and radiation use efficiency under prevailing environmental conditions (Hebbar et al. 2008; Loison et al. 2016; Rahman et al. 2017; Nagender et al. 2017). Cotton growth models might useful for assisting in synthesis of researchers understanding regarding relationships of genetics, physiological besides environment interactions, and among various disciplines besides organization of growth data of crop. In the most recent decade, cotton models have been applied broadly in cotton belt regions for simulation of crop responses to various abiotic factors. In recent decade, a lot of cotton mechanistic simulations have been reported on both the growth, development as well as seed cotton yield from planting to physiological maturity in response to both nonspecific and specific location environment circumstances (Sommer et al. 2008; Zamora et al. 2009; Kamali et al. 2011; Reddy et al. 2016; Qian et al. 2017).

21.2 Crop Management Practices

Table 21.2 showed that the GOSSYM model was well evaluated and validated in Arizona, America. Model performance was well for higher nitrogen levels. Model predicted that farmers should apply 56 kg ha^{-1} besides predicted increased yield of lint 224 kg ha^{-1} (Reddy et al. 2016). The COTTON2K model was good performed under pruning and topping effect in cotton. Strong agreement between recorded and simulation lint yield was obtained, with 10% error in China (Yang et al. 2008). CROPGRO-Cotton model simulated biomass accumulation for leaves and stems very good, excluding during reproductive growth, when model simulated leaf and stem weights were under estimated for rainfed and irrigated circumstances. Model performance for the final SCY was very good, with RMSE value of 312 kg ha^{-1} in America (Guerra et al. 2005). The InfoCrop model accurateness was 86% and 89% for total dry matter and seed yield, correspondingly under different sowing dates effect in India (Hebbar et al. 2008). The evaluation results of CropSyst model showed that early cotton growth and LAI development were predicted with satisfactory accuracy. Nevertheless, final total dry matter was somewhat overvalued by CropSyst model, due to the reason that several unaccounted cotton plant stresses at the studied locations reduced genuine aboveground total dry matter, leading to RMSE value approximately 2.00 Mg ha^{-1}. Few characteristics of cotton crop like indeterminate growth pattern could not be amalgamated in detail in CropSyst model (Sommer et al. 2008). CROPGRO model results presented that mechanism (i.e., SLA, LAI, LFMAX, and biomass apportioning) influencing cotton productivity in shaded environmental conditions affect the model behavior. The model showed a close agreement between field observed besides model simulated total dry matter during first besides second year of study with $R^2 = 0.95$; $R^2 = 0.92$, respectively (Zamora et al. 2009). In Pakistan, CROPGRO model performance was good; that results showed that RMSE values for LAI approached more than 1 in several of nitrogen treatments, RMSE range values for biomass, and SCY were reasonably well in acceptable range from 367 to 497 kg ha^{-1} and 122 to 227 kg ha^{-1}, respectively (Wajid et al. 2014). CROPGRO-Cotton model was effectively calibrated and validated in field conditions of Cameroon. Indeed, the RMSE of maximum LAI and SCY were 28.01% and 25.74%, correspondingly (Loison et al. 2016). Seed cotton production was predicted with acceptable RMSE range from 137 to 382 kg ha^{-1} at final harvesting, whereas dynamics of time series for seed cotton yield were simulated reasonably well with higher values of *d*-index ranging from 0.95 to 0.99 for all varieties during evaluation. Sowing date analyses by using historic weather data during 1980 to 2013, close agreed with field recorded SCY trend, showed reduction of 27% for late sowing from 20th April to 21st June, whereas, first too advance sowing date 10th March had also faced 8% decline for the entire varieties in Pakistan (Rahman et al. 2017). The DNDC model can be useful research tool for simulation of soil and root respiration in oasis cotton cropping system. The model analysis showed that air temperature, rainfall, soil organic carbon contents, fertilization, and irrigation have a positively impact on soil respiration. Root respiration was higher sensitive to

Table 21.2 Various cotton models used for crop management practices worldwide

Country	Types of applications	Cotton Model name	Parameters studied	References
Pakistan	Cultivars and sowing date management	CROPGRO	Phenology, LAI, biomass, SCY	Rahman et al. (2017)
America	Farmers field test	CROPGRO	SCY, biomass, boll and stem weight	Guerra et al. (2005)
Cameroon	Cultivar selection	CROPGRO	LAI, SCY	Loison et al. (2016)
Pakistan	Optimum phosphorus fertilizer	CROPGRO	Biomass, LAI, SCY	Amin et al. (2017)
Pakistan	Nitrogen rates and planting dates	CROPGRO	LAI, biomass, SCY	Arshad et al. (2017)
China	Sowing date	COSIM	Phenology, lint yield	
India	Plant density and nitrogen levels	CROPGRO	Phenology, biomass, SCY	Nagender et al. (2017)
China	Phosphorus management	Olsen-P	Phosphorus quantity	Wang et al. (2015)
Pakistan	Nitrogen levels and planting dates	CROPGRO	Phenology, LAI, biomass	Wajid et al. (2014)
China	Fertilization, soil organic C	DNDC	Root respiration	Yu and Zhao (2015)
India	Sowing date	Info crop	Phenology, LAI, biomass, SCY	Hebbar et al. (2008)
China	Aeration stress	CROPR	SCY	Qian et al. (2017)
Uzbekistan	Nitrogen fertilizer	CropSyst	Phenology, LAI, biomass, SCY	Sommer et al. (2008)
America	Light levels or shading in alley cropping	CROPGRO	SCY, biomass, LAI	Zamora et al. (2009)
China	Pruning and topping effect	COTTON2K	SCY, biomass, boll weight	Yang et al. (2008)
China	Sowing date	COSIM	SCY	Liting et al. (2018)
America	Nitrogen management	GOSSYM	Lint yield LAI, biomass	Reddy et al. (2016)
South Africa	Cultivar selection	AMMI	Boll weight, SCY	Pretorius et al. (2015)
India	Yield gap	CROPGRO	SCY	Sadanshiv et al. (2012)
Iran	Cultivar	AMMI	Biomass, SCY	Kamali et al. (2011)
America	Soil type effect	GOSSYM	Seedcotton production	Iqbal and Whisler (2000)
India	Sowing date	InfoCrop	Boll weight, LAI, biomass, SCY	Tak (2014)

climate conditions and crop management practices as compared to soil physical and chemical properties (Yu and Zhao 2015). Simulating precision of COSIM cotton model for LAI and biomass was more than 87% in comparison to the field observed data and more than 83% as compared with the field observed values for the lint yield under different sowing dates in China. CROPGRO-Cotton model performance was well, because there was a close well agreement between observed and predicted seed cotton production across different planting densities and nitrogen rates with root mean square error value of 300 kg ha^{-1}, whereas, the positive coefficient residual mass (CRM) value showed the tendency of the cotton model to inderpredict the seed cotton yield via 5%. With respect to NRMSE value, the model prediction was well with 9% for variety MRC-7201 as compared to other varieties (Nagender et al. 2017). With improved CROPR-Cotton model, the RMSE value for seed cotton yield was 686.22 kg ha^{-1} with an NRMSE value of 14.87%. With the original CROPR model, the RMSE value was 1019.02 kg ha^{-1} with an NRMSE of 22.08% under soil aeration stress conditions. The performance of improved model was well as compared to original model (Qian et al. 2017). Under different planting dates and nitrogen levels, CROPGRO model performance was good in Pakistan, because the RMSE range values from 278 to 573 kg ha^{-1} and from 237 to 422 kg ha^{-1} were attained for TDM and SCY, correspondingly (Arshad et al. 2017).

21.3 Irrigation Management

Table 21.3 showed that the modified OZCOT cotton model provided reasonable predictions of yield for solid planted and skips row cotton crop under irrigated and rainfed conditions in Australia (Milroy et al. 2004). AquaCrop model did accurate simulation of evapo-transpiration (error value $<13\%$). Model predicted SCY value was inside 10% of measurements under various irrigation regimes in first year of study, with errors up to 32%. The model well simulated the tendency in entire soil water contents; however, deviation exists for individual soil layers (Farahani et al. 2009). The predicted results validate that EPIC cotton model can be applied as a research tool for cotton crop in case of fully irrigated and deficit circumstances in southern cotton growing areas of Texas. Model seems to be successful for long-term and pre-season choices for managing irrigation for cotton crop in America (Ko et al. 2009). AquaCrop model was evaluated well for drip-irrigated cotton, and its performance was well in hot and dry climate in the eastern Mediterranean region. Well-predicted consequences of evapo-transpiration, total dry matter, seed cotton production, and soil moisture contents under four irrigation levels are promising considering the simplicity owing to its demanded lowest input data (Hussein et al. 2011). Attia et al. 2016 reported that performance of Cotton2K model was well under different irrigation regimes. Field recorded and model predicted seed cotton yield was not significantly different from each other. The relationship between average seed cotton production and evapo-transpiration replacement showed that irrigation at 112% evapo-transpiration maximize seed cotton production at 1406 kg ha^{-1} in

Table 21.3 Various cotton models used for irrigation management worldwide

Country	Cotton model name	Parameters studied	References
America	SALUS	LAI, biomass, SCY	Dzotsi et al. (2015)
Greece	AquaCrop	Soil water contents, canopy cover, biomass, SCY	Tsakmakis et al. (2019)
Australia	OZCOT	WUE	McCarthy et al. (2014)
America	EPIC	WUE, lint yield	Ko et al. (2009)
China	SWAT	Soil water contents, biomass, lint yield	Han et al. (2015)
Greece	AquaCrop	SCY	Linker et al. (2016)
China	AquaCrop	WUE, canopy cover	Tan et al. (2018)
China	CROPGRO	Soil water contents	Li et al. (2019a, b)
Iran	AquaCrop	Phenology, SCY	Heidariniya et al. (2012)
America	CROPGRO	LAI, SCY	Modala et al. (2015)
Syria	AquaCrop	Canopy cover, biomass, SCY, WUE	Hussein et al. (2011)
America	GOSSYM	WUE, SCY	Liang et al. (2012)
China	CROPGRO	Evapo-transpiration	Lifeng et al. (2015)
America	CROPGRO	LAI, biomass yield	Thorp et al. (2017)
America	Cotton2K	Lint yield	Attia et al. (2016)
America	CROPGRO	Soil moisture contents, evapo-transpiration	Adhikari et al. (2017)
China	AquaCrop	SCY	Li et al. (2019a, b)
Syria	AquaCrop	Canopy cover, evapotranspiration, biomass, SCY	Farahani et al. (2009)
China	EPIC	Soil water contents, LAI, biomass, SCY	Wang et al. (2017a, b)
Syria	CROPWAT	Crop water use, SCY	Ibrahim and Yacoub (2009)
America	CROPGRO	Phenology, LAI, SCY	Pathak et al. (2007)
Australia	OZCOT	Lint yield	Milroy et al. (2004)

America (Yu and Zhao 2015). The SWAT model showed that volumetric soil water contents, leaf area indices, aboveground TDM, and SCY simulated were in well agreement with the observed data. Additionally, model simulation results also showed that underground water is an important source of water for cotton. As compared to control simulation which had underground water, 23% of transpiration is supplied through capillary rise, with an enhancement of seed cotton production by 20% in China (Han et al. 2015). The model performance was good, while changes in irrigation quantities during validation, as showed by a good agreement between models predicted besides field, observed LAI, plant height, and seed cotton yield under all irrigation scheduling treatments. Simulated SCY was within acceptable error range from −24% to 13% (Modala et al. 2015). The EPIC model performed well with the normalized root mean square error ranging from of 2.6% to 22.6% for soil moisture content, from 6.3% to 14.1% for LAI, from 4.9% to 7.2% for aboveground dry matter as well as with the error ranging from 2.5% to 8.2% for seed cotton production in China (Wang et al. 2017a, b). The evaluated and validated CROPGRO-Cotton model was well capable for simulation of SCY besides seasonal ET during a long-term period from 1924 to 2012 historic period in America under irrigated circumstances (Adhikari et al. 2017). Root zone soil moisture depletion was better simulated by CROPGRO-Cotton model as compared to FAO-56 irrigation scheduling spreadsheet, possibly due to its more multifaceted soil profile simulation module in America (Thorp et al. 2017). Well calibrated and evaluated of AquaCrop Model results showed that highest water productivity could be achieved through applying one irrigation (at seedling), two irrigations (at seedling plus squaring), and three irrigations (at seedling, squaring plus flowering) during wet, normal, and dry year, respectively, in China (Li et al. 2019a, b). AquaCrop model can be applied as a tool to simulate growth response of cotton to water under film-mulched drip irrigation. Based upon principle of high productivity besides WUE, predicted outcomes depicted that right irrigation quantities were mentioned at 358–457 mm and 406–462 mm for silty loam and sandy loam soils in China (Tan et al. 2018). When the AquaCrop model run in growing degree days mode, accuracy of simulated biomass besides soil water contents was fair for all proposed sets ($R^2 \geq 0.93$, EF ≥ 0.91), but SCY was simulated adequately. Similarly, in calendar days mode, the soil water contents were simulated accurately ($R^2 \geq 0.75$, EF ≥ 0.72) in Greece (Tsakmakis et al. 2019).

21.4 Phenology

Koukouli and Georgiou (2018) reported that InfoCrop model results that the phenology of cotton crop will somewhat earlier in the future scenario due to the climate thermal trend in spring season. The projections results indicated that the anthesis and physiological maturity date were advanced by 3 and 2 days, correspondingly during from 2020 to 2050 as well as 3 to 5 days advance during from 2070 to 2100 in comparison to the baseline period in Greece. InfoCrop model performance was well for anthesis and physiological maturity with RMSE value of 3 and 5 days, correspondingly in India (Hebbar et al. 2008). The CROPGRO model predicted values of cotton crop phenological stages, and phases were reliable with the field observed data, with RMSE value under 2 days for the period of both years of study in Pakistan (Wajid et al. 2014). DSSAT-CROPGRO-Cotton model was successfully calibrated and evaluated in field conditions of Cameroon, because the RRMSE value of flowering date and boll opening date validated dataset were 5.3% and 4.3%, correspondingly (Loison et al. 2016). CROPGRO model predicted flowering and maturity days were very close with field recorded data of phenological stages by slope = 0.82, $R^2 = 0.71$, $P < 0.01$ as well as slope = 0.82, $R^2 = 0.73$, $P < 0.01$, correspondingly. After that, good calibrated and evaluated cotton model was applied to prediction of crop phenology in 1980–2015 in Punjab, Pakistan. Model-simulated phenological stages were accelerated more, and phases reductions were more due to climate thermal trends in comparison to field recorded phenological stages and phases.

Table 21.4 Various cotton models used for assessment of climate change worldwide

Country	Cotton model name	Adaptation strategies	References
Pakistan	CROPGRO	Phosphorus fertilizer application	Amin et al. (2018)
Benin	CROPGRO	Improve soil fertility	Amouzou et al. (2018)
Pakistan	CROPGRO	Sowing date, cultivars	Habib ur Rahman et al. (2018)
Pakistan	GP-MCMC-copula	Heat resistant cultivars	Ali et al. (2018)
Greece	AquaCrop	Sowing date	Voloudakis et al. (2015)
America	CROPGRO	Efficient irrigation strategies	Adhikari et al. (2016)
India	CROPGRO	Sowing date	Kumar et al. (2017)
India	ARIMA	Resources conservation strategies	Poyyamozhi and Mohideen (2016)
America	EPIC	Enhance RUE of cultivars	Ko et al. (2009)
America	CROPGRO	Sowing date	Thorp et al. (2014)
India	InfoCrop	Cultivars longer duration	Venugopalan et al. (2014)
America	GOSSYM	Heat and drought tolerant varieties	Richardson et al. (2002)
Greece	CropSyst	Planting date	Koukouli and Georgiou (2018)

Nevertheless, 30.21% of the negative impact of the climate warming trend was minimized by growing of latest cotton varieties, which required higher growing degree days (Ahmad et al. 2017). Model predicted values for anthesis stages by CROPGRO-Cotton model for all cotton cultivars were well agreed with observed data, with RMSE value of 1.0 days, coefficient residual mass (CRM) value of −1, and NRMSE value of 1%. Cultivars predicted values for days to maturity were well agreed with field recorded data, with RMSE value of 2 days, CRM value of 2, and NRMSE value of 2% (Nagender et al. 2017). Relationship between field-recorded and model-predicted flowering days and boll maturity days of all studied treatments showed that overall performance of CROPGRO model in Pakistan was well (range value of RMSE from 1.58 to 1.82 days, range value of d from 0.94 to 0.95) and (range value of RMSE from 1.68 to 2.90 days, range value of d from 0.91 to 0.97), correspondingly (Rahman et al. 2017).

21.5 Climate Change

Table 21.4 showed that the CropSyst model simulation under climate change scenario of SRES B1 resulted in an enhancement through 6% during 2020 to 2050 and a reduction by almost 15% in seed cotton production during 2070 to 2100 in Greece (Koukouli and Georgiou 2018). The GOSSYM cotton model predicted that maximum obtainable cotton fruit production efficiency was increased under changing climatic conditions through an adaptation strategy of heat- and drought-tolerant cultivars in America (Richardson et al. 2002). When the parameter radiation use efficiency (RUE) was used in EPIC model for simulation of the climate variability in SCY, then a significantly well-correlated coefficient of 0.86 and RMSE of 0.22 t ha^{-1} was gained, presenting insignificant difference between model simulated and field observed yield data (Ko et al. 2009). In Pakistan, DSSAT-CROPGRO-model simulated that variety FH-142 would give a more percentage seed cotton production as compared to MNH-886 during 2025–2050 and the less percentage seed cotton production would be for MNH-886 at physiological maturity date under climatic warming scenarios in Punjab. Farming community might have to apply a moderate level of phosphorous fertilizer 57 kg ha^{-1} as an adaptation strategy for reduce the harmful influences of climate change during future scenarios (Amin et al. 2018). The CROPGRO model-simulated SCY with RMSE ranged from 105 to 1107 kg ha^{-1} (3% to 28% of average value) as well as cumulative seasonal evapo-transpiration was predicted with RMSE ranged from 12 to 42 mm (1% to 5% of average value). Variability in seed cotton production and evapo-transpiration due to the applied experiment treatments were predicted well ($d \geq 0.95$), independent of the year to year changes caused by the seasonal factors in America (Thorp et al. 2014). Prior to application of AquaCrop model, model was well validated with RMSE value for SCY, and total dry matter was 0.17 and 0.49 t ha^{-1}, correspondingly, whereas, *d* value (Index of Agreement value) was 0.93 and 0.94, respectively. When model was applied after evaluation, model results indicated that a trend toward enhancing SCY through the ending of the current century was detected under the majority of the global circulation models (GCMs), particularly in cotton growing belts of western Greece and northern Greece (Voloudakis et al. 2015). CROPGRO model results indicated that during calibration and evaluation of model, the predicted SCY showed very good agreement with field observed yield ($r^2 \geq 0.94$; $d \geq 0.83$). On an average basis, model-simulated seed cotton production across cotton growing belt reduced within, ranging from 4% to 17% during 2041 to 2070 climate change scenarios in comparison to the baseline weather data from 1971 to 2000 period in America (Adhikari et al. 2016). The ARIMA cotton model showed that total growing area and seed cotton production can be enhanced in future scenario, if adaptation strategies like proper land reclamation and conservation measures are adopted in India (Poyyamozhi and Mohideen 2016). The CROPGRO model results indicated that SCY are projected to reduction by 8% on an averaged basis during 2039s scenario and 20% reduction during 2069s scenario under the RCP 4.5 in comparison to the historic weather data from 1980 to 2010. Average seed cotton production is projected to decline through 12% and 30%, respectively, on an averaged basis under the RCP 8.5 scenario in Pakistan (Habib ur Rahman et al. 2018). The CROPGRO-Cotton model simulated higher SCY with planting dates in month of June in comparison to month of July in Benin. In the climate change scenario of RCP 2.6, model simulated a reduction in WUE by 20% without any soil amendment, and by 4% with an integrated soil crop management as compared baseline period, but an enhancing of cotton-seed yield by 2% with a higher use of mineral fertilizer amount in Benin (Amouzou et al. 2018).

21.6 Economics and Policy Making

Cotton models are useful tool for cotton crop economic analysis. Policy makers used cotton models results for making appropriate strategies for obtain maximum achievable yield under almost all major agroecological conditions. In spite of some limitations, the cotton growth modeling approach remains helpful for the economic assessment under the impacts of future global climate change; therefore, it provides assistance in the formulation of national policies for adaptation strategies purpose. Planning in cotton industry, managing operations, and concerns of management decisions upon environmental concerns are also well supported through employing modeling approach worldwide (Wang et al. 2017a, b; Li et al. 2019a, b).

21.7 Conclusion

Cotton growth models are useful research tools worldwide. Mostly cotton models were applied for assessment of climate change, cotton phenology, cotton crop management practices, and irrigation management options worldwide. All cotton models were successfully used at local, regional, and national levels worldwide, but among all cotton growth models, CROPGRO-Cotton model was mostly employed by cotton scientists. For irrigation management strategies, mostly AquaCrop model was used by researchers.

Acknowledgments The author acknowledged the funding by the Higher Education Commission (HEC), Islamabad (HEC-NRPU-4511), and Bahauddin Zakariya University, Multan.

References

Abbas Q, Ahmad S (2018) Effect of different sowing times and cultivars on cotton fiber quality under stable cotton-wheat cropping system in southern Punjab, Pakistan. Pak J Life Soc Sci 16:77–84

Adhikari P, Ale S, Bordovsky JP, Thorp KR, Modala NR, Rajan N, Barnes EM (2016) Simulating future climate change impacts on seed cotton yield in the Texas High Plains using the CSM-CROPGRO-Cotton model. Agric Water Manag 164:317–330

Adhikari P, Gowda PH, Marek GW, Brauer DK, Kisekka I, Northup B, Rocateli A (2017) Calibration and validation of CSM-CROPGRO-Cotton model using lysimeter data in the Texas High Plains. J Contemp Water Res Educ 162(1):61–78

Ahmad S, Raza I (2014) Optimization of management practices to improve cotton fiber quality under irrigated arid environment. J Food Agric Environ 2(2):609–613

Ahmad S, Raza I, Ali H, Shahzad AN, Atiq-ur-Rehman, Sarwar N (2014) Response of cotton crop to exogenous application of glycinebetaine under sufficient and scarce water conditions. Braz J Bot 37(4):407–415

Ahmad S, Abbas Q, Abbas G, Fatima Z, Atique-ur-Rehman, Naz S, Younis H, Khan RJ, Nasim W, Habib ur Rehman M, Ahmad A, Rasul G, Khan MA, Hasanuzzaman M (2017) Quantification of climate warming and crop management impacts on cotton phenology. Plants 6(7):1–16

Ahmad S, Iqbal M, Muhammad T, Mehmood A, Ahmad S, Hasanuzzaman M (2018) Cotton productivity enhanced through transplanting and early sowing. Acta Sci Biol Sci 40:e34610

Ali H, Afzal MN, Ahmad F, Ahmad S, Akhtar M, Atif R (2011) Effect of sowing dates, plant spacing and nitrogen application on growth and productivity on cotton crop. Int J Sci Eng Res 2 (9):1–6

Ali H, Abid SA, Ahmad S, Sarwar N, Arooj M, Mahmood A, Shahzad AN (2013a) Integrated weed management in cotton cultivated in the alternate-furrow planting system. J Food Agric Environ 11(3&4):1664–1669

Ali H, Abid SA, Ahmad S, Sarwar N, Arooj M, Mahmood A, Shahzad AN (2013b) Impact of integrated weed management on flat-sown cotton (*Gossypium hirsutum* L.). J Anim Plant Sci 23 (4):1185–1192

Ali H, Hameed RA, Ahmad S, Shahzad AN, Sarwar N (2014a) Efficacy of different techniques of nitrogen application on American cotton under semi-arid conditions. J Food Agric Environ 12 (1):157–160

Ali H, Hussain GS, Hussain S, Shahzad AN, Ahmad S, Javeed HMR, Sarwar N (2014b) Early sowing reduces cotton leaf curl virus occurrence and improves cotton productivity. Cer Agron Moldova XLVII(4):71–81

Ali M, Deo RC, Downs NJ, Maraseni T (2018) Cotton yield prediction with Markov Chain Monte Carlo-based simulation model integrated with genetic programing algorithm: a new hybrid copula-driven approach. Agric For Meteorol 263:428–448

Amin A, Nasim W, Mubeen M, Nadeem M, Ali L, Hammad HM, Sultana SR, Jabran K, Habib ur Rehman M, Ahmad S, Awais M, Rasool A, Fahad S, Saud S, Shah AN, Ihsan Z, Ali S, Bajwa AA, Hakeem KR, Ameen A, Amanullah, Rehman HU, Alghabar F, Jatoi GH, Akram M, Khan A, Islam F, Ata-Ul-Karim ST, Rehmani MIA, Hussain S, Razaq M, Fathi A (2017) Optimizing the phosphorus use in cotton by using CSM-CROPGRO-cotton model for semi-arid climate of Vehari-Punjab, Pakistan. Environ Sci Pollut Res 24(6):5811–5823

Amin A, Nasim W, Mubeen M, Ahmad A, Nadeem M, Urich P, Fahad S, Ahmad S, Wajid A, Tabassum F, Hammad HM, Sultana SR, Anwar S, Baloch SK, Wahid A, Wilkerson CJ, Hoogenboom G (2018) Simulated CSM-CROPGRO-cotton yield under projected future climate by SimCLIM for southern Punjab, Pakistan. Agric Syst 167:213–222

Amouzou KA, Naab JB, Lamers JP, Borgemeister C, Becker M, Vlek PL (2018) CROPGRO-Cotton model for determining climate change impacts on yield, water-and N-use efficiencies of cotton in the Dry Savanna of West Africa. Agric Syst 165:85–96

Arshad MN, Ahmad A, Wajid SA, Cheema MJM, Schwartz MW (2017) Adapting DSSAT model for simulation of cotton yield for nitrogen levels and planting dates. Agron J 109(6):2639–2648

Attia A, Rajan N, Nair SS, DeLaune PB, Xue Q, Ibrahim AMH, Hays DB (2016) Modeling cotton lint yield and water use efficiency responses to irrigation scheduling using Cotton2K. Agron J 108(4):1614

Baker DN, Lambert JR, McKinion JM (1983) GOSSYM: a simulator of cotton crop growth and yield. South Carolina Agric Stn Bull 1089:134

Brown LG, Jones JW, Hesketh JD, Hartsog JD, Whisler FD, Harris FS (1985) COTCROP: computer simulation of growth and yield. Information Bulletin 69. Mississippi Agricultural and Forestry Experiment Station, Mississippi State, MS

Duncan WG (1972) SIMCOT: a simulator of cotton growth and yield. In: Proc. workshop on tree growth dynamics and modeling. Duke university, Oct 11–12, 1971, pp 115–118

Duncan WG, Loomis RS, Williams WA, Hanau R (1967) A model for simulating photosynthesis in plant communities. Hilgardia J Agric Sci 38:181–205

Dzotsi KA, Basso B, Jones JW (2015) Parameter and uncertainty estimation for maize, peanut and cotton using the SALUS crop model. Agric Syst 135:31–47

Farahani HJ, Izzi G, Oweis TY (2009) Parameterization and evaluation of the AquaCrop model for full and deficit irrigated cotton. Agron J 101(3):469–476

Guerra LC, Garcia AG, Hoogenboom G, Bednarz CW, Jones JW (2005) Evaluation of a new model to simulate growth and development of cotton. In: ASA, CSSA, and SSSA annual meetings, Salt Lake City, UT, pp 6–10

Han M, Zhao C, Simunek J, Feng G (2015) Evaluating the impact of groundwater on cotton growth and root zone water balance using Hydrus-1D coupled with a crop growth model. Agric Water Manag 160:64–75

Hearn AB (1994) OZCOT: a simulation model for cotton crop management. Agric Syst 44:257–259

Hearn AB, da Roza GD (1985) A simple model for crop management applications for cotton 2 (*Gossypium hirsutum* L). Field Crop Res 12:49–69

Hebbar KB, Venugopalan MV, Seshasai MVR, Rao KV, Patil BC, Prakash AH, Kumar V, Hebbar KR, Jeyakumar P, Bandhopadhyay KK, Rao MRK (2008) Predicting cotton production using Infocrop-cotton simulation model, remote sensing and spatial agro-climatic data. Curr Sci 95 (11):1570–1579

Heidariniya M, Naseri AA, Boroumandnasab S, Moshkabadi BS, Nasrolahi AH (2012) Evalution of AquaCrop model application in irrigation management of cotton. World Rural Obs 4:55–59

Hodges HF, Whisler FD, Bridges SM, Reddy KR, McKinnion JM (1998) Simulation in crop management-GOSSYM/COMAX. In: Peart RM, Curry RB (eds) Agricultural systems modeling and simulation. Marcel Dekker, Inc., New York

Hussein F, Janat M, Yakoub A (2011) Simulating cotton yield response to deficit irrigation with the FAO AquaCrop model. Span J Agric Res 9(4):1319–1330

Ibrahim B, Yacoub A (2009) Using CROPWAT model in scheduling irrigation and determining water use of cotton in two areas in Syria. Arab Gulf J Sci Res 27(1):70–77

Iqbal J, Whisler FD (2000) Spatial variability of soil physical properties and validation of Gossym on selected growth parameters of a cotton crop. In: 2000 proceedings Beltwide cotton conferences, San Antonio, USA, 4–8 January, 2000, vol 2. National Cotton Council, pp 1430–1431

Jallas E, Martin P, Sequeira R, Turner S, Cretenet MGE (2000) Virtual COTONS1, the firstborn of the next generation of simulation model. Virtual Worlds 1834:235–244

Kamali SD, Jeloudar NB, Alishah O (2011) The assessment of adaptability and stability of yield on cotton cultivars by using uniparametric, non-parametric methods and AMMI model. Iran J Field Crop Sci 42(2):397–407

Khan MB, Khaliq A, Ahmad S (2004) Performance of mashbean intercropped in cotton planted in different planting patterns. J Res (Sci) 15(2):191–197

Ko J, Piccinni G, Steglich E (2009) Using EPIC model to manage irrigated cotton and maize. Agric Water Manag 96(9):1323–1331

Koukouli P, Georgiou P (2018) Evaluation of climate change impacts on cotton yield using Cropsyst and regression models. J Adv Agric 8(1):1433–1451

Kumar S, Niwas R, Khichar ML, Kumar Y, Premdeep AS (2017) Sensitivity analysis of DSSAT CROPGRO-Cotton model for cotton under different growing environments. Indian J Ecol 44 (4):237–241

Ladewig H, Thomas JK (1992) A follow-up evaluation of the GOSSYM/COMAX cotton program. Texas Agric Ext Serv 14:1–47

Lambert JR, Baker DN (1984) RHIZOS, a simulator of root growth and soil processes: model descriptions. South Carolina Agric Expt Tech Bull 1080

Landivar JA, Reddy KR, Hodges HF (2010) Physiological simulation of cotton growth and yield. In: Physiology of cotton. Springer, Dordrecht, pp 318–331

Lemmon H (1986) COMAX: an expert system for cotton crop management. Science 233:29–33

Li W, Zhou Z, Meng Y, Xu N, Fok M (2009) Modeling boll maturation period, seed growth, protein, and oil content of cotton (*Gossypium hirsutum* L.) in China. Field Crop Res 112 (2–3):131–140

Li F, Yu D, Zhao Y (2019a) Irrigation scheduling optimization for cotton based on the AquaCrop model. Water Resour Manag 33(1):39–55

Li M, Du Y, Zhang F, Bai Y, Fan J, Zhang J, Chen S (2019b) Simulation of cotton growth and soil water content under film-mulched drip irrigation using modified CSM-CROPGRO-cotton model. Agric Water Manag 218:124–138

Liang XZ, Xu M, Gao W, Reddy KR, Kunkel K, Schmoldt DL, Samel AN (2012) A distributed cotton growth model developed from GOSSYM and its parameter determination. Agron J 104 (3):661–674

Lifeng W, Zhang F, Fan J, Zhou H, Xing Y, Qiang S (2015) Sensitivity and uncertainty analysis for CROPGRO-cotton model at different irrigation levels. Transactions of the Chinese society of. Agric Eng 31(15):55–64

Linker R, Ioslovich I, Sylaios G, Plauborg F, Battilani A (2016) Optimal model-based deficit irrigation scheduling using AquaCrop: a simulation study with cotton, potato and tomato. Agric Water Manag 163:236–243

Liting H, Xuebiao P, Xuejiao W, Qi H (2018) Dynamic prediction of cotton regional yield based on the COSIM model-a case study of Akesu City, Xinjiang. Cotton Sci 1:10–19

Loison R, Audebert A, Debaeke P, Gérardeaux E (2016) Design of African rainfed cotton ideotypes using DSSAT CROPGRO-Cotton. In: International Crop Modelling Symposium, 15–17 Mar 2016, Berlin

Marani A, Cardon GE, Phene CJ (1992a) CALGOS, a version of GOSSYM adapted for irrigated cotton. I. Drip irrigation, soil water transport, and root growth. In: Herber DJ, Richter DA (eds) Proc Beltwide cotton Conf. National Cotton Council of America, Memphis, TN, pp 1352–1357

Marani A, Phene CJ, Cardon GE (1992b) CALGOS, a version of GOSSYM adapted for irrigated cotton. III. Leaf an boll growth routines. In: Herber DJ, Richter DA (eds) Proc. Beltwide cotton Conf. National Cotton Council of America, Memphis, TN, pp 1361–1364

McCarthy AC, Hancock NH, Raine SR (2014) Simulation of irrigation control strategies for cotton using model predictive control within the VARIwise simulation framework. Comput Electron Agric 101:135–147

McKinion JM, Baker DN, Hesketh JD, Jones JW (1975) SIMCOT II: a simulation of cotton growth and yield. ARS-S-52, USDA, pp 27–82

Milroy SP, Bange MP, Hearn AB (2004) Row configuration in rainfed cotton systems: modification of the OZCOT simulation model. Agric Syst 82(1):1–16

Modala NR, Ale S, Rajan N, Munster CL, DeLaune PB, Thorp KR, Nair SS, Barnes EM (2015) Evaluation of the CSM-CROPGRO-Cotton model for the Texas rolling plains region and simulation of deficit irrigation strategies for increasing water use efficiency. Trans ASABE 58 (3):685–696

Mutsaers HJW (1984) KUTUN: a morphogenetic model for cotton (*Gossypium hirsutum* L.). Agric Syst 14:229–257

Nagender T, Reddy DR, Sreenivas G, Rani PL, Surekha K, Gupta A, Sreekanth PD (2017) Evaluation of CSM-CROPGRO-Cotton model under varied plant densities and nitrogen levels for simulating crop growth, development and seed cotton yield. Int J Pure Appl Biosci 5 (4):1738–1754

Pan X, Han X, Shi Y (1996) A cotton growth and development simulation model for culture management-COTGROW. Sci Agric Sin 29:94–96. (In Chinese, with English abstract)

Pathak TB, Fraisse CW, Jones JW, Messina CD, Hoogenboom G (2007) Use of global sensitivity analysis for CROPGRO cotton model development. Trans ASABE 50(6):2295–2302

Poyyamozhi S, Mohideen AK (2016) Forecasting of cotton production in India using ARIMA model. Asia Pac J Oper Res 1:70–74

Pretorius MM, Allemann J, Smith MF (2015) Use of the AMMI model to analyse cultivar-environment interaction in cotton under irrigation in South Africa. Afr J Agric Res 2:76–80

Qian L, Wang XG, Luo WB, Qi ZM, Sun HW, Luo YY (2017) An improved CROPR model for estimating cotton yield under soil aeration stress. Crop Pasture Sci 68(4):366–377

Rahman MH, Ahmad A, Wajid A, Hussain M, Rasul F, Ishaque W, Islam MA, Shelia V, Awais M, Ullah A, Wahid A (2017) Application of CSM-CROPGRO-Cotton model for cultivars and

optimum planting dates: evaluation in changing semi-arid climate. Field Crop Res. https://doi.org/10.1016/j.fcr.2017.007
Rahman MH, Ahmad A, Wang X, Wajid A, Nasim W, Hussain M, Ahmad B, Ahmad I, Ali Z, Ishaque W, Awais M, Shelia V, Ahmad S, Fahad S, Alam M, Ullah H, Hoogenboom G (2018) Multi-model projections of future climate and climate change impacts uncertainty assessment for cotton production in Pakistan. Agric For Meteorol 253–254:94–113
Reddy KR, Kakani VG, McKinion JM, Baker DN (2016) Applications of a cotton simulation model, GOSSYM, for crop management, economic, and policy decisions. In: Agricultural system models in field research and technology transfer. CRC Press, Boca Raton, FL, pp 44–65
Richardson AG, Reddy KR, Boone ML (2002) Sensitivity analysis of climate change impacts on cotton production using the GOSSYM crop model. Int J Biotron 31:25–49
Ritchie JT (1972) Model for predicting evaporation from a row crop with incomplete cover. Water Resour Res 8(5):1204–1213
Sadanshiv NS, Chatterji S, Sen TK, Venugopalan MV, Tiwary P, Wagh NS, Arun C (2012) Application of a crop simulation model for quantification of yield gap of cotton in Wardha district, Maharashtra. Agropedology 22(2):74–79
Shuli F, Jarwar AH, Wang X, Wang L, Ma Q (2018) Overview of the cotton in Pakistan and its future prospects. Pak J Agric Res 31(4):396–407
Sommer R, Kienzler K, Conrad C, Ibragimov N, Lamers J, Martius C, Vlek P (2008) Evaluation of the CropSyst model for simulating the potential yield of cotton. Agron Sustain Dev 28 (2):345–354
Tak V (2014) Heat unit requirement and validation of infocrop model for Bt-cotton in middle Gujarat agroclimatic region. Doctoral dissertation, AAU, Anand
Tan S, Wang Q, Zhang J, Chen Y, Shan Y, Xu D (2018) Performance of AquaCrop model for cotton growth simulation under film-mulched drip irrigation in southern Xinjiang, China. Agric Water Manag 196:99–113
Tariq M, Yasmeen A, Ahmad S, Hussain N, Afzal MN, Hasanuzzaman M (2017) Shedding of fruiting structures in cotton: factors, compensation and prevention. Trop Subtrop Agroecosyst 20(2):251–262
Tariq M, Afzal MN, Muhammad D, Ahmad S, Shahzad AN, Kiran A, Wakeel A (2018) Relationship of tissue potassium content with yield and fiber quality components of Bt cotton as influenced by potassium application methods. Field Crop Res 229:37–43
Thorp KR, Barnes EM, Hunsaker DJ, Kimball BA, White JW, Nazareth VJ, Hoogenboom G (2014) Evaluation of CSM-CROPGRO-Cotton for simulating effects of management and climate change on cotton growth and evapotranspiration in an arid environment. Trans ASABE 57 (6):1627–1642
Thorp KR, Hunsaker DJ, Bronson KF, Andrade-Sanchez P, Barnes EM (2017) Cotton irrigation scheduling using a crop growth model and FAO-56 methods: field and simulation studies. Trans ASABE 60(6):2023–2039
Tsakmakis ID, Kokkos NP, Gikas GD, Pisinaras V, Hatzigiannakis E, Arampatzis G, Sylaios GK (2019) Evaluation of AquaCrop model simulations of cotton growth under deficit irrigation with an emphasis on root growth and water extraction patterns. Agric Water Manag 213:419–432
Usman M, Ahmad A, Ahmad S, Irshad M, Khaliq T, Wajid A, Hussain K, Nasim W, Chattha TM, Trethowan R, Hoogenboom G (2009) Development and application of crop water stress index for scheduling irrigation in cotton (*Gossypium hirsutum* L.) under semiarid environment. J Food Agric Environ 7(3&4):386–391
Venugopalan MV, Tiwary P, Ray SK, Chatterji K, Bhattacharyya T, Bandhopadhyay KK, Sarkar D, Chandran P, Pal DK, Mandal DK (2014) InfoCrop-cotton simulation model–its application in land quality assessment for cotton cultivation. Curr Sci 107(9):1512–1518
Voloudakis D, Karamanos A, Economou G, Kalivas D, Vahamidis P, Kotoulas V, Kapsomenakis J, Zerefos C (2015) Prediction of climate change impacts on cotton yields in Greece under eight climatic models using the AquaCrop crop simulation model and discriminant function analysis. Agric Water Manag 147:116–128

Wajid A, Ahmad A, Hussain M, ur Rahman MH, Khaliq T, Mubeen M, Rasul F, Bashir U, Awais M, Iqbal J, Sultana SR (2014) Modeling growth, development and seed-cotton yield for varying nitrogen increments and planting dates using DSSAT. Pak J Agric Sci 51:641–650

Wall GW, Amthor JS, Kimball BA (1994) COTCO2: a cotton growth simulation model for global change. Agric For Meteorol 70:289–342

Wang B, Liu H, Wang XH, Li JM, Ma YB, Ma XW (2015) Soil phosphorus accumulation model for an arid area of North-Western China with 3-year rotation of wheat, maize and cotton. J Agric Sci 153(7):1247–1256

Wang J, Li J, Guan H (2017a) Evaluation of drip irrigation system uniformity on cotton yield in an arid region using a two-dimensional soil water transport and crop growth coupling model. Irrig Drain 66(3):351–364

Wang X, Pan X, Wang S, Hu L, Guo Y, Li X (2017b) Dynamic prediction method for cotton yield based on COSIM model in Xinjiang. Trans Chin Soc Agric Eng 33(8):160–165

Yang Y, Ouyang Z, Yang Y, Liu X (2008) Simulation of the effect of pruning and topping on cotton growth using COTTON2K model. Field Crop Res 106(2):126–137

Yu Y, Zhao C (2015) Modelling soil and root respiration in a cotton field using the DNDC model. J Plant Nutr Soil Sci 178(5):787–791

Zamora DS, Jose S, Jones JW, Cropper WP (2009) Modeling cotton production response to shading in a pecan alleycropping system using CROPGRO. Agrofor Syst 76(2):423–435

Zhang JP, Tugwell NP, Cochran MJ, Bourland FM, Oosterhuis DM, Klein CD (1994) COTMAN: a computer-aided cotton management system for late-season practices. In: Herber DJ, Richter DA (eds) Proc. Beltwide cotton Conf. National Cotton Council of America, Memphis, TN, pp 1286–1287

Chapter 22
Climate Resilient Cotton Production System: A Case Study in Pakistan

Muhammad Habib ur Rahman, Ishfaq Ahmad, Abdul Ghaffar, Ghulam Haider, Ashfaq Ahmad, Burhan Ahmad, Muhammad Tariq, Wajid Nasim, Ghulam Rasul, Shah Fahad, Shakeel Ahmad, and Gerrit Hoogenboom

---

M. H. ur. Rahman (✉)
Department of Agronomy, Muhammad Nawaz Shareef University of Agriculture, Multan, Pakistan

Institute of Crop Science and Resource Conservation (INRES) Crop Science Group, University Bonn, Bonn, Germany
e-mail: habib.rahman@mnsuam.edu.pk

I. Ahmad
Centre for Climate Research and Development, COMSATS University, Islamabad, Pakistan

A. Ghaffar · G. Haider
Department of Agronomy, Muhammad Nawaz Shareef University of Agriculture, Multan, Pakistan

A. Ahmad
Climate Change, US-Pakistan Centre for Advanced Studies in Agriculture and Food Security, Department of Agronomy, University of Agriculture, Faisalabad, Pakistan

B. Ahmad
Pakistan Meteorological Department, Islamabad, Pakistan

M. Tariq
Central Cotton Research Institute, Multan, Pakistan

W. Nasim
Department of Agronomy, University College of Agriculture and Environmental Sciences, Islamia University of Bahawalpur, Bahawalpur, Pakistan

G. Rasul
Pakistan Meteorological Department, Islamabad, Pakistan

International Center for Integrated Mountain Development, Kathmandu, Nepal

S. Fahad
Department of Agriculture, University of Swabi, Swabi, Pakistan

College of Plant Science and Technology, Huazhong Agricultural University, Wuhan, People's Republic of China

S. Ahmad, M. Hasanuzzaman (eds.), *Cotton Production and Uses*,
https://doi.org/10.1007/978-981-15-1472-2_22

Abstract Cotton production is most vulnerable to climate change particularly in Pakistan, and sustainable cotton yield is critical to accomplish the future demand of the country. Climate change has negative impact on cotton production in major parts of the cotton-growing regions. It hampers not only the yield but also quality of fiber and has negative impact on socioeconomic conditions of farmers. Climate, crop, and economic multidisciplinary modeling approach are being used to assess the impact of climate change and development of adaptation strategies for sustainable cotton production. Climate change scenarios revealed the increase in both maximum and minimum temperature and uncertain rainfall patterns throughout the world and especially in dry and arid areas of the world like Pakistan. Rainfall would increase and decrease as projected by multi-GCMs and RCPs, and it is fact that these changes in climate would lead to negative effect on cotton crop production, and sustainable cotton production in the future is under threat due to climate variablity. Generally, mostly general circulation model (GCM) scenario projected the reduction in cotton yield as compared with the baseline during both timer periods and RCPs tested. Adaptation strategies can minimize the negative impact of climate change. So, changes in crop management practices (sowing, planting density, irrigation, and plant protection) may be good adaptation strategies for sustainable cotton production under changing climate scenarios of the world. Climate resilient cotton production system has potential to minimize the negative impacts of climate change on cotton crop by developing heat and drought resilient germplasm, mitigation technology to reduce GHG emission, and application of decision support system (DSS) and use of ICT-based technologies for sustainable cotton crop production. It is time to adopt climate, energy, and water smart cotton production technologies and practices for sustainable cotton production in the future.

Keywords Climate change · Phenology · Adaptation · Resilient · Sustainable · DSSAT

Abbreviations

AEZ	Arid irrigated zone
BVT	Bee vectoring technology
CO_2	Carbon dioxide
DSSAT	Decision Support System for Agrotechnology Transfer
GCMs	General circulation models
GHG	Greenhouse gas
GIS	Geographical information system

S. Ahmad
Department of Agronomy, Faculty of Agricultural Sciences and Technology, Bahauddin Zakariya University, Multan, Pakistan

G. Hoogenboom
Agricultural and Biological Engineering Department, Institute for Sustainable Food Systems (ISFS), University of Florida, Gainesville, FL, USA

ICT	Information and communication technologies
IPM	Integrated pest management
RCPs	Representative concentration pathways
SR	Solar radiation

22.1 Introduction

22.1.1 Significance of Cotton Crop

Cotton, being a queen of the fiber, enjoys itself a predominant position among all other cash crops in the world (Usman et al. 2009; Ahmad et al. 2014, 2017a, 2018; Abbas and Ahmad 2018; Ahmad and Raza 2014; Ali et al. 2011, 2013a, b, 2014a, b). Cotton is the mostly used fiber across the globe and had significant contribution in textile industry of the world; therefore, its demand and consumption has been projected to increase then its production (USDA 2018). It can provide not only fiber for the textile industry but also contributes a major role in oil industries by providing oil-rich seed and protein. Total global cotton production worldwide in 2018 was 124 million bales. The United States, China, India, and Pakistan were the largest producers across the world and contribute more than half of the total production volume. Among the top exporter the United States was ranked first, exported 15.5 million bales around the world. Cotton production varied across the world due to differences in production potential mainly due to environmental and climatic factors (Fig. 22.1). Cotton production supports the world largest fiber and textile industry which includes textile mills, spindles, looms, knitwear units, garment units, ginners, and oil expellers. It is by all measures the world's most important economic sector with enormous forward and

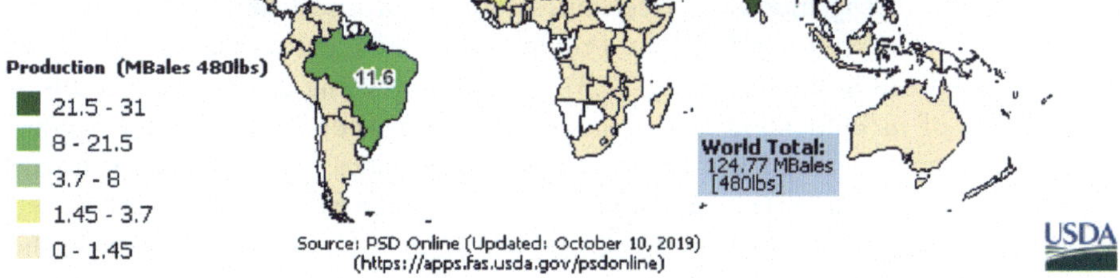

Fig. 22.1 Cotton production and areas across the world; regions have differences in production potential due to variations in environmental and climatic conditions

backward linkages. Cotton yarn production from fiber is clustered in major cotton-growing countries including the USA, China, Pakistan, India, and Russia (3/4 yarn production is done in less developing countries of the world); yarn production in different regions and countries can be seen in the map of yarn production (Fig. 22.2). It is estimated in the 94th annual report "Cotton Outlook" of USDA that world's cotton production will decline in 2018–2019 by 3.6% due to yield decline in cotton-producing countries (USDA 2018).

Cotton is already occupying about less than 3% of world's agricultural land which is 36% of 10.3 billion hectare land suitable for agriculture on planet earth. Thus, according to an estimate, if the area occupied under cotton is 2.5% of total arable land, then it will be contributing 0.1–0.3% of global greenhouse gas emission (Cotton incorporated 2009; ITC 2011). It indicates that cotton is not the principal source of greenhouse gas emission. However, in-depth analysis of cotton production suggests that cotton at same time is a contributor and victim of global warming and climate change (ITC 2011; Amin et al. 2017, 2018; Khan et al. 2004; Rahman et al. 2018; Tariq et al. 2017, 2018). For instance, according to the estimates of different case studies, production-related GHG emission was estimated about 0.38–0.92 tons of carbon dioxide (CO_2) equivalents per bale of cotton. On the other hand, only ozone pollution out of several other environmental constraints for cotton was found to cause 23% yield loss in upland cotton (Grantz 2003).

Pakistan stands at fourth, third, and second position in terms of cotton production, largest producer of yarn, and largest exporter, respectively, across the world. It is fifth largest consumer of cotton yarn. Cotton and cotton products pay 1.6% in GDP while 55% contribution is country earning with respect to foreign exchange (GOP 2018). Cotton crop is being cultivated on an area of 3 million hectares which covers almost 15% area of the total area cultivated while almost 1.3 million farmers out of 5 million total farmers grow cotton crop. Cotton crop-growing areas and production in Pakistan (Punjab and Sindh) can be seen in cotton production map of Pakistan (Fig. 22.3) while Punjab is the major cotton-producing (74.7%) province in Pakistan. Cotton production is highly variable over the past years due to climatic variability, high climate extremes, weaker adaptive capacity of cotton growers to these climate extremes, high insect pest infestation, poor crop management practices, and policy challenges (Ashraf and Iftikhar 2013; Batool and Saeed 2017). Recently during 2015–2016 cotton production was roughly dropped by 38% in Punjab than the last years (6.3 million bales only as compared with the 10.3 million bales); it was mainly due to erratic and intense rainfall during monsoon, weed infestation, high insect pest infestation, and low-quality cotton seed (GOP 2016). Province-wise area and production of cotton crop is highlighted in the map (Fig. 22.3). Cotton crop being an ancient crop has a long history of its cultivation across the world. After wheat crop, it is the second largest cash crop on area basis in the country. It provides raw material for textile industry and it is an important source of foreign exchange earnings (GOP 2018).

Although Pakistan ranked fourth position in terms of cotton production in the world, still cotton crop is climate sensitive and suffered most due to environmental and climatic factors in the country. As it is depicted from several studies that cotton is the most vulnerable crop and its production would be reduced due to climate change (Rahman et al. 2018), it is clearly depicted from the ground facts that its

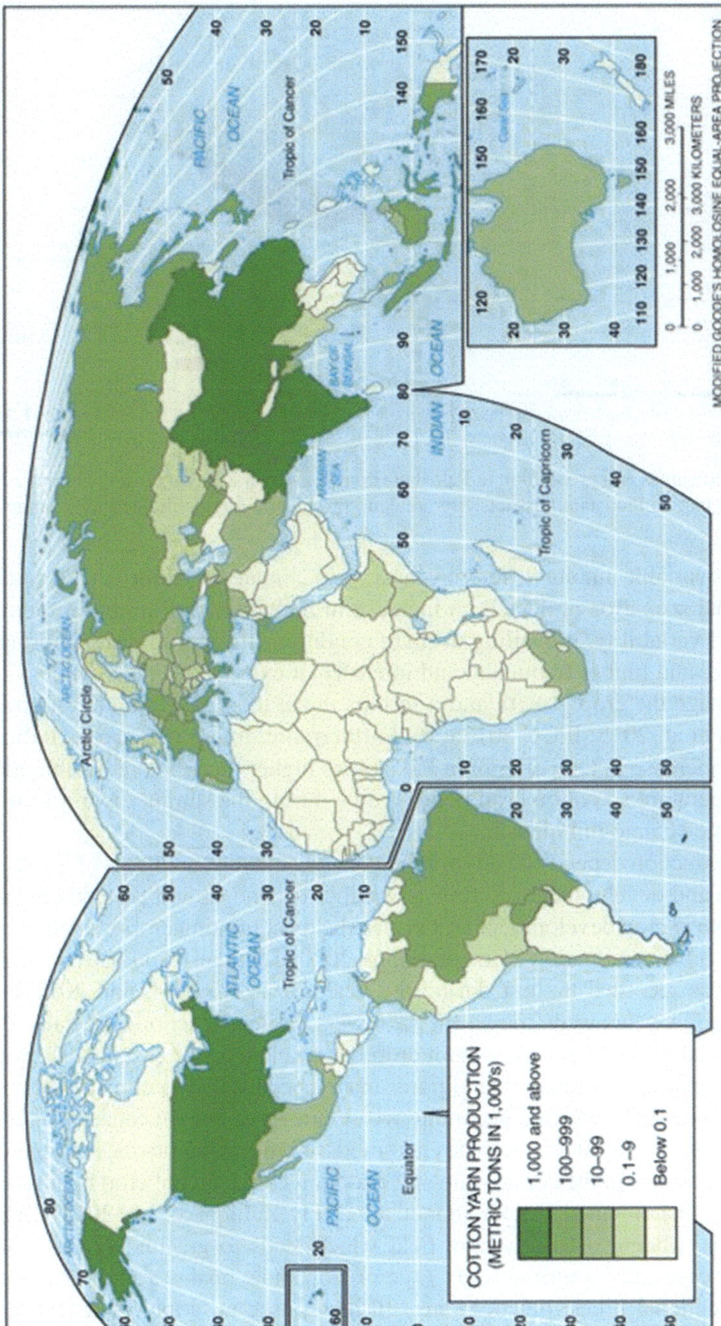

Fig. 22.2 Cotton yarn production (metric tons) in different areas of the world; cotton yarn production from fiber is clustered in major cotton-growing countries including the USA, China, Pakistan, India, and Russia (3/4 yarn production is done in less developing countries of the world)

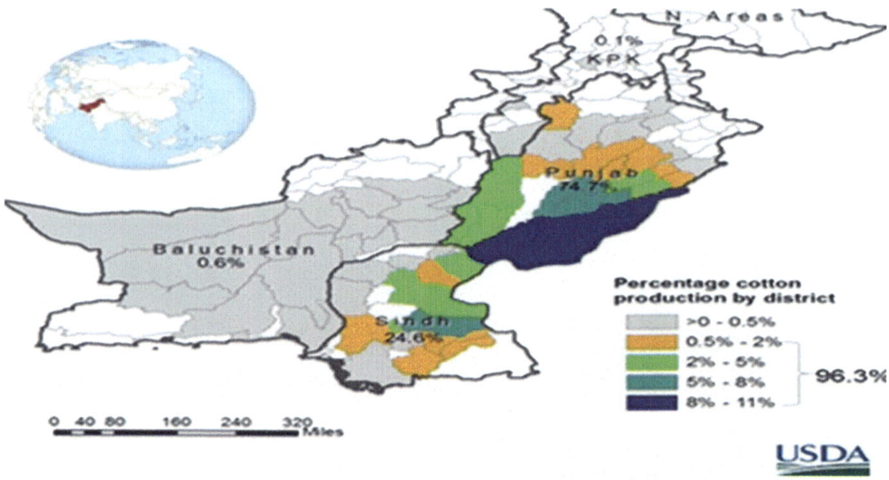

Fig. 22.3 Cotton areas and production in Pakistan (Punjab and Sindh); Punjab is the major cotton-producing province in Pakistan. District-wise area and production are highlighted in the map

production is variable among different years due to climate variability in the country. Future climate scenarios projected an increase in maximum and minimum temperature, rainfall variability, prevailing drought conditions, irrigation water shortage at farmer's field, and higher frequency and intensity of extreme weather events would severely damage the crop growth and a serious threat to sustainable cotton production (Ahmad et al. 2015; IPCC 2013a, b). Furthermore, future projections indicated that average increase in temperature in Pakistan is higher than the predictable global average temperature so cotton crop being grown during the summer is most vulnerable to climate change in Pakistan.

Cotton production depends on environmental and climatic conditions of the region. Crop growth and development of crops is weather dependent and temperature has a significant role in crop development as it is also the most limiting crop-reducing factor especially in crop production (Hoogenboom 2000; Rahman et al. 2016). Cotton production was recorded highest during the years of 1991, 2004, and 2011 in the country while few years were known for the lower cotton production especially 1976, 1983, 1993, and 2015. Generally cotton crop has challenges for sustainable production due to changing in climatic conditions and especially due to climatic extremes. Major issues and challenges are becoming worse due to lacking in capacity of cotton growers to adopt the latest technologies and modern production practices as majority of the cotton growers in the country are still adopting conventional crop management practices and technologies (Makhdum et al. 2011; Zulfiqar et al. 2017). Higher fertilizer dose is being used at cotton field which leads to greenhouse gas (GHG) emission and especially contributes to CO_2 emission and global warming (Rasul et al. 2016). Conventional tillage practices are still being used at farmer field which also lead to more GHG emission. Lower use efficiency of water, fertilizer, radiation, and pesticides is the major challenge in current cotton cropping system and conventional cotton production leads to higher cost of production and lower resource use efficiencies (Raza and Ahmad 2015; Zulfiqar et al. 2017). Cotton production system is badly affected by climate change and is most vulnerable to climate extremes due to weaker adaptive capacity. Situation becomes worse as excessive irrigation water is being used along with intensive pesticide and fertilizer application to cotton crop in current cropping system (Rasul et al. 2012; Nazli et al. 2012; Zulfiqar et al. 2017). Current cropping system is most vulnerable to climate change especially in Pakistan and sustainable crop production is crucial to fulfill the future demand of the country (Rahman et al. 2018; Zhu et al. 2013).

22.1.2 Climate Change and Cotton Production

Changing climate is likely to upsurge the susceptibility of agriculture especially the cropping system (Ahmad et al. 2015) due to unexpected and extreme changes in climate especially increase in temperature and variability in rainfall patterns in the world (Rosenzweig et al. 2014; IPCC 2014a, b). There is explicit change in weather pattern of Pakistan (Ahmad et al. 2015). These changes have been amplified from last decades due to more emission of greenhouse gasses. Mean temperature is increasing 0.2–0.6 °C per decade in Pakistan (Ahmad et al. 2015) and night temperature is increasing more than day temperature. Climate change and variability has a severe impact on crop production and could also be the reason for shift in cropping system in different regions of the Punjab province (Ahmad et al. 2019). Climate being a critical factor in crop productivity across the globe as it effects the growth, development, phenological phases, and cotton yield, it has a significant role than other factors that contribute for crop production (Bradow and Davidonis 2000). Increase in temperature at any cotton growth phase may lead to a negative impact on cotton productivity, while cotton being a C_3 crop can get benefits from elevated CO_2 as fertilization.

However, combined effect of elevated temperature and CO_2 might have adverse effect on cotton production as elevated temperature has a prominent role (Singh et al. 2007). Expected variation in climate especially elevated temperature is estimated to have harmful effects on cotton crop and especially to cotton production (Ahmad et al. 2015; Rahman et al. 2016, 2017, 2018). Night temperature and especially elevated night temperature is more harmful for cotton crop as the specific difference between day and night temperature is crucial for optimum cotton production (Singh et al. 2007). Temperature and rainfall patterns affect the crop choice in different zones (Ghazala and Rasul 2009) and for sustainable cotton production, current spatial and temporal changes in rainfall and temperature need to be considered.

Climate change and variability has a severe impact on cotton production and is the major cause of shift in sowing time of cotton crop (Rahman et al. 2018). Variability in rainfall pattern and fluctuation in temperature have a negative impact on cotton crop growth, development, and production (Bange and Milroy 2004). Cotton crop is specific for temperature requirements at different growth phases (Ahmad et al. 2017a), physiological process, growth development and yield is dependent on climatic conditions (Singh et al. 2007). Temperature variation above and below normal conditions had a significant negative impact on cotton growth and ultimately cotton yield (Luo et al. 2014). Temperature stress at any crop growth phase leads to variability in cotton yield; higher day temperature followed by high night temperature may aggravate the harmful effect on cotton growth and yield. High temperature than normal leads to shorten growth phases, resulting in stunted growth and smaller boll which ultimately lead to lower cotton production (Reddy and Zhao 2005). Cotton yield was decreased up to 51% by increasing temperature from 28 to 32 °C in Indian Punjab which was assessed with the aid of crop models; cotton crop was severely damaged from sowing to anthesis phase (Wajid et al. 2014). Increase in summer temperature especially when exceeds to 45 °C in Punjab and Sindh provinces lead to sever reduction in seed cotton yield as boll development phases are most sensitive to heat stress (Rahman et al. 2004, 2018). As temperature is a critical factor in cotton production, especially boll setting and boll development phases are damaged significantly when maximum temperature during the day exceeds to 30 °C for only a period of 13 h (Reddy et al. 2002). Quality of cotton fiber is also affected by environmental factors and especially temperature and rainfall (Reddy et al. 2007). All quality parameters are negatively affected by environmental factors especially temperature and erratic and intense rainfall.

Elevated CO_2 alone has a positive impact on cotton crop growth, development, and yield due to CO_2 fertilization being a C_3 plant (Reddy et al. 2004). Increase in seed cotton yield of 5.5% by doubling the CO_2 at current temperature but 56% yield decrease only due to increase in maximum temperature while 28% reduction in seed cotton yield observed by the increment in minimum (night) temperature (Jalota et al. 2006). Elevated CO_2 has a positive effect on growth, development, biomass accumulation, and ultimately cotton yield because elevated CO_2 enhances photosynthesis efficiency by reducing stomatal conductance and transpiration (Reddy et al. 2007; Luo et al. 2016). As it is evident from various studies that interactive effect of increase in temperature, elevated CO_2, and variable rainfall patterns might possibly offset the optimistic effect of doubled concentration of carbon dioxide (Williams et al. 2015; Rahman et al. 2018). Intensive cotton production plans by changing in current production and cropping system through adopting the adaptation technologies/strategies could have potential to reduce the adverse effect of climate variability and sustainable cotton production would only be possible.

22.1.3 Climate of Cotton Zones in Pakistan

Climate is a prime factor that exerts major influence on vegetation, soil type, and water resources. Climate is the major driving force for crop production while cotton crop is very specific to its environmental requirements. Climate of cotton zone in Pakistan lies in arid to semiarid region in both provinces Sindh and Punjab. Cotton crop is suitable to this climatic region as it requires normally 4–5 months of uniform high temperature (28–43 °C) while optimum air temperature for its proper vegetative growth and development required 21–29 °C and reproductive growth phases required 27–31 °C temperature ranges (Singh et al. 2007; Bange et al. 2008; Rahman et al. 2018). Cotton crop requires minimum rainfall during its early growing periods while dry months are desirable for good-quality cotton production (Bange et al. 2008). Excessive and erratic rainfall at later growth stages/phases causes damages to crop and especially reproductive growth phases are severely damaged to excessive rainfall (Singh et al. 2007; Rahman et al. 2018). Drought conditions are already prevailed in this climatic zone while climatic conditions become worse in the future due to climate change and climate shift in the region. Irrigation water availability is more threaten and exacerbated in the cotton region due to being arid climate and cotton crop water requirement depends on irrigation water supply. Irrigation water availability in the cotton zones through the Indus basin would be unpredictable due to climatic variability in the country (IPCC 2014a, b). In the past few decades, cotton zone has experienced severe droughts and followed by devastating floods especially flood in 2010 which have contributed to low crop yield and especially the cotton crop being the main crop of this zone. Moreover, the mean temperature across the country has already been increased by 0.65 °C in the past 30 years (Ahmad et al. 2015). Future projections of climate change in cotton zone of Punjab showed the expected changes and increase of 1.2–1.8 °C during near term while higher increase (2.2–3.1 °C) of mean temperature is expected during second time period which is known as mid-century (2040–2069) under emission scenarios of RCP 4.5. Bigger changes and higher variations in climatic variables are expected under the GHG emission scenarios of RCP 8.5, as results showed that increase in mean temperature of 1.4–2.2 °C and 3.0–3.9 °C is projected for the time periods of near term and mid-century, respectively. Similarly, changes and variation in rainfall are projected ranged from −8% to 15% during the time period of near term and −5% to 17% change is projected under RCP 4.5 scenario during mid-century time period, while bigger changes and higher variations are projected for RCP 8.5, as it is expected that −8% to 22% and −2% to 20% ranges are anticipated for the time periods of near term and mid-century, respectively. Results of future projections about cotton yield and production revealed the reduction in seed cotton yield of 8% till 2039 while 20% decrease on an average basis is expected by 2069 under the lower emission scenario of RCP 4.5 as compared with the baseline yield. More variability is expected as likely higher increase in temperature and rainfall variability projected under RCP 8.5 scenario, results revealed that a reduction of 12% and 30% in mean seed yield on an average basis under RCP 8.5 (Rahman et al. 2018). Temperatures are also expected to increase more in winter than in summer which is more dangerous to crops and heat waves during cotton season are the threats to sustainable cotton production. Rainfall projections are more uncertain and less clear due to uncertainties in the model structure (Collins et al. 2013; Rahman et al. 2018). Impact of climate change on cotton yield may vary across different regions in the country and it will be significant without proper adaptation technologies/strategies for cotton crop.

Future projections across Pakistan (a) and especially in Punjab (b) and cotton zones represent the increase in the average annual mean temperature (Fig. 22.4), by 2050, revealed the threat to sustainable cotton production in the country. Similarly projected changes in annual precipitation across Pakistan (a) and Punjab (b) by 2050 represented the variability in rainfall and uncertainty especially in cotton zone.

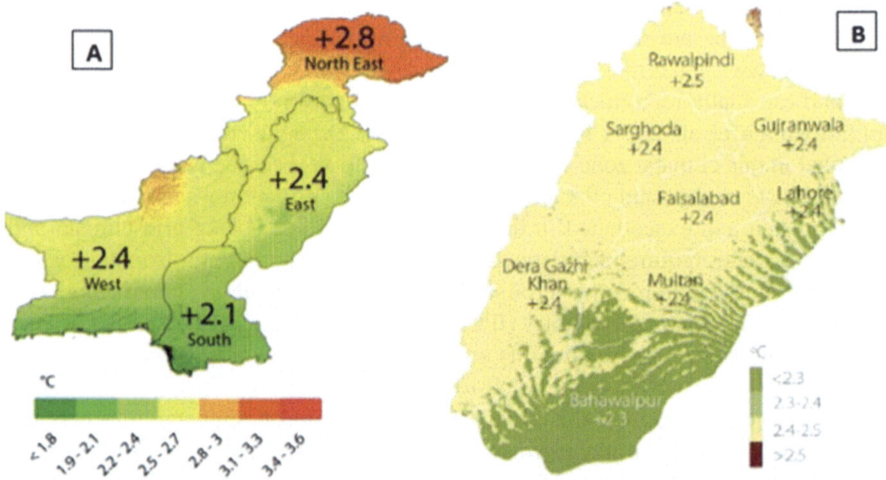

Fig. 22.4 Projected changes in temperature across Pakistan (**a**) and Punjab (**b**) by 2050 represent the increase in the average annual mean temperature especially cotton zone in Pakistan

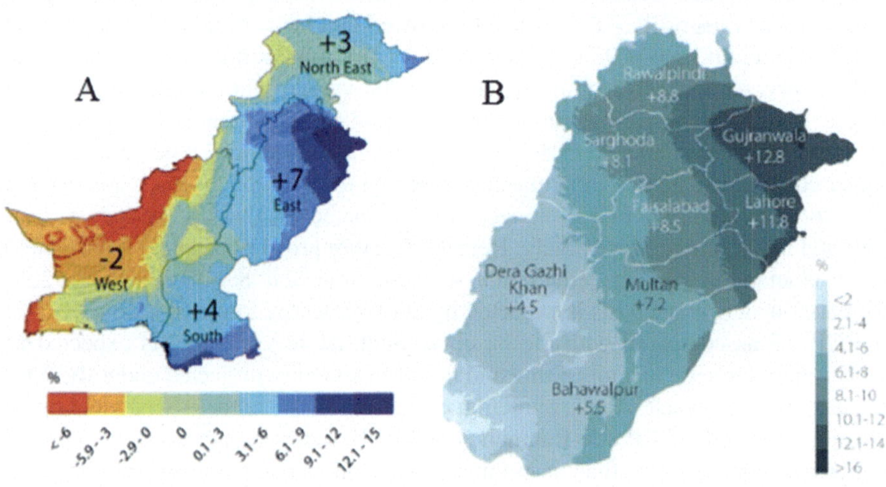

Fig. 22.5 Projected changes in annual precipitation across Pakistan (**a**) and Punjab (**b**) by 2050 represent the variability in rainfall and uncertainty especially in cotton zone. Generally maps depict both increase and decrease in annual rainfall across Pakistan

Generally maps depict both increase and decrease in annual rainfall across Pakistan and Punjab (Fig. 22.5). Variation in maximum and minimum temperature and ranges of Punjab in Kharif season (30-year mean), both temperature ranges are quite high in cotton zone 37–40 °C and 25–29 °C (Fig. 22.6). Similarly, rainfall variability is depicted clearly from the map while rainfall amount is less than 350 mm as it is in arid region (Fig. 22.7).

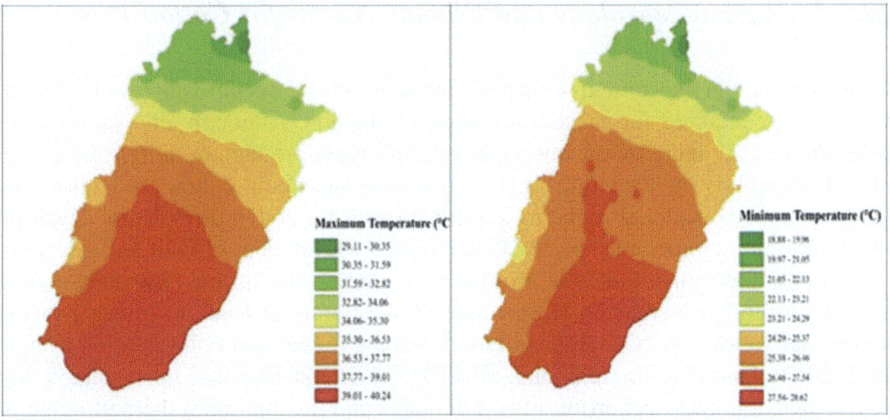

Fig. 22.6 Maximum and minimum temperature variation of Punjab in Kharif season (30-year mean); both temperature ranges are quite high in cotton zone

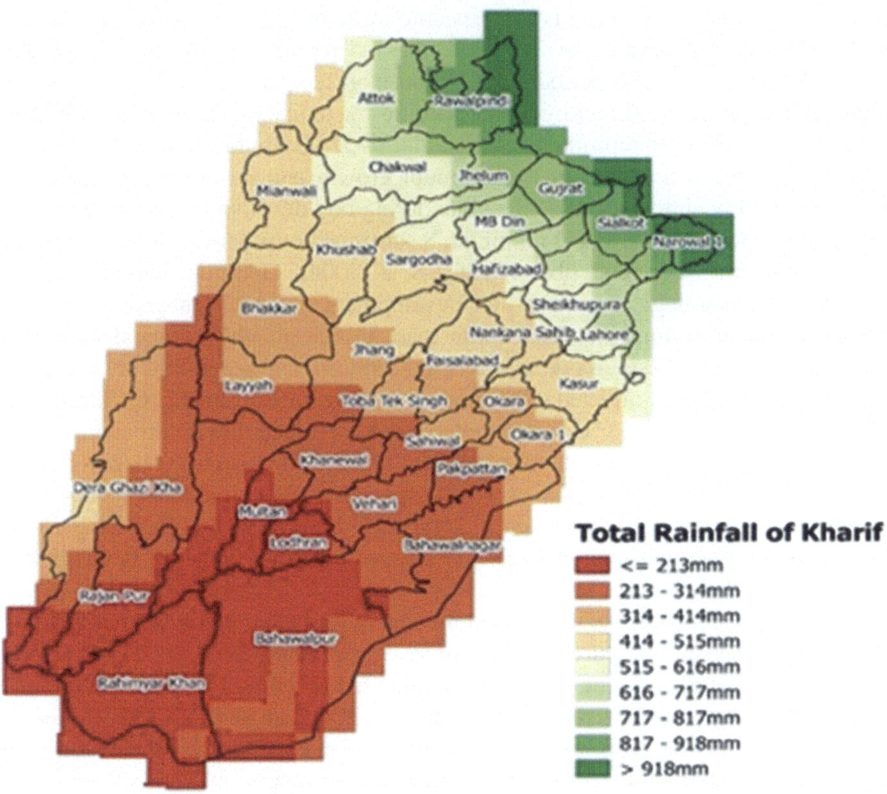

Fig. 22.7 Average (2000–2017) seasonal rainfall distribution across the Punjab during cotton growing seasons

22.1.4 Agrometeorology and Climate Norms for Cotton Crop

Cotton crop is very specific in terms of agrometeorology; specific ranges of weather variables are required for successful cotton production. Although all-weather variables are crucial and play an important role in cotton production, temperature and rainfall variability have significant contribution in successful cotton production. As temperature is considered as the most critical factor for cotton growth and development, optimum temperature of 33 °C is required for efficient growth and development as it is the temperature that promotes the photosynthesis and ultimately cotton growth. Although increasing temperature above optimum leads to lower growth, temperature above 36 °C significantly reduced the flower and boll retention (Singh et al. 2007). Optimum thermal range of 23–32 °C is considered as most suitable for physiological process occurring in cotton plants and this optimum thermal range is also considered as optimal range for growth and development of cotton (Cottee et al. 2010). Temperature has a major role to determine the phenological phases especially initiation and ending of different growth phases and generally whole growth of cotton crop is totally dependent on temperature for being thermo period sensitive (Luo et al. 2014). Cotton crop is very specific for temperature ranges, as it requires uniformly high temperature of 28–43 °C for 4–5 months during its crop growth period. Vegetative growth period required mean air temperature of 21–29 °C while reproductive phases also required very specific for temperature; optimum air temperature for majority of reproductive phases is 27–32 °C (Reddy et al. 2007). Cotton requires 8–9 mean hours daily for optimum growth and development and mean relative humidity almost about 70%, while during boll development relative humidity should be less than 70% and minimum 8 sunshine hours per day are required for the development of a good fiber (Reddy et al. 2004). Optimum temperature for majority of physiological growth and development phases is ranged 25–33 °C; however as temperature increased more than 33 °C, all physiological processes especially photosynthesis in cotton are being reduced and generally cotton growth and development badly affected when temperature increased from 42 °C (Singh et al. 2007). A temperature of 33 °C is reported as optimal temperature for efficient growth and development of cotton plants while above 36 °C temperature is reported as detrimental for reproductive phases which especially cause significant reduction in flower and boll retention (Reddy et al. 2004). Minimum temperature below 15 °C retards the crop growth and reduces the square formation and boll development (Bange and Milroy 2004).

Cotton crop is specific for its temperature ranges for the initiation and competition of each phase and crop requires specific thermal time (Reddy et al. 2007) and critical growth phases of cotton crop are adversely affected by climate extremes especially high temperature stress at reproductive phases (Rahman et al. 2016). Cotton is photoperiod-sensitive crop while it is perennial and indeterminate in terms of morphology (Bange et al. 2008). Accumulated degree days above a specific temperature threshold are deliberated as a good appraisal to see the temperature effect on specific growth phases of cotton crop (Wajid et al. 2014; Rahman et al. 2016).

Sowing time of cotton crop and even each growth phase initiation and completion total depends on temperature. Temperature is a critical factor that defines the sowing of a crop while climatic attributes describe the optimum sowing time in any region and it is crucial for the sustainable cotton production specially the current environmental and climatic challenges to achieve the potential (Singh et al. 2007). Cotton crop seedlings are also temperature sensitive as cold temperature stress results in poor crop stand and delay in developmental and phenological phases, poor physiological growth attributes, and less productivity (Constable and Bange 2006; Conaty et al. 2012). Cold shock due to chilling temperature at early growth phases leads to slow down the growth process and late developmental phases showed that temperature has a significant role at any stage of cotton crop development. A significant difference in day and night temperature is also crucial for optimum cotton growth and lower night (>20 °C) temperature particularly during boll development phases leads to stunted boll growth and poor cotton yield (Bange and Milroy 2004). Studies on sowing time and cotton crop performance at different growth phases are well documented in Rahman et al. (2016). Sowing of cotton crop at optimum time can only harvest peak solar radiation and optimal cotton crop norms can be exposed for better production, while late sowing has to face the sub- and super-optimal climatic conditions at critical growth phases especially heat stress and drought conditions prevails at reproductive phases leads to flower and boll shedding and ultimately lower cotton production (Arshad et al. 2007; Rahman et al. 2004, 2016, 2017, 2018). Late sowing has also faced the high-temperature stress while optimal temperature ranges for physiological and metabolic processes from 23 to 32 °C, and super-optimal temperature leads to short reproductive phases than the early or normal sowing and ultimately resulted in poor production (Pettigrew and Johnson 2005; Conaty et al. 2012). Sowing at right time produced more yield due to better boll retention and more growing period for reproductive phases and timely completion of each phenophase (Rahman et al. 2016).

Although cotton crop requires irrigation amount of 650–750 mm depending on sowing time, as early sowing cotton needs more irrigation water, cotton crop requires minimum rainfall and it is adopted in arid climatic zones just like arid region of Punjab and Sindh provinces. Heavy rainfall especially erratic and intense is more dangerous to cotton crop during either vegetative or reproductive phases. Dry growing season is good for the production of quality fiber as rainfall during later growth phases may cause squares and boll shedding while opened boll is dangerously affected by erratic and intense rainfall especially in terms of cotton quality (Vara Prasad et al. 2005; Singh et al. 2007). High rainfall leads to high humidity which provides the favorable climatic conditions for many of the insect pest and diseases. Excessive rainfall promotes the top growth, more biomass, and lower yield as it is an indeterminate crop so balanced fertilizer and irrigation management are crucial. Best suitable areas for cotton on the basis of environmental conditions like soil (texture, EC, and fertility), climate (minimum and maximum temperature and rainfall), and irrigation water availability and moisture index were assessed and found that cotton crop is most suitable in the lower areas of Punjab province like Khanewal, Vehari, Multan, Muzaffargarh, Bahawalpur, Bahawalnagar, etc.

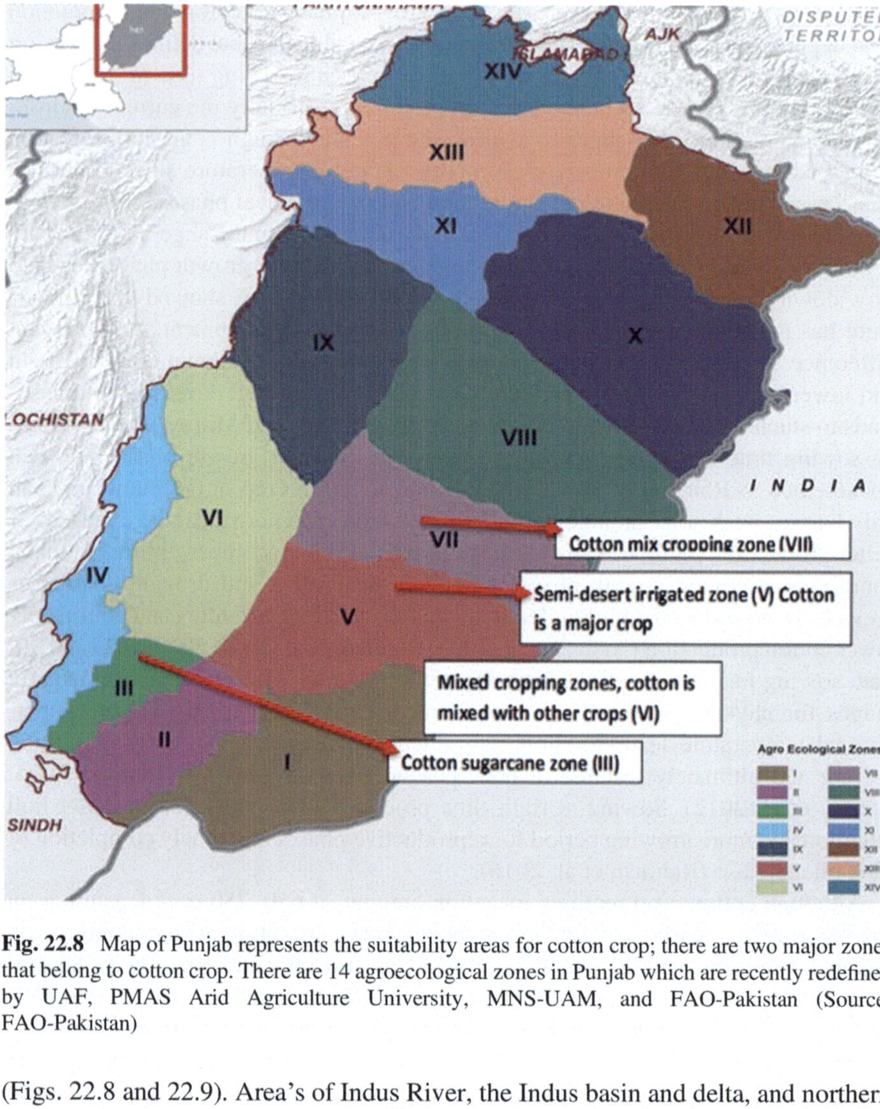

Fig. 22.8 Map of Punjab represents the suitability areas for cotton crop; there are two major zones that belong to cotton crop. There are 14 agroecological zones in Punjab which are recently redefined by UAF, PMAS Arid Agriculture University, MNS-UAM, and FAO-Pakistan (Source: FAO-Pakistan)

(Figs. 22.8 and 22.9). Area's of Indus River, the Indus basin and delta, and northern and southern irrigated plains where cotton is the main and dominant crop while recently developed AEZs showed that highest yield areas in cotton zone are arid irrigated zone (AEZs-II), cotton-sugarcane zone (III), and cotton mix cropping zone (VII) in Punjab-Pakistan. These zones are found in four divisions (Sahiwal, Multan, D.G. Khan, and Bahawalpur). The soil in these AEZ is sandy loamy and the temperature for crop production ranges from maximum 30.7 °C to minimum 17.7 °C. Based on economic suitability, Rajanpur is highly suitable for cotton production with a net return of Rs. 13,487 per hectare, followed by Rahim Yar Khan and Bahawalpur with net returns of Rs. 13,089 and Rs. 12,905 per hectare, respectively.

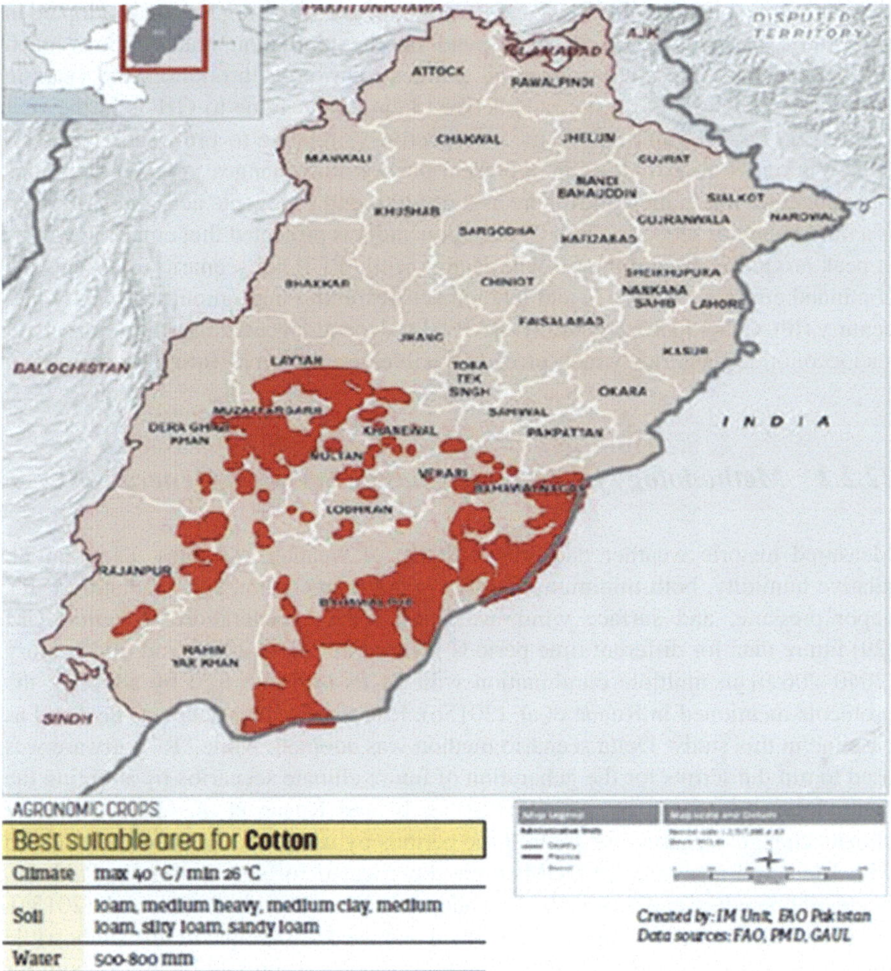

Fig. 22.9 Cotton crop suitability map in Punjab on the basis of crop norms especially maximum temperature, minimum temperature, rainfall, soil texture, soil EC, and moisture index. Crop suitability map represents that south Punjab is only suitable for cotton crop in Punjab; South Punjab is the main contributor region in cotton production (Source: FAO-Pakistan)

22.2 Climate Change Scenarios for Cotton Season in Pakistan

22.2.1 General Circulation Models (GCMs) and Representative Concentration Pathways (RCPs)

Representative concentration pathways (RCPs) basically represent the concentration of greenhouse gasses as it is adopted in fifth assessment report by IPCC. Two concentration pathways (RCP 4.5 and RCP 8.5) of greenhouse gases were adopted in this study; basically these RCPs depend on the prediction data of how much greenhouse gases are emitted from all sources/contributor throughout the years in the future. RCPs represent the radioactive forcing values due to GHGs in the year 2100 (2.6, 4.5, 6.0, and 8.5 W m^{-2}, respectively) relative to pre-industrial levels (Van Vuuren et al. 2011). RCPs represent the possible changes in GHG emissions and concentration in the atmosphere as a result of anthropogenic activities. RCP 4.5 is a mild behavior scenario of GHG emission and it is projected that emission will be at peak around 2040 and then it will decline while RCP 8.5 scenario represents the continued emission of GHGs and its radioactive forcing throughout the twenty first century (IPCC 2014a, b). Generally, RCPs also depend on certain assumptions about socioeconomic scenarios which provide flexible description of future scenarios.

22.2.2 Methodology for Climate Change Scenario Generation

Measured historic weather data (1980–2010) of weather variables, i.e., rainfall, relative humidity, both minimum and maximum temperature, solar radiation (SR), vapor pressure, and surface wind, was used for the generation of multi-GCM (29) future data for different time periods [near term (2010–2039) and mid-century (2040–2069)] in multiple combination with RCPs (4.5 and 8.5) by adopting the protocols mentioned in Ruane et al. (2015b). Historic weather data was declared as baseline in this study. Delta scenario method was adopted, while "R" software was used to run the scripts for the generation of future climate scenarios by adopting the methodology proposed by AgMIP (2013a, b) and Ruane et al. (2013). Data of climate change scenarios for future time periods by using two RCPs of 4.5 and 8.5 with latest and available 29 GCMs were downscaled for cotton zone in Pakistan. Standard procedure and protocols were adopted as described in Ruane et al. (2015a). CO_2 concentration of 390 ppm was adopted and used as baseline in this study while for future climate CO_2 was used as follows: during near term 423 and 432 ppm for RCP 4.5 and RCP 8.5, respectively, and during mid-century time period 499 and 571 ppm for RCP 4.5 and RCP 8.5, respectively (Rosenzweig et al. 2014). Comprehensive explanation about RCPs, GCMs, and CO_2 concentrations can be seen in AgMIP (2014) and Rahman et al. (2018).

22.2.3 Climate Change Scenarios in Near Term (2010–2039) and Mid-century (2040–2069)

22.2.3.1 Future Climate Scenarios During Near Term (2010–2039)

General circulation models (GCMs) were categorized into different groups (hot dry, hot wet, moderate, moderate dry, cool dry, cool wet, moderate wet, very hot, and dry scenarios) due to variation in mean temperature and rainfall during cotton growing season. Classification of GCMs generally represents the behavior of each GCM on the basis of changes and variation in temperature and precipitation. Baseline seasonal mean temperature (April to October) and rainfall are 31.92 °C and 165 mm, respectively, while changes are computed as variation and differences between baseline and future temperature and precipitation. Generally, there would be increase in seasonal mean temperature and variability in precipitation patterns that are projected by all GCMs and RCPs during all-studied time periods in the cotton zone of Pakistan. Mean ensemble of 29 GCMs showed that RCP 4.5 scenario is relatively moderate while RCP 8.5 scenario showed harsh behavior as higher increase in seasonal mean temperature of 2.01 and 3.86 °C (relative to baseline) is projected than RCP 4.5 (1.80 and 3.41 °C) for the time period of 2040 and 2069, respectively (Figs. 22.10 and 22.11). Mean increase of 4.3% and reduction of 3.9% relative to baseline is projected in seasonal rainfall during cotton growing season for RCP 4.5 scenario for the time period of 2040 and 2069, respectively (Fig. 22.12). Likewise, mean ensemble of 29 GCMs showed the reduction of 5% and 7% for the time period of 2040 and 2069, respectively, in emission scenario of RCP 8.5; largely RCP 8.5 scenario is related with significant variation both in precipitation and temperature than RCP 4.5 scenario (Fig. 22.13).

Different GCMs showed variation due to differences in behavior while GCMs have been categorized into different groups on the basis of variation in temperature and precipitation. Generally smaller changes relative to baseline during cotton growing season would range from 0.65 °C (HADGEM2-CC) to 3.46 °C (MIROC-ESM), during near-term time period, whereas the significant higher deviation/changes are observed in INMCM4 (1.82 °C) and MIROC-ESM (5.89 °C) under RCP 4.5 emission scenario than baseline for mid-century time periods (Fig. 22.10). Changes in mean seasonal temperature ranged from 0.72 °C (HADGEM2-ES) to 2.82 °C (MIROC-ESM) in near-term time period while higher changes ranged from 2.40 °C (INMCM4) to 6.06 °C (MIROC-ESM) in RCP 8.5 till 2069 time period (Fig. 22.11). Major and significant changes/variation under RCP 8.5 scenario revealed the harsh behavior. Individual response of GCM and division into different categories on the basis of variation and deviation from baseline can be observed in Figs. 22.10 and 22.11 with respect to changes in mean seasonal temperature. It is projected clearly from the response of each GCM that there would be more increase in minimum temperature as compared with the maximum temperature for the studied time periods which is more detrimental to cotton growth and development and especially cotton production.

Maximum consensus of GCMs about changes in seasonal temperature, mild scenarios, worse and worst scenarios, and hotter GCMs is deliberated in details in Figs. 22.10 and 22.11. Results of GCMs for maximum consensus under RCP 4.5 projected that there would be an increase in seasonal average temperature of 1.4–1.86 °C and 2.8–3.4 °C for 2040 and 2069 time spans, respectively (Fig. 22.10). MIROC-ESM found the hottest one among all GCMs under studied time periods (2040 and 2069 time spans) for the RCP 4.5 emission scenario. Similarly, maximum consensus about increase in mean maximum temperature in

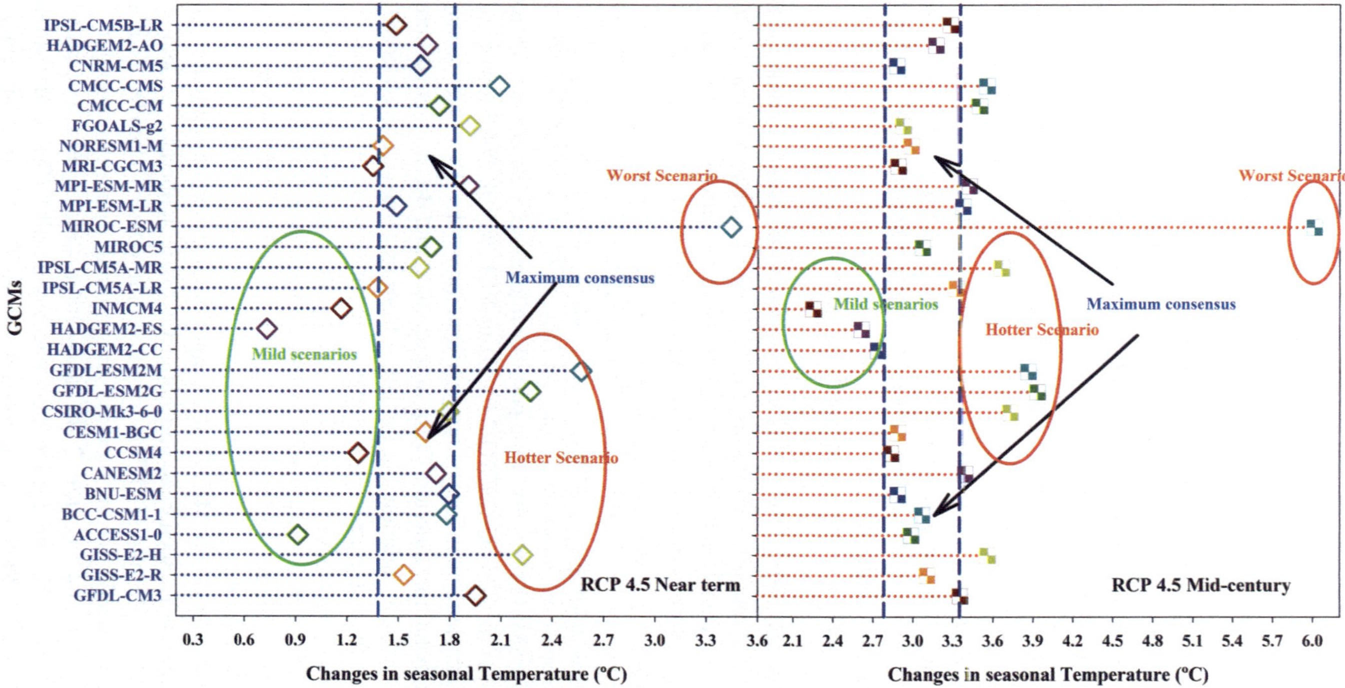

Fig. 22.10 GCM evaluation and categorization on the basis of changes in mean seasonal temperature (°C) during cotton growing season as projected by climate models (GCMs) relative to baseline period (1980–2010) under GHG emission scenario of *RCP 4.5* during *near term (2010–2039)* and *mid-century (2040–2069)* time periods, respectively. Maximum consensus developed by the climate models (GCMs) in both time periods is shown by the vertical blue dotted lines. Green and red circles represent the mild and hotter scenarios, respectively, while hottest and worst scenario is shown by MIROC-ESM in both time periods (near term and mid-century)

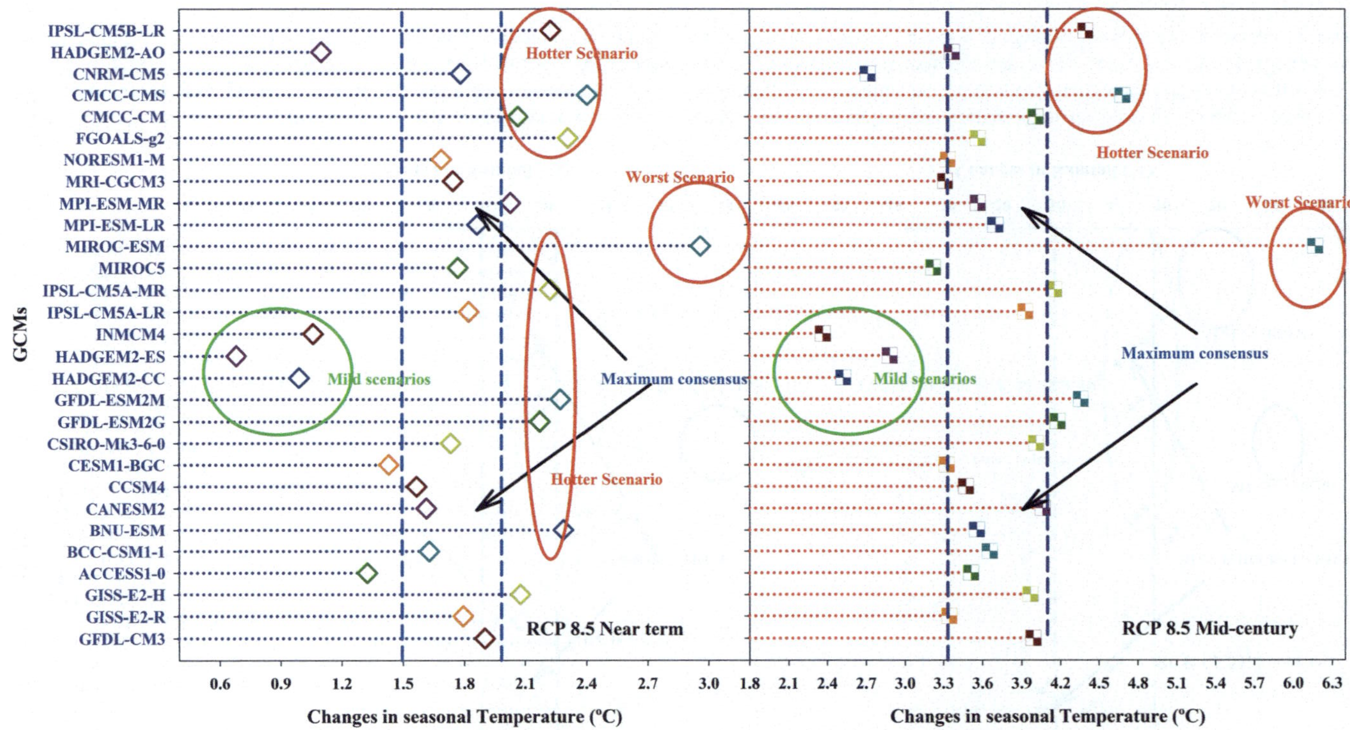

Fig. 22.11 GCM evaluation and categorization on the basis of changes in mean seasonal temperature (°C) during cotton growing season as projected by climate models (GCMs) relative to baseline period (1980–2010) under GHG emission scenario of *RCP 8.5* during *near term (2010–2039)* and *mid-century (2040–2069)* time periods, respectively. Maximum consensus developed by the climate models (GCMs) in both time periods is shown by the vertical blue dotted lines. Green and red circles represent the mild and hotter scenarios, respectively, while hottest and worst scenario is shown by MIROC-ESM in both time periods (near term and mid-century)

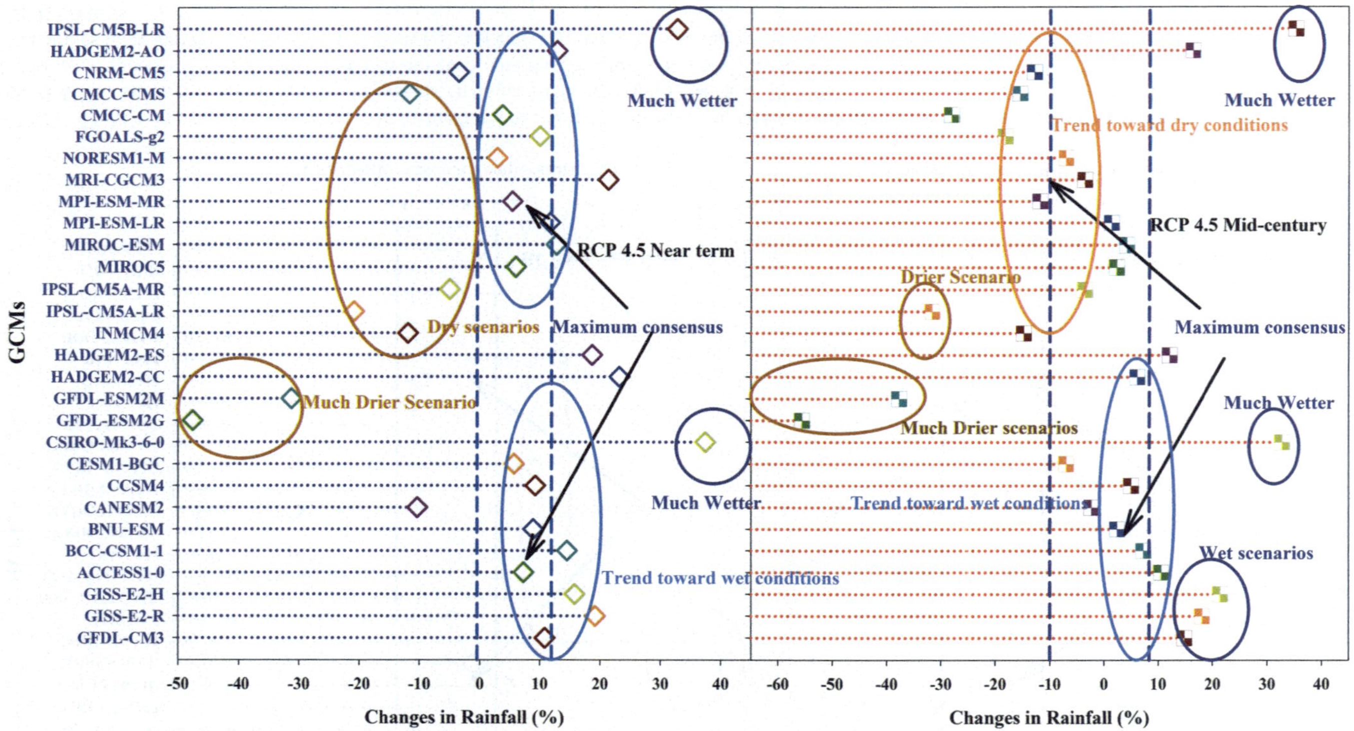

Fig. 22.12 GCM evaluation and categorization (dry, much drier, maximum consensus, wet, and much wetter scenarios) on the basis of changes in total seasonal rainfall (mm) during cotton growing season as projected by climate models (GCMs) relative to baseline period (1980–2010) under GHG emission scenario of *RCP 4.5* during *near term (2010–2039)* and *mid-century (2040–2069)* time periods, respectively. Maximum consensus developed by the climate models (GCMs) in both time periods is shown by the vertical blue dotted lines. Different color circles represent the dry, much drier, wet, and much wetter scenarios in both time periods (near term and mid-century)

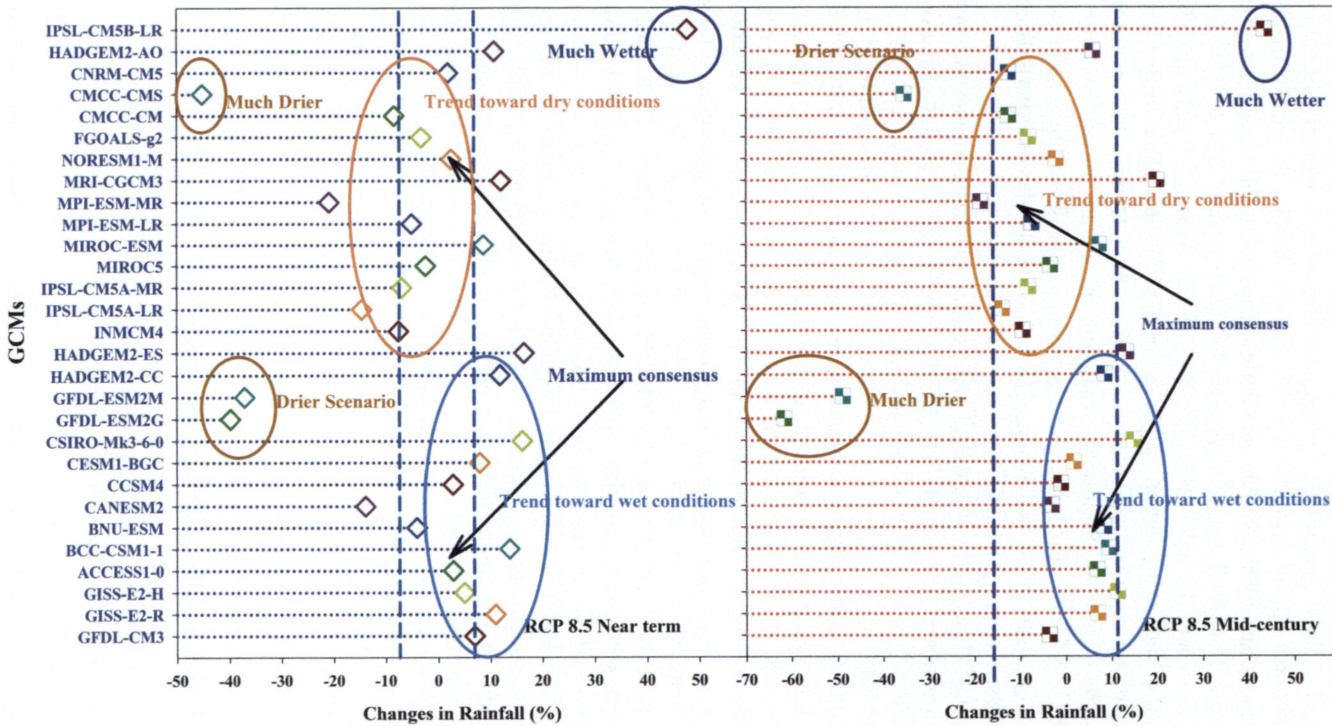

Fig. 22.13 GCM evaluation and categorization (dry, much drier, maximum consensus, wet, and much wetter scenarios) on the basis of changes in total seasonal rainfall (mm) during cotton growing season as projected by climate models (GCMs) relative to baseline period (1980–2010) under GHG emission scenario of *RCP 8.5* during *near term (2010–2039)* and *mid-century (2040–2069)* time periods, respectively. Maximum consensus developed by the climate models (GCMs) in both time periods is shown by the vertical blue dotted lines. Different color circles represent the dry, much drier, wet, and much wetter scenarios in both time periods (near term and mid-century)

RCP 8.5 scenario would range from 1.5 to 2 °C and 3.3 to 4.1 °C relative to baseline for 2040 and 2069 time spans, respectively (Fig. 22.11).

Climate scenarios about precipitation revealed that seasonal changes and variation in rainfall patterns during cotton season are more uncertain. Results of maximum consensuses for rainfall showed the smaller increase in rainfall amount relative to baseline. Changes in rainfall would range from 0% to 10% and −10% to 8.5% for 2040 and 2069 time periods under RCP 4.5 (Fig. 22.12). Mostly, climate models revealed the trend of increasing rainfall while fewer models showed the decreasing tendency of rainfall in under RCP 4.5 scenario during cotton growing season but it seems most uncertain patterns. Significant reduction in rainfall amount is clearly depicted under RCP 8.5 scenario for both time periods; it is projected that change in rainfall would range from −9% to 8% and −18% to 11% for the time periods of 2040 and 2069, respectively (Fig. 22.13).

There are few GCMs as well that are declared as drier than seasonal baseline but highest reduction is observed for the GFDL-ESM2G under both emission scenarios (RCP 4.5 and RCP 8.5) in both time periods of 2050 and 2069. There are few GCMs like IPSL-CM5B-LR that was found wet and hot in both timer periods and RCPs as compared with other GCMs. There are some worst-case scenarios as well where higher increase in temperature and reduction in precipitation are projected. Furthermore few models were also found wetter and much wetter and some may declare as dry and drier but one thing is common among all GCMs and studied RCP scenarios that rainfall patterns are more uncertain and cotton crop is very specific for its irrigation water demand. Some GCMs showed less uncertain and showed stable behavior during both time periods under both RCPs, namely, CCSM4, HadGEM2-CC, HadGEM2-ES, INMCM4, CanESM2, CNRM-CM5, ACCESS1-0, BNU-ESM, and MIROC5. These GCMs are recommended and can be applied for the assessment of climate change impacts for other crops in Pakistan. Generally climate models are uncertain, climate change impact assessment studies about crops especially for cotton crop should rely on more than one climate models as uncertainties are common in emission scenarios (Rahman et al. 2018). It is recommended across the globe about modeling studies to use multi-models for climate change as it can provide more consistent result about decision management especially in agriculture sector as it is more risky than others (Asseng et al. 2013; Rosenzweig et al. 2014; Rahman et al. 2018).

22.3 Impact of Climate Change on Cotton Production

22.3.1 Climate Change Impact Assessment for Cotton Crop During Near Term (2010–2039)

Cotton crop is very sensitive for its climatic requirements and slight variation may lead to changes in growth and development and ultimately lead to reduction in production. Weather variables especially temperature and rainfall have a significant role for cotton production and variation or change in these factors may lead to severe reduction in cotton production. High temperature above normal which is required for optimum growth and development of cotton crop speeds up the phenological phases and had a negative impact on vegetative and especially reproductive growth phases of cotton and ultimately reduces the cotton production. Sustainable cotton production becomes more sensitive especially in arid climate where rainfall variability and high temperature are already a severe threat to cotton crop. Generally, mostly GCM scenario projected the reduction in cotton yield as compared with the baseline during both time periods and RCPs tested.

Seed cotton yield would reduce but an important contribution of variation existed in yield (−2.7% to −49%) among studied GCMs under RCP 4.5 during near term; similarly yield reduction ranges from −3% to −32% under scenario of RCP 8.5 during near term although less uncertainty was found among GCMs under RCP 8.5 scenario. Generally, less seed cotton yield reduction is observed under RCP 4.5 but few GCMs are more uncertain (Figs. 22.14 and 22.15). Although all GCMs showed the reduction in cotton crop yield in Pakistan during near term, there are few worse-case scenarios as well that showed the higher reduction in cotton yield because those are found hotter and variable in term of rainfall. Drier GCMs and hottest at the same time are found worst-case scenarios for cotton production as heat and drought conditions are more detrimental to cotton crop. Overall mean ensemble of 29 GCMs showed the reduction of 10% and 17% in seed cotton yield in Pakistan under GHG emission scenarios of RCP 4.5 and RCP 8.5, respectively, during near term. More reduction in cotton yield is expected due to higher increase in mean seasonal temperature and significant amount of rainfall variability during cotton growing season in Pakistan. Drier conditions with below potential evapotranspiration and especially high night temperature lead to more reduction in cotton yield in worst-case scenarios of GCMs during near term (Figs. 22.14 and 22.15).

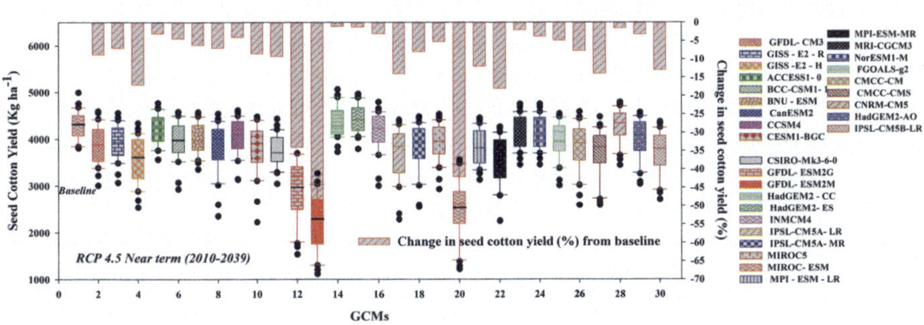

Fig. 22.14 Mean seed cotton yield of cultivars and yield gained/lost as compared with baseline (historic) simulated by CSM-CROPGRO-Cotton based on 29 GCMs under RCP 4.5 for near term of the twenty first century (2010–2039). Black dots at both ends of box plots represent the lowest and highest SCY points in different growing years. Changes in mean seed cotton yield (%) of each GCM depict by the bar graphs against each GCM

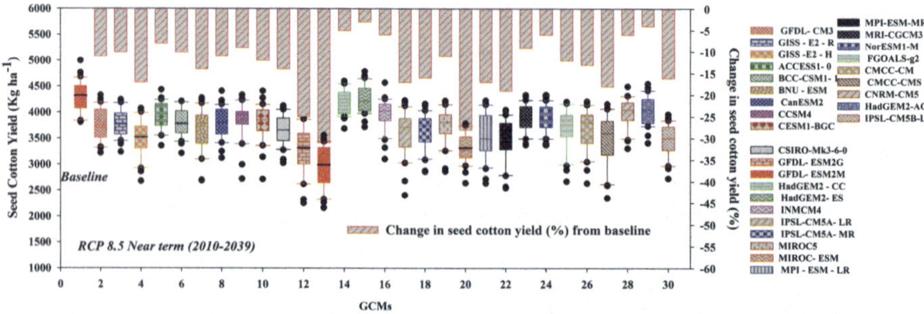

Fig. 22.15 Mean seed cotton yield of cultivars and yield gained/lost as compared with baseline (historic) simulated by CSM-CROPGRO-Cotton based on 29 GCMs under RCP 8.5 for near term of the twenty first century (2010–2039). Black dots at both ends of box plots represent the lowest and highest SCY points in different growing years. Changes in mean seed cotton yield (%) of each GCM depict by the bar graphs against each GCM

22.3.2 Climate Change Impact Assessment for Cotton Crop During Mid-century (2040–2069)

Significantly higher losses of cotton yield are projected for the time period of 2040–2060 for both RCP 4.5 and RCP 8.5 but especially higher reduction is expected in RCP 8.5 due to worst scenarios of increasing temperature both maximum and due to more increase in minimum temperature accompanied by higher rainfall variability and especially the drier conditions during cotton growth especially at reproductive phases. It is projected that there would be a decrease of 25% and 39% (mean ensemble of 29 GCMs) in cotton yield under RCP 4.5 and RCP 8.5, respectively, during mid-century as compared with the seasonal average baseline cotton yield (3919 kg ha^{-1}). Variation in the yield of cotton crop would range from −5% to −57% and −7% to −72% under RCP 4.5 and 8.5 scenarios, respectively, during mid-century in Pakistan (Figs. 22.16 and 22.17). Major reduction in cotton yield is attributed for the same GCMs as observed during the near-term time period for both RCP 4.5 and RCP 8.5 emission scenarios. Similar causes as mentioned and discussed in previous section like higher increase in night temperature accompanied with drier conditions, heat stress during reproductive growth phases of cotton, and rainfall variability may be the significant contributor in lower cotton yield production under future climate scenarios of Pakistan.

More seed cotton yield reduction is projected for CMCC-CMS, IPSL-CM5B-LR, GISS-E2-H, GFDL-ESM2M, and GFDL-ESM2G and GFDL-ESM2M and GFDL-ESM2G are hotter and drier while HADGEM2-ES and HADGEM2-CC are milder and wet which leads to lower reduction in cotton yield. High variation was found among GCMs for cotton yield but normally ten GCMs, including MIROC-ESM, GFDL-ESM2G, GFDL-ESM2M, IPSL-CM5A-LR, CMCC-CMS, GFDL-CM, CSIRO-MK3-6-0, MPI-ESM-MR, CANESM2, and IPSL-CM5A-MR, projected significant variation in climatic variables especially temperature and rainfall, while mild behavior is projected by the four GCMs, namely, HADGEM2-CC, INMCM4,

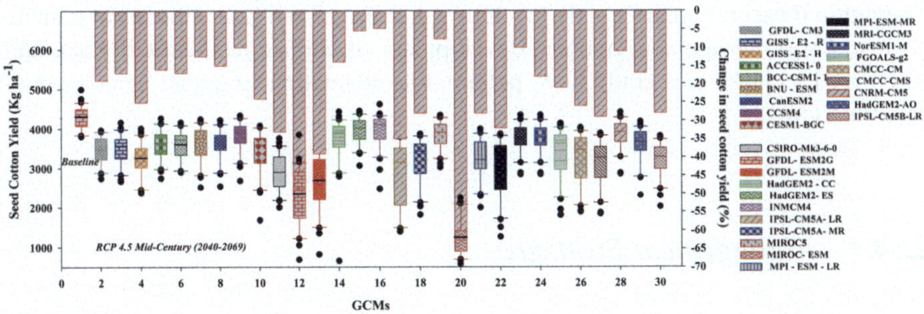

Fig. 22.16 Mean seed cotton yield of cultivars and yield lost as compared with baseline (historic) simulated by CSM-CROPGRO-Cotton based on 29 GCMs under RCP 4.5 for mid of the twenty first century (2040–2069). Black dots at both ends of box plots represent the lowest and highest SCY points in different growing years. Changes in mean seed cotton yield (%) of each GCM depict by the bar graphs against each GCM

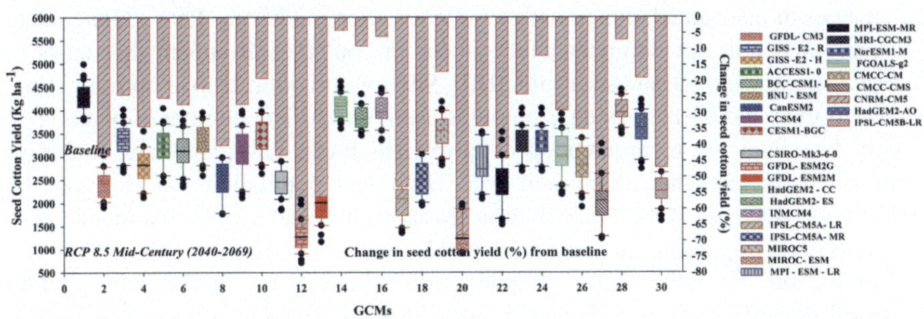

Fig. 22.17 Mean seed cotton yield of cultivars and yield lost as compared with baseline (historic) simulated by CSM-CROPGRO-Cotton based on 29 GCMs under RCP 8.5 for mid of the twenty first century (2040–2069). Black dots at both ends of box plots represent the lowest and highest SCY points in different growing years. Changes in mean seed cotton yield (%) of each GCM depict by the bar graphs against each GCM

HADGEM2-ES, and CNRM-CM5, than all other studied GCMs and generally speaking changes overall are moderate (Figs. 22.14, 22.15, 22.16, and 22.17). Few GCMs are found hotter and drier as well like GFDL-ESM2G and GFDL-ESM2M but few others (HADGEM2-ES, HADGEM2-CC, and INMCM4) are, although hot, have wet conditions during cotton growing seasons which lead to lower reduction of cotton yield in these GCMs during both studied time periods.

22.4 Adaptation Technology Development for Sustainable Cotton Production Under Climate Change Scenarios

Climate change has negative effects on growth and yield of cotton. Extreme temperatures and uncertain rainfall patterns decrease the seed cotton yield by reducing the length of growing periods. To ensure the high yield under changing climate, adaptation measures are necessary to address the current and future threats of climate change. Crop management options, development of climate resilient cultivars, and ICT-based technologies could have potential as adaptation strategies in context of changing climate.

22.4.1 Management Strategies

Modification in production technology of cotton is important for confronting the risk associated with climate change. Different management strategies as shown in Table 22.1 could be useful for mitigating the negative effects of climate change on cotton.

Adjustment of sowing dates is very important and associated with growing season length. Too early and late sowing of cotton reduced the yield by increasing the chances of inset attack due to unfavorable weather conditions. Late sowing delay flowering and boll development which occur in cooler environment that leads to decrease in the yield (Braunack et al. 2012). Sowing of cotton crop at optimum time increased the yield by increasing growing season length. The number of plants m^{-2} is very less than the recommended at farm field level; few plants are unable to germinate that causes the reduction in yield. High plant population also decreases the yield due to decrease in boll size and increases fruit shedding. So, maintenance of optimum plant population increased the yield.

Application of slow release (coated) and balanced use of fertilizers (NPK) increased the efficiency and meet the nutritional requirement of crop that lead to increase in the yield. Future scenarios of irrigation water showed that per capita availability of water would be reduced in the future due to climate change and less water would be available for crop production (Ahmad et al. 2019). Efficient methods such as drip irrigation in cotton reduced the losses and provide water directly to root

Table 22.1 Crop management strategies as adaptation for cotton under climate change

Adaptations	References
Adjustment of sowing dates	Huang (2016)
Optimum plant population	Wrather et al. (2008)
Precise application of nutrients (balanced use of fertilizers, slow released fertilizers)	Wrather et al. (2008)
Efficient water technologies (drip irrigation)	Bhaskar et al. (2005)
Integrated pest and disease management	Hillocks (2005)
Drainage of water during excess rainfall	Manik et al. (2019)
Reducing post-harvest losses (mechanical picking)	Muthamilselvan et al. (2007)

zone of crop which prevent from drought or severe water stress. Insect pest and disease attack on cotton causes the huge reduction in yield of cotton.

Integrated pest management (IPM) eliminated the use of pesticides and minimized the toxicity of chemical due to the use of cultural and biological methods for pest control. Insect pest and disease will become more problematic due to climate change (Petzoldt and Seaman 2006). IPM would be useful strategy for control of insets pest and diseases. Some practices adopted in IPM method are timely removal of weeds and removal of alternate host plants and cotton sticks from the field. Animals like sheep and goats are allowed to feed after last picking. Bee vectoring technology (BVT) is part of IPM which is a biological control of pest and leads to heather and stronger plants (Arora et al. 2011). As far as water is concerned, cotton is sensitive to stagnant water. Excess water promotes the fruit shedding in cotton due to inadequate aeration in the root zone. Removal of excess water from the field by making drain at low end of the field is a useful strategy for cotton especially during monsoon season (Muthamilselvan et al. 2007; Perumal et al. 2006). Cotton is an indeterminate crop and picking of cotton is done 3–5 times at final stages. Manual picking of cotton is labor intensive and cost accounts for 30–35% of total cultivation. It also increased the post-harvest losses due to picking of immature boll. Mechanical picking of cotton reduced the total cost and post-harvest losses.

22.4.2 Heat and Drought Resilient Germplasm Development

Heat and drought stress caused the physiological and chemical changes in cotton which affected the growth and yield. Generally, heat and drought conditions restricted the root growth, plant height, boll development, and fiber quality. However, photosynthetic activity, stomatal conductance, and water potential are also decreased (Chastain et al. 2016). Therefore, it is necessary to develop heat- and drought-tolerant mechanism in cotton to avoid multiple stresses and to survive under harsh environment.

Future rise in temperature would reduce the phenological events such as days to flowering and maturity. Regaining the phenological event under high temperature through breeding could be best strategy in developing heat-tolerant cultivars (Ahmad et al. 2017a). For example, a study was conducted for the development of climate resilient cultivars of cotton in Pakistan using Decision Support System for Agrotechnology Transfer (DSSAT). Future rise in maximum temperature of 3.6 °C and minimum temperature of 3.8 °C was used in a model. Impact of rise in temperature decreased the days to anthesis by 10%, maturity by 20%, and yield by 60%. Genetic coefficients of cotton in DSSAT were adjusted to regain the phenology and yield. The time required for the cultivar to reach a final pod was decreased by 30% and fraction of daily growth which is portioned to seed and shell was increased by 10%. Other phenology-related traits were also adjusted to get the climate resilient cultivar of cotton (Ahmad et al. 2017b). Drought-tolerant cultivars could be developed through selection of potential traits, which can enhance the performance under water deficit condition. Several traits exist that are associated with drought tolerance such as leaf conductance, leaf water potential, rolling, osmotic adjustment, and extraction of soil water (Turner 1979; Woodfin et al. 1979). However, plant breeder can use different techniques for the evaluation of these traits to develop drought-tolerant cultivars. The development of heat and drought tolerant can offset the yield losses and sustain the productivity under changing climate scenarios.

22.4.3 Application of Decision Support System for Sustainable Cotton Production

Decision support system (DSS) is computer-based system that helps in decision-making. DSS addresses the issues related to efficient agronomic practices, conservation, nutrient management, insect pest, and disease management which would be useful in adapting the environmental vulnerabilities. Nutrients could be managed through DSS. Fertilizers are very expensive and huge losses were observed during application at farm level, which reduced the yield. In the future due to rise in temperature, fertilizer losses would be increased through volatilization. So DSS has been designed for the recommendation of optimum doses of fertilizer to crops. For example, CROPGRO simulates the yield of cotton by the integrating soil, water, genetics, and environment system (Hoogenboom 2000; Rahman et al. 2017). It helps in simulating the nitrogen balance and optimization of fertilizer in cotton (Wang et al. 2013). Haifa Nutri-Net is another DSS that assists the growers in nutrient and irrigation management under changing climate (Achilea et al. 2005). Planning Land Applications of Nutrients for Efficiency and the Environment (PLANET) is another system which provides the best management practices for crops and recommends the fertilizer application based on previous crop (Gibbons et al. 2005).

Insects and pests can be managed through DSS to increase the production under changing climate. For example, CLIMEX model is used to examine the insect and pest distribution. It has been used by more than 20 countries for insect and pest management (Walker and White 2001). Another model SOPRA is also used for monitoring and management of pest population (Mir and Quadri 2009). DSS has been used for water management and timely estimation of drought for cotton (Loi and Tangtham 2005). For climate change adaptations, climate forecasting is necessary because it provides the scientific bases. Decision Support System for Agrotechnology Transfer (DSSAT) has been designed to assess the climate impacts (Jones et al. 2000); another model on geographical information system (GIS) and remote sensing based has been developed to estimate the risk associated with climate. Forecasting of climate based on DSS helps in adjustment of crops in different climatic conditions.

22.4.4 Use of ICT for Better Cotton Production Under Climate Change

Information and communication technologies (ICT) play a key role for adapting climate change. The main steps for the development of adaptations through ICT are given in Fig. 22.18. Similar steps were described by Sala (2009). The first step of adaptation is the observation on how the climatic variations are occurring on a specific site. Observation could be carried out by different tools like sensor-based networks and remote sensing techniques. The collected data on climatic impacts on cotton are stored and digitized for communication of different institutes. After the data is analyzed for planning, for this purpose computer-based models are used for decision-making. GIS can facilitate in development of adaptation measures based on observation for stakeholders. The developed adaptation is implemented to farmers for decision-making as shown in step 3 in Fig. 22.18. The tools useful for implementation and management are forecasting and early warning system. Capacity

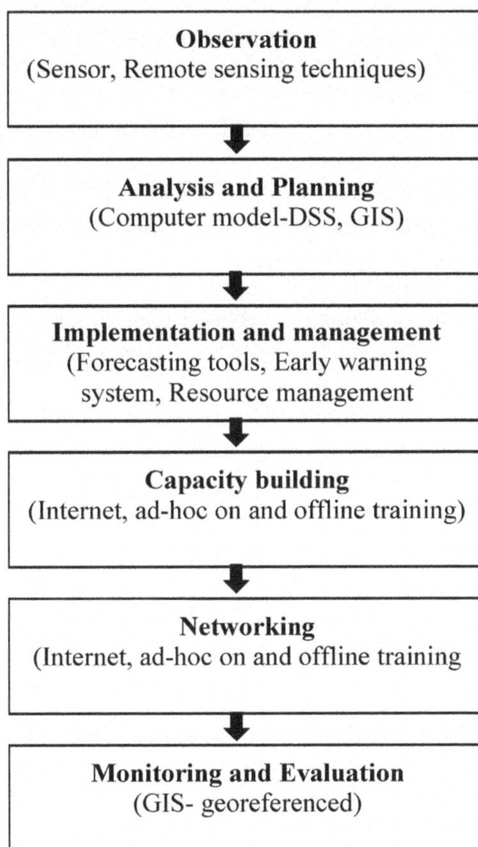

Fig. 22.18 ICT-based steps for the development of adaptation strategies for sustainable cotton production to mitigate the negative impact of climate change

building on climate adaptation can be employed for the awareness of farmers in cotton zones. Training, seminars, and workshops could be conducted using ad hoc on and offline system. The final step is networking in which data is stored and retried for comparing the information with other knowledge partner in different areas for precise decision. The final stage of adaptation through ICT is monitoring and evaluation, which could be done using GIS tools. It allows geo-reference information and support for monitoring and evaluation of developed adaptations strategies.

22.4.5 Potential Options of Climate Resilient Cotton Production

These are the few options that may have potential to develop climate resilient cotton production:

1. Improving resource use efficiency (water, fertilizers, agrochemicals).
2. Changing from conventional to good agricultural practices.
3. Carbon sequestration as a climate change mitigation strategy.
4. Breeding heat and drought climate resilient varieties with high resource use efficiency.
5. Optimizing sowing dates to harvest the maximum solar radiation and optimum climate norms.
6. Reduced tillage operations for the low emission of GHGs.
7. Prolonging soil cover to control weeds and enhance water use efficiency and water productivity.
8. Increasing plant density.
9. Introduction of carbon pricing policy.
10. Developing energy efficient technologies.
11. Improving soil organic matter.
12. Promotion of biodiversity.
13. Improvement in agriculture extension services (farmer education).
14. Climate smart village development for the promotion of climate resilient cropping system.

22.4.6 Strategies/Technologies for Climate Smart Cotton Production

Climate smart technologies and practices are the need of the time to develop climate smart cotton production system as climate smart agriculture (CSA) tackles with three important pillars as: (1) sustainably increasing agricultural productivity and income to meet national food security, (2) adapting and building resilience in agricultural systems to climate change, and (3) reducing and/or removing greenhouse gas emissions or increasing carbon sequestration (FAO 2013, 2017). Practices and technologies in agriculture are deliberated as smart especially in terms of climate that have potential and assist to achieve at least one component of CSA. Climate smart agriculture system generally improves the resource use efficiency and increased resilience and productivity by reducing greenhouse gas emissions. Climate smart cotton production system has technologies and strategies to increase sustainable cotton production, farm income, and livelihood, improves water and fertilizer use efficiency and develops resilience to climate variability, and generally lowers the emission of GHG emission due to different poor management practices (FAO 2010, 2012, 2016). Vulnerability of cotton production system due to climate extremes can be minimized significantly by adopting CSA technologies and practices and hence it revealed the potential (FAO 2010). Few studies on climate smart cotton showed that new practices and technologies and especially adaption in current cotton system will strengthen the sustainable cotton production by improving resource use efficiency (Ashraf and Iftikhar 2013; Pasha 2015; Imran et al. 2018; Rahman et al. 2018).

Carbon, energy, water, and knowledge smart climate agricultural technologies and practices are now being adopted and need to be adopt for cotton production system as well because cotton is most sensitive to any stress (Imran et al. 2018). CSA practices and technologies are also being adopted by even small landholders (Khatri-Chhetri et al. 2017) in Punjab especially in cotton belt which will improve cotton production, WUE, and NUE and ultimately improve the efficiency of key inputs in cotton-based cropping system. Cotton production can be improved by adopting CSA practices like sowing on raised bed or ridges as has potential to save water and high efficiency and it enhances nutrients uptake and transport (Khatri-Chhetri et al. 2016; Imran et al. 2018). Sowing on beds also ensures good crop stand and optimum plant population as planting density is the key contributor in cotton yield. It also improves germination and crop stand under unfavorable environmental conditions and it also saves crop for waterlogging situation in case of erratic and intense rainfall during monsoon. Climate resilient genotype development which can withstand at high temperature and drought conditions is also among top agenda of CSA, as climate smart varieties have potential to minimize the negative impact of climate change and survive even under unfavorable climatic conditions (Rane and Nagarajan 2004). Currently genotypes being cultivated at farmer's field have lack of sustainability and most vulnerable to climate extremes and only survive for few years and lose potential quickly and cause cotton yield stagnation (Rahman et al. 2016). Water-saving technologies especially high irrigation efficiency system have potential and especially drip irrigation is being adopted as CSA practice for cotton crop to save irrigation water by reducing surface water losses and it has also potential to improve nutrient use efficiency (Watto and Mugera 2015; Dağdelen et al. 2009; Manpreet et al. 2007). Cotton crop yield is improved by adopting drip irrigation through saving water and good crop stand which ultimately lead to better crop production by avoiding environmental stress. It has been observed that application of climate smart practices and technologies considerably reduces the adverse effects of climate change and improves cotton yield and production and ultimately better livelihood of farmers at grassroots scale. There is a need of the time to adopt climate smart cotton production system for the sustainable cotton production under climate change scenarios.

22.5 Conclusions

Current agriculture production systems are most vulnerable to climate change; climate extremes and climate variability are threats to sustainable crop production across the world to fulfill the needs of ever-increasing population. Food and fiber security is under threat due to environmental challenges and especially current cotton-based cropping system and cotton crop is most sensitive to climate extremes and environmental stress. Cotton crop is a significant cash crop, it has a prominent role in cropping system and support the GDP. Climate change is expected to enhance the vulnerability of cotton crop as there is significant shift in seasons and increase in the number of climate extreme events across the world. Climate change scenarios revealed the increase in both maximum and minimum temperature and uncertain rainfall patterns throughout the world and especially in dry and arid areas of the world. Rainfall would increase and decrease as projected by multi-GCMs and RCPs and it is fact that these changes in climate would lead to negative effect on cotton crop production and sustainable cotton production in the future is under threat. Weather plays a crucial role as it determines the initiation and ending period of phenological stages during crop growing cycle. Climate change has a negative impact on cotton production in major parts of the cotton-growing regions. It not only hampers the yield but quality of fiber and has a negative impact on socioeconomic conditions of farmers. Climate, crop, and economic multidisciplinary modeling approach are being used to assess the impact of climate change and adaptation strategy development for sustainable cotton production. Changes in crop management practices (sowing, planting density, irrigation, plant protection) may be good adaptation strategies for sustainable cotton production under changing climate scenarios of the world. Climate resilient cotton production system has potential to cope with the negative impacts on cotton crop by developing heat and drought resilient germplasm, mitigation technology to reduce GHG emission, and application of decision support system and use of ICT-based technologies for sustainable cotton crop production. It is time to adopt climate, energy, and water smart cotton production technologies and practices for sustainable cotton production in the future.

References

Abbas Q, Ahmad S (2018) Effect of different sowing times and cultivars on cotton fiber quality under stable cotton-wheat cropping system in southern Punjab, Pakistan. Pak J Life Soc Sci 16:77–84

Achilea O, Ronen E, Elharrar G (2005) Haifa Nurti-Net–a comprehensive crop Nutrition software, operated over the web. EFITA/WCCA, pp 25–28

AgMIP (2013a) Guide for running AgMIP climate scenario generation tools with R in windows. AgMIP. http://www.agmip.org/wp-content/uploads/2013/10/Guide-forRunning-AgMIP-Climate-Scenario-Generation-with-R-v2.3.pdf

AgMIP (2013b) The coordinated climate-crop modeling project C3MP: an initiative of the agricultural model inter comparison and improvement project. C3MP protocols and procedures. AgMIP. http://research.agmip.org/download/attachments/1998899/C3MP+Protocols+v2.pdf

AgMIP (2014) Guide for regional integrated assessments: handbook of methods and procedures, Version 5.1. AgMIP. http://www.agmip.org/wpcontent/uploads/2013/06/AgMIPRegional%20Research-Team-Handbook-v4.2.pdf

Ahmad S, Raza I (2014) Optimization of management practices to improve cotton fiber quality under irrigated arid environment. J Food Agric Environ 2(2):609–613

Ahmad S, Raza I, Ali H, Shahzad AN, Atiq-ur-Rehman, Sarwar N (2014) Response of cotton crop to exogenous application of glycinebetaine under sufficient and scarce water conditions. Braz J Bot 37(4):407–415

Ahmad A, Ashfaq M, Rasul G, Wajid SA, Khaliq T, Rasul F, Saeed U, Rahman MH, Hussain J, Baig IA, Naqvi SAA, Bokhari SAA, Ahmad S, Naseem W, Hoogenboom G, Valdivia R (2015) Impact of climate change on the rice–wheat cropping system of Pakistan. In: Rosenzweig C, Hillel D (eds) Handbook of climate change and agro ecosystems: the agricultural model inter comparison and improvement project integrated crop and economic assessments, Part 2. Imperial College Press, London

Ahmad S, Abbas Q, Abbas G, Fatima Z, Atique-ur-Rehman, Naz S, Younis H, Khan RJ, Nasim W, Habib ur Rehman M, Ahmad A, Rasul G, Khan MA, Hasanuzzaman M (2017a) Quantification of climate warming and crop management impacts on cotton phenology. Plants 6(1):E7. https://doi.org/10.3390/plants6010007

Ahmad A, Ashfaq M, Rasul G, Wajid SA, Khaliq T, Rasul F, Saeed U, Ahmad I, Nasir J, Baig IA (2017b) AgMIP-Pakistan RRT final report. Agric Model Intercomp Improv Proj. 1–76

Ahmad S, Iqbal M, Muhammad T, Mehmood A, Ahmad S, Hasanuzzaman M (2018) Cotton productivity enhanced through transplanting and early sowing. Acta Sci Biol Sci 40:e34610

Ahmad I, Wajid SA, Ahmad A, Cheema MJM, Judge J (2019) Optimizing irrigation and nitrogen requirements for maize through empirical modeling in semi-arid environment. Environ Sci Pollut Res Int 26(2):1227–1237

Ali H, Afzal MN, Ahmad F, Ahmad S, Akhtar M, Atif R (2011) Effect of sowing dates, plant spacing and nitrogen application on growth and productivity on cotton crop. Int J Sci Eng Res 2 (9):1–6

Ali H, Abid SA, Ahmad S, Sarwar N, Arooj M, Mahmood A, Shahzad AN (2013a) Integrated weed management in cotton cultivated in the alternate-furrow planting system. J Food Agric Environ 11(3&4):1664–1669

Ali H, Abid SA, Ahmad S, Sarwar N, Arooj M, Mahmood A, Shahzad AN (2013b) Impact of integrated weed management on flat-sown cotton (*Gossypium hirsutum* L.). J Anim Plant Sci 23 (4):1185–1192

Ali H, Hameed RA, Ahmad S, Shahzad AN, Sarwar N (2014a) Efficacy of different techniques of nitrogen application on American cotton under semi-arid conditions. J Food Agric Environ 12 (1):157–160

Ali H, Hussain GS, Hussain S, Shahzad AN, Ahmad S, Javeed HMR, Sarwar N (2014b) Early sowing reduces cotton leaf curl virus occurrence and improves cotton productivity. Cer Agron Moldova XLVII(4):71–81

Amin A, Nasim W, Mubeen M, Nadeem M, Ali L, Hammad HM, Sultana SR, Jabran K, Habib ur Rehman M, Ahmad S, Awais M, Rasool A, Fahad S, Saud S, Shah AN, Ihsan Z, Ali S, Bajwa AA, Hakeem KR, Ameen A, Amanullah, Rehman HU, Alghabar F, Jatoi GH, Akram M, Khan A, Islam F, Ata-Ul-Karim ST, Rehmani MIA, Hussain S, Razaq M, Fathi A (2017) Optimizing the phosphorus use in cotton by using CSM-CROPGRO-cotton model for semi-arid climate of Vehari-Punjab, Pakistan. Environ Sci Pollut Res 24(6):5811–5823

Amin A, Nasim W, Mubeen M, Ahmad A, Nadeem M, Urich P, Fahad S, Ahmad S, Wajid A, Tabassum F, Hammad HM, Sultana SR, Anwar S, Baloch SK, Wahid A, Wilkerson CJ, Hoogenboom G (2018) Simulated CSM-CROPGRO-cotton yield under projected future climate by SimCLIM for southern Punjab. Pak Agric Syst 167:213–222

Arora R, Jindal V, Rathore P, Kumar R, Singh V, Bajaj L (2011) Integrated pest management of cotton in Punjab, India. Radcliffe's IPM world Textb., St. Paul University, Minnesota. http://www.ipmworld.umn.edu/chapters/Arora.html

Arshad M, Wajid A, Maqsood M, Hussain K, Aslam M, Ibrahim M (2007) Response of growth, yield and quality of different cotton cultivars to sowing dates. Pak J Agric Sci 44:208–212

Ashraf S, Iftikhar M (2013) Mitigation and adaptation strategies for climate variability: a case of cotton growers in the Punjab, Pakistan. Int J Agric Ext 1(1):30–35

Asseng S, Ewert F, Rosenzweig C, Jones JW, Hatfield JL, Ruane AC, Boote KJ, Thorburn PJ, Rötter RP, Cammarano D, Brisson N, Basso B, Martre P, Aggarwal PK, Angulo C, Bertuzzi P, Biernath C, Challinor AJ, Doltra J, Gayler S, Goldberg R, Grant R, Heng L, Hooker J, Hunt LA, Ingwersen J, Izaurralde RC, Kersebaum KC, Müller C, Kumar NS, Nendel C, O'Leary G, Olesen JE, Osborne TM, Palosuo T, Priesack E, Ripoche D, Semenov MA, Shcherbak I, Steduto P, Stöckle C, Stratonovitch P, Streck T, Supit I, Tao F, Travasso M, Waha K, Wallach D, White JW, Williams JR, Wolf J (2013) Uncertainty insimulating wheat yields under climate change. Nat Clim Chang 3:827–832

Bange MP, Milroy SP (2004) Impact of short term exposure to cold temperatures on early development of cotton (*Gossypium hirsutum* L.). Aust J Agric Res 55:655–664

Bange MP, Caton SJ, Milroy SP (2008) Managing yields of high fruit retention in transgenic cotton (*Gossypium hirsutum* L.) using sowing date. Aust J Agric Res 59:733–741

Batool S, Saeed F (2017) Mapping the cotton value chain in Pakistan: a preliminary assessment for identification of climate vulnerabilities & pathways to adaptation. Sustainable Development Policy Institute, Islamabad, Pakistan, pp 1–60

Bhaskar KS, Rao MRK, Mendhe PN, Suryavanshi MR (2005) Micro irrigation management in cotton. CICR Technical Bulletin No. 31. Cent. Inst. Cott. Res., Nagpur, India

Bradow JM, Davidonis GH (2000) Quantitation of fiber quality and the cotton production-processing interface: a physiologist's perspective. J Cotton Sci 4:34–64

Braunack MV, Bange MP, Johnston DB (2012) Can planting date and cultivar selection improve resource use efficiency of cotton systems? Field Crop Res 137:1–11

Chastain DR, Snider JL, Choinski JS, Collins GD, Perry CD, Whitaker J, Grey TL, Sorensen RB, van Iersel M, Byrd SA, Porter W (2016) Leaf ontogeny strongly influences photosynthetic tolerance to drought and high temperature in Gossypium hirsutum. J Plant Physiol 199:18–28

Collins M, Knutti R, Arblaster J, Dufresne JL, Fichefet T, Friedlingstein P, Gao X, Gutowski WJ, Johns T, Krinner G, Shongwe M, Tebaldi C, Weaver AJ, Wehner M (2013) Long term climate change: projections, commitments and irreversibility. In: Stocker TF, Qin D, Plattner GK, Tignor M, Allen SK, Boschung J, Nauels A, Xia Y, Bex V, Midgley PM (eds) Climate change. The physical science basis. Contribution of working group I to the fifth assessment report of the Intergovernmental Panel on Climate Change. Cambridge University Press, Cambridge, UK, pp 1029–1036. https://doi.org/10.1017/CBO9781107415324.024

Conaty WC, Burke JJ, Mahan JR, Neilsen JE, Sutton BG (2012) Determining the optimum plant temperature of cotton physiology and yield to improve plant-based irrigation scheduling. Crop Sci 52:1828–1836

Constable GA, Bange MP (2006) What is cotton's sustainable yield potential? Aust Cotton Grower 26:6–10

Cottee NS, Tan DKY, Bange MP, Cothren JT, Campbell LC (2010) Multi-level determination of heat tolerance in cotton (*Gossypium hirsutum* L.) under field conditions. Crop Sci 50:2553–2564

Cotton Incorporated (2009) Summary of life cycle inventory data for cotton (field to bale – version 1.12 July 2009). Cary, NC, Cotton Incorporated, p 31

Dağdelen N, Başal H, Yılmaz E, Gürbüz T, Akcay S (2009) Different drip irrigation regimes affect cotton yield, water use efficiency and fiber quality in western Turkey. Agric Water Manag 96 (1):111–120

FAO (2010) "Climate-smart" agriculture policies, practices and financing for food security, adaptation and mitigation. Food and Agriculture Organization of the United State of America (FAO),

Rome, Italy, pp 1–49. http://www.fao.org/docrep/013/i1881e/i1881e00.htm. Accessed 21 Dec 2018

FAO (2013) Climate smart agriculture sourcebook. Food and Agriculture Organization of the United Nations (FAO), Rome. http://www.fao.org/docrep/018/i3325e/i3325e.pdf

FAO (2017) FAOSTAT Pakistan. Food and Agriculture Organization of the United Nations (FAO), Rome. http://www.fao.org/faostat/en

FAO (Food and Agriculture Organisation of the United Nations) (2012) The state of food and agriculture - investing in agriculture for a better future. FAO, Rome, Italy

FAO (Food and Agriculture Organisation of the United Nations) (2016) AQUASTAT website. http://www.fao.org/nr/water/aquastat/main/index.stm

Ghazala N, Rasul G (2009) Water requirement of wheat crop in Pakistan. Pak J Meteorol 6 (11):89–97

Gibbons MM, Fawcett CP, Warings RJ, Dearne K, Dampney P, Richardson SJ (2005) PLANET nutrient management decision support system–a standard approach to fertilizer recommendations. EFITA/WCCA, pp 25–28

GOP (2016) Final crop estimates. 2016, Crop Reporting Service, Governmement of Punjab, Lahore. http://crs.agripunjab.gov.pk

GOP (Government of Punjab) (2018) Cotton under drip irrigation system - a case study. Directorate General Agriculture (Water Managment), Government of Punjab, Lahore. http://ofwm.agripunjab.gov.pk

Grantz DA (2003) Ozone impacts on cotton: towards an integrated mechanism. Environ Pollut 126:331–344

Hillocks RJ (2005) Integrated management of insect pests, diseases and weeds of cotton in Africa. Integr Pest Manag Rev 1:31–47

Hoogenboom G (2000) Contribution of agrometeorology to the simulation of crop production and its applications. Agric For Meteorol 103:137–157

Huang J (2016) Different sowing dates affected cotton yield and yield components. Int J Plant Prod 10:63–84

Imran MA, Ali A, Ashfaq M, Hassan S, Culas R, Ma C (2018) Impact of climate smart agriculture (CSA) practices on cotton production and livelihood of farmers in Punjab. Pak Sustain 10:2101. https://doi.org/10.3390/su10062101

IPCC (2013a) Summary for policymakers. In: Climate change 2013. The physical science basis. Contribution of working group I to the fifth assessment report of the Intergovernmental Panel on Climate Change, p 33. https://doi.org/10.1017/CBO9781107415324

IPCC (2013b) Summary for policymakers. In: Stocker TF, Qin D, Plattner G-K, Tignor M, Allen SK, Boschung J, Nauels A, Xia Y, Bex V, Midgley PM (eds) Climate change 2013: the physical science basis. Contribution of working group I to the fifth assessment report of the Intergovernmental Panel on Climate Change. IPCC, Cambridge, UK. http://www.ipcc.ch/pdf/assessment-report/ar5/wg1/WGIAR5_SPM_brochure_en.pdf

IPCC (2014a) Climate change 2014. Synthesis report summary chapter for policymakers 31. IPCC. https://doi.org/10.1017/CBO9781107415324

IPCC (2014b) Summary for policymakers. In: C.B. Field, V.R. Barros, D.J. Dokken, K.J. Mach, M.D. Mastrandrea, T.E. Bilir, M. Chatterjee, K.L. Ebi, Y.O. Estrada, R.C. Genova, B. Girma, E.S. Kissel, A.N. Levy, S. MacCracken, P.R. Mastrandrea, and L.L. White Editor. Climate change 2014: impacts, adaptation, and vulnerability. Part A: Global and sectoral aspects. Contribution of working group II to the fifth assessment report of the Intergovernmental Panel on Climate Change, 2014, IPCC: Cambridge, UK. 1–32. http://ipcc-wg2.gov/AR5/images/uploads/WG2AR5_SPM_FINAL.pdf

ITC (2011) Trade map. International Trade Centre, Switzerland. http://www.trademap.org/

Jalota SK, Sood A, Chahal GBS, Choudhary BU (2006) Crop water productivity of cotton (*Gossypium hirsutum* L.) – wheat (*Triticum aestivum* L.) system as influenced by deficit irrigation, soil texture and precipitation. Agric Water Manag 84:137–146

Jones JW, Hansen JW, Royce FS, Messina CD (2000) Potential benefits of climate forecasting to agriculture. Agric Ecosyst Environ 82:169–184

Khan MB, Khaliq A, Ahmad S (2004) Performance of mashbean intercropped in cotton planted in different planting patterns. J Res (Sci) 15(2):191–197

Khatri-Chhetri A, Aryal JP, Sapkota TB, Khurana R (2016) Economic benefits of climate-smart agricultural practices to smallholder farmers in the Indo-Gangetic Plains of India. Curr Sci 110:1251–1256

Khatri-Chhetri A, Aggarwal PK, Joshi PK, Vyas S (2017) Farmers' prioritization of climate-smart agriculture (CSA) technologies. Agric Syst 151:184–191

Loi NK, Tangtham N (2005) Decision support system (DSS) for sustainable watershed management in Dong Nai Watershed-Vietnam. In: Synergistic approach to appropriate forestry technology for sustaining rainforest ecosystems: proceedings International Forestry Seminar. Universiti Putra, Malaysia

Luo Q, Bange M, Clancy L (2014) Cotton crop phenology in a new temperature regime. Ecol Model 285:22–29

Luo Q, Bange M, Braunack M, Johnston D (2016) Effectiveness of agronomic practices in dealing with climate change impacts in the Australian cotton industry–a simulation study. Agric Syst 147:1–9

Makhdum AH, Khan HN, Ahmad S (2011) Reducing cotton footprints through implementation of better management practices in cotton production; a step towards better cotton initiative. In: Proceedings of the fifth meeting of the Asian Cotton Research and Development Network, Lahore, Pakistan, 23–26 Feb 2011, pp 1–18. https://www.icac.org/tis/regional_networks/asian_network/meeting_5/. Accessed 4 Jan 2018

Manik SN, Pengilley G, Dean G, Field B, Shabala S, Zhou M (2019) Soil and crop management practices to minimize the impact of waterlogging on crop productivity. Front Plant Sci 10:140. https://doi.org/10.3389/fpls.2019.00140

Manpreet S, Gumber RK, Brar AS, Mukesh S (2007) Performance of drip irrigation systems in cotton in relation to lateral patterns and irrigation levels. In: World Cotton Research Conference-4, Lubbock, Texas, USA, 10–14 Sept 2007. International Cotton Advisory Committee (ICAC)

Mir SA, Quadri SMK (2009) Decision support systems: concepts, progress and issues–a review. In: Lichtfouse E (ed) Climate change, intercropping, pest control and beneficial microorganisms, Sustainable agricultural reviews, vol 2. Springer, Dordrecht, pp 373–399

Muthamilselvan M, Rangasamy K, Ananthakrishnan D, Manian R (2007) Mechanical picking of cotton: a review. Agric Rev 28:118–126

Nazli H, Orden D, Sarker R, Meilke K (2012) Bt cotton adoption and wellbeing of farmers in Pakistan. In: Proceedings of the International Association of Agricultural Economists (IAAE) Triennial Conference, Foz do Iguaçu, Brazil, 18–24 Aug 2012, pp 1–26. ageconsearch.umn.edu/bitstream/126172/

Pasha HA (2015) Growth of the provincial economies. Institute for Policy Reforms (IPR), Pakistan. http://ipr.org.pk/wp-content/uploads/2016/04/GROWTH-OF-PROVINCIAL-ECONOMICS-.pdf

Perumal NK, Hebba KB, Rao MRK, Singh P (2006) Physiological disorders in cotton. Central Institute for Cotton Research, Nagpur, pp 1–25

Pettigrew WT, Johnson JT (2005) Effects of different seeding rates and plant growth regulators on earlyplanted cotton. J Cotton Sci 9:189–198

Petzoldt C, Seaman A (2006) Climate change effects on insects and pathogens. Clim Change Agric: Promoting Practical Profitable Responses 3:6–16

Rahman H, Malik SA, Saleem M (2004) Heat tolerance of upland cotton during the fruiting stage evaluated using cellular membrane thermos stability. Field Crop Res 85:149–158

Rahman MH, Ahmad A, Wajid A, Hussain M, Akhtar J, Hoogenboom G (2016) Estimation of temporal variation resilience in cotton varieties using statistical models. Pak J Agric Sci 53:787–807

Rahman MH, Ahmad A, Wajid A, Hussain M, Rasul F, Ishaque W, Islam MA, Shelia V, Awais M, Ullah A, Wahid A, Sultana SR, Saud S, Khan S, Fahad S, Hussain M, Hussain S, Nasim W (2017) Application of CSMCROPGRO-Cotton model for cultivars and optimum planting dates: evaluation in changing semi-arid climate. Field Crop Res. https://doi.org/10.1016/j.fcr.2017.007

Rahman MH, Ahmad A, Wang X, Wajid A, Nasim W, Hussain M, Ahmad B, Ahmad I, Ali Z, Ishaque W, Awais M, Shelia V, Ahmad S, Fahad S, Alam M, Ullah H, Hoogenboom G (2018) Multi-model projections of future climate and climate change impacts uncertainty assessment for cotton production in Pakistan. Agric For Meteorol 253–254:94–113

Rane J, Nagarajan S (2004) High temperature index—for field evaluation of heat tolerance in wheat varieties. Agric Syst 79(2):243–255

Rasul G, Mahmood A, Sadiq A, Khan SI (2012) Vulnerability of the Indus Delta to climate change in Pakistan. Pak J Meteorol 8(16):89–107

Rasul F, Gull U, Rahman MH, Hussain Q, Chaudhary HJ, Matloob A, Shahzad S, Iqbal S, Shelia V, Masood S, Bajwa HM (2016) Biochar an emerging technology for climate change mitigation. J Environ Agric Sci 9:37–43

Raza A, Ahmad M (2015) Analysing the impact of climate change on cotton productivity in Punjab and Sindh, Pakistan. Pakistan Institute of Development Economics (PIDE), Islamabad, Pakistan, pp 1–32

Reddy KR, Zhao D (2005) Interactive effects of elevated CO_2 and potassium deficiency on photosynthesis, growth, and biomass partitioning of cotton. Field Crop Res 94:201–213

Reddy KR, Doma PR, Mearns LO, Boone MYL, Hodges HF, Richardson AG, Kakani VG (2002) Simulating the impacts of climate change on cotton production in the Mississippi delta. Clim Res 22:271–281

Reddy KR, Koti S, Davidonis GH, Reddy VR (2004) Interactive effects of carbon dioxide and nitrogen nutrition on cotton growth, development, yield, and fiber quality. Agronomy 96:1148–1157

Reddy KC, Malik RK, Reddy SS, Nayakatawa EZ (2007) Cotton growth and yield response to nitrogen applied through fresh and composted poultry litter. J Cotton Sci 11:26–34

Rosenzweig C, Elliott J, Deryng D, Ruane AC, Müller C, Arneth A, Boote KJ, Folberth C, Glotter M, Khabarov N, Neumann K, Piontek F, Pugh TAM, Schmid E, Stehfest E, Yang H, Jones JW (2014) Assessing agricultural risks of climate change in the 21st century in a global gridded crop model inter comparison. Proc Natl Acad Sci U S A 111:3268–3273

Ruane AC, Cecil LD, Horton RM, Gordón R, McCollum R, Brown D, Killough B, Goldberg R, Greeley AP, Rosenzweig C (2013) Climate change impact uncertainties for maize in Panama: farm information, climate projections, and yield sensitivities. Agric For Meteorol 170:132–145

Ruane AC, Goldberg R, Chryssanthacopoulos J (2015a) Climate forcing datasets for agricultural modeling: merged products for gap-filling and historical climate series estimation. Agric For Meteorol 200:233–248

Ruane AC, Winter JM, McDermid SP, Hudson NI (2015b) AgMIP climate datasets and scenarios for integrated assessment. In: Rosenzweig C, Hillel D (eds) Handbook of climate change and agroecosystems: the agricultural model inter comparison and improvement project (AgMIP) integrated crop and economic assessments, Part 1, ICP series on climate change impacts, adaptation, and mitigation, vol 3. Imperial College Press, London, pp 45–78

Sala S (2009) Information and communication technologies for climate change adaptation, with a focus on the agricultural sector. In: Thinkpiece for CGIAR science forum workshop on "ICTs transforming agricultural science, research and technology generation" Wageningen, Netherlands, pp 16–17

Singh RP, Vara Prasad PV, Sunita K, Giri SN, Reddy KR (2007) Influence of high temperature and breeding for heat tolerance in cotton: a review. Adv Agron 93:313–385

Tariq M, Yasmeen A, Ahmad S, Hussain N, Afzal MN, Hasanuzzaman M (2017) Shedding of fruiting structures in cotton: factors, compensation and prevention. Trop Subtrop Agroecosyst 20(2):251–262

Tariq M, Afzal MN, Muhammad D, Ahmad S, Shahzad AN, Kiran A, Wakeel A (2018) Relationship of tissue potassium content with yield and fiber quality components of Bt cotton as influenced by potassium application methods. Field Crop Res 229:37–43

Turner NC (1979) Drought resistance and adaptation to water deficits in crop plants. In: Mussell H, Staples RC (eds) Stress physiology in crop plants. Wiley, New York, pp 343–372

USDA (United States Department of Agriculture) (2018) Pakistan sugar annual report. Global Agricultural Information Network. https://gain.fas.usda.gov

Usman M, Ahmad A, Ahmad S, Irshad M, Khaliq T, Wajid A, Hussain K, Nasim W, Chattha TM, Trethowan R, Hoogenboom G (2009) Development and application of crop water stress index for scheduling irrigation in cotton (*Gossypium hirsutum* L.) under semiarid environment. J Food Agric Environ 7(3&4):386–391

Van Vuuren DP, Edmonds J, Kainuma M et al (2011) The representative concentration pathways: an overview. Clim Chang 109:5. https://doi.org/10.1007/s10584-011-0148-z

Vara Prasad PV, Allen LH, Boote KJ (2005) Crop responses to elevated carbon dioxide and interactions with temperature. J Crop Improv 13:113–155

Wajid A, Ahmad A, Hussain M, Rahman MH, Khaliq T, Mubeen M, Rasul F, Bashir U, Awais M, Iqbal J, Sultana SR, Hoogenboom G (2014) Modeling growth, development and seed-cotton yield for varying nitrogen increments and planting dates using DSSAT. Pak J Agric Sci 51:639–647

Walker P, White DH (2001) INSIGHT: a framework for determining impacts of changes in land and water policies. Environ Int 27:127–132

Wang S, Li X, Lu J, Hong J, Chen G, Xue X, Li J, Wei Y, Zou J, Liu G (2013) Effects of controlled-release urea application on the growth, yield and nitrogen recovery efficiency of cotton. Agric Sci 4:33–38

Watto MA, Mugera AW (2015) Econometric estimation of groundwater irrigation efficiency of cotton cultivation farms in Pakistan. J Hydrol Reg Stud 4:193–211

Williams A, White N, Mushtaq S, Cockfield G, Power B, Kouadio L (2015) Quantifying the response of cotton production in Eastern Australia to climate change. Clim Chang 129:183–196

Woodfin CA, Rosenow DT, Clark LE, Johnson JW (1979) Differential response of sorghum cultivars to drought stress. Am Soc Agron Abstr, p 82

Wrather JA, Phipps BJ, Stevens WE, Phillips AS, Vories ED (2008) Cotton planting date and plant population effects on yield and fiber quality in the Mississippi Delta. J Cotton Sci 12:1–7

Zhu T, Ringler C, Iqbal MM, Sulser TB, Goheer MA (2013) Climate change impacts and adaptation options for water and food in Pakistan: scenario analysis using an integrated global water and food projections model. Water Int 38:651–669

Zulfiqar F, Datta A, Thapa GB (2017) Determinants and resource use efficiency of "better cotton": an innovative cleaner production alternative. J Clean Prod 166:1372–1380

Chapter 23
Cotton Ontogeny

Muhammad Tariq, Ghulam Abbas, Azra Yasmeen, and Shakeel Ahmad

Abstract The chapter deals with fundamental transformation in cotton during various developmental phases from germinating seeds to formation of mature seeds. Initially, the complex metabolic changes occur in seeds followed by rehydration. The storage lipids and protein act as a source of energy during germination process and following energy requirement is fulfilled from photosynthesis. The cotton crop enters a reproductive growth following a 4–5 weeks of germination. The flavonoids are responsible for appearance of flower's colour. The nutrient uptake increases from flowering to fruiting and declines at maturity. Depending upon the mobility, the nutrients accumulated in leaves are translocated to developing bolls. The developing seeds start accumulation of oil and protein. Among three boll components, the fibres and seed coat are rich in cellulose and starch, respectively. Oil and protein are accumulated in the developing embryos. The oil contents increase with boll age and starch initially increases and then drops down. However, protein contents are initially high and drop down during second week of anthesis and then followed by steady increase. The auxin and ethylene production in developing bolls decreases with age.

Keywords Germination · Establishment · Seed · Lipids · Protein · Nutrients

---

M. Tariq (✉)
Agronomy Section, Central Cotton Research Institute, Multan, Pakistan
e-mail: mtariq131@ccri.gov.pk

G. Abbas · A. Yasmeen
Department of Agronomy, Bahauddin Zakariya University, Multan, Pakistan

S. Ahmad
Department of Agronomy, Faculty of Agricultural Sciences and Technology, Bahauddin Zakariya University, Multan, Pakistan

S. Ahmad, M. Hasanuzzaman (eds.), *Cotton Production and Uses*,
https://doi.org/10.1007/978-981-15-1472-2_23

23.1 Introduction

The word "ontogeny" has Greek origin which means mode of production and it depicts the history of structural changes. Because it deals with studies on developmental aspects of organism, it is also called as ontogenesis or morphogenesis. Morphogenesis is the basic process for appearance of various plant structures and every developmental stage is characterized with specific process (Usman et al. 2009; Ahmad et al. 2014, 2018; Abbas and Ahmad 2018; Ahmad and Raza 2014; Ali et al. 2011, 2013a, b, 2014a, b). Therefore, it covers the important aspects of development during the life span of organisms. The term differs with phylogeny in a sense that phylogeny depicts the evolutionary history and ontogeny reflects the developmental changes from fertilized eggs to mature organisms. The ontogeny is an important aspect of plant development where plants go a sequence of development and transition from simple to more complex structure. Various structures, metabolic events and functions appeared and diminished during the course of development starting from germination seeds to formation of mature seeds and fibres.

The apparent developmental changes in cotton include seedling establishment from seeds, formation of squares and flowers and further boll development processes. These visible changes are accompanied with degradation of previous compounds and formation of some new biomolecules. For example, lipids and protein synthesized during seed development are degraded during germination (Thakare et al. 2014), synthesis of flavonoids for flower colour during flowering (Tan et al. 2013) and again accumulation of oil, protein, starch and cellulose during seed development (Leffler 1986b; Ruan 2005). Besides these, the nutrient uptake and their further compartmentation in various plant parts also changes with crop age (Leffler 1986a; Mullins and Burmester 2010). The main objective of the chapter was to give a basic overview of processes being carried out during various development phases of cotton.

23.2 Germination and Seedling Ontogeny

The metabolic events during seed germination are the most complex phenomenon in the plant life cycle. The stored products are mobilized and transported and new compounds are developed. The water uptake in cotton seed takes place from chalazal aperture. The seed germination starts with rehydration which generally completes within 4–6 h. It is derived from material potential difference between cell wall and cellular contents of the seed. The rate of respiration is relatively high during this phase. The water uptake is reduced later on and accompanied with higher metabolic activities in the seed. Again, water uptake and respiration rate increase during the protrusion of radicle (true germination). The root growth is faster at initial stages than shoot growth. At very initial stages, the stored lipids and proteins act as a source of energy for germinating seeds. The amylase, protease and lipase are three hydrolytic enzymes involved in the cotton seed germination. Amylase and protease degrade the starch and protein, respectively, while lipase hydrolyses the triacylglycerol into fatty acid and glycerol (Thakare et al. 2014). The lipid reservoirs are mobilized following the radicle protrusion and utilized by the emerging seedlings within a period of week. Glyoxylate cycle increases when developing seedlings depend upon the reservoirs of the cotyledon and photosynthesis is going to start. The lipid metabolism came to an end with appearance of true leaves. The major portion of starch is present in stems and roots prior to reproductive growth (Taliercio et al. 2010). The significant portion of carbohydrates is translocated to developing roots; hence, initially aboveground growth is slower than roots. The cotton seed primarily develops roots, stems and leaves which continue to grow along with reproductive developments. The root extension continues up to first bloom and declines later on.

23.3 Flowering Ontogeny

The apical meristem of the main stem received signal from leaves to the primary axis to shift from vegetative to reproductive branch. On account of indeterminate growth habit, the reproductive life span of cotton may last for several months provided that genotypes and environment are supportive. Therefore, the cotton observes a range of environmental conditions through variable blooming dates.

The leaves accumulate dry matter with the square development and reach at peak when square transformed into white flower. The dry weight of square bract also increases with square development. The initial 15-day dry matter accumulation in floral bud is slow which increases three times during the last week of development. Initially, the hexose, sucrose and starch (non-structural carbohydrates) concentration in floral buds do not change with time which is followed by significant increase at anthesis. The leaves are rich in starch during floral bud development (Zhao and Oosterhuis 2000). The increasing concentration of such assimilates with square aging actually depicts the translocation out of leaves.

The dominant colour scheme in various cotton germplasm are creamy, yellow and red in order of abundance. The cream and yellow colour are most common in *Gossypium hirsutum* L. followed by white. The flower colour change after anthesis is a common feature of cotton species and generally visible on the second day. The appearance of various flower colours and post-anthesis dramatic colour changes are due to flavonoid (flavonol and anthocyanin) contents (Tan et al. 2013). Hence, biosynthesis and accumulation of anthocyanin is the major cause of dramatic flower colour changes, although anthocyanin accumulation is genetically controlled with the limited effect from light.

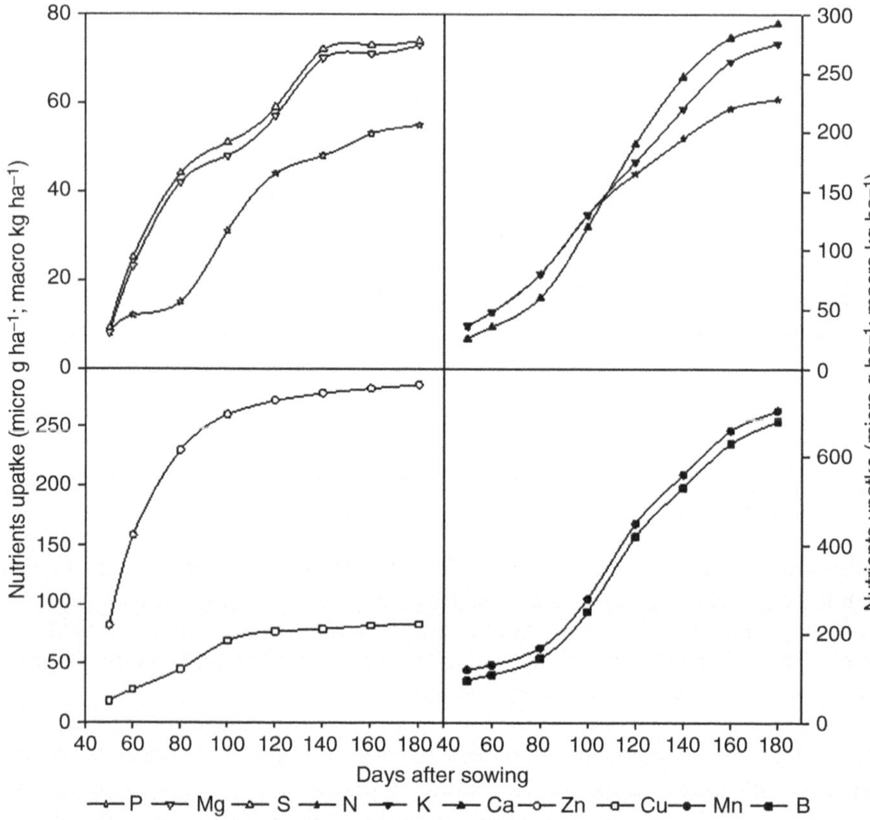

Fig. 23.1 Macronutrient (N, P, K, Ca, S, Mg) and micronutrient (Mn, B, Zn and Cu) seasonal uptake pattern for cotton crop. (Source: Modified and adapted from Rochester et al. 2012)

23.3.1 Ontogeny of Nutrients and Dry Matter

The rate of nutrient uptake varies with crop age and mainly dictated by biomass accumulation and fruit load (Ahmad et al. 2017; Amin et al. 2017, 2018; Khan et al. 2004; Rahman et al. 2018; Tariq et al. 2017, 2018). The nutrient uptake remains slow during initial growth stage and accelerated from flowering to fruiting followed by decline at maturity (Mullins and Burmester 2010). The typical uptake curve of various nutrients is given in Fig. 23.1. The leaf nitrogen, phosphorus, potassium, ferric, copper and zinc are negatively affected by crop age and calcium, magnesium, sodium, manganese, sulphur and boron increase with age (Rochester et al. 2012). The decreasing nutrient concentration in the leaves may indicate the translocation to developing bolls. The nutrient compartments among boll components (seed, lint and walls) revealed that the significant portion of nitrogen, phosphorus, magnesium, ferric, copper and zinc is present in seeds during flower transformation to bolls. The wall of bolls accumulates more potassium, calcium and sulphur during boll development. The lint is rich in cellulose and only minor quantity of these nutrients is translocated to lint (Rochester et al. 2012). However, the translocation of the nutrients in plants is also affected by mobility of nutrients. The increasing concentration of nutrients in developing seeds indicates the translocation out of bur and fibre (Leffler 1986a).

The contribution of various plant parts (leaves, stem and fruiting parts) in total dry matter production varies with crop age. Initially, the leaves account much more for dry matter which decline during crop maturity. The fruiting parts serve as a principal component of dry matter after 120 days of crop emergence. Similarly, stem dry matter also increases with age without any reduction during crop cycle. The accumulation of dry matter is slow for initial 2 months and then gets accelerated.

23.4 Seed and Fibre Ontogeny

The boll development starts following pollination which is classified into enlargement, filling and maturation phase. The enlargement phase is comparatively a longer phase which lasts about 3 weeks and fibre produced on seed gets longer. As the boll size and fibre elongation slow down, the filling stage gets started. The filling period requires relatively less time and ceased about 10–15 days before boll opening. The boll components also accumulate dry weight and secondary wall of fibre is formed during this phase. The developing seed also accumulates oil and protein during filling stage. No significant changes in dry matter of boll components occur during maturation phase. The nutrient concentration initially drops down and then increases during filling period (Leffler and Tubertini 1976). The oil and protein concentration of developing seeds increases with age, while starch concentration initially increases after anthesis and declines later on (Fig. 23.2). Among the soluble sugar, maltose remains dominant sugar during seed development (Leffler 1986b). Cellulose is accumulated in the fibre and starch in the seed coat, while oil and protein are accumulated in the embryo during the process of seed development (Fig. 23.3). The bolls attain its maximum weight and size during maturation phase. The fibre mass accounts for 40–45% of total seed weight on dry weight basis. During secondary wall thickening of fibre, two carbon storage products are actively synthesized, i.e. cellulose in the fibre and oil in the embryo. The sucrose translocated from leaves serves as the energy source to carry out biosynthesis of these products. The sucrose moves in two directions from the seed coat, inward for the inner seed coat and embryo, outward for fibres. The inward translocation is used for synthesis of lipids and protein in the embryo (Ruan 2005).

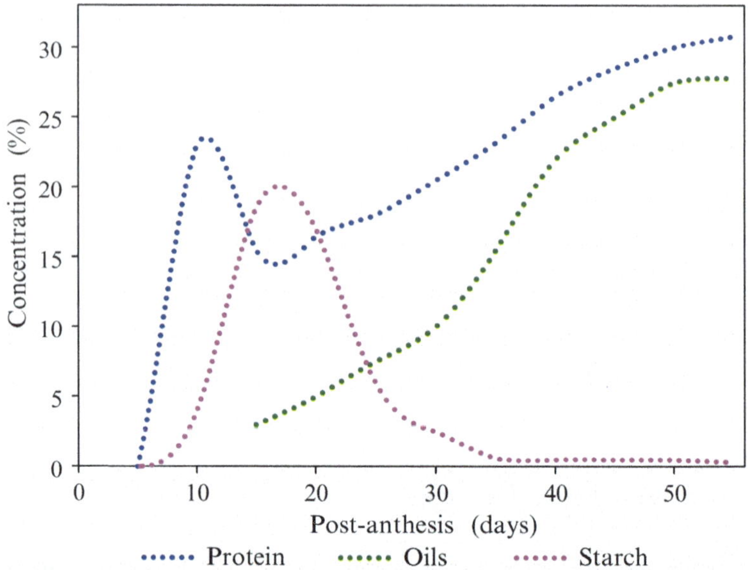

Fig. 23.2 Protein, oil and starch accumulation trends in developing cotton bolls. (Source: Modified and adapted from Leffler 1986b)

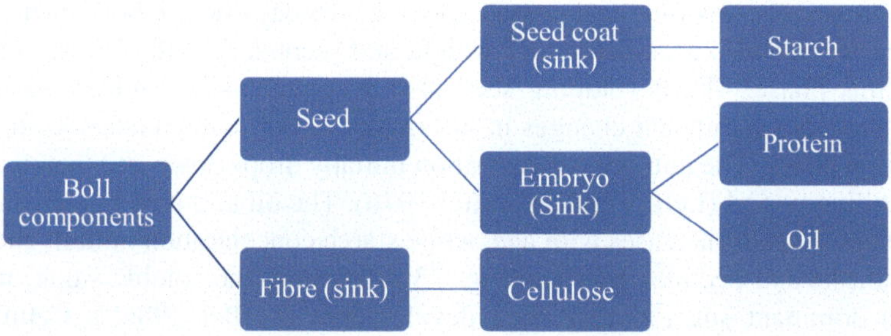

Fig. 23.3 The major sink for cellulose, starch, protein and oil in developing boll

23.4.1 Carbohydrate Dynamics During Seed Ontogeny

The carbohydrates are translocated from leaves to the seed coat and endosperm of developing bolls at very early stages. The starch present in the seed coat and endosperm is also translocated to the embryo. Later on, the starch present in the embryo is also declined and metabolized into complex sugar. The import of carbohydrates is ceased at maturity (about 45 days post-anthesis) and non-structural carbohydrates are only rearranged (Benedict et al. 1976). The oil is another important component of cotton seed which is mainly composed of triacylglycerol molecules. The biosynthesis of triacylglycerol containing fatty acids is carried out in the stroma of plastids. The detail information has been mentioned in various literatures (Browse and Ohlrogge 1995; Xu et al. 2016). Two types of protein including globulins and albumins are present in cotton seed. These are synthesized and stored in protein storage vacuoles (Dure III and Chlan 1981). Gossypol is an important phenolic compound in the cotton seed which may be free or in bonded form with protein. The free form of gossypol is toxic, particularly for non-ruminant animals.

23.4.2 Hormonal Dynamics During Seed and Fibre Ontogeny

The fibres are originated from ovule epidermal cells and initiation takes place from the day of anthesis and completed in 5 days post-anthesis. The major steps in fibre development include initiation, elongation, secondary wall thickening and maturation. Indole acetic acid, gibberellic acid and brassinosteroids are required for initial fibre development (Yang et al. 2006). The positive impact of indole acetic acid on fibre initiation (Zhang et al. 2011), gibberellic acid both for fibre initiation and elongation (Liao et al. 2009) and brassinosteroids for fibre development (Sun and Allen 2005) has been reported. The abscisic acid decreased fibre elongation (Dasani and Thaker 2006). Although reactive oxygen species (ROS) are produced during stress in plants, the role in fibre initiation has been witnessed in literature (Zhang et al. 2010). Ca ions are also accumulated during fibre elongation (Stiff and Haigler 2012). The turgor pressure and callose synthesis is increased during elongation. The aquaporin protein also facilitates the turgor-driven elongation. An increase in fibre length up to 12.7% has been reported through application of brassinolide (Sun and Allen 2005). The auxin concentration in developing bolls decreases with age from second week of anthesis (Rodgers 1981). The ethylene production is negatively affected by age of developing bolls (Guinn 1982).

23.5 Conclusion

The series of metabolic events are carried out along with phenological transition during cotton life span. Some biomolecules are degraded, translocated and converted to other forms. Therefore, the profile of biomolecules changes with the age of crop. The leaves act as major sink for nutrients prior to onset of reproductive development and later on developing bolls become major sink. The major metabolic changes occur during seed germination and seed development process. The biomolecules accumulated during seed development are degraded during seed germination. The seed and fibre are important boll components which act as sink for cellulose and starch, protein and oil, respectively. The concentration of these biomolecules varies along with the age of developing bolls.

References

Abbas Q, Ahmad S (2018) Effect of different sowing times and cultivars on cotton fiber quality under stable cotton-wheat cropping system in southern Punjab, Pakistan. Pak J Life Soc Sci 16:77–84

Ahmad S, Raza I (2014) Optimization of management practices to improve cotton fiber quality under irrigated arid environment. J Food Agric Environ 2(2):609–613

Ahmad S, Raza I, Ali H, Shahzad AN, Atiq-ur-Rehman, Sarwar N (2014) Response of cotton crop to exogenous application of glycinebetaine under sufficient and scarce water conditions. Braz J Bot 37(4):407–415

Ahmad S, Abbas Q, Abbas G, Fatima Z, Atique-ur-Rehman, Naz S, Younis H, Khan RJ, Nasim W, Habib urRehman M, Ahmad A, Rasul G, Khan MA, Hasanuzzaman M (2017) Quantification of climate warming and crop management impacts on cotton phenology. Plants 6(7):1–16

Ahmad S, Iqbal M, Muhammad T, Mehmood A, Ahmad S, Hasanuzzaman M (2018) Cotton productivity enhanced through transplanting and early sowing. Acta Sci Biol Sci 40:e34610

Ali II, Afzal MN, Ahmad F, Ahmad S, Akhtar M, Atif R (2011) Effect of sowing dates, plant spacing and nitrogen application on growth and productivity on cotton crop. Int J Sci Eng Res 2 (9):1–6

Ali H, Abid SA, Ahmad S, Sarwar N, Arooj M, Mahmood A, Shahzad AN (2013a) Integrated weed management in cotton cultivated in the alternate-furrow planting system. J Food Agric Environ 11(3&4):1664–1669

Ali H, Abid SA, Ahmad S, Sarwar N, Arooj M, Mahmood A, Shahzad AN (2013b) Impact of integrated weed management on flat-sown cotton (*Gossypium hirsutum* L.). J Anim Plant Sci 23 (4):1185–1192

Ali H, Hameed RA, Ahmad S, Shahzad AN, Sarwar N (2014a) Efficacy of different techniques of nitrogen application on American cotton under semi-arid conditions. J Food Agric Environ 12 (1):157–160

Ali H, Hussain GS, Hussain S, Shahzad AN, Ahmad S, Javeed HMR, Sarwar N (2014b) Early sowing reduces cotton leaf curl virus occurrence and improves cotton productivity. Cer Agron Moldova XLVII(4):71–81

Amin A, Nasim W, Mubeen M, Nadeem M, Ali L, Hammad HM, Sultana SR, Jabran K, Habib urRehman M, Ahmad S, Awais M, Rasool A, Fahad S, Saud S, Shah AN, Ihsan Z, Ali S, Bajwa AA, Hakeem KR, Ameen A, Amanullah, Rehman HU, Alghabar F, Jatoi GH, Akram M, Khan A, Islam F, Ata-Ul-Karim ST, Rehmani MIA, Hussain S, Razaq M, Fathi A (2017) Optimizing the phosphorus use in cotton by using CSM-CROPGRO-cotton model for semi-arid climate of Vehari-Punjab, Pakistan. Environ Sci Pollut Res 24(6):5811–5823

Amin A, Nasim W, Mubeen M, Ahmad A, Nadeem M, Urich P, Fahad S, Ahmad S, Wajid A, Tabassum F, Hammad HM, Sultana SR, Anwar S, Baloch SK, Wahid A, Wilkerson CJ, Hoogenboom G (2018) Simulated CSM-CROPGRO-cotton yield under projected future climate by SimCLIM for southern Punjab, Pakistan. Agric Syst 167:213–222

Benedict CK, Kohel RJ, Schubert AM (1976) Transport of 14C-assimilates to cottonseed: integrity of funiculus during seed filling stage. Crop Sci 16:23–21

Browse J, Ohlrogge JB (1995) Lipid biosynthesis. Plant Cell 7:957–970

Dasani SH, Thaker VS (2006) Role of abscisic acid in cotton fiber development. Russ J Plant Physiol 53:62–67

Dure L III, Chlan C (1981) Developmental biochemistry of cottonseed embryogenesis and germination. XII. Purification and properties of principal storage proteins. Plant Physiol 68:180–186

Guinn G (1982) Fruit age and changes in abscisic acid content, ethylene production, and abscission rate of cotton fruits. Plant Physiol 69:349–352

Khan MB, Khaliq A, Ahmad S (2004) Performance of mashbean intercropped in cotton planted in different planting patterns. J Res (Sci) 15(2):191–197

Leffler HR (1986a) Mineral compartmentation within the boll. USDA-ARS Stoneville, Mississippi. In: Mauney JR, Stewart JMD (eds) Cotton physiology, The Cotton Foundation reference book series, vol 1. The Cotton Foundation, Memphis, TN

Leffler HR (1986b) Developmental aspects of planting seed quality. USDA-ARS Stoneville, Mississippi. In: Mauney JR, Stewart JMD (eds) Cotton physiology, The Cotton Foundation reference book series, vol 1. The Cotton Foundation, Memphis, TN

Leffler HR, Tubertini BS (1976) Development of cotton fruit. II. Accumulation and distribution of mineral nutrients. Agron J 68:858–861

Liao W, Ruan M, Cui B, Xu N, Lu J, Peng M (2009) Isolation and characterization of aGAI/RGA-like gene from *Gossypium hirsutum* L. Plant Growth Regul 58:35–45

Mullins GL, Burmester CH (2010) Relation of growth and development to mineral nutrition. In: Stewart JM, Oosterhuis DM, Heitholt JM, Mauney JR (eds) Physiology of cotton. Springer, New York. http://www.springerlink.com/index/u88838767k45m032.pdf

Rahman MH, Ahmad A, Wang X, Wajid A, Nasim W, Hussain M, Ahmad B, Ahmad I, Ali Z, Ishaque W, Awais M, Shelia V, Ahmad S, Fahad S, Alam M, Ullah H, Hoogenboom G (2018) Multi-model projections of future climate and climate change impacts uncertainty assessment for cotton production in Pakistan. Agric For Meteorol 253–254:94–113

Rochester IJ, Constable GA, Oosterhuis DM, Errington M (2012) Nutritional requirements of cotton during flowering and fruiting: flowering and fruiting in cotton. In: Oosterhuis DM, Cothren JT (eds) Flowering and fruiting in cotton. The Cotton Foundation, Cordova, TN

Rodgers JP (1981) Cotton fruit development and abscission variations in the level of auxins. Trop Agr (Trin) 58:63–72

Ruan YL (2005) Recent advances in understanding cotton fibre and seed development. Seed Sci Res 15:269–280

Stiff MR, Haigler CH (2012) Recent advances in cotton fiber development. In: Oosterhuis DM, Cothren JT (eds) Flowering and fruiting in cotton. Cordova, TN, The Cotton Foundation

Sun Y, Allen RD (2005) Functional analysis of the BIN2 genes of cotton. Mol Genet Genomics 274:51–59

Taliercio E, Kwanyen P, Scheffler J (2010) Nitrogen metabolism in cotton stems and roots during reproductive development. J Cotton Sci 14:107–112

Tan J, Wang M, Tu L, Nie Y, Lin Y, Zhang X (2013) The flavonoid pathway regulates the petal colors of cotton flower. PLoS One 8(8):e72364

Tariq M, Yasmeen A, Ahmad S, Hussain N, Afzal MN, Hasanuzzaman M (2017) Shedding of fruiting structures in cotton: factors, compensation and prevention. Trop Subtrop Agroecosyst 20(2):251–262

Tariq M, Afzal MN, Muhammad D, Ahmad S, Shahzad AN, Kiran A, Wakeel A (2018) Relationship of tissue potassium content with yield and fiber quality components of Bt cotton as influenced by potassium application methods. Field Crop Res 229:37–43

Thakare HS, Kumar V, Singh C (2014) Effect of different hydrolytic enzymes on germination of inter and intra specific cotton hybrids and parents. Bioscan 9(3):943–946

Usman M, Ahmad A, Ahmad S, Irshad M, Khaliq T, Wajid A, Hussain K, Nasim W, Chattha TM, Trethowan R, Hoogenboom G (2009) Development and application of crop water stress index for scheduling irrigation in cotton (*Gossypium hirsutum* L.) under semiarid environment. J Food Agric Environ 7(3&4):386–391

Xu Z, Li J, Guo X, Jin S, Zhang X (2016) Metabolic engineering of cottonseed oil biosynthesis pathway via RNA interference. Sci Rep 6:1–14

Yang SS, Cheung F, Lee JJ, Ha M, Wei NE, Sze SH, Stelly DM, Thaxton P, Triplett B, Town CD, Chen ZJ (2006) Accumulation of genome-specific transcripts, transcription factors and phytohormonal regulators during early stages of fiber cell development in allotetraploid cotton. Plant J 47:761–775

Zhang D, Zhang T, Guo W (2010) Effect of H_2O_2 on fiber initiation using fiber retardation initiation mutants in cotton (*Gossypium hirsutum* L.). J Plant Physiol 167:393–399

Zhang M, Xheng X, Song S, Zeng Q, Hou L, Li D, Zhao J, Wei Y, Li X, Luo M, Xiao Y, Luo X, Zhang J, Xiang C, Pei Y (2011) Spatiotemporal manipulation of auxin biosynthesis in cotton ovule epidermal cells enhances fiber yield and quality. Nat Biotechnol 29:453–458

Zhao D, Oosterhuis DM (2000) Dynamics of non-structural carbohydrates in developing leaves, bracts and floral buds of cotton. Environ Exp Bot 43:185–195

Chapter 24
Molecular Breeding of Cotton for Drought Stress Tolerance

Muhammad Asif Saleem, Abdul Qayyum, Waqas Malik, and Muhammad Waqas Amjid

Abstract Availability of freshwater to crops is declining year by year, so there is a need to exploit genetic mechanism of drought tolerance in crops including cotton. Drought-tolerant cultivars may be tailored if generated data regarding inheritance of drought-related traits to tolerance is practically used. The complexity in inheritance of drought tolerance has been a main reason of slow progress. Although a lot of conventional and nonconventional research work has been conducted for the traits related to abiotic stress in cotton, fruitful field results have not been obtained. There is a need to understand drought stress and mechanisms adopted by cotton against drought stress. These include morphological, physiological, biochemical, and genetic responses in cotton. Identification of the important genes related to drought tolerance would also be a major contribution. The impactful genes and major QTLs could be stacked in a single cotton plant, using gene pyramiding, and this may produce the future cotton plant for upcoming adverse environment.

Keywords Gene pyramiding · QTL · Upland cotton

Abbreviations

AA	Ascorbic acid
ABA	Abscisic acid
APX	Ascorbate peroxidase
CAT	Catalase
EL	Electrolyte leakage
GA	Gibberellic acid
GPX	Guaiacol peroxidase
GR	Glutathione reductase
GWAS	Genome-wide association study

M. A. Saleem (✉) · A. Qayyum · W. Malik
Department of Plant Breeding and Genetics, Bahauddin Zakariya University, Multan, Pakistan
e-mail: drasifsaleem@bzu.edu.pk

M. W. Amjid
Department of Agriculture, Bacha Khan University, KPK, Pakistan

S. Ahmad, M. Hasanuzzaman (eds.), *Cotton Production and Uses*,
https://doi.org/10.1007/978-981-15-1472-2_24

H_2O_2	Hydrogen peroxide
$HO^•$	Hydroxyl radical
IPT	Isopentenyltransferase
JA	Jasmonic acid
LEA	Late embryogenesis abundant
MAS	Marker-assisted selection
$NO^•$	Nitric oxide
$^1O_2$	Singlet oxygen
$O_2^{•-}$	Superoxide anion radical
ROS	Reactive oxygen species
SA	Salicylic acid

24.1 Introduction

Origin of cotton is one of the most incredible stories in archives of crop origin, evolution and domestication. *Gossypium* comprises of more than 50 species (45 diploid, 5 tetraploid) distributed to tropical, subtropical, and semiarid regions worldwide with eight genomes (A, B, C, D, E, F, G, and K) with four cultivated species from family Malvaceae. *Gossypium hirsutum* and *Gossypium barbadense* are tetraploid (AADD) while *Gossypium herbaceum* and *Gossypium arboreum* are diploid (AA) cultivated species of cotton (Wendel and Cronn 2003).

The key purpose for cultivating cotton is fiber and oil (Ahmad et al. 2014, 2017, 2018; Abbas and Ahmad 2018; Ahmad and Raza 2014; Ali et al. 2011, 2013a, b, 2014a, b; Usman et al. 2009). Seed cotton is ginned to separate the fiber from seed. Separated fiber is a spinnable resource of yarn that is knitted into fabrics. Textile industry depends on fabrics for clothes, towels, and additional households. Cottonseed is composed of 45% meal, 28% hull, 17% crude oil, and 10% short fibers (Smith 1995). Cotton meal comprises of high content of protein (41%) and being used in animal feed. Cotton oil is used as edible oil, in plastics, and soaps.

Upland cotton is an important cash crop with a remarkable impact on economy of China, India, United State, Brazil, and Pakistan as these are top cotton-producing countries (Statista 2018). There are certain limitations to higher production of cotton because cotton is highly sensitive to abiotic stresses like salt, drought, and heat and to biotic issues such as diseases and pests. Among abiotic stresses drought and heat stress are causing frequent economic losses worldwide. Climate is changing due to global warming which may increase the atmosphere temperature by 1.8–4.0 °C in the next decade and may drastically affect cotton fiber parameters (Dai et al. 2017; Amin et al. 2017, 2018; Khan et al. 2004; Rahman et al. 2018; Tariq et al. 2017, 2018). Future climate change escalation will bring heat stress along with drought episodes in most parts of the world. Climate fluctuation accompanied with emissions of greenhouse gases in the atmosphere is predominantly responsible occurrence of abiotic stresses especially drought and heat stress. Among abiotic stresses, drought is more limiting factor to crop productivity. Drought can adversely reduce yield and productivity depending upon severity, growth stage, and extent of occurrence of drought. Cotton plant handles drought stress by different built-in morphological, physiological, biochemical, and molecular responses to minimize lint yield reduction (Khan et al. 2018). Reduced fresh water availability and upcoming climate change effects are making condition more adverse for cotton around the world.

24.2 Drought Stress

As the moisture level in soil decreases the tissue water level in plants is also decreased. Drought stress begins with insufficiency of fresh water in plant to fulfill transpirational demands. Drought stress affects the physiology of plants by cellular and molecular mechanisms. Water contents in plants are linked to moisture contents in soil. Lesser water content and less leaf turgidity resulted in augmented stomatal resistance that lowers photosynthesis affecting growth of plants.

Tissue water plays a vital role in biochemical processes and standard functioning is extremely disturbed under drought (Burke and O'Mahony 2001). Under drought, because of lesser stomatal conductance, CO_2 uptake is decreased. This decrease in intercellular CO_2 concentration leads to reduced photosynthesis. Under severe water shortage, photosystem activity is affected leading to lower photosynthetic rate combined with photorespiration. The process makes way for the production of ROS that cause damage to ATP synthase, which reduces RuBP production in Calvin cycle and production of sucrose synthesis. Impaired photosynthesis hinders plants from normal growth (Lawlor and Tezara 2009). Drought affects productivity particularly when it happens at critical development stages (Salehi-Lisar and Bakhshayeshan-Agdam 2016).

Plants have natural ability to respond drought stress by some built-in physiological and molecular changes. These changes depend upon severity and extent of drought stress along stage of plant development. In general, plants adopt mechanisms (escape, avoidance, resistance, and tolerance) to endure in drought (Mitra 2001). Mechanism in plants which enables them to complete their life cycle in promising moisture circumstances before onset of meteorological drought is called drought escape. When the amount of water through transpiration exceeds than absorption, it causes dehydration. Plants can avoid dehydration (drought avoidance) by some physiological and morphological mechanisms to maintain balance between loss and uptake of water. These mechanisms are stomatal closure, leaf rolling, and changes in leaf angle, shedding of older leaves, cuticle thickness, and small canopy leaf area by reducing growth. However when drought prolongs, avoidance becomes difficult for plants. Plant can also resist drought stress by root enhancement and osmotic adjustment. Capability of crops to endure low water contents is termed drought tolerance (Ullah et al. 2017).

24.3 Impact on Cotton

Cotton is relatively more sensitive to drought compared to other field crops. In cotton, reduced water level in cellular tissues reduces photosynthesis rate due to increased production of reactive oxygen species. Amount of chlorophyll, carotenoids, proteins, and starch contents significantly reduced in water shortage condition (Parida et al. 2008). Severe drought stress increases the levels of glutathione-*S*-transferase, superoxide dismutase, and proline increase along with their gene expression while that of malondialdehyde content decreased (Singh et al. 2015). In cotton, drought stress affects photosynthesis due to stomatal resistance as well as causes non-stomatal limitations to photosynthesis in case of moderate to severe stress of water (Ennahli and Earl 2005).

Drought stress decreases leaf water content, cell membrane permeability, chlorophyll a, chlorophyll b, accumulation of biomass, leaf water content, and actual quantum yield of photosystem II (Wang et al. 2007). Low leaf water contents affects cell expansion (Schonfeld et al. 1988) and retards leaf, stem, root elongation as well as reduces the number of floral buds. The overall development of cotton plant, either vegetative or non-vegetative, halted under drought stress.

The affected plant may be recognized by reduced plant height, shedding of bolls, reduced number of nodes and leaf size. Drought stress at cotton boll development stage may seriously affect yield of cotton as yield has direct dependency on number of bolls per plant (Niu et al. 2018). However under drought stress the rate of root extension may be similar as in controlled conditions.

24.4 Morphological Response of Cotton Under Drought

Development and productivity of cotton is adversely affected due to drought stress. Plants respond to drought in different ways; it adjusts their growth period to mature early and avoid seasonal drought stress. Cotton tolerates, escapes, or recovers drought stress by adapting several morphological approaches (Khan et al. 2018). Cotton has a certain level of adaptations to drought stress owing to its perennial nature (Singh et al. 2018). In cotton drought stress limits root development, plant height, leaf growth, shoot growth, leaf area, vegetative growth, quality, and yield of fiber (Hasan et al. 2018). It has been found that water deficit limits 50% dry matter accumulation of *Gossypium barbadense* (Ullah et al. 2017). Under drought treatment, 60% reduction was observed in leaf area when compared to the control (Singh et al. 2018). Drought affects the root growth which in turn may leads to reduced biomass accumulation in cotton. Cotton undergoing water deficit explores moisture and nutrients by deeper root penetration (Fang and Xiong 2015).

Cotton showed some adaptations toward drought stress effect on roots. It enables increased root length and decreased shoot length; the enhanced root/shoot ratio indicates water assimilation and enhanced drought tolerance (Hasan et al. 2018).

The increased lateral roots enhance the root surface absorption area and increased root proliferation which are desirable traits for drought adaptation (Khan et al. 2018). Drought effect can be seen on cotton leaves which depict turgor loss, drooping, wilting, discoloration, yellowing, and premature senescence. Cotton exhibits leaf rolling, smaller and thicker leaves, abundant trichomes, smaller stomata and thick cuticle, and well-developed vascular bundle sheaths as stress adaptive traits (Fang and Xiong 2015). Drought stress at reproductive stage induces reproductive part abscission and boll size reduction. Reduced carbon assimilation leads to less biomass accumulation in cotton which causes large yield losses due to drought stress (Jawdat et al. 2018). Water deficit at flowering stage has been found to be most detrimental on seed cotton yield. Drought stress during fiber cell development affects the quality of the lint. Hence identification of more traits for drought tolerance under irrigated and water deficit agro-environment is needed to assist molecular breeding of for enhanced fiber quality and yield traits in cotton.

24.5 Physiological Response of Cotton Under Drought

Cotton maintains stomatal regulation and osmotic adjustment to tolerate stress (Ullah et al. 2017). Cotton leaves showed limited photosynthesis, decreased transpiration, low stomatal conductance and water potential when tested under drought conditions (Fang and Xiong 2015; Hasan et al. 2018). It has been established that 90% transpiration in plants occurs through stomata; hence stomatal regulation is an important mechanism through which plants maintain cellular function (Liang et al. 2016). In cotton, stomatal closure during water deficit helps in reducing water loss by limiting high transpiration rates. Negative correlation exists between stomatal conductance and drought tolerance which suggest stomatal conductance as a prospective indicator for drought tolerance in cotton (Fang and Xiong 2015).

Drought stress affects cellular structures by inducing osmotic stress and oxidative damage. This oxidative damage results from stomatal closure and decreased rates of gaseous exchange (Soomro et al. 2011). Damage to cell membranes due to drought stress causes electrolyte leakage (EL %). Electrolyte leakage is used as gauge to assess integrity and permeability of cell membranes and the subsequent leaking of intracellular contents. Sugars and amino acids generally termed as membrane compatible solutes protect the plasma membrane from desiccation-induced damage. Thus, osmotic adjustment and degree of membrane protection are linked; drought stress decreases membrane stability through lipid peroxidation caused by active oxygen species.

Drought stress affects cell turgidity and cellular homeostasis; hence osmotic adjustment is a crucial adaptation in crop plants against drought-induced damage (Wang et al. 2016b). Drought-tolerant genotype maintained higher relative water content in their leaves under water stress (Parida et al. 2007). Accumulation of osmolyte or osmoprotectants is a key strategy in plants to maintain cell homeostasis to withstand drought (Soomro et al. 2011). Proline, sugars, glycine betaine, alkaloids, amino acids, and inorganic ions are the compatible osmolytes that regulate cell homeostasis and alleviate stress in cell membranes (Singh et al. 2018). These osmolytes also act as free radical scavengers and protective agent for enzyme function (Wang et al. 2016b). Drought-tolerant species of cotton accumulate higher proline content; a positive link between proline levels in leaves and drought stress resistance has been demonstrated by many researchers (Khan et al. 2018; Wang et al. 2016b). These solutes protect vital proteins, enzymes, and membranes from the damage due to oxidative damage and higher inorganic ion concentrations under drought stress. The exogenous application of phytohormones, osmoprotectants (proline plus glycine betaine), and inorganic substances has been proved to be effective in mitigating injurious effects of drought stress in cotton (Fang and Xiong 2015). Drought stress increases accumulation of osmolytes in cotton such as prolines, soluble proteins, soluble sugars, and betaines (Hasan et al. 2018; Parida et al. 2007). Transgenic cotton plants showed better tolerance to drought stress due to enhanced accumulation of glycine; these plants exhibited higher photosynthetic rate, enhanced osmotic adjustment, increased relative water content, less lipid membrane peroxidation, and less electrolyte leakage (Pilon et al. 2015).

Drought stress damages photosynthetic apparatus and alters chlorophyll content, thus impacting photosynthesis in plants (Fang and Xiong 2015). Chlorophyll a, b, total chlorophyll content, and a/b ratio are declined by drought stress in cotton (Soomro et al. 2011). A net 66% photosynthesis decline has been observed in mature leaves of cotton under drought stress (Ullah et al. 2017). Drought stress decreases protein, starch, chlorophylls, and carotenoid contents in cotton and an increasing tendency was observed upon recovery from stress (Parida et al. 2007; Soomro et al. 2011). Hence chlorophyll content, electrolyte leakage, membrane stability, and relative water content act as an effective criterion to screen cotton genotypes and select most efficient genotypes.

24.6 Biochemical Response of Cotton Under Drought

Cotton tolerates external stresses through certain adaptations. Drought tolerance mechanism is connected to numerous biochemical processes. These processes are controlled by hormonal interactions in plant body. Plant growth regulators, for example, abscisic acid (ABA), jasmonic acid (JA), gibberellic acid (GA), and salicylic acid (SA), play an important role in various physiological and biochemical processes in plant life cycle; specially during plant development, reproduction, and stress signaling (Pandey et al. 2003). ABA synthesis is promoted by osmotic stress which induces adaptive physiological changes and activates expression of stress-related genes. Signal perception occurs by the plasma membranes; this can be through ABA-dependent or ABA-independent pathways. ABA-dependent signaling plays a crucial role in the expression of stress-responsive genes (Mittal et al. 2014).

Drought treatments reduced the GA content of roots; upon rewatering GA content and CAT activity increases (Niu et al. 2018). JA and its active derivatives jasmonates play significant part in plant responses to drought stress; JA is associated in stomatal closure, root elongation, fruit development, and viable pollen production (Ullah et al. 2017). Drought stress breaks cellular physiological homeostasis by inducing accumulation of reactive oxygen species (ROS) in plants. These include superoxide anion radical ($O_2^{\bullet-}$), hydrogen peroxide (H_2O_2), singlet oxygen ($1O_2$), hydroxyl radical ($HO^{\bullet}$), and nitric oxide ($NO^{\bullet}$) (Wang et al. 2016b). These reactive oxygen species oxidize photosynthetic pigments, proteins, lipids, DNA, and RNA and increase the damaging processes in the cell (Pilon et al. 2015). Drought conditions enable the photorespiration and enhance RuBP oxygenation due to reduced CO_2 fixation (Wang et al. 2016b). Plants have antioxidant defense mechanism against this oxidative damage, which controls cellular ROS levels during stress conditions (Fang and Xiong 2015). The antioxidant machinery has two arms: (1) enzymatic components, like catalase (CAT), SOD, ascorbate peroxidase (APX), glutathione reductase (GR), and guaiacol peroxidase (GPX), and (2) nonenzymatic antioxidants, for example, reduced glutathione (GSH), ascorbic acid (AA), a-tocopherol, flavonoids, carotenoids, and proline. These two components work together to scavenge ROS (Pilon et al. 2015). Under drought stress, ROS production increases and reduces the activity of the photosynthetic apparatus. Superoxide dismutase (SOD) converts O_2^{-} into H_2O_2, which is further converted to water by ascorbate peroxidase (APX). Antioxidant response of cotton cultivars determines their resistance capability to drought stress. Drought-tolerant cultivar has active antioxidative enzyme mechanism, which decreases the oxidative stress induced by lipid peroxidation (Fang and Xiong 2015). Decreased antioxidant enzyme activities in transgenic *G. barbadense* resulted in increased oxidative stress under drought conditions. These results show the importance of antioxidant defense mechanisms. Genes and certain factors are involved in improving the antioxidant machinery of cotton plants, such as Zn, that further need to be explored (Khan et al. 2018).

24.7 Molecular Response of Cotton Under Drought

Modern tools of biotechnology and genetic engineering now holds leading role to dissect complex nature of abiotic stresses tolerance. Genetic tolerance to abiotic stresses are quantitative in nature, involving many genes and QTLs (Fang and Xiong 2015). High-density genome-wide association study by using SNP in cotton is being used to study quantitative traits in cotton. Five QTLs were identified on A-genome and nine QTLs on D-genome by using SNP markers for abiotic stress along with 12 putative key genes through GBS-SNP-based high-density genetic mapping (Diouf et al. 2017). Twenty QTLs associated with drought tolerance were identified along with number of candidate genes. Four important drought tolerance-related genes, PP2C encoding a protein phosphatase 2C, HAT22 encoding a homeobox-leucine zipper protein, RD2 encoding a response to desiccation 2 protein, and PIP2 encoding a plasma membrane intrinsic protein 2, were potentially candidate genes for conferring drought tolerance in cotton (Hou et al. 2018). Transcriptomics and functional genomics have unveiled plenty of genes for drought resistance which include transcription factors, protein kinases, and some structural genes. Some NAC proteins regulate plant defense mechanism under drought stress. NAC protein *GhNAC2* reduced wilting and leaf abscission in cotton under drought stress. Overexpression of *GhNAC2* suppressed the ethylene pathway and activated the ABA/JA pathway which leads to longer roots, larger leaves, and hence higher yield in cotton under drought (Gunapati et al. 2016). Transcription factors are involved in biological processes. R2R3-MYB is the biggest family of transcription factors. The number R2R3-MYB transcription factor varies in different plant species. Cotton has narrow genetic pool due to constant breeding of same species. Other cotton species could be used to broad the genetic base of cotton. For this purpose, R2R3-MYB transcription factor was investigated through genome-wide characterization and 205 putative R2R3-MYB genes were identified on D-genome in *Gossypium raimondii* which were distributed across 13 chromosomes in various densities. MYB genes were found to be expressed under drought stress at seedling stage in cotton (He et al. 2016). Transcription factors (TF) help plants to cope with abiotic stresses. The bZIP (basic leucine zipper) is important TF in plants to activate ABA accumulation. GhABF2 encodes for bZIP TF in cotton. Overexpression of GhABF2 improves drought resistance in cotton plant (Liang et al. 2016). Sixteen putative genes were identified by using SNPs linked with drought resistance and were found to be associated with ghr-miR169a and ghr-miR164 (Magwanga et al. 2018a). Late embryogenesis abundant (LEA) proteins play a key part in the mechanism of drought stress. LEA genes were identified in *Gossypium arboreum*, *Gossypium hirsutum*, and *Gossypium raimondii*. All the LEA genes contained W-Box, MBS, ABRE, and TAC elements in their promoters which are known to be functionally involved to confer drought stress in crop plants (Magwanga et al. 2018b). Overexpression of RAV TFs and bZIP helps cotton to cope drought (Mittal et al. 2014). ABP9 gene was introduced into *Gossypium hirsutum* L. Overexpression of ABP9 confers drought tolerance in cotton by better root systems, higher germination, reduced stomatal aperture, and stomatal density (Wang et al. 2017). Mitogen-activated protein kinase kinase kinase (MAPKKK) is associated in plant stress response. MAPKKK genes were identified in *Gossypium raimondii* genome and *Gossypium hirsutum*. Gene expression arrays discovered that MAPKKKs intricate in abiotic stresses (Zhang et al. 2018). The isopentenyltransferase gene (IPT) confers resistance against water deficit. The expression of IPT is critical about the time at which drought occurs. Occurrence of drought stress before flowering stage enhances cotton yield (Zhu et al. 2018a). Some other important genes against abiotic stresses are listed in Table 24.1.

Table 24.1 Some important cotton genes for abiotic stress tolerance

Sr. No.	Gene	Trait	References
1	HSPCB	Peptide synthesis activated in drought-tolerant genotypes	Voloudakis et al. (2002)
2	GHSP26	Regulates cell metabolism, improves drought tolerance	(Maqbool et al. 2007)
3	KC3	Improves drought tolerance	(Selvam et al. 2009)
4	DREB 1 AND 2	Improves stress tolerance in crop plants	(Liu et al. 1998)
5	TPS (trehalose-6-phosphate synthase)	Protects proteins under water stress	(Kosmas et al. 2006)
6	*GhABF2* (bZIP)	Regulates genes related to ABA and increases the activities of SOD and CAT	(Liang et al. 2016)
7	*GhNAC2*	Lengthens roots, enhances drought tolerance	(Gunapati et al. 2016)
8	*GbMYB5*	Reduces stomatal size and rate of opening of stomata	(Chen et al. 2015)
9	*GhWRKY41*	Induces stomatal closure and higher antioxidant activity and lower malondialdehyde content	Chu et al. (2015)
10	*GhMKK3*	Enhances drought tolerance	Wang et al. (2016a)
11	*GhMAP3K40*	Enhances drought tolerance at the germination stage	(Chen et al. 2015)
12	*GhMPK17*	Increases salt and osmotic stress tolerance	Zhang et al. (2014)
13	*GbMPK3*	Increases drought and oxidative stress tolerance	Long et al. (2014)
14	*GhMKK1*	Increases drought and salt tolerance	Lu et al. (2013)
15	*GhMPK2*	Enhances drought and salt tolerance	Zhang et al. (2011)

24.8 What Is Next? Gene Pyramiding?

In cotton, drought being the major abiotic stress is significantly reducing the crop yield. Although scientists have made a lot of progress in developing the drought-tolerant varieties/cultivars through conventional breeding techniques, it is a tedious work which requires a lot of time, labor, and money. Alternatively, marker-assisted selection (MAS) is an efficient technique through which the genomic regions influenced by stress conditions are identified. Many QTLs have been identified for drought stress tolerance in different crops. Pyramiding the desirable genes from different sources into single genotype using the MAS technique may lead toward the development of drought-tolerant varieties/cultivars.

To develop future plant, resistant against drought stress, gene pyramiding could play pivotal role. The technique involves stacking important genes related to a particular trait, from different sources into a single genotype. The simultaneous expression of stacked genes may provide sufficient resilience against adverse conditions. Principally, the stacked genes are fixed into homozygous states that make plant with predictable performance. A number of examples are settled such as in maize; nine genes from different sources of two categories, plant defense response genes, and anti-apoptotic genes have been pyramided into a single maize line (Zhu et al. 2018b). Many of abiotic stress tolerance traits are quantitative in nature, confining breeders for conventional breeding approaches of plant improvement. However, scientists have reported many important genes for abiotic stress tolerance that could be stacked in a single genotype to make future plant resistant against drought stress. The cumulative expression of important genes in a single cotton plant may provide "stay green" type of ability under drought stress condition. The variety of genes would enable cotton to withstand adverse environmental conditions.

24.9 Conclusion and Future Perspective

Climate change is the consistent part of green planet but now its acceleration and adverse effect to agriculture has been phenomenal. Future breeding of cotton, a sensitive crop against drought, must imply modern techniques and knowledge to develop plant with some extra traits that could make cotton withstand unfavorable conditions. The mechanism of drought stress tolerance is complex in nature in cotton. QTL identification for drought tolerance in cotton is an ongoing strategy but the results have not been translated into a meaningful product. Many genes have also been identified that can overcome drought stress. A combination of stacked genes in cotton plant may provide sufficient empowerment against drought stress. Gene pyramiding could be future breeding of cotton for drought stress tolerance.

Abiotic and biotic tolerance, high yield, quality improvement, and production of special purpose cotton (long staple, medium long staple, color, and organic cotton) are new requirements in cotton production of the world. Currently genetic base of cotton is becoming narrow; that is why different exotic germplasm should be incorporated into different breeding programs to improve the cotton. Cotton genotypes from diverse origins should be assessed for the genomic regions conferring resistance through genome-wide association study (GWAS) to identify QTNs and genes by applying SNPs. Different genotypes conferring resistance may be incorporated in breeding programs. Identified genes may also be further validated using transcriptomics analysis to identify the expression of the said genes. For this purpose, the phenotypic data for abiotic stress-related trait can be investigated, and genome-wide SNP polymorphic loci can be studied using SNP chip. Based on phenotypic and genomic data, GWAS analysis is performed to study the genetic diversity across different cotton varieties/genotypes. The technique is usful to identify the elite alleles associated with the target traits, and to further investigate the transfer of variation. These selected candidate genes will be functional confirmation and may provide genetic modification of traits in the future. The process of identifying useful genes followed by gene pyramiding, may lead to develop future cotton plant.

References

Abbas Q, Ahmad S (2018) Effect of different sowing times and cultivars on cotton fiber quality under stable cotton-wheat cropping system in southern Punjab, Pakistan. Pak J Life Soc Sci 16:77–84

Ahmad S, Raza I (2014) Optimization of management practices to improve cotton fiber quality under irrigated arid environment. J Food Agric Environ 2(2):609–613

Ahmad S, Raza I, Ali H, Shahzad AN, Atiq-ur-Rehman, Sarwar N (2014) Response of cotton crop to exogenous application of glycinebetaine under sufficient and scarce water conditions. Braz J Bot 37(4):407–415

Ahmad S, Abbas Q, Abbas G, Fatima Z, Atique-ur-Rehman, Naz S, Younis H, Khan RJ, Nasim W, Habib ur Rehman M, Ahmad A, Rasul G, Khan MA, Hasanuzzaman M (2017) Quantification of climate warming and crop management impacts on cotton phenology. Plants 6(7):1–16

Ahmad S, Iqbal M, Muhammad T, Mehmood A, Ahmad S, Hasanuzzaman M (2018) Cotton productivity enhanced through transplanting and early sowing. Acta Sci Biol Sci 40:e34610

Ali H, Afzal MN, Ahmad F, Ahmad S, Akhtar M, Atif R (2011) Effect of sowing dates, plant spacing and nitrogen application on growth and productivity on cotton crop. Int J Sci Eng Res 2 (9):1–6

Ali H, Abid SA, Ahmad S, Sarwar N, Arooj M, Mahmood A, Shahzad AN (2013a) Integrated weed management in cotton cultivated in the alternate-furrow planting system. J Food Agric Environ 11(3&4):1664–1669

Ali H, Abid SA, Ahmad S, Sarwar N, Arooj M, Mahmood A, Shahzad AN (2013b) Impact of integrated weed management on flat-sown cotton (*Gossypium hirsutum* L.). J Anim Plant Sci 23 (4):1185–1192

Ali H, Hameed RA, Ahmad S, Shahzad AN, Sarwar N (2014a) Efficacy of different techniques of nitrogen application on American cotton under semi-arid conditions. J Food Agric Environ 12 (1):157–160

Ali H, Hussain GS, Hussain S, Shahzad AN, Ahmad S, Javeed HMR, Sarwar N (2014b) Early sowing reduces cotton leaf curl virus occurrence and improves cotton productivity. Cer Agron Moldova XLVII(4):71–81

Amin A, Nasim W, Mubeen M, Nadeem M, Ali L, Hammad HM, Sultana SR, Jabran K, Habib ur Rehman M, Ahmad S, Awais M, Rasool A, Fahad S, Saud S, Shah AN, Ihsan Z, Ali S, Bajwa AA, Hakeem KR, Ameen A, Amanullah, Rehman HU, Alghabar F, Jatoi GH, Akram M, Khan A, Islam F, Ata-Ul-Karim ST, Rehmani MIA, Hussain S, Razaq M, Fathi A (2017) Optimizing the phosphorus use in cotton by using CSM-CROPGRO-cotton model for semi-arid climate of Vehari-Punjab, Pakistan. Environ Sci Pollut Res 24(6):5811–5823

Amin A, Nasim W, Mubeen M, Ahmad A, Nadeem M, Urich P, Fahad S, Ahmad S, Wajid A, Tabassum F, Hammad HM, Sultana SR, Anwar S, Baloch SK, Wahid A, Wilkerson CJ, Hoogenboom G (2018) Simulated CSM-CROPGRO-cotton yield under projected future climate by SimCLIM for southern Punjab, Pakistan. Agric Syst 167:213–222

Burke JJ, O'Mahony PJ (2001) Protective role in acquired thermotolerance of developmentally regulated heat shock proteins in cotton seeds. J Cotton Sci 5:174–183

Chen X, Wang J, Zhu M, Jia H, Liu D, Hao L, Guo X (2015) A cotton Raf-like MAP3K gene, GhMAP3K40, mediates reduced tolerance to biotic and abiotic stress in Nicotiana benthamiana by negatively regulating growth and development. Plant Sci 240:10–24

Chu X, Wang C, Chen X, Lu W, Li H, Wang X, Hao L, Guo X (2015) The cotton WRKY gene GhWRKY41 positively regulates salt and drought stress tolerance in transgenic Nicotiana benthamiana. PLoS One 10:e0143022

Dai Y, Yang J, Hu W, Zahoor R, Chen B, Zhao W, Meng Y, Zhou Z (2017) Simulative global warming negatively affects cotton fiber length through shortening fiber rapid elongation duration. Sci Rep 7:9264

Diouf L, Pan Z, He SP, Gong WF, Jia YH, Magwanga RO, Romy KRE, Or Rashid H, Kirungu JN, Du X (2017) High-density linkage map construction and mapping of salt-tolerant QTLs at seedling stage in upland cotton using genotyping by sequencing (GBS). Int J Mol Sci 18:E2622

Ennahli S, Earl HJ (2005) Physiological limitations to photosynthetic carbon assimilation in cotton under water stress. Crop Sci 45:2374–2382

Fang Y, Xiong L (2015) General mechanisms of drought response and their application in drought resistance improvement in plants. Cell Mol Life Sci 72:673–689

Gunapati S, Naresh R, Ranjan S, Nigam D, Hans A, Verma PC, Gadre R, Pathre UV, Sane AP, Sane VA (2016) Expression of GhNAC2 from G. herbaceum, improves root growth and imparts tolerance to drought in transgenic cotton and Arabidopsis. Sci Rep 6:24978

Hasan M, Ma F, Prodhan Z, Li F, Shen H, Chen Y, Wang X (2018) Molecular and physio-biochemical characterization of cotton species for assessing drought stress tolerance. Int J Mol Sci 19:2636

He Q, Jones DC, Li W, Xie F, Ma J, Sun R, Wang Q, Zhu S, Zhang B (2016) Genome-wide identification of R2R3-MYB genes and expression analyses during abiotic stress in Gossypium raimondii. Sci Rep 6:22980

Hou S, Zhu G, Li Y, Li W, Fu J, Niu E, Li L, Zhang D, Guo W (2018) Genome-wide association studies reveal genetic variation and candidate genes of drought stress related traits in cotton (*Gossypium hirsutum* L.). Front Plant Sci 9:1276

Jawdat D, Allaf A, Taher N, Al-Zier A, Morsel N, Ajii Z, Al-Safadi B (2018) Cotton flowering behavior, fiber traits and gene expression under water-shortage stress. Adv Hortic Sci 32:79–92

Khan MB, Khaliq A, Ahmad S (2004) Performance of mashbean intercropped in cotton planted in different planting patterns. J Res (Sci) 15(2):191–197

Khan A, Pan X, Najeeb U, Tan DKY, Fahad S, Zahoor R, Luo H (2018) Coping with drought: stress and adaptive mechanisms, and management through cultural and molecular alternatives in cotton as vital constituents for plant stress resilience and fitness. Biol Res 51:47

Kosmas SA, Argyrokastritis A, Loukas MG, Eliopoulos E, Tsakas S, Kaltsikes PJ (2006) Isolation and characterization of drought-related trehalose 6-phosphate-synthase gene from cultivated cotton (Gossypium hirsutum L.). Planta 223(2):329–339

Lawlor DW, Tezara W (2009) Causes of decreased photosynthetic rate and metabolic capacity in water-deficient leaf cells: a critical evaluation of mechanisms and integration of processes. Ann Bot 103:561–579

Liang C, Meng Z, Meng Z, Malik W, Yan R, Lwin KM, Lin F, Wang Y, Sun G, Zhou T, Zhu T, Li J, Jin S, Guo S, Zhang R (2016) GhABF2, a bZIP transcription factor, confers drought and salinity tolerance in cotton (*Gossypium hirsutum* L.). Sci Rep 6:35040

Liu Q, Kasuga M, Sakuma Y, Abe H, Miura S, Yamaguchi-Shinozaki K, Shinozaki K (1998) Two transcription factors, DREB1 and DREB2, with an EREBP/AP2 DNA binding domain separate two cellular signal transduction pathways in drought- and low-temperature-responsive gene expression, respectively, in Arabidopsis. Plant Cell 10(8):1391–1406

Long L, Gao W, Xu L, Liu M, Luo X, He X, Yang X, Zhang X, Zhu L (2014) GbMPK3, a mitogen-activated protein kinase from cotton, enhances drought and oxidative stress tolerance in tobacco. Plant Cell Tissue Organ Cult 116:153–162

Lu W, Chu X, Li Y, Wang C, Guo X (2013) Cotton GhMKK1 induces the tolerance of salt and drought stress, and mediates defence responses to pathogen infection in transgenic Nicotiana benthamiana. PLoS One 8:e68503

Magwanga RO, Lu P, Kirungu JN, Diouf L, Dong Q, Hu Y, Cai X, Xu Y, Hou Y, Zhou Z, Wang X, Wang K, Liu F (2018a) GBS mapping and analysis of genes conserved between Gossypium tomentosum and Gossypium hirsutum cotton cultivars that respond to drought stress at the seedling stage of the BC(2)F(2) generation. Int J Mol Sci 19:E1614

Magwanga RO, Lu P, Kirungu JN, Lu H, Wang X, Cai X, Zhou Z, Zhang Z, Salih H, Wang K, Liu F (2018b) Characterization of the late embryogenesis abundant (LEA) proteins family and their role in drought stress tolerance in upland cotton. BMC Genet 19:6

Maqbool A, Zahur M, Irfan M, Qaiser U, Rashid B, Husnain T, Riazuddin S (2007) Identification, characterization and expression of drought related alpha-crystalline heat shock protein gene (from Desi cotton). Crop Sci 47(6):2437

Mitra J (2001) Genetics and genetic improvement of drought resistance in crop plants. Curr Sci 80:758–763

Mittal A, Gampala SS, Ritchie GL, Payton P, Burke JJ, Rock CD (2014) Related to ABA-Insensitive3(ABI3)/Viviparous1 and AtABI5 transcription factor coexpression in cotton enhances drought stress adaptation. Plant Biotechnol J 12:578–589

Niu J, Zhang S, Liu S, Ma H, Chen J, Shen Q, Ge C, Zhang X, Pang C, Zhao X (2018) The compensation effects of physiology and yield in cotton after drought stress. J Plant Physiol 224:30–48

Pandey D, Goswami C, Kumar B (2003) Physiological effects of plant hormones in cotton under drought. Biol Plantarum 47:535–540

Parida AK, Dagaonkar VS, Phalak MS, Umalkar G, Aurangabadkar LP (2007) Alterations in photosynthetic pigments, protein and osmotic components in cotton genotypes subjected to short-term drought stress followed by recovery. Plant Biotechnol Rep 1:37–48

Parida AK, Dagaonkar VS, Phalak MS, Aurangabadkar LP (2008) Differential responses of the enzymes involved in proline biosynthesis and degradation in drought tolerant and sensitive cotton genotypes during drought stress and recovery. Acta Physiol Plant 30:619–627

Pilon C, Oosterhuis DM, Ritchie GL, Paiva EA (2015) Photosynthetic efficiency and antioxidant activity of cotton under drought stress during early floral bud development. Am J Exp Agric 9. http://www.journalrepository.org/media/journals/AJEA_2/2015/Sep/Pilon962015AJEA21477.pdf

Rahman MH, Ahmad A, Wang X, Wajid A, Nasim W, Hussain M, Ahmad B, Ahmad I, Ali Z, Ishaque W, Awais M, Shelia V, Ahmad S, Fahad S, Alam M, Ullah H, Hoogenboom G (2018) Multi-model projections of future climate and climate change impacts uncertainty assessment for cotton production in Pakistan. Agric For Meteorol 253–254:94–113

Salehi-Lisar SY, Bakhshayeshan-Agdam H (2016) Drought stress in plants: causes, consequences, and tolerance. In: Drought stress tolerance in plants, vol 1. Springer, New York, pp 1–16. https://www.springer.com/gp/book/9783319288970

Schonfeld MA, Johnson RC, Carver BF, Mornhinweg DW (1988) Water relations in winter wheat as drought resistance indicators. Crop Sci 28:526–531

Selvam JN, Kumaraeadibel N, Gopikrishnan A, Kumar BK, Ravikesavan R, Boopathi MN (2009) Identification of a noval drought tolerance gene in (Gossypium hirsutum L.cv) KC3. Communications in biometry and crop. Sciences 4(1):9–13

Singh R, Pandey N, Naskar J, Shirke PA (2015) Physiological performance and differential expression profiling of genes associated with drought tolerance in contrasting varieties of two Gossypium species. Protoplasma 252:423–438

Singh B, Norvell E, Wijewardana C, Wallace T, Chastain D, Reddy K (2018) Assessing morphological characteristics of elite cotton lines from different breeding programmes for low temperature and drought tolerance. J Agron Crop Sci 204:467–476

Smith CW (1995) Cotton (Gossypium hirsutum L.). In: Crop production, evolution, history and technology. John Wiley and Sons. Inc., New York, USA

Soomro MH, Markhand GS, Soomro BA (2011) Screening Pakistani cotton for drought tolerance. Pak J Bot 44:383–388

Statista (2018). http://www.statista.com/statistics/263055/cotton-productionworldwide-by-top-countries/

Tariq M, Yasmeen A, Ahmad S, Hussain N, Afzal MN, Hasanuzzaman M (2017) Shedding of fruiting structures in cotton: factors, compensation and prevention. Trop Subtrop Agroecosyst 20(2):251–262

Tariq M, Afzal MN, Muhammad D, Ahmad S, Shahzad AN, Kiran A, Wakeel A (2018) Relationship of tissue potassium content with yield and fiber quality components of Bt cotton as influenced by potassium application methods. Field Crop Res 229:37–43

Ullah A, Sun H, Yang X, Zhang X (2017) Drought coping strategies in cotton: increased crop per drop. Plant Biotechnol J 15:271–284

Usman M, Ahmad A, Ahmad S, Irshad M, Khaliq T, Wajid A, Hussain K, Nasim W, Chattha TM, Trethowan R, Hoogenboom G (2009) Development and application of crop water stress index for scheduling irrigation in cotton (*Gossypium hirsutum* L.) under semiarid environment. J Food Agric Environ 7(3&4):386–391

Voloudakis AE, Kosmas SA, Tsakas S, Eliopoulos E, Loukas M, Kosmidou K (2002) Expression of selected drought-related genes and physiological response of Greek cotton varieties. Funct Plant Biol 29:1237–1245

Wang CY, Isoda A, Li MS, Wang DL (2007) Growth and eco-physiological performance of cotton under water stress conditions. Agric Sci China 6:949–955

Wang C, Lu W, He X, Wang F, Zhou Y, Guo X, Guo X (2016a) The cotton mitogen-activated protein kinase kinase 3 functions in drought tolerance by regulating stomatal responses and root growth. Plant Cell Physiol 57:1629–1642

Wang R, Gao M, Ji S, Wang S, Meng Y, Zhou Z (2016b) Carbon allocation, osmotic adjustment, antioxidant capacity and growth in cotton under long-term soil drought during flowering and boll-forming period. Plant Physiol Biochem 107:137–146

Wang C, Lu G, Hao Y, Guo H, Guo Y, Zhao J, Cheng H (2017) ABP9, a maize bZIP transcription factor, enhances tolerance to salt and drought in transgenic cotton. Planta 246:453–469

Wendel JF, Cronn RC (2003) Polyploidy and the evolutionary history of cotton. Adv Agron 78:139–186

Zhang H, Shen G, Kuppu S, Gaxiola R, Payton P (2011) Creating drought-and salt-tolerant cotton by overexpressing a vacuolar pyrophosphatase gene. Plant Signal Behav 6:861–863

Zhang J, Zou D, Li Y, Sun X, Wang NN, Gong SY, Zheng Y, Li XB (2014) GhMPK17, a cotton mitogen-activated protein kinase, is involved in plant response to high salinity and osmotic stresses and ABA signaling. PLoS One 9:e95642

Zhang X, Mi X, Chen C, Wang H, Guo W (2018) Identification on mitogen-activated protein kinase signaling cascades by integrating protein interaction with transcriptional profiling analysis in cotton. Sci Rep 8:8178

Zhu X, Sun L, Kuppu S, Hu R, Mishra N, Smith J, Esmaeili N, Herath M, Gore MA, Payton P, Shen G, Zhang H (2018a) The yield difference between wild-type cotton and transgenic cotton that expresses IPT depends on when water-deficit stress is applied. Sci Rep 8:2538

Zhu X, Zhao J, Abbas HMK, Liu Y, Cheng M, Huang J, Cheng W, Wang B, Bai C, Wang G (2018b) Pyramiding of nine transgenes in maize generates high-level resistance against necrotrophic maize pathogens. Theor Appl Genet 131:2145–2156

Chapter 25
Biotechnology for Cotton Improvement

Khezir Hayat, Adem Bardak, Dony Parlak, Farzana Ashraf, Hafiz Muhammad Imran, Hafiz Abdul Haq, Muhammad Azam Mian, Zahid Mehmood, and Muhammad Naeem Akhtar

Abstract Cotton is a natural fiber crop in the world. The ever-increasing demands of the fast-growing population for food, feed, fiber, and fuel, which is estimated to be 11 billion all over the world in 2050, urge to enhance food production 2–3 times. But limitations in conventional breeding program for genetic upgrading are due to limited knowledge about yield and fiber traits. Use of molecular markers and exploitation of DNA polymorphism is one of the noteworthy developments in the field of molecular genetics. Availability of reference genome of *G. raimondii* L., *G. arboreum* L., and next-generation sequencing routed it on fast track for exploring variability among genotypes of cotton. Genomic research could be quantitative trait loci mapping, genome-wide associations, and next-generation sequencing approaches.

Keywords Marker-assisted selection · Quantitative trait locus · Deoxyribonucleic acid · Simple sequence repeat

Abbreviations

AFLP	Amplified fragment length polymorphism
CAPS	Cleaved amplified polymorphic sequence
DNA	Deoxyribonucleic acid
EST-SSRs	Expressed sequence tags

K. Hayat (✉) · F. Ashraf · H. M. Imran · H. A. Haq · M. A. Mian · Z. Mehmood
Central Cotton Research Institute, Multan, Pakistan

A. Bardak
Department of Agricultural Biotechnology, Kahramanmaras Sutcu Imam University Kahramanmaras, Kahramanmaras, Turkey

D. Parlak
Kahramanmaras Sutcu Imam University Kahramanmaras, Kahramanmaras, Turkey

M. N. Akhtar
Department of Soil and Environmental Sciences, MNS University of Agriculture, Multan, Pakistan

Pesticide Laboratory, Multan, Pakistan

S. Ahmad, M. Hasanuzzaman (eds.), *Cotton Production and Uses*,
https://doi.org/10.1007/978-981-15-1472-2_25

GBS	Genotyping by sequencing
GM	Genetically modified
GMO	Genetically modified organisms
GWAS	Genome-wide association
ICAC	International Cotton Advisory Committee
ISAAA	International Service for the Acquisition of Agri-biotech Applications
ISSR	Inter-simple sequence repeat
MAS	Marker-assisted selection
MT	Million metric tons
PCR	Polymerase chain reaction
QTL	Quantitative trait locus
RAPDs	Random amplified polymorphic DNAs
RFLP	Restriction fragment length polymorphism
SCAR	Sequence characterized amplified region
SNP	Single nucleotide polymorphism
SSR	Simple sequence repeat
STS	Sequence-tagged site
USDA	United State Department of Agriculture

25.1 Introduction

The ultimate source of natural fiber in the world is cotton (*Gossypium* spp.) which contributes a lot for global economy (Cuming et al. 2015), as well as major oilseed crop. Cotton is sown in over 100 countries across the world (ICAC 2018) on an area of 33.4 million hectare with total production of about 120 million bales (USDA 2013–2014).

It is the need of the day to increase food production 2–3 times as it is speculated that world population might be 11 billion up to 2050 which will produce alarming competition for food, feed, and fiber (Ahmad et al. 2014, 2017, 2018; Abbas and Ahmad 2018; Ahmad and Raza 2014; Ali et al. 2011, 2013a, b, 2014; Usman et al. 2009). Moreover, recently the abrupt fluctuation in food items warns to develop innovations in order to fulfill requirements of the people. Biotechnology can make a noteworthy role for efforts as demonstrated in cotton and further crops. New advances in biotechnology have made it possible to develop transgenic plants having genes that were not likely to be developed sexually. Cotton has been a leader in using biotechnology. Meanwhile, limiting resources and various biotic and abiotic stresses are responsible for decrease in production (Usman et al. 2009; Amin et al. 2017, 2018; Rahman et al. 2018; Tariq et al. 2017, 2018). These issues bring an urgent necessity for enhancing productivity of crops. The effective utilization of available genetic diversity is a key for combating such scarce resources (Hu et al. 2014).

Hence, it has to be considered the value of classical breeding to modern biotechnological era. There would be a need of fewer resources with sound genetic pattern. The present plant breeding is a joint venture of various subjects like genetics, pathology, soil, water, and nutrition as well as has many strategies invented more than century ago. The development of well-established traditional plant breeding system is a way to overcome the merits of biotechnology. A cultivar is known as the product of transference of many genes. For obtaining a target of 27.5 million bales, it is necessary to manipulate genetic variation more effectively via plant breeding and biotechnology (ICAC 2018).

Biotechnology means technology used to resolve any problem via any living organism (Jost 1941): It can also be defined as diverse biological methods which involve any parts of the cells for the betterment of bacteria, animals, and plants. It is also used to shuffle genes from one organism to another via genetic engineering being a popular tool in current era. Genetically modified crops (GMO) are sown on an area of 189.9 million hectare in contrast to 1.7 million hectare during 1996, clearly showing the popularity of biotech products all over the globe. In addition, GMO crops shared US$ 17.2 billion all over the world with more than 30% seed value (ISAA 2017).

It is highly popular among the different shareholders to develop biotech crops for overcoming the requirements of the market. Biosafety measures are kept in forward in GMO crops as compared to classical breeding as a number of scientists were not allowed to sell such product due to noncompliance of such measures (Kim et al. 2009; Lee 2011). And a number of other essential things have to be kept in mind like desirable characters and demand in the market property rights of the genes utilized in GMOs. Park et al. (2018) revealed that about 13 years were spent for the use of public casted about US$ 15 million with proper intellectual rights.

25.2 Agricultural Biotechnology Products and Their Value

Agricultural biotechnology is implemented to enhance production of plants, principally by decreasing the production cost mostly in crops sown in the temperate zones. It is the capability of this technology which permits better yield with better adaptability, resistance against various diseases, improved soil conservation using limited resources, better nutritive value and storage, and ultimately low cost to the market. GMO crops are sown in 24 countries all over the world on an area of 189.8 million hectare with uplift of 3% area in contrast to 2016 with economic value of US$ 186.1 billion (ISAAA 2017). The top countries which are going to adopt this technology to highest content include the USA (94.5%), Brazil (94%), Argentina (about 100%), Canada (95%), and India (95%), and the improvement of biotech products in these regions is due to precise and fast-track approval of GMO crops keeping in view the occurrence of recent diseases and pest as well as the changing global environment. The GMO crops have broadened to the crops other than the four best crops, i.e., maize, soybeans, cotton, and canola, for furnishing diverse choice to the consumers. These crops consist of eggplants, papaya, alfalfa, sugar beets, potatoes, etc., about all available in the market. The GMO outputs obtained from 2.15 billion hectare greatly fulfill the needs of 7.6 billion people for food, feed, fiber, and fuel. Therefore, it is compulsory to use such products on the globe for population needs which is speculated to be 9.8 billion till 2050 (UN 2017). Currently, biotech crops share US$ 17.2 billion in the world and it is projected that it will be about 8.3% in 2022 to 10.5% in 2025 from joint contribution of biotech crops and seeds (ISSA 2017). GMO cotton is grown on an area of 0.34 billion hectare all over the world while in Pakistan covered an area of 0.3 million hectare. The merits from these crops include insect resistance, herbicide resistance, stacked insect resistance/herbicide resistance, and other stresses which have been done since 1996. Traits have been transferred which had the ability to increase the quantity of oils, proteins, carbohydrates, and starch in food and root crops. The advancements in next-generation sequencing may fasten breeding programs as large number of functional genes can be searched for drought and salt tolerance and other quantitative characters. This will be of high value to the farmers which use marginal lands all over the globe as the improvement of such characters with traditional methods is less fruitful in crops of immense importance.

25.3 Molecular Breeding and Marker-Assisted Selection

Development of cultivars involves many steps in classical breeding and it takes 10–15 years depending upon the crop species. With the innovations in biotechnology, within 7–10 years a new variety with required traits can be released for general cultivation. One of the methods which is applied for speeding up the breeding is called molecular breeding which involves selection based upon molecular level also known as marker-assisted selection (MAS) (Bolek et al. 2016). It is the way to use biotechnology for the development of new genotypes which combines MAS and transgenics (Moose and Mumm 2008). MAS has an important role in agriculture as there is an issue of biosafety roles in many countries of the world for the GMO crops (Nicolia et al. 2014). It is a type of breeding in which selection is based upon DNA level for required characters in crop species. MAS has been used broadly for the improvement of crops with high better yield, quality, and tolerant to other biotic and abiotic stresses as being an efficient and important way.

It is a prerequisite to learn about the development and utilization of molecular markers for improving yield of crops from genetic perspectives; as a whole the goal of this chapter is to learn advancement of DNA markers and their role in cotton breeding.

DNA marker is a precise sequence of nucleotides with a designated place on a chromosome (Kumar 1999), or it can be defined as a unit of inheritance with precise location within a chromosome which can be differentiated phenotypically (Kingm and Stansfield 1990; Schulmann 2007). Molecular markers are portioned into three categories: (1) philological markers that have morphological characters, (2) biochemical markers in which isozymes are utilized for variation, and (3) genetic markers, which describe the position of change in DNA sequence (Joshi and Nguyen 1993; Gupta et al. 1999). Molecular markers have the capability of homoplasy for the identification of homo- and heterozygotes (Roychowhury et al. 2014). DNA markers have high reliability for restoration of fertility than phonological markers (Shanti et al. 2001); molecular markers with more recombination in gene pool entries are more favorable for MAS (Bolek 2003).

Many scientists have described that MAS has many merits over conventional breeding methods (Collard and Mackill 2008; Waqas et al. 2014). Selection is made on DNA for required characters in contrast to morphological basis (Helentjaris et al. 1986), allowing the foundation for MAS (Welsh and McClelland 1990; Struss and Plieske 1998). DNA markers are favored for refining of major characters in a number of plants like maize (Stuber et al. 1999; Tuberosa et al. 2003), rice (Mackill et al. 1999), wheat (Koebner and Summers 2003), and barley (Thomas 2003; Williams 2003). *Gossypium* genus is the source of natural fiber and MAS has not achieved success to the desired level due to limited genetic diversity and compatibility problems during domestication as well as less recombination extent (Rahman et al. 2005; Abdurakhmonov et al. 2008).

Several agronomic and quality characters as well as diseases are found to be governed by many genes which are known as mutagenic characters (Polygenic characters). For improving such characters, breeders have been forced to adopt marker-assisted selection by using mutations in genes (Bolek et al. 2005). Variations in such traits with particular quantitative trait locus (QTL) are efficient and economical than classical methods (Collard et al. 2005). Introgression of desired characters in upland cotton (*Gossypium hirsutum*) is carried by MAS with least deterioration of genes via genetic maps and QTLs (Tanksley et al. 1989; Abdurakhmonov et al. 2011). With innovations in next-generation sequencing, a number of markers can be used throughout different species with genome-wide association studies (Schuster 2011).

In order to get desirable outcomes from such diverse techniques, it is of vital value to know about development and uses of such markers in the plants.

25.3.1 DNA Markers in Cotton

Characterization of germplasm using DNA in plants is carried for analyzing genetic variation and identifying desired QTLs for required characters in genetic stock. Germplasm maintenance relies directly on degree of variation among the genotypes (Jubrael et al. 2005). DNA markers can be designed within a short spell due to availability of bioinformatics (Andersen and Lubberstedt 2003) and such markers are more authentic for the scientists as being the base for detection and conservation of diversity, MAS, and sequencing (Kalia et al. 2011).

The evolution of DNA markers relies upon its ability, recombination, and expenditure (Bernardo 2008). Molecular markers are differentiated into categories using their method of observing DNA variation like (a) hybridization based, (b) polymerase chain reaction (PCR) based, and (c) DNA sequence based.

25.3.1.1 Restriction Fragment Length Polymorphism (RFLP)

Foremost popular kind of hybridization-based markers utilized in plants for linkage mapping during 1975 was restriction fragment length polymorphism (RFLP) (Helentjaris et al. 1986). This type of marker was found to be broadly realistic for determining genetic diversity and DNA-based germplasm maintenance. A number of researchers have applied RFLP for genetic mapping in cotton (Ulloa and Meredith Jr 2000). The credibility of RFLP was observed for bacterial blight resistance in gene pool and QTLs were detected for this disease and verified in the germplasm (Wright et al. 1998). RFLPs have been used widely in cotton and played significant part in breeding of cultivars (Rahman et al. 2009). The designing of such markers needs expensive chemicals and also more duration is required for analysis of marker data which hinders its exploitation in MAS (Agarwal et al. 2008).

25.3.1.2 Random Amplified Polymorphic DNAs (RAPDs)

RAPDs involve PCR in which DNA sequence of 10 bp is amplified with primer combination. Genotypes are characterized with RAPDs after applying primers which manifest reproducibility about change in genetic variation. RAPDs have been used in cotton for a number of traits (Rahman et al. 2002; Hussain et al. 2005; Khan et al. 2000). Introgression was studied and then genetic diversity was observed using RAPDs and endorsed that this methodology is highly suitable (Sheidail et al. 2007). Resistance against jassids, mites, and aphids was studied with RAPDs and also collations were observed (Geng et al. 1995). Genetic diversity, phylogenetic studies, and genetic mapping have been done via RAPDs (Zhang et al. 2009). Drawbacks of RAPDs are that for maximizing reproducibility; it needs to follow PCR conditions but in true sense such bands are difficult to amplify.

25.3.1.3 Amplified Fragment Length Polymorphism (AFLP)

The marker which is used for joining the PCR-based markers with RFLP is AFLP which attaches known sequence of DNA with the portion of DNA evolved from restriction enzymes in RFLP (Lynch and Walsh 1998). Peculiarity of AFLP is its capability to "representation of genome" as DNA points spread throughout the genome simultaneously. AFLPs have been applied for characterization of germplasm and their ancestor's pattern (Lacape et al. 2003; Abdalla et al. 2001) as well as for linkage mapping in cotton (Zhang et al. 2005). The pros of AFLPs are authentic and polymorphic (Jones et al. 1997), do not require analysis of sequence and are highly informative, and have nature to analyze huge quantity of reproducible loci with individual primer on gel electrophoresis in contrast to RFLPs and single sequence repeats (Russell et al. 1997). Moderately degraded DNA and good-quality

DNA can be utilized for annealing; however it should be free of restriction enzyme and PCR inhibitors.

25.3.1.4 SSR (Simple Sequence Repeat)

Reproducible pattern observed in DNA with 1–6 bp replicating units is known as short tandem repeats or microsatellites or simple sequence repeats (SSRs) (Bidichandani et al. 1998). Such type of molecular markers has been used extensively for conservation of germplasm, comparison of genotypes, population analyses, and genetic mapping. Microsatellites are the most popular DNA markers for determining ancestral pattern (Zhang et al. 2005). These are extensively used for DNA profiling as forensic, genetic diversity study while also used in conservation biology and in variety development (Coetes and Byrne 2005). International Cotton Genome Initiative was initiated that will pave the way for development of highly condensed genetic map of cotton which will ultimately be a road map for the cotton breeders according to prevailing problems (Yu et al. 2005).

Microsatellites have been used for determining genetic variation among related species and cultivated cotton (Liu et al. 2006). Fiber quality traits have been improved using SSRs by a number of cotton breeders (Ulloa et al. 2002; Zhang et al. 2013). A lots of studies for disease resistance/tolerance like verticillium wilt (*Verticillium dahliae* Kleb.) (Bolek et al. 2005), bacterial blight, and cotton leaf curl virus (Aslam et al. 1999). SSRs were used for association studies in a germplasm collection consisting of 241 genotypes of upland cotton (Qin et al. 2015). They found that SSRs can be useful for improving agronomic and economic traits which will ultimately contribute toward marker-assisted selection. Zhang et al. (2016) observed fiber quality QTLs using introgressed lines (ILs) using SSRs. They observed multiple QTLs and suggested that such QTLs can be used for determining origin of the lines.

25.3.1.5 Inter-Simple Sequence Repeat (ISSR)

The DNA marker in which DNA sequences are magnified from polymorphic point among two same SSRs and those that are present at contagious point are known as inter-simple sequence repeat (ISSR). AFLP and SSR benefits are combined in ISSR (Bornet and Branchard 2001). ISSR has been applied largely for advancement of economical traits in cotton as well as for characterization of germplasm and linkage mapping (Bornet and Branchard 2001; Sica et al. 2005). Genetic variation for agronomic characters was studied in F_2 population of upland cotton using ISSR. Farahani et al. (2018) determined genetic diversity in germplasm using ISSR. They showed that such information is useful for devising any strategy relating to improvement in economic traits.

25.3.1.6 Sequence Characterized Amplified Region (SCAR)

PCR-based DNA marker by which DNA fragment is identified via pair of specific primers is recognized as sequence characterized amplified region (Paran and Michelmore 1993). SCAR has merit over RAPDs as being able to locate just an individual point. Such type of markers has better choice as being codominant in contrast to RAPDs and also can be applied for genetic mapping in cloned libraries. Fiber quality (Guo et al. 2003) and fertility restoration (Wang et al. 2007; Zhang et al. 2012) have been observed in cotton (Nair et al. 1995; Liu et al. 1999). The peculiarity of SCAR is its economic value for development and high reproducibility for comparing huge populations (Guo et al. 2006).

25.3.1.7 Sequence-Tagged Sites (STS)

PCR-based DNA markers which through precise single nucleotides evolves a sequence connected to a required character known as sequence-tagged site (STS). STS are described with a set of primers that sequence a RFLP fragment having minute copy number (Blake et al. 1996). Being easy to apply, more polymorphism, codominant nature, and reliable for sequencing, STS have been used in cotton improvement. Feng et al. (2005) evolved restorer parental material for hybrid cotton using STS.

25.3.1.8 Expressed Sequence Tags (EST-SSRs)

Expressed sequence tags (EST-SSRs) are the highly prevailing points in the species in contrast to genic SSRs from non-translated points (Cuadrado and Schwardzacher 1998). Diverse sequence determination ways allow the identification of SSRs from the ESTs. And a number of methods have been applied for development of ESTs in *Gossypium* species (Qureshi et al. 2004), ancestral pattern. ESTs have been applied for gene identification (Hughes and Friedman 2005) and marker detection (Michalek et al. 2002). ESTs are better than SSRs as are found in transcribed regions and fastly searched by bioinformatics (Varshney et al. 2005). Highly condensed genetic map was developed in cotton using ESTs (Wang et al. 2015). They showed ESTs were found to best for genomic studies and also deduced that ESTs can be applied for searching desirable traits in cotton.

25.3.1.9 Cleaved Amplified Polymorphic Sequence (CAPS)

The CAPS combines RFLP in PCR by which DNA fragment is magnified succeeded by digestion of restriction enzyme. EST-derived CAPS oligonucleotides have more pros than the genomic SSRs for comparison mapping in MAS. CAPS have been applied with success in devising patents in cotton as well as conservation of germplasm, analysis of genetic variation in breeding, and genomic studies.

25.3.1.10 Single Nucleotide Polymorphism (SNP)

DNA markers which are differentiated by an individual base among two genotypes and also have specific position in the genome are known as single nucleotide polymorphism (SNP) (Ayeh 2008). SNPs were first discovered in humans and found to be mostly available type of hereditary variation among individuals of two species. SNPs are considered to be greatly reliable as help to identify phenotype quickly (Batley and Edwards 2007). SNPs are the best markers for the scientific community in breeding and MAS as these are abundantly available in the genome with good polymorphism whether within or between individuals as compared to SSRs (Berard et al. 2009). Evolution of SNPs is not an easy task as it involves detection of SNP and its validation; both of these steps are highly interfered by intrinsic genome structure and mostly nonavailability of prescribed genome. The availability of high-throughput sequencing technologies like 454 Life Sciences (Roche Applied Science), HiSeq (Illumina, San Diego, CA), and Ion Torrent (Life Technologies Corporation, Carlsbad, CA) has allowed to detect SNPs with economic value at whole genome-wide association mapping. These have been identified in number of species like humans (Sachidanandam et al. 2001), *Arabidopsis thaliana* (Jander et al. 2002), maize (Ching et al. 2002), wheat, and cotton.

SNPs have been used for analyzing genetic variation, ancestral pattern, and genetic mapping in cotton (Van Deynze et al. 2009). Affymetrix has evolved "gene chip" and has been released commercially. GenBank, dbEST, and RefSeq are used for the identification of SNPs for sequence verification. Byers et al. (2012) developed enormous number of SNPs by reduced representation of the genome using Roche 454 pyro-sequencing in polyploidy cotton. The popularity of SNPs is that these are being used in genomics economically for gene identification and genetic mapping for improving traits of interest and also being applied in cloning. Cotton 63K chip has been used for the construction of the most dense genetic map of cotton in mapping population of inter- and intraspecies (Hulse-Kemp et al. 2015). This map has been used in MAS as a base for the improvement of economic traits and agronomic characters. Palanga et al. (2017) used SNPs for identification of QTLs related to verticillium disease resistance in an introgressed recombinant inbred line (RIL) population. As a whole 119 QTLs were detected but 40 QTLs found to be versatile which can be utilized in future breeding for devising strategy to verticillium resistance.

25.3.2 Genotyping by Sequencing (GBS)

Genotyping by sequencing or genotyping by synthesis is a methodology of next-generation sequencing in which libraries are developed by reduced representation of the genome with restriction enzymes and DNA adaptors (Elshire et al. 2011; Poland et al. 2012). The peculiarity of this method is to evolve enormous number of markers in a single step without the reference genome in any species. Moreover, it has been refined since its evolution in conjunction with restriction endonucleases for genome reduction via high-throughput sequencing technologies (Baird et al. 2008). GBS also allows analysis of mapping population in an easy way using association mapping at whole genome (Poland and Trevor 2012). It has been applied successfully in lots of crop species like *Zea mays* (Elshire et al. 2011), *Sorghum bicolor* with slight change (Poland et al. 2012), and cotton (Van Deynze et al. 2009; Fang et al. 2014; Islam et al. 2015). GBS has been used in cotton for the advancement of many yield and yield attributes and fiber traits but studies are highly interfered due to complex nature of cotton genome (Li et al. 2014). Bhatti (2018) identified 32 QTLs in upland cotton using GBS and also revealed that most of the QTLs were stable in multiple locations.

25.3.3 Genome-Wide Association (GWAS)

Biparental populations have been used in genomic studies for observing genetic variation in the gene pool, development of linkage maps, and identification of QTLs for traits of interest from economical perspectives (Chen et al. 2007). It is a point of concern to identify closely linked DNA markers due to limited crossing over for marker-assisted selection. And such populations have a few minor QTLs which are unable to evolve high magnitude of polymorphism.

The alternative and more reliable way to detect QTLs of interest is called linkage disequilibrium or LD mapping in which accessions preserved in gene bank or genetic stock are used with high recombination (Zhao et al. 2014). Association analysis depends upon connection of alleles among trait and DNA marker. Genome-wide association mapping (GWAS) is an authentic way to identify reliable parents for plant breeding program. It also overcomes problems of populations derived from hybridizing parents due to more extent of polymorphism. It is directly related with magnitude of linkage disequilibrium. The population structure analysis is the backbone for such type of mapping as it permits to observe individual genotype ancestral pattern. As a whole it allows selection of parents using magnitude of ancestral recombination and linkage disequilibrium (Lu et al. 2009).

Association mapping has been used for the refinement of various agronomic and fiber quality characters (Abdurakhmonov et al. 2008, 2009). It was shown that QTLs can be used for improving traits using germplasm entries. Waqas et al. (2014) revealed that with the availability of SNPs, association mapping can be conducted at whole genome for the development of highly saturated maps and detection of stable QTLs. Jia et al. (2014) used GWAS to determine QTLs for salt and drought tolerance in a germplasm collection of cotton. They revealed that associations among markers and phenotypic data can be used for devising molecular breeding. Li et al. (2016) used association mapping for determining QTLs for fiber quality in a mapping population consisting 188 individuals and found 71 QTLs for yield and fiber traits but 12 QTLs were highly reliable as were observed in multiple locations which can be useful for MAS. Mei et al. (2017) applied this mapping for assessing morphology parameters like fruiting branch in upland cotton. They screened 39 lines and their F_1 consisting of 178 and assumed that branching pattern can be applied as a selection parameter as the genetic variation varied exclusively between the genotypes.

25.4 Conclusion

Tremendous progress has been made in cotton biotechnology and GMO crops since the release of transgenic. GMO crops covered an area of 189.8 million hectare all over the globe in 24 countries during 2017 for food, feed, and processing. DNA markers have immense value for devising any breeding strategy. These are considered to be reliable source of tracing genetic diversity. QTLs can be detected and applied for reliable selection. Cotton production is stagnant due to limited genetic diversity in the germplasm which is also affected by many biotic and abiotic problems. Molecular markers have capability to tackle these obstacles more efficiently via next-generation sequencing techniques for transferring of economical characters from wild species to upland cotton. Primarily, SSRs are the predominant DNA markers being used for crop improvement in number of species for a long time. The abundant availability of SNPs with innovations in high-throughput sequencing has allowed breeders to analyze genetic diversity in a precise way. As a whole by using proper biotechnology methods and molecular breeding, the production of cotton can be increased to fulfill requirements of burgeoning population which is thought to be 11 billion till 2050.

References

Abbas Q, Ahmad S (2018) Effect of different sowing times and cultivars on cotton fiber quality under stable cotton-wheat cropping system in southern Punjab, Pakistan. Pak J Life Soc Sci 16:77–84

Abdalla AM, Reddy OUK, El-Zik KM, Pepper AE (2001) Genetic diversity and relationships of diploid and tetraploid cottons revealed using AFLP. Theor Appl Genet 102(2–3):222–229

Abdurakhmonov IY, Kohel RJ, Yu JZ, Pepper AE, Abdullaev AA, Kushanov FN, Salakhutdinov IB, Buriev ZT, Saha S, Scheffler BE, Jenkins JN, Abdukarimov A (2008) Molecular diversity and association mapping of fiber quality traits in exotic *G. hirsutum* L. germplasm. Genomics 92 (6):478–487

Abdurakhmonov IY, Saha S, Jenkins JN, Buriev ZT, Shermatov SE, Scheffler BE, Pepper AE, Yu JZ, Kohel RJ, Abdukarimov A (2009) Linkage disequilibrium based association mapping of fiber quality traits in *G. hirsutum* L. variety germplasm. Genetica 136:401–417

Abdurakhmonov IY, Buriev ZT, Shermatov SE (2011) Marker-assisted selection for complex fiber traits in cotton. In: 5th World Cotton Research Conference, Special session of ICGI, Mumbai, India 7–12 November, 2011

Agarwal M, Shrivastava N, Padh H (2008) Advances in molecular marker techniques and their applications in plant sciences. Plant Cell Rep 27(4):617–631

Ahmad S, Raza I (2014) Optimization of management practices to improve cotton fiber quality under irrigated arid environment. J Food Agric Environ 2(2):609–613

Ahmad S, Raza I, Ali H, Shahzad AN, Atiq-ur-Rehman, Sarwar N (2014) Response of cotton crop to exogenous application of glycinebetaine under sufficient and scarce water conditions. Braz J Bot 37(4):407–415

Ahmad S, Abbas Q, Abbas G, Fatima Z, Atique-ur-Rehman, Naz S, Younis H, Khan RJ, Nasim W, Habib ur Rehman M, Ahmad A, Rasul G, Khan MA, Hasanuzzaman M (2017) Quantification of climate warming and crop management impacts on cotton phenology. Plants 6(7):1–16

Ahmad S, Iqbal M, Muhammad T, Mehmood A, Ahmad S, Hasanuzzaman M (2018) Cotton productivity enhanced through transplanting and early sowing. Acta Sci Biol Sci 40:e34610

Ali H, Afzal MN, Ahmad F, Ahmad S, Akhtar M, Atif R (2011) Effect of sowing dates, plant spacing and nitrogen application on growth and productivity on cotton crop. Int J Sci Eng Res 2 (9):1–6

Ali H, Abid SA, Ahmad S, Sarwar N, Arooj M, Mahmood A, Shahzad AN (2013a) Integrated weed management in cotton cultivated in the alternate-furrow planting system. J Food Agric Environ 11(3–4):1664–1669

Ali H, Abid SA, Ahmad S, Sarwar N, Arooj M, Mahmood A, Shahzad AN (2013b) Impact of integrated weed management on flat-sown cotton (*Gossypium hirsutum* L.). J Anim Plant Sci 23 (4):1185–1192

Ali H, Hameed RA, Ahmad S, Shahzad AN, Sarwar N (2014) Efficacy of different techniques of nitrogen application on American cotton under semi-arid conditions. J Food Agric Environ 12 (1):157–160

Amin A, Nasim W, Mubeen M, Nadeem M, Ali L, Hammad HM, Sultana SR, Jabran K, Habib ur Rehman M, Ahmad S, Awais M, Rasool A, Fahad S, Saud S, Shah AN, Ihsan Z, Ali S, Bajwa AA, Hakeem KR, Ameen A, Amanullah, Rehman HU, Alghabar F, Jatoi GH, Akram M, Khan A, Islam F, Ata-Ul-Karim ST, Rehmani MIA, Hussain S, Razaq M, Fathi A (2017) Optimizing the phosphorus use in cotton by using CSM-CROPGRO-cotton model for semi-arid climate of Vehari-Punjab, Pakistan. Environ Sci Pollut Res 24(6):5811–5823

Amin A, Nasim W, Mubeen M, Ahmad A, Nadeem M, Urich P, Fahad S, Ahmad S, Wajid A, Tabassum F, Hammad HM, Sultana SR, Anwar S, Baloch SK, Wahid A, Wilkerson CJ, Hoogenboom G (2018) Simulated CSM-CROPGRO-cotton yield under projected future climate by SimCLIM for southern Punjab, Pakistan. Agr Syst 167:213–222

Andersen JR, Lubberstedt T (2003) Functional markers in plants. Trends Plant Sci 8(11):554–560

Aslam M, Jiang C, Wright R, Paterson AH (1999) Identification of molecular markers linked to leaf curl virus disease resistance in cotton. Pak J Biol Sci 2(1):124–126

Ayeh KO (2008) Expressed sequence tags (ESTs) and single nucleotide polymorphisms (SNPs): emerging molecular marker tools for improving agronomic traits in plant biotechnology. Afr J Biotechnol 7(4):331–341

Baird NA, Etter PD, Atwood TS, Currey MC, Shiver AL, Lewis ZA, Selker EU, Cresko WA, Johnson EA (2008) Rapid SNP discovery and genetic mapping using sequenced RAD markers. PLoS One 3:e3376

Batley J, Edwards D (2007) SNP applications in plants. In: Oraguzie NC, Rikkerink EHA, Gardiner SE, de Silva HN (eds) Association mapping in plants Springer, New York, NY, pp 95–102

Berard A, Le Paslier MC, Dardevet M, Exbrayat VF, Bonnin I, Cenci A, Haudry A, Brunel D, Ravel C (2009) High-throughput single nucleotide polymorphism genotyping in wheat (*Triticum* spp.). Plant Biotechnol J 7:364–374

Bernardo R (2008) Molecular markers and selection for complex traits in plants: learning from the last 20 years. Crop Sci 48(5):1649–1664

Bhatti (2018) Association analysis and mapping of fiber quality in cotton. Ph. D Thesis submitted to Kahramanmaras Sutcu Imam University, Turkey

Bidichandani SI, Ashizawa T, Patel PI (1998) The GAA triplet-repeat expansion in Friedreich ataxia interferes with transcription and may be associated with an unusual DNA structure. Am J Hum Genet 62:111–121

Blake TK, Kadyrzhanova D, Shepherd KW, Islam AKMR, Langridge PL, McDonald CL, Talbert LE (1996) STS PCR markers appropriate for wheat-barley introgression. Theor Appl Genet 82:715–721

Bolek Y (2003) Status of genome mapping and use in cotton improvement. KSÜ Fen ve Mühendislik Dergisi 6(2):72–79

Bolek Y, El-Zik KM, Pepper AE, Bell AA, Magill CW, Thaxton PM, Reddy OUK (2005) Mapping of verticillium wilt resistance genes in cotton. Plant Sci 168:1581–1590

Bolek Y, Hayat K, Bardak A, Azhar MT (2016) Molecular breeding of cotton. In: Adurakhmonov IY (ed) Cotton research. Intech Publishers, pp 123–166

Bornet B, Branchard M (2001) Nonanchored inter simple sequence repeat (ISSR) markers: reproducible and specific tools for genome fingerprinting. Plant Mol Biol Rep 19(3):209–215

Byers RL, Harker DB, Yourstone SM, Maughan PJ, Udall JA (2012) Development and mapping of SNP assays in allotetraploid cotton. Theor Appl Genet 124:1201–1214

Chen ZJ et al (2007) Toward sequencing cotton (*Gossypium*) genomes. Plant Physiol 145:1303–1310

Ching A, Caldwell KS, Jung M, Dolan M, Smith OS, Tingey S, Morgante M, Rafalski AJ (2002) SNP frequency, haplotype structure and linkage disequilibrium in elite maize inbred lines. BMC Genet 3:19

Coetes DJ, Byrne M (2005) Genetic variation in plant populations. Chapter 9. In: Henry RJ (ed) Plant diversity and evolutions-genotypic and phenotypic variation in higher plants. CABI Publishing, Wallingford, pp 139–164

Collard BC, Mackill DJ (2008) Marker assisted selection: an approach for precision plant breeding in the twenty-first century. Philos Trans R Soc B 363:557–572

Collard BCY, Jahufer MZZ, Brouwer JB, Pang ECK (2005) An introduction to markers, quantitative trait loci (QTL) mapping and marker-assisted selection for crop improvement: the basic concepts. Euphytica 142:169–196

Cuadrado A, Schwardzacher T (1998) The chromosomal organization of simple sequence repeats in wheat and rye genomes. Chromosoma 107:587–594

Cuming DS, Altan F, Akdemir H, Tosun M, Gurel A, Tanyolac B (2015) QTL analysis of fiber color and fiber quality in naturally green colored cotton (*Gossypium hirsutum* L.). Turk J Field Crops 20(1):49–58

Elshire RJ, Glaubitz JC, Sun Q, Poland JA, Kawamoto K, Buckler ES, Mitchell SE (2011) A robust, simple genotyping-by-sequencing (GBS) approach for high diversity species. PLoS One 6(5): e19379

Fang DD, Jenkins JN, Deng DD, McCarty JC, Li P, Wu J (2014) Quantitative trait loci analysis of fiber quality traits using a randommated recombinant inbred population in upland cotton (*Gossypium hirsutum* L.). BMC Genomics 15:397

Farahani F, Sheidai M, Koohdar F (2018) Genetic finger printing of cotton cultivars by ISSR molecular markers. Genetika 50(2):627–634

Feng CD, Stewart JMD, Zhang JF (2005) STS markers linked to the Rf1 fertility restore gene of cotton. Theor Appl Genet 110(2):237–243

Geng C, Gong Z, Huang JQ, Zhang ZC (1995) Identification of difference between cotton cultivars (*G. hirsutum*) using the RAPD method. Jiangsu J Agric Sci 11:21–24

Guo G, Shen X, Zhang T, Yu JZ, Kohel RJ (2003) Development of scar marker linked to a major qtl for high fiber strength and its usage in molecular-marker assisted selection in upland cotton. Crop Sci 43:2252–2256

Guo W, Wang W, Zhou B, Zhang T (2006) Cross species transferability of *G. arboreum*- derived EST-SSRs in the diploid species of *Gossypium*. Theor Appl Genet 112:1573–1581

Gupta PK, Varshney RK, Sharma PC, Ramesh B (1999) Molecular markers and their applications in wheat breeding. Plant Breed 118:369–390

Helentjaris T, Slocum M, Wright S, Schaefer A, Nienhuis J (1986) Construction of genetic linkage maps in maize and tomato using restriction fragment length polymorphisms. Theor Appl Genet 61:761–769

Hughes AL, Friedman R (2005) Expression patterns of duplicate genes in the developing root in *Arabidopsis thaliana*. J Mol Evol 60:247–256

Hu S, Wu S, Wang Y, Zhao H, Yuanyan Z (2014) Genetic diversity and genetic structure of different types of natural populations in *Osmanthus fragrans* Lour. and the relationships with sex ratio, population structure, and geographic isolation. Sci World J. Article ID 817080

Hulse-Kemp AM, Lemm J, Plieske J, Ashrafi H, Buyyarapu R, Fang DD et al (2015) Development of a 63K SNP array for cotton and high-density mapping of intraspecific and interspecific populations of *Gossypium* spp. G3 (Bethesda) 5:1187–1209

Hussain A, Ibrahim M, Tajammal MA, Naz MA (2005) Cotton varieties of Pakistan. Federal Seed Certification and Registration Department. Ministry of Food, Agriculture & Livestock (MINFAL), Government of Pakistan

International Cotton Advisory Committee (2018). http://www.icac.org

International Service for the Acquisition of Agri-biotech (ISAA) (2017). http://www.isaaa.org

Islam MS, Thyssen GN, Jenkins JN et al (2015) Detection, validation, and application of genotyping-by-sequencing based single nucleotide polymorphisms in upland cotton. Plant Genome 8(1):1–10

Jander G, Norris SR, Rounsley SD, Bush DF, Levin IM, Last RL (2002) Arabidopsis mapbased cloning in the post-genome era. Plant Physiol 129:440–450

Jia Y, Sun J, Wang X, Zhou Z, Pan Z, He S et al (2014) Molecular diversity and association analysis of drought and salt tolerance in *Gossypium hirsutum* L. germplasm. J Integr Agric 13:1845–1853

Jones CJ, Edwards KJ, Castaglione S, Winfield MO, Sala F, van de Wiel C, Bredemeijer G, Vosman B, Matthes M, Daly A, Brettschneider R, Bettini P, Buiatti M, Maestri E, Malcevschi A, Marmiroli N, Aert R, Volckaert G, Rueda J, Linacero R, Vasquez A, Karp A (1997) Reproducibility testing of RAPIDs, AFLP and SSR markers in plants by a network of European laboratories. Mol Breed 3:381–390

Joshi CP, Nguyen HT (1993) RAPD (random amplified polymorphic DNA) analysis based inter varietal genetic relationships among hexaploid wheats. Plant Sci 93:95–103

Jost A (1941) Sexual reproduction in yeast. Lecture at Technical Institute Lwow, Poland.

Jubrael JM, Udupa SM, Baum M (2005) Assessment of AFLP-based genetic relationships among date palm (*Phoenix dactylifera* L.) varieties of Iraq. J Am Soc Hortic Sci 130(3):442–447

Kalia RK, Rai MK, Kalia S, Singh R, Dhawan AK (2011) Microsatellite markers: an overview of the recent progress in plants. Euphytica 177:309–334

Khan SA, Hussain D, Askari E, Stewart JMD, Malik KA, Zafar Y (2000) Molecular phylogeny of *Gossypium* species by DNA fingerprinting. Theor Appl Genet 101(5–6):931–938

Kim CG, Jeong SC, Yoon WK, Park KW, Choi KH, Kim HM (2009) Development of genetically modified crops based on considerations of risk assessment and management. J Plant Biotechnol 36:360–365

Kingm RC, Stansfield WD (1990) A dictionary of genetics, 4th edn. Oxford University Press, New York, NY, p 188

Koebner RM, Summers RW (2003) 21st century wheat breeding: plot selection or plate detection? Trends Biotechnol 21:59–63

Kumar LS (1999) DNA markers in plant improvement: an overview. Biotechnol Adv 17:143–182

Lacape JM, Nguyen TB, Thibivilliers S, Bojinov B, Courtois B, Cantrell RG, Burr B, Hau B (2003) A combined RFLP-SSR-AFLP map of tetraploid cotton based on a *Gossypium hirsutum* x *Gossypium barbadense* backcross population. Genome 46:612–626

Lee SW (2011) Strategies for the development of GM crops in accordance with the environmental risk assessment (I). J Plant Biotechnol 38:125–129

Li F, Fan G, Wang K, Sun F, Yuan Y, Song G, Li Q, Ma Z, Lu C, Zou C, Chen W, Liang X, Shang H, Liu W, Shi C, Xiao G, Gou C, Ye W, Xu X, Zhang X, Wei H, Li Z, Zhang G, Wang J, Liu K, Kohel RJ, Percy RG, Yu JZ, Zhu YX, Wang J, Yu S (2014) Genome sequence of the cultivated cotton *Gossypium arboreum*. Nat Genet 46:567–572

Li C, Dong Y, Zhao T, Li L, Li C, Yu E, Mei L, Daud MK, He Q, Chen J, Zhu S (2016) Genome-wide SNP linkage mapping and QTL analysis for fiber quality and yield traits in the upland cotton recombinant inbred lines population. Front Plant Sci 7:1356

Liu D, Guo X, Lin Z, Nie Y, Zhang X (2006) Genetic diversity of Asian cotton (*Gossypium arboreum* L.) in China evaluated by microsatellite analysis. Genet Resour Crop Evol 53 (6):1145–1152

Liu Z, Sun Q, Ni Z, Yang T, McIntosh RA (1999) Development of SCAR markers linked to the Pm21 gene conferring resistance to powdery mildew in common wheat. Plant Breed 118:215–219

Lu YL, Yan JB, Guimarães CT, Taba S, Hao ZF, Gao SB et al (2009) Molecular characterization of global maize breeding germplasm based on genome-wide single nucleotide polymorphisms. Theor Appl Genet 120:93–115

Lynch M, Walsh B (1998) Genetics and analysis of quantitative traits. Sinauer, Sunderland, MA

Mackill DJ, Nguyen HT, Zhang J (1999) Use of molecular markers in plant improvement programs for rainfed lowland rice. Field Crops Res 64:177–185

Mei Y, Yu J, Xue A, Fan S, Song M, Pang C, Pei W, Yu S, Zhu J (2017) Dissecting genetic network of fruit branch traits in upland cotton by association mapping using SSR markers. PLoS One 12 (1):e0162815

Michalek W, Weschke W, Pleissner KP, Graner A (2002) EST analysis in barley defines a unigene set comprising 4,000 genes. Theor Appl Genet 104:97–103

Moose SP, Mumm RH (2008) Molecular plant breeding as the foundation for 21st century crop improvement. Plant Physiol 147:969–977

Nair S, Bentur JS, Rao UP, Mohan M (1995) DNA markers tightly linked to a gall midge resistance gene (Gm2) are potentially useful for marker-aided selection in rice breeding. Theor Appl Genet 91:68–73

Nicolia A, Manzo A, Veronesi F, Rosellini D (2014) An overview of the last 10 years of genetically engineered crop safety research. Crit Rev Biotechnol 34(1):77–88

Palanga KK, Jamshed M, Rashid MHO, Gong J, Li J, Iqbal MS, Liu A, Shang H, Shi Y, Chen T, Ge Q, Zhang Z, Dilnur T, Li W, Li P, Gong W, Yuan Y (2017) Quantitative trait locus mapping for verticillium wilt resistance in an upland cotton recombinant inbred line using snp-based high density genetic map. Front Plant Sci 8:382

Paran I, Michelmore RW (1993) Development of reliable PCR based markers linked to downy mildew resistance genes in lettuce. Theor Appl Genet 85:985–993

Park SH, Cho J, Kim YC, Kim SM, Lim SM, Lee GS (2018) National program for developing biotech crops in Korea. Plant Breed Biotechnol 6(3):171–176

Poland JA, Trevor WR (2012) Genotyping-by-sequencing for plant breeding and genetics. Plant Genome 5:92–102

Poland JA, Brown PJ, Sorrells ME, Jannink JL (2012) Development of high-density genetic maps for barley and wheat using a novel two enzyme genotyping-by-sequencing approach. PLoS One 7:e32253

Qin H, Chen M, Yi X, Bie S, Zhang C, Zhang Y, Lan J, Meng Y, Jiao C (2015) Identification of associated SSR markers for yield component and fiber quality traits based on frame map and upland cotton collections. PLoS One 10(1):e0118073

Qureshi SN, Saha S, Kantety RV, Jenkins JN (2004) EST-SSR: a new class of genetic markers in cotton. J Cotton Sci 8:112–123
Rahman M, Hussain D, Zafar Y (2002) Estimation of genetic divergence among elite cotton (*Gossypium hirsutum* L.) cultivars/genotypes by DNA fingerprinting technology. Crop Sci 42:2137–2144
Rahman M, Asif M, Ullah I, Malik KA, Zafar Y (2005) Overview of cotton genomic studies in Pakistan. In: Plant & Animal Genome Conference XIII. San Diego, CA. USA
Rahman M, Zafar Y, Paterson AH (2009) Gossypium DNA markers: types, numbers and uses. In: Paterson AH (ed) Genetics and genomics of cotton. Springer, Dordrecht, pp 101–139
Rahman MH, Ahmad A, Wang X, Wajid A, Nasim W, Hussain M, Ahmad B, Ahmad I, Ali Z, Ishaque W, Awais M, Shelia V, Ahmad S, Fahad S, Alam M, Ullah H, Hoogenboom G (2018) Multi-model projections of future climate and climate change impacts uncertainty assessment for cotton production in Pakistan. Agric For Meteorol 253-254:94–113
Roychowhury R, Taoutaou A, Khalid RK, Mohamed RAG, Jagatpati T (2014) Crop improvement in the ear of climate change. In: Roychowhury (ed) I.K International Publication house Ltd
Russell JR, Fuller JD, Macaulay M, Hatz BG, Jahoor A, Powell W, Waugh R (1997) Direct comparison of levels of genetic variation among barley accessions detected by RFLPs, AFLPs, SSRs and RAPDs. Theor Appl Genet 95:714–722
Sachidanandam R, Weissman D, Schmidt SC, Kakol JM et al (2001) A map of human genome sequence variation containing 1.42 million single nucleotide polymorphisms. Nature 409:928–933
Schulmann AH (2007) Molecular markers to assess genetic diversity. Euphytica 158(3):313–321
Schuster I (2011) Marker-assisted selection for quantitative traits. Crop Breed Appl Biotechnol S1:50–55
Shanti ML, George MLC, Cruz CMV, Bernardo MA, Nelson RJ, Reddy JN, Leung H, Sridhar R (2001) Identification of resistance genes effective against rice bacterial blight pathogen in eastern India. Plant Dis 85:506–512
Sheidail M, Shahriari ZH, Rokinzadeh H, Nourmohammadi Z (2007) RAPD and cytogenetic study of some tetraploid cotton (*Gossypium hirsutum* L) cultivars and their hybrids. Cytologia 72:77–82
Sica M, Gamba G, Montieri S, Gaudio L, Aceto S (2005) ISSR markers show differentiation among Italian populations of *Asparagus acutifolius* L. BMC Genet 6:17
Struss D, Plieske J (1998) The use of microsatellite markers for detection of genetic diversity in barley populations. Theor Appl Genet 97:308–315
Stuber CW, Polacco M, Senior ML (1999) Synergy of empirical breeding, marker-assisted selection, and genomics to increase crop yield potential. Crop Sci 39:1571–1583
Tanksley SD, Young ND, Paterson AH, Bonierbale MW (1989) RFLP mapping in plant breeding: new tools for an old science. Nat Biotechnol 7:257–264
Tariq M, Yasmeen A, Ahmad S, Hussain N, Afzal MN, Hasanuzzaman M (2017) Shedding of fruiting structures in cotton: factors, compensation and prevention. Trop Subtrop Agroecosyst 20(2):251–262
Tariq M, Afzal MN, Muhammad D, Ahmad S, Shahzad AN, Kiran A, Wakeel A (2018) Relationship of tissue potassium content with yield and fiber quality components of Bt cotton as influenced by potassium application methods. Field Crop Res 229:37–43
Thomas W (2003) Prospects for molecular breeding of barley. Ann Appl Biol 142:1–12
Tuberosa R, Salvi S, Sanguineti MC, Maccaferri M, Giuliani S, Landi P (2003) Searching for QTLs controlling root traits in maize: a critical appraisal. Plant and Soil 255:35–54
Ulloa M, Meredith WR Jr (2000) Genetic linkage map and QTL analysis of agronomic and fiber traits in an intraspecific population. J Cotton Sci 4(3):161–170
Ulloa M, Meredith WR Jr, Shappley ZW, Kahler AL (2002) RFLP genetic linkage maps from four F2.3 populations and a joinmap of *Gossypium hirsutum* L. Theor Appl Genet 104:200–208
United nation organization. https://www.un.org

United States Department of Agriculture (USDA) (2013) Cotton and wool outlook. Available at: http://www.ers.usda.gov/publications/cws-cotton-and-wool-outlook/cws-13e.aspx#. UjhSRRzCBmg

Usman M, Ahmad A, Ahmad S, Irshad M, Khaliq T, Wajid A, Hussain K, Nasim W, Chattha TM, Trethowan R, Hoogenboom G (2009) Development and application of crop water stress index for scheduling irrigation in cotton (*Gossypium hirsutum* L.) under semiarid environment. J Food Agric Environ 7(3–4):386–391

Van Deynze V, Stoffel AK, Lee M, Wilkins TA, Kozik A, Cantrell RG, Yu JZ, Kohel RJ, Stelly DM (2009) Sampling nucleotide diversity in cotton. BMC Plant Biol 9:125

Varshney RK, Graner A, Sorrells ME (2005) Genic microsatellite markers in plants: features and applications. Trends Biotechnol 23:48–55

Wang F, Stewart JMD, Zhang J (2007) Molecular markers linked to the Rf2 fertility restorer gene in cotton. Genome 50(9):818–824

Wang Q et al (2015) Genome-wide mining, characterization, and development of microsatellite markers in *Gossypium* species. Sci Rep 5:10638

Waqas M, Khan AA, Ashraf J, Qayyum A, Ahmad MQ, Iqbal MZ, Abbasi GH (2014) Molecular markers and cotton genetic improvement: current status and future prospects. Scientific World Journal 2014:607091

Welsh J, McClelland M (1990) Fingerprinting genomes using PCR with arbitrary primers. Nucleic Acids Res 18:7213–7218

Williams KJ (2003) The molecular genetics of disease resistance in barley. Aust J Agr Res 54:1065–1079

Wright RJ, Thaxton PM, El-Zik KM, Paterson AH (1998) D-subgenome bias of Xcm resistance genes in tetraploid *Gossypium* (cotton) suggests that polyploid formation has created novel avenues for evolution. Genetics 149(4):1987–1996

Yu J, Wang J, Lin W, Li S, Li H et al (2005) The genomes of *Oryza sativa*: a history of duplications. PLoS Biol 3(2):e38

Zhang ZS, Xiao YH, Luo M, Li XB, Luo XY, Hou L, Li DM, Pei Y (2005) Construction of a genetic linkage map and QTL analysis of fiber-related traits in upland cotton (*Gossypium hirsutum* L.). Euphytica 144(1-2):91–99

Zhang ZS, Hu MC, Zhang J, Liu DJ, Zheng J, Zhang K, Wang W, Wan Q (2009) Construction of a comprehensive PCR-based marker linkage map and QTL mapping for fiber quality traits in upland cotton (*Gossypium hirsutum* L.). Mol Breed 24:49–61

Zhang X, Meng Z, Zhou T, Sun G, Shi J, Yu Y, Zhang R, Guo S (2012) Mitochondrial SCAR and SSR markers for distinguishing cytoplasmic male sterile lines from their isogenic maintainer lines in cotton. Plant Breed 131(4):563–570

Zhang T, Qian N, Zhu X et al (2013) Variations and transmission of QTL alleles for yield and fiber qualities in upland cotton cultivars developed in China. PLoS One 8(2). Article ID e57220

Zhang SW, Zhu XF, Feng LC, Gao X, Yang B, Zhang TZ, Zhou BL (2016) Mapping of fiber quality QTLs reveals useful variation and footprints of cotton domestication using introgression lines. Sci Rep 6:31954

Zhao Y, Wang H, Chen W, Li Y (2014) Genetic structure, linkage disequilibrium and association mapping of verticillium wilt resistance in elite cotton (*Gossypium hirsutum* L.) germplasm population. PLoS One 9(1):e86308

Chapter 26
Development of Transgenic Cotton for Combating Biotic and Abiotic Stresses

Babar Hussain and Sultan Mahmood

Abstract Cotton is an important crop as it produces valuable textile fibre and contributes to economy of various textile producing countries. However, cotton production is under great threat due to persistent climate change as more frequent drought spells, soil salinization and heat stress, plant pathogens and insect pests significantly reduce its growth, development, fibre and seed yield. Cotton yield has shown up to 50–60% yield losses due to drought and biotic stresses. However, crop improvement through classical breeding takes almost 7–8 years making classical breeding a tedious, laborious, time-consuming and expensive process. On the other hand, DNA recombinant technology ensures the transfer and integration of target gene into cotton genome within a cropping season, thus shortening the breeding cycle significantly. Therefore, transgenic cotton with improved drought, heat and salinity tolerance and resistance against herbicides, diseases and insects has been developed extensively in relatively shorter time. Transgenic method has also helped in improvement of the traits which are considered difficult or impossible to be improved through classical breeding. Another success of transgenic cotton is its great acceptance to farming community due to economic benefits resulting from biotic and abiotic stress tolerance and reduced costs of insect pest sprays. Furthermore, transgenic cotton is under cultivation in various regions and countries; thus we have reviewed the progress in development for biotic and abiotic stress tolerance in cotton through transgenic method. This information will be a valuable resource for cotton breeders and biotechnologists for planning of their future research.

Keywords Transgenic cotton · GMOs · Salinity tolerance · Drought tolerance · Heat tolerance · Herbicide resistance · Insect resistance · Disease resistance

The original version of this chapter was revised. A correction to this chapter can be found at https://doi.org/10.1007/978-981-15-1472-2_31.

B. Hussain (✉)
Faculty of Engineering and Natural Sciences, Sabanci University, Istanbul, Turkey

Faculty of Life Sciences, University of Central Punjab, Lahore, Pakistan
e-mail: babar.hussain@ucp.edu.pk

S. Mahmood
Department of Plant Breeding and Genetics, Bahauddin Zakariya University, Multan, Pakistan

S. Ahmad, M. Hasanuzzaman (eds.), *Cotton Production and Uses*,
https://doi.org/10.1007/978-981-15-1472-2_26

26.1 Introduction

Cotton is an important crop as it produces valuable textile fibre and contributes to economy of various textile producing countries such as Pakistan (Khan et al. 2004; Ahmad et al. 2014, 2018; Abbas and Ahmad 2018; Ahmad and Raza 2014; Ali et al. 2011a, b, 2013a, b, 2014a, b; Usman et al. 2009). However, its production is under great threat due to persistent climate change as it is expected that global temperature may rise another 1.5 °C in 2030 instead of speculated 2040 (Xu et al. 2018; Ahmad et al. 2017; Amin et al. 2017, 2018; Rahman et al. 2018; Tariq et al. 2017, 2018). Therefore, global warming is resulting in higher temperature and heat stress is the most eminent threat to global agriculture production as it increases the intensity of other abiotic stresses; e.g. drought spells are becoming more intense and frequent, soil salinization is resulting in soil degradation on large scale, and unexpected rains, plant pathogens and insect pest population are on the rise (Budak et al. 2015; Hussain 2015; Hussain et al. 2017). Heat stress increases the cell injury in cotton (Khan et al. 2008) and decreases the number of sympodial branches, number of seeds and seed cotton yield (Khan et al. 2008). Heat stress also increases the drought spells that severely reduces cotton growth, development, fibre quality and yield; e.g. drought reduced the annual cotton production by 34% in Pakistan (Ullah et al. 2017). Similarly, soil salinity also severely affects the plant growth and development, thus reducing the root and shoot lengths and fresh and dry tissue biomass in cotton (Farooq et al. 2019).

Cotton seed and fibre yield is also decreased by various diseases and insect pest so the breeding for biotic and abiotic stress tolerance in cotton for sustainable production under climate change scenario is vital. However, crop improvement through classical breeding methods such as pedigree method takes almost 7–8 years thus making classical breeding a tedious, laborious, time-consuming and expensive process (Hussain et al. 2012; Minhas et al. 2018). However, DNA recombinant technology or transgenic method that involves the transfer of stress responsive genes from other sources to cotton reduces the breeding cycle significantly and confers stress tolerance. Therefore, this genetically modified organism (GMO) technology has shown great potential of varietal development with anticipated traits across the globe. It is greatly accepted by plant researchers and breeders because of its ability for robust development of superior varieties that result in higher crop production. The GMOs provide the benefits of higher production and more biotic and abiotic stress tolerance. Transgenic method is routinely being used to transfer biotic and abiotic stress responsive genes in cotton and has shown great success (Zhang et al. 2017, 2009; Mishra et al. 2017; Awan et al. 2015; Wang et al. 2015; Chen et al. 2016; Farooq et al. 2019; Hao et al. 2018). Another reason behind the success of GMO cotton is its great acceptance to farming community due to economic benefits resulting from biotic and abiotic stress tolerance and reduced costs of insect pest sprays. Therefore, transgenic Bt cotton and other GMO crops are widely grown in countries like the USA, Canada, India and Pakistan (Hussain 2015). Keeping in view the success of GMO cotton, this book chapter focusses on use of GMO technology for improvement of drought, heat, salinity tolerance and resistance against herbicides, diseases and insect pests. We hope that this chapter will be a great source for breeders and biotechnologists to plan their future research keeping in mind the genes and methods discussed in this chapter.

26.2 Transgenic Cotton with Improved Drought Tolerance

In recent years, drought spells have increased due the effect of climate change or global warming (Ali et al. 2011a, b; Budak et al. 2015). Drought severely affects growth, development, fibre quality and yield in cotton. For example, annual 34% decline in cotton production was reported in Pakistan (from 14.4 to 9.68 million bales) due to increased drought spells and high temperature (Ullah et al. 2017). Similarly, crop losses amounted to as high as 50–67% over a period of 50 years in the United States and other parts of the world, thus incurring losses to farming community and agro-based industry (Comas et al. 2013; Budak et al. 2015). Therefore, identification of drought tolerance germplasm and development of drought resilience is key for sustainable cotton yield in climate change scenario (Table 26.1).

Breeding for drought is not easy as drought tolerance is a complex trait and plants respond to drought through various genes, hormones, signalling molecules, transcription factors (TFs), proteins, microRNAs (miRNAs), cofactors, metabolites and ions (Budak et al. 2015). For example, proline content, abscisic acid (ABA) content, soluble sugar content, peroxidase (POD) and superoxide dismutase (SOD) enzymatic activity increased in cotton under drought. However, gibberellic acid (GA) content and catalase (CAT) activity decreased under drought stress while the interaction between GA and ABA signalling played a vital role in compensatory root growth after drought was withdrawn (Niu et al. 2018). This complexity in drought response hinders the rapid development of drought-tolerant cotton cultivars through classical breeding; e.g. development of a drought-tolerant cotton variety, BH-167, took seven cropping seasons (Minhas et al. 2018). This makes the improvement process through classical breeding tedious, laborious, time consuming and expensive (Hussain 2015). However, with the advances in molecular biology and sequencing techniques (Hussain 2015; Budak et al. 2015), it has become much easier to identify any major drought tolerance gene/genes in cotton wild relatives or other crops, transfer them to cotton through transformation techniques and improve the tolerance in cultivated cotton.

Various genes have been transferred into cotton for improving drought tolerance. Deeper and dense root system is an indicator of drought tolerance in plants. The *Agrobacterium tumefaciens*-mediated transformation of a rice TF gene, SNAC1, into cotton cultivar YZ1 enhanced the plant growth and particularly root development as compared to the wild-type plants under drought and salinity. The proline content in transgenic plants was enhanced but the malondialdehyde (MDA) content was reduced under both stresses conferring the tolerance to salinity and drought. The

Table 26.1 Summary of transgenes transformed in cotton for improvement of drought tolerance and related traits whereas ↑ means increase/improvement in a trait and ↓ shows decrease or reduction in a trait

Transgene	Donor	Traits improved in transgenic cotton plants	Reference
Beta	*Escherichia coli*	↑ drought tolerance; ↑ glycine betaine content; ↑ relative water content (RWC); ↓ ion leakage;↑ photosynthesis; ↑ osmotic adjustment; ↓ lipid membrane peroxidation	Lv et al. (2007)
AnnBj1	Mustard	↑ drought tolerance;↑ oxidative tolerance; ↑RWC; ↑ sucrose; ↑ peroxidase (POD) activity; ↑ proline content; ↑ fresh weight; ↑ dry weight; ↑ total chlorophyll; ↓ hydrogen peroxide	Divya et al. (2010)
SNAC1	*O. Sativa*	↑ drought tolerances; ↑ root development; ↑ proline content; ↑ boll number; ↓ transpiration rate; ↓ malondialdehyde (MDA) content	Liu et al. (2014)
AtEDT1 Or*HDG11*	*A. Thaliana*	↑ drought tolerance; ↑ root system; ↑ soluble sugar content; ↑ proline content; ↑ antioxidant enzymes; ↑ stomatal and leaf epidermal cell size; ↑ cotton yield; ↓ stomatal density	Yu et al. (2016)
ScALDH21	*Syntrichia caninervis*	↑ drought tolerance; ↑ proline content; ↑ POD activity; ↑ photosynthetic rate; ↑ boll size; ↑ plant height; ↑ fibre yield; ↓ lipid peroxidation	Yang et al. (2016)
OsSIZ1	*O. Sativa*	↑ drought tolerance; ↑ growth; ↑ net photosynthesis rate; ↑ fibre yield	Mishra et al. (2017)
AmDUF1517	*Ammopiptanthus mongolicus*	↑ drought tolerance; ↑ cold tolerance; ↑ antioxidant enzymes, i.e. glutathione S-transferase (GST), POD, superoxide dismutase (SOD), catalase (CAT) activity; reactive oxygen species (ROS) homeostasis; ↓ cell membrane injury	Hao et al. (2018)
IPT	*A. tumefaciens*	↑ vegetative stage drought tolerance	Zhu et al. (2018)
StDREB2	Potato	↑ drought tolerance; ↑ boll number; ↑ plant biomass; ↑ boll number; ↑ soluble sugars content; ↑ RWC; ↑ soluble protein content; ↑ proline content; ↑ chlorophyll content; ↑ ROS scavenging; ↑ CAT, GST, POD, SOD activity; ↑ *GhERF2*, *GhNAC3*, *GhRD22*, *GhDREB1A*, *GhDREB1B*, *GhDREB1C* expression; ↓ hydrogen peroxide; ↓ MDA content	El-Esawi et al. (2019)

drought tolerance in transgenic cotton was also measured in terms of lower transpiration rate and higher boll number (Liu et al. 2014). Similarly, transgenic cotton plants transformed with betA gene showed increased glycine betaine (GB) accumulation and drought tolerance in comparison to the control plants. The transgenics showed higher relative water content (RWC), photosynthesis and osmotic adjustment while ion leakage and lipid membrane peroxidation were reduced (Lv et al. 2007). In another study, transgenic cotton with mustard annexin (AnnBj1) transgene yielded salinity, oxidative and drought tolerance. The drought tolerance was associated with reduced hydrogen peroxide in guard cells and increased POD activity, sucrose, proline, fresh weight, RWC, dry weights and total chlorophyll in transformed plants (Divya et al. 2010).

Transformation of *Arabidopsis* enhanced drought tolerance1/homeodomain glabrous11 gene (AtEDT1/HDG11) conferred drought and salinity tolerance in cotton by improving root system, soluble sugar content, proline content and activities of antioxidant enzymes as compared with wild-type plants. Transformed plants showed reduced stomatal density and increased stomatal and leaf epidermal cell size. Under field conditions, transgenic cotton demonstrated greatly improved drought tolerance and improved cotton yield under drought and normal conditions (Yu et al. 2016). Similarly, transformation of a *Syntrichia caninervis* gene, ScALDH21, into cotton conferred drought tolerance by enhanced proline content, photosynthetic rate and POD activity while lipid peroxidation was reduced. Transgenic plants showed larger bolls and greater plant height and fibre yield in field conditions (Yang et al. 2016). In another study, overexpression of SUMO E3 ligase gene (OsSIZ1) in cotton conferred drought tolerance through better growth, higher net photosynthesis rate and higher fibre yield in greenhouse and field conditions when compared with non-transgenic plants in both growth chamber and greenhouse conditions (Mishra et al. 2017).

Transformation of cotton with an *Ammopiptanthus mongolicus* gene, AmDUF1517, enhanced the drought, salt and cold tolerance in transgenic plants by higher production and activity of antioxidant enzyme such as glutathione S-transferase (GST), POD, SOD and CAT, reactive oxygen species (ROS) homeostasis and avoiding the cell membrane injury (Hao et al. 2018). Similarly, overexpression of potato TF, StDREB2, in cotton conferred drought tolerance by enhanced boll number, plant biomass, boll number, soluble sugar content, RWC, soluble protein content, proline content, chlorophyll content, ROS scavenging and antioxidant enzyme (CAT, GST, POD, SOD) activity in transgenic plants than wild-type plants. Contrarily, hydrogen peroxide and malondialdehyde contents were greatly reduced in transgenic plants. Additionally, transgenic plants had higher expression levels of stress-responsive TFs such as GhERF2, GhNAC3, GhRD22, GhDREB1A, GhDREB1B, GhDREB1C and antioxidant enzyme genes than wild-type plants (El-Esawi et al. 2019). However, transgenic cotton with isopentenyltransferase gene (IPT) responded differently to drought stress at vegetative and reproductive stage; i.e. transgenic plants outperformed control plants only when drought occurred at vegetative stage (Zhu et al. 2018). Therefore, selection of the transgenes that express at the required growth stage in plants is vital and should be decided according to occurrence of stress.

26.3 Transgenic Cotton with Improved Salinity Tolerance

More than one billion hectares of land worldwide is affected by soil salinity (Wicke et al. 2011) and this marginal land is expected to increase due to soil salinization that is caused by human activities (land clearing and saline irrigation) and natural processes. Soil salinity severely affects the plant growth and development (Farooq et al. 2019; Hussain et al. 2015) and it reduced the root and shoot lengths and fresh and dry tissue biomass in cotton (Farooq et al. 2019) as Na+ influx disrupts leaf function and nutrient balance and even causes leaf/plant mortality due to sodium toxicity in non-halophytes like wheat, cotton, rice, etc. (Hussain et al. 2017). On the other hand, global demand for food and fibre is increasing due to growing population. Therefore, transgenic method for crop improvement comes into play due to its robust results (Budak et al. 2015). Several genes have been transformed in cotton to improve salt tolerance; e.g. transformation of cotton with mustard annexin (AnnBj1) gene conferred salt tolerance through reduced production of H_2O_2, increased POD activity, proline, sucrose, RWC, dry weights and total chlorophyll. Transgenic cotton treated with salinity showed enhanced expression levels of genes coding for Δ-pyrroline-5-carboxylate synthetase in leaves and sucrose synthase, sucrose phosphate synthase and cellulose synthase A in the fibre and leaves. There were no adverse effects on fibre quality, normal seed development and cellulose content under salinity (Divya et al. 2010). Similarly, overexpression of SNAC1 TF in cotton conferred salt tolerance through enhanced root development and proline content and reduced MDA content as compared to the wild types under salinity (Liu et al. 2014). In another study, transformation of cotton with HDG11 gene conferred salinity tolerance by improving root system, proline content, and activities of antioxidant enzymes as compared with wild-type plants (Yu et al. 2016).

GB is a drought and salinity responsive osmoprotectant and choline monooxygenase (CMO) is a major catalyst in its biosynthesis. A CMO gene got from *Atriplex hortensis*, AhCMO, was transferred to improve salt tolerance. The expression of AhCMO gene was confirmed through molecular analysis. In greenhouse, transgenic plants accumulated 26% and 131% more GB than control plants under normal and saline conditions, respectively. On the other hand, electrolyte leakage, MDA content and osmotic potential were greatly reduced in leaves of the transgenic plants than wild-type plants under saline condition. Additionally, Fv/Fm and net photosynthesis rates were not affected significantly in transgenic lines as compared to non-transgenic plants, thus making transgenic cotton more salt tolerant than the non-transgenic ones. This is due to elevated GB that provides protection to the cell membrane and photosynthetic machinery. Under field condition, transgenic plants had significantly higher seed cotton yield under saline field conditions (Zhang et al. 2009). An *Apocynum venetum* DEAD-box helicase 1 (AvDH1) gene is expressed under salinity and its overexpression in cotton conferred salinity tolerance. The integration and expression of AvDH1 gene was confirmed through Southern and Northern blotting. In growth chamber, transgenic cotton was much more salt tolerant than control plants. Transgenic plants showed reduced membrane ion leakage and enhanced SOD activity that resulted in lower membrane damage and higher plant survival rates. In saline field condition, the transgenic plants had enhanced number of bolls, seed cotton yield and boll weight than control plants. The difference between transgenic and non-transgenic plants increased under high soil salinity levels (Chen et al. 2016). This study indicates that transgenic cotton expressing AvDH1 is a promising option for increasing crop productivity in saline fields. The above-mentioned studies show the increased cotton yield and fibre quality in transgenic cotton lines under field conditions as compared to non-transgenic plants, thus providing a promising opportunity to develop salt resilient cotton cultivars for farmers' field (Table 26.2).

26.4 Transgenic Cotton with Improved Heat Tolerance

The accumulation of carbon dioxide and greenhouse gasses in the atmosphere is causing global warming or constant increase in worldwide temperature. There is a dire need to stop global warming at 1.5 °C above pre-industrial levels, the limit set in the 2015 Paris climate agreement, and if the globe warms up by 2 °C, the number of people facing water scarcity will double. This extra 0.5 °C warming will make over 1.5 billion people vulnerable to deadly heatwaves in addition to increased disease risk for hundreds of millions. There are good chances that we may breach 1.5 °C in 2030 instead of speculated 2040 (Xu et al. 2018). Among the abiotic stresses faced by plants, heat stress is the most eminent threat to global agriculture production and food security as it increases the occurrence of other abiotic stresses (Budak et al. 2015; Hussain 2015; Hussain et al. 2017) such as more frequent drought spells, soil salinization, unexpected rains, plant pathogens and insect pests.

Heat stress in cotton increases the cell injury (Khan et al. 2008) and decreases the number of sympodial branches, number of seeds and seed cotton yield (Khan et al. 2008). The evaluation of square Bt cotton under heat stress found that temperature higher than 38 °C significantly reduced the square insecticidal protein contents after 24 h of elevated temperature. However, temperature below 38 °C had little effect on the square amino acid concentrations, glutamic-pyruvic transaminase (GPT) activity, soluble protein contents, glutamic-oxalacetic transaminase (GOT) and peptidase activities but temperature above 38 °C decreased the soluble protein contents and GPT and GOT activities and increased the amino acid concentrations and peptidase activities. Thus higher temperature led to unstable insect resistance during square stage (Wang et al. 2015). Therefore, transgenic method can be used to enhance the heat tolerance in cotton in relatively shorter time than classical breeding to pace with rapid climate change.

Table 26.2 Summary of transgenes transformed in cotton for improvement of salinity, heat and herbicide resistance/tolerance and related traits whereas ↑ means increase/improvement in a trait and ↓ shows decrease or reduction in a trait

Transgene	Donor	Traits improved in transgenic cotton plants	Reference
AhCMO	*Atriplex hortensis*	↑ salt tolerance; ↑ glycine betaine content; ↑ Fv/Fm; ↑ cell membrane stability; ↑ net photosynthesis rate; ↑ osmotic adjustment; ↑ seed cotton yield; ↓ malondialdehyde (MDA) content; ↓ osmotic potential; ↓ electrolyte leakage	Zhang et al. (2009)
AnnBj1	Mustard	↑ salt tolerance; ↑ oxidative tolerance; ↑ drought tolerance; ↑ RWC; ↑ sucrose; ↑ peroxidase (POD) activity; ↑ proline content; ↑ fresh weight; ↑ dry weight; ↑ total chlorophyll; ↓ hydrogen peroxide	Divya et al. (2010)
SNAC1	*O. Sativa*	↑ salt tolerances; ↑ drought tolerance; ↑ root development; ↑ proline content; ↑ boll number; ↓ transpiration rate; ↓ MDA content	Liu et al. (2014)
AtEDT1 Or*HDG11*	*A. Thaliana*	↑ salt tolerance; ↑ drought tolerance; ↑ root system; ↑ soluble sugar content; ↑proline content; ↑ antioxidant enzymes; ↑ stomatal and leaf epidermal cell size; ↑ cotton yield; ↓ stomatal density	Yu et al. (2016)
AvDH1	*Apocynum venetum*	↑ salt tolerance; ↑ plant survival rates ↑ SOD activity; ↑ number of bolls; ↑ boll weights; ↑ seed cotton yield; ↓ membrane ion leakage	Chen et al. (2016)
AtHSP101	*A. thaliana*	↑ heat tolerance; ↑ pollen tube elongation; ↑ pollen germination rate; ↑ boll set; ↑ number of seeds; ↑ fibre yield	Burke and Chen (2015)
OsSIZ1	*O. Sativa*	↑ heat tolerance; ↑ drought tolerance; ↑ growth; ↑ net photosynthesis rate; ↑ fibre yield	Mishra et al. (2017)
Phos-Phinothric	*S. hygroscopicus*	↑ herbicide (Basta @ 15,000 ppm) tolerance	Keller et al. (1997)
GTGene Or*EPSPS*	*S. typhimurium*, *E. coli*	↑ herbicide (glyphosate @ 1900 ml/acre) tolerance	Awan et al. (2015)
EPSPS	*S. typhimurium*, *E. coli*	↑ herbicide (glyphosate @ 1900 ml/acre) tolerance	Latif et al. (2015)
G2-aroA Or*EPSPS*	*S. typhimurium*, *E. coli*	↑ herbicide (glyphosate @ 45.0 mmol/L) tolerance	Zhang et al. (2017)

An *Arabidopsis thaliana* heat shock protein 101 (AtHSP101) known for its important role in heat tolerance at vegetative stage was overexpressed in cotton Coker 312 through *Agrobacterium*-mediated transformation. The pollens from cotton lines transformed with AtHSP101 exhibited significantly greater pollen tube elongation and higher germination rate under higher temperatures or post-heat exposure when compared with non-transgenic plants. Under both the field and greenhouse condition, transgenic cotton lines exhibited significantly higher boll set and number of seeds under elevated day and night temperatures as compared with control plants (Burke and Chen 2015). Therefore, enhanced heat tolerance of reproductive machinery in transgenic cotton is pivotal for higher and sustainable yield in climate change scenario. Similarly, the SUMO E3 ligase gene (AtSIZ1) is an important gene for abiotic stress response in plants as the plants with loss of function mutation in AtSIZ1 lead to enhanced sensitive to abiotic stress such as drought, salinity and heat stresses. Overexpression of its rice homolog, OsSIZ1, in cotton conferred drought and heat tolerance through better growth, development and higher net photosynthesis rate when compared with non-transgenic plants in both growth chamber and greenhouse conditions (Mishra et al. 2017). The heat and drought tolerance of transgenic cotton lines was correlated with higher fibre yield in greenhouse and field condition trials in rainfed and reduced irrigation conditions. Therefore, net photosynthesis rate and stable reproductive machinery under heat stress are important traits that can be exploited for breeding heat-tolerant cotton (Table 26.2).

26.5 Transgenic Cotton with Improved Herbicide Resistance

Herbicide-resistant plants are those who can tolerate the herbicides; thus spray of herbicides just kills the weeds/unnecessary herbs without harming the main crop in the field. For example, cotton injury in glyphosate-resistant cotton was only 5% (Burke et al. 2005). However, this efficiency has improved in recent resistant types. Resistant against herbicides may possibly be increased by using different methods. Different herbicide enzymes were introduced into plants that have excellent potential to resist the herbicides action against plants. Moreover, plants can be genetically modified to prevent access of herbicide to target site of plants. Glyphosate is an active ingredient for broad-spectrum weedicides and has been focused in most of such researchers for development of herbicide-tolerant crops (Gianessi 2013). Different genes have been transferred into cotton for improving herbicide tolerance.

Transformation of a *Streptomyces hygroscopicus* gene coding for phosphinothric was performed in four cotton varieties through particle bombardment. The transgenic plants were recovered based on expression of marker gene β-glucuronidase (gus) and integration and expression of transgene in cotton genome was confirmed by Southern and Northern blot analyses. Transgenic plants showed tolerance to Basta® herbicide up to 15,000 ppm in greenhouse. The herbicide tolerance was inherited to next generations in a Mendelian fashion, thus showing the stable integration of transgene (Keller et al. 1997). Similarly, *Agrobacterium*-mediated transformation of cotton cultivar MNH-786 with GT gene (with 1.05%

transformation efficiency) was confirmed by ELISA and gene-specific PCR. Transgenic cotton lines survived 1900 ml/acre dose of glyphosate spray (Awan et al. 2015) while all herbs/weeds growing along the crop burned completely 5 days after the glyphosate spray.

Agrobacterium-mediated transformation of transgenic cotton variety, CEMB-02 (having Cry1Ac and Cry2A genes), with a 5-enolpyruvylshikimate-3-phosphate synthase gene (EPSPS) conferred herbicide tolerance and transgenic plants survived 1900 ml/acre dose of glyphosate spray (41%) while herbs/weeds burned completely. Additionally, insects feeding on transgenic plants showed 100% mortality rate due to presence of Cry1Ac and Cry2A genes (Latif et al. 2015). Using the same transformation method, a novel gene G2-aroA coding for EPSPS was transferred into cotton cultivar K312 and its integration and expression was confirmed by PCR and Southern and Western blot analyses. The transgenic plants showed significantly higher resistance to glyphosate as compared to wild-type plants and transgenics kept growth and bloom and produced seeds even after being applied with 45.0 mmol/L of glyphosate (Zhang et al. 2017). Thus G2-aroA or EPSPS gene is a strong candidate for future breeding for herbicide tolerance in cotton and other crops (Table 26.2).

26.6 Transgenic Cotton with Improved Disease Resistance

Plant diseases pose great threat to international food security as they cause 10% losses to global crop production (Hussain 2015). The potato blight in Ireland, maize leaf bight in the USA and coffee rust in Brazil are examples of complete crop failure to plant diseases. Similarly, Great Bengal Famine of 1943 also highlights that crop failure could put human life at risk. The fast-evolving pathogens and emergence of new pathogen races also make the breeding for disease resistance a challenging task (Hussain 2015). The transgenic method is one of the most prevalent methods for breeding disease resistance plants as it shortens the breeding cycle. Various genes from different resources have been introduced in cotton for improving its disease resistance that are listed in Table 26.3. Recently, it has been noted that chitinase-encoding transgenes from different fungi have the ability to increase plant resistance against fungal and bacterial pathogens as a broad range level (Tohidfar et al. 2012). Chitinase-encoding gene has been identified from diverse group of fungi. There are two types of diseases that pose serious threat to cotton production worldwide: viral and fungal disease.

Cotton leaf curl virus (CLCuV), a *Begomovirus*, belonging to *Geminiviridae* family, is an eminent threat to cotton production as it causes the leaf curl disease (CLCuD) and reduces the cotton yield up to 80% in India and Pakistan. Gemini viruses mainly occur in tropical and warm areas and cause significant economic losses to farmers as CLCuD cause growth stunting, thickened leaf veins, leaf curling and darkening and leaf enations on lower leaf. The plants infected with CLCuD have stunted growth, flowering and boll setting, thus lowering the cotton yield (Sattar et al. 2013). The resistance against CLCuV tends to collapse in resistance varieties

Table 26.3 Summary of transgenes transformed in cotton for improvement of disease resistance and related traits whereas ↑ means increase/improvement in a trait and ↓ shows decrease or reduction in a trait

Transgene	Donor	Improved traits in transgenic cotton	Reference
Tv-ech1	*Trichoderma virens*	↑ elevated endochitinase activities ↑ resistance against *Rhizoctonia solani* fungus ↑ resistance against *Alternaria alternate* fungus	Emani et al. (2003)
AV2	CLCKV	↑ cotton leaf curl virus (CLCuD) resistance	Sanjaya et al. (2005)
D4E1	Synthetic antimicrobial peptide	↑ resistance against *Fusarium verticillioides* fungus ↑ resistance against *Verticillium dahliae* fungus ↑ resistance against *Thielaviopsis basicola* fungus	Rajasekaran et al. (2005)
chitinase (*chi*) gene	Common bean	↑ resistance against *Verticillium dahlia* fungus	Tohidfa et al. (2009)
Chi II	*O. Sativa*	↑ resistance against *Fusarium oxysporum* fungus ↑ resistance against *Alternaria macrospora* fungus ↑ survival rate; ↓ number and length of lesions	Ganesan et al. (2009)
AC1	Geminiviruses	↑ cotton leaf curl virus (CLCuD) resistance	Amudha et al. (2010)
AC1	Geminiviruses	↑ cotton leaf curl virus (CLCuD) resistance ↓ and delayed disease symptoms	Hashmi et al. (2011)
AV1	Geminiviruses	↑ cotton leaf curl virus (CLCuD) resistance	Amudha et al. (2011)
Endochitinase	Phaseolus	↑ resistance against *Verticillium dahlia* fungus ↓ foliar disease symptoms, vascular discoloration ↑ plant height	Tohidfar et al. (2012)
βC1	CLCV antisense gene	↑ cotton leaf curl virus (CLCuD) resistance ↓ disease viral symptoms	Sohrab et al. (2016)
SG1, SG4	*Gastrodia elata*	↑ resistance against *Verticillium dahlia* wilt	Wang et al. (2016)

under mild or severe infection at initial growth stages. This resistance leaves cotton under constant threat of infection, infestation and cultivars developed against the virus through classical breeding shown low to moderate resistance and highly resistance varieties are not present. Therefore, an effective way of developing virus-resistant plants is pathogen-derived resistance (PDR) that uses antisense and cross-protection approach. For begomoviruses, overexpression of viral coat protein

(CP) in plants has been a promising tool for developing resistance against viruses in plants. Several antisense genes and CPs from geminiviruses such as *AC1*, *AV1*, *AV2* and *βC1* have been overexpressed in cotton that conferred enhanced resistance against CLCuV (Sanjaya et al. 2005; Amudha et al. 2010, 2011; Hashmi et al. 2011; Sohrab et al. 2016).

Fungi are one of the most devastating pathogens that causes significant yield losses in cotton; e.g. cotton *Verticillium* wilt caused by *Verticillium dahlia* fungus is a notorious disease that attacks cotton seedlings in most cotton-growing areas. *V. dahliae* is a soil-borne pathogen, which infects the plants through root system causing stunted growth, wilting and defoliation, thus incurring 15–70% yield losses (Tohidfa et al. 2009; Wang et al. 2016). However, the remedial against the fungal pathogens is hidden in another group of fungi that produce endochitinase enzyme-coding genes. Mycoparasitic fungi are a rich source of genes that could be transferred to crop plants for enhancing resistance against insect pests and fungal plant pathogens. Many chitinase-encoding genes from some fungi have been transformed in cotton to confer resistance against a broad range of fungal pathogens as chitinase has ability to degrade chitin that makes up 3–60% of the fungi cell wall (Collinge et al. 1993). For example, overexpression of chitinase ending *Tv-ech1* gene from *Trichoderma virens* fungus in cotton conferred broad-spectrum resistance against fungal pathogens *Rhizoctonia solani* and *Alternaria alternate* (Emani et al. 2003). Similarly, overexpression of *D4E1* in cotton greatly enhanced resistance against three fungi, i.e. *Fusarium verticillioides*, *Thielaviopsis basicola* and *Verticillium dahlia* (Rajasekaran et al. 2005). In another study, overexpression of chitinase-encoding genes *chi* and *Chi II* from common bean and rice, respectively, conferred tolerance against *Verticillium dahliae*, *Fusarium oxysporum* and *Alternaria macrospora* measured in terms of higher survival rate, less number and length of disease lesions (Tohidfa et al. 2009; Ganesan et al. 2009).

The overexpression of an *Endochitinase*-encoding gene from fungus *Phaseolus* in cotton conferred resistance against *Verticillium dahlia* fungus that was measured in terms of decreased foliar disease symptoms and vascular discoloration. On the other hand, transgenic plants had higher plant height as compared with non-transgenic plants (Tohidfar et al. 2012). Enhanced resistance against *Verticillium dahlia* was also observed in transgenic cotton transformed with *SG1* and *SG4* gene transferred from *Gastrodia elata* fungus (Wang et al. 2016). Thus, chitinase-encoding genes are useful resource for engineering plants resistant to plant pathogens particularly fungal plant pathogens.

26.7 Transgenic Cotton with Improved Insect Resistance

Insects particularly Lepidoptera are serious threat to global crop production as they chew almost all parts of the plants (Shabbir et al. 2014). Several insects affect the cotton yield but the cotton bollworm (*Helicoverpa armigera*) is one of the most devastating pests that causes great yield losses; e.g. its outbreak caused losses of over

1.5 billion US dollars in China in 1992. Other important cotton insects are fall armyworm (*Spodoptera frugiperda*), cotton boll weevil (*Anthonomus grandis*), cotton bollworm (*Heliothis zea*), pink bollworm (*Pectinophora gossypiella*), etc. (Perlak et al. 1990; Wu et al. 2011; de Oliveira et al. 2016; Yue et al. 2016; Wan et al. 2017). The transgenic cotton overexpressing a *Bacillus thuringiensis* (Bt) bacterium-origin *Cry1Ab* and *Cry1Ac* genes is an effective approach for controlling the chewing insects in cotton since 1990s (Perlak et al. 1990; Liu et al. 2017; Chen et al. 2017). The significant success of Bt cotton is associated with production of *Cry1Ac* toxin in Bt cotton which causes insect mortality. For example, when *Cry1Ac*-resistant and *Cry1Ac*-susceptible cotton bollworm (*Helicoverpa armigera*) was fed with transgenic cotton overexpressing synthetic *Vip3AcAa* and Bt-*Cry1Ac* genes, it caused swelling and breakage in the microvilli and endoplasmic reticulum of the midgut cells of bollworm larvae. Additionally, the inner cristae of the midgut cells' mitochondria became indistinct and disruption in the boundaries of karyotheca in the nucleus of midgut cells was also observed. These changes in microvilli and goblet cell cavity of the midgut epithelial cells of the bollworm eventually killed the insect (Chen et al. 2018). Therefore, this insect mortality without toxic insecticide spays provides a unique opportunity to get rid of insects in a more environmentally safe way.

The transgenic Bt cotton helps to decrease the number of pesticide sprays on cotton to control the chewing/feeding insects, thus reduces the production cost and increases the cotton yield, thus increasing the profit for farmers. Therefore, Bt cotton has widely been accepted by farming community and being grown in several countries including Australia, China, India, Pakistan and the USA. During last couple of decades, the growing area under cotton has continuously increased; e.g. the area under *Bt* cotton cultivation has increased significantly in the Cotton Belt of the Punjab province of Pakistan (Bahawalpur, DG Khan, Faisalabad, Jhang, Multan and Vehari districts) and majority of wheat farmers were satisfied with growing transgenic cotton (Hussain 2015). Several Bt genes such as *Cry1Aa*, *Cry1Ab*, *Cry1Ac*, *Cry1Ba*, *Cry2A*, *Cry3A*, etc., have been identified as insect-controlling transgenic genes and many Cry and other genes have been transformed in cotton for making them resistant against insects (Table 26.4). For example, cotton overexpressing *CryIAb* and *CryIAc* genes from Bt showed increased resistance against lepidopteran insects: cabbage looper (*Trichoplusia ni*), beet armyworm (*Spodoptera exigua*) and cotton bollworm (*Heliothis zea*) (Perlak et al. 1990). Similarly, transformation of Pakistani cotton cultivar MNH-786 with *Cry1Ac* and *Cry2A* genes conferred resistance against *Helicoverpa armigera* and all its larvae feeding on recombinant cotton were found dead while the ones on non-transgenic cotton cultivar kept growing (Awan et al. 2015). In another study, overexpression of *Cry1Ia12* Bt gene in cotton conferred resistance against fall armyworm (*Spodoptera frugiperda*) and cotton boll weevil (*Anthonomus grandis*) (de Oliveira et al. 2016) and *Cry1Ac* overexpressed in cotton incurred pink bollworm (*Pectinophora gossypiella*) resistance (Wan et al. 2017).

However, Bt-origin *Cry* genes are not the only resistance transgenes against insects in cotton and other genes have also been transformed in cotton for the

Table 26.4 Summary of transgenes transformed in cotton for improvement of insect resistance and related traits whereas ↑ means increase/improvement in a trait and ↓ shows decrease or reduction in a trait

Transgene	Donor	Improved traits in transgenic cotton	Reference
cryIAb+ *cryIAc*	*Bacillus thuringiensis*	↑ resistance against lepidopteran insects ↑ resistance against cabbage looper (*Trichoplusia ni*) ↑ resistance against beet armyworm (*Spodoptera exigua*) ↑ resistance against cotton bollworm (*Heliothis zea*)	Perlak et al. (1990)
vip3A	Synthetic	↑ resistance against fall armyworm (*S. frugiperda*) ↑ resistance against beet armyworm (*Spodoptera exigua*) ↑ resistance against cotton bollworm (*Helicoverpa zea*)	Wu et al. (2011)
Cry1Ia12	*Bacillus thuringiensis*	↑ resistance against fall armyworm (*Spodoptera frugiperda*) ↑ resistance against cotton boll weevil (*Anthonomus grandis*)	de Oliveira et al. (2016)
NPF	*H. armigera*	↑ resistance against cotton bollworm (*Helicoverpa armigera*) ↑ knock down of larval npf in transgenic plants ↓ food consumption, body size and body weight of insect	Yue et al. (2016)
Vip3AcAa + *Cry1Ac*	Synthetic+ *Bacillus thuringiensis*	↑ resistance against cotton bollworm (*H. armigera*) ↑ resistance against *H. armigera* resistance against *Cry1Ac* ↑ pyramid resistance by *Vip3AcAa* gene	Chen et al. (2017)
Cry1Ac	*Bacillus thuringiensis*	↑ resistance against pink bollworm (*Pectinophora gossypiella*) ↑ resistance against pink bollworm by crossing Bt and non-Bt cotton to slow the resistance against *Cry1Ac*	Wan et al. (2017)
dsHaHR3	Cotton bollworm	↑ resistance against cotton bollworm (*H. armigera*) ↑ larval mortality and deformities of pupation and adult eclosion ↓ *HaHR3* expression in bollworm by RNAi reaction of *dsHaHR3* ↑ cotton yield	Han et al. (2017)
Vip3AcAa + *Cry1Ac*	Synthetic+ *Bacillus thuringiensis*	↑ resistance against cotton bollworm (*H. armigera*) ↑ swelling and breakage in the microvilli of midgut cells of larvae ↑ indistinctness of inner cristae of the midgut cells' mitochondria ↑ swelling and fractures in midgut cells' endoplasmic reticulum ↑ disruption in boundaries of karyotheca in the nucleus of midgut cells ↑ mortality of larvae	Chen et al. (2018)

purpose. For example, overexpression of chimeric *Vip3A* gene in cotton cv. Zhongmiansuo35 increased resistance against fall armyworm (*S. frugiperda*), beet armyworm (*Spodoptera exigua*) and cotton bollworm (*Helicoverpa zea*) (Wu et al. 2011). Similarly, transformation of cotton with a novel *H. armigera* gene coding for neuropeptide F (*NPF*) produced two mature peptides, NPF1 and NPF2, which regulated the insect feeding behaviour. The transformed gene knocked down the larval *npf* by RNA interference (RNAi) that resulted in reduced food consumption, size and body weight of insect in addition to reduced glycogen and increased trehalose as compared to controls, thus controlling insect population (Yue et al. 2016). In another study, overexpression of a novel cotton bollworm-origin gene, *dsHaHR3*, in cotton knocked down the bollworm *HaHR3* gene by RNAi reaction in insects feeding on transgenic plants. This enhanced the larval mortality and deformities of pupation and adult eclosion, thus conferring resistance against the *H. armigera* (Han et al. 2017).

The insects may develop resistance against Bt genes with the passage of time, a phenomenon known as resistance breakdown. However, expression of chimeric *Vip3AcAa* gene along with *Cry1Ac* in cotton resulted in stable resistance against *H. armigera* having resistance against *Cry1Ac* (Chen et al. 2017, 2018); thus the gene pyramiding helps to cope with resistance in insects against Bt toxin. Another method for stable resistance against insects is crossing Bt and non-Bt cotton (Wan et al. 2017) to slow the resistance against *Cry1Ac* in pink bollworm (*Pectinophora gossypiella*).

26.8 Prospects

Considerable improvement in cotton for tolerance against biotic and abiotic stresses has been made in recent years. However, there is the need to focus on improving transformation efficiency to make the method more acceptable and efficient. Similarly, there is the need to focus on improving the nutrient mixture of cotton to address the malnutrition in people using its oil. Malnutrition has also been addressed through these GMO foods, which is an excellent source of nutrients for our coming generations. Undernourishment may possibly be controlled through these genetically modified food and plants. It has been reported that half a million children are going to face either complete or partial blindness due to nutrient and vitamin A deficiencies. Therefore, it is demand of time to increase GM technology that will control different constrains of human health. Recently, clustered regularly interspaced short palindromic repeats/CRISPR-associated protein 9 (CRISPR/Cas9) system has been successfully employed to develop resistance against drought, heat, salinity, herbicides and diseases in cotton and other crops (Hussain et al. 2018). However, it has not been used to breed for resistance against insects but the power and ease which CRISPR/Cas9 system brings in will be very crucial for the future landscape of future farming fields.

References

Abbas Q, Ahmad S (2018) Effect of different sowing times and cultivars on cotton fiber quality under stable cotton-wheat cropping system in southern Punjab, Pakistan. Pak J Life Soc Sci 16:77–84

Ahmad S, Raza I (2014) Optimization of management practices to improve cotton fiber quality under irrigated arid environment. J Food Agric Environ 2(2):609–613

Ahmad S, Raza I, Ali H, Shahzad AN, Atiq-ur-Rehman, Sarwar N (2014) Response of cotton crop to exogenous application of glycinebetaine under sufficient and scarce water conditions. Braz J Bot 37(4):407–415

Ahmad S, Abbas Q, Abbas G, Fatima Z, Atique-ur-Rehman, Naz S, Younis H, Khan RJ, Nasim W, Habib ur Rehman M, Ahmad A, Rasul G, Khan MA, Hasanuzzaman M (2017) Quantification of climate warming and crop management impacts on cotton phenology. Plants 6(7):1–16

Ahmad S, Iqbal M, Muhammad T, Mehmood A, Ahmad S, Hasanuzzaman M (2018) Cotton productivity enhanced through transplanting and early sowing. Acta Sci Biol Sci 40:e34610

Ali H, Afzal MN, Ahmad F, Ahmad S, Akhtar M, Atif R (2011a) Effect of sowing dates, plant spacing and nitrogen application on growth and productivity on cotton crop. Int J Sci Eng Res 2 (9):1–6

Ali Q, Elahi M, Hussain B, Hussain N, Fawad K (2011b) Genetic improvement of maize (*Zea mays* L.) against drought stress: an overview. Agric Sci Res J 1:228–237

Ali H, Abid SA, Ahmad S, Sarwar N, Arooj M, Mahmood A, Shahzad AN (2013a) Integrated weed management in cotton cultivated in the alternate-furrow planting system. J Food Agric Environ 11(3–4):1664–1669

Ali H, Abid SA, Ahmad S, Sarwar N, Arooj M, Mahmood A, Shahzad AN (2013b) Impact of integrated weed management on flat-sown cotton (*Gossypium hirsutum* L.). J Anim Plant Sci 23 (4):1185–1192

Ali H, Hameed RA, Ahmad S, Shahzad AN, Sarwar N (2014a) Efficacy of different techniques of nitrogen application on American cotton under semi-arid conditions. J Food Agric Environ 12 (1):157–160

Ali H, Hussain GS, Hussain S, Shahzad AN, Ahmad S, Javeed HMR, Sarwar N (2014b) Early sowing reduces cotton leaf curl virus occurrence and improves cotton productivity. Cercetări Agronomice în Moldova XLVII(4):71–81

Amin A, Nasim W, Mubeen M, Nadeem M, Ali L, Hammad HM, Sultana SR, Jabran K, Habib ur Rehman M, Ahmad S, Awais M, Rasool A, Fahad S, Saud S, Shah AN, Ihsan Z, Ali S, Bajwa AA, Hakeem KR, Ameen A, Amanullah, Rehman HU, Alghabar F, Jatoi GH, Akram M, Khan A, Islam F, Ata-Ul-Karim ST, Rehmani MIA, Hussain S, Razaq M, Fathi A (2017) Optimizing the phosphorus use in cotton by using CSM-CROPGRO-cotton model for semi-arid climate of Vehari-Punjab, Pakistan. Environ Sci Pollut Res 24(6):5811–5823

Amin A, Nasim W, Mubeen M, Ahmad A, Nadeem M, Urich P, Fahad S, Ahmad S, Wajid A, Tabassum F, Hammad HM, Sultana SR, Anwar S, Baloch SK, Wahid A, Wilkerson CJ, Hoogenboom G (2018) Simulated CSM-CROPGRO-cotton yield under projected future climate by SimCLIM for southern Punjab, Pakistan. Agr Syst 167:213–222

Amudha J, Balasubramani G, Malathi VG, Monga D, Bansal KC, Kranthi KR (2010) Cotton transgenics with antisense AC1 gene for resistance against cotton leaf curl virus. Electron J Plant Breed 1(4):360–369. Available at: https://core.ac.uk/download/pdf/25864638.pdf. Accessed 22 Jun 2019

Amudha J, Balasubramani G, Malathi VG, Monga D, Kranthi KR (2011) Cotton leaf curl virus resistance transgenics with antisense coat protein gene (AV1). Curr Sci 101:300–307

Awan MF, Abbas MA, Muzaffar A, Ali A, Tabassum B, Rao AQ, Nasir IA, Husnain T (2015) Transformation of insect and herbicide resistance genes in cotton (*Gossypium hirsutum* L.). J Agric Sci Technol 17:275–285

Budak H, Hussain B, Khan Z, Ozturk NZ, Ullah N (2015) From genetics to functional genomics: improvement in drought signaling and tolerance in wheat. Front Plant Sci 6:1012

Burke JJ, Chen J (2015) Enhancement of reproductive heat tolerance in plants. PLoS One 10: e0122933

Burke IC, Troxler SC, Askew SD, Wilcut JW, Smith WD (2005) Weed management systems in glyphosate-resistant cotton. Weed Technol 19:422–429

Chen J, Wan S, Liu H, Fan S, Zhang Y, Wang W, Xia M, Yuan R, Deng F, Shen F (2016) Overexpression of an Apocynumvenetum DEAD-Box Helicase gene (AvDH1) in cotton confers salinity tolerance and increases yield in a saline field. Front Plant Sci 6:1227

Chen W, Lu G, Cheng H, Liu C, Xiao Y, Xu C, Shen Z, Soberón M, Bravo A, Wu K (2017) Transgenic cotton co-expressing chimeric Vip3AcAa and Cry1Ac confers effective protection against Cry1Ac-resistant cotton bollworm. Transgenic Res 26:763–774

Chen W, Liu C, Lu G, Cheng H, Shen Z, Wu K (2018) Effects of Vip3AcAa+Cry1Ac cotton on midgut tissue in Helicoverpa armigera (Lepidoptera: Noctuidae). J Insect Sci 18:13

Collinge DB, Kragh KM, Mikkelsen JD, Nielsen KK, Rasmussen U, Vad K (1993) Plant chitinases. Plant J 3:31–40

Comas LH, Becker SR, Cruz VMV, Byrne PF, Dierig DA (2013) Root traits contributing to plant productivity under drought. Front Plant Sci 4:442

Divya K, Jami SK, Kirti PB (2010) Constitutive expression of mustard annexin, AnnBj1 enhances abiotic stress tolerance and fiber quality in cotton under stress. Plant Mol Biol 73:293–308

El-Esawi M, Alayafi A, El-Esawi MA, Alayafi AA (2019) Overexpression of StDREB2 transcription factor enhances drought stress tolerance in cotton (*Gossypium barbadense* L.). Genes (Basel) 10:142

Emani C, Garcia JM, Lopata-Finch E, Pozo MJ, Uribe P, Kim D-J, Sunilkumar G, Cook DR, Kenerley CM, Rathore KS (2003) Enhanced fungal resistance in transgenic cotton expressing an endochitinase gene from Trichoderma virens. Plant Biotechnol J 1(5):321–336. Available at: https://www2.eez.csic.es/~pozo/cotton1.pdf. Accessed 18 Jun 2019

Farooq MA, Shakeel A, Atif RM, Saleem MF (2019) Genotypic variations in salinity tolerance among Bt cotton. Pak J Bot 51:1945

Ganesan M, Bhanumathi P, Kumari KG, Prabha AL, Song P-S, Jayabalan N, Ganesan M, Bhanumathi P, Kumari KG, Prabha AL et al (2009) Transgenic Indian cotton (*Gossypium hirsutum*) harboring rice Chitinase gene (Chi II) confers resistance to two fungal pathogens. Am J Biochem Biotechnol 5:63–74

Gianessi LP (2013) The increasing importance of herbicides in worldwide crop production. Pest Manag Sci 69:1099–1105

Han Q, Wang Z, He Y, Xiong Y, Lv S, Li S, Zhang Z, Qiu D, Zeng H (2017) Transgenic cotton plants expressing the HaHR3 gene conferred enhanced resistance to Helicoverpa armigera and improved cotton yield. Int J Mol Sci 18:1874

Hao Y, Lu G, Wang L, Wang C, Guo H, Li Y, Cheng H (2018) Overexpression of AmDUF1517 enhanced tolerance to salinity, drought, and cold stress in transgenic cotton. J Integr Agric 17:2204–2214

Hashmi JA, Zafar Y, Arshad M, Mansoor S, Asad S (2011) Engineering cotton (*Gossypium hirsutum* L.) for resistance to cotton leaf curl disease using viral truncated AC1 DNA sequences. Virus Genes 42:286–296

Hussain B (2015) Modernization in plant breeding approaches for improving biotic stress resistance in crop plants. Turk J Agric For 39:515–530

Hussain B, Khan MA, Ali Q, Shaukat S (2012) Double haploid production is the best method for genetic improvement and genetic studies of wheat. IJAVMS 6:216–228

Hussain B, Khan AS, Ali Z (2015) Genetic variation in wheat germplasm for salinity tolerance at seedling stage: improved statistical inference. Turk J Agric For 39:182–192

Hussain B, Lucas SJ, Ozturk L, Budak H (2017) Mapping QTLs conferring salt tolerance and micronutrient concentrations at seedling stage in wheat. Sci Rep 7:15662

Hussain B, Lucas SJ, Budak H (2018) CRISPR/Cas9 in plants: at play in the genome and at work for crop improvement. Brief Funct Genomics 17:319–328

Keller G, Spatola L, McCabe D, Martinell B, Swain W, John ME (1997) Transgenic cotton resistant to herbicide bialaphos. Transgenic Res 6:385–392

Khan MB, Khaliq A, Ahmad S (2004) Performance of mashbean intercropped in cotton planted in different planting patterns. J Res (Sci) 15(2):191–197

Khan AI, Khan IA, Sadaqat HA (2008) Heat tolerance is variable in cotton (*Gossypium hirsutumn* L.) and can be exploited for breeding of better yielding cultivars under high temperature regimes. Pak J Bot 40:2053–2058

Latif A, Rao AQ, Khan MAU, Shahid N, Bajwa KS, Ashraf MA, Abbas MA, Azam M, Shahid AA, Nasir IA, Husnain T (2015) Herbicide-resistant cotton (*Gossypium hirsutum*) plants: an alternative way of manual weed removal. BMC Res Notes 8:453

Liu G, Li X, Jin S, Liu X, Zhu L, Nie Y, Zhang X (2014) Overexpression of rice NAC gene SNAC1 improves drought and salt tolerance by enhancing root development and reducing transpiration rate in transgenic cotton. PLoS One 9:e86895

Liu J, He Z, Rasheed A, Wen W, Yan J, Zhang P, Wan Y (2017) Genome-wide association mapping of black point reaction in common wheat (*Triticum aestivum* L.). BMC Plant Biol 17:220

Lv S, Yang A, Zhang K, Wang L, Zhang J (2007) Increase of glycinebetaine synthesis improves drought tolerance in cotton. Mol Breed 20:233–248

Minhas R, Shah SM, Akhtar LH, Awais S, Shah S (2018) Development of a new drought tolerant cotton variety "BH-167" by using pedigree method. J Environ Agric Sci 14:54–62

Mishra N, Sun L, Zhu X, Smith J, Prakash Srivastava A, Yang X, Pehlivan N, Esmaeili N, Luo H, Shen G, Jones D, Auld D, Burke J, Payton P, Zhang H (2017) Overexpression of the rice SUMO E3 Ligase gene OsSIZ1 in cotton enhances drought and heat tolerance, and substantially improves fiber yields in the field under reduced irrigation and rainfed conditions. Plant Cell Physiol 58:735–746

Niu J, Zhang S, Liu S, Ma H, Chen J, Shen Q, Ge C, Zhang X, Pang C, Zhao X (2018) The compensation effects of physiology and yield in cotton after drought stress. J Plant Physiol 224–225:30–48

de Oliveira RS, Oliveira-Neto OB, Moura HFN, de Macedo LLP, Arraes FBM, Lucena WA, Lourenço-Tessutti IT, de Deus Barbosa AA, da Silva MCM, Grossi-de-Sa MF (2016) Transgenic cotton plants expressing Cry1Ia12 toxin confer resistance to fall Armyworm (Spodoptera frugiperda) and cotton boll weevil (Anthonomus grandis). Front Plant Sci 7:165

Perlak FJ, Deaton RW, Armstrong TA, Fuchs RL, Sims SR, Greenplate JT, Fischhoff DA (1990) Insect resistant cotton plants. Nat Biotechnol 8:939–943

Rahman MH, Ahmad A, Wang X, Wajid A, Nasim W, Hussain M, Ahmad B, Ahmad I, Ali Z, Ishaque W, Awais M, Shelia V, Ahmad S, Fahad S, Alam M, Ullah H, Hoogenboom G (2018) Multi-model projections of future climate and climate change impacts uncertainty assessment for cotton production in Pakistan. Agric For Meteorol 253-254:94–113

Rajasekaran K, Cary JW, Jaynes JM, Cleveland TE (2005) Disease resistance conferred by the expression of a gene encoding a synthetic peptide in transgenic cotton (*Gossypium hirsutum* L.) plants. Plant Biotechnol J 3:545–554

Sanjaya, Satyavathi VV, Prasad V, Kirthi N, Maiya SP, Savithri HS, Sita GL (2005) Development of cotton transgenics with antisense AV2 gene for resistance against cotton leaf curl virus (CLCuD) via Agrobacterium tumefaciens. Plant Cell Tiss Org Cult 81:55–63

Sattar MN, Kvarnheden A, Saeed M, Briddon RW (2013) Cotton leaf curl disease - an emerging threat to cotton production worldwide. J Gen Virol 94:695–710

Shabbir MZ, Arshad M, Hussain B, Nadeem I, Ali S, Abbasi A, Ali Q (2014) Genotypic response of chickpea (*Cicer arietinum* L.) for resistance against gram pod borer (Helicoverpa Armigera). Adv Life Sci 2:23–30

Sohrab SS, Kamal MA, Ilah A, Husen A, Bhattacharya PS, Rana D (2016) Development of Cotton leaf curl virus resistant transgenic cotton using antisense βC1 gene. Saudi J Biol Sci 23:358–362

Tariq M, Yasmeen A, Ahmad S, Hussain N, Afzal MN, Hasanuzzaman M (2017) Shedding of fruiting structures in cotton: factors, compensation and prevention. Trop Subtrop Agroecosyst 20(2):251–262

Tariq M, Afzal MN, Muhammad D, Ahmad S, Shahzad AN, Kiran A, Wakeel A (2018) Relationship of tissue potassium content with yield and fiber quality components of Bt cotton as influenced by potassium application methods. Field Crop Res 229:37–43

Tohidfa M, Rassouli H, Ghareyazie B, Najafi J (2009) Evaluation of stability of Chitinase gene in transgenic offspring of cotton *(Gossypium hirsutum)*. Iran J Biotechnol 7(1):45–50. National Institute of Genetic Engineering and Biotechnology. Available at: http://www.ijbiotech.com/article_7073_0.html. Accessed 21 Jun 2019

Tohidfar M, Hossaini R, Bashir NS, Meisam T (2012) Enhanced resistance to Verticillium dahliae in transgenic cotton expressing an Endochitinase gene from Phaseolus vulgaris. Czech J Genet Plant Breed 48:33–41

Ullah A, Sun H, Yang X, Zhang X (2017) Drought coping strategies in cotton: increased crop per drop. Plant Biotechnol J 15:271–284

Usman M, Ahmad A, Ahmad S, Irshad M, Khaliq T, Wajid A, Hussain K, Nasim W, Chattha TM, Trethowan R, Hoogenboom G (2009) Development and application of crop water stress index for scheduling irrigation in cotton (*Gossypium hirsutum* L.) under semiarid environment. J Food Agric Environ 7(3–4):386–391

Wan P, Xu D, Cong S, Jiang Y, Huang Y, Wang J, Wu H, Wang L, Wu K, Carrière Y, Mathias A, Tabashnik BE (2017) Hybridizing transgenic Bt cotton with non-Bt cotton counters resistance in pink bollworm. Proc Natl Acad Sci 114:5413–5418

Wang J, Chen Y, Yao M, Li Y, Wen Y, Chen Y, Zhang X, Chen D (2015) The effects of high temperature level on square Bt protein concentration of Bt cotton. J Integr Agric 14:1971–1979

Wang Y, Liang C, Wu S, Zhang X, Tang J, Jian G, Jiao G, Li F, Chu C (2016) Significant improvement of cotton Verticillium wilt resistance by manipulating the expression of Gastrodia antifungal proteins. Mol Plant 9:1436–1439

Wicke B, Smeets E, Dornburg V, Vashev B, Gaiser T, Turkenburg W, Faaij A (2011) The global technical and economic potential of bioenergy from salt-affected soils. Energ Environ Sci 4:2669

Wu J, Luo X, Zhang X, Shi Y, Tian Y (2011) Development of insect-resistant transgenic cotton with chimeric TVip3A∗ accumulating in chloroplasts. Transgenic Res 20:963–973

Xu Y, Ramanathan V, Victor DG (2018) Global warming will happen faster than we think. Nature 564:30–32

Yang H, Zhang D, Li X, Li H, Zhang D, Lan H, Wood AJ, Wang J (2016) Overexpression of ScALDH21 gene in cotton improves drought tolerance and growth in greenhouse and field conditions. Mol Breed 36:34

Yu LH, Wu SJ, Peng YS, Liu RN, Chen X, Zhao P, Xu P, Zhu JB, Jiao GL, Pei Y, Xiang CB (2016) Arabidopsis EDT1/HDG11 improves drought and salt tolerance in cotton and poplar and increases cotton yield in the field. Plant Biotechnol J 14:72–84

Yue Z, Liu X, Zhou Z, Hou G, Hua J, Zhao Z (2016) Development of a novel-type transgenic cotton plant for control of cotton bollworm. Plant Biotechnol J 14:1747–1755

Zhang H, Dong H, Li W, Sun Y, Chen S, Kong X (2009) Increased glycine betaine synthesis and salinity tolerance in AhCMO transgenic cotton lines. Mol Breed 23:289–298

Zhang X, Tang Q, Wang X, Wang Z (2017) Development of glyphosate-tolerant transgenic cotton plants harboring the G2-aroA gene. J Integr Agric 16:551–558

Zhu X, Sun L, Kuppu S, Hu R, Mishra N, Smith J, Esmaeili N, Herath M, Gore MA, Payton P, Shen G, Zhang H (2018) The yield difference between wild-type cotton and transgenic cotton that expresses IPT depends on when water-deficit stress is applied. Sci Rep 8:2538

Chapter 27
Production and Processing of Quality Cotton Seed

Atique-ur-Rehman, Muhammad Kamran, and Irfan Afzal

Abstract Quality seed is a prerequisite for even stand establishment which fabricates a pathway toward higher productivity; however non-availability of better-quality cotton seed at planting is the prime reason of poor germination and crop stand failure in developing countries. Cotton seed accounts for 15–25% of the value of crop, and it is important to maintain its quality after harvest and storage to avoid any spoilage. Early and uniform stand establishment enables crop to utilize input resources for longer time and accumulate enough dry matter during vegetative and reproductive stages. The decision of picking highly relates with physical and physiological maturity of crop during upcoming dry and hot season. Right after harvesting, seed cotton needs to be dried below 12% to avoid physical damages on seed coat during ginning. Furthermore, the cotton seed should be dried to 8% moisture by adopting a quick and efficient drying method. Precise delinting with commercial sulfuric acid removes fuzz from seed and facilitates seed grading and delivery at the time of sowing. After delinting, the physiological quality of cotton seed can be preserved by placing seed at 8% moisture content in proper storage conditions. Thus, maintenance of seed quality during production and storage ascertains a milestone toward seed security.

Keywords Seed quality · Post-harvest management · Cotton · Picking · Crop yield

Abbreviations

B	Boron
CAP	Cold atmospheric pressure plasma
eqMC	Equilibrium moisture content
ISTA	International Seed Testing Association
PICS	Purdue Improved Crop Storage

Atique-ur-Rehman (✉)
Department of Agronomy, Bahauddin Zakariya University, Multan, Pakistan
e-mail: dr.atique@bzu.edu.pk

M. Kamran · I. Afzal
Department of Agronomy, University of Agriculture, Faisalabad, Pakistan

S. Ahmad, M. Hasanuzzaman (eds.), *Cotton Production and Uses*,
https://doi.org/10.1007/978-981-15-1472-2_27

RH Relative humidity
TZ Tetrazolium chloride

27.1 Introduction

Cotton is known as white gold as it is the primary source of fiber and oil to support the biggest textile, oil, and animal feed industries throughout the globe and contributes economic impact of at least $600 billion worldwide (Ashraf et al. 2018). The upland cotton comprises more than 50 cultivars which provides the opportunity for understanding the breeding with wild species to achieve higher productivity of lint and seed as well as stress-resilient germplasm (Rahman et al. 2009). On production basis more than 80% of cotton is cultivated in Brazil, China, India, Pakistan, Turkey, the USA, and Uzbekistan (Solomon 2007). In spite of this prominent status in oil, textile, and foreign exchange, however, tragically, its yield from unit area is stagnant and much less than potential (Ahmad et al. 2014, 2017, 2018; Abbas and Ahmad 2018; Ahmad and Raza 2014; Ali et al. 2011, 2013a, b, 2014a, b; Usman et al. 2009). There are multifarious reasons, which can be attributed to numerous factors that imitate on cotton growth and yield (Amin et al. 2017, 2018; Khan et al. 2004; Rahman et al. 2018; Tariq et al. 2017, 2018). Among these, lack of quality cotton seed is important. Although there is greatly advanced technological package for cotton production, lack of quality cotton seed has been perceived as a serious issue. Good-quality seed of improved varieties is one of the key inputs for attaining high cotton yield with more economic benefits.

High-quality seeds emerge quickly and establish uniform stand which enables crop to grow perfectly in field and ultimately ensures higher fiber and seed production. Similarly, genetically uniform crop established plants (Rahman et al. 2008) perform phenological events simultaneously and exhibit maturity at same time. Furthermore, harvesting time and post-harvest processing and storage under optimum conditions significantly contribute to quality maintenance till next planting seasons (Arah et al. 2016). Harvesting of fully mature cotton from more than 80% boll during hot and dry weather guarantees the high quality of lint and seed production (Shuli et al. 2018). After harvesting, preparation of seed cotton for ginning by immediate drying below 12% moisture content separates lint of good staple length and strength without affecting the seed coat.

Cotton seed is further subjected to drying up to 8% moisture content and processing (Usberti et al. 2006). The remaining fibers are cleaned from seed surface by delinting with commercial sulfuric acid, and healthy seeds are separated from broken and empty seeds to ensure physically good-quality seed. Subsequently, the seeds are stored under dry and cold storage conditions to preserve the physiological quality of seed. The physical, physiological, and genetic purity of seed should be assessed during storage and before planting by subjecting the sample to various tests drafted by ISTA.

After storage, performance of seed in field can be enhanced by applying seed invigoration techniques including priming, heat treatment, and seed surface treatment. Planting of quality seeds under optimum field conditions transforms maximum seeds to seedlings quickly and displays uniform stand (Finch-Savage and Bassel 2016). The ideal transformation of vegetative growth to reproductive phase produces good-quality lint and seeds that ensures higher economic returns (Kerby et al. 1997).

This chapter covers the importance of quality seed along with biotic and abiotic stresses during production. Moreover, the importance of picking, handling, processing, and storage of cotton seed to preserve quality is also discussed briefly. The chapter also elaborates some seed invigoration techniques to enhance seed quality, which equip the crop with good characteristics during growth and development and ensure higher economic returns.

27.2 Cotton Seed Quality

Like all dicot seeds a mature cotton seed contains all the necessary parts that successfully transform into normal seedlings after providing optimal conditions for germination after sowing. Physically cotton seed has dark brown color and oval shape with a pointed end called the micropyle which is the point of radicle protrusion during germination and a circular side called the chalaza which encloses cotyledons above the hypocotyl that will nourish the newly emerged seedlings through stored food. The gossypol tissues are also significantly visible within seeds and on growing tissues (Ritchie et al. 2007).

Physically good-quality mature seed looks uniform in shape, weight, and color and is free from insect damage and other forms of damage, inert matter, disease and other crop seeds. Physical seed quality involves the percentage of other than seed materials (Singh et al. 2015). Physiologically good-quality seed shows expression in further generation and multiplication under optimal conditions. Seeds that show expression of successful germination are alive or viable, while seeds able to produce seedlings with normal shoot and normal root under given conditions are vigorous. Seed performance in production of elite seedlings can be classified as high-, medium-, or low-vigor seeds (França et al. 2018).

27.2.1 Determining Cotton Seed Quality

Several characteristics are determined to find the quality of seed including the genetic and physical purity, health or physical condition of seed, ability to germinate or vigor, relation to specific type and extent of microorganisms that can cause a disease/problem after germination, and to some extent the nature and consistency of seed treatment if any. Maximum germination percentage, successful transformation in normal seedlings, and eventually establishment of uniform stand are virtues of good-quality seed.

27.2.1.1 Genetic Purity

Genetic purity is much imperative to achieve good-quality produce of same kind. It is often hard to find a specific seed from a lot of seed of different varieties after harvesting. Mixture of different genotypes results in a number of problems such as difficulty in harvesting and post-harvest handling due to variation in maturity. Furthermore, if unwanted seed is not separated, it will multiply and produce more seed. A careful inspection and removal of such mixing is mandatory during the growing period to keep genetic purity of cotton seed. Vigor of seed determines the productivity of crop, which is possible from a quality seed (Rathinavel 2014). There is great variation among cotton genotypes for different morphological and physiological characteristics both under normal and stress conditions (Bakhsh et al. 2019). Moreover, response of different cotton genotypes varies under different stresses (Abbas et al. 2008; Latif et al. 2014). Selection of appropriate genotype suited to particular ecological conditions can improve cotton yield (Abbas et al. 2008). The contribution of genetic purity of cotton seed has been well recognized by a number of scientists (Cherry and Leffler 1984; Dowd et al. 2010; Lukonge et al. 2007; Pettigrew and Dowd 2012). Despite the big genetic inconsistency among different cotton cultivars, little effort is done in improving the composition of seed. However, in recent past, cotton breeding was focused in improving fiber yield and quality, while little was done to improve cotton seed for its oil and composition. Multifarious uses of cotton seed oil such as food, feed, and biofuel have convinced breeders for uplifting its oil percentage and quality. Efforts for improvement in cotton seed are intensified for oil and its composition without forfeiting yield and quality of lint (Paterson 2009). In cotton, genetic purity is certain by vigilant and methodical maintenance of single variety, multiplication of seed and specific production practices, and sound recognized quality promising procedures like performing regular field inspections, having specific farms of variety, and maintaining gins and storehouses.

27.2.1.2 Physical Qualities

Physical qualities mention the freedom of desired seed from other materials and its health, moisture content, appearance, size, color, etc. For physical purity of cotton seeds, these are cleaned to remove all distant materials like dirt, dust, chaff, and stones. For size determination of cotton seeds, groups of samples comprising 100 parts are selected randomly. From each group, ten seeds are taken for linear dimensions and the size of seed is assessed (Calısır et al. 2005; Yalcın and Ozarslan 2004). Normally, after the ginning process, the cotton seed separated from fiber has a residual covering of lint. This lint can change the physical appearance/characteristics

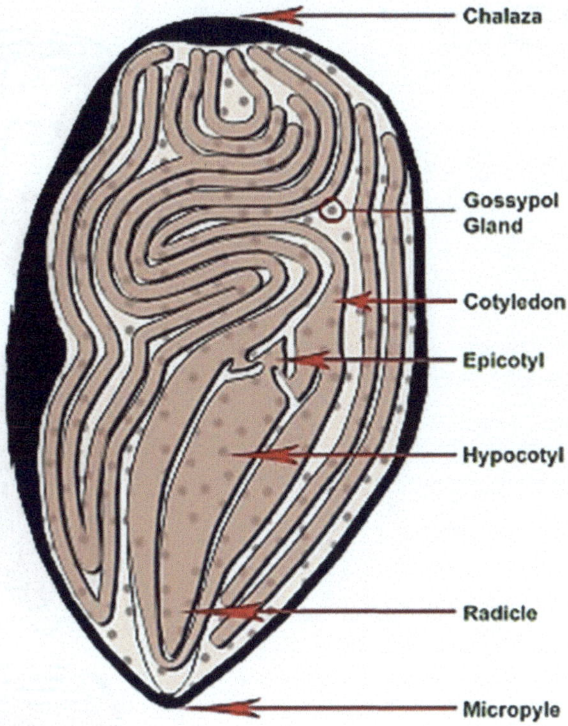

Fig. 27.1 The structure of mature cotton seed includes all the necessary components which transform it into a seedling under favorable conditions

and can obstruct with subsequent processes of handling, grading, and sowing of seed into the field. To perform these processes, seed must be of free-flowing nature and without fuzz (Figs. 27.1 and 27.2). The removal of external hairs from seed covering is done by the process of delinting. There are several types of procedures adequate for conditioning of linted or fuzzy seeds to free the product that can be uniformly treated and planted with considerable precision. In this process of delinting, fuzzy cotton seed is taken in a container, and concentration of H_2SO_4 is added @ 100 ml/kg of seed. With the addition of acid, seed is constantly stirred for 2–3 min for uniform coverage and treatment. After a few minutes, when seeds turn brown, they are thoroughly washed with cold water to remove acid. Care should be taken while washing as improper washing will affect the viability of the seed. After washing, seed is placed in water with a ratio of 1:10 for removal of floaters. For a complete wash to remove acid, seed is immersed in 0.5% $CaCl_2$ solution for 10–15 min. For delinting of bulk of cotton seed, a delinting machine is used.

Different cleaning equipment and density separators can remove physical contaminants and impurities or immature seeds. Likewise, seed treaters are available for seed treatments with fungicides or insecticides that can treat seed with more than one treatment simultaneously with controlled dosages and in sequence. Physical quality is easy to maintain owing to the presence of quality assurance procedures with updated processing technology and facilities currently accessible for the processing

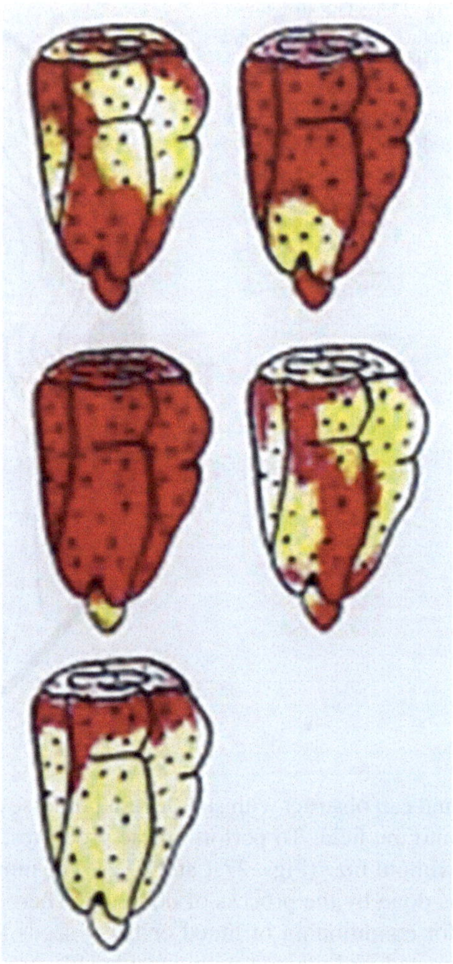

Fig. 27.2 Examples of non-viable seeds

and preparing of cotton seed. Cotton seed quality problems relating to genetic and physical purity, physical disorder, and somehow seed-borne microorganisms are not a big subject to address. These problems arise usually due to lapses in quality control, mismanagement, and/or lack of facilities and can be solved through proper curative arrangements. However, poor viability, less germination, and vigor with meager production potential are problems that happen most commonly. Indeed, technological advancement in production, harvesting, and conditioning of huge quantities of cotton seed is becoming obligatory which has worsen the germination/vigor problems. Cotton seed germination ability and vigor are deemed vital for seed quality. Earlier canopy development and light interception by cotton plant during lateral growth stages depend upon seed germination and vigor, which consequently determine cotton development (Reddy et al. 2017). However, these can be affected by indeterminate habit, climatic conditions during the growing of plant, mechanization used for harvesting and post-harvesting, and primacy given to lint as compared to seed. Efforts are being made for the improvement in seed germination and vigor to make crop stand better under prevailing conditions. Adequate techniques have been developed to evaluate the physical purity of cotton seed lots. Physical purity of 96% is recommended for cotton seed for uniform rapid germination. Infestation of disease and insect pest may occur during seed storage under different environmental factors, which are important to keep under control. Analysis of seed for its viability must be checked before sowing it into the field and quantity of seed can be decided accordingly.

Physiological observations like seed viability and vigor ensure normal seedlings followed by even stand establishment prior to sowing with aim to identify the seed lot which ultimately gives expected returns. The International Seed Testing Association (ISTA) provides facilities to test the quality of seed throughout the world. The viability of seed lot can be predicted by using standard germination test and tetrazolium chloride test.

27.2.2 Germination and Vigor

High germination percentage is the prime factor for a good crop stand. Standard procedures/tests are available for determination of cotton seed germination. Samples of 100 seeds are grown in controlled media such as blotters and moist paper towels or in sand with a temperature of 20–30 °C. Seed germination count is made after a few days and final counting is done when germination becomes constant. Special closed containers can be used for samples not responding under usual conditions (Toole and Drummond 1924). Normal developed seedlings counted at final stage are expressed as germination percentage. Nonetheless, there may be difficulties in obtaining germination percentage from different tests owing to inherent characters of cotton seed and conditions contributing to germination. For example, cotton seed with more than 10% moisture content is subjected to mold during testing, while cotton seed with 5–6% moisture content is found with hard covering (Toole and Drummond 1924). Germination of cotton seed is affected by temperature and percentage of germinated seeds is greater at relatively high temperature, i.e., 30 °C as compared to 20–30 °C. Furthermore, the process of seed germination is completed earlier at higher temperature. Optimum germination temperature for cotton seed is found between 30 and 33 °C (Krzyzanowski and Delouche 2011). Likewise, Camp and Walker (1972) reported that cotton seed can germinate at a maximum temperature of 39–40 °C, while minimum temperature at which cotton seed can germinate is about 12 °C (Ludwig 1932), whereas Pereira et al. (2005) reported that cotton seed could not germinate at 14 °C. However, tolerance to low temperature (16–22 °C) is found to be better in high-quality cotton seed as compared to medium-quality seed (Krzyzanowski and Delouche 2011). Moreover, high temperature range (36–38 °C) helps medium-quality cotton seed to germinate consistently (Krzyzanowski and Delouche 2011).

Immediate assessment of quality is required during acceptance and storage of cotton seed. Several methods are available to quickly determine moisture content, mechanical damage, and contaminants. Nonetheless, germination, the most crucial, takes at least 4–5 days. Some quick tests are available to estimate germination in a short time. In this method, 50–100 cotton seeds are kept in a container that allows quick longitudinal bisection of cotton seed. Germinated embryos are visually assessed by their "fullness" and color, which takes almost 15 min.

Tetrazolium chloride (TZ) for seed viability is widely used in the cotton seed industry. TZ has the ability to react with oxygen utilized during respiration and staining. The seed is cut transversely one-third from the distal end and stained in 1.0% TZ solution at 30 °C for 18 h. The staining on embryo area expresses the respiration and liveliness of seed. For evaluation, embryo is extracted from seed coat and stained or unstained tissues are observed (ISTA 2015).

Seed germination only reveals the sharp variation in degree of seed deterioration, while the performance of seed under adverse conditions shows correlation with seed vigor. The potential of seed to transform into normal seedlings even under suboptimal conditions can be assessed by performing a vigor test (Finch-Savage and Bassel 2016). This test is performed to assess the narrow variation of seed deterioration stages in different seed lots (Baalbaki et al. 2009) during storage and in the field environment. Cotton seeds are gorgeous in oil (25–40%) which needs exceptional care in storage. During storage, seed vigor determines the conditions adequate to preserve seed quality. Accelerated aging test is performed by placing the seeds on perforated plate over the germination tray, which has distilled water at the bottom. The plates are placed in an incubator or special accelerated chamber which creates the stressful environment in chamber by 100% humidity and 45 °C temperature for 72 h. Afterward, the standard germination test is performed by placing 50 seeds in four replicates between triple layers of moist filter paper in germinator at 25 °C. The observation of radicle protrusion was taken after 4 days and 7 days, respectively. The germinated seeds showed vigor after episode of high humidity and heat stresses (TeKrony and Egli 1997).

Some quick tests based on EC of seed exudates from proplasts are used for estimation of seed viability. This approach demonstrates a close relationship between seed quality and "protoplast" permeability. The current flow or EC methods are most reliable and accurate in determining the high- or low-quality seed; however, medium-quality seeds are difficult to judge. A relatively different method for estimation of germination is soaking of cotton seed in water that is done at 65–70 °C for 90 min. A refractometer is used to read the leachates and good seeds are considered to have readings below 0.2, while poor-quality seeds have readings above 0.6. Free fatty acid contents in cotton seed are found to be closely related with germination of seeds and thus are greatly used as a rough index for quality of cotton planting seed. With increase in fatty acids, germination percentage of cotton seed decreases, and seeds having more than 1% free fatty acids or 3% extracted oil cannot germinate. On the basis of this, it is recommended by scientists to save cotton seeds with free fatty acid contents of not more than 1%. However, there is contradiction for considering free fatty acids as criteria for cotton seed quality. For instance, there are some specific fatty acid contents that are used as parameters for quality of cotton seed.

During seed storage, seed viability is maintained by using different packing materials. However, it is difficult to maintain seed viability even under properly maintained storage conditions. Inherent high vigor quality of seed is thus important to maintain proper seed viability. Maintenance of seed viability of different species under different storage conditions with controlled environment in storage is the most important to research. During storage, cotton seeds have significant genotypic differences in a number of physiological traits, and storing in airtight polythene bag is better for maintaining the physiological quality of cotton seeds as compared to cloth bag (Salam et al. 2017). Cotton seeds of different varieties may have different vigor indexes, and seed treatments used to keep vigor have no or very little effect on them.

27.2.2.1 Improving Germination and Vigor of Cotton Seed

High-quality cotton seeds are aimed to get more cotton yield. Cotton seed quality may change with germination rate and environment and soil conditions (Shaheen et al. 2012). Furthermore, during storage, various factors like initial seed moisture contents, high temperature, and ambient relative humidity can influence seed viability. Poor-quality cotton seed deteriorates more quickly than good-quality seed under particular conditions (Iqbal et al. 2002). Owing to the importance of seed quality and germination potential for cotton production, different methods are developed to foster rapid germination under various environmental conditions. Different seed treatments were applied including priming, heat, and cold treatments as well as seed surface treatment to improve the physiological quality of cotton seeds (Bolek et al. 2013).

Priming is the pre-sowing seed treatment in which the necessary metabolic processes for germination take place without actual germination. Priming is done by soaking seeds in aerated osmotic solution of different concentration to provide them enough water to prolong the lag phase but prevent radicle protrusion (Taylor et al. 1998; Afzal et al. 2017). Priming has positive impact on cotton seed emergence even in stress conditions of uneven temperature by accelerating the emergence in field thus reducing emergence time (Casenave and Toselli 2007). Primed cotton seed is less sensitive to moisture stress and thus priming increases the germination rate in cotton (Murungu et al. 2005). The seedlings emerged from primed seed develop a root system faster and proliferate the roots to moisture and nutrients quicker and have improved growth and development against biotic and abiotic stresses (Mustafa et al. 2017). Priming not only improves the germination of cotton seed under thermal and water stress conditions (Casenave and Toselli 2007), but also primed seed could also be stored at least 6–12 months under ambient conditions (Toselli and Casenave 2014).

For improving the performance of seeds under stress, different treatments like hormones, growth regulators, and different chemicals are frequently used. For example, at the time of sowing of cotton, temperature is very critical and emergence of cotton seedling may be delayed under unfavorable temperature (Krzyzanowski and Delouche 2011). Cotton seed priming is found effective to apply different agents. Polyethylene glycol, potassium (K_3PO_4, KH_2PO_4, KNO_3), and sodium (NaCl) are generally used in priming (Ghassemi-Golezani et al. 2013; Arif et al. 2014; Hameed et al. 2013). Significant improvements in seed vigor, germination rate, and seedling growth occur by these priming techniques. Cotton requires high temperatures for good stand establishment, and low temperature can hinder germination and emergence. Cotton seed priming with KNO_3, 5-aminolevulinic acid (ALA), gibberellic acid (GA_3), methyl jasmonate, acetylsalicylic acid (ASA), or kinetin improves the germination rate under cold temperature. Different growth regulators could not improve germination rate. However, cotton seed priming with KNO_3 improves germination and vigor of seedlings and growth of cotton under low temperature (Çokkizgin and Bölek 2015). Earlicr Suzuki and Khan (2000) recorded improved seed vigor, germination, and seedling growth of snap bean by humidification. Contrarily, Basra et al. (2003) tested cotton seed humidification and found that it has no significant effect on vigor and germination of cotton seeds. Similarly, Thomas and Christiansen (1971) also described that humidification or preconditioning of good-quality seed had no improvement in the emergence of cotton seed. They concluded that preconditioning is more effective in low-quality seeds as compared to high-quality cotton seeds.

The advancements in agriculture introduce cold atmospheric pressure plasma (CAP) to enhance the cotton seed performance against temperature stress in field. CAP treatment utilizes ionized gas which has the ability to react and excite biological molecules and form reactive oxygen and nitrogen species to improve the rate of water absorption in germinating seeds and develop chilling tolerance (Gerard et al. 2018). Cotton seed exposed to pulsed electromagnetic field for 15 or 30 min performed better in terms of root shoot growth, transpiration rate, stomatal conductance, nutrient accumulation, photosynthetic rate, and ultimately yield (Bilalis et al. 2013). Heat treatment at 60 °C for 8 h before sowing is a promising technique to enhance seed vigor of upland cotton and ultimately showed better seedling performance in the field (Basra et al. 2004). Moreover, the exposure of cotton seed to saturated humidity for 24 h also influenced the quality of seed as depicted by minimum value of seed leachate peroxide in treated seeds than untreated ones (Basra et al. 2003).

27.3 Cotton Seed Quality During Production

Cotton has an indeterminate growth pattern; therefore, it has continuous vegetative growth and reproductive stages. During field appraisal cotton is the most responsive crop against biotic and abiotic environments. Thus, the selection of high-quality seed at sowing puts significant impact on future performance of cotton crop in field. High-quality seed quickly and efficiently transforms into seedlings and establishes uniform stand quickly and strengthens the plant to respond to stresses effectively. However, various factors are responsible for the maintenance of its quality during production.

27.3.1 Planting Time

The journey of the plant in field critically depends upon time of sowing. Optimum sowing time provides optimum conditions, which enable maximum seeds for germination and uniform stand establishment in field. Early or late sowing checks the germination and stand establishment by cold and heat spell. Cotton seed emerges well and quick from warm and moist soil, while temperatures below 65 °F cause slower water absorption and metabolic processes and delayed germination in seed (Larson and Mapp 1997).

Secondly planting time is also important to fit the critical stages of cotton in suitable conditions as early or late blooming directly influences boll retention, seed and lint development, and ultimately economic returns. After sowing the episode of 25–28 °C enables cotton seed to imbibe well and prepared and utilized reserves to emerge radicle. Radicle emerged through micropyle grows deeper in soil, through a tap root system supplying water and nutrients (Kamran et al. 2017).

A rule of thumb for cotton-growing belt is that the cotton sowing should be done when the temperature of soil at depth of 4 in. reaches above 65 °F for 3 consecutive days with warm temperature in forecast (Oosterhuis et al. 1992).

27.3.2 Seedbed Environment

Proper germination and seedling establishment are only ensured when the seedbed is finely prepared. These processes determine uniformity, crop density, and efficient use of inputs and free them from weed infestation which ultimately contribute to higher yield and quality of the crop (Hadas et al. 1985). Right after germination, the root system of cotton develops faster than shoot system as at the time of emergence of cotyledons the root can go deeper up to 10 in. and root development is critical step to further growth of cotton plant. The cotton develops tap root and lateral roots which collectively develop a basal root system to absorb moisture and necessary nutrients in well-drained and perforated soil, but cold, acidic, water-deficient, hard soils hinder the root growth.

Any hindrance in root development during early stages leads to unsatisfactory production season and economic returns. Besides, cotton follows an epigeal emergence pattern; cotyledons need well-pulverized soil to emerge out from soil. Any impedance at soil surface leads to delay in emergence, which causes development of extra-thick hypocotyl and big shank or thick-legged cotton. Soil crusting and compaction restrict gas exchange and in turn seedling emergence (Hadas et al.

1985). Adopting management strategies during field preparation can minimize the stresses during root development.

27.3.3 Nutrient Application

The amount of nutrient uptake by cotton plants depends upon the potential of the plant to assimilate the nutrient as biomass in vegetative and reproductive parts. The cotton uptakes nitrogen and potassium in greater amount at about 200 kg/hectare (Hodges 1992), while some nutrients are required in lesser quantity so-called micronutrients like Ca which is taken by plant at less than 30 g/hectare in a season. The uptake and requirement of nutrient in cotton is dependent upon variety, pattern of nutrient accumulation, and growth stage of plant as cotton uptakes maximum nutrients from day 101 to 130 days after sowing and younger plants uptake more nutrients than younger ones (Mullins and Bumester 2010).

The deficiency of essential nutrients affects both vegetative and reproductive phases of cotton. Nitrogen controls the structural and functional integrity of protein to control growth and prevents abscission of square, bolls, and photosystems (Reddy et al. 1996). Similarly, phosphorus also has significant importance during cell division and transport and is assimilated by the economic parts of plants like boll and seeds. Potassium has a significant role in transportation of carbohydrate metabolites from vegetative parts to reproductive parts (Sangakkara et al. 2000) and activates more than 60 enzymes (Taiz and Zeiger 2012).

Boron (B) is an essential micronutrient that is significantly important during seed production. Cotton plant containing less than 20 mg/kg B in fully expanded leaf blades expressed deficiency symptoms (Zhao and Oosterhuis 2002) like abnormal fertilization during seed development. Foliar application of B on plant during reproductive phase improved boll weight and lint and seed yield and quality like seed germination and vigor by 17% and 25%, respectively (Dordas 2006).

27.3.4 Abiotic Factors

Temperature and moisture are the critical environmental factors that affect the performance of cotton plant throughout the growing period. Cotton accumulates a specific amount of heat units to transform from one developmental stage to a further stage. High or low temperature from optimum level checks the growth and development of cotton crop. Photoperiod and temperature are the fundamental factors, which determine the growth of cotton plant during all critical stages both vegetative and reproductive. Cotton plant grows well at temperature ranges from 20 to 30 °C during life cycle (Reddy et al. 1996).

High temperature affects growth stages of cotton including germination, seedling development, stand establishment (Burke and Wanjura 2010), reproduction, maturity, and picking in field as well as after harvesting during seed handling, processing, and storage. Cotton seed germination is very sensitive to temperature. After germination, heat stress during early vegetative development resulted in stunted morphological characteristics (Reddy et al. 1997). Temperature above 30 °C during reproduction directly affects the blooms by flower shedding (Oosterhuis 2002) and impairs growth of pollen tube (Burke et al. 2004) which ultimately leads to reduced boll size and number of bolls and seed cotton during reproductive phase.

27.3.5 Biotic Factors

Throughout the life cycle, cotton plant remains susceptible to insect pest attack. Pest attacks cotton plant's root, stem, leaves, and fruits and decreases the cotton production from 35% to 45% yearly (Masood et al. 2011). Conventionally cotton pest has been controlled by heavy dose of chemicals as more than 60% of total chemical is being sprayed on cotton for controlling pests.

Dusky cotton bug (*Oxycarenus laetus* Kirby) has become a major pest which affects the quality of seed and lint during development (Abbas et al. 2015). Cotton bugs feed on bolls and seed by injecting piercing mouth parts inside and leave a greasy spot and females lay eggs on lint and feed on seed. The induction of Bt cotton reinforces dusky bug attack on seed during development. The maximum loss was recorded in quantitative factors, i.e., seed cotton weight (42.92%), seed weight (40.84%), and oil content (35.16%) by cotton bug (Srinivas and Patil 2004). Dusky bug deteriorates seed quality by reducing seed productivity and oil content, and its population sharply increases in humid environment prior to picking (Iqbal et al. 2018).

Similarly, red cotton bug (*D. koenigii*) also affects both lint and seed quality by injecting piercing mouth parts in fruits and sucking the sap of seed and fruits. The nymphs and adults of red cotton bug contaminate the lint color or quality during processing; that is why this pest is also known as cotton stainer (Jaleel et al. 2013).

27.4 Cotton Seed Quality During Harvesting and Post-harvesting

Picking at proper time and maintenance of dryness of cotton seed during processing and handling ensure a high-quality seed. Soon after picking, the quality of cotton seed begins to decline and environment plays a key role in determining seed quality (Suzuki et al. 2014; Deho et al. 2017).

27.4.1 Picking

Picking is the critical factor for seed quality (Deho et al. 2017). Cotton is a perennial plant and has indeterminate growth habit so time of maturity is based on the capacity of plant to carry maximum fruiting branches, as the plant does not have sufficient resource supply to initiate new fruiting branches (cutout). The fully developed seeds and lint come out from bolls and are ready to be picked out indicating the physical maturity of the plant (Bange and Milroy 2000).

The maturity of cotton depends on genotypes and environmental conditions during production seasons. The genotypes bred for early maturity develop canopy quicker and intercept more light and produce more dry matter to partition in fruits and mature early to accomplish reproductive phase, while long-season cultivars spend a long time in field to develop and mature. Thus, early maturation helps to enhance productivity by relieving late season risks linked with biotic and abiotic stresses (Suzuki et al. 2014).

The maturity can also be calculated on the basis of node positions on the main stem above the uppermost white flower. The fifth node above the white flower represents physiological cutout date because subsequent flowers have low availability of plant reserves to produce bolls of reasonable size and quality (Oosterhuis et al. 1992). Physiological cutout is the best indication of crop maturity, which is useful to plan the picking in for coming warm and dry weather.

At physical and physiological maturity, the cotton should be picked on forecast of dry and warm weather. Humid and cold weather at the time of picking is detrimental for cotton picking as it increases the moisture percentage in lint and affects the quality of both lint and seed during handling and processing. The time of picking is also critical to avoid the moisture gain in seed cotton; early morning picking, along with late evening picking, introduces moisture in seed cotton which later on enhances the risk of quality loss due to fungus infestation and rapid deterioration. Cotton should be harvested at minimum seed cotton moisture content (12%). Higher percentage of moisture increases temperature that accelerates the deterioration of cotton seed prior to ginning.

27.4.2 Ginning

Cotton seed quality is affected during ginning due to mechanical injuries. The efficiency of ginning is defined by amount of fiber separated from seed (ginning out-turn), fiber length, and amount of seed without slight injury on seed coat. The seed cotton moisture and speed of machine especially saw gin are important factors, which determine the efficiency of ginning.

Inappropriate handling after picking can damage the seed mechanically during ginning. Minimum seed damage was found when seed cotton moisture fell below 12%, and an increase in moisture content is directly associated with seed damage. If ginning is done at above 12% moisture content, seed coat damage increases directly with increase in the moisture content (Anthony et al. 1994). Moreover, the overspeed of ginning machine leaves the fiber on seed surface and causes fiber damage, as broken-off fibers reduce and shorten the staple length and create naps in cotton (Gourlot et al. 2018).

27.4.3 Delinting

In cotton the ginning produces two valuable products, i.e., fiber for textile industries and seed for sowing. After ginning some fibers remain attached on seed coat which results in production of fuzzy seeds. The fuzzy seeds should be cleaned by process of removing lint from seed surface called delinting. This process facilitates seed grading, cleaning, surface treatment, and seed placement in field through planter during mechanized farming.

Delinting of cotton seed with 98% commercial sulfuric acid is the most common and effective method to remove fuzz from cotton seed surface. The commercial sulfuric acid and fuzzy seed were mixed in a ratio of 1 L to 10 kg, respectively. The seeds are continuously stirred for 10 min for long staple length and 30 min for short staple length to burn the fuzz. Afterward, the seeds are washed immediately in running water till whole seeds are neutralized and then dried under shade (Gravanis et al. 2004).

During mechanical process, physical abrasion removes fuzz from the seed surface. The machine contained seven drum linings with one or two roller nylon or steel bristle brushes used for scrubbing the lint from the seed surface twice for 10 min. The process removes lint at 95% and ensures the germination of cotton seed at up to 88.5%. But this machine is still used for experimentation, and further investigation and research is needed to design the machine and protocol for delinting of seed on commercial scales (Holt et al. 2017).

27.4.4 Cotton Seed Drying

Drying is one of the most critical processes in sustaining cotton seed quality. Seed is a hygroscopic material; depending upon the external environment, the translocation of water between seed and surrounding continuously takes place (Bradford et al. 2018). Water molecules always exist in a state of vibration; the molecule located near the seed surface evaporates in less humid surroundings (desorption) and gets dry. Similarly, the water molecule in humid surroundings when contacted with the seed surface is absorbed inside (absorption) and moisture is gained. The trade of moisture between the seed and surroundings continues until both processes, i.e., desorption and absorption, develop a state of equilibrium. At this state the moisture in seed is called equilibrium moisture content, while the humidity of the surroundings is called equilibrium relative humidity (Soysal and Öztekin 1999). The moisture inside the seed can be desorbed by lowering the relative humidity from the surroundings. The relative humidity can be reduced by removing the moisture from the surroundings or elevating the temperature of air.

27.4.4.1 Sun-Drying

The drying of cotton seed in the sun is a cost-effective approach in an area that receives prolonged sunshine at daytime. The sun radiation elevates the temperature of surroundings and reduces the relative humidity thus evaporating moisture from the seed getting it dry. But this process has many limitations in that it is an uncontrolled drying process, the sun warms the air and seed simultaneously and the seed gets irreparable quality loss, and fluctuations in surrounding weather directly influence the seed quality (Adam et al. 2018). Cotton seed stored for a week with initial viability of 95%, with 10% moisture content at 35 °C temperature, lost viability by 1.5% (Ellis and Roberts 1980).

27.4.4.2 Forced Air-Drying

During forced air-drying, natural air or air supplemented with heat at suitable temperature is blown through a layer of seed till it dries (Ashley et al. 2018). This means of cotton seed drying is practiced by seed industries. The seed is placed over the sieve in chamber and hot air is passed through the seed. The heat expands the air volume and reduces the relative humidity of air and dry air expels out the moisture from the seed. The temperature and air flow are controlled by industries to dry the seed with minimum effect on the seed quality. Cotton seed gets dried in 72 h by flowing hot air of 45 °C. The drying temperature should be safe and not above the thermal tolerance of the seed.

27.4.4.3 Chemical Drying, Desiccants, and Sorption Dryers

The relative humidity (RH) of the surroundings is reduced by absorbing the moisture from air keeping the temperature uninterrupted. The dry air at lower RH is passed through seeds and moisture evaporates from the seed getting it dry. The best way is to dry the seed up to the required moisture level quickly without disturbing the seed quality. Recent research has reported zeolite as an effective tool for drying seeds which is required for long-term storage (Bradford et al. 2018). Drying beads have hygroscopic property as microscopic pores in beads are equal to the size of water molecules; only water molecules are bound tightly within pores. However, in contrast to conventional drying, when beads are placed along with seeds in a hermetic container, they absorb moisture and lower RH without heating air and disturbing equilibrium between RH and seed moisture contents. Kamran et al.

(2017) corroborated that beads and sun-drying took 1 h and 6 h, respectively, to reduce cotton seed moisture content from 11% to 8%.

27.4.5 Storage

Seed longevity during storage depends upon genetically regulated aging character of seed that is influenced by initial quality of seed at time of storage; history of seed before storage, i.e., environment during picking and processing; equilibrium moisture content (eqMC) of seed; temperature; and storage duration (Biabani et al. 2011). Damage of seed during storage is inevitable (Mosavi et al. 2011). An ideal storage module maintains the good quality of seed for a longer time by keeping the environment dry and cool (Bradford et al. 2018).

The principle for seed longevity emphasized on seed moisture contents, temperature, and oxygen (Ellis and Roberts 1980). Seed adjusts its moisture contents with RH of storage environment. Dry chain principle is quick drying of seed followed by storage in dry module to preserve seed viability for a long time (Bradford et al. 2018). Dry chain concept is analog to cold chain concept; however, when seed is dried and hermetically stored at low moisture, seed can maintain good quality without further energy input for cooling.

Storage conditions in favor of seed quality can be explained by the principle proposed by Harrington (1972) which stated that each 1% reduction of seed moisture or each 5 °C decrease in temperature doubled the storage life of the seed (within the moisture range of 5–14%). Previous studies reveal that cotton seed adjusts at lowest moisture content as compared to cereals having higher moisture contents when RH level is exceeded in the surroundings (Bakhtavar et al. 2019). Seed moisture isotherm clearly indicates that low RH provides the best storage module by maintaining low seed moisture content but higher RH elevates seed moisture contents and accelerates seed deterioration.

27.4.5.1 Conventional Storage

In rural areas of developing countries, farmers still store cotton seeds on open ground or pack them in clothes or jute bags and keep them in godowns. As these bags are not airtight, so moisture from the seed will fluctuate with respect to external RH and temperature and seed quality remains at risk (Karthikeyan et al. 2009).

Soon after ginning, producers spread the seeds generally on the ground. An advantage of this technique is that it is not expensive along with not accomplishing safe storage conditions to seeds due to weather uncertainties. Fluctuation is straightly influenced by conditions like RH, temperature, and aeration that ultimately lead toward seed deterioration.

27.4.5.2 Cold Storage

The main biochemical reactions (lipid peroxidation and non-enzymatic protein glycosylation with reducing sugars) are responsible to expedite the aging process in seed during storage (Murthy et al. 2003) because enzymes of these reactions accelerate the metabolic process when moisture and temperature are exceeded. Under such environment, seeds typically quickly lose viability within a few days or weeks, while seeds with low moisture contents under cold storage are likely to be in a glassy state. The low molecular mobility of the seed cytoplasm with high viscosity could inhibit many detrimental processes in the seed (Murthy et al. 2003).

There is no doubt that cold storage keeps the seeds in good quality because low temperature arrests the catabolic and physiological processes in the seeds (Das et al. 1998), but despite this cold storage has many limitations in that having a large quantity of seed requires walk-in refrigerators and some seeds show negative performance regarding germination and vigor after annual storage at freezing temperature. Cryogenic storage is one of the advanced but expensive forms of cold storage in which germplasm/seeds are suspended in liquid nitrogen (−196 °C) in unique tanks.

27.4.5.3 Hermetically Sealed Storage

Hermetic storage is a pest- and chemical-free storage technology which is the most commonly adopted technology among farmers in the developing world (Afzal et al. 2017; Bakhtavar et al. 2019). Hermetic storage creates an environment, which is free from oxygen and has a higher rate of carbon dioxide that hinders insect growth. Seed storage at elevated RH and moisture leads to fungal and insect growth along with aflatoxin production.

Hermetic bags are being used by farmers for preservation of seed quality in developing countries. Among these bags, Super Bags by GrainPro Inc. and Purdue Improved Crop Storage (PICS) bags are extensively used for storage of cereals, pulses, and oilseeds. These bags prevent the exchange of moisture and air and have very low vapor transmission rate (Afzal et al. 2017). Such conditions become lethal to insect and fungal growth. There is no doubt that these bags are costly than the traditional bags; however, the need for insecticide use is eliminated.

The efficiency of hermetic storage is determined by efficiency of drying and packaging material. After drying at 8% moisture content, the cotton seed should immediately be stored in an airtight container, which maintained the low moisture content for a longer time. Bakhtavar et al. (2019) revealed that cotton seed stored in hermetic Super Bag at 8% initial seed moisture content gave maximum germination (68%) after 18 months. This process of storage needs limitation as the seed must be dried to lower moisture content; otherwise, the seed with higher moisture content will be deteriorated with great extent than any other storage method (Lane and Woloshuk 2017).

27.5 Conclusions

It is very difficult task to produce good-quality cotton seed in the developing world, as the prime focus of farmers is always lint not seed. The maintenance of cotton seed quality during production is risky due to climate change and variability such as temperature and unpredicted rainfalls. Picking of fully opened mature bolls, in hot and dry weather, delivers dry seed cotton, which shields the seed from physical damages during ginning. Right after picking seed moisture contents and temperature are important factors to conserve the seed quality during processing and storage. An integrated approach is needed to maintain seed dryness during harvesting and post-harvest handling.

References

Abbas Q, Ahmad S (2018) Effect of different sowing times and cultivars on cotton fiber quality under stable cotton-wheat cropping system in southern Punjab, Pakistan. Pak J Life Soc Sci 16:77–84

Abbas A, Ali MA, Khan TM (2008) Studies on gene effects of seed cotton yield and its attributes in five American cotton cultivars. J Agric Soc Sci 4:147–152

Abbas M, Hafeez F, Farooq M, Ali A (2015) Dusky cotton bug *Oxycarenus* spp. (Hemiptera: Lygaeidae): hibernating sites and management by using plant extracts under laboratory conditions. Polish J Entomol 84(3):127–136

Adam O, Abisoye O, Joseph A, Julius O, Ayooluwa O (2018) The effect of traditional shade drying method on seed germination and vigour of two varieties of tomato. Asian Res J Agric 10(1):1–6

Afzal I, Bakhtavar MA, Ashfaq M, Sagheer M, Baributsa D (2017) Maintaining dryness during storage contributes to higher quality of maize seeds. J Stored Prod Res 72:49–53

Ahmad S, Raza I (2014) Optimization of management practices to improve cotton fiber quality under irrigated arid environment. J Food Agric Environ 2(2):609–613

Ahmad S, Raza I, Ali H, Shahzad AN, Atiq-ur-Rehman, Sarwar N (2014) Response of cotton crop to exogenous application of glycinebetaine under sufficient and scarce water conditions. Braz J Bot 37(4):407–415

Ahmad S, Abbas Q, Abbas G, Fatima Z, Atique-ur-Rehman, Naz S, Younis H, Khan RJ, Nasim W, Habib ur Rehman M, Ahmad A, Rasul G, Khan MA, Hasanuzzaman M (2017) Quantification of climate warming and crop management impacts on cotton phenology. Plants 6(7):1–16

Ahmad S, Iqbal M, Muhammad T, Mehmood A, Ahmad S, Hasanuzzaman M (2018) Cotton productivity enhanced through transplanting and early sowing. Acta Sci Biol Sci 40:e34610

Ali H, Afzal MN, Ahmad F, Ahmad S, Akhtar M, Atif R (2011) Effect of sowing dates, plant spacing and nitrogen application on growth and productivity on cotton crop. Int J Sci Eng Res 2 (9):1–6

Ali H, Abid SA, Ahmad S, Sarwar N, Arooj M, Mahmood A, Shahzad AN (2013a) Integrated weed management in cotton cultivated in the alternate-furrow planting system. J Food Agric Environ 11(3–4):1664–1669

Ali H, Abid SA, Ahmad S, Sarwar N, Arooj M, Mahmood A, Shahzad AN (2013b) Impact of integrated weed management on flat-sown cotton (*Gossypium hirsutum* L.). J Anim Plant Sci 23 (4):1185–1192

Ali H, Hameed RA, Ahmad S, Shahzad AN, Sarwar N (2014a) Efficacy of different techniques of nitrogen application on American cotton under semi-arid conditions. J Food Agric Environ 12 (1):157–160

Ali H, Hussain GS, Hussain S, Shahzad AN, Ahmad S, Javeed HMR, Sarwar N (2014b) Early sowing reduces cotton leaf curl virus occurrence and improves cotton productivity. Cercetări Agronomice în Moldova XLVII(4):71–81

Amin A, Nasim W, Mubeen M, Nadeem M, Ali L, Hammad HM, Sultana SR, Jabran K, Habib ur Rehman M, Ahmad S, Awais M, Rasool A, Fahad S, Saud S, Shah AN, Ihsan Z, Ali S, Bajwa AA, Hakeem KR, Ameen A, Amanullah, Rehman HU, Alghabar F, Jatoi GH, Akram M, Khan A, Islam F, Ata-Ul-Karim ST, Rehmani MIA, Hussain S, Razaq M, Fathi A (2017) Optimizing the phosphorus use in cotton by using CSM-CROPGRO-cotton model for semi-arid climate of Vehari-Punjab, Pakistan. Environ Sci Pollut Res 24(6):5811–5823

Amin A, Nasim W, Mubeen M, Ahmad A, Nadeem M, Urich P, Fahad S, Ahmad S, Wajid A, Tabassum F, Hammad HM, Sultana SR, Anwar S, Baloch SK, Wahid A, Wilkerson CJ, Hoogenboom G (2018) Simulated CSM-CROPGRO-cotton yield under projected future climate by SimCLIM for southern Punjab, Pakistan. Agr Syst 167:213–222

Anthony WS, William D, Mayfield D (1994) Cotton Ginners Handbook, rev. U.S. Department of Agriculture, Agricultural Handbook, vol 503. U.S. Dept. of Agriculture, Washington, DC. 348 p

Arah IK, Ahorbo GK, Anku EK, Kumah EK, Amaglo H (2016) Postharvest handling practices and treatment methods for tomato handlers in developing countries. Adv Agric 2016:6436945. https://doi.org/10.1155/2016/6436945. 8 p

Arif M, Jan MT, Mian IA, Khan SA, Hollington P, Harris D (2014) Evaluating the impact of osmopriming varying with polyethylene glycol concentrations and durations on soybean. Int J Agric Biol 16:2

Ashley H, Thomas J, Holt G, Valco T (2018) Cottonseed air-handling and storage requirement. J Cotton Sci 22:47–59

Ashraf J, Zuo D, Wang Q, Malik W, Zhang Y, Abid MA, Song G (2018) Recent insights into cotton functional genomics: progress and future perspectives. Plant Biotechnol 16(3):699–713

Baalbaki R, Elias S, Marcos-Filho J, Mcdonald MB (2009) Seed vigor testing handbook. Association of Official Seed Analysts, Ithaca, NY. 341 p

Bakhsh A, Rehman M, Salman S, Ullah R (2019) Evaluation of cotton genotypes for seed cotton yield and fiber quality traits under water stress and non-stress conditions. Sarhad J Agric 35:161–170

Bakhtavar MA, Afzal I, Basra SMA (2019) Moisture adsorption isotherms and quality of seeds stored in conventional packaging materials and hermetic Super Bag. PLoS One 14(2):e0207569

Bange M, Milroy SP (2000) Timing of crop maturity in cotton: impact of dry matter production and partitioning. Field Crop Res 68(2):143–155

Basra SMA, Asraf M, Ahmed R (2003) Effect of pre-sowing humidification treatment on cottonseed vigour. Pak J Life Soc Sci 1(1):37–41

Basra SMA, Ashraf M, Iqba N, Khaliq A, Ahmad R (2004) Physiological and biochemical aspects of pre-sowing heat stress on cottonseed. Seed Sci Technol 32:837–846

Biabani A, Boggs LC, Katozi M, Sabouri H (2011) Effects of seed deterioration and inoculation with *Mesorhizobium ciceri* on yield and plant performance of chickpea. Aust J Crop Sci 5 (1):66–70

Bilalis DJ, Katsenios N, Efthimiandou A, Karkanis A, Khan EM, Mitsis T (2013) Magnetic field pre-sowing treatment as an organic friendly technique to promote plant growth and chemical elements accumulation in early stages of cotton. Aust J Crop Sci 79(1):46–50

Bolek Y, Nas MN, Cokkizgini H (2013) Hydropriming and hot water induced heat shock increase cotton seed germination and seedling emergence at low temperature. Turk J Agric For 37:300–306

Bradford KJ, Dahal P, Van Asbrouck J, Kunusoth K, Bello P, Thompson J (2018) The dry chain: reducing postharvest losses and improving food safety in humid climates. Trends Food Sci Technol 71:84–93

Burke JJ, Wanjura DF (2010) Plant responses to temperature extremes. In: Physiology of cotton. Springer, Dordrecht, pp 123–128

Burke JJ, Velten J, Oliver MJ (2004) *In vitro* analysis of cotton pollen germination. Agron J 96:359–368

Calısır S, Marakoglu T, Ogut H, Ozturk O (2005) Physical properties of rapeseed (*Brassica napus oleifera* L.). J Food Eng 69:61–66

Camp AF, Walker MN (1972) Soil temperature studies with cotton. II. The relation of soil temperature to germination and growth of cotton. Florida Agric Exp Station Bull 189:17–32

Casenave EC, Toselli ME (2007) Hydropriming as pre-treatment of cotton germination under thermal water stress conditions. Seed Sci Technol 35(1):88–98

Cherry JP, Leffler HR (1984) Seed. In: Ko-hel RJ, Lewis CF (eds) Cotton. ASA, CSSA, and SSSA, Madison, WI, pp 511–569

Çokkizgin H, Bölek Y (2015) Priming treatments for improvement of germination and emergence of cotton seeds at low temperature. Plant Breed Seed Sci 71:121–134

Das BK, Barua IC, Dey SC (1998) Effect of packing material, storage condition and duration of storage on seed viability, vigour and seedling survivability in Rajmah (*Phaseolus vulgaris* L.). Legume Res 21(2):91–95

Deho ZA, Laghari S, Abro S (2017) Impact of irrigation frequencies and picking timings on fiber quality and seed germination of cotton varieties. Pak J Bot 49:2347–2352

Dordas C (2006) Foliar boron application affects lint and seed yield and improves seed quality of cotton grown on calcareous soils. Nutr Cycl Agroecosyst 76:19. https://doi.org/10.1007/s10705-006-9037-7

Dowd MK, Boykin DL, Meredith WR Jr, Campbell BT, Bourland FM, Gannaway JR, Glass KM, Zhang J (2010) Fatty acid profiles of cottonseed genotypes from the National Cotton Variety Trials. J Cotton Sci 14:64–73

Ellis RH, Roberts EH (1980) Improved equations for the prediction of seed longevity. Ann Bot 45:13–30

Finch-Savage WE, Bassel GW (2016) Seed vigour and crop establishment: extending performance beyond adaptation. J Exp Bot 67(3):567–591

França PRC, Almeida FAC, Silva ORRF, Jerônimo JF (2018) Physiological quality of Gossypium hirsutum L. seeds after the thermal delinting process. Comunicata Scientiae 9(3):465–474

Gerard JJB, Groot D, Jundt A, Anthony B, Murphy B, Bange MP, Prochnow AM (2018) Cold plasma treatment of cotton seed germination improvement. Sci Rep 8:14372

Ghassemi-Golezani K, Dastborhan S, Zehtab-Salmasi S (2013) Seed priming and field performance of borage (*Borago officinalis* L.) under different irrigation treatments. Int J Agron Plant Prod 4:82–87

Gourlot JP, Togola M, Goze E, Bachelier B, Coulibaly M, Traore AK (2018) Ginning: a way of measuring its specific impact on fiber quality. In: International cotton conference Bremen. Available at: https://baumwollboerse.de/wp-content/uploads/2018/03/Gourlot-2018_Bremen_IntlConf_V03.pdf

Gravanis FT, Chouliarars NA, Xifilidou S, Gougoulias N, Vagelas IK, Gemtos TA (2004) Effect of cotton seed acid delinting product and gin trash on soil nitrification and the possibility in disseminating cotton pathogen. Acta Horticult 698(698):273–278

Hadas A, Wolf D, Rawitz E (1985) Residual compaction effects on cotton stand and yield. Trans Am Soc Agric Eng 28:691–696

Hameed A, Sheikh MA, Jamil A, Basra SMA (2013) Seed priming with sodium silicate enhances seed germination and seedling growth in wheat (*Triticum aestivum* L.) under water deficit stress induced by polyethylene glycol. Pak J Life Soc Sci 11:19–24

Harrington JF (1972) Seed storage and longevity. In: Kozlowski TT (ed) Seed biology, vol 3. Academic Press, New York, NY, pp 145–245

Hodges SC (1992) Nutrient deficiency disorders. In: Hillocks RJ (ed) Cotton diseases. CABI Publishing, Wellingford, pp 355–403

Holt G, Wedegaertner T, Wanjura J, Pelletier M, Delhom C, Duke S (2017) Engineering and ginning development and evaluation of a novel bench-top mechanical cotton seed delinter for cotton breeders. J Cotton Sci 21:18–28

Iqbal N, Basra SMA, Khalil-ur-Rehman (2002) Evaluation of vigour and oil quality in cottonseed during accelerated ageing. Int J Agric Biol 4:318–322

Iqbal J, Butta SA, Alqarni AS, Owayss AA, Ansari MJ (2018) Seasonal population dynamics of dusky cotton bug (*Oxycarenus* spp.) in transgenic cotton varieties under field conditions. Saudi J Biol Sci 25:1122–1127

ISTA (2015) International rules for seed testing. International Seed Testing Association, Switzerland

Jaleel W, Saeed S, Naqqash MN (2013) Biology and bionomic of *Dysdercus koenigii* F. (Hemiptera: Pyrrhocoridae) under laboratory conditions. Pak J Agric Sci 50:373–378

Kamran M, Afzal I, Basra SMA, Khan SHU, Mahmood A (2017) Improvement of cotton crop performance by estimating optimum sowing and picking time. Int J Agric Biol 19:241–247

Karthikeyan C, Veeraragavathatham D, Karpagam D, Firdouse SA (2009) Traditional storage practices. Indian J Trad Know 8(4):564–568

Kerby TA, Plant RE, Horrocks RD (1997) Height to node ratio as an index of early season cotton growth. J Prod Agric 10:80–83

Khan MB, Ahmad S, Khaliq A (2004) Performance of mashbean intercropped in cotton planted in different planting patterns. J Res (Sci) 15(2):191–197

Krzyzanowski FC, Delouche JC (2011) Germination of cotton seed in relation to temperature. Revista Brasileira de Sementes 33:543–548

Lane B, Woloshuk C (2017) Impact of storage environment on the efficacy of hermetic storage bags. J Stored Prod Res 72:83–89

Larson JA, Mapp HP (1997) Cotton cultivar, planting, irrigating and harvesting decision under risk. J Agric Resour Econ 22(1):157–173

Latif A, Ahmad T, Hayat S, Sarwar G, Ehsan MZ, Raza M, Sarwer M, Khan IA (2014) Genetics of yield and some yield contributing traits in upland cotton (*Gossypium hirsutum* L.). J Plant Breed Crop Sci 6:57–63

Ludwig CA (1932) The germination of cotton seed at low temperature. J Agric Res 44:367–380

Lukonge E, Labuschagne MT, Hugo A (2007) The evaluation of oil and fatty acid composition in seed of cotton accessions from various countries. J Sci Food Agric 87:340–347

Masood A, Arif MJ, Hamed M, Talpur MA (2011) Field performance of *Trichogrammachilonis* against cotton bollworms IPM approach. Pak J Agric Agric Eng Vet Sci 27:176–184

Mosavi NSM, Tilebeni HG, Kordfirouzjae GH, Sadeghi M, Sedighi E (2011) Free fatty acid and electrical conductivity changes in cotton seed (*Gossypium hirsutum* L.) under seed deteriorating conditions. Int J Agric Sci 1(2):62–66

Mullins GL, Bumester CH (2010) Relation of growth and development to mineral nutrition. In: Stewart JM, Oosterhuis DM, Heitholt JM, Mauney JR (eds) Physiology of cotton. Springer, New York, NY, pp 97–105. Available at: http://www.springerlink.com/index/u88838767k45m032.pdf

Murthy UMN, Kumar PP, Sun WQ (2003) Mechanisms of seed ageing under different storage conditions for *Vigna radiata* L. Wilczek: lipid peroxidation, sugar hydrolysis, Maillard reactions and their relationship to glass state transition. J Exp Bot 54(384):1057–1106

Murungu FS, Nyamugafata P, Chiduza C, Clark LJ, Whalley WR (2005) Effects of seed priming and water potential on germination of cotton (*Gossypium hirsutum* L.) and maize (*Zea mays* L.) in laboratory assays. Afr J Plant Soil 22:64–70

Mustafa HSB, Mahmood T, Ullah A, Sharif A, Bhatti AN, Nadeem M, Ali R (2017) Role of seed priming to enhance growth and development of crop plants against biotic and abiotic stresses. Bull Bio All Sci Res 2(2):1–11

Oosterhuis DM (2002) Day or night high temperature: a major cause of yield variability. Cotton Grower 46:8–9

Oosterhuis DM, Bourland FM, Tugwell NP (1992) Basis for the nodes above white flower cotton monitoring system. Ark Farm Res 41:3–6

Paterson A (2009) Genetics and genomics of cotton. Springer, New York, NY

Pereira JR, Alves MC, Lima SV, Souza RN, Santos JW, Bezerra JR (2005) Germinação de sementes de algodãoherbaceio, cultivadas sob diferentestratamentos de irrigação, emdiferentescondiçõesambientais de plantio. In: Congressobrasileiro de algodão, Salvador.

Resumo Salvador: Embrapa Algodão, 2005. Available at: http://www.cnpa.embrapa.br/produtos/algodao/publicacoes/trabalhos_cba5/212.pdf

Pettigrew WT, Dowd MK (2012) Interactions between irrigation regimes and varieties result in altered cotton- seed composition. J Cotton Sci 16:42–52

Rahman M, Tabassum N, Ullah I, Asif M, Zafar Y (2008) Studying the extent of genetic diversity among *Gossypium arboreum* L. genotypes/cultivars using DNA fingerprinting. Genet Resour Crop Evol 55:331–339

Rahman M, Zafar Y, Paterson AH (2009) *Gossypium* DNA markers types, number and uses. In: Paterson AH (ed) Genomics of cotton. Springer, Dordrecht. https://doi.org/10.1007/978-0-387-70810-2_5

Rahman MH, Ahmad A, Wang X, Wajid A, Nasim W, Hussain M, Ahmad B, Ahmad I, Ali Z, Ishaque W, Awais M, Shelia V, Ahmad S, Fahad S, Alam M, Ullah H, Hoogenboom G (2018) Multi-model projections of future climate and climate change impacts uncertainty assessment for cotton production in Pakistan. Agric For Meteorol 253-254:94–113

Rathinavel K (2014) Influence of storage temperature and seed treatments on viability of cotton seed (*Gossypium hirsutum* L.). Cotton Res J 6:1–6

Reddy AR, Reddy KR, Padjung R, Hodges HF (1996) Nitrogen nutrition and photosynthesis in leaves of Pima cotton. J Plant Nutr 19:755–770

Reddy KR, Hodges HF, Mckinion JM (1997) A comparison of scenarios for the effect of global climate change on cotton growth and yield. Funct Plant Biol 24:707–713

Reddy KR, Brand D, Wijewardana C, Gao W (2017) Temperature effects on cotton seedling emergence, growth, and development. Agron J 109:1379–1387

Ritchie GL, Bednarz CW, Jost PH, Brown SM (2007) Cotton growth and development. Ext Agron Bull 1252. Revised June 2007

Salam MA, Haque MM, Islam MO, Uddin MN, Haque MN (2017) Genotypic variation in yield and fiber quality traits of cotton grown from seeds packed in different packaging materials. Sarhad J Agric 33:255–262

Sangakkara UR, Frehner M, Nosberger J (2000) Effect of soil moisture and potassium fertilizer on shoot water potential, photosynthesis and partitioning of carbon in mungbean and cowpea. J Agron Crop Sci 185:201–207

Shaheen HL, Shahbaz M, Ullah I, Iqbal MZ (2012) Morpho-physiological responses of cotton (*Gossypium hirsutum*) to salt stress. Int J Agric Biol 14:980–984

Shuli F, Jarwar AH, Wang X, Wang L, Ma Q (2018) Overview of the cotton in Pakistan and its future prospects. Pak J Agric Res 31(4):396–407

Singh N, Yadav A, Varma A (2015) Effect of plant growth promoting activity of rhizobacteria on cluster bean (*Cyamopsis tetragonoloba* L.) plant growth and biochemical constituents. Int J Curr Microbiol App Sci 4:1071–1082

Solomon S (2007) Contribution of working group I contribution to the fourth assessment report of the IPCC. In: Solomon S, Qin D, Manning M, Chen Z, Marquis M, Averyt KB (eds) Climate Change 2007-the physical science basis. Cambridge University Press, Cambridge, NY

Soysal Y, Öztekin S (1999) Equilibrium moisture content equations for some medicinal and aromatic plants. J Agric Eng Res 74:317–324

Srinivas M, Patil BV (2004) Quantitative and qualitative loss caused by dusky cotton bug, Oxyacarenus laetus kirby on cotton. Karnataka J Agric Sci 17(3):487–490

Suzuki H, Khan AA (2000) Effective temperatures and duration for seed humidification in snap bean (*Phaseolus vulgaris* L.). Seed Sci Technol 28:381–389

Suzuki N, Rivero RM, Shulaev V, Blumwald E, Mittler R (2014) Abiotic and biotic stress combinations. New Phytol 203:32–43

Taiz L, Zeiger E (2012) Plant physiology, 5th edn. Sinauer Associates, Inc., Sunderland, MA

Tariq M, Yasmeen A, Ahmad S, Hussain N, Afzal MN, Hasanuzzaman M (2017) Shedding of fruiting structures in cotton: factors, compensation and prevention. Trop Subtrop Agroecosyst 20(2):251–262

Tariq M, Afzal MN, Muhammad D, Ahmad S, Shahzad AN, Kiran A, Wakeel A (2018) Relationship of tissue potassium content with yield and fiber quality components of Bt cotton as influenced by potassium application methods. Field Crop Res 229:37–43

Taylor AG, Allen PS, Bennett MA, Bradford KJ, Burrisand JS, Misra MK (1998) Seed enhancements. Seed Sci Res 8:245–256

TeKrony DM, Egli DB (1997) Relationship between standard germination, accelerated ageing germination and field emergence in soybean. In: Ellis RH, Black M, Murdoch AJ, Hong TD (eds) Basic and applied aspects of seed biology. Kluwer Academic, London, pp 593–600

Thomas RQ, Christiansen MN (1971) Seed hydration- chilling treatment effects on germination and subsequent growth and fruiting of cotton. Crop Sci 2:454–456

Toole EH, Drummond PL (1924) The germination of cotton seed. J Agric Res 28:285–292

Toselli ME, Casenave EC (2014) Is the enhancement produced by priming in cottonseeds maintained during storage? Bragantia Campinas 73:372–376

Usberti R, Roberts EH, Ellis RH (2006) Prediction of cotton longevity. Pesq Agropec Bras Brasília 41(9):1435–1441

Usman M, Ahmad A, Ahmad S, Irshad M, Khaliq T, Wajid A, Hussain K, Nasim W, Chattha TM, Trethowan R, Hoogenboom G (2009) Development and application of crop water stress index for scheduling irrigation in cotton (*Gossypium hirsutum* L.) under semiarid environment. J Food Agric Environ 7(3–4):386–391

Yalcın I, Ozarslan C (2004) Physical properties of vetch seed. Biosyst Eng 88:507–512

Zhao DL, Oosterhuis DM (2002) Cotton carbon exchange, nonstructural carbohydrates, and boron distribution in tissues during development of boron deficiency. Field Crop Res 78:75–87

Chapter 28
Quality Aspects of Cotton Lint

Muhammad Ilyas Sarwar and Danish Iqbal

Abstract Cotton is the key natural fibre used in the textile industry, the quality of which is extremely critical for successful processing. Many factors influence its fibre quality. The fibre quality of the open boll is affected by pre-harvesting and post-harvesting practices. Mature seed cotton is harvested and transferred to the ginning industry. The separation of fibres from the seed is the main goal of ginning. Besides varietal and environmental factors, the pre- and post-ginning practices decide the quality. The quality of cotton is determined by multiple measurements like fibre length, strength, micronaire, colour, trash, etc. These parameters play a necessary role in marketing.

This chapter deliberates all post-harvesting factors that affect quality like handling, storage, cleaning, ginning, operational health and safety and classification of cotton.

Keywords Fibre quality · Seed cotton · Ginning

Abbreviations

AFIS	Advanced fibre information system
FEC	Feeder-extractor-cleaner
GI	Galvanized iron
GOT	Ginning out-turn
HVI	High-volume instrument
LVI	Low-volume instrument
PCSI	Pakistan Cotton Standard Institute
USDA	United States Department of Agriculture

M. I. Sarwar (✉) · D. Iqbal
Fibre Technology Section, Central Cotton Research Institute, Multan, Pakistan
e-mail: mianilyas222@ccri.gov.pk

S. Ahmad, M. Hasanuzzaman (eds.), *Cotton Production and Uses*,
https://doi.org/10.1007/978-981-15-1472-2_28

28.1 Concept of Cotton Fibre Quality

Cotton is the backbone of the world's textile trade. Most of our everyday fabrics are made from cotton. It is a natural fibre mainly composed of cellulose which is the most abundant natural material present in wood (Usman et al. 2009; Ahmad et al. 2014, 2017, 2018; Abbas and Ahmad 2018; Ahmad and Raza 2014; Ali et al. 2011, 2013a, b, 2014a, b). It is seed hair, a single hyper-extended cell rising from the protodermal cells of the external integument layer of the seed coat. Cotton fibre development responds independently to variations in the micro- and macro-environments likewise with all living plant cells (Amin et al. 2017, 2018; Khan et al. 2004; Rahman et al. 2018; Tariq et al. 2017, 2018). Consequently, the fibres on a sole seed constitute diversities of cell wall thickness, fibre length, shape and physical maturity. Environmental variations inside the canopy of the plant and amongst distinct plants and within and amongst fields confirm that the population of fibre in each boll, definitely on each seed, incorporates an extensive range of fibre traits and that every single bale of cotton holds a highly variable fibre population. The cellulose contents of raw cotton vary from 88% to 96% of the dry weight. Scoured, bleached or dry cotton fabric is approximately 99% cellulose. The original quality and traits of the fibre are dependent primarily on the variety, environmental conditions during development and agronomical practices. When there is no weathering effect, the highest-quality fibre is obtained from fully mature and newly opened cotton bolls. The successful processing of lint depends on the proper management of highly variable fibre properties that have been proven to affect the quality and productivity of the finished product during and after the harvest. If fibre blending strategies and subsequent spinning and dyeing processes are to be optimized for specific end uses and profitability, textile mill managers need accurate and effective methods for descriptive and predictive quantitative measurement of these highly variable fibre properties. Fibre quality means approximately entirely changed for cotton growers and cotton processors. Growers or processors do not have contact to post-harvest mechanisms that improve the intrinsic fibre quality. Cotton quality has historically employed both visual and mechanical methods (Anthony and Mayfield 1994). Cotton quality characteristics, e.g. fibre length, fineness, micronaire value, strength and colour, have ever been influential in defining the values of fibre as discussed below.

28.1.1 Fibre Length

It is "the mean length by number of the longer one-half of the fibres, by weight, in sample". It is typically the measurement of longer fibres or most frequently occurring fibres. Fibre length differs with variety. The fibre length and distribution of length on seed are also influenced by the stresses throughout development of fibre as well as the mechanical processes at harvest and post-harvest. The length of the fibre

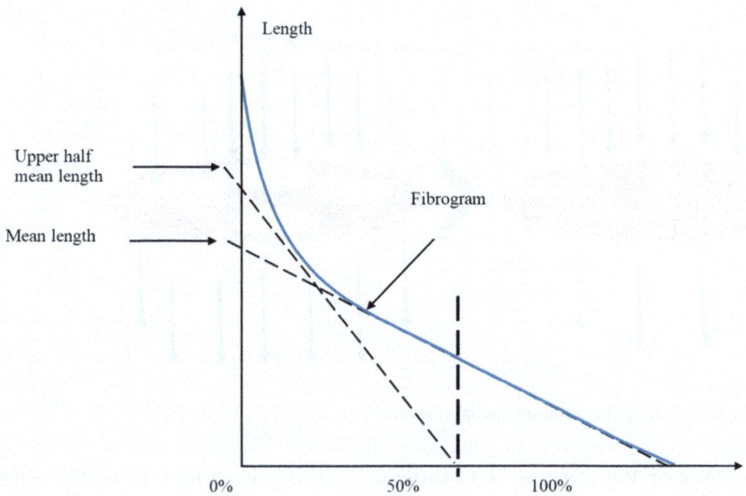

Fig. 28.1 Cotton fibre length measured by HVI

determines the spinning machine setting (Fig. 28.1). Longer fibres can take lower twist levels, have higher yarn strength and can be spun at higher processing speeds for more finer counts of yarn.

28.1.2 *Fibre Strength*

It is "the amount of force required to break a bundle of fibres, reported in grams force per tex". A tex is equivalent to mass in grams of one thousand metres of fibres. The capacity of cotton to survive tensile forces plays a vital role in spinning. Fabric and yarn strength are related to fibre strength.

28.1.3 *Length Uniformity*

It is "the ratio between the mean length and the upper half mean length of the fibres, expressed as a percentage". Variations in length result in increased processing waste and decline in process ability and quality of yarn.

28.1.4 *Micronaire*

"It is measurement of fibre fineness and maturity"; principle of air flow is used for the measurement. The linear density determines the amount of fibres required for the

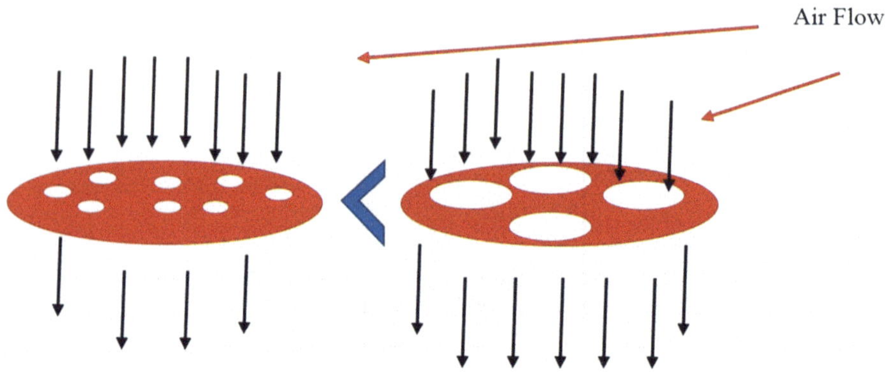

Fig. 28.2 Principle of micronaire measurement

cross section of the yarn to determine the spinnable count. Immature fibres are obtained in cotton with a low micronaire value. Cotton with a high micronaire value is taken as coarse (high linear density) which gives less fibres in yarn cross section (Fig. 28.2).

28.1.5 Colour Grade

The colour grade is ascertained by the degree of whiteness (Rd) and yellowness (+b). Reflectance specifies how dull or bright the cotton is and yellowness designates the degree of colour pigmentation. In addition to severe staining, cotton colour and "preparation" levels have no direct effect on processing capacity. Significant differences in colour can create problems in dyeing.

28.1.6 Trash %

It is the quantity of the non-lint constituents in cotton, such as bark and leaves from the cotton plant. Trash denotes to the portion of the plant that is added during the harvest and is then broken down into smaller portions during ginning process. Although it is easy to remove large particles, excessive trash leads to an increase in spinning waste. High dust levels can affect the rotor spinning efficiency and quality of the product. Grass and bark are tough to isolate from cotton fibres in mill.

28.2 Post-harvesting and Storage Management

Cotton is Pakistan's main cash income crop and its products are obtained in the shape of cottonseed and lint. The higher the market interest rate, the more the growers will be sought to maintain their quality and high standards. In fact, the grower has achieved the production target but forgot the quality steps. Pakistani cotton is genetically sustained and healthy. However, it belongs to the unclassified classification, including the use of incorrect methods for picking, processing and ginning, and there are always pricing issues depending on the quality of the seed cotton. This is why low-standard cotton enters the market.

Pakistani cotton is hand-picked, which should be done after 60% bolls are fully opened. The picking of cotton should be done after 10:00 am in the presence of sunlight when crop is fully dry from humidity and dew. The picking should be done variety-wise and it is necessary that picked seed cotton be labelled and reserved variety-wise and/or be sent to the ginning factory to sustain cottonseed and lint variety-wise.

The work of cotton picking is assigned mostly to female sects in the population, and they can easily complete work in a shorter period of time than male counterparts. Women have better picking skills, which is the main reason for saving time. Female cotton-picking personnel for training purposes should be carried out under the supervision of an expert or trained grower whilst standing in the crop to obtain clean cotton and maintain its standards. Women must be using cotton cloth to collect seed cotton, and materials made of any synthetic fibres must be avoided. However, in order to avoid impurities such as dust in the seed cotton, it should be sorted from the lower side to the top. This step leads to clean cotton picking.

Cotton picking should be carried out after the crop has been dry from dew. It is reported that during the picking period, the cotton crop may show dew due to the persistence of climatic conditions. Therefore, picking in early time may increase the probabilities of seed cotton adding humidity and dew. Dew-mixed cotton or lint that remains wet or damp when stored in a place causes damage to nearby lint and cottonseeds; lint quality and seed germination may be getting deteriorated. The ginning factories do not have facilities for cottonseed drying.

In the case of precipitation, seed cotton would not be picked instantly, but crops and seed cotton should be kept dry. Cotton should be picked when the seed cotton or lint is dried after exposure to air and sunlight. There is a risk of shower during picking, and in the case of rain, the fibres turn black and will get a low price. In fact, spoilage and impurities in seed cotton may occur due to serious damage caused by pests and diseases. This damage is largely determined by the production of cotton plants, such as cotton bolls, squares and flowers, which may rot immediately. Dark and rotten seed cotton appears when the boll is open. Therefore, seed cotton should be picked at the time of picking. Otherwise it will open and the empty boll will be picked and mixed with the lint as it is picked quickly. On several occasions, due to unskilled picking of cotton crop leaves, empty cotton bolls, leaves and stems, flowers, immature bolls, branches and weeds were mute by lint. These materials must not be left behind and mixed with cotton, but care should be taken to remove these impurities. The picking, storage and sale of such cotton or lint is at risk.

Weeds can cause difficulties when picking cotton from cotton bolls, and easy mixing can damage the quality of the lint. Cotton must be harvested in a timely manner and a second pick will be maintained after the first picking interval of 15–20. If picked before time, raw materials or immature fibres can be picked, which will be sold at a low price on the market. Agricultural experts recommend first picking on each planting. This seed can be highly germinated and does not contain many pests (including insects and diseases). However, cotton should not be picked early or late. In the case of early picking, a small fibre length with shrinkage quality will be obtained which will result in immature fibres and substandard fabrics; also fibres obtained from early-picked cotton bolls will immediately darken. Seeds obtained from early picking did not have any good quality in terms of low seed germination and low edible oil content. Irrigation water should be limited during picking to properly use the cotton boll, which helps to easily pick cotton and increase the strength of the cotton fibre.

For delayed picking, the quality of lint may be lost. Seed cotton may be lost for a long time, and continuous dew and air blowing through the dust on the cotton will change the colour of the lint. Fast winds may shed seed cotton or lint from open cotton bolls and may be one of the sources of reduced yield per acre. The cotton fibres selected in the later stage made the appearance dirty, reducing the fineness and lustre of the clothes. Because of lack of fibre strength in late-picked cotton, low-quality fibres may often be produced in textile and garment factories.

There is a growing trend for impurities or contamination during cotton picking, storage and transportation to the factory. Rural female folks engaged in cotton picking began picking before dew and humidity were dry. The reason behind this is to increase the weight and volume of the picked cotton in the case of dew or humidity, in order to achieve higher wages. There is a risk of directly affecting the quality of lint. This effect of dew or humidity can be traced when filling bales and storing lint. The second increasing trend in impurities and pollution in cotton lint may be rural women who use plastic bags and silk scarves to pick seed cotton. This type of contaminating impurities can occur during spinning and dyeing of fibres, threads and clothing.

The mixing of human hair during picking cotton creates problems in the ginning, spinning and weaving processes, which can cause losses to the textile industry. It is a drawback that the women put water into the harvested cotton at different time intervals to gain weight to get higher wages. This practice may undermine cotton and its quality, leading to low market prices.

Therefore, cotton must be saved from impurities and pollution when picking cotton after dew drying. This harvested cotton should not be stored on damp soil, but placed in a cloth bag and dry floor before being sent to the market or factory. If cotton is not treated immediately to the factory, then it is lying in an exposed place; in this case the cotton must be covered with cloth at night to avoid deterioration due to open air and moisture, but during the day the large cloth must be removed for proper sun exposure.

Many attributes affect the quality of seed cotton during storage. The most significant is the moisture content. Further these include time of storage, green matter, initial temperature of seed cotton, temperature of seed cotton, weather parameters (relative humidity, rainfall temperature) and protection from wet floor and rain. The findings provide useful guidance on how storage variables affect quality. These results clearly show that whilst it is not possible to accurately predict how stored variables affect quality, guidelines for safe, efficient storage are useful.

28.3 Handling and Heap Formation at Gin Yard

The collection of seed cotton from diverse regions is channeled by a cotton dealer to the ginning unit through different means of transportation, e.g. trolleys, trucks and animal carts. Cotton transport should be carried out in open trolleys covered with protective cotton, avoiding the use of polyproline, polyethylene and jute bags for cotton transport. The weighing of seed cotton is done initially on a factory-based weighing scale, and it is then passed through the factory gate and stored in the seed cotton storage yard. Some progressive units maintain internal laboratories to measure changes in the amount of lint (ginning out-turn, GOT), moisture percentage and impurities. Mostly the seed cotton quality is based on the experience of cotton selectors that do on-site qualitative valuation.

Batches of seed cotton from different places/sources are mixed/blend with one another for making cotton heap in the cotton storage yard. The heap is usually made to maintain the daily ginning capacity of the unit. The practice of manual labor is often used to make and mix cotton heaps. Due to unavailability of labors and also for time saving, some ginning units use mobile conveyors and/or tractors for making heaps. The main quality parameters considered for heap making are trash, moisture content percentage and colour.

28.4 Transportation of Seed Cotton Towards Gin Machine

There are actually two common alternative systems used for the feeding of seed cotton from the heap of cotton to the separator. First one is a pneumatic transmission system from cotton heap to the separator, in which a suction fan draws seed cotton and carries it over a pipe to bring it to the separator. A stone catcher in the suction line is also installed to isolate heavy elements such as brick pieces, heavy iron pieces, stones and particles.

A substitute system uses a conveyor belt for the feeding of cotton to the separator. Normally, two-stage conveyors are mounted in series with a fan between them for the extraction of heavy particles. The air flow produced by the fan is to throw the cotton bolls onto the second-stage conveyor belt, although the heavy particles fall into the trash container, placed directly below.

28.5 Cotton Contamination

The cotton lint quality is ascertained by the characteristics of the fibre. The most important thing is the degree of pollution, which greatly affects its price. Even a single foreign fibre can cause yarn or fabric degradation and is responsible for an estimated yearly loss in export earnings of US$1.4 billion to US$3 billion. Pakistani cotton is contaminated between 18 and 19 g/bale, whilst international standards necessitate up to 2.5 g/bale of cotton.

Most of the contamination comes from impurities that are incorporated into the bales during harvesting, ginning and baling and packaging due to human contact. The contamination of the seed cotton can be carried out at each step, from picking to ginning. Since cotton is picked by hand-picked methods by Pakistan's rural women, the main cause of contamination is human hair and any excess fabric. In addition, jute bags of pickers, sticks and weeds, leaves, immature bolls, dust, toffee lids and plastic bags are another key source. Furthermore, by adding water by the picker, picking the cotton early in the morning before the dew is dried and storing the cotton on moist soil to increase its weight, its quality is destroyed.

Measures to reduce contamination:

- Pickers should use cotton cloth bags instead of jute or polypropylene bags.
- The head of the picker should be covered by cotton fabric as hair or any other fabric fibre should not be mixed with seed cotton.
- Picking must be done at the appropriate period when air and sunshine have dried out the dew and moisture completely.
- Cotton must be stored on brick floors and covered with cotton sheets.
- Metallic body open trolleys would be for rapid conveyance of seed cotton from cotton field to factory or market.
- Eight percent moisture has to be sustained during pressing.

The lack of awareness is the main cause of this contamination. Achieving the zero contamination target is not possible, but dropping it to a lower level is considerably feasible. In order to improve the quality of cotton, the government has introduced a gradual scale of premium.

28.6 Pre-cleaning System

The pre-cleaning phase during the ginning process is very important. The effectiveness and success rate of this stage can be attained by effectively using the available contact area. A feed bucket is fitted out with a feed roller mounted in front of the beater and cleaner. This will adjust the feeding of seed cotton to these machines.

The productivity of cleaning of seed cotton is dependent on numerous factors, including cotton moisture content, machine design, degree of treatment, adjustment of speed and machine condition, quantity and nature of waste, cotton distribution throughout the machine and the varieties of cotton. One recommendation is to use a SMEDA-approved standard instrument to adjust the cleaning machine and optimize the speed of the roller based on the machine's physical conditions, the machine manufacturer's recommendations and the seed cotton quality. Four hundred to 600 rpm is the standard speed of the cleaner rolls.

In order to sustain the productivity and efficiency of the machines, preventive maintenance is also necessary. As this process does not contain material other than cotton, so it is very efficient at this stage. This is attained by the effective action of the stone catcher in the pneumatic transportation system and/or an effective air-flow supply in the two phases of conveyor system.

The spikes length must be from 1.50 to 2.00 in., and the spacing of 6.34 mm should be between the spikes pair. The 6–8 mm holes should have mesh/screen. Replace the broken and bent spikes with new even spikes during maintenance. The grid should be monitored and replaced by the new grid immediately after wearing. Also, keep in mind, when choosing a new or modifying a present beater, that an inclined beater along with grid bar/rod has a prominent efficiency of cleaning than a horizontal beater because of gravitational force.

28.7 Mechanized Cleaning

This step involves the cleaning of pre-cleaned seed cotton which is then beaten by the beater. The beater contains rollers about four to six with spikes on the external surface of the roller. The fibres are opened up and further trash is removed as the seed cotton passes through the rollers. These impurities are passed to the dust chamber by a suction blower (commonly referred to as a selection fan). The cotton is conveyed by a belt conveyor or pneumatically to the beater and then to the gin machine.

28.8 Parts of Gin Stand and Their Impact on Production and Quality

Ginning is the main process to execute on the gin stand. A saw-type gin stand is generally installed in Pakistan, usually mentioned as a saw gin. The unit for ginning consists of five to six serrated gin stands that are fitted side-by-side.

After mechanized cleaning, a feeding worm is installed above from the gin stand to carry the seed cotton to the feed bucket of the ginning machine. Saw gins generally consist of a pair of feed rolls, a set of beater rolls, a saw roll and a hull roll. The main part of saw ginning machine is the saw roll. It has a precise length, and its circular saws are of exact diameter with a number of circumferential teeth. The saw rolls are mounted on the iron shaft that is driven by an electrical motor. Also, some ginning units are installed with feeder-extractor-cleaner (FEC) before the beater of the gin machine for advanced cleaning.

The saw ginning machine separates the lint from the seed and other undesirable matters such as neps and motes. One or two transportation fans are installed accompanied by a network of ducts to carry the lint cotton from the ginning machine to the condenser's top. The delivery air is inserted into the saw roller edges through the nozzle and carries the lint along with it. The seed detached from lint is accumulated for further processing.

The ginning has a high impact on the cotton quality, the regulation of cotton moisture throughout ginning and cleaning and the degree of cleanliness. It has large influence on some quality traits but very little impact on others. Fibre length and the relevant parameters and quality traits like short fibre index and length uniformity are mostly affected by the ginning process. Fibre damage is higher when damp or dry cotton is processed. Ginning affects grade in regard to preparation and foreign matter but has little effect on fibre colour except in extreme cases (Abdullah et al. 2014).

28.9 Lint Cleaning, Conditioning and Bale Pressing

The lint that is obtained from saw gin is transmitted into the condenser. The condenser is basically a roller filter that rotates at a controlled speed having a galvanized iron (GI) screen on its edges. The lint is converted into a thick plate, called a bat, as it passes through the condenser. Any dust remaining in the cotton lint will also be discharged to the chamber of dust chamber during this process. The bat is made at the channel of the condenser and is moved by gravity to the bale press. Generally, the slide is equipped with a sprinkler that enhances moisture to the lint. There is not any standard system to adjust the sprayer and install it before packing on the slide, except for the blow and test methods to maintain the lowest level of moisture in the cotton bale at 9%. The moisture metre is used to measure moisture.

There are two machines that are used for lint bat conversion into a lint bale, placed in order, called press and tramper. Trampers are mainly composed of a container and a power-driven plunger. The lint is filled up and compacted by repetitive stroking by the plunger. The belt pressure is driven by a hydraulic machine to a preset pressure and then by an oil-based hydraulic pump after passing through tramper. The wire is wrapped up around the bale after pressing. Ultimately, the bales are sent to the bale yard after being weighed on a scale.

28.10 Estimation of Cotton Fibre Quality

The cotton fibre quality can be estimated through different techniques and instruments. The main equipment are high-volume instrument (HVI), advanced fibre information system (AFIS), fibrograph, etc., which are available in the market.

Fig. 28.3 High-volume instrument (HVI)

28.10.1 High-Volume Instrument (HVI)

High-volume instrument is the major cotton testing instrument that is used throughout the globe for main cotton research operations with maximum accuracy and precision (Fig. 28.3). Fibre strength, length, micronaire value, uniformity, colour, impurities, fibre elongation, cotton maturity, short fibre index and moisture content are the standard measurements for assessing a particular cotton type. All of these characteristics are important for cotton research and new strains development. Approximately, a complete test on the HVI requires 30 g of lint sample.

28.10.1.1 Length/Strength Module (910)

The length/strength module evaluates two samples at the same time; placing around 8–10 g of cotton lint sample in each bucket, it automatically makes the combs from the bucket material. The comb slides along the comb track until the first comb is positioned in front of the brusher. The function of brusher is to automatically clean and align fibres in the beards and remove the loose fibres. After that both combs with fibre beard are moved along the comb track to the module scanner, where the fibre beards are scanned by lens and jaw system from the tip to base for measuring length, short fibre index and length uniformity. The ratio between the average length of the fibres and the average length of the upper half expressed as a percentage is the length uniformity. Fibre strength and elongation are determined by settling the clamps on the 3.2 mm (1/8 in.) spacing to break the tapered beard.

28.10.1.2 Micronaire Module (920)

Micronaire value is a measurement of fibre fineness and maturity. The air-flow equipment is used to quantify the air permeability of the lint weighing 8.5–11.5 g compacted into a fixed chamber.

28.10.2 Low-Volume Instrument (LVI)

Low-volume instrument measures uniformity, length and short fibre index. The working principle of LVI is photoelectric. The test sample is scanned by an optical system that is highly stable and sensitive. The light beam is convergent on single side of the beard, over the lens placement. The specialized light beam from the source passes by the beard and into the photocell. The passing of light beam by the beard measures the amount of fibres contained in the sample.

28.10.3 Advanced Fibre Information System (AFIS)

Fibre length, neps, fibre diameter and trash of individual fibre are measured by AFIS which delivers data on distribution of assessed properties. The method of splitting up a sample into individual fibres by the instrument is similar to the processing conditions in up-to-date textile technology.

28.11 The Cotton Classification

Cotton marketing is incomparable in all field and fruit crops. Cotton quality can be represented by numerous measurements made by the cotton classers and described in various grades. The term "cotton classification" brings up the covering of standardized measures to quantify the physical properties of raw cotton that affect the quality of finished products or process efficiency. Classification is developed by the United States Department of Agriculture (USDA) on international level based on HVI classification, and the Pakistan Cotton Standard Institute (PCSI) classifies Pakistani cotton for local market based on the classer's grade composed of three components: trash, preparation and colour.

Cotton fibre quality is primarily classified by its length, micronaire value, strength, maturity, uniformity index, colour grade and short fibre index. After the lint bale is prepared after the ginning process, samples taken from each bale are classified accordingly using a high-volume instrument (HVI) and with the assistance of a professional called a classer. A scientific quality control verification (calibration) is established periodically to settle if the instrument accuracy is maintained. To maintain the instrument accuracy, calibration (scientific quality control check) is done periodically as recommended by the manufacturer.

28.11.1 USDA Cotton Classifications

Cotton as a natural product usually contains non-lint contents, such as leaves and bark from cotton plants, called trash. Bark and grass may be more difficult to move out because they align with the fibres and cause major problems during the spinning action. The trash amount also affects the value of cotton because of the spinning mill requirement to remove the trash before processing. The naturally varied differences in the quality of fibre, in combination with differences in end-use requirements, result in substantial inconsistency in the cost of the cotton lint to the processor. Hence, a classification of discounts and premiums has been well-known to denote a stated base quality. Generally, cotton fibre value increases as the bulk-averaged fibres increase in length, whiteness (+Rd), micronaire and strength, and discounts are made for equally low MIC (less than 3.5) and high MIC (more than 4.9). The USDA has established grading standards in collaboration with the whole cotton industry.

28.11.1.1 Fibre Length

Fibre length is associated with variety, but water stress, nutrient deficiency or cotton plant exposure to heat stress may shorten it. Cotton of a particular variety grows fibres of nearly identical length. Over-cleaning or low moisture content of seed cotton ginned at the plant also damages the length. Length influences yarn strength, evenness, yarn count and the efficiency of spinning. Other quality factors are also important. Fibre length rating is shown in Table 28.1.

28.11.1.2 Length Uniformity

It is the proportion of the average length to the average length of the upper half, stated as percentage. However, there is a variability in the fibre length naturally, so the length uniformity is lesser than 100. The fibres possibly will vary inside the bale,

Table 28.1 Rating of fibre length

Description	Rating (mm)
Short staple	20.64–23.81
Medium staple	24.61–27.78
Medium to long staple	28.58–30.96
Long staple	31.75–35.72
Extra-long staple	Above 36.51

Table 28.2 Rating of length uniformity

Description	Rating (%)
Very high	Above 85
High	83–85
Intermediate	80–82
Low	77–79
Very low	Below 77

Table 28.3 Rating of micronaire value

Description	Rating
Very fine	Below 3.0
Fine	3.0–3.9
Average	4.0–4.9
Coarse	5.0–5.9
Very coarse	6.0 and above

and length uniformity permits for determination of inconsistency within the bale. It affects yarn strength and evenness; also cotton with a low uniformity has high percentage of short fibres. The rating of length uniformity is shown in Table 28.2 (USDA 2001).

28.11.1.3 Fibre Micronaire

The fineness is important to determine the type of yarn that can be made from the fibres. The finer the fibres, the finer the yarn. During growth, the micronaire is affected by environmental conditions such as humidity, temperature, plant nutrition, and sunlight extremes in cotton bolls or plant populations. Fibre fineness influences the final product in a number of ways. To prevent fibre damage during opening, cleaning and discarding of low-micronaire-value cotton, processing speeds must be slow. Finer fibres have high yarn strength. Dye absorbency and retention are dependent on fineness and maturity; the greater the maturity, the greater will be the absorbency and retention as described in Table 28.3 (USDA 2001).

28.11.1.4 Fibre Strength

Fibre strength depends mainly on the variety. It can be influenced by deficiency of plant nutrient and climatic conditions. Yarn strength and fibre strength are highly correlated. Cotton with higher fibre strength can hold up breakage during processing. The strength of the fibres ultimately affects the fabric made from these fibres.

Fibre strength rating is given in Table 28.4.

Table 28.4 Rating of fibre strength

Description	Rating (g/tex)
Very strong	31 and above
Strong	29–30
Average	26–28
Intermediate	24–25
Weak	23 and below

Table 28.5 Colour grades of cotton

Description	White	Light spotted	Spotted	Tinged	Yellow stained
Good middling	11[a]	12	13	–	–
Strict middling	21[a]	22	23[a]	24	25
Middling	31[a]	32	33[a]	34[a]	35
Strict low middling	41[a]	42	43[a]	44[a]	–
Low middling	51[a]	52	53[a]	54[a]	–
Strict good ordinary	61[a]	62	63[a]	–	–
Good ordinary	71[a]	–	–	–	–
Below grade	81	82	83	84	85

Note: The sources for rating tables of fibre properties are from ICA Bremen and USDA cotton classification

[a]Physical standards: All others are descriptive

28.11.1.5 Colour Grade

The colour of fibres is also important. The colour grade is quantified by reflectance (Rd) and yellowness (+b). Very white cotton is usually more valuable, and the colour of the cotton may turn yellow when exposed to elements before harvest. When the boll is opened, the mature cotton is white and clean. Yellowing may be an important factor in frost, drought or early harvesting to aid in the early termination of growth. Grey is mainly the result of contact with moisture and weathering on site. Weathering can be controlled, but the risk of weathering damage can be reduced by minimizing the time between the first and last opening. Honeydew from mites, fungal growth or sugar on the lint can also produce grey cotton, but this can be managed by controlling the mites before they produce significant amounts of honeydew. Under certain conditions, such as drought stress, rain-grown cotton produces more spots than irrigated cotton. Otherwise the colours tend to be similar. Colour deterioration affects the capability of cotton fibres to retain and absorb dyes and finishes.

There are 25 official colour grades of cotton, plus five grades below the grade, as shown in the table below. The USDA maintains physical standards for 15 colour grades. Others are descriptive standards (Table 28.5).

Table 28.6 PCSI grading

Grade	Seed cotton	Lint
Super	• Open and healthy seed cotton • Colour white • Leaf particles in a little bit amount	• Colour white (Rd = 77.58%) • Trash % is less than 3% • A-index pricing value in international market
Grade-1	• Open and healthy seed cotton • Colour white • Leaf particles in a little amount	• Colour white (Rd = 75.40%) • Trash %age is 3–4% • A-index pricing value in International Market
Grade-2	• Open seed cotton with some un-opened bolls • Colour white • Leaf particles in a little amount	• Colour white (Rd = 73.16%) • Trash %age is 4–5% • B-index pricing value in international market
Grade-3	• Open seed cotton with some un-opened and yellow bolls • Colour average white • Leaf particles with little bit of stems of the plant	• Colour white (Rd = 70.22%) • Trash %age is 5–6% • Base grade with no deduction
Grade-4	• Immature seed cotton in great quantity • Colour yellowish white • Leaf particles with little bit of stems of the plant in great extent	• Used for coarser counts • Colour white (Rd = 67.00%) • Trash %age is 7–8%
Grade-5	• Immature seed cotton in great quantity • Colour yellowish grey • Leaf particles with stems of the plant in great extent and diseased bolls	• Used for coarser counts • Colour white (Rd = 63.89%) • Trash %age is 8–9%

28.11.2 PCSI Cotton Classification

Cotton trade around the world is based on grade, length of fibre and other fibre properties. PCSI recommends proper picking and better ginning practices, grading of seed cotton and lint classification as well as marketing system based on quality. PCSI classifies the cotton in the form of seed cotton and lint. There are six official grades based on classer's classification as illustrated in Table 28.6.

28.12 Occupational Safety and Health

The science of providing, implementing, assessing, and controlling hazards in the workplace that may endanger workers' health considers possible impacts on nearby communities and the overall environment.

The standards of occupational safety and health are significantly diverse amongst countries, economic sectors and firm sizes. The rate of workplace incidence of death differs in some countries, and there seems to be an important difference between developing and developed nations; for example, Pakistani factory workers are eight times more likely to die at work than French factory workers. Occupational safety and health performance rates vary widely amongst economic sectors in various countries. Statistics show that the world's highest occupational mortality rates are in construction, forestry, agriculture and mining. In general, large industries have good safety record than small enterprises. It seems that the death and severe injury rate in a small workplace (up to 50 workers) is two times that of a large workplace (more than 200 workers).

Like other processing industries, Pakistan's cotton ginning industry has many hazards. The gin causes the highest damage on the hands, followed by injuries to the spine, eyes, feet, arms, shoulders, legs, head and chest. The overall financial costs of injury and illness consist of direct costs (medicinal and other recompense) and indirect costs (loss of worktime, downtime, loss in revenue, costs of insurance, loss in productivity and numerous other loss aspects). It is easy to determine direct costs as they are lower than indirect costs.

The health syndromes in ginning enterprises are listed below:

1. Air quality
2. Noise exposure
3. Illumination
4. Physical safety
5. Firefighting system
6. Medical emergency procedure

References

Abbas Q, Ahmad S (2018) Effect of different sowing times and cultivars on cotton fiber quality under stable cotton-wheat cropping system in southern Punjab, Pakistan. Pak J Life Soc Sci 16:77–84

Abdullah M, Abbas G, Hussain M (2014) Sustainable cotton production in Pakistan's cotton ginning SMEs. Better ginning practices (BGPs) manual. WWF-Pakistan, Karachi, p 138. Available at: https://docplayer.net/54960502-European-union-pak-sustainable-cotton-production-in-pakistan-s-cotton-ginning-smes-spring-better-ginning-practices-bgps-manual.html

Ahmad S, Raza I (2014) Optimization of management practices to improve cotton fiber quality under irrigated arid environment. J Food Agric Environ 2(2):609–613

Ahmad S, Raza I, Ali H, Shahzad AN, Atiq-ur-Rehman, Sarwar N (2014) Response of cotton crop to exogenous application of glycinebetaine under sufficient and scarce water conditions. Braz J Bot 37(4):407–415

Ahmad S, Abbas Q, Abbas G, Fatima Z, Atique-ur-Rehman, Naz S, Younis H, Khan RJ, Nasim W, Habib ur Rehman M, Ahmad A, Rasul G, Khan MA, Hasanuzzaman M (2017) Quantification of climate warming and crop management impacts on cotton phenology. Plants 6(7):1–16

Ahmad S, Iqbal M, Muhammad T, Mehmood A, Ahmad S, Hasanuzzaman M (2018) Cotton productivity enhanced through transplanting and early sowing. Acta Sci Biol Sci e34610:40

Ali H, Afzal MN, Ahmad F, Ahmad S, Akhtar M, Atif R (2011) Effect of sowing dates, plant spacing and nitrogen application on growth and productivity on cotton crop. Int J Sci Eng Res 2 (9):1–6

Ali H, Abid SA, Ahmad S, Sarwar N, Arooj M, Mahmood A, Shahzad AN (2013a) Integrated weed management in cotton cultivated in the alternate-furrow planting system. J Food Agric Environ 11(3–4):1664–1669

Ali H, Abid SA, Ahmad S, Sarwar N, Arooj M, Mahmood A, Shahzad AN (2013b) Impact of integrated weed management on flat-sown cotton (*Gossypium hirsutum* L.). J Anim Plant Sci 23 (4):1185–1192

Ali H, Hameed RA, Ahmad S, Shahzad AN, Sarwar N (2014a) Efficacy of different techniques of nitrogen application on American cotton under semi-arid conditions. J Food Agric Environ 12 (1):157–160

Ali H, Hussain GS, Hussain S, Shahzad AN, Ahmad S, Javeed HMR, Sarwar N (2014b) Early sowing reduces cotton leaf curl virus occurrence and improves cotton productivity. Cercetări Agronomice în Moldova XLVII(4):71–81

Amin A, Nasim W, Mubeen M, Nadeem M, Ali L, Hammad HM, Sultana SR, Jabran K, Habib ur Rehman M, Ahmad S, Awais M, Rasool A, Fahad S, Saud S, Shah AN, Ihsan Z, Ali S, Bajwa AA, Hakeem KR, Ameen A, Amanullah, Rehman HU, Alghabar F, Jatoi GH, Akram M, Khan A, Islam F, Ata-Ul-Karim ST, Rehmani MIA, Hussain S, Razaq M, Fathi A (2017) Optimizing the phosphorus use in cotton by using CSM-CROPGRO-cotton model for semi-arid climate of Vehari-Punjab, Pakistan. Environ Sci Pollut Res 24(6):5811–5823

Amin A, Nasim W, Mubeen M, Ahmad A, Nadeem M, Urich P, Fahad S, Ahmad S, Wajid A, Tabassum F, Hammad HM, Sultana SR, Anwar S, Baloch SK, Wahid A, Wilkerson CJ, Hoogenboom G (2018) Simulated CSM-CROPGRO-cotton yield under projected future climate by SimCLIM for southern Punjab, Pakistan. Agr Syst 167:213–222

Anthony WS, Mayfield WD (eds) (1994) Cotton Ginners Handbook, rev. U.S. Department of Agriculture, Agricultural Handbook, vol 503. U.S. Dept. of Agriculture, Washington, DC. 348 p

Khan MB, Khaliq A, Ahmad S (2004) Performance of mashbean intercropped in cotton planted in different planting patterns. J Res (Sci) 15(2):191–197

Rahman MH, Ahmad A, Wang X, Wajid A, Nasim W, Hussain M, Ahmad B, Ahmad I, Ali Z, Ishaque W, Awais M, Shelia V, Ahmad S, Fahad S, Alam M, Ullah H, Hoogenboom G (2018) Multi-model projections of future climate and climate change impacts uncertainty assessment for cotton production in Pakistan. Agric For Meteorol 253-254:94–113

Tariq M, Yasmeen A, Ahmad S, Hussain N, Afzal MN, Hasanuzzaman M (2017) Shedding of fruiting structures in cotton: factors, compensation and prevention. Trop Subtrop Agroecosyst 20(2):251–262

Tariq M, Afzal MN, Muhammad D, Ahmad S, Shahzad AN, Kiran A, Wakeel A (2018) Relationship of tissue potassium content with yield and fiber quality components of Bt cotton as influenced by potassium application methods. Field Crop Res 229:37–43

USDA (2001) The classification of cotton. US Department of Agriculture, Agricultural Marketing Service, Agricultural Handbook Number 566. U.S. Dept. of Agriculture, Washington, DC, p 23

Usman M, Ahmad A, Ahmad S, Irshad M, Khaliq T, Wajid A, Hussain K, Nasim W, Chattha TM, Trethowan R, Hoogenboom G (2009) Development and application of crop water stress index for scheduling irrigation in cotton (*Gossypium hirsutum* L.) under semiarid environment. J Food Agric Environ 7(3–4):386–391

Chapter 29
Modern Concepts and Techniques for Better Cotton Production

Abdul Ghaffar, Muhammad Habib ur Rahman, Hafiz Rizwan Ali, Ghulam Haider, Saeed Ahmad, Shah Fahad, and Shakeel Ahmad

Abstract Sustainable cotton production in current environmental conditions is under threat due to climatic variability and shortage of ever-decreasing resources for agricultural crops. There is dire need to improve the cotton production to fulfill increasing demands of the ever increasing world population which will rise up to nine billion till 2050. Poor soil health, poor water quality and water shortage, insect pest complex, and unpredictable climatic patterns are predominant problems to cotton production. Hence, there is a great challenge to manage cotton crop in a sustainable fashion without the degradation of soil, water, and environment due to climate variability. There are several factors associated with low production of cotton including improper sowing and picking, poor pesticide spraying approaches, inappropriate amount and time of irrigation, processing and ginning through inappropriate and primitive procedures, heat stress, lack of disease- and pest-tolerant varieties, improper nutrient management, improper disease management, and improper weed management. It is the need of the hour to adopt the modern technologies and applications for sustainable cotton production. There are several modern technologies which can increase the production of cotton and make the idea of sustainability feasible because of their site-specific management of

---

A. Ghaffar · H. R. Ali · G. Haider · S. Ahmad
Department of Agronomy, Muhammad Nawaz Shareef University of Agriculture, Multan, Pakistan

M. Habib ur Rahman (✉)
Department of Agronomy, Muhammad Nawaz Shareef University of Agriculture, Multan, Pakistan

Institute of Crop Science and Resource Conservation (INRES) Crop Science Group, University Bonn, Bonn, Germany
e-mail: habib.rahman@mnsuam.edu.pk

S. Fahad
Department of Agriculture, University of Swabi, Swabi, Khyber Pakhtunkhwa, Pakistan

College of Plant Science and Technology, Huazhong Agricultural University, Wuhan, P.R. China

S. Ahmad
Department of Agronomy, Faculty of Agricultural Sciences and Technology, Bahauddin Zakariya University, Multan, Pakistan

S. Ahmad, M. Hasanuzzaman (eds.), *Cotton Production and Uses*,
https://doi.org/10.1007/978-981-15-1472-2_29

all agricultural inputs. GPS, GIS, and remote sensing technologies make the precise seeding of cotton seed, fertilizers, and pesticides. IPM, IWM, and INM are the well-developed modern concepts which not only reduce the cost of production but also mitigate the emission of greenhouse gases. For sustainable cotton production, implementation of these modern concepts is crucial so that the human beings will get benefits in the future. Therefore, this chapter will be focused on the recently developed technologies which can be sustainably utilized for the better management of cotton crop across the world. This chapter will explore the importance of Decision Support system (DSS) for sustainable cotton production; role of GPS, GIS, and remote sensing for identifying site-specific factors such as soil quality indicators; importance of transgenic cotton; impact of mechanical sowing and picking on sustainable cotton production; use of UAVs for nutrient and pesticide management; and impacts of modern concepts on increasing agronomic production and advancing global fiber and oil security.

Keywords Sustainable cotton production · GIS · GPS · Remote sensing · Fiber security

Abbreviations

ARIMA	Autoregressive integrated moving average
ARMA	Autoregressive moving average
CSM	Cropping system model
DSS	Decision support system
EC	Electrical conductivity
ET	Evapotranspiration
FDR	Frequency domain reflectometry
GIS	Geographic information system
GPS	Global positioning system
GSM	Global system for mobile communication
IPM	Integrated pest management
IRS	Information retrieval system
IWM	Integrated weed management
LAI	Leaf area index
MARS	Marker-assisted recurrent selection
MAS	Marker-assisted selection
NDVI	Normalized difference vegetation index
NMR	Nuclear magnetic resonance
PA	Precision agriculture
RS	Remote sensing
SCY	Seed cotton yield
SEBAL	Surface energy balance algorithm for land
UAV	Unmanned aerial vehicle
VRA	Variable rate application
VWC	Volumetric water content
WHO	World Health Organization
WUE	Water use efficiency

29.1 Introduction

29.1.1 Significance of Modern/Advanced Technology for Sustainable Cotton Production

Modern technology plays an important role in agricultural productivity. Today with the implementation of modern technology, the farmer is able to grow crops in different soil conditions by minimum application of input resources. Inventions of modern machinery reduce farmer effort which are used for manual working in field and also reduce production time. By using this machinery, farmers can improve soil fertility, sowing method, fertilizer application, and crop protection during the whole season. In sustainable cotton production, some procedures such as integrated nutrient, weed, and pest management are involved. By using modern technology and integrated management together, it will increase the overall SCY with minimum inputs and the farmer gets maximum profit. These increases in production will boost the country's GDP. Including all of above it is also found that precision agriculture reduces the fossil fuel consumption which pollutes the environment.

29.1.2 Recent Advancements in Cotton Production

Modern technology implementation in cotton production improves its yield and reduces human efforts (Usman et al. 2009; Ahmad et al. 2014; Abbas and Ahmad 2018; Ahmad and Raza 2014; Ali et al. 2011, 2013a, b, 2014a, b). Application of information technology enables the farmer to think innovatively and make a decision at the right time (Ahmad et al. 2017, 2018; Amin et al. 2017, 2018; Khan et al. 2004; Rahman et al. 2017, 2018; Tariq et al. 2017, 2018). Other technologies including robots, satellite imagery, GPS, GIS, and RS are used in precision agriculture. These technologies provide information about crop health, nutrient status, irrigation management, and cotton yield (biological, SCY). There are also new resistant and high-yielding cotton varieties (transgenic cotton) (Rocha-munive et al. 2018) and variety selection methods such as MAS (Lema 2018) and MARS (Ribaut et al. 2010). Nano-biotechnology is also used in agriculture to protect plants against pathogens and monitor crop growth. It is also used to study the role and regulation of plant hormones. For fertilizer management, UAVs (Daughtry et al. 2018), sensors (Jia et al. 2014), and variable rate technology have recently been adopted for precise fertilizer application. Sprinkler and drip irrigation method increases the WUE in crops as well as in orchards.

29.1.3 Application of DSS for Sustainable Cotton Production

DSS is an interactive computer program that contains a wide range of information needed by decision makers to solve unstructured and difficult problems about uncertain and complex situations by accessing the data processes and analytical reasoning. These computer-based programs are being used for solving problems regarding multiple disciplines in agriculture to get specified objectives (Mir et al. 2015; Malik et al. 2018). For instance, farmers frequently need information to make decisions about which quantity of manure is needed to spread in cotton crop. For this purpose there are many tools for decision support such as mobile applications that quantify the nutrient amount present within manures applied in crop at varying amounts (Rose et al. 2018). Fertilizer management is adopted for managing and optimizing fertilizer use in crop production to enhance crop yield and decrease fertilizer input cost. Another technique, IrriSAT, which has been devised based on weather for irrigation scheduling, quantifies the amount of water required for cotton crop. This program was developed by the Commonwealth Scientific and Industrial Research Organisation (CSIRO) for the Cotton Research and Development Corporation (CRDC) (Vleeshouwer et al. 2015). Many studies reported that farmer engagement is noticeably very low with computer-based applications for decision support (Rahman et al. 2016; Kerselaers et al. 2015; Lindblom et al. 2017; Ogunti et al. 2018). To resolve this problem, various researches have evaluated how DSS tools can be updated and supplied to increase their use (Kerselaers et al. 2015).

29.2 Soil Sampling and Analysis Using Advanced Technologies for Better Cotton Production

Soil is the natural and valuable nonrenewable resource which contributes in sustainability of the ecosystem throughout the world, but soil productivity and fertility is under threat due to climate change across the world (Incrocci et al. 2019). According to the Soil Science Society of America (2013), soil plays a vital role in (1) formation of basic ecosystem, (2) provision of essential nutrients to crops and forests, (3) biomass production in forestry as well as in agriculture, (4) growth medium, (5) carbon pool formation, (6) filtering of our water, and (7) regulation of Earth's temperature. In an agricultural point of view, soil is a key factor for crop production since it depends on soil fertility and soil health. Soil fertility could be determined by soil testing to evaluate the fertilizer status to achieve maximum crop yield and minimize the potential losses to the environment (Dawson and Knowles 2018). There are advanced technologies (GIS and GPS) for soil testing to evaluate the soil topography, soil health, and soil management; soil mapping can be done by adopting the latest technologies available.

29.2.1 Use of GIS and GPS for Cotton Crop Production

The application of GIS in agriculture was started in the mid-1990s, with the invention and broader use of GPS for precision agriculture. GIS is able to manage large data of soil. The data of organic matter content and nutrient status of soil are very important for sustainable production (Senthurpandian et al. 2010). GIS produced soil fertility status spatially as well as temporarily for site-specific management of resources (nutrients, water) for sustainable crop production (Patil et al. 2011). For this purpose GIS technology can simplify as well as help in dealing with soil variability (Sarmah et al. 2018). The geographic information system enables the researcher to study the soil properties by spatial visualization in the procedure of vector and raster maps (Glowienka et al. 2016).

GPS is a form of satellite network revolving around the globe. This system operates by determining reflecting GPS signals from the Earth's surface. It requires satellites of GPS which provide radar transmitters and GPS receiver. The position of radar transmitters and GPS receiver should be co-located, but the angle of incidence may vary over the selected points of soil, but in general it is above 30° (Privette et al. 2011). Mapping and determining variations in soil properties of a field needs accurate information of position from where samples were taken. In GPS soil sampling, grid method of sampling is used, and location of sampling point should be accurate to develop soil data layer.

29.2.2 Application of Remote Sensing for Soil Sampling in Cotton Field

Remote sensing is a method of data collection that records the quantity of electromagnetic radiation emitted or reflected from substances on Earth at different wavelengths. These radiations travel from substance/source directly through space vacuum or are reflected indirectly and are captured by specific sensors (Jensen 2009). Different objects have different reflectance properties such as rock, soils, water, vegetation, etc. Hence, RS could help in characterization of soil attributes (Grunwald et al. 2015). Application of remote sensing in soil is based on soil spectral reflectance. The nature and shape of spectral reflectance curve of a soil depends on chemical and physical properties of soil (Wadodkar et al. 2014). Application of remote sensing in soil sampling is mainly in spatial valuation of soil fertility in the sense of nutrient deficiency by soil fertility map preparation. The remote sensing has been applied for soil classification based on soil crusting, salinization, texture, moisture, and mineralogy.

29.2.3 Advanced Techniques in Soil Analysis

29.2.3.1 X-ray Spectroscopy

X-ray spectroscopic techniques are widely used in soil science, environmental sciences, and agronomy. Techniques utilizing diffuse reflectance of near- and mid-infrared rays have been effectively used to evaluate the characteristics of a wide range of soils (Minasny et al. 2009; Kamau-Rewe et al. 2011; Towett et al. 2013), and they have also been used at wide scale coupled with enhanced geostatic procedure. It is a quick way for estimating composition of soil samples. For sample analysis it does not require acid digestion of the samples; thus it may be used as a screening tool.

29.2.3.2 Phosphorus-31 NMR Spectroscopy

Evaluating the dynamics and form of soil phosphorus is important to sustain agricultural productivity with minimum environmental risks. An emerging technique, phosphorus-31 NMR spectroscopy, was used by Newman and Tate for the first time in 1980 on extracts of soil to evaluate the soil phosphorus (P) and its forms in soil, preferably organic P; nevertheless, it must be operated accurately to give meaningful results (Cade-Menun and Liu 2014; Cade-Menun 2017). Total P of soil is expressed generally in mg/kg ranges; as concentration of soil extracts increases, the total P concentration increases as well in every NMR sample, significantly enhancing NMR response (Cade-Menun 2017). In this technique, for characterization of organic P, a mixture of ethylenediaminetetraacetic acid and sodium hydroxide is used with the soil extract (Doolette and Smernik 2015).

29.3 Modern Technologies for Cotton Genotype/Cultivar Development

29.3.1 Use of Modern Technology for Selection of Genotype/ Cultivars for Different Ecological Zones Under Contrasting Climate

Plant population is a key factor in contributing to overall cotton yield. Selection of cultivar under contrasting environment is very important, and breeders also have developed environment-specific varieties. Some techniques are also used for selection of varieties such as marker-assisted selection, marker-assisted recurrent selection, seed health test, and germination potential.

29.3.1.1 Marker-Assisted Selection (MAS)

Introducing a new variety by traditional and conventional breeding method could approximately take 8–10 years to develop it. Breeders are interested in a new method to speed up this process. By the invention of marker-assisted selection, a large number of genetic material can be identified (Lema 2018). MAS is known as a breeding technique in which selection and detection of DNA marker are combined into a traditional and conventional breeding program (Jiang 2013). The efficiency and effectiveness of traditional and conventional breeding for selecting genotype can be improved by a large range of molecular markers (Kumar et al. 2011). Breeders would be able to bypass the traditional method of phenotype-based selection by wide use of molecular markers that are included in growing plants to physiological and harvesting maturity and also observe physical characteristics to assume fundamental makeup. MAS can be used at seedling stage for selection which provides high efficiency and precision at a low cost (Farokhzadeh and Alifakheri 2014).

29.3.1.2 Marker-Assisted Recurrent Selection (MARS)

Genotype selection under different ecological zones is also possible by MARS, and it is a technique in which molecular markers are used to identify and select various genomic regions that are involved in complex trait expressions to collect the best genotype within populations (Ribaut et al. 2010). The efficiency of complex traits by MARS is possible, but a disadvantage is that the selection cycle is long which restricts the practical use of this new breeding method (Lema 2018) because it takes 5–6 years for genotype selection. Molecular markers working under genotypic assays may be cheaper, faster, and accurate as compared to conventional phenotypic assays while depending on traits and conditions; thus these techniques may result in high efficiency in time, effort, and resource saving (Jiang 2013).

29.3.2 Transgenic Cotton

Biotechnology is used for genetic modification (GM) in which the genetic material of various living organisms is manipulated, modifying them to perform specific functions. In 1996, transgenic cotton was for the first time sown (James 2016; Rocha-munive et al. 2018), because non-*Bt* cotton failed due to pest pressure in cotton-producing areas (Terán-Vargas et al. 2005). Since the last two decades, transgenic cotton adoption has been increased globally resulting to a rise of about 42% in area in 2017–2018. The major problem of non-transgenic cotton was bollworms causing severe loss in SCY (Singh 2018). By using transgenic cotton, yield has been increased manifold around the world (Asghar et al. 2016) through controlling bollworms. Transgenic cotton, in the start of it, became popular through reduction in insecticide sprays against bollworm pests (Singh 2018). It is reported that *Bt* cotton is resistant to lepidopteran pests due to the presence of *cry* genes extracted from *Bacillus thuringiensis* (Bt). It showed resistance to larval stage of lepidopteran pests including *Helicoverpa zea*, *Pectinophora gossypiella*, *Spodoptera exigua*, and *Heliothis virescens* (Benedict et al. 1993; James 2016; Rocha-Munive et al. 2018).

29.3.3 Modern Concepts in Seed Testing and Viability

Seed is a very small and fragile embryonic plant and it is also the foundation of agriculture. Seed viability and quality are important factors in modern agriculture for optimum population and maximum yield (Tsedaley 2015). Seed viability is measured in which it is tested how seeds grow and develop into plants. Seed quality is an important characteristic for crop productivity as well as for food security under changing climate (Finch-Savage and Bassel 2016). Caverzan et al. (2018) reported that there are various factors which limit the growth and productivity of crops. Thus, the physiological quality of seed is more important for uniform establishment of crops. Seed vigor reflects several properties which determine the seed viability and quality and emergence potential of crops under variable climatic conditions (Finch-Savage and Bassel 2016).

29.4 New Concepts of Cotton Planting

29.4.1 New Concepts in Tillage for Seed Bed Preparation to Ensure Low GHG Emission

Soil tillage has a pronounced effect on soil tilth and balance between soil and greenhouse gases (GHGs) (nitrous oxide, carbon dioxide, and methane). The excessive emission of these GHGs leads to climate change and global warming (Mangalassery et al. 2014; Campbell et al. 2014). Krištof et al. (2014) reported that when tillage intensity increases, it results in high amounts of carbon dioxide emission from the soil into the atmosphere. Abdalla et al. (2016) concluded that emission of carbon dioxide is 27% more in tillage soils than in no-tillage soils in arid climate regions, while in humid climate regions, emission of carbon dioxide is 16% more in tillage soils than in no-tillage soils. To reduce the greenhouse gas emission from the soil into the atmosphere, conservation tillage (reduced, minimum, zero tillage, direct seeding) has been reported to be adopted as it decreases fossil fuel usage for field preparation and also enhances carbon sequestration in soil. Maraseni and Cockfield (2011) found that we can improve yield and reduce cost of production and greenhouse gas emission from soil through conservation tillage.

29.4.2 Mechanical Sowing of Cotton

The main objective of mechanical sowing is the uniform distribution of seeds at the optimum spaces and depth. The effects of optimal sowing are maximum germination, sprouting, and no competition for space, water, nutrients, and light resulting in improved yield (Turan et al. 2015). Mechanical sowing of cotton is possible by automated seed-cum-fertilizer drill, tractor-drawn planter. Mechanical sowing reduces the pressure of labor cost and time consumption. In automated seed-cum-fertilizer drill, seed and fertilizer are fitted together and placed simultaneously at specific line spacing. It is used in arid regions for cotton sowing. The advantages of this drill are that (1) fertilizer and seed are in separate compartments, (2) it opens the furrows at calibrated uniform depth, and (3) seed and fertilizer are covered and soil around the seed becomes compacted. This mechanical planter can sow 10–24 rows at once. The planter opens a furrow or trench in each row, drops the seeds in right amount, covers these seeds, and presses the soil on them (National Cotton Council of America).

29.4.3 Advanced/Modern Concept in Cotton Planting

Cotton is a source of raw material that mediates the quality of end product such as fiber. Cotton cultivation has been made obvious by the following unsustainable methods: extensive irrigation, mono-cropping, and excessive use of synthetic fertilizers. But now the trend has been changed toward sustainability (Radhakrishnan 2017). Many practices and programs have been started around the world to promote cotton production. Modern techniques have been developed regarding cotton sowing. Different sowing methods have been created for different environmental conditions such as environment-specific cultivar selection and invention of different implements for sowing (planter which maintains plant-to-plant and line-to-line distance, seed-cum-fertilizer drill).

29.4.4 Sowing Techniques Under Different Cotton-Based Cropping Systems

Cropping systems are defined as the sequences and management techniques of crops on a specific field over time (Goglio et al. 2018). Wheat-cotton-based cropping system is used in South Punjab, Pakistan; due to climate change wheat yield suffers after cotton harvest in wheat-cotton systems. Wheat relay cropping in cotton is an option to get maximum system profitability and wheat yield (Shah et al. 2016; Sajjad et al. 2018). Zero-till drill and manual broadcasting methods are used for relay cropping of wheat in cotton (Buttar et al. 2013; Nasrullah et al. 2017).

29.5 Modern Concepts in Nutrient Management

29.5.1 Soil Test-Based Nutrient Application

Maintaining soil fertility for intensive systems is important besides improving crop yield. The crops of intensive cropping systems use large amount of nutrients from soil during growing season (Kumar et al. 2017a, b, c). To overcome nutrient deficiency in the next cropping season, optimum fertilizer application is required. Fertilizers are the most costly inputs for nutrient application in agriculture, and their application should be rational and accurate (Kashyap et al. 2018). Soil testing is a unique technique to evaluate the soil health and fertility status for balanced fertilizer application. Soil test result and fertilizer recommendation interpretation is important to improve soil fertility status and crop yield.

29.5.2 Use of UAVs for Fertilizer Management

The use of UAVs/drones has been adopted in the recent decade consisting of high-resolution sensors and image processing unit which are reliable and cost effective for fertilizer prediction in agriculture (Daughtry et al. 2018). These drones have the capability to fly automatically to survey the crops from above for detailed growth observation (Wang et al. 2013; Guan et al. 2019). Schut et al. (2018) studied the fertilizer and yield response with UAVs and proposed that high-resolution images of UAVs can be useful to evaluate temporal and spatial variability of crop yield and crop yield response to fertilizers. Hunt et al. (2018) used a parafoil-wing UAV for colored-infrared images to assess the nitrogen status of potato (*Solanum tuberosum* L.). They reported that each treatment of applied nitrogen could be distinguished precisely in that image. Ballester et al. (2017) worked on assessing in-season nitrogen status and lint yield of cotton by using UAV images. They proposed that UAVs can monitor cotton nitrogen status variability at commercial farm and elaborate the challenges about information based on high-resolution images for fertilizer recommendation in in-season cotton crop at early growth stages.

29.5.3 Application of Sensors for Cotton Crop Management

Crop observations near the ground are an effective method which contributes a significant role in precision agriculture and agricultural production (Jia et al. 2014). These ground-based crop observations provide fast, nondestructive, real-time, and inexpensive crop growth information. On the basis of this information, the grower can increase yield by optimum and proper irrigation, fertilizer application, pest control, and harvesting on time (Li et al. 2009). Jia et al. (2014) worked on monitoring cotton growth and nitrogen status by using GreenSeeker handheld sensor to measure the canopy cover. They concluded that the relationships between above-ground total nitrogen content and canopy cover were precise and had an R^2 value of 0.926 and an RMSE value of 1.631 g/m^2. Sui and Thomasson (2006) reported that plant height and spectral information observed by sensing systems had significant correlation with N concentration of cotton leaf. Proximal sensor is also used for N status assessment. Commonly three proximal optical sensors (canopy reflectance sensor, fluorescence-based flavonol meters, and chlorophyll meter) are used for assessing N status in different crops (Padilla et al. 2018).

29.5.4 Vermicomposting

Vermicomposting is the procedure in which mesophilic bio-oxidation and stabilization processes of organic materials are introduced by microorganisms and earthworms. This will increase decomposition rate by changing stabilization of that organic materials and also modifying physical and bio-chemical properties (Lim et al. 2015). Lim et al. (2015) reported that vermicomposting can improve soil properties (physical, chemical, and biological). Soil treated with vermicompost has good porosity, aeration, water retention, and bulk density. Vermicompost contains macro- and micronutrients, enzymes, vitamins, antibiotics, and plant growth hormones. Nutrient balance depends on the C/N ratio present in vermicompost. The sources of vermicomposts are sewage sludge (Ludibeth et al. 2012), water weeds (Najar and Khan 2013), cotton waste of hospitals (Pramanik and Chung 2010), and animal manures. Rekha et al. (2018) studied the comparison between plant growth enhancer and vermicompost in *Capsicum annuum* L. crop. They concluded that plants treated with vermicompost exhibited significant growth than plant growth enhancer (IAA, GA)-treated plants.

29.5.5 Integrated Nutrient Management (INM)

The INM is a strategy for improving and maintaining soil fertility for sustainable crop production for a long period of time and reducing inorganic fertilizer cost

(Sindhi et al. 2018). Agriculture based on soil health and fertility ensures balanced and adequate nutrient supply to the plants. Vora et al. (2015) worked on INM effect on cotton and soil fertility after harvesting. They used ten different sources of nutrients which were fertilizers (organic, inorganic), gypsum, compost, castor cake, and vermicompost in their studies. They recorded that treatment under combined use of inorganic fertilizer, compost, castor cake, and bio-fertilizer returns highest yield of cotton as compared to control.

Reddy et al. (2017) studied the evaluation of INM interventions by using poultry manure and phospho-compost in cotton. They reported that lower yield is obtained from poultry manure-treated plots than phospho-compost-treated plots and also concluded that nutrient uptake by plants was improved in all INM interventions due to soil nutrient improvement. The soil organic matter and nutrient status continuously decreased due to intensive farming. The integrated utilization of organic and inorganic nutrient sources improves the crop productivity and fertility status of the soil.

29.6 Water Management Practices to Enhance WUE and Productivity in Cotton Crop

29.6.1 Consumptive Use of Irrigation Water

Over-exploitation of fresh water has been increased by twofold from the last decade to increase crop production and food security for the ever-increasing world population (Liu et al. 2009; Yuan and Shen 2013). This over-exploitation declining the ground water table will reduce the fresh water supply for crop production, which puts emphasis on using precious water resources efficiently (Pawar and Khanna 2018). Several strategies have been devised to enhance the consumptive use of irrigation water or WUE including varying sowing methods (Ali et al. 2017), lateral-move machines and center pivots (Roth et al. 2014), and models. Ali et al. (2017) performed field experiment on improving cotton yield by enhancing the WUE under different sowing techniques including flat sowing, ridge sowing, and bed planting at different locations in the province of Punjab. They found that maximum WUE (6.79 kg/ha/ mm) and SCY (3432.50 kg/ha) were obtained from flat sowing with earthing up on alternative rows.

Difallah et al. (2017) presented a linear programming model to determine the optimum amount of water required by the crop and the experiment was conducted in Algeria. They reported that this model could decrease about 28.5% of water consumption. Irrigating the crops with surface irrigation causes uneven water distribution, loss of water in the form of deep percolation, and seepage and also gives rise to weed growth. They suggested that sprinkler irrigation has high WUE and one can save water at about 30–60%.

29.6.2 Remote Sensing for Command Area Management

Remote sensing technique can provide automatic, continuous, and nondestructive information, which is easily implemented in data transmission systems to have real-time access to data collected from smartphone or a remote computer (Marino and Alvino 2018). Digital analysis is used to obtain information from satellite images; remote sensing is a dynamic, highly accurate data source to estimate irrigated areas as well as to monitor their evaluation based on time. Quebrajo et al. (2018) studied the irrigation management by using soil remote sensing and thermal imaging in sugar beet crop. Images were captured by thermal imaging camera on UAV for the evaluation of water and soil moisture status in sugar beet plants. They observed that sugar content and fresh root mass tended to be low when thermal imaging detected the water stress at a higher level.

Quantification of WUE in the field needs precise and high spatiotemporal resolution approximates of ET from the surface (Wu et al. 2009; Al Zayed et al. 2015; Ma et al. 2018). The irrigation requirement of crop can be determined by these factors: crop water requirement, soil water status, amount of rainfall, and efficiency of irrigation systems (Khanal et al. 2017). Studies explored the use of thermal remote sensing images in detecting the soil moisture status to facilitate irrigation scheduling (Khanal et al. 2017). Hassan-Esfahani et al. (2015) reported that use of thermal images can estimate accurately the spatial distribution of soil moisture at the surface in a Utah oats and alfalfa farm.

29.6.3 Irrigated Crop Area Monitoring

Generally 80% of fresh water is used in agriculture and it will be increased because of the ever-increasing population, and it is necessary to devise systems for sustainable water use (Pathan and Hate 2016). Identification of irrigated crop area is important for status monitoring, water management, and yield estimation (Wu and De Pauw 2011). For monitoring irrigated crop area, Internet of Things (IOT) techniques have been developed based on sensor information and they calculate the amount of water required. This system runs with two sensors to acquire data of soil temperature and humidity, temperature, and sunshine hours per day. This system is based on sensor values and estimates the amount of water needed for irrigation (Rao and Sridhar 2018).

Lepage et al. (2009) evaluated the SAMIR software used to estimate evapotranspiration and budget of irrigation water on a large scale, working on basis of satellite images. Remote sensing provides a synoptic view of crop development basic information for computing reliable evapotranspiration (ET) obtained from FAO method. Eunice (2013) introduced Zigbee network to monitor irrigation in paddy field. The architecture of this network is that different nodes are placed in the crop field. It calculates physical values such as pressure, temperature, water level, and humidity that can be observed in the field of paddy crop. The data observed from different places of the crop field is transmitted to GSM node, and then this data will be transferred to a personal computer through a gateway. A server which is connected to a database stores extreme values of water level, humidity, and temperature. If the sensors give data of extreme threshold level, the alarm unit produces a specific alarm sound to inform the farmer about the crop field.

29.6.4 GIS and Sensor-Based Irrigation

The reliable and timely monitoring and observation of water resources and exploration systematically and developing new techniques for irrigation water saving. To do this, it is important to use modern techniques of assessment, surveying, design, investigation, and implementation. GIS is considered as an effective tool for the management for irrigation water saving (Ali 2011). GIS is a system used to observe, store, evaluate, manage, and present data linked with different locations (Panwar 2015). Kamal and Amin (2010) studied the irrigation water management in rice by using GIS and developed a GIS-based water management model for scheduling irrigation water deliveries daily and regularly assessing irrigation water performance. The "scheduling" database computes the amount of water required by crop, and the "monitoring" database gives information about uniform distribution of water and deficit or excess. They suggested that GIS is an efficient tool to monitor irrigation water management in precision farming. Taghvaeian et al. (2018) assessed the irrigation performance in Southern California by using GIS and remote sensing. They implemented SEBAL to calculate actual ET in one irrigation and potential ET was assessed by Priestley-Taylor method. Remotely sensed data was integrated with ground data in GIS to compute several drainage and irrigation performance indicators. They concluded based on assessed data through GIS and RS that water consumption across the district was uniform and relative ET was high, showing that irrigation water supply was adequate.

Soil moisture sensors are being widely used for irrigation management in the last decade. A large number of affordable and reliable sensors are available working under the principle of FDR to calculate EC and VWC in soil (Montesano et al. 2015). These sensors when installed at different locations and multiple depths can give water status in soil profile. Computing the soil water tension as an active point for scheduling irrigation is an approach often used. Leib et al. recommended 30 to −40 kPa tension for irrigation threshold in sandy loam soils, and in silt loam soils, it was between −40 and −60 kPa. He also noted the important factors including site selection for sensor installation, sensor depths, and crop growth stages. Masseroni et al. (2018) studied the surface irrigation management by sensors in rice field. They used water-level sensor that provides enough data to regulate the inflow of water in real time and to cut off the water flow rate at a specific time.

29.6.4.1 Application of NDVI for Stress Management in Cotton Crop

Vegetation indices are used for the estimation of vegetation cover, and these indices are derived from satellite sensor data; among these indices, NDVI is commonly used to assess the vegetation spread area and quality (Vani and Mandla 2017). It was introduced in 1973 by Rousil; vegetation phenology is used in NDVI algorithm because green vegetation reflects more in NIR and less in visible light, while in sparse vegetation, reflection of NIR is less and a large portion of visible light is reflected. NDVI values are derived by combining these reflectance ratios (Vani and Mandla 2017). Stone and Bauer studied the irrigation management at variable rates by using NDVI to evaluate NDVI potential to estimate various crop coefficients for making spatial irrigation recommendations. They conducted a study on cotton irrigation under central pivot irrigation systems that compared the checkbook method (irrigation amount applied according to crop age and total weekly precipitation) with irrigation precipitation based on NDVI. Irrigation events were initiated through soil sensors. It was observed that irrigation amounts were different in the NDVI-based method from rates recommended by the checkbook method up to 70 days after sowing when NDVI value differences among field area and plant density were nonexistent. The results suggested that irrigation prescription based on NDVI method is warranted. Despite all of these, NDVI has some limitations to its application such as NDVI having same values for two different crop water conditions. Another fact is that NDVI readings may be influenced by errors inherent to the measuring technique; irrigation scheduling cannot be recommended at the moment based on NDVI value in one season study because it requires continuous study (Vani and Mandla 2017).

29.7 Improved Weed Management

The excessive weed proliferation in cotton decreases the cotton seed yield because weed would compete with cotton for nutrients, space, and moisture. Pre and post-sowing weed management is very important part among cultural practices, so it cannot be neglected. There are many improved methods that have been devised to control weeds including chemical and mechanical weed control, crop rotation, stale bed preparation, and many others.

29.7.1 Pre-sowing Weed Management

29.7.1.1 Use of Glyphosate

Glyphosate is the most extensively used herbicide around the world. The feature of broad-spectrum application makes glyphosate a prominent herbicide, and it can be applied at different times during the whole season. The positive impact of glyphosate herbicide has been observed in transforming agricultural practices regarding weed management in glyphosate-tolerant crops. The combined effect of a broad-spectrum herbicide such as glyphosate and crops tolerant to this specific herbicide allowed the use of efficient and simplified weed management that reduced the utilization of alternative technologies like hand and tillage weeding. Use of glyphosate herbicide before emergence of cotton reduces the weed management cost through chemical and labor costs. It is also promoted in the conservation tillage technique because it reduces mechanical weeding in cotton (Held et al. 2016).

29.7.2 Stale Seed Bed Preparation (SSB)

In stale bed preparation, a seed bed is prepared some days, weeks, or months before sowing of crop. Stale seed bed method allows the weeds to grow extensively before crop planting or sowing. After the extensive growth of weeds, they are controlled through early soil tillage or non-selective broad-spectrum herbicides. Tillage plays a role in controlling weeds by destroying the emerging weed seedlings, burying the seeds, and also delaying the growth of perennial weeds. This technique decreases weed emergence and early crop competition with weeds.

29.7.3 Mechanical Weeding

Mechanical weeding is the best reported non-selective method that is most effective in controlling annual weeds. It is the physical removal of weeds by using different tools including disks, hoes, rotary weeders, cultivators, and mechanical choppers. These implements are designed to uproot, cut, or cover weed seedlings. Mechanical weeding in cotton field can be done with primary or secondary tillage. Primary tillage such as moldboard plow is used for deep soil cultivation which leads to uprooting and shredding of weed flora growing in the cotton field before sowing. It also helps in burying a large quantity of weed seeds in a deeper layer of soil. Secondary tillage, for example, harrowing and disking, plays a role in shredding weed biomass. Both of this primary and secondary tillage destroy the weed biomass quickly in field, and follow-up with cultivator use may be important to uproot and dislodge the weed flora after tillage. Mechanical weed management in cotton field after sowing is more effective when weeds are smaller than cotton plants. Despite these, frequent use of tillage for mechanical weeding may lead to soil degradation, weed seed germination present near the soil surface, and perennial weed dispersal by breaking the vegetative structures present underground (Ashigh et al. 2015).

29.7.4 Transgenic Cotton

Application of various chemicals is needed for insect and disease management during the growing season of cotton. The SCY was severely affected by weed infestation in field crop. High weed infestation period is a few weeks after planting. When using herbicide, it is necessary to introduce herbicide-resistant cotton varieties. In 1997, a glyphosate-resistant cotton variety was used commercially resulting in a high yield of cotton (Latif et al. 2015). This new tool is also environment-friendly because it reduces the quantity and number of sprays during the whole growing season. Use of glyphosate in glyphosate-resistant cotton has non-residual, non-mobile, and less environment toxicity effect compared with residual herbicide. In glyphosate-tolerant cotton, weed management is easier than conventional method. But continuous use of same herbicide increases the resistance of weeds against this herbicide, and preventive methods should be adopted to reduce this rapid resistance in weeds. Farmers should also use some alternative herbicide for controlling weeds (Holtzapffel et al. 2008).

29.7.5 Use of Satellite Imagery for Specific Weed Management

Temporal weed monitoring is the first and primary step in site-specific weed management in cotton; using GIS, RS, and UAV, it is a technique in which careful mapping and monitoring of weed infestation in early growth stages is performed (Papadopoulos et al. 2018; López-Granados 2011). When the remote sensing system collects the reflectance from ground level, reflectance values from specific and individual features are used for average over the entire remote sensor pixel area. RS consists of low- and high-resolution pixels. In high resolution, images are explained through small pixels, while in low resolution, large pixels describe the reflected image (Clay et al. 2004). In weed monitoring strategy, two steps are involved: (1) weed map creation on the basis of frequent collected data and (2) weed detection in real time, integrating it with sensors, processing method, and actuation system. Despite these, this system also has some limitations including difficulty in weed detection at a very initial growth stage (Fernández-Quintanilla et al. 2018).

29.8 Modern Concepts in Pest and Disease Management

29.8.1 Plowing Practices

Reduced tillage practices can enhance insect pest issues because of previous crop residues left on the soil surface (Johnson et al. 2001). These crop residues provide food and shelter as host crops. Aphids and certain cotton insects might increase in areas where preceding crop residues are not incorporated at planting. Infestation risks can be reduced by employing herbicides before the planting of the next crop. When soil is disturbed for the purpose of preparing soil for planting, insect pests are exposed which are killed by natural enemies and/or because of lack of food. Reduced tillage practices increase the number of predators which plays a significant role in biological control (Gencsoylu and Yalcin 2004). Thus, plowing practices are considered the best strategies for controlling insect pests across the world.

29.8.2 Trap Cropping

Trap cropping involves planting minor areas nearby the protected crop with a crop that is susceptible to specific pests that will serve as an attractant and can be economically destroyed (Shelton and Badenes-Perez 2006). Specific plant species can influence associated insects by attracting them (Altieri and Nicholls 2004). The attractiveness of the trap crop causes insects to move to the trap crop and stay away from the major crop, leaving the major crop undisturbed.

Insects are major biological constraints in successful production of crops on a worldwide basis (Croft et al. 1985; Attia et al. 2013). Trap crops have been successfully used for suppression of insects across the world. Trap cropping has been used in China and the USA to conserve and enhance natural enemies of cotton aphid (Xia et al. 1998). In Egypt, basil crop is intercropped into cotton field as trap crop for the suppression of insect pests (Schader et al. 2005). In New Zealand, Rea et al. (2002) have studied and reported that trap crop, e.g., black mustard, was an effective control strategy against *N. viridula* in sweet corn. Trap cropping might be an effective approach to manage *N. viridula*. Sorghum has been used as trap crop against *N. viridula* adults positively in cotton fields (Tillman 2006).

The application of insecticides has played an important role in controlling insects, but their excessive use has many environmental issues (Ecobichon 2001). Trap crops can be an effective strategy for managing lygus bugs (*Lygus hesperus*) in the cotton field. Alfalfa planted in the cotton field can draw bugs out of the cotton field which will reduce the damage to cotton blossoms (Pedigo 1989). Inter-crops are very useful for suppression of insects as these crops provide a complex environment (Lal 2016). Thus, trap crops and inter-crops can be used for the suppression of insect pests in cotton crop production as sustainable approach.

29.8.3 Pheromone Traps

A *pheromone* is a chemical that is produced by an animal that changes the behavior of another animal of same species. The pheromones are used for the monitoring and control of insects because they perform important functions, e.g., mating disruption, mass trapping, attract-and-kill, and push-pull (Tewari et al. 2014).

Cotton is considered a well-known crop which is often attacked by many insect pests and hence consumes a large number of plant protection products (Deguine et al. 2008). Monitoring and control of American bollworm, spotted bollworm, pink bollworm, and *Spodoptera* can be done through the use of pheromone traps. Pheromones should be installed at a distance of 50 m from the field. Trapped moths are removed on a daily basis. Yellow pans and sticky traps have been used for the monitoring and control of whitefly in the field of cotton (mostly 25 yellow pans or sticky traps per one hectare).

Different types of sex pheromones have been used for different insect pests, and the most used sex pheromone in the adult moth of cotton bollworm (*Helicoverpa armigera*) is "D. Z11-16AL (97), Z9-16AL hexadecenal." It is also reported that sex pheromones (10E 12E)-10, 12 hexadecadienal for cotton spotted bollworm and Z7 Z11-16AC (50), Z7 E11-16AC (50) hexadecadienyl acetate for pink bollworm have been successfully used (Shah et al. 2011).

Gossyplure is a synthetic sex pheromone which has been significantly used against pink bollworm in monitoring and control of cotton fields. This pheromone reduced the population of larvae infesting cotton bolls because of disruption of premating pheromone communication between male and female moths (Henneberry 2007).

29.8.4 Application of Drones

Agriculture drones have been successfully used for spraying fertilizer and pesticides across the world (Kale et al. 2015). Crop spraying with the help of drones is up to five times faster than spraying with regular machinery (MIT 2016). As a drone can scan the ground and apply liquids quickly with great precision, this sophisticated equipment is considered well for the spraying of cotton crop. A cotton farmer can use drones to monitor a field for insects, disease, and other pests more efficiently compared to traditional scouting methods. Drones have a heat sensor which can detect hot spots in fields and quickly treat disease before the crop is lost.

The WHO estimated that one million cases were affected by manual spraying of pesticides in the crop field. The UAV aircrafts are used to spray pesticides to prevent health issues in humans. This laid the foundation to develop such technologies (drones) which reduce the wastage of chemicals into the atmosphere and water (Mogili and Deepak 2018).

29.8.5 *Improved IPM Techniques*

29.8.5.1 Cultural Methods

Cultural control is used to suppress pest population by making the environment less favorable for survival, growth, and reproduction of pest species. It includes the selection of cotton field, variety selection, crop rotation, deep plowing, and the time of planting. This method has been successfully used across the world. In cultural control, crop rotation is one of the best strategies for the suppression of soil-borne diseases and pests such as root-knot nematodes which are a well-known constraint in cotton production across the world (Starr et al. 2005). Use of resistant/tolerant varieties, avoiding excess nitrogen application, and destruction of crop residues are adopted strategies which can reduce the pest population into the cotton field (Singh et al. 2008). Early planting of cotton in the field is recommended to escape pink bollworm and American bollworm infestation (Draz 2009).

Sanitation of the field can be carried out through the cleaning of field equipment which is considered an important preventive control strategy of IPM. Different equipment (often ridger and drill) are used in the production of cotton. These equipment may carry many insect pests into the field. Use of clean and certified seed may prevent the introduction of pests into the field.

29.8.5.2 Biological Control

Biological control of insect pests is considered a sustainable approach in the suppression of multiple pests of cotton across the world (Tomson et al. 2017). *Trichogramma* species have been significantly used in reducing the bollworm population because *Trichogramma* species feed on 37–40% of the cotton bollworm eggs (Ba et al. 2008). In China, whitefly (*B. tabaci*) has been satisfactorily controlled through natural enemies (Shen et al. 2005; Naranjo 2009). In India, the use of neem products/neem-based pesticides and release of *Trichoderma* species have been used as a sustainable strategy in reducing the population of insect pests (Singh et al. 2008). *B. bassiana* has reduced the population of cotton plant bugs when it was sprayed on the cotton field (Tong et al. 2010). El-wakeil et al. (2006) stated that combined use of neem products and *Trichogramma* species is considered as the best biocontrol agent.

29.8.5.3 Mechanical Practices

The insect management techniques across the world are progressively leaning toward a more environment-friendly agriculture without upsetting the balance of an ecosystem (Tilman et al. 2002). Sex pheromone traps and light traps have been successfully used as mechanical methods in insect pest management across the world. Different types of traps have been widely used for the monitoring of insect pests such as yellow traps for aphid and white traps for thrips.

29.8.5.4 Chemical Control

Seed treatment shields seed or young seedling against pests and diseases transmitted by seed or soil and stimulates germination and plant growth. Seed treatment is an effective and economical control method for early-season insect pests in the cultivation of cotton. It can shield cotton seed and young seedling from insect pests, nematodes, diseases, and other threats to health and productivity of cotton crop. Various studies have shown that seed treatment is a sustainable strategy for controlling insect pests and diseases across the world. In the Mid-South, neonicotinoid seed treatments have provided significant harvest and economic profits in cotton production (North et al. 2017).

The effectiveness of seed treatment may vary from fungicide to fungicide, but treated cotton seed provides greater cotton yields (Copeland et al. 2016). Cotton seeds treated with the combination of polymer and fungicide give significantly higher seed quality parameters (Vijaya et al. 2017). In Brazil, seed treatment has been successfully used against bacterial blight and damping-off diseases in cotton production (de Medeiros et al. 2015). In India, seed treatment showed better results in the control of leaf blight of cotton (Sangeetha et al. 2018). Thus, there is a need to explore the significances of seed treatment at farm level because only a few use seed treatment at farm level due to lack of awareness.

29.9 Cotton Crop Yield Estimation/Forecasting

29.9.1 Use of GIS and Remote Sensing for Cotton Yield Estimation and Forecasting

29.9.1.1 Need for GIS and RS

It is very difficult to determine sample locations and measure a sufficient number of samples for the estimation of yield/biomass of cotton across the world. Moreover, the results obtained through surveying in the fields were not accurate (Zang 1998). Similarly, obtaining information about yield and biomass of cotton crop is a labor-intensive and slow process. Hence, there is a need to introduce new technologies which are cost effective and provide timely information.

29.9.1.2 Methodology

Remote sensing has gained much significance across the world as it has the potential and capability to provide spatial information at a global level on features and phenomena on Earth on a real-time basis. It is a very useful technology used for the estimation of crop acreage and production and is cost effective and timely (Gitonga 1995; Dalezios et al. 2001). RS techniques can offer a complete study of the surface of the Earth on a daily basis. It can be used for the estimation of net primary agricultural production over time and space. This can be attained by using vegetation indices; however, they are not a direct measure of productivity/biomass, but they are correlated with LAI and biomass (Todd et al. 1998). Remote sensing can detect the biophysical functions in the plant which provides a platform for the estimation of cotton yield (Dalezios et al. 2001).

There are four main ways by which remote sensing forecasts the crop yield and biomass. These ways are divided into four categories:

Remote Sensing Methods Based on Empirical Methods

In this method, spectral indices are calculated from satellite images such as NDVI which are indirectly used for the estimation of cotton yield and biomass.

Remote Sensing Methods Based on Water Consumption Balance Method

In this method, the whole growth period of cotton crop can be divided into different sets whose evaporation fraction is measured. This method estimates productivity as a function of evaporation fraction during crop growth and uses the water consumption balance model to estimate evaporation fraction.

Crop Growth Models

Crop growth models focus on complex interaction of physiological processes with the environment. Crop growth models are combined with spectral observation provided by satellite data. In this way, cotton yield and biomass are estimated.

Monteith Model

Remote sensing with the integration of Monteith model can forecast crop yield by using biomass. It is a simple and useful paradigm for modeling crop yield and biomass. This method can be successfully employed for cotton yield estimation across the world.

Table 29.1 Overview of some crop models and their impact on the study of cotton yield forecasting across the world

Crop model	Study location	Study impact	Sources
CROPGRO-Cotton	Pakistan	Positive effects	Wajid et al. (2014); Rahman et al. (2017, 2018)
Info Crop Growth	India	Results satisfactory	Hebbar et al. (2008)
CROPGRO-Cotton	India	Positive effects	Kumar et al. (2017a, b, c)
ARIMA	India	Positive effects	Debnath et al. (2013)
CROPGRO-Cotton	Georgia	Positive effects	Ortiz et al. (2009)
Semi-empirical	China	Positive effects	–
CROPGRO-Cotton	Georgia	Positive effects	Pathak et al. (2009)
CROPGRO-Cotton	Pakistan	Positive effects	Arshad et al. (2017)
CROPGRO-Cotton	Georgia	Positive effects	Hoogenboom et al. (2004)
CSM-DSSAT	USA	Positive effects	Jones et al. (2003)
AquaCrop	Spain	Positive effects	Garcia-Vila et al. (2009)

29.9.2 Application of Crop Models for Cotton Yield Forecasting

Being a major cash crop of Pakistan and due to its significant contribution toward the agrarian economy of a country, it is useful to know about production and productivity status of cotton in the future. One can utilize historical available data of production and productivity for predicting future production and productivity of cotton by using forecasting models (Table 29.1).

The CROPGRO-Cotton model can be used for simulation of growth and SCY for different weather, soil, and husbandry practices (Ortiz et al. 2009; Rahman et al. 2017). Several studies have been carried out regarding application of crop models across the world. Semi-empirical models were used for the prediction of cotton growth and SCY in response to various nitrogen fertilizers (Lie et al. 2009). Now many researches have been done on application of crop models for prediction of SCY in Pakistan (Wajid et al. 2014; Arshad et al. 2017; Rahman et al. 2018). Crop models have been used for the estimation of cotton yield under future climate to develop site-specific adaptation strategies for adjustment of sowing dates, irrigation, and fertilizer (Arshad et al. 2017).

CSM-CROPGRO-Cotton can be used for simulation of growth and yield of cotton for different weather and soil conditions and managing practices (Jones et al. 2003; Hoogenboom et al. 2017). However, a model was found to have the ability to evaluate cotton production for climate change situations (Murthy 2004). For prediction of crop development and yield, all models take different sets of parameters into consideration like cultivar characteristics, minimum and maximum temperature, solar radiation, and crop management's elements (Hoogenboom et al. 2017). Kumar et al. (2017a, b, c) have reported that CROPGRO-Cotton model can be run for the estimation of SCY, biomass, and LAI of cotton crop. Wajid et al. (2014) have used CSM-CROPGRO-Cotton model for the prediction of SCY and total dry matter. They have got satisfactory results for the different parameters of the cotton. Crop models have been satisfactorily used for the field-scale forecasting of crop yields and biomass for wheat, rice, and cotton in Pakistan (Bastiaanssen and Ali 2003).

Papageorgiou et al. (2011) have studied and reported that fuzzy cognitive map might be the convenient tool in the prediction of cotton yield and improvement of crop management. The ARMA and ARIMA models of forecasting of cotton and sugarcane yield have been used (Ali et al. 2015).

Crop area and productivity appraisal are an essential procedure in supporting policy decision regarding land use allocation, food security, and environmental issues. The ARIMA model has been successfully used in Asia. Debnath et al. (2013) used the ARIMA model for the forecasting of cotton production and yield in India. An AquaCrop model is a useful tool which is used for the estimation of cotton yield/biomass in response to water management. This model has been used to assist managers for making decisions in cotton irrigation under water-restricted conditions (Garcia-Vila et al. 2009). In India, the Info Crop growth models have been satisfactorily used in combination with RS and GIS for the estimation of cotton production in irrigated areas (Hebbar et al. 2008).

Future projections of climate change in cotton zone of Punjab, Pakistan, showed that there would be increases in temperature from 1.2 to 1.8 °C and 2.2–3.1 °C in RCP 4.5 scenario, while 1.4–2.2 °C and 3.0–3.9 °C increases are anticipated in RCP 8.5 scenario, for near-term (2010–2039) and mid-century (2040–2069), respectively. Similarly, rainfall variations are anticipated at −8% to 15% and −5% to 17% for RCP 4.5, while −8% to 22% and −2% to 20% variations are anticipated for RCP 8.5, in near-term and mid-century, respectively. SCYs are projected to decrease by 8% on average by 2039 and 20% by 2069 for RCP 4.5 relative to baseline (1980–2010). Mean SCYs are projected to decrease by 12% and 30% on average for RCP 8.5 (Rahman et al. 2018).

29.10 Modern Techniques in Cotton Picking and Storage

29.10.1 Mechanical Picking

Cotton mechanization is playing an important role in the life of farmers as millions of farmers are dependent on cotton crop directly or indirectly across the world (Deshmukh and Mohanty 2016). In the world, Australia, the USA, and Israel are the countries where all the cotton is harvested mechanically (Muthamilselvan et al. 2007). In Asia, this is somehow well adopted in India because it is the world's largest producer of cotton. Mechanical picking of cotton is a newly developed technology in Pakistan. Mechanical picking machines have become necessary to minimize the hard work involved in hand picking and save cost of production. Mechanical picking will enhance the production of cleaner grade of seed cotton. Machine picking increases fiber length and fiber strength when it is compared with hand picking (Tian et al. 2017). Further, the mechanical cotton picking system will also be helpful in

Fig. 29.1 Mechanical picking of cotton in Pakistan (**Source**: Central Cotton Research Institute (CCRI), Multan, Pakistan)

achieving timeliness of operation for the next crop. However, there are some issues to mechanical picking of cotton such as initial large cost of the imported mechanical pickers, management of homogeneous height of cotton plants, and lack of credit facilities. Mechanical picking of cotton improves the fiber quality and seed cotton uniformity (Khalifa et al. 2009).

29.10.2 Need for Mechanical Picking

There are many issues associated with manual picking such as it being tedious and costlier than other agricultural operations, non-availability of labor, and delayed picking causing yield loss and affecting the overall quality of lint. Farmers are finding difficulties to complete picking operation on time even after spending more money across the world. In Pakistan, late wheat sowing due to late picking of cotton is a major issue in cotton-wheat cropping system which causes significant yield reduction of wheat (Tahir et al. 2009). This indicates that there is urgent need of mechanical picking in the world to increase the economic value of the field. Pakistan has been importing cotton picking machines from Uzbekistan. Mechanical picker (Fig. 29.1). Cotton picking machines have been firstly imported by the CCRI to encourage innovation and mechanized farming in the country and enhance per-acre crop output to alleviate poverty in rural areas of the country.

29.10.3 Significances of Mechanical Cotton Pickers

Mechanical cotton pickers have a lot of significances:

1. These mechanical cotton pickers reduce the loss of lint yield left on the plant during manual picking.
2. These also reduce the cost of production of cotton crop and improve the quality of lint.
3. These cotton pickers have also reduced the yield reduction of wheat crop through timely picking of cotton lint.

Across the world, there is an urgent need to promote mechanical picking of cotton because it not only reduces the cost of production but it also fulfills the gap of labor shortage. Similarly, it reduces the risk of negative climate impacts on the cotton crop.

29.11 Emerging Technologies

29.11.1 Robots

Agricultural robots have been successfully used for seeding, harvesting, weed control, and chemical applications (Cariou et al. 2009) and for improving productivity and efficiency across the world (Foglia and Reina 2006). Agricultural robots have the ability to scout weeds present in the field, find open bolls in the field, pick cotton in the lab, improve fiber and seed quality, increase the value of markdown cotton, reduce weather risk, increase yield, lower harvest cost, and enhance cotton sustainability (Wang et al. 2008).

Various types of agricultural robots have been developed which perform different tasks in the field. Cotton is a crop which is more attacked by insect pests and diseases. Robots can be used to monitor and identify diseases in field during early stage. Newly developed robot (eAGROBOT) is used to detect diseases in cotton at an early stage using image processing techniques (Pilli et al. 2015). Similarly, soil-sensing-survey robots that use an electronic nose to determine chemical soil properties have been developed (Pobkrut and Kerdcharoen 2014). These robots can timely provide variability present in the soil fertility of a cotton field which is the most important approach toward increasing resource use efficiency. Haibo et al. (2010) established a wheat-seeder robot which utilizes an air suction precision seeding mechanism to precisely drop seeds using RTK-GNSS module. This robot can also be adjusted in seeding of cotton.

Robots can reduce fuel consumption and air pollution (Gonzalez-de-Soto et al. 2016). The conventional cultivation in which crops are conducted and managed manually can be improved by using intelligent machines such as robots (Xia et al. 2015). Many agricultural robotics based on visual guidance are developed for automatic operations in agriculture, like micro-dosing, de-leafing, and weed and insect management (Slaughter et al. 2008; Ota et al. 2007; Sogaard and Lund 2007). Labor shortage is the largest problem in the agriculture sector and very costly. Hence this sector demands for robots that are efficiently working (Bechar et al. 2015).

29.11.2 Computer and IT Applications

Information technologies, like radars, mobile/telephones, FAX machines, computers, satellites, etc., in the world contribute to numerous forms of information systems like IRS that help us in solving problems or making decisions. An IRS is an environment of people, technologies, and procedures that help us to find data, information, and knowledge resources that can be found in a specific library, for that particular matter, wherever they exist. These IT-based technologies are useful for cultivation of cotton. Amount of water sprinkled in a balanced quantity is also computerized. The production capacity of the farm has been increased owing to use of IT in agriculture. There are fewer losses as now work can be monitored by computer in a traditional field like agriculture, wherein we can boost yield and reduce the chances of errors.

29.11.3 Decision Support Systems (DSS)

Computer systems that offer users with provision to investigate complex information and help in making decisions are called DSS. These are information systems with a specific task to help people in the problem-solving and decision-making process. These systems consist of a collection of people, procedures, software, and databases with a purpose. In such systems, computer is the most important technology.

29.11.4 Precision Agriculture (PA)

PA is abstracted by a system approach to re-organize agricultural system toward a low-input, high-efficiency, sustainable agriculture (Cook and Bramley 1998). It has been used for the evaluation of how the small areas or plots have different production environments and yields in Asia (Farid et al. 2013; Chandel et al. 2014; Shivanna and Nagendrappa 2014). Precision farming of cotton has become the requisite of agriculture because of global warming, climate change, reduction in water resources, health hazards due to more application of insecticides and pesticides, and use of excessive amounts of fertilizers to meet the needs of the growing population. These problems can be handled by using fewer inputs, eliminating health hazards, and making the environment safer by lowering the use of insecticides, pesticides, and fertilizers. It becomes possible when we know about the potential of a specific land

and the composition of all nutrients necessary for better yield and soil water requirements.

This modern technology can be successfully applied in cotton farming by using computers, GPS, telecommunications, farm implements, GIS, machine guidance, and RS (Zhang et al. 2002). All these equipment coordinate with one another and make agricultural practices very easy and more input responsive. With the help of these modern technologies, precision farming is capable of collecting data regarding production variability in both space and time. This variability in the cotton field can be responded positively in proper time which is considered a major property of precision farming.

The VRA in precision agriculture is an area of technology that focuses on automated application of materials in a given field. Technology in which materials (fertilizers, chemicals, and seeds) are applied is based on data that is collected by sensors, maps, and GPS; hence this technology helps in accurate application of these materials into the field.

Map-based VRA adjusts application rate based on an electronic map. With the help of GPS and prescription maps, variable rate technology can measure the field position, and hence the concentration of agricultural inputs required by the cotton crop is changed as the applicator moves through the cotton field. There is no need for maps and positioning system in sensor-based VRAs. Sensors on applicator quantify soil properties or crop characteristics "on the go"; based on this continuous availability of information, a control system estimates input requirements of soil or plants and transfers information to a controller which makes sure there is availability of inputs to the location measured by sensors. This variable rate technology can be a sustainable approach toward enhancing the cotton production and decreasing the cost of production due to accurate application of all agricultural inputs into the cotton fields.

29.11.4.1 Significances of Precision Farming in Cotton Production

1. It decreases the bad environmental impacts of inputs used like fertilizers, pesticides, insecticides, etc. and the cost of production of cotton (Hedley 2015).
2. There is precise seeding in the fields by the seed drill controlled by the GPS system.
3. Spray nozzles having sensors shut down when there is no vegetation under the nozzles thus reducing the wastage of fertilizers and pesticides.
4. Auto-steer and sensors with GPS system ensure the automatic application of agricultural practices so farmers can work at night in the fields also.
5. Maps of the soil give all the information about the inputs to be supplied in the soil; hence farmers can apply nutrients more precisely and timely and can reduce wastage of nutrients (Srinivasan 2006; McBratney et al. 2005).
6. Satellite imagery provided by remote sensing can be used for the evaluation of requirements of different inputs.

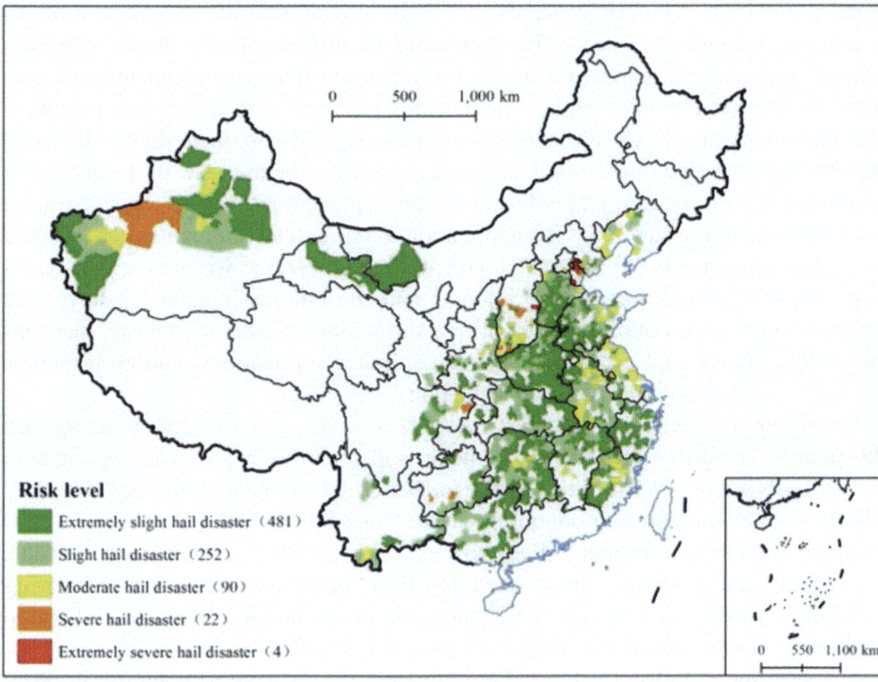

Fig. 29.2 GIS-based risk assessment of hail disasters affecting cotton and its spatiotemporal evolution in China (**Source**: Wang et al. 2016)

GIS technology has been successfully used for the management of cotton crop in China and many other countries of the world. In China, GIS was used for evaluation of hail disaster risk management to the cotton crop as shown in Fig. 29.2. According to this figure, Wang et al. (2016) have reported that hail catastrophe risk is low in China except for North China Plain and cotton-planting areas in Xinjiang Uygur Autonomous Region. Similarly, this technology can be used for many other special GIS-based risk assessments across the world.

29.12 Conclusion

The maintenance and improvement of soil quality, soil health, water quality, adequate water availability, and environment quality are the main challenges associated with sustainable cotton production. Hence modern concepts and technology applications in cotton production are crucial for the optimization of productivity and profitability and enhancement of resource use efficiency and sustainability with reduced environmental impacts. Innovations in cotton production are the need of the hour to meet the ever-increasing food and fiber demand of the ever-increasing world population. Modern concepts and technologies provide the basic tools to assess and manage the variability prevailing in different cotton-based cropping systems. Cotton crop is most sensitive to weather and other environmental stresses. Sustainable cotton production by efficient utilization of ever-decreasing resources and implementation of good management practices will maximize the productivity and profitability of cotton-based cropping systems. Precise use of resources in combination with modern technology is ideally appropriate to play a major role in sustainable cotton production. The application of modern technologies such as DSSs for crop management; the use of RS, GIS, GPS, and UAVs; wireless sensor-based crop monitoring and forecasting system; and agricultural machinery have key potential to enhance cotton production in a sustainable manner. Similarly, mechanical cotton sowing and picking, IT-based computer applications, and crop models can be used for sustainable cotton production.

Moreover, modern and innovative concepts such as X-ray spectroscopy and phosphorus-31 NMR spectroscopy for better soil sampling analysis and application of new tools such as MAS and MARS for better selection of genotypes have key roles in sustainable cotton production. On the other hand, modern concepts in seed testing and viability; minimum tillage concept for low GHG emission; new concepts in nutrient management; sensor-based fertilizer, crop, and irrigation monitoring system; and modern concepts in weed management can be used for successful cotton production across the world. Various types of agricultural robots can be used for precise seeding of cotton seed, weed control, and pest management and for improving resource use efficiency and cotton productivity. PA is a requisite of sustainable cotton production which can be used for the improvement of soil health, water quality, and environmental quality under cotton-based cropping systems. In conclusion, there is a need to adopt all the above-discussed modern concepts and technologies for the promotion of sustainable cotton production; conservation of soil, water, and other agricultural resources; and improving the environmental quality.

References

Abbas Q, Ahmad S (2018) Effect of different sowing times and cultivars on cotton fiber quality under stable cotton-wheat cropping system in southern Punjab, Pakistan. Pak J Life Soc Sci 16:77–84

Abdalla K, Chivenge P, Ciais P, Chaplot V (2016) No-tillage lessens soil CO_2 emissions the most under arid and sandy soil conditions: results from a meta-analysis. Biogeosciences 13 (12):3619–3633

Ahmad S, Raza I (2014) Optimization of management practices to improve cotton fiber quality under irrigated arid environment. J Food Agric Environ 2(2):609–613

Ahmad S, Raza I, Ali H, Shahzad AN, Atiq-ur-Rehman, Sarwar N (2014) Response of cotton crop to exogenous application of glycinebetaine under sufficient and scarce water conditions. Braz J Bot 37(4):407–415

Ahmad S, Abbas Q, Abbas G, Fatima Z, Atique-ur-Rehman, Naz S, Younis H, Khan RJ, Nasim W, Habib ur Rehman M, Ahmad A, Rasul G, Khan MA, Hasanuzzaman M (2017) Quantification of climate warming and crop management impacts on cotton phenology. Plants 6(7):1–16

Ahmad S, Iqbal M, Muhammad T, Mehmood A, Ahmad S, Hasanuzzaman M (2018) Cotton productivity enhanced through transplanting and early sowing. Acta Sci Biol Sci 40:e34610

Al Zayed IS, Elagib NA, Ribbe L, Heinrich J (2015) Spatio-temporal performance of large-scale Gezira irrigation scheme, Sudan. Agr Syst 133:131–142

Ali MH (2011) GIS in irrigation and water management. In: Practices of irrigation & on-farm water management, vol 2. Springer, New York, NY

Ali H, Afzal MN, Ahmad F, Ahmad S, Akhtar M, Atif R (2011) Effect of sowing dates, plant spacing and nitrogen application on growth and productivity on cotton crop. Int J Sci Eng Res 2 (9):1–6

Ali H, Abid SA, Ahmad S, Sarwar N, Arooj M, Mahmood A, Shahzad AN (2013a) Integrated weed management in cotton cultivated in the alternate-furrow planting system. J Food Agric Environ 11(3–4):1664–1669

Ali H, Abid SA, Ahmad S, Sarwar N, Arooj M, Mahmood A, Shahzad AN (2013b) Impact of integrated weed management on flat-sown cotton (*Gossypium hirsutum* L.). J Anim Plant Sci 23 (4):1185–1192

Ali H, Hameed RA, Ahmad S, Shahzad AN, Sarwar N (2014a) Efficacy of different techniques of nitrogen application on American cotton under semi-arid conditions. J Food Agric Environ 12 (1):157–160

Ali H, Hussain GS, Hussain S, Shahzad AN, Ahmad S, Javeed HMR, Sarwar N (2014b) Early sowing reduces cotton leaf curl virus occurrence and improves cotton productivity. Cercetări Agronomice în Moldova XLVII(4):71–81

Ali S, Badar N, Fatima H (2015) Forecasting production and yield of sugar cane and cotton crops of Pakistan for 2013-2030. Sarhad J Agric 31(1):1–9

Ali L, Anum W, Hussain G, Shahid MI (2017) Enhancement in cotton (*Gossypium hirsutum* L.) crop yield by water use efficiency under various planting techniques. Environ Earth Ecol 1 (2):6–16

Altieri M, Nicholls C (2004) Biodiversity and pest management in agroecosystems, 2nd edn. CRC Press, Boca Raton, FL, pp 1–252

Amin A, Nasim W, Mubeen M, Nadeem M, Ali L, Hammad HM, Sultana SR, Jabran K, Habib ur Rehman M, Ahmad S, Awais M, Rasool A, Fahad S, Saud S, Shah AN, Ihsan Z, Ali S, Bajwa AA, Hakeem KR, Ameen A, Amanullah, Rehman HU, Alghabar F, Jatoi GH, Akram M, Khan A, Islam F, Ata-Ul-Karim ST, Rehmani MIA, Hussain S, Razaq M, Fathi A (2017) Optimizing the phosphorus use in cotton by using CSM-CROPGRO-cotton model for semi-arid climate of Vehari-Punjab, Pakistan. Environ Sci Pollut Res 24(6):5811–5823

Amin A, Nasim W, Mubeen M, Ahmad A, Nadeem M, Urich P, Fahad S, Ahmad S, Wajid A, Tabassum F, Hammad HM, Sultana SR, Anwar S, Baloch SK, Wahid A, Wilkerson CJ, Hoogenboom G (2018) Simulated CSM-CROPGRO-cotton yield under projected future climate by SimCLIM for southern Punjab, Pakistan. Agr Syst 167:213–222

Arshad MN, Ahmad A, Wajid SA, Cheema MJM, Schwartz MW (2017) Adapting DSSAT model for simulation of cotton yield for nitrogen levels and planting dates. Agron J 109(6):2639–2648

Asghar M, Farooq M, Hussain M (2016) Productivity and profitability of cotton – wheat system as influenced by relay intercropping of insect resistant transgenic cotton in bed planted wheat. Eur J Agron 75:33–41

Ashigh J, Mohseni-Moghadam M, Idowu J, Hamilton C (2015) Weed management in cotton. NMSU, Las Cruces, NM

Attia S, Grissa KL, Lognay G, Bitume E, Hance T, Mailleux AC (2013) A review of the major biological approaches to control the worldwide pest Tetranychus urticae (Acari: Tetranychidae) with special reference to natural pesticides. J Pest Sci 86(3):361–386

Ba HGL, Ai HMT, Wu PE (2008) Preliminary study on biological control of cotton bollworm, Helicoverpa armigera (Hubner) with Trichogramma in Tulufan in Xinjiang autonomous region. China Cotton 35:17–18

Ballester C, Hornbuckle J, Brinkhoff J, Smith J, Quayle W (2017) Assessment of in-season cotton nitrogen status and lint yield prediction from unmanned aerial system imagery. Remote Sens (Basel) 9(11):1149

Bastiaanssen WGM, Ali S (2003) A new crop yield forecasting model based on satellite measurements applied across the Indus Basin, Pakistan. Agric Ecosyst Environ 94(3):321–340

Bechar A, Nof SY, Wachs JP (2015) A review and framework of laser-based collaboration support. Ann Rev Control 39:30–45

Benedict JH, Sachs ES, Altman DW, Ring DR, Stone TB, Sims SR (1993) Impact of δ-endotoxin-producing transgenic cotton on insect–plant interactions with Heliothis virescens and Helicoverpa zea (Lepidoptera: Noctuidae). Environ Entomol 22(1):1–9

Buttar GS, Sidhu HS, Singh V, Jat ML, Gupta R, Singh Y, Singh B (2013) Relay planting of wheat in cotton: an innovative technology for enhancing productivity and profitability of wheat in cotton–wheat production system of South Asia. Exp Agric 49(1):19–30

Cade-Menun BJ (2017) Characterizing phosphorus forms in cropland soils with solution 31 P-NMR: past studies and future research needs. Chem Biol Technol Agric 4(1):19

Cade-Menun B, Liu CW (2014) Solution phosphorus-31 nuclear magnetic resonance spectroscopy of soils from 2005 to 2013: a review of sample preparation and experimental parameters. Soil Sci Soc Am J 78(1):19–37

Campbell B, Chen L, Dygert C, Dick W (2014) Tillage and crop rotation impacts on greenhouse gas fluxes from soil at two long-term agronomic experimental sites in Ohio. J Soil Water Conserv 69 (6):543–552

Cariou C, Lenain R, Thuilot B, Berducat M (2009) Automatic guidance of a four-wheel-steering mobile robot of accurate field operations. J Field Robot 26(6-7):504–518

Caverzan A, Giacomin R, Müller M, Biazus C, Lângaro NC, Chavarria G (2018) How does seed vigor affect soybean yield components? Agron J 110(4):1318–1327

Chandel NS, Agrawal KN, Tripathi H, Garg SK (2014) Development of yield maps in wheat using yield monitor. Bhartiya Krishi Anusandhan Patrika 29(3):111–115

Clay SA, Chang J, Clay DE, Reede CL, Dalsted K (2004) Using remote sensing to develop weed management zones in soybeans. Site Spec Manag Guide 42:1–4

Cook SE, Bramley RG (1998) Precision agriculture – opportunities, benefits and pitfalls. Aust J Exp Agric 38:753–763

Copeland JD, Dodds DM, Catchot AL, Gore J, Wilson DG Jr (2016) Evaluation of PRE herbicides and seed treatment on thrips infestation and cotton growth, development, and yield. Agron J 108 (6):2355–2364

Croft BA, Cook RJ, MacKenzie DR (1985) Biological constraints. In: Gibbs M, Carlson C (eds) Crop productivity--research imperatives revisited: an international conference held at Boyne Highlands Inn, Harbor Springs, Michigan, pp 177–195

Dalezios NR, Domenikiotis C, Loukas A, Tzortzios ST, Kalaitzidis C (2001) Cotton yield estimation based on NOAA/AVHRR produced NDVI. Phys Chem Earth B Hydrol Ocean Atmos 26 (3):247–251

Daughtry DW, Porter WM, Harris GH, Noland RL, Snider JL, Virk S (2018) Correlating plant nitrogen status in cotton with UAV based multispectral imagery. In: A paper in 14th International Conference on Precision Agriculture 24-27 June, 2018, Montreal, Quebec, Canada. ISPA, Monticello, IL, pp 1–9

Dawson A, Knowles O (2018) To grid or not to grid – a review of soil sampling strategies. In: Currie LD, Christensen CL (eds) Farm environmental planning – science, policy and practice. Fertilizer and Lime Research Centre, Massey University, Palmerston North. Available at: http://flrc.massey.ac.nz/publications.html

Debnath MK, Bera K, Mishra P (2013) Forecasting area, production and yield of cotton in India using ARIMA model. J Space Sci Technol 2(1):16–20

Deguine JP, Ferron P, Russell D (2008) Sustainable pest management for cotton production. A review. Agron Sustain Dev 28:113–137

Deshmukh AS, Mohanty A (2016) Cotton mechanisation in India and across globe: a review. Int J Adv Res Eng Sci Technol 3(1):66

Difallah W, Benahmed K, Draoui B, Bounaama F (2017) Linear optimization model for efficient use of irrigation water. Int J Agron 2017:5353648

Doolette AL, Smernik RJ (2015) Quantitative analysis of 31P NMR spectra of soil extracts–dealing with overlap of broad and sharp signals. Magn Reson Chem 53(9):679–685

Draz KA (2009) Cotton pests. Faculty of Agriculture, Alexandria University, Damanhour

Ecobichon DJ (2001) Pesticide use in developing countries. Toxicology 160(1-3):27–33

El-Wakeil NE, Gaafar NM, Vidal S (2006) Side effect of some neem products on natural enemies of Helicoverpa (Trichogramma spp.) and Chrysoperla carnea. Arch Phytopathol Plant Protect 39 (6):445–455

Eunice MA (2013) Real time paddy crop field monitoring using Zigbee Network. Int J Eng Sci Res 4(1):1208–1213

Farid HU, Bakhsh A, Ahmad N, Ahmad A (2013) Evaluation of management zones for site-specific application of crop inputs. Pak J Life Soc Sci 11(1):29–35

Farokhzadeh S, Alifakheri B (2014) Marker-assisted selection for disease resistance: applications in breeding (Review). Int J Agric Crop Sci 7:1392–1405

Fernández-Quintanilla C, Peña JM, Andújar D, Dorado J, Ribeiro A, López-Granados F (2018) Is the current state of the art of weed monitoring suitable for site-specific weed management in arable crops? Weed Res 58(4):259–272

Finch-Savage WE, Bassel GW (2016) Seed vigour and crop establishment: extending performance beyond adaptation. J Exp Bot 67(3):567–591

Foglia MM, Reina G (2006) Agricultural robot for radicchio harvesting. J Field Robot 23 (6-7):363–377

Garcia-Vila M, Fereres E, Mateos L, Orgaz F, Steduto P (2009) Deficit irrigation optimization of cotton with AquaCrop. Agron J 101(3):477–487

Gencsoylu I, Yalcin I (2004) The effect of different tillage systems on cotton pests and predators in cotton fields. Asian J Plant Sci 3(1):39–44

Gitonga GW (1995) Forecasting maize crop yield in Kenya using meteorological satellite data. In: Proceedings of the EUMETSAT Meteorological Satellite Data Users' Conference: Polar Orbiting Systems, Winchester, UK, 1995, pp 93–100

Glowienka E, Michalowska K, Pekala A, Hejmanowska B (2016) Application of GIS and remote sensing techniques in multitemporal analyses of soil properties in the foreland of the carpathians. IOP Conf Ser Earth Environ Sci 44(5):052044

Goglio P, Brankatschk G, Knudsen MT, Williams AG, Nemecek T (2018) Addressing crop interactions within cropping systems in LCA. Int J Life Cycle Assess 23(9):1735–1743

Gonzalez-de-Soto M, Emmi L, Benavides C, Garcia I, Gonzalez-de-Santos P (2016) Reducing air pollution with hybrid-powered robotic tractors for precision agriculture. Biosyst Eng 143:79–94

Grunwald S, Vasques GM, Rivero RG (2015) Fusion of soil and remote sensing data to model soil properties. Adv Agron 131:1–109

Guan S, Fukami K, Matsunaka H, Okami M, Tanaka R, Nakano H, Sakai T, Nakano K, Ohdan H, Takahashi K (2019) Assessing correlation of high-resolution NDVI with fertilizer application level and yield of rice and wheat crops using small UAVs. Remote Sens (Basel) 11(2):112

Haibo L, Qing L, Yufeng X, Chuijje Y (2010) Research and development on the key technology of wheat single seed robot. In: IEEE World Automation Congress. IEEE, Washington, DC, pp 339–343

Hassan-Esfahani L, Torres-Rua A, Jensen A, McKee M (2015) Assessment of surface soil moisture using high-resolution multi-spectral imagery and artificial neural networks. Remote Sens (Basel) 7(3):2627–2646

Hebbar KB, Venugopalan MV, Seshasai MVR, Rao KV, Patil BC, Prakash AH, Kumar V, Hebbar KR, Jeyakumar P, Bandhopadhyay KK, Rao MRK, Khadi BM, Aggarwal PK (2008) Predicting cotton production using Info crop-cotton simulation model, remote sensing and spatial agro-climatic data. Curr Sci 95:1570–1579

Hedley C (2015) The role of precision agriculture for improved nutrient management on farms. J Sci Food Agric 95(1):12–19

Held A, Hudson J, Martin L, Reeves W (2016) Benefits and safety of glyphosate. MONSANTO, St. Louis, MO

Henneberry TJ (2007) Integrated systems for control of the pink bollworm Pectinophora gossypiella in cotton. In: Vreysen MJB, Robinson AS, Hendrichs J (eds) Area-wide control of insect pests. Springer, Dordrecht, pp 567–579

Holtzapffel R, Mewett O, Wesley V, Hattersley P (2008) Genetically modified crops: tools for insect pest and weed control in cotton and canola. Australian Government Bureau of Rural Sciences, Canberra, ACT

Hoogenboom G, Jones JW, Wilkens PW, Porter CH, Batchelor WD, Hunt LA, Boote KJ, Singh U, Uryasev O, Bowen WT, Gijsman AJ, Du Toit AS, White JW, Tsuji GY (2004) Decision support simulating cotton yield using CROPGRO model system for agrotechnology transfer. ver. 4.0. University of Hawaii, Honolulu, Hawaii

Hoogenboom G, Porter CH, Shelia V, Boote KJ, Singh U, White JW, Hunt LA, Ogoshi R, Lizaso JI, Koo J, Asseng S, Singels A, Moreno LP, Jones JW (2017) Decision Support System for Agrotechnology Transfer (DSSAT), Version 4.7. DSSAT Foundation, Gainesville, FL. Available at: http://dssat.net

Hunt ER, Horneck DA, Spinelli CB, Turner RW, Bruce AE, Gadler DJ, Brungardt JJ, Hamm PB (2018) Monitoring nitrogen status of potatoes using small unmanned aerial vehicles. Precis Agric 19(2):314–333

Incrocci L, Marzialetti P, Incrocci G, Di A, Balendonck J, Bibbiani C, Bibbiani C, Spagnol S, Pardossi A (2019) Sensor-based management of container nursery crops irrigated with fresh or saline water. Agric Water Manag 213:49–61

James C (2016) Global status of commercialized biotech/GM crops: 2016: ISAAA brief No. 52. ISAAA, Ithaca, NY

Jensen JR (2009) Remote sensing of the environment: an earth resource perspective, 2nd edn. Pearson Education India, Noida

Jia B, He H, Ma F, Diao M, Jiang G, Zheng Z, Cui J, Fan H (2014) Use of a digital camera to monitor the growth and nitrogen status of cotton. Scientific World Journal 2014:602647

Jiang GL (2013) Molecular markers and marker-assisted breeding in plants. Chapter 3. In: Plant breeding from laboratories to fields. InTech, Rijeka, pp 45–83

Johnson WC, Brenneman TB, Baker SH, Johnson AW, Sumner DR, Mullinix BG (2001) Tillage and pest management considerations in a peanut–cotton rotation in the southeastern Coastal Plain. Agron J 93(3):570–576

Jones JW, Hoogenboom G, Porter CH, Boote KJ, Batchelor WD, Hunt LA, Wilkens PW, Singh U, Gijsman AJ, Ritchie JT (2003) The DSSAT cropping system model. Eur J Agron 18 (3–4):235–265

Kale S, Khandagale S, Gaikwad S, Narve S, Gangal P (2015) Agriculture drone for spraying fertilizer and pesticides. Int J Adv Res Comp Sci Softw Eng 5(12):804–807

Kamal RM, Amin MSM (2010) GIS-based irrigation water management for precision farming of rice. Int J Agric Biol Eng 3(3):27–35

Kamau-Rewe M, Rasche F, Cobo JG, Dercon G, Shepherd KD, Cadisch G (2011) Generic prediction of soil organic carbon in alfisols using diffuse reflectance Fourier-transform mid-infrared spectroscopy. Soil Sci Soc Am J 75:2358–2360

Kashyap N, Das KN, Deka B, Dutta M (2018) Soil test based fertilizer prescriptions under integrated plant nutrient supply for hybrid rice (cv. US–382) in alluvial soils of Jorhat District of Assam, India. Int J Curr Microbiol App Sci 7(5):3570–3576

Kerselaers E, Rogge L, Lauwers G, Van Huylenbroeck G (2015) Decision support for prioritizing of land to be presented for agriculture: can participatory tool development help? Comp Electron Agric 110:208–220

Khalifa EM, Kholief RM, Eitawil MA, Neamatallah MA (2009) Influence of harvesting methods on ginning operation and fiber qualities for Egyptian cotton. Mansoura Uni J Agric Sci 34:1431

Khan MB, Khaliq A, Ahmad S (2004) Performance of mashbean intercropped in cotton planted in different planting patterns. J Res (Sci) 15(2):191–197

Khanal S, Fulton J, Shearer S (2017) An overview of current and potential applications of thermal remote sensing in precision agriculture. Comp Electron Agric 139:22–32

Krištof K, Šima T, Nozdrovický L, Findura P (2014) The effect of soil tillage intensity on carbon dioxide emissions released from soil into the atmosphere. Agron Res 12(1):115–120

Kumar J, Choudhary AK, Solanki RK, Pratap A (2011) Towards marker-assisted selection in pulses: a review. Plant Breed 130(3):297–313

Kumar A, Brar NS, Pal S, Singh P (2017a) Available soil macro and micro-nutrients under rice-wheat cropping system in District Tarn Taran of Punjab, India. Ecol Environ Conserv 23 (1):201–206

Kumar S, Niwas R, Khichar ML, Singh A, Badal P, Kumar Y, Chauthan AS (2017b) Genetic coefficient and validation of DSSAT model for cotton under different growing environments. Int J Curr Microbiol App Sci 6(4):1031–1041

Kumar U, Shahid M, Tripathi R, Mohanty S, Kumar A, Bhattacharyya P, Lal B, Gautam P, Raja R, Panda BB, Jambhulkar NN, Shukla AK, Nayak AK (2017c) Variation of functional diversity of soil microbial community in sub-humid tropical rice-rice cropping system under long-term organic and inorganic fertilization. Ecol Indic 73:536–543

Lal S (2016) Impact of plant diversity on the insect pest complex of maize. Doctoral Dissertation, MPUAT, Udaipur

Latif A, Rao AQ, Khan MAU, Shahid N, Bajwa KS, Ashraf MA, Abbas MA, Azam M, Shahid AA, Nasir IA, Husnain T (2015) Herbicide-resistant cotton (*Gossypium hirsutum*) plants: an alternative way of manual weed removal. BMC Res Notes 8(1):453

Lema M (2018) Marker assisted selection in comparison to conventional plant breeding: review article. Agric Res Tech Open Access J 14(2):555914. https://doi.org/10.19080/ARTOAJ.2018.14.555914

Lepage M, Simonneaux V, Thomas S, Metral J, Duchemin B, Kharrou H, Cherkaoui M, Chehbouni A (2009) SAMIR a tool for irrigation monitoring using remote sensing for evapotranspiration estimate. MELIA, Marrakech

Li W, Zhou Z, Meng Y, Xu N, Fok M (2009) Modeling boll maturation period, seed growth, protein, and oil content of cotton (*Gossypium hirsutum* L.) in China. Field Crop Res 112:131–140

Lim SL, Wu TY, Lim PN, Shak KP (2015) The use of vermicompost in organic farming: overview, effects on soil and economics. J Sci Food Agric 95(6):1143–1156

Lindblom J, Lundstrom C, Ljung M, Jonsson A (2017) Promoting sustainable intensification in precision agriculture: review of decision support system development and strategies. Precis Agric 18:309–331

Liu J, Zehnder AJB, Yang H (2009) Global consumptive water use for crop production: the importance of green water and virtual water. Water Resour Res 45(5):1–15

López-Granados F (2011) Weed detection for site-specific weed management: mapping and real-time approaches. Weed Res 51(1):1–11

Ludibeth SM, Marina IE, Vicenta EM (2012) Vermicomposting of sewage sludge: earthworm population and agronomic advantages. Compost Sci Util 20(1):11–17

Ma Y, Liu S, Song L, Xu Z, Liu Y, Xu T, Zhu Z (2018) Estimation of daily evapotranspiration and irrigation water efficiency at a Landsat-like scale for an arid irrigation area using multi-source remote sensing data. Remote Sens Environ 216:715–734

Malik M, Sehgal M, Kanojia AK, Singh RV (2018) A review paper on decision support system/ expert system developed on mango. Int J Plant Protec 11(1):119–123

Mangalassery S, Sjögersten S, Sparkes DL, Sturrock CJ, Craigon J, Mooney SJ (2014) To what extent can zero tillage lead to a reduction in greenhouse gas emissions from temperate soils? Sci Rep 4:4586

Maraseni TN, Cockfield G (2011) Does the adoption of zero tillage reduce greenhouse gas emissions? An assessment for the grains industry in Australia. Agr Syst 104(6):451–458

Marino S, Alvino A (2018) Detection of homogeneous wheat areas using multi-temporal UAS images and ground truth data analyzed by cluster analysis. Eur J Remote Sens 51(1):266–275
Masseroni D, Moller P, Tyrell R, Romani M, Lasagna A, Sali G, Facchi A, Gandolfi C (2018) Evaluating performances of the first automatic system for paddy irrigation in Europe. Agric Water Manag 201:58–69
McBratney A, Whelan B, Ancev T, Bouma J (2005) Future directions of precision agriculture. Precis Agric 6:7–23
de Medeiros FHV, de Souza RM, Ferro HM, Zanotto E, Machado JDC, de Medeiros FCL (2015) Screening of endospore-forming bacteria for cotton seed treatment against bacterial blight and damping-off. Adv Plants Agric Res 2(4):167–172
Minasny B, Tranter G, McBratney AB, Brough DM, Murphy BW (2009) Regional transferability of mid-infrared diffuse reflectance spectroscopic prediction for soil chemical properties. Geoderma 153(1–2):155–162
Mir SA, Qasim M, Arfat Y, Mubarak T, Bhat ZA, Bhat JA, Bangroo SA, Sofi TA (2015) Decision support systems in a global agricultural perspective - a comprehensive review. Int J Agric Sci 7 (1):403–415
MIT (2016) Six ways drones are revolutionizing agriculture. MIT Technol Rev
Mogili UMR, Deepak BBVL (2018) Review on application of drone systems in precision agriculture. Proc Comp Sci 133:502–509
Montesano FF, Serio F, Mininni C, Signore A, Parente A, Santamaria P (2015) Tensiometer-based irrigation management of subirrigated soilless tomato: effects of substrate matric potential control on crop performance. Front Plant Sci 6:1150
Murthy VRK (2004) Crop growth modeling and its applications in agricultural meteorology. In: Sivakumar MVK, Roy PS, Harsen K, Saha SK (eds) Satellite Remote Sensing and GIS Applications in Agricultural Meteorology Workshop, Dehra Dun, India. World Meteorological Organisation, Geneva, pp 235–261
Muthamilselvan M, Rangasamyt K, Ananthakrishnan D, Manian R (2007) Mechanical picking of cotton - a review. Agric Rev 28(2):118–126
Najar IA, Khan AB (2013) Management of fresh water weeds (macrophytes) by vermicomposting using Eisenia fetida. Environ Sci Pollut Res 20(9):6406–6417
Naranjo SE (2009) Impacts of Bt crops on non-target invertebrates and insecticide use patterns. CAB Rev Perspect Agric Veterin Sci Nutr Nat Resour 4(011):1
Nasrullah HM, Aslam M, Akhtar M, Ali B, Majid A, Akram M, Farooq U (2017) Relay cropping of cotton in standing wheat: an innovative approach for enhancing the productivity and income of small farm. Roman Agric Res 34:187–195
North JH, Gore J, Catchot AL, Stewart SD, Lorenz GM, Musser FR, Cook DR, Kerns DL, Dodds DM (2017) Value of neonicotinoid insecticide seed treatments in mid-south cotton (Gossypium hirsutum [Malvales: Malvaceae]) production systems. J Econ Entomol 111(1):10–15
Ogunti EO, Akingbade FK, Segun A, Oladimeji O (2018) Decision support system using mobile applications in the provision of day to day information about farm status to improve crop yield. Periodic Eng Nat Sci 6:89–99
Ortiz BV, Hoogenboom G, Vellidis G, Boote KJ, Davis RF, Perry C (2009) Adapting the CROPGRO cotton model to simulate cotton biomass and yield under southern root-knot nematode parasitism. Trans ASABE 52:2129–2140
Ota T, Bontsema J, Hayashi S, Kubota K, van Henten E, van Os EA, Ajiki K (2007) Development of a cucumber leaf picking device for greenhouse production. Biosyst Eng 98:381–390
Padilla FM, Gallardo M, Peña-Fleitas MT, de Souza R, Thompson RB (2018) Proximal optical sensors for nitrogen management of vegetable crops: a review. Sensors (Basel) 18(7):E2083
Panwar R (2015) GIS and remote sensing applications in natural resources management. Int J Innov Res Adv Stud 2(4)
Papadopoulos AV, Kati V, Chachalis D, Kotoulas V, Stamatiadis S (2018) Weed mapping in cotton using ground-based sensors and GIS. Environ Monit Assess 190(10):622

Papageorgiou EI, Markinos AT, Gemtos TA (2011) Fuzzy cognitive map based approach for predicting yield in cotton crop production as a basis for decision support system in precision agriculture application. Appl Soft Comput 11(4):3643–3657

Pathak TB, Jones JW, Fraisse C, Wright D, Hoogenboom G, Judge J (2009) Uncertainty analysis of CROPGRO-cotton model. In: American Geophysical Union, Fall Meeting. AGU, Washington, DC

Pathan SA, Hate SG (2016) Automated Irrigation system using Wireless Sensor Network. Int J Eng Tech Res 5(6):6–9

Patil SS, Patil VC, Al-Gaadi KA (2011) Spatial variability in fertility status of surface soils. World Appl Sci J 14:1020–1024

Pawar J, Khanna R (2018) More crop per drop: ways to increase water use efficiency for crop production: a review. Int J Chem Stud 6(3):3573–3578

Pedigo LP (1989) Entomology and pest management. McMillan Publishing Company, New York, NY, p 646

Pilli SK, Nallathambi B, George SJ, Diwanji V (2015) Eagrobot-a robot for early crop disease detection using image processing. In: 2nd International Conference on Electronics and Communication Systems. IEEE, Washington, DC, pp 1684–1689

Pobkrut T, Kerdcharoen T (2014) Soil sensing survey robots based on electronic nose. In: 2014 14th International Conference on Control, Automation and Systems (ICCAS 2014). IEEE, Washington, DC, pp 1604–1609

Pramanik P, Chung YR (2010) Efficacy of vermicomposting for recycling organic portion of hospital wastes using Eisenia fetida: standardization of cow manure proportion to increase enzymatic activities and fungal biomass. Environmentalist 30(3):267–272

Privette CV, Khalilian A, Torres O, Katzberg S (2011) Remote sensing of environment utilizing space-based GPS technology to determine hydrological properties of soils. Remote Sens Environ 115(12):3582–3586

Quebrajo L, Perez-Ruiz M, Pérez-Urrestarazu L, Martínez G, Egea G (2018) Linking thermal imaging and soil remote sensing to enhance irrigation management of sugar beet. Biosyst Eng 165:77–87

Radhakrishnan S (2017) Sustainable cotton production. In: Sustainable fibres and textiles. Woodhead Publishing, Cambridge, pp 21–67

Rahman MH, Ahmad A, Wajid A, Hussain M, Akhtar J, Hoogenboom G (2016) Estimation of temporal variation resilience in cotton varieties using statistical models. Pak J Agric Sci 53:787

Rahman MHU, Ahmad A, Wajid A, Hussain M, Rasul F, Ishaque W, Islam MA, Shelia V, Awais M, Ullah A, Wahid A, Sultana SR, Saud S, Khan S, Fahad S, Hussain M, Hussain S, Nasim W (2017) Application of CSM-CROPGRO-cotton model for cultivars and optimum planting dates: evaluation in changing semi-arid climate. Field Crop Res 238: 139–152. https://doi.org/10.1016/j.fcr.2017.07.007

Rahman MH, Ahmad A, Wang X, Wajid A, Nasim W, Hussain M, Ahmad B, Ahmad I, Ali Z, Ishaque W, Awais M, Muddasir M, Shelia V, Ahmad S, Fahad S, Alam M, Ullah H, Hoogenboom G (2018) Multi-model projections of future climate and climate change impacts uncertainty assessment for cotton production in Pakistan. Agric For Meteorol 253-254:94–113

Rao RN, Sridhar B (2018) IoT based smart crop-field monitoring and automation irrigation system. In: 2018 2nd International Conference on Inventive Systems and Control (ICISC). IEEE, Washington, DC, pp 478–483

Rea JH, Wratten SD, Sedcole R, Cameron PJ, Davis SI, Chapman RB (2002) Trap cropping to manage green vegetable bug Nezara viridula (L.) (Heteroptera: Pentatomidae) in sweet corn in New Zealand. Agric For Entomol 4:101–107

Reddy DD, Blaise D, Kumrawat B, Singh AK (2017) Evaluation of integrated nutrient management interventions for cotton (*Gossypium hirsutum*) on a Vertisol in central India. Commun Soil Sci Plan 48(4):469–475

Rekha GS, Kaleena PK, Elumalai D, Srikumaran MP, Maheswari VN (2018) Effects of vermicompost and plant growth enhancers on the exo-morphological features of Capsicum annum (Linn.) Hepper. Int J Recycl Organ Waste Agric 7(1):83–88

Ribaut JM, de Vicente MC, Delannay X (2010) Molecular breeding in developing countries: challenges and perspectives. Curr Opin Plant Biol 13(2):213–218

Rocha-Munive MG, Soberón M, Castañeda S, Niaves E, Scheinvar E, Eguiarte LE, Mota-Sánchez D, Rosales-Robles E, Nava-Camberos U, Martínez-Carrillo JL, Blanco CA, Bravo A, Souza V (2018) Evaluation of the impact of genetically modified cotton after 20 years of cultivation in Mexico. Front Bioeng Biotechnol 6:82

Rose DC, Morris C, Lobley M, Winter M, Sutherland WJ, Dicks LV (2018) Exploring the spatialities of technological and user re-scripting: the case of decision support tools in UK agriculture. Geoforum 89:11–18

Roth G, Harris G, Gillies M, Montgomery J, Wigginton D (2014) Water-use efficiency and productivity trends in Australian irrigated cotton: a review. Crop Pasture Sci 64(12):1033–1048

Sajjad A, Anjum SA, Ahmad R, Waraich EA (2018) Relay cropping of wheat (*Triticum aestivum* L.) in cotton (*Gossypium hirsutum* L.) improves the profitability of cotton-wheat cropping system in Punjab, Pakistan. Environ Sci Pollut Res 25(1):782–789

Sangeetha KD, Ashtaputre SA, Ramya TS, Kavyashree MC, Anil GH (2018) Exploration of fungicides against *Alternaria* leaf blight of cotton in Northern parts of Karnataka, India. Int J Chem Stud 6(3):2127–2129

Sarmah K, Deka CR, Sharma U, Sarma R (2018) Role of GIS based technologies in sustainable agriculture resource planning & management using spatial decision support approach. Int J Innov Res Eng Manag 5(1):30–34

Schader C, Zaller JG, Kopke U (2005) Cotton-basil intercropping, effects on pests, yields and economical parameters in an organic field in Fayoum, Egypt. Biol Agric Hortic 23:59–72

Schut AGT, Traore PCS, Blaes X, de By RA (2018) Assessing yield and fertilizer response in heterogeneous smallholder fields with UAVs and satellites. Field Crop Res 221:98–107

Senthurpandian VK, Jayaganesh S, Srinivas S, Palani N, Muraleedharan N (2010) Application of geographic information system to fertility management of tea soils of Anamallais. Asian J Earth Sci 3(3):136–141

Shah MA, Memon N, Baloch AA (2011) Use of sex pheromones and light traps for monitoring the population of adult moths of cotton bollworms in Hyderabad, Sindh, Pakistan. Sarhad J Agric 27(3):435–442

Shah MA, Farooq M, Hussain M (2016) Productivity and profitability of cotton–wheat system as influenced by relay intercropping of insect resistant transgenic cotton in bed planted wheat. Eur J Agron 75:33–41

Shelton AM, Badenes-Perez FR (2006) Concepts and applications of trap cropping in pest management. Annu Rev Entomol 51:285–308

Shen M, Yang X, Cox-Foster D, Cui L (2005) The role of varroa mites in infections of Kashmir bee virus (KBV) and deformed wing virus (DWV) in honey bees. Virology 342(1):141–149

Shivanna AM, Nagendrappa G (2014) Chemical analysis of soil samples to evaluate the soil fertility status of selected command areas of three tanks in Tiptur Taluk of Karnataka, India. IOSR J App Chem 7(11):1–5

Sindhi SJ, Thanki JD, Desai LJ (2018) A review on integrated nutrient management (INM) approach for maize. J Pharmacogn Phytochem 7(4):3266–3269

Singh S (2018) Transgenic cotton-its adoption, threats and challenges ahead: a review. J Entomol Zool Stud 6(5):1989–1997

Singh A, Vasisht AK, Kumar R, Das DK (2008) Adoption of integrated pest management practices in paddy and cotton: a case study in Haryana and Punjab. Agric Econ Res Rev 21(2):1–6

Slaughter DC, Giles DK, Downey D (2008) Autonomous robotic weed control systems: a review. Comp Electron Agric 61:63–78

Sogaard HT, Lund I (2007) Application accuracy of a machine vision-controlled robotic micro-dosing system. Biosyst Eng 96:315–322

Soil Science Society of America (2013) 2013 Annual Meeting. Available at: https://www.soils.org/meetings/2013-annual-meeting

Srinivasan A (2006) Handbook of precision agriculture: principles and applications. Food Products Press, Binghamton, NY, pp 3–18

Starr JL, Carneiro RG, Ruano O (2005) Nematode parasites of cotton and other tropical fibre crops. In: Luc M, Sikora RA, Bridge J (eds) Plant parasitic nematodes in sub-tropical and tropical agriculture, 2nd edn. CAB International, Wallingford, pp 539–556

Sui R, Thomasson JA (2006) Ground-based sensing system for cotton nitrogen status determination. Trans ASABE 49(6):1983–1991

Taghvaeian S, Neale CMU, Osterberg JC, Sritharan S, Watts DR (2018) Remote sensing and GIS techniques for assessing irrigation performance: case study in Southern California. J Irrig Drain Eng 144(6):05018002

Tahir M, Ali A, Nadeem MA, Hussain A, Khalid F (2009) Effect of different sowing dates on growth and yield of wheat (*Triticum aestivum* L.) varieties in district Jhang, Pakistan. Pak J Life Soc Sci 7:66–69

Tariq M, Yasmeen A, Ahmad S, Hussain N, Afzal MN, Hasanuzzaman M (2017) Shedding of fruiting structures in cotton: factors, compensation and prevention. Trop Subtrop Agroecosyst 20(2):251–262

Tariq M, Afzal MN, Muhammad D, Ahmad S, Shahzad AN, Kiran A, Wakeel A (2018) Relationship of tissue potassium content with yield and fiber quality components of Bt cotton as influenced by potassium application methods. Field Crop Res 229:37–43

Terán-Vargas AP, Rodríguez JC, Blanco CA, Martínez-Carrillo JL, Cibrián-Tovar J, Sánchez-Arroyo H, Rodríguez-del-Bosque LA, Stanley D (2005) Bollgard cotton and resistance of tobacco budworm (*Lepidoptera: Noctuidae*) to conventional insecticides in southern Tamaulipas, Mexico. J Econ Entomol 98:2203–2209

Tewari S, Leskey TC, Nielsen AL, Pinero JC, Rodriguez-Saona CR (2014) Use of pheromones in insect pest management, with special attention to weevil pheromones. In: Integrated pest management. Academic Press, London, pp 141–168

Tian J, Zhang X, Yang Y, Yang C, Xu S, Zuo W, Zhang W, Dong H, Jiu X, Yu Y, Zhao Z (2017) How to reduce cotton fiber damage in the Xinjiang China. Ind Crop Prod 109:803–811

Tillman PG (2006) Sorghum as a trap crop for *Nezara viridula* L. (Heteroptera: Pentatomidae) in cotton in the southern United States. Environ Entomol 35(3):771–783

Tilman D, Cassman KG, Matson PA, Naylor R, Polasky S (2002) Agricultural sustainability and intensive production practices. Nature 418:671–677

Todd SW, Hoffer RM, Milchunas DG (1998) Biomass estimation on grazed and ungrazed rangelands using spectral indices. Int J Remote Sens 19(3):427–438

Tomson M, Sahayaraj K, Kumar V, Avery PB, McKenzie CL, Osborne LS (2017) Mass rearing and augmentative biological control evaluation of Rhynocoris fuscipes (Hemiptera: Reduviidae) against multiple pests of cotton. Pest Manag Sci 73(8):743–1752

Tong YJ, Wu KM, Lu YH, Gao XW (2010) Pathogenicity of Beauveria spp. Strains to three species of mirids, Apolygus lucorum, Adelphocoris suturalis and Adelphocoris lineolatus. Acta Phytophylacica Sinca 37:172–176

Towett EK, Shepherd KD, Cadisch G (2013) Quantification of total element concentrations in soils using total X-ray fluorescence spectroscopy (TXRF). Sci Total Environ 463-464:374–388

Tsedaley B (2015) Review on seed health tests and detection methods of seedborne diseases. J Biol Agric Healthcare 5(5):176–185

Turan J, Višacki V, Mehandžić S, Findura P, Burg P, Sedlar A (2015) Sowing quality indicators for a seed drill with overpressure. Acta Univ Agric Silvic Mendel Brun 62(6):1487–1492

Usman M, Ahmad A, Ahmad S, Irshad M, Khaliq T, Wajid A, Hussain K, Nasim W, Chattha TM, Trethowan R, Hoogenboom G (2009) Development and application of crop water stress index for scheduling irrigation in cotton (*Gossypium hirsutum* L.) under semiarid environment. J Food Agric Environ 7(3–4):386–391

Vani V, Mandla VR (2017) Comparative study of NDVI and SAVI vegetation Indices in Anantapur district semi-arid areas. Int J Civil Eng Technol 8(4):559–566

Vijaya MBN, Rai PK, Srivastava DK, Bara BM, Kumar R (2017) Effects of polymer seed coating, fungicide seed treatment and storage duration on seedling characteristics of cotton (*Gossypium hirsutum*) seeds. J Pharma Phytochem 6(4):534–536

Vleeshouwer J, Car NJ, Hornbuckle J (2015) A cotton irrigator's decision support system and benchmarking tool using national, regional and local data. In: Int. Symposium on Environ. Software Systems. Springer, Cham, pp 187–195

Vora VD, Rakholiya KD, Rupapara KV, Sutaria GS, Akbari KN (2015) Effect of integrated nutrient management on Bt cotton and post-harvest soil fertility under dry farming agriculture. Asian J Agric Res 9(6):350–356

Wadodkar MR, Ravishankar T, Joshi AK (2014) Application of remote sensing techniques for soil fertility assessment. Available at: https://www.academia.edu/11077591/APPLICATION_OF_REMOTE_SENSING_TECHNIQUES_FOR_SOIL_FERTILITY_ASSESSMENT

Wajid A, Ahmad A, Hussain M, Rahman MH, Khaliq T, Mubeen M, Rasul F, Bashir U, Awais M, Iqbal J, Sultana SR (2014) Modeling growth, development and seed-cotton yield for varying nitrogen increments and planting dates using DSSAT. Pak J Agric Sci 51:641–650

Wang M, Wei J, Yuan J, Xu K (2008) A research for intelligent cotton picking robot based on machine vision. In: International Conference on Information and Automation. Zhangjiajie, China. IEEE, Washington, DC

Wang S, Li X, Lu J, Hong J, Chen G, Xue X, Li J, Wei Y, Zou J, Liu G (2013) Effects of controlled-release urea application on the growth, yield and nitrogen recovery efficiency of cotton. Agric Sci 4(12):33–38

Wang L, Hu G, Yue Y, Ye X, Li M, Zhao J, Wan J (2016) GIS-based risk assessment of hail disasters affecting cotton and its spatiotemporal evolution in China. Sustainability 8(3):1–20

Wu W, De Pauw E (2011) A simple algorithm to identify irrigated croplands by remote sensing. In: Proceedings of the 34th International Symposium on Remote Sensing of Environment (ISRSE), Sydney, Australia. Arinex, Sydney, NSW, pp 10–15

Wu K, Lu Y, Wang Z (2009) Advance in integrated pest management of crops in China. Chin Bull Entomol 46(6):831–836

Xia J, Cui J, Ma L, Dong S, Cui X (1998) The role of transgenic BT cotton in integrated insect pest management. Acta Gossypii Sin 11:57–64

Xia C, Wang L, Chung BK, Lee JM (2015) In situ 3D segmentation of individual plant leaves using a RGB-D camera for agricultural automation. Sensors (Basel) 15(8):20463–20479

Yuan Z, Shen Y (2013) Estimation of agricultural water consumption from meteorological and yield data: a case study of Hebei, North China. PLoS One 8(3):e58685

Zang X (1998) On the estimation of biomass of submerged vegetation using Landsat thematic mapper (TM) imagery: a case study of the Honghu Lake, PR China. Int J Remote Sens 19 (1):11–20

Zhang N, Wang M, Wang N (2002) Precision agriculture – a worldwide overview. Comp Electron Agric 36(2-3):113–132

Chapter 30
Diverse Uses of Cotton: From Products to Byproducts

Hassan Munir, Fahd Rasul, Ashfaq Ahmad, Muhammad Sajid, Salman Ayub, Muhammad Arif, Pakeeza Iqbal, Amna Khan, Zartash Fatima, Shakeel Ahmad, and Muhammad Azam Khan

Abstract Cotton is a multifaceted crop of which wholesome or all parts individually can be used for their byproducts in addition to their domestic or economic uses. It provides lint raw material to an ever-increasing textile industry, cotton seed oil for culinary purposes, and edible oil and protein-rich oil cake residue for livestock. Cotton can benefit human being through its sticks, fibers, seed, and oil as the primary products, whereas several secondary products are manufactured by utilizing these

H. Munir (✉) · F. Rasul · M. Sajid · S. Ayub
Department of Agronomy, University of Agriculture, Faisalabad, Pakistan
e-mail: hmbajwa@uaf.edu.pk

A. Ahmad
Department of Agronomy, University of Agriculture, Faisalabad, Pakistan

US.-Pakistan Centre for Advanced Studies in Agriculture and Food Security, University of Agriculture, Faisalabad, Pakistan

M. Arif
Department of Agronomy, University of Agriculture Peshawar, Peshawar, Pakistan

P. Iqbal
Department of Botany, University of Agriculture, Faisalabad, Pakistan

A. Khan
Department of Agronomy, Bahauddin Zakariya University, Multan, Pakistan

Department of Agronomy, University College of Agriculture, University of Sargodha, Sargodha, Pakistan

Z. Fatima
Department of Agronomy, Bahauddin Zakariya University, Multan, Pakistan

S. Ahmad
Department of Agronomy, Faculty of Agricultural Sciences and Technology, Bahauddin Zakariya University, Multan, Pakistan
e-mail: shakeelahmad@bzu.edu.pk

M. A. Khan
In-Service Agriculture Training Institute, Sargodha, Sargodha, Pakistan

S. Ahmad, M. Hasanuzzaman (eds.), *Cotton Production and Uses*,
https://doi.org/10.1007/978-981-15-1472-2_30

components of cotton. Cash crop appraisal is, however, based on its multifaceted uses, and a significant proportion in value added in agriculture is well established through cotton textile, hosiery made-up, and raw and fine products of cotton apart from its use in many surgical products.

Keywords Textile · Fuel · Fiber · Lint · Cash crop · Seed

30.1 Introduction

Cotton is distinguished as a source of natural fiber since the very early human civilization due to its need to provide basic clothing and hence was identified as an important plant species for mankind (Fryxell 1979; Ahmad et al. 2014, 2017, 2018; Abbas and Ahmad 2018; Ahmad and Raza 2014; Ali et al. 2011, 2013a, b, 2014a, b; Usman et al. 2009). At present, cotton is among the top fiber crops that are being grown on a vast area belonging to nearly 50 nations across the globe. The regions of cotton cultivation comprise of temperate to tropical conditions with efficient commercial cotton production (Smith and Hirth 1988; Amin et al. 2017, 2018; Khan et al. 2004; Rahman et al. 2018; Tariq et al. 2017, 2018; Usman et al. 2009). Cotton is presently considered among the top 20 important crops where its appraisal is different from other crops as it is not a staple crop (Wendel et al. 2010). Cotton ranks first among fiber crops across the globe. Its fiber is known as "king of the fibers" which has the ability for being naturally absorbent, permeable, soft, and viable as well as durable with no skin rashes and allergies when meeting human skin. For being a natural thread, cotton lint-based fabric is never undesired for its origination from waste material as compare to chemically made threads in which a lot of waste material is involved during different steps of manufacturing. Apart from its fiber appraisal, cotton is among the top ten crops for its appraisal as oilseed, a rich source of edible oil. In addition, cotton is being modified in terms of its genetics, and transgenic cotton is counted among the crops with greatest area under cultivation across the globe (FAOSTAT 2012). Fiber staple of cotton is worth to be knitted or woven into a variety of fabrics, for example, corduroy, velvet, jersey, velour, and chambray. Furthermore, cotton is used at a large scale in making fishnets, archival papers, in binding of books and stationery, food packaging, feed seedcake, pillows, and quilts. After the extraction of oil from cotton seed, the remaining part which is known as "cotton seed cake" is being fed to cattle and is especially a rich source of energy for ruminants. Cotton seed left over considered as waste after extraction of oil is named "banola (Urdu), wareva (Punjabi), and seed cake (English)." Cotton seed oil is used in cooking after blending it with different other edible oils in a variety of goods like margarine, cosmetics, pharmaceuticals, rubber, soap, emulsifier, fattening, vanaspati ghee, and plastics. The short fiber residue of cotton seed named linters left as waste after ginning are used to produce products like bank notes, X-ray films, swabs, cotton buds, bandages, etc. In this chapter, an attempt has been made to colligate the major uses of products of cotton and its byproducts covering all portions of the cotton plant, i.e., raw, refined, and finished goods.

30.2 Raw Uses

30.2.1 Cotton Sticks

Cotton was cultivated on approximately 2,489,000 hectares in Pakistan, and the production was 10,671,000 bales during the fiscal year 2016–2017 (Economic Survey of Pakistan 2017). Cotton stick is the major residue that contributes threefold of the lint weight obtained from cotton plant. In rural areas of Pakistan and many agrarian economies of the world, cotton sticks are mostly appraised for fuel. The amount of cotton crop residues, called with different names, i.e., cotton straw, cotton sticks, cottonwood, etc., usually volumes from 5 to 7 tons/hectare (Silanikove and Levanon 1986). Hence, cotton sticks are referred as the biofuel for domestic energy consumption starting from the stove to kiln; however, bioenergy reactors or gasification chambers can be the source of utilization of these sticks in this energy scarce arena.

30.2.2 Baskets

Making baskets from different parts of cotton benefits not only at domestic level, but it is also being utilized for marketing a number of vegetables, fruits, and other industrial products. Because of the strength of cotton plant sticks, their rigid parts are used to make strong baskets. Good and fine quality fibers can also be used in making enormous art products like wall hangings, gift packings, and closet boxes.

30.2.3 Fences and Cages

Making fences is another efficient use of cotton stick. Semi-dried cotton sticks can be flexible for being knitted like a woven web, or it can be moistened for the said purpose to make it flexible. Such fences are cost-effective for the small households in addition to keeping pets and birds by making cages with same sticks.

30.2.4 Cotton Bolls

Cotton lint grows in a fluffy round and sound clumps (Locales) and is called boll. These fluffy fiber clusters or bolls are living cells with outer walls surrounding the cytoplasm. The outer wall of cotton fiber is made up of several layers of microscopic fibers mainly composed of cellulose and microfibrils. In raw conditions, the fiber is an almost pure cellulose sheath with a fine coat of wax which makes it hydrophobic or "waterproof." The top-notch utility of cotton is to make fine fiber and knit the textile. Linters are undersized fibers called fuzz on the seed coat; those can be used as source of cellulose in order to make plastics, explosive material, and many other products. These fuzzy fibers are not being utilized for making cloth but help make lining used for making furniture in addition to quality paper manufacturing. Furthermore, bumpers, dashboards, and other plastic-oriented car or vehicle parts and mattresses are made from this fuzzy material. Cotton fiber is naturally either white or colored. There are varieties that produce brown, khaki, yellow, and greyish-green fibers. The American Indian natives usually utilized colored cotton, while in modern and industrialized countries, the development of chemical dyes increased demand for colorless lint.

30.2.5 Cotton Lint/Fiber

High durability has made cotton renewable; hence, its fiber can be used in a variety of man's ways (Chen et al. 2007). Lint from cotton plant is twirled and then woven into fabrics, i.e., chambray, velvet, velour, corduroy, jersey, and flannel. Sixty percent of the world's net cotton harvest is utilized in making cloth, while the rest is used in industrial products and home furnishings. Most renowned cotton products include t-shirts, denim jeans, towels, socks, bedsheets, and underwear etc. Fiber is also used in tentage making, car tires, ropes, fishnets, and book bindings. Its fiber is used in different home textiles, i.e., rugs, blankets, diapers and tampons, window coverings, and face masks etc. Fine fiber can be used in the production of currency, papers, stationery items, art papers, degrees, and document folders etc.

30.2.6 Cotton Locule

Locule means a cavity of the ovary in which ovules develop or preserve. Locules carry the cotton seeds in them which can be utilized in numerous ways. Three products including meal, hulls, and oil were obtained after the crushing and islolation of cotton seed. Oil is used for salad dressing and cooking and as vegetable shortening. The meal and hulls are used as feed for poultry, fish meals, and feed for livestock as well as fertilizer separately or in combinations. Cotton trash in the form of leaves and stalks can be mixed with soil to trigger its fertility status for being a source of organic matter. Cotton seed is also used in high protein concentration for making baked food products. As far as the aesthetic purposes of locules are concerned, these can be used in interior decoration and can be made scented and colored using dyes and fragrances.

30.3 Cotton Seed

More than half of the cotton (cotton seed) consists of seed, a beneficial byproduct gained during production of fiber. One ton of cotton seed can contribute to 1/3 of seed coat, while nearly 20% oil can be extracted from it, and half of this seed after pressing can even be used as animal meal (Cotton Australia 2018). Cotton seed meets the protein need of half of a billion people and many billions of animals across the globe. The most common utilizations of cotton seed are oil for cooking and feeding the livestock animals. Cotton seed is crushed or pressed to get cotton seed oil with a number of utilizations starting from domestic up to industrial avenues. While processing the seed of cotton, cleaning is done as first step, and this continues even up to the extraction of pharmaceutical or industrially important compounds such as gossypol. Removing dirt, inert matter, plant debris, and short fine fibers is among the cleaning procedures to get fine cotton important for making quality paper. Technically, paper composed of both small and long staple having fibers is more durable as it resists mishandlings easily. Even printing currency is worth based on certain blends of cotton and linen, such as ¼ linen and ¾ cotton blend is used to print dollar bills that costs 9.6 cents/dollar size bill (Kavilanz 2011). Cotton seed can be cooked as meal and is commonly used to feed livestock and cattle as it's a rich source of energy. Seed oil can also be used in making of various industrial goods, i.e., margarine, soap, emulsifiers, pharmaceuticals, cosmetics, rubber, candles, water proofing, etc. Cotton seed oil has a profound quality that it is free of cholesterol; it is having high amounts of polyunsaturated fats and has massive amount of antioxidant (vitamin E) which increases the shelf life of chickens, pigs, and rabbits, among others.

30.4 Lint Uses

30.4.1 Raw Fabric Making

Fabric initially woven from the cotton thread is named grey cloth. Such weaving is highly dependent on spinning process in which staples of cotton are spun and cotton thread is being made (Smith and Hirth 1988). Process involved in fiber making comprising of conversion of raw cotton to thread is called "spinning," and the other where the thread is changed into fabric is called "weaving." Textile manufacturing is based on basic steps such as spinning and weaving of cotton. These processes have become technically advanced and are now done by blending precious materials such as the addition of feathers, beads, bird feathers, etc. Such processed products are valued in the elite community of the society since long (Follensbee 2008). Raw cotton is used in medical field for making medical swabs and bandages and in other related products and surgical affairs, because it can easily be sterilized and can also be used to cover or wrap the wounds.

30.4.2 Fine Cotton (or Chintz)

Fine cotton is 100% cotton fabric as well as fine and basic cotton. Fine cotton has a plain weave, but as compared to the basic cotton, it is lower in weight and softer too as it is generated from fine yarns. Various types of fine cotton are produced, i.e., dyed, bleached, and printed. When it is at a temperature which is not more than 60 °C, it ensures the long-lasting or permanent color.

Glazed calicos are termed chintz and those were imported from India subcontinent for use as clothing stuff. These glazed chintz are fine enough to be printed and decorated and can be stuffed with material from bird feathers and can also be used for making drawings, flowers, etc., with half tone designing. A number of uses of chintz are found in households like bedlinen, window decorations and shades, and installation with delicate clothing having resistance to stains of different origin particularly from dirt. Furthermore, at times these chintz are parts of a dress of feminine gender and kid garments particularly during summers.

30.4.3 Cotton Rope and Twine

Cotton ropes are an important commodity for domestic as well as for industrial uses. Ropes are being used from knitting charpais (Indian woven bed) to making anchor supporting ropes. From tying to knotting and from binding to making spirals, every flexible binding is based on use of cotton ropes. Twine is a light weight string or strong thread composed of two or more than two smaller strands lashed and then got twisted together. Natural twine is used in various forms like sewing upholstery, making rugs, jewelry making, packaging, weaving nets, wrapping things, critical household applications, food services, meat packing, advertising, and assorted industrial uses. It is also used to tie up a turkey, roast, and various poultry or meats prepared for rotisserie, grilling, or oven. Just keep it away from the open flame. Furthermore, it is used to tie up stuffed chicken breasts and meat roll ups, holding up the delicious ingredients inside. One can tie parcels or herb bouquets and gift tags or presents using a twin.

30.4.4 Textiles

Textile products are used by all of us and not just for aesthetic reasoning. We use them for protection against hazardous elements, to improve our quality of life, for safety purposes, and even for technical reasons. Textiles are highly esteemed commodities like bullion and emerald and were signified when used by status quo in the old era.

Textiles are classified as described by Bergan (1987).

30.4.4.1 Clothing

For being rich, elite community of the society used cotton garments in the ages when it was real-time scarce due to low production and less yield potential of cotton varieties. Still its fineness in the merit of its worth and cost of cloth varies with it; hence, consider it well regarded among the fibrous crops such as yucca and maguey. Clothing made from cotton is having additional uses in the household, such as kitchen napkin, hand gloves, and covers.

30.4.4.2 Armor Manufacture

Textiles were an important element of armors natively called "ichcahuipilli" in Mexican civilization from where upland cotton originated. In many civilizations, battles and wars were ranked among the most substantial sacramental and sacred inferences, and combatants were exceptionally statused. Battle armory was designed with ample use of cotton clothing in combination with bird fluffs so that combatant could carry them easily for being lighter in weight.

30.5 Sociopolitical Significance

In Asian environment, silk route was also named textile route and was based on long travel associated with business activities as well as supply of the basic needs to humans in far flung, less privileged areas by crossing Pamir heights and through the Himalayas. A trajectory of wearing precious, fine, and attractive clothes was highly respected since the dark ages, and people were dealt on the basis of what they had worn. Even today, clothes are the symbol of recognition when the world has become a small village and people are interacting across the borders frequently. Textiles and fine cloth were part of the decorative clothings worn by idols and gods in the old civilizations of Euphrates or in the Mesoamerican Myths.

30.6 Economic Commodity

As far as the bartered trade system of the past is concerned, textiles remained among the top precious commodities to be bartered. Such exchanges were historically reported against spices in India before Christ. For being a light weight commodity, cotton clothings were carried for longer distances for ease in travel and were marketed with high prices in order to get grains and other precious commodities in many folds by weight. Even today, cotton made-ups are being exported as costly commodities and are a source of heavy foreign exchange earnings in the matter of financial status of any nation or country across the world.

30.6.1 Bags

In daily routine life, cotton bag is an important and handy product that if made from natural cotton fiber will completely serve as eco-friendly; hence, cotton is completely biodegradable in nature. The cotton bag material is free of any environmental hazard and made of pure fiber. These bags can easily be washed when they are dirty and can be used again and again for a long time. The surface of cotton bags is very suitable for designing or printing and, therefore, can be used as promotional material. Furthermore, these bags can be used for packing, shopping, etc. Customers like to get a cotton bag in order to carry lightweight items like gifts, garments, etc., due to the smooth and light nature of the fabric.

30.6.2 Rugs and Carpets

Central and Western Asians developed carpets first as coverings for beaten/broken earth floors. From old times, carpets covering the floors of houses and tents as well as mosques and palaces were made from cotton thread. In the homes of elite eastern families, floor coverings serve an aesthetic as well as a pragmatic function. Rugs are often arranged in a traditional arrangement, partially to allow for simultaneous display; the size of carpet and shape are determined by the selected place within that arrangement.

30.6.3 Towels

Towels are a basic everyday need to sponge out the wetness. These are necessarily required from their need while working in kitchen to taking shower in the bathrooms in addition to other cleaning purposes. Towels are made up of good quality cotton material with high absorbance and durability. Cotton-made towels are destined for their softness and their property for not infecting skins with any allergy. However, they are quickly washable as well as their cleanliness is near to excellence too. They are good on skin because they are not made from any petrochemicals.

30.6.4 Paper

While manufacturing paper, durability, resilience, and price affordability are always kept in mind. Hence, making paper for longer durations, in superior quality for a variety of objectives such as printing, dying, painting, etc., and avoiding age-based deterioration are among the targets. Cotton-based paper luckily has all these characters; however, variation is found from lot to lot and patch to patch. Many of the important documents are being written on paper made from cotton since the very past. Today, important documents, registries, agreements, and govt. sector archives are being executed using quality cotton paper; even academic degrees and manuscripts are being printed on such papers with cotton as constituent with varying compositions from 25% to 100%. Light penetration and crossing is one useful method to assess the quality of paper; hence, holding paper in the direction of light can help us understand how good that paper is. Market value in paper industry is ruled by acid-free paper across the world for being the most precious one; hence, offset paper is a term used for getting white, colorable, printable, and durable paper for official and durable needs.

30.6.5 Hosiery

Hosiery is a stuff which is worn by human as an expounded outfit, as vests, linings, and undergarments. These are the products which are categorized on the basis of denier or opacity. Five to 15 lower denier measurements explain a tendency to become absorbent that can be fragile as per its look, where forty and above denier stuff is impenetrable to light. Hosiery articles are among the largest cotton commodities being consumed regarding dressing, sports, and aerobics purposes.

30.6.6 Surgical Uses

In addition to cotton dressing and clothing uses as garments, surgical appraisal of cotton is one important avenue of cotton utilization. In old times natural fiber is used in many processes. One of the earliest examples is the use of fibers in healthcare facilities for wooden dentures. Cotton fiber has been used in healthcare due to its softness, purity, hypoallergenic purity, and absorbency properties. Cotton can also be sterilized by all three major sterilization methods such as steam, gamma radiation, and ethylene oxide. Undocumented use of cotton seed as food remained for long in the Mesoamerican civilizations. Roasting of seed for edible purpose is still part of rural households in scarce areas of the region. In addition, cotton seed use for abortion of men semen is also known. The families with edible use of cotton oil were having less birth rate that later proved in the form of sterility and retarded potassium absorption to a serious extent. Certain studies were also found that birth control can be ensured by dietary use of cotton seed or oil in reduced concentrations for those men who wish no more children in the days to come (Coutinho et al. 2000).

Cotton seed has a gland termed as "gossypol gland" responsible for secretion of an alkaloid substance "gossypol" with high anti-nutritive significance. During late 1980, human immunodeficiency virus (HIV) has emerged as real time threat to mankind for which a number of treatments were considered for their potential impact on reduced epidemiology of this virus disease. Gossypol was found to be effective for both traditional mean of cure as well as its significance as substitute medicine (Polsky et al. 1989; Ratsula et al. 1983).

Injuries based on unlucky incidence are part of human and animal life for being warm blooded in nature. Saving blood by stitching the wounds is one empirical solution to avoid excessive blood loss. Statures used for this purpose at once help life sustain under any time of injury or surgery. Cotton thread used for the purpose is one most suitable as it has the tendency to be used in pure or in blended form related to the kind, nature, and extent of injury. Many thread types are in practice today with dissolvability or non-dissolvability in the human or animal body. Thus, pure cotton such as original catguts (polydioxanone) or cotton-blended threads such as caprolactone, proline polyglycolic acid, polylactic acid, etc., for saturating the wounds are key tools of almost all surgeons today. In addition, a number of polymer-based threads are also used for surgery purposes with absorbable character.

Blood choking and coagulation is one needed attribute to immediately relieve the patient from hazard. For the purpose, cotton gauze or sponges are being employed by the surgeons. Being biodegradable and absorbent to slurry type secretion and blood, such gauzes help to act like sponges and dry wounds to initiate the healing process. During surgery sponges called gauze sponges with different sizes. It is also worth to mention that the nonwoven material or gauzes help remediate more efficiently than the ones being made with woven lint of cotton. Another advantage of cotton-based gauzes is that we can make them sterile with little effort as compare to other materials. Latest technology has suggested use of some blended materials for the same purpose; however, the significance of cotton-based gauzes is still valid and superseded them.

Healing bigger wounds or injuries are highly dependent on cotton-based bandages particularly in case of major operative procedures. Bandages are mainly designed to restrict the movement of injured part of the body. Bandages are part of the dressing and hence support the procedures undertaken for improving skin tissues and abrasions. Other uses of bandage are based on their types such as "elastic bandages;" those are used to provide support or to reduce the swelling on an uneven part of human body. Bandages used tightly can help reduce the flow of blood to the suburbs of a crack or wound and hence help give tissue a chance to recover in case of heavy bleeding conditions. A wide range and types of bandages such cloth slips and flexible bandages are available in the market; those can be improvised keeping in view the conditions of the wound as well as the wound location and purpose of improvision. Absorbent cotton is usually a bleached, non-lubricated fine piece of sheath available in different dimensions. Absorbent type of bandages is often in direct contact of human body; hence, these should be free from any risk of human health and meet the pharmaceutical parameter. Absorbent cotton is also termed as wool cotton. Excessive use of it in surgical traumas named it as "surgical cotton" too. Cotton wool is used mainly in nursing home, dispensaries, hospitals, etc., because it has high absorbency power, due to absorbent. Today, baby diapers and other absorbent type of cotton commodities are popular for daily life use and consume a lot of capital.

30.6.7 Biodegradable Packaging

The step which combine the fungicide inside a cast and gin waste called a "tool" where the two-component combine to result in spongy material same in the appearance of thermopore sheet. Such thermopore or polystyrene type spongy material obtained from cotton provides a cheap and environment-friendly alternative to get things packed. Volume of this type of intervention is as larger as $2 billion in the market.

30.6.8 Embroidery

Cotton is spun manually (by hand) for its fine-tuned utilization as stuff for embroidery of dresses. In Pakistani environment, needed work-based embroidery is very popular; hence, the use of this sort of six-stranded, twisted thread is highly demanded here. Quality stich is the base of such popularity in such embroidery, and their use is dated back to sixteen century. Few more uses are also mentioned below:

- Matte-Finish cotton is a French type used on borders of the cloth often named as French border cotton.
- Two-ply and spiral, twisted cotton thread is named pearl cotton with a variety of thickness starting from 3 to 16 finca, i.e., from heaviest to finest, respectively.

30.7 Conclusion

Cotton is a multifaceted crop of which wholesome or all parts individually can be used for their byproducts in addition to their domestic or economic uses. Cotton can benefit human being through its sticks, fibers, seed, and oil as the primary products, whereas several secondary products are manufactured by utilizing these components of cotton. Cash crop appraisal is, however, based on its such multifaceted uses, and a significant proportion in value added in agriculture is well established through cotton textile, hosiery made-up, and raw and fine products of cotton apart from its use in many surgical products as mentioned in above paragraphs. The team of authors foresee that the use of this silver line thread can, however, be a savior of human life in future climate change scenario from high temperature, skin dryness, and respiratory infections, a need of today and the days to come.

References

Abbas Q, Ahmad S (2018) Effect of different sowing times and cultivars on cotton fiber quality under stable cotton-wheat cropping system in southern Punjab, Pakistan. Pak J Life Soc Sci 16:77–84

Ahmad S, Raza I (2014) Optimization of management practices to improve cotton fiber quality under irrigated arid environment. J Food Agric Environ 2(2):609–613

Ahmad S, Raza I, Ali H, Shahzad AN, Atiq-ur-Rehman, Sarwar N (2014) Response of cotton crop to exogenous application of glycinebetaine under sufficient and scarce water conditions. Braz J Bot 37(4):407–415

Ahmad S, Abbas Q, Abbas G, Fatima Z, Atique-ur-Rehman, Naz S, Younis H, Khan RJ, Nasim W, Habib ur Rehman M, Ahmad A, Rasul G, Khan MA, Hasanuzzaman M (2017) Quantification of climate warming and crop management impacts on cotton phenology. Plants 6(7):1–16

Ahmad S, Iqbal M, Muhammad T, Mehmood A, Ahmad S, Hasanuzzaman M (2018) Cotton productivity enhanced through transplanting and early sowing. Acta Sci Biol Sci 40:e34610

Ali H, Afzal MN, Ahmad F, Ahmad S, Akhtar M, Atif R (2011) Effect of sowing dates, plant spacing and nitrogen application on growth and productivity on cotton crop. Int J Sci Eng Res 2 (9):1–6

Ali H, Abid SA, Ahmad S, Sarwar N, Arooj M, Mahmood A, Shahzad AN (2013a) Integrated weed management in cotton cultivated in the alternate-furrow planting system. J Food Agric Environ 11(3–4):1664–1669

Ali H, Abid SA, Ahmad S, Sarwar N, Arooj M, Mahmood A, Shahzad AN (2013b) Impact of integrated weed management on flat-sown cotton (*Gossypium hirsutum* L.). J Anim Plant Sci 23 (4):1185–1192

Ali H, Hameed RA, Ahmad S, Shahzad AN, Sarwar N (2014a) Efficacy of different techniques of nitrogen application on American cotton under semi-arid conditions. J Food Agric Environ 12 (1):157–160

Ali H, Hussain GS, Hussain S, Shahzad AN, Ahmad S, Javeed HMR, Sarwar N (2014b) Early sowing reduces cotton leaf curl virus occurrence and improves cotton productivity. Cercetări Agronomice în Moldova XLVII(4):71–81

Amin A, Nasim W, Mubeen M, Nadeem M, Ali L, Hammad HM, Sultana SR, Jabran K, Habib ur Rehman M, Ahmad S, Awais M, Rasool A, Fahad S, Saud S, Shah AN, Ihsan Z, Ali S, Bajwa AA, Hakeem KR, Ameen A, Amanullah, Rehman HU, Alghabar F, Jatoi GH, Akram M,

Khan A, Islam F, Ata-Ul-Karim ST, Rehmani MIA, Hussain S, Razaq M, Fathi A (2017) Optimizing the phosphorus use in cotton by using CSM-CROPGRO-cotton model for semi-arid climate of Vehari-Punjab, Pakistan. Environ Sci Pollut Res 24(6):5811–5823

Amin A, Nasim W, Mubeen M, Ahmad A, Nadeem M, Urich P, Fahad S, Ahmad S, Wajid A, Tabassum F, Hammad HM, Sultana SR, Anwar S, Baloch SK, Wahid A, Wilkerson CJ, Hoogenboom G (2018) Simulated CSM-CROPGRO-cotton yield under projected future climate by SimCLIM for southern Punjab, Pakistan. Agr Syst 167:213–222

Bergan FF (1987) Cotton in Aztec Mexico: production, distribution and uses. Mexic Stud 3 (2):235–262

Chen Y et al. (2007) Potential of agricultural residues and hay for bioethanol production. Appl Biochem Biotechnol 142(3):276–290

Cotton Australia (2018) Cotton Australia Annual 2018. Australian Cotton Industry Statistics, New South Wales

Coutinho EM, Athayde C, Atta G (2000) Gossypol blood levels and inhibition of spermatogenesis in men taking gossypol as a contraceptive. A multicenter, international, dose-finding study. Contraception 61(1):61–67

Economic Survey of Pakistan (2017) Cabinet Division. Ministry of Finance, Govt. of Pakistan, Islamabad

FAOSTAT (2012) Searchable online statistical database. Food and Agriculture Division of the United Nations, Rome. Available at: http://faostat.fao.org/. Accessed Apr 2019

Follensbee BJA (2008) Fiber technology and weaving in formative-period gulf coast cultures. Anc Meso 19(1):87–110

Fryxell PA (1979) The natural history of the cotton tribe (Malvaceae, tribe Gossypieae). Texas A&M University Press, College Station, TX

Kavilanz P (2011) Dollar bills are made of cotton. CNN-Money. Available at: http://money.cnn.com/2011/03/08/news/economy/dollar_cotton_prices/. Accessed Mar 2019

Khan MB, Khaliq A, Ahmad S (2004) Performance of mashbean intercropped in cotton planted in different planting patterns. J Res (Sci) 15(2):191–197

Polsky B, Segal SJ, Baron PA, Gold JW, Ueno H, Armstrong D (1989) Inactivation of human immune-deficiency virus in vitro by gossypol. Contraception 39(6):579–587

Rahman MH, Ahmad A, Wang X, Wajid A, Nasim W, Hussain M, Ahmad B, Ahmad I, Ali Z, Ishaque W, Awais M, Shelia V, Ahmad S, Fahad S, Alam M, Ullah H, Hoogenboom G (2018) Multi-model projections of future climate and climate change impacts uncertainty assessment for cotton production in Pakistan. Agric For Meteorol 253-254:94–113

Ratsula K, Haukkamaa M, Wichmann K, Luukkainen T (1983) Vaginal contraception with gossypol: a clinical study. Contraception 27(6):571–576

Silanikove N, Levanon D (1986) Cotton straw: composition, variability and effect of anaerobic preservation. Biomass 9:101–112

Smith ME, Hirth KG (1988) The development of prehispanic cotton-spinning technology in western Morelos, Mexico. J Field Archaeol 15(3):349–358

Tariq M, Yasmeen A, Ahmad S, Hussain N, Afzal MN, Hasanuzzaman M (2017) Shedding of fruiting structures in cotton: factors, compensation and prevention. Trop Subtrop Agroecosyst 20(2):251–262

Tariq M, Afzal MN, Muhammad D, Ahmad S, Shahzad AN, Kiran A, Wakeel A (2018) Relationship of tissue potassium content with yield and fiber quality components of Bt cotton as influenced by potassium application methods. Field Crop Res 229:37–43

Usman M, Ahmad A, Ahmad S, Irshad M, Khaliq T, Wajid A, Hussain K, Nasim W, Chattha TM, Trethowan R, Hoogenboom G (2009) Development and application of crop water stress index for scheduling irrigation in cotton (*Gossypium hirsutum* L.) under semiarid environment. J Food Agric Environ 7(3–4):386–391

Wendel JF, Brubaker CL, Seelanan T (2010) The origin and evolution of *Gossypium*. In: Stewart JM, Oosterhuis DM, Heitholt JJ, Mauney JR (eds) Physiology of cotton. Springer, Dordrecht, pp 1–18

Correction to: Development of Transgenic Cotton for Combating Biotic and Abiotic Stresses

Babar Hussain and Sultan Mahmood

Correction to:
Chapter 26 in: S. Ahmad, M. Hasanuzzaman (eds.), *Cotton Production and Uses*, https://doi.org/10.1007/978-981-15-1472-2_26

The original version of the chapter was inadvertently published with incorrect sequence of the authors. The correct sequence is

Babar Hussain and Sultan Mahmood

In addition, Dr. Babar Hussain's email was incorrectly linked with his coauthor. These errors have now been corrected with this erratum.

The updated online version of this chapter can be found at https://doi.org/10.1007/978-981-15-1472-2_26

S. Ahmad, M. Hasanuzzaman (eds.), *Cotton Production and Uses*,
https://doi.org/10.1007/978-981-15-1472-2_31

The manufacturer's authorised representative in the EU is Springer Nature Customer Service Centre GmbH, Europaplatz 3, 69115 Heidelberg, Germany. If you have any concerns regarding our products, please contact ProductSafety@springernature.com

Printed and bound by CPI Group (UK) Ltd, Croydon, CR0 4YY
25/09/2026
02222882-0009